Fossil Vertebrates of Greece Vol. 2

Evangelos Vlachos
Editor

Fossil Vertebrates of Greece Vol. 2

Laurasiatherians, Artiodactyles, Perissodactyles, Carnivorans, and Island Endemics

Springer

Editor
Evangelos Vlachos
Museo Paleontólogico Egidio Feruglio
CONICET
Trelew, Chubut, Argentina

ISBN 978-3-030-68441-9 ISBN 978-3-030-68442-6 (eBook)
https://doi.org/10.1007/978-3-030-68442-6

This Springer imprint is published by the registered company Springer Nature Switzerland AG
The registered company address is: Gewerbestrasse 11, 6330 Cham, Switzerland

This book is dedicated
to the memory of
our friend and colleague
Katerina Vasileiadou

Foreword

When I first started working on fossil mammal material from Greece, the one thing I wanted most was something that did not exist: a comprehensive guide to the region's fossil vertebrates. *Fossil Vertebrates of Greece* fills this gap admirably, marking a major and timely milestone in the synthesis of Greece's natural heritage. For the first time, an account of almost all of the fossil vertebrate species known from Greece—from the smallest conodonts to the Titans of the Proboscidea—can be found in one place. By bringing these data and knowledge together, *Fossil Vertebrates of Greece* is more than a regional record of past biodiversity. It makes the clearest statement yet that Greece is a treasure trove of vertebrate fossil history, and—thanks to its synthetic nature and the inclusion of vital geographical, geological, and historical context as well as collections information—it is also a tool that will help to unlock a new wave of high-quality, globally important research.

With the exception of one Oligocene Anthracothere (chapter "The Fossil Record of Anthracotheres (Mammalia: Artiodactyla: Hippopotamoidea) in Greece"), the material covered in Volume 2 is Neogene onwards, further emphasizing the richness of Greece's fossil record from these periods. Iconic sites like Pikermi, yielding equally iconic beasts, such as the extraordinary Chalicotheres (chapter "The Fossil Record of Felids (Mammalia: Carnivora: Felidae) in Greece"), of course stand out, as does the richness of the bovid fossil record (chapter "The Fossil Record of Bovids (Mammalia: Artiodactyla: Ruminantia: Pecora: Bovidae) in Greece"). But so too do the rare taxa—especially those scarce Pliocene records of tapirs (chapter "The Fossil Record of Tapirs (Mammalia: Perissodactyla: Tapiridae) in Greece"), felids (chapter "The Fossil Record of Felids (Mammalia: Carnivora: Felidae) in Greece"), and canids (chapter "The Fossil Record of Canids (Mammalia: Carnivora: Canidae) in Greece"), so tantalizing in their scarcity and paleoenvironmental promise. The Quaternary record of *Hippopotamus* (chapter "The Fossil Record of Insular Endemic Mammals from Greece") provides a rich basis for paleoenvironmental reconstruction, but this too flags one of the biggest challenges for Quaternary mammal research in Greece: a robust and independent chronology. Greece's plentiful karst makes for excellent taphonomic conditions for the preservation of Quaternary fauna, but brings with it the complexity of dating cave deposits. This, however, is something that is worth attempting, given the richness of material highlighted in Volume 2, all hailing from a time and place so important to our understanding of the impact of large climatic shifts and sea level changes on humans and their environment.

The impact of Quaternary sea level change can be expected to be felt first, and most strongly, on islands (given the direct link between sea level change and island area and isolation, which are themselves inextricably linked with island biodiversity), and Greece's islands have acted as natural laboratories for these effects for the whole of the Quaternary. The resulting endemic faunas (chapter "The Fossil Record of Insular Endemic Mammals from Greece") are, and will continue to be, vital areas for paleoenvironmental and evolutionary research—but they can only be understood in the wider framework of mainland Faunas, as the preceding chapters of *Fossil Vertebrates of Greece* make clear (and why I so longed for just such a synthetic overview at the start of my Ph.D.).

Works like this are a way for a discipline to showcase both its treasures and the community that studies them. Our colleagues are saying, "We have incredible material, and we want you to come and work with us on it." In this context, the forthright look at the history and wider colonial influence on Vertebrate Paleontology in Greece taken by Vlachos and colleagues in these volumes provide powerful lessons for new collaborations. The accompanying unflinching but vital assessment of the current status of scholarship and research in Greece shows that there is much to be proud of, but also much that could be improved. Together *Fossil Vertebrates of Greece* becomes more than a major work of scholarship; it is a catalyst for change.

The birth and development of our discipline coincided with the birth and development of the modern Greek nation, and the influence of the Western powers that shaped and buffeted nineteenth and early twentieth century Greek politics is apparent in the flow of fossil vertebrates from Greece to collections in London, Paris, Vienna, and Munich. The international scattering of these specimens—many of them illegally, like the fossils of *Mammuthus creticus* from Cape Maléka that Dorothea Bate deliberately and knowingly smuggled out of Crete (they are now in the NHM, London)—is emblematic of the struggle Greek paleontologists have faced from the outset. Vlachos highlights the lack of both a national museum of Natural History and a national paleontological society as key hurdles to the progression of Greek paleontology, as evidenced by the lag in scientific milestones for first Greek men and then Greek women. He does not need to join the dots between this and the historical legacy of foreign research in Greece; it is obvious.

I am proud to have a research connection to Dorothea Bate, the first woman to have described a Greek vertebrate fossil species. I have benefitted not only from easy access to her Greek specimens but also the strong personal connection to Greek paleontology that Bate's story gave me. This helped engender a feeling of "belonging" to a place I had never visited, giving me the confidence and desire to pursue research there. I am so grateful for this last aspect especially, and the adventures, discoveries, and friendships it spurred. But by not troubling myself to learn about, let alone face up to, the full colonial legacy on the history of Greek paleontology, I did not fully appreciate just how generous, kind, and open my Greek colleagues have been toward collaboration, when they must surely have felt some degree of chagrin at yet another foreigner trying her luck in their backyard. My gratitude to them is now even more profound.

Science is at its best when it is collaborative, open, and international, and this is exactly what synthetic works like *Fossil Vertebrates of Greece* foster as they facilitate access to knowledge, sites, and material. But we should all be alert to the imbalance in advantage that the hand of history has dealt, and ensure we don't repeat the same mistakes. I can't help wondering, for example, how the course of Greek paleontological history might have run differently had the pioneering discoverer of *Mammuthus creticus* been someone whose surname ended in –ακη? Might she have inspired and supported a whole host of other Greek female pioneers? These connections do matter and are wonderful to trace—like the pioneering women linked through Petralona Cave—from Anna Petrochilou's first exploration of the site; through the first description of a new Greek fossil species—a horse from Petralona Cave—by a Greek woman, Evangelia Tsoukala; to Katerina Harvati's pioneering work on the Petralona hominin skull. And rather neatly, Spyridoula Pappa's work with Tsoukala on the ursids from Petralona Cave (chapter "The Fossil Record of Bears (Carnivora: Ursidae) in Greece") joins these pioneers back to Dorothea Bate: Pappa is now Curator of Fossil Mammals at the NHM, London, with Bate's fossils (including *M. creticus*) under her care.

Looking to the future, *Fossil Vertebrates of Greece* will be the benchmark for both the state of knowledge and the demographics of Vertebrate Paleontology in Greece, and for Greek vertebrate paleontologists globally. It will be the baseline from which progress can be measured, and even greater strides can be made. The Who's-Who of authors in these volumes, from emeritus professor to up-and-coming student, is testament to a time of unprecedented activity, scholarship. The outstanding effort to reflect the current diversity of Greek vertebrate paleontology researchers, especially women active in the field, in the author list regardless of seniority, sews the seed for change and the hope for a more equitable future.

It is always impossible in a foreword to do justice to a work of scholarship, especially one of such breadth at this, encompassing a fossil record that stretches from the Paleozoic to the Holocene, and that showcases the discoveries and achievements of so many dear colleagues. But I predict that, in 10 years time, there will be groundbreaking projects, research networks, and collaborations—and the research positions that stem from them—that will trace their origins back to these two volumes, and the names therein. In that sense, *Fossil Vertebrates of Greece* will speak for itself in the legacy it leaves.

Department of Life Sciences
Natural History Museum
London, UK

Victoria Herridge

Preface

Each young student of vertebrate paleontology in Greece quickly learns about two important dreams of every Greek paleontologist: the creation of a National Museum of Natural History and the necessity of a volume that summarizes the fossil record of vertebrates from Greece. My case was not different. Fueled by the dreams and ideas of my former supervisor, Prof. Evangelia Tsoukala, those dreams became my dreams as well. To my mind, these two important goals are connected: a National Museum of Natural History would lead to the publication of such a book eventually. Similarly, the existence of a book about the fossil record of Greek vertebrates would highlight the necessity of a National Natural History Museum and would assist in its creation.

The fossil record of Greek vertebrates preserves important information that helps formulating several ideas on the evolution of life in the area, especially for the last 25 million years in the case of terrestrial environments. In the case of marine environments, however, Greek fossils trace back to the Mesozoic and Paleozoic as well. Some Greek localities such as Pikermi near Athens, Samos Island in the Aegean, and Petralona Cave near Thessaloniki are among the most famous fossil localities in the world. Greek fossils form important parts of the collections and exhibitions of numerous international museums of natural history.

As soon as I was given the chance, I decided to pursue one of these two dreams, the publication of a high-quality volume that would review the fossil record of Greek vertebrates. It was February 2017 when the original plan for this project was made, and immediately the challenges of this task were clear. The need to gather most of the active paleontologists with first-hand experience and research on these fossils, organize the chapters, design numerous high-quality figures depicting the most important fossils, go through an extensive literature search, and bring all this heterogeneous information together into a book. Fortunately, my teachers, professors, colleagues, and friends answered the call and blessed me with their trust to organize this project. Almost 4 years later, this project grew into a two-volume book with a combined length of more than 1300 pages with more than 300 figures and maps, arranged in 42 chapters written by 31 different scientists.

This project is divided into two volumes: the first one covering basal vertebrates, amphibians, reptiles, afrotherians, glires, and primates, whereas the second deals with laurasiatherians, artiodactyles, perissodactyles, carnivorans, and island endemics. Each volume is organized in chapters per taxonomic group (in most cases per family), containing the most up-to-date information on the systematics, taxonomy, distribution, and evolution of each clade. Each chapter contains maps and tables with all known vertebrate occurrences from Greece. The most important and well-preserved fossils are illustrated. Our aim is to present detailed information of all valid vertebrate species with special emphasis on taxa whose type locality is in Greece. Some chapters are quite extensive, whereas others are quite brief, reflecting the alpha diversity of the given group in the fossil record of the country. Even so, this project is not complete: for different reasons, chapters on some important clades are missing. But given the circumstances during the last 4 years, we have done our utmost best to present the most complete and inclusive possible result.

The documented vertebrate assemblage has a long and complex evolutionary history. Since its appearance in the geological record, Greece stands in between three continents, Europe, Asia, and Africa, playing an important role in our understanding of the events that shaped the biota in the Eastern Mediterranean through time. As such, Greek vertebrates have multiple origins that can be inferred by known dispersals from the surrounding areas, as well as some vertebrates that represent native, endemic, forms of life. The geological and tectonic complexity of the region has further affected the evolution of Greek vertebrates leading to its unique extant fauna.

This book primarily addresses all those who are looking for a detailed account of the evolution of life in Greece: the scientific community, paleontologists and neontologists, teachers and professors, students and researchers from Greece and abroad. However, because of the numerous figures and illustrations, this book could be useful for a wider audience as well. This book summarizes what has been done in Greece by many generations of Greek and foreign paleontologists, providing a state of the art of the taxonomy and systematics of fossil vertebrates from Greece. This monumental task also has pointed out several areas of potential future research. As such, I hope that this book will prove to be a useful reference and research tool for the future generations of paleontologists working with the amazing fossils from the country. I do not pretend that this project will be the final word on the fossil record of vertebrates in Greece. To the contrary, I view this project as the starting point for similar efforts in the future, which will fill the gaps, correct the mistakes, and expand the information presented herein.

Trelew, Chubut, Argentina Evangelos Vlachos

Spellings of Toponyms and Personal Names

The reader should understand that the spelling of most of the Greek localities with fossil vertebrates varies in the literature and might vary in the chapters of this volume as well. In many cases, some localities are known from two or more completely different names. Whereas we opted for the most uniform result, in some cases the choice between these different spellings and/or names depends on personal preferences of the authors and historical usage, which are respected herein. To avoid confusion and ambiguity, the great majority of the localities in the tables herein is accompanied by the collection number in the Paleobiology Database. To some extent, the same problem exists with some Greek personal names.

Acknowledgments

First and foremost I would like to express my sincere thanks to all contributing authors of this two-volume project. They are beloved teachers and professors, esteemed collaborators and colleagues, respected scientists and dear friends who entrusted me to edit and coordinate this project. I deeply thank them for their collaboration and for giving their best scientific efforts to meet the goals of this huge effort. Among them, I feel obligated to specially mention E. Tsoukala, G.D. Koufos, D.S. Kostopoulos, A. Athanassiou, and of course K. Vasileiadou for their crucial help and support in matters exceeding their respective chapters, including helping to organize the list of chapters and their main authors. The completion of this project would not have been possible without the help, support, and understanding of Springer-Nature and especially the Senior Publishing Editors Zachary Romano, Aaron Schiller, Dr. Johanna Schwarz, and Claudia Mannsperger. I am indebted to many wonderful scientists that reviewed the chapters, namely (in alphabetical order): S. Bartolini, G. Baryshnikov, G. Carnevale, M. Cherin, A. Collareta, M. Coombs, R. Croitor, J. de Vos, E. Delson, D. Geraads, L. Hristova, Ch. Kevrekidis, J. Klietmann, T. Lehmann, F. Lihoreau, V. Popov, R. Rabinovich, M. Roksandic, G. Roessner, C. Romano, M. Salesa, R. Sardella, G.L. Stringer, W. Schwarzhans, E. Syromyatnikova, M. Uhen, L. van den Hoek Ostende, J. van der Made, A. Villa, A. Xafis, D. Youlatos, and three anonymous reviewers; most of the chapters have been additionally reviewed by other contributors of this project. Juliana Sterli accompanied me along this journey in any way possible.

Contents

Abbreviations

Symbols

★ The presence of a star next to a taxon name denotes that the type of this taxon (e.g., type specimen in the case of species-group name and type species in the case of a genus-group name) is from Greece. These type localities are marked also with boldface in the tables of the various chapters of this book.

⊙ The presence of the two concentric circles next to a taxon name denotes that it is an extant species. Extant species are also followed by their vernacular common name.

Institutional Abbreviations

AMNH	American Museum of Natural History, New York
AMPG	Department of Palaeontology and Historical Geology, University of Athens, Greece
BMNH	British Museum of Natural History (= Natural History Museum), London
BSPM	Bayrische Staatssamlung für Paläontologie und historische Geologie, Munich, Germany
CM	Carnegie Museum of Natural History, Pittsburgh, Pennsylvania
FSL	Univ. Claude-Bernard, Lyon, France
GPIH	Geologisch-Paläontologisches Institut der Universität Hamburg (Centrum für Naturkunde, Hamburg)
GPMM	Geologisch-Paläontologisches Museum (Geomuseum) der Westfälischen Wilhelms-Universität Münster
HLMD	Hessisches Landesmuseum, Darmstadt, Germany

IGF	Museum of Natural History of the University of Florence, Italy
IPS	Institut de Paleontologia Miquel Crusafont de Sabadell, Barcelona, Spain
IPUW	Institut für Paläontologie der Universität Wien, Austria
ISERB	Institute of Speleology "Emil Racoviţă," Bucharest, Romania
LAY	Lucas Mallada Institute, Spain
LGPUT	Laboratory of Geology and Palaeontology, School of Geology, Aristotle University of Thessaloniki, Greece
MAFI	Magyar Állami Földtani Intézet, Budapest
MCGL	Musée Cantonal de Géologie Lausanne
MCM	Montevarchi Paleontological Museum, Tuscany, Italy
MGPP	Museo di Geologia e Paleontologia dell'Università di Padova
MGPT-PU	Museum of Geology and Paleontology of Torino University, Italy
MHNO	Musée des Sciences Naturelles, Orleans
MNH.MG	Museum of Natural History of Milia, Grevena, Greece
MNHB	Museum der Naturkunde für Humboldt Universität zu Berlin
MNHN	Museum National d'Histoire Naturelle, Paris, France
MNM	Museum für Naturkunde, Magdeburg, Germany
MPUR	Museum of Palaeontology, Department of Earth Sciences of Sapienza University, Rome, Italy
MSZ	Strasbourg University Zoological Museum
NHMA	Natural History Museum of Aegean, Mytilinii, Samos
NHMB	Naturhistorisches Museum, Basel
NHMC	Natural History Museum of Crete, Herakleion, Crete, Greece
NHMUK or NHML	Natural History Museum, London, UK
NHMW	Naturhistorisches Museum Wien, Austria
NKUA	National and Kapodistrian University of Athens, Greece
NMBE	Naturhistorisches Museum Bern
NMP	Natural History Museum of Nostiko, Kastoria, Greece
NRM	Naturhistoriska Riksmuseet Stockholm
PIN	Palaeontological Institute of the Russian Academy of Sciences
PMMS	Paleontological Museum of Mytilinii, Samos
RIM	collection of B. von Freyberg, Erlangen, Germany
SAM	Natural History Museum of Stuttgart, Germany
SMF	Senckenbergisches Naturhistorisches Museum-Frankfurt, Germany
SMNK	Staatliches Museum für Naturkunde Karlsruhe
SMNS	Staatlisches Museum für Naturkunde, Stuttgart, Germany
STIPB	Steinmann-Institut für Geologie, Mineralogie und Paläontologie (Goldfuss Museum), Rheinische Friedrich-Wilhelms-Universität Bonn

UCMP	University of California Museum of Paleontology (Vertebrate Collection)
UGR	Ungarische Geologische Reichsanstalt, Budapest, Hungary
UMSL	University of Montpellier, France
UU	Department of Earth Sciences of the University of Utrecht, The Netherlands
YPM	Yale Peabody Museum of Natural History

Other Abbreviations

GPTS	Geomagnetic Polarity Time Scale
Ma	million years ago
MN	Mammal Neogene Zones
MNQ	Quaternary Mammal Zones

Introduction to the Study of Greek Fossil Vertebrates

Evangelos Vlachos

1 Vertebrates

Vertebrates (Chordata: Vertebrata) are animals closely related to other animals with a notochord, namely the lancelets (Cephalochordata) and ascidians (Urochordata). Although traditionally considered as subphyla of the phylum Chordata, a recent review by Irie et al. (2018) proposes the recognition of Vertebrata as a separate phylum, because of new evidence from comparative genomic, gene expression, and morphological studies. Chordates form a natural or monophyletic group defined by several shared derived characters (or synapomorphies) which separate them from other deuterostomians: the presence of an endostyle, the presence of a dorsal and hollow nerve cord, the presence of a notochord, myomeres, and the presence of a postanal tail (de Iuliis and Pulerà 2019). Vertebrates are, of course, animals that have a backbone; however, this simplistic definition does not hold against modern definitions of this clade. During the last two decades, extensive discussions on the definition of vertebrates took place, especially considering the relationships between vertebrates without jaws (e.g., lampreys and hagfishes) and vertebrates with jaws (i.e., gnathostomatans). Lampreys possess only rudimentary vertebrae, termed arcualia, considered as precursors of true vertebrae (de Iuliis and Pulerà 2019). Solving the relationships between these three groups was crucial to define Vertebrata: the currently accepted scheme groups lampreys and hagfishes (= cyclostomatans) which form the sister group of jawed vertebrates (= gnathostomatans) (de Iuliis and Pulerà 2019 and references therein). Vertebrates are defined by several synapomorphies: two rounds of genome-wide duplication, full repertoire of neurogenic placodes, a neural crest, the formation of tripartite brain, at least one set of semicircular canals in the ear, and the presence of arcualia (de Iuliis and Pulerà 2019 and

E. Vlachos (✉)
Museo Paleontológico Egidio Feruglio, CONICET, Trelew, Chubut, Argentina
e-mail: evlacho@mef.org.ar

E. Vlachos (ed.), *Fossil Vertebrates of Greece Vol. 2*,
https://doi.org/10.1007/978-3-030-68442-6_1

references therein)—the last character is the modern expression of the classical phrase “vertebrates are animals with backbone.”

As vertebrates are composed of several and many hard parts, they have an important potential for fossilization in terrestrial and marine environments. There are more than 50,000 collections with more than 245,000 occurrences of vertebrates around the world, covering the entire Phanerozoic (Paleobiology Database 2020). Greece holds more than 3000 occurrences from more than 500 collections of fossils. The actual number of fossil specimens is impossible to calculate precisely at the moment. Considering each individual as a single specimen, even my most conservative and preliminary estimations suggest that 50,000 fossil individuals is a realistic starting point—the actual number of fossil specimens is higher as many of these individuals are represented by several specimens of their skeleton. Considering that Solounias (1981) calculates that only from Samos more than 30,000 specimens are dispersed around the world, my estimation herein is certainly a quite conservative one.

Data Statement

The data and information used to perform the basic analyses that support the statements in the following sections are based on the list of references of various chapters in this book and the Paleobiology Database.

2 The Fossil Treasures of Greece

The history of vertebrate fossils in Greece starts in the 1830s with Pikermi (Attica, near Athens), arguably the most famous and one of the most important fossiliferous localities of the country. The discovery of the locality is credited to the Scottish historian G. Finley who spotted bone accumulations in the Pikermi ravine, during prospection of the area in 1836. Together with the help of the German ornithologist A. von Lindermayer, Finley continued exploring the area and made the first excavations there (Symeonidis and Theodorou 1989; Roussiakis et al. 2019). The next part of the story is a classic one: the diamonds of the Bavarian soldier! The tale goes that in 1838 a Bavarian soldier, who presumably participated in the excavations of Finley and Lindermayer, took to Munich some fossils from Pikermi (Woodward 1901). Some of them contained calcite crystals in their cavities, which he thought were diamonds (Gaudry 1862–1867). The zoologist A. Wagner immediately refuted the treasure find of the soldier, only to suggest that the real “treasure” was something else: amazing fossils of extinct animals, including the maxilla of the cercopithecid *Mesopithecus* (Wagner 1839).

The publication of Wagner (1839) should be considered as the official beginning of vertebrate paleontological research in Greece. There are some earlier mentions of fossil discoveries from Greece by Finley, Jameson, and Goldfuβ (for further information, see Athanassiou this volume and Giaourtsakis this volume), but these works contained simple mentions without any actual study of the fossils.

Almost immediately, several excavations took place in Pikermi, enriching collections in the country and abroad: J. R. Roth in 1852–1853, H. Mitzopoulos in 1853, A. Chairetis in 1854, A. Gaudry in the winter of 1855–1860 and summer of 1860, W. Dames in 1880s and 1890s, M. Neumayer and L. von Tausch in 1885, the Prince of Orleans in 1888, M. from Dijon in 1895, A. S. Woodward and Th. Skouphos in the turn of the century, and O. Abel in 1912 (Roussiakis et al. 2019).

Around the same time, another famous paleontological area from Greece is discovered: the Adrianos ravine, in Mytilinii basin of Samos Island. The discovery of these localities is credited to C. J. Forsyth Major who collected specimens in 1885–1887 and also returned later in 1887 and 1889 to search for more fossils (Forsyth Major 1894; Koufos 2009). However, fossil mammals from Samos were already known by accidental findings given to Italian travelers, now housed in Padova (Koufos 2009 and references therein). In the case of Samos, it is likely that the fossils were also known much earlier, hidden inside the ancient Greek myths of monsters, gods, and amazons (Mayor 2000; Koufos 2009; and references therein). Just like in Pikermi, many more excavations took place afterwards and by numerous people, also enriching collections in European and North American museums: T. Stutzel in the end of the nineteenth century, B. Sturtz in 1889–1900, A. Hentschel in 1901–1902, E. Fraas in 1901, K. Acker, Th. Skouphos in 1903, and B. Brown in 1921–1924 (see Koufos 2009 and references therein).

At the beginning of the nineteenth century, the trio of the classical vertebrate paleontological localities from Greece is completed. During the World War I (and particularly in 1915–1916), the French naturalist C. Arambourg was serving in the 156th French Infantry Division and discovered fossils in various localities of the Axios valley and Thessaloniki area, joined under the term “Salonique”: Ravin X, Ravin des Zouaves, Ravin R, Ravin C, Ravin de Vatilük (= Vathylakkos), Ravin de Konikovo (= Dytiko), and Falaise de Karabouroun (= Megalo Emvolon). Additional collections were made later in the area by Dr. Puyhaubert. These fossils were packed and sent to the MNHN of Paris (via Algeria) and formed the subject of the monumental monograph of Arambourg and Piveteau (1929).

These three, in the broad sense, localities (Pikermi, Samos, and “Salonique”) form the basis not only of the mammalian paleontology in Greece but also that of other vertebrate groups: the legendary Owen (1857) named the first snake from Greece, based on some vertebrae from Megalo Emvolon; Gaudry (1862) named the first tortoise from Greece, based on some shells from Pikermi; Forsyth Major (1891) named the first avian species from Greece and Samos; Woodward (1901) mentioned the first fossils of giant tortoises, from Pikermi.

The leading Greek figure of that time is, without any doubt, Th. Skouphos, who not only excavated in both Pikermi and Samos, but also initiated the investigations in another important paleontological area: the Megalopolis basin in Peloponnesus peninsula (see Athanassiou n.d., volume 1). Skouphos also started the first university position on paleontology in 1906, in the Faculty of Natural Sciences of the University of Athens. At the same time, he founded the Laboratory and Museum of Geology and Paleontology, which are the “ancestors” of the present-day paleontological institutions of the National and Kapodistrian University of Athens.

During this *Early Stage* of vertebrate paleontological research in Greece—and amidst the destruction of the World Wars—paleontological research is mostly focused on Pikermi, Samos, and "Salonique," although fossils from other localities from Greece begin to appear. Perhaps the most notable cases include the early investigations of the endemic vertebrates from Crete (e.g., Simonelli 1894, 1907; Bate 1907, 1912) and the early Miocene fossils from Chios (e.g., Paraskevaidis 1940). The great majority of the published research during this stage (which contains roughly 90 published papers, books, and monographs; Fig. 1) is made by foreign scientists and deals with Greek fossils housed in foreign institutions. The most important collections are located in BSPM (Germany), MNHN (France), NHMUK (UK), and NHMW/IPUW (Austria), including more than 75 holotypes of species named from Greek type localities.

Following the end of the World War II, the *Intermediate Stage* of vertebrate paleontological research in Greece commences. This period is marked by the specific focus of enriching or creating fossil collections in national institutions, mostly Greek universities and their associated museums. During the *Intermediate Stage*, the number of trained Greek paleontologists with permanent positions gradually increases and expands geographically. As a result, Greek paleontologists gradually emerge as important and leading authorities on several topics, with an important component of foreign collaboration as well. Paleontological research continued in

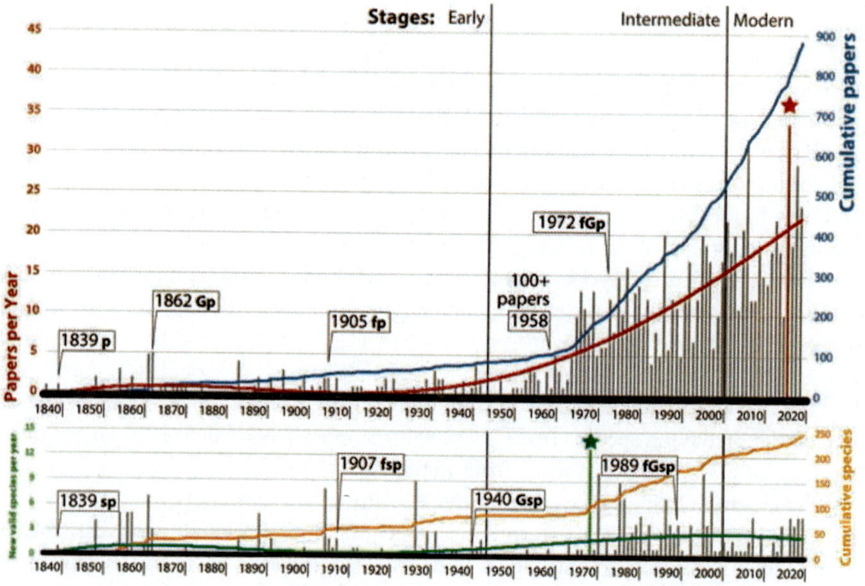

Fig. 1 Counts of published papers (red line) and new species (green line) based on Greek fossils per year vs. their cumulative growth of knowledge on Greek fossils (blue and orange lines, respectively). The diagram is divided into the three stages that are defined in the text. Some milestones are indicated, whereas stars indicate the most prolific years. *f* female, *G* Greek author, *p* published paper, *sp* new species (considered as valid at present)

Athens after the times of Skouphos, with P. Psarianos, M. Mitzopoulos, G. Marinos, and N. Symeonidis. During this period, J. Melentis, who started working in Athens, moves to Thessaloniki and the School of Geology (former Faculty of Natural Sciences) and impulses paleontological research there. The Museum of Geology and Paleontology of the University of Thessaloniki opened in 1940 and operates officially since the 1990. Melentis and Symeonidis are, without any doubt, two leading figures of the first part of the *Intermediate Stage*, forming two poles of paleontological research in Thessaloniki and Athens, respectively. They carried out numerous excavations in areas like Samos, Attica, Megalopolis, Axios valley, and Crete that have been accompanied in many cases by the creation of museums and exhibitions, and numerous key publications that remain relevant to date. Also, both professors invested in training new paleontologists, creating a solid basis for paleontological research in the country. As a result, during the last third of the twentieth century, additional vertebrate paleontologists joined, both in Athens (M. Dermitzakis, G. Theodorou, C. Doukas, P. Pavlakis) and Thessaloniki (G. Koufos, E. Tsoukala) with permanent positions of teaching and research. They continued this legacy, with excavations, development of museums and exhibitions, publications and studies, formation of new students, and national and international collaboration, causing a sharp increase and subsequent accelerated growth in the paleontological knowledge of the country after the 1970s (Fig. 1).

During the *Intermediate Stage*, the classic, now, localities of Pikermi, Samos, and "Salonique" are re-visited, with new excavations, revisions of the previously published materials, and discovery of new materials and applications of new techniques and methodologies (see the relevant chapters of this book). New paleontological sites are discovered, focusing the interest in adjacent areas (e.g., Western Macedonia, Chalkidiki peninsula, Mygdonia basin, Peloponnese, Thessaly, Euboea Island, other Aegean Islands). Numerous new exhibitions emerge, holding and preserving important new collections of fossils (see the next section). The core of permanent paleontologists and researchers train new doctoral students who will lead paleontological research during the transition to the *Modern Stage*; there is no need to mention their names, as they are many of the authors of the various chapters in this book.

During the second half of the twentieth century, the scientific production has soared compared to the Early Stage, with more than 411 additional publications (Fig. 1). This accumulative information led to amazing new discoveries and the descriptions of many new taxa of vertebrates from Greece. These taxa and the localities are extensively covered in the following chapters, and they will not be repeated herein.

It is difficult to define a boundary between the *Intermediate* and the *Modern Stage* of paleontological research, which I consider to be defined by the presence of permanent working paleontological teams in Greece and leading studies on vertebrate paleontology. Also, this varies per taxonomic group and locality. For most mammalian families, this transition was achieved already in the 1980s, whereas in reptiles, for example, this is a recent phenomenon and occurred in the twenty-first century. The Geological Department of the Athens University, containing the oldest paleontological research position in the country, has a head-start of almost half a century over the School of Geology of the University of Thessaloniki; the Department of Geology of the University of Patras only recently has gained a member specializing on vertebrate paleontology.

The *Modern Stage* of paleontological research in Greece—roughly defined herein from the beginning of the twenty-first century—finds an unprecedented group of highly qualified Greek vertebrate paleontologists, who lead the scene in the national and international levels. There are at least 20 active researchers with a Ph.D. and approximately ten Ph.D. candidates; more than a third of them are women, which have been significantly underrepresented in the previous stages. Adding to the names mentioned before, more Greek paleontologists hold now permanent positions in Greek universities and abroad (e.g., D. S. Kostopoulos, S. R. Roussiakis, N. Solounias, K. Harvati, G. Iliopoulos, G. Lyras). They have formed teams with researchers, permanent staff members, postdocs, and PhD/Master's candidates. Additionally, permanent or long-term research positions are occupied by vertebrate paleontologists in Museums (e.g., K. Vasileiadou) and other national governmental institutions (e.g., A. Athanassiou). This pool of scientists is actively discovering, preserving, and describing the Greek fossil record of vertebrate animals. Almost every year, new fossil localities are discovered. During the last decades, even more museums and exhibitions have been developed (see below), bringing paleontology closer to the public. And only in the last two decades, more than 381 new scientific works have been published (Fig. 1); this is nearly as much as the published record during the *Intermediate Stage*, especially if we consider that additional works published in 2020 are not counted herein. In several years during the *Modern Stage*, more than 20 new papers per year have been consistently published by Greek vertebrate paleontologists, marking unprecedented scientific production records. These last two decades also saw an increased national collaboration, both intra- and inter-institutionally, producing comprehensive and integrated works of global interest and outreach. This modern community of Greek paleontologists is much more specialized and diverse than before, with many scientists specializing in single family-group taxa and/or to new analytical techniques.

As a result of the growth during the *Intermediate* and *Modern Stages*, the fossil collections in Greek institutions have grown, now hosting more than 100 different holotypes of vertebrate species. LGPUT (Thessaloniki; 52+ holotypes) and AMPG (Athens; 41+ holotypes) have been transformed as the two most important institutions in terms of the number of type specimens preserved.

3 Milestones

The purpose of this short section is to mention and draw attention to some important milestones related to the published record of the fossil vertebrates from Greece (Fig. 1). As it has been mentioned above, the first publication to include studies on vertebrate fossils from Greece is that of Wagner (1839), presenting fossils from

Pikermi. In the same publication, the first ever genus and species is named from Greek fossils: *Mesopithecus pentelicus* Wagner, 1839, from the late Miocene of Pikermi. Thus, the famous type specimen (BS and valid PM-PIK-AS II.11), should be considered—objectively, based on the published record—as the first vertebrate fossil to be formally described and published from Greece.

The first publication written (at least partly) by Greek authors and dealing with some vertebrate fossils is the one of Mitzopoulos et al. (1862), dealing with a general report of the specimens in the Physiographic Museum of Athens. Another alternative is the work of Skouphos (1905), presenting important elephant fossils from Megalopolis. The first new species erected by a Greek, though, came much later by Paraskevaidis (1940): *Georgiomeryx georgalasi*. The first publication dealing with vertebrate fossils from Greece and written by a female author is by the British pioneer Dorothy Bate (Bate 1905), who also named few years later the first species from Greece named by a female author: *Elephas* (now *Mammuthus*) *creticus* from Crete (Bate 1907). The first paleontological publication by a Greek female author came 60 years later, when Helene Sakellariou-Mane published some gomphothere remains from Axios valley, near Thessaloniki (Sakellariou-Mane 1972). However, a special mention must be made to Anna Petrochilou, one of the founders of speleological research and exploration in Greece. Although she never studied or published fossilized remains in detail, she mentioned fossil vertebrates found in many caves that she explored, presented in a series of scientific works in the *Bulletin of the Hellenic Speleological Society* in the 1960s and 1970s. Some of these caves described by her would become some of the most prolific fossil vertebrate localities in the country. For example, she remarked that the study of the paleontological remains from Petralona Cave is mandatory, immediately after the first explorations in the cave (Petrochilou 1964; Petrochilos 1965). The first new species named from Greece by a Greek female author came even later, when E. Tsoukala (1989) named *Equus petraloniensis* from Petralona Cave. During the last few decades, the participation of female paleontologists in Greece has been rising, as is clearly evident in the list of authors in this volume.

Based on the cited publications herein, the top ten of authors—in terms of total number of contributions—have participated in more than half of the available literature dealing with Greek vertebrate fossils. This list includes (in descending order based on the number of their contributions): G. D. Koufos, D. S. Kostopoulos, E. Tsoukala, N. Symeonidis, A. Athanassiou, J. K. Melentis, L. de Bonis, G. Theodorou, S. Roussiakis, and G. Bouvrain.

In these nearly 900 publications published so far dealing primarily with Greek vertebrate fossils, paleontologists have identified and described more than 1100 different species of vertebrates preserved as fossils in the sediments and rocks of Greece, collected and excavated in more than 500 different localities in the country (Fig. 2). One out of four of these species (at least 254 and counting only those considered valid herein) were new to science and have been named based on unique type fossil specimens from Greece. The extreme majority of these new species (at least 219 spp.) were mammals; mammalian species represent half of the identified species (at least 538 spp.) in the country.

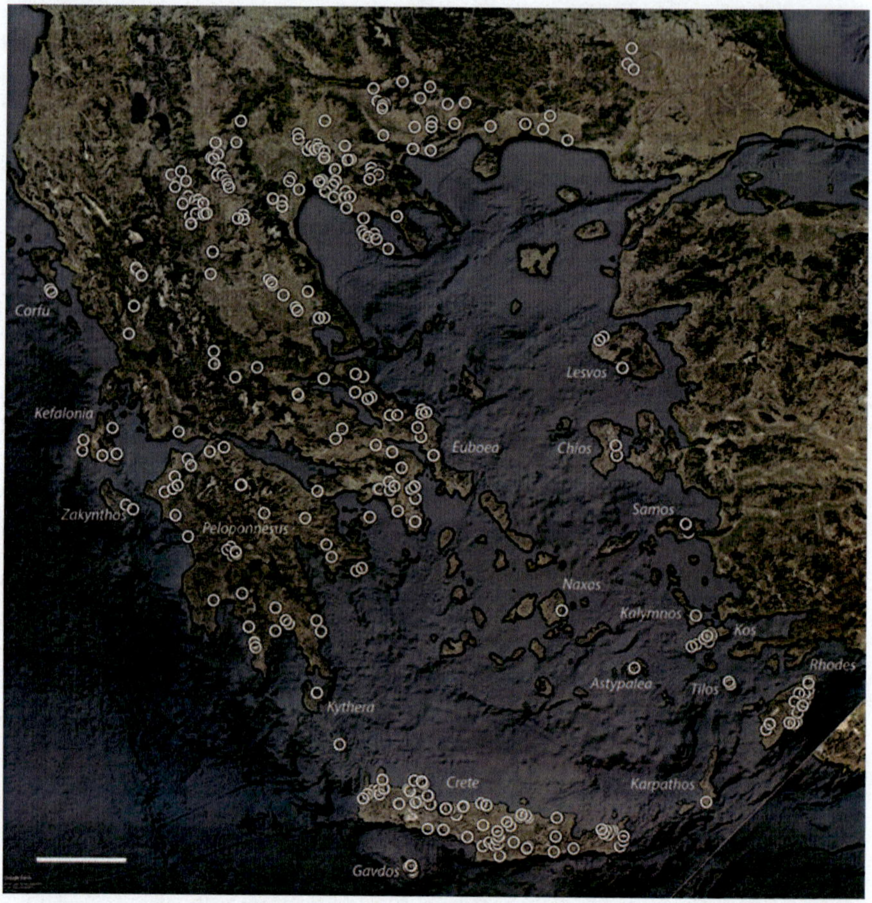

Fig. 2 Map of Greece showing the geographic distribution of the most important localities with fossils of vertebrates. Data from various chapters of this book and the Paleobiology Database. Image exported from Google Earth Pro © 2019, map data from the US Department of State Geographer, SIO, NOAA, U.S. Navy, NGA, GEBCO, image from Landsat/Copernicus. Scale bar equals 80 km, North faces upward

4 Future Perspectives

The current status is most certainly positive and allows projecting the future of paleontology in the country with optimism. But we should not forget to evaluate the context in which the modern paleontological community has evolved; if anything, this has been a tough struggle against extinction. We need to keep in mind, and compared to other countries with similarly important vertebrate fossil record, that paleontological research in Greece lacks many of the advantages that are frequently found abroad. For example, there is no specialized National Paleontological Society, which could define and preserve our rights and trigger communication, divulgation, and growth of the community. This Society could also assist in the creation of laws specific to fossils and excavations that are placed outside the archaeological context of existing legislation; these laws would cover and protect the extreme majority of the fossils presented herein. The lack of a National Paleontological Society is also partially responsible for the lack of a specialized scientific journal of paleontology with a concrete regional focus. In many cases, important works and discoveries are rejected from top international journals because of their regional interest, whereas at the same time, this publishing niche is empty in Greece.

For a long time, and especially during the *Intermediate Stage*, this niche was partly filled with the "*Annales Géologiques des Pays Hélléniques*" (1942–2004), a journal founded by M. Mitzopoulos and succeeded by the "*Hellenic Journal of Geosciences*"—the last volume of this journal was published in 2010. To place it in context, and although *Annales* had a broader Earth Sciences scope, more than 90 papers related to Greek fossil vertebrates have been published there, with a publishing frequency that diminished through time, in stark contrast to the overall publishing record that experienced an accelerated growth. It seems to me that a modern, open access, regional paleontological journal is essential for the continuous scientific growth of the Greek paleontological community. Finally, a major obstacle on the development of paleontological research in Greece is the lack of a clear growth plan and associated funding. Only very few of the completed PhD studies have been supported by adequate scholarships. Very few of the paleontological excavations and field collections received funding from local, national, or international entities. Permanent research positions are almost exclusively linked to teaching positions in universities. Additional research positions that will not be linked with teaching need to be created, otherwise all this recent growth in human resources remains unused. Proper and professional preparation laboratories are virtually nonexistent, and the preparation of fossils is mainly done through volunteering work by the researchers themselves and students, without the assistance of professional preparators. Collections are seriously understaffed and poorly financed, and researchers devote a lot of their time for curatorial work; the preservation of important type specimens and other unique fossils is left to the responsibility of the scientists alone. At least, many regional museums and exhibitions have been developed (Table 1) and, in several cases, provide high-quality exhibits and fulfill an important educational and touristic role. Even so, in many cases, they lack a clear plan of viable development which would permit the protection of the fossils and the sustainability of these institutions. But in numerous cases, these facilities are understaffed as well as their development has been done with poor funding and limited long-term planning. All these famous fossiliferous localities and many new ones that have been discovered along the way remain unprotected as well. At the same time, Greece still lacks a National Museum of Natural History, a prerequisite for paleontological growth in many countries of the world. And last but not least, during the *Modern Stage* (so far the most fruitful period from a scientific point of view), research has been carried out under a severe financial crisis and brain drain to other countries.

It would not be a stretch to claim that the current status of paleontological research in the country is based on two pillars: the quality and importance of the

Table 1 List of major museums, institutions, and collections containing fossils of vertebrates from Greece

Name	City	Acronym
Museum of Paleontology and Geology of the National and Kapodistrian University of Athens	Athens	AMPG
Palaeontological Exhibition of Pikermi	Pikermi	–
Palaeontological Museum of Megalopolis	Issoma Karyon	–
Palaeontological Museum of Rethymnon, Crete	Rethymnon, Crete	–
Vrisa Natural History Museum	Vrisa, Lesvos	AMPG-VM
Dwarf Elephant's Exhibition	Tilos	
Museum of Fossil Mammals of Kerasia	Kerasia, Euboea	AMPG-K
Museum of Geology – Palaeontology – Palaeonthropology, School of Geology, Aristotle University of Thessaloniki	Thessaloniki	LGPUT
Aristotle Museum of Natural History	Thessaloniki	ΑΜΦΙΘ
Palaeontological Exhibition of Milia, Grevena	Milia, Grevena	LGPUT-MIL
Palaeontological Exhibition of Kalamoto	Kalamoto, Langadas	LGPUT-KAL
Museum of Loutra Almopias Cave	Aridea	LGPUT-LAC
Palaeontological Exhibition of Agios Georgios Hill	Kilkis	LGPUT-SGK
Palaeontological Exhibition of Thermopigi	Thermopigi	LGPUT-SIT
Historical and Palaeontological Collection of Siatista and Kaloneri Elephant	Siatista	LGPUT-SIA LGPUT-AKL
Palaeontological Exhibition of Neokaissareia	Neokaissareia, Pieria	LGPUT-NKP
Aegean Museum of the Natural History	Samos	PMMS
Natural History Museum of Crete	Herakleion, Crete	NHMC
Goulandris Natural History Museum	Athens	GNHM
Natural History Museum of the Lesvos Petrified Forest	Sigri, Lesvos	–
Petralona Museum	Petralona, Chalkidiki	–
Palaeontological and Historical Museum of Ptolemaida	Ptolemaida	PHP
Natural History Museum of Volos	Volos	–
Perdikkas' Elephant Exhibition	Perdikkas, Eordaia	–
Museum of Palaeontology and Palaeobotany of Nostimo	Nostimo, Kastoria	NMP
"Anna Petrochilou Center," Perama Cave	Perama, Ioannina	–
Historical – Folklore and Natural History Museum of Kozani	Kozani	–

Greek fossils, and the selfless work of all the professors and researchers during the *Intermediate Stage*. Nearly two centuries after the discoveries in Pikermi, many more "diamonds" have been discovered in the country, treated beyond doubt as scientific "treasures" by the Greek and foreign paleontologists who discovered, preserved, and studied them. But if we are to build upon this momentum and treat this amazing paleontological heritage accordingly, serious investment and changes need to be made.

Acknowledgments I first and foremost thank E. Tsoukala for the continuous and selfless guidance and inspiration, and for comments that improved this short account. I also deeply thank G.D. Koufos, G.E. Theodorou, D.S. Kostopoulos, K. Vasileiadou, A. Athanassiou, and S.R. Roussiakis for all the help in completing this project and this manuscript. Finally, I deeply thank all the authors of this book who provided excellent and detailed accounts, allowing the preparation of this short summary herein.

References

Arambourg C, Piveteau J (1929) Les vertébrés du Pontien de Salonique. Ann Paleontol 18:57–139

Athanassiou A (n.d.) The fossil record of non-endemic elephants and mammoths (Mammalia: Proboscidea: Elephantidae) in Greece. In: Vlachos E (ed) The fossil vertebrates of Greece. Vol. 1 – basal vertebrates, amphibians, reptiles, afrotherians, glires, and primates. Springer – Nature Publishing Group, Cham. (Volume 1)

Athanassiou A (this volume) The fossil record of non-endemic fossil deer (Mammalia: Artiodactyla: Cervidae) in Greece. In: Vlachos E (ed) The fossil vertebrates of Greece Vol. 2 – laurasiatherians, artiodactyles, perissodactyles, carnivorans, and island endemics. Springer – Nature Publishing Group, Cham

Bate DMA (1905) Four and a half months in Crete in search of Pleistocene mammalian remains. Geol Mag 2:193–202

Bate DMA (1907) On elephant remains from Crete, with description of *Elephas creticus* sp. n. Proc Zool Soc London 1907:238–250

Bate DMA (1912) On a New Species of Mouse and other Rodent Remains from Crete. Geol Mag 9:4–6

De Iuliis G, Pulerà D (2019) The dissection of vertebrates. Academic Press, Cambridge, MA

Forsyth Major CI (1891) Considérations nouvelles sur la faune des vertébrés du Miocène supérieur dans l'île de Samos. C R Hebd Séances Acad Sci 113:608–610

Forsyth Major CI (1894) Le gisement ossifère de Mytilini et catalogue d'ossemens fossiles recueillis à Mitylini, île de Samos, et déposés au Collège Galliard, à Lausanne. G Bridel et Cie (éditeurs), Lausanne

Gaudry A (1862) Résultats des fouilles exécutées en Gréce sous les auspices de l'Académie. C R Acad Sci Paris 52:297–300

Gaudry A (1862–1867) Animaux fossiles et géologie de l'Attique. Savy, Paris

Giaourtsakis IX (this volume) The fossil record of rhinocerotids (Mammalia: Perissodactyla: Rhinocerotidae) in Greece. In: Vlachos E (ed) The fossil vertebrates of Greece vol. 2 – laurasiatherians, artiodactyles, perissodactyles, carnivorans, and island endemics. Springer – Nature Publishing Group, Cham

Irie N, Satoh N, Kuratani S (2018) The phylum Vertebrata: a case for zoological recognition. Zool Lett 4:32. https://doi.org/10.1186/s40851-018-0114-y

Koufos GD (2009) The Late Miocene mammal faunas of the Mytilinii Basin, Samos Island, Greece: new collection. History of the Samos fossil mammals. Beit Paläontol 31:1–12

Mayor A (2000) The first fossil hunters: paleontology in Greek and Roman times. Princeton University Press, Princeton, NJ

Mitzopoulos H, Orphanides TG, von Heldreich T (1862) [General report on the accessions in the collections of the physiographic museum in Athens during the academic year 1860–61]. Athens (in Greek)

Owen R (1857) On the fossil vertebræ of a Serpent (*Laophis crotaloïdes*, Ow.) discovered by Capt. Spratt, R.N., in a tertiary formation at Salonica. Quart J Geol Soc Lond 13:196–199

Paleobiology Database (2020) Navigator. Accessed 7 Jan 2000

Paraskevaidis I (1940) Eine obermiocäne Fauna von Chios. N Jb Min Geol Paläont 83:363–442

Petrochilos J (1965) Découverte de restes de Mammifères du Quaternaire moyen dans la région de Petralona en Chalkidiki. Troisième Congr Int Spéléol Wien

Petrochilou A (1964) Spilaio Kokkinon Petron Petralonon Chalkidikis Arith. 1044. Bull Hellenic Speleol Soc 7:160–166

Roussiakis S, Filis P, Sklavounou P, Giaourtsakis I, Kargopoulos N, Theodorou G (2019) Pikermi: a classical European fossil mammal geotope in the spotlight. Eur Geol 48:28–32

Sakellariou-Mane H (1972) Presence d'*Anancus arvernensis* dans les couches supérieures du Pliocène de la Valée du Vardar. Folia Biochim Biol Graeca 9:31–36

Simonelli V (1894) Appunti sopra i terreni neogenici e quaternari dell'Isola di Candia. Atti Accad Naz Lincei Cl Sci Fis 2:265–268

Skouphos TG (1905) Über die palæontologischen Ausgrabungen in Griechenland in Beziehung auf das Vorhandensein des Menschen. Comp Rendus Congr Int Archéol Athènes 1905:231–236

Simonelli V (1907) Mammiferi quaternari dell'isola di Candia I. Memorie, Accademia delle Scienze. Ist Bologna, Cl Sc Fis 4:455–471

Solounias N (1981) The Turolian fauna from the island of Samos, Greece. Contr Vertebr Evol 6:1–232

Symeonidis N, Theodorou G (1989) History of fossil life in Pikermi, Attica. Attiko Topio kai Perivallon, Athina

Tsoukala ES (1989) Contribution to the study of the Pleistocene fauna of large mammals (Carnivora, Perissodactyla, Artiodactyla) from Petralona cave Chalkidiki (N. Greece). Ph.D. thesis. Aristotle University of Thessaloniki, Thessaloniki

Wagner A (1839) Fossile Ueberreste von einem Affenschädel und andern Säugethierreste aus Griechenland. Gel Anz Bayer Akad Wiss 38:301–312

Woodward AS (1901) On the bone-beds of Pikermi, Attica and on similar deposits in Northern Euboea. Geol Mag 4:481–486

The Sedimentary Basins from the Miocene to the Present in Greece: Examples for the Most Studied Basins from North Greece

Angelos G. Maravelis, Nicolina Bourli, Evangelos Vlachos, and Avraam Zelilidis

1 Introduction

There are many post-Oligocene basins in Greece, where both terrestrial, lacustrine-lagoonal, or marine deposits have been accumulated (Fig. 1). These correspond to the main areas of targeted field research to find fossil vertebrates, although in some exceptional cases, such fossils could be found in older rocks as well. For the purpose of this book, which concerns the fossil vertebrates from Greece ranging mainly from the Miocene to present time, we present an introduction on three major basins, situated in central and north Greece: Mesohellenic, Axios-Thermaikos, and Strymonikos basins, and their associated minor ones. These are some of the basins where many vertebrate fossils have been discovered and include also more detailed studies on the stratigraphy and paleoenvironment (Fig. 1).

These basins are situated in central and north Greece and have been developed on the Internal Hellenic Units (Fig. 2). The Mesohellenic basin is situated in the contact between External and Internal Hellenides, the Axios-Thermaikos basin on Axios-Vardar unit, whereas the Strymonikos basin on the contact between Serbomacedonian with Rhodope Massif.

It seems that these three major basins were active from late Eocene to the present, during which clastic deposits were accumulated. However, all of them were

A. G. Maravelis
Department of Geology, Aristotle University of Thessaloniki, Thessaloniki, Greece
e-mail: angmar@geo.auth.gr

N. Bourli · A. Zelilidis (✉)
Department of Geology, University of Patras, Patra, Greece
e-mail: n_bourli@upnet.gr; a.zelilidis@upatras.gr

E. Vlachos
Museo Paleontológico Egidio Feruglio, CONICET, Trelew, Chubut, Argentina
e-mail: evlacho@mef.org.ar

E. Vlachos (ed.), *Fossil Vertebrates of Greece Vol. 2*,
https://doi.org/10.1007/978-3-030-68442-6_2

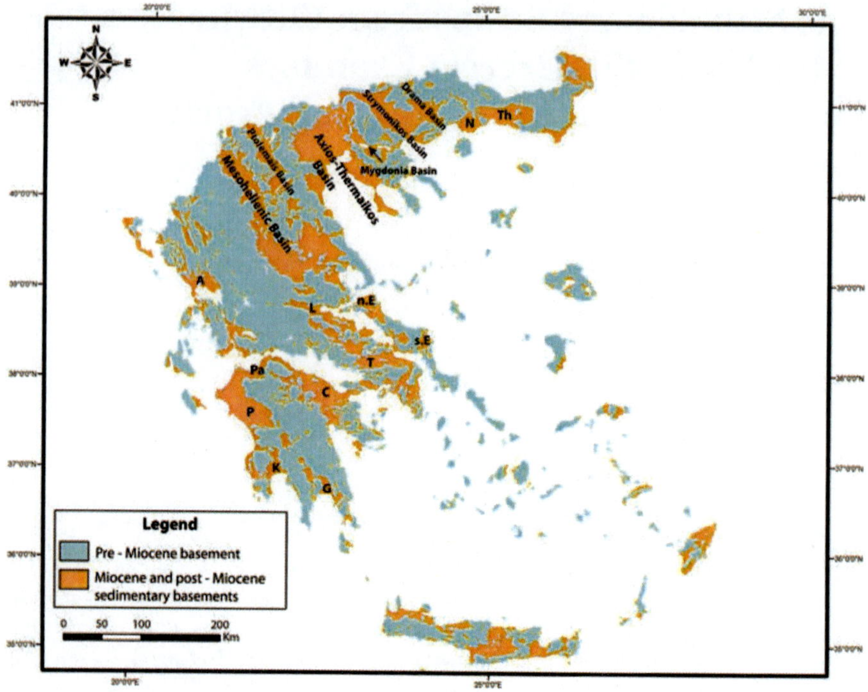

Fig. 1 The post-Oligocene basins of Greece, where the studied Mesohellenic, Ptolemais, Axios-Thermaikos, Mygdonia, and Strymonikos basins were showed. Additional, smaller, basins not presented in detail in the text: K, Kalamata; G, Gytheion; P, Pyrgos; Pa, Patras; C, Corinth; T, Thiva; L, Lamia; n.E, North Evia; s.E, South Evia; A, Amvrakikos Gulf; N, Nestos; and Th, Thrace

accompanied by three minor, and younger, basins, which were developed east of the major basins from the Miocene onwards. The Mesohellenic major basin is accompanied by the Ptolemais minor basin, the Axios-Thermaikos major basin is accompanied by the Mygdonia minor basin, and finally, the Strymonikos major basin is accompanied by the Drama minor basin (see Figs. 3, 4, and 5).

The remaining post-Alpine deposits from Greece are located in other, smaller basins, which nevertheless show important fossil vertebrate potential. These basins were strongly influenced by normal faults producing asymmetrical grabens.

At least four basins can be distinguished in Crete Island (Kasteli-Chania, Rethymnon, Irakleion-Messara, Ag. Nikolaos-Siteia basins), being characterized by strong synsedimentary tectonic influence with normal faults, where sedimentation started during late Miocene time (Tortonian) and continues with some interruptions as far as during the Pleistocene (Kontopoulos et al. 1996; Pasadakis et al. 2012; Moforis et al. 2013; Zidianakis et al. 2015; Maravelis et al. 2016; Zelilidis et al. 2016). Compared to other post-Oligocene basins from Greece that are predominantly terrestrial, the Neogene deposits from Crete contain several marine sediments. As such, they play an important role to our knowledge of the fossil record of

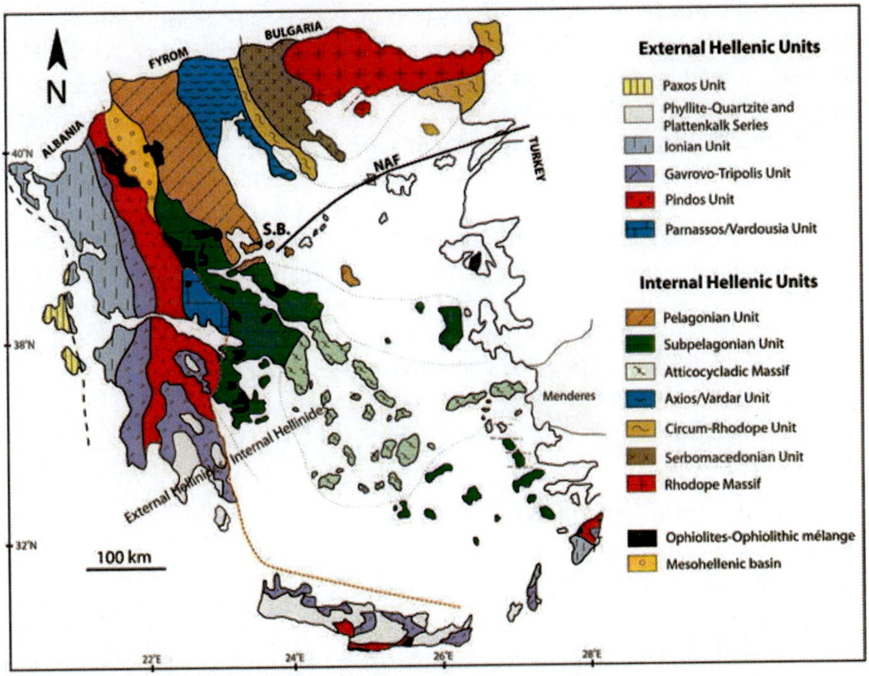

Fig. 2 Geological map of Greece showing the External and Internal Hellenides with their Units, Series, and Massifs; S.B., Sporades Basin (modified from Koukouvelas 2019)

vertebrates adapted to marine niches, including a diverse record of ray-finned fishes (see Argyriou volume 1, and references therein) and the entire Greek fossil record of sea cows (see Iliopoulos et al. volume 1, and references therein).

The remaining Aegean islands also contain smaller scale Neogene deposits. The most interesting are situated in Rhodes Island, where sedimentation during the Neogene covered a very high percentage of the present-day island. These deposits have not been extensively explored for fossil vertebrates, but they most certainly show an important potential, as evidenced by the rich diverse micromammalian faunas of the island (see Vasileiadou and Sylvestrou volume 1, Vasileiadou and Doukas this volume; and references therein).

In Peloponnesus, also there are many quite large basins, like these of Kalamata, Gytheion, Pyrgos, Patras, and Corinth (see Fig. 1), where mostly sedimentation took place in deltaic environments from early Pliocene to present, with many cycles of sedimentation, introducing regression and transgression events, owed mostly to tectonic activity and less to eustatic sea-level changes (Kontopoulos and Zelilidis 1992; Poulimenos et al. 1993; Zelilidis and Kontopoulos 1994; Zelilidis 2000; Zelilidis and Kontopoulos 2001). Many fossil vertebrates have been discovered in these deposits, including mainly large mammals like elephants and hippos (Athanassiou volume 1, this volume) or turtles (Vlachos volume 1).

There are also smaller basins, characterized by Neogene deposits, like these of Thiva, Lamia, north Evia, central Evia in eastern part of central Greece, Amvrakikos gulf in the western Greece, and Nestos and Thrace basins in North Greece (see Fig. 1). With the exception of Evia island, the other basins mentioned above contain only some few known localities with fossil vertebrates.

Although there are many basins with great interest, as mentioned above (Fig. 1), we focused in three synthetic basins (a major with an accompanied minor) as their complicated evolution seems to present the higher interest. Moreover, as this chapter represent the introductory chapter for all other chapters, with these three-basin analyses, a detailed description for post-Oligocene evolution in Greece is presented. Finally, all detailed maps were organized with unified basement, different lithologies and environments with chronologically divisions and not per formation, in order readers to understand and follow the remaining chapters.

2 The Mesohellenic Basin (MHB)

Geological Setting

The Mesohellenic Basin (MHB) corresponds to a Late Eocene/Oligocene–Middle Miocene thrust-top basin that was developed because of the westwards progradation of the Pindos Orogen (Avramidis et al. 2000; Zelilidis et al. 2002). The MHB is approximately 150 km long and 30 km wide (Figs. 1 and 3) and is positioned at the boundary between the Apulian Platform to the west and the Pelagonian Microplate to the east (Doutsos et al. 1994).

The basement of the MHB is represented by the ophiolite complex, remnants of the older subducted oceanic crust (of the so-called Pindos Ocean) that was subsequently emplaced over the margin of the continental crust (Apulian Platform) during compression (Moores 1969), along with Upper Cretaceous limestones of the western Pelagonian margin (Killias et al. 2015). The dip directions of the sedimentary succession at the western margin of the MHB (toward the ENE), and the eastern margin (toward the WSW) define a large-scale asymmetrical syncline (Killias et al. 2015), whereas to the north, the basin is further subdivided into two narrower synclines, separated by a region of uplift (Doutsos et al. 1994). Structural analyses reveal that during the Middle Eocene to Quaternary, the MHB displays a complex tectonic pattern (Vamvaka et al. 2010) that occurred in semi-ductile to brittle conditions (Killias et al. 2015).

Contrasting scenarios have been proposed to account for the origin of the MHB, with the most possible being: (1) a multi-story strike-slip and piggy-back-type basin from Middle Eocene to the present, above the westward emplacing Neotethyan ophiolites and Pelagonian units on the cold Hellenic accretionary prism (Killias et al. 2015); (2) a foreland-type basin developed in response to eastwards back-

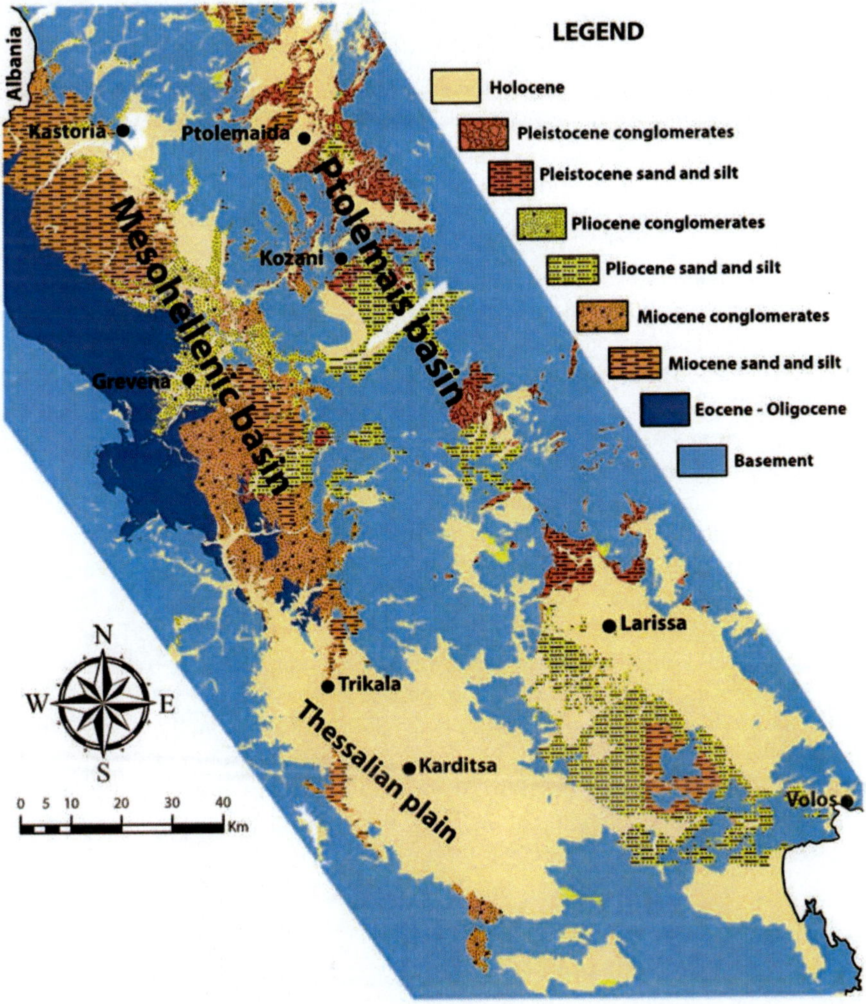

Fig. 3 Geological map of the Mesohellenic and accompanied Ptolemais basins

thrusting from Eocene to Miocene, because of the growth of the Pindos Orogen (Doutsos et al. 1994); (3) a forearc-type basin that developed during the early stages of a Mid–Late Eocene subduction (Pindos Ocean) and turned into a piggy-back basin as a result of Oligocene underthrusting of the Gavrovo-Tripolis domain (Ferriere et al. 2004).

Since the Late Miocene, the MHB exhibits gentle strike-slip deformation along major faults, uplift, and erosion, with the oldest rocks being exposed along the western basin margin (Doutsos et al. 1994).

Stratigraphic Evolution

The MHB receives sediments that have been accumulated in a variety of depositional environments from the Early Oligocene to the Middle Miocene. The stratigraphic record of the MHB initiates with the Eptahori Formation (1000-m thick, Lower Oligocene) that comprises conglomerates and interbedded sandstones that accumulated on a fan-delta depositional environment (Zelilidis and Kontopoulos 1996; Zelilidis et al. 2002; Avramidis et al. 2002). These deposits evolve abruptly upward into fine-grained submarine fan sediments. At the southwestern margin of the MHB, the Eptahori Formation unconformably overlays the Middle to Upper Eocene shales and sandstones of the Kranea Formation (Bizon et al. 1968). The Kranea Formation (<2 km thick) evolves up-sequence from fan-delta conglomerates and shales to turbidite sandstones and shales, and finally to deltaic and flood plain deposits (Zelilidis and Kontopoulos 1996; Zelilidis et al. 2002). During the Late Oligocene to Middle Miocene, the depositional environments are represented by fan-delta deposits that evolve into shelf deposits around the uplifted areas to the north and south of the MHB. Submarine fan deposits were continuously deposited in the central part of the MHB. Over the Eptahori Formation, the Pentalofos Formation (2500 m thick, Upper Oligocene–Lower Miocene) comprises submarine fan deposits that evolve to the southeast into shelf deposits. The overlying Tsotili Formation (1500 m thick, Lower to Middle Miocene) is also composed of submarine fan deposits, which in the northern part of the basin are overlain by Middle Miocene sandy shelf deposits (Ondria beds, 350 m thick). Finally, in the eastern parts of the basin, fluvial deposits have been accumulated. South of the MHB, the Thessalian plain is characterized by coarse-grained shallow water deltaic deposits that formed after the separation of the MHB into three-distinct sub-basins (according to Zelilidis 2003), during early Miocene.

Vertebrate Fossils in the MHB Basin

The majority of the vertebrate fossils from the MHB basin area are known actually from the more recent, Plio–Pleistocene deposits. These sediments contain perhaps the most important records of derived proboscidean specimens, including elephants and mammoths (see Athanassiou volume 1, and references therein), and mastodons (see Konidaris and Tsoukala volume 1, and references therein). The Miocene deposits are much poorer in fossil vertebrate remains, given that they mostly represent more coastal or littoral depositional environments; they are, however, quite rich in invertebrate fossil remains. As such, Miocene vertebrate localities from the MHB basin are few, including some aquatic turtle remains (see Vlachos volume 1, and references therein). These few occurrences, however, are of great importance for their respective clades and highlight the important potential of vertebrate fossil findings in the Miocene and older deposits of the MHB.

3 The Ptolemais Basin (PTB)

Geological Setting

The Ptolemais Basin (PTB) is a northwest-southeast trending sedimentary basin and is located in the Pelagonian Zone (Figs. 1 and 3). The study region was affected by at least six deformational stages from the Carboniferous to the Miocene (Mountrakis 1983). This tectonic activity triggered uplift and deformation of the basement rocks, along with the development of large-scale synclines and anticlines (Mountrakis 1983). The PTB basin was developed along a preexisting syncline, located within the regional mountain ranges. The study area is regarded as a rift-type basin developed because of extensional tectonic activity, which followed the Alpine orogenesis in Greece (Anastasopoulos and Koukouzas 1972; Koukouzas et al. 1979, 1981, 1984, 1985). During the Pliocene, NE-SW to NNE-SSW directed normal faults affected the Neogene sedimentary succession of the PTB and controlled topography, forming grabens and horsts. Tectonic analysis studies suggest at the early stages of basin development, the PTB was controlled by NW-SE directed normal faults, whereas the later stages are impacted by NE-SW normal faults (Pavlides 1985; Pavlides and Mountrakis 1987). The fault activity is subdivided into two stages that were active during the Late Miocene-Pleistocene and post-Early Pleistocene, respectively (Pavlides and Mountrakis 1987).

Stratigraphic Evolution

In the PTB, Neogene in age sediments (Miocene to Pliocene) are deposited, which come from the erosion of the preexisting basement rocks. These basement rocks belong to the Pelagonian Zone and consist of Pre-Cambrian to Paleozoic metamorphic rocks, Carboniferous granites, Triassic to Lower Jurassic limestones, and Middle Jurassic ophiolites that are unconformably overlaid by Upper Cretaceous limestones and submarine fan deposits (Mountrakis 1983). The Neogene (Late Miocene to Early Pleistocene) deposits have been interpreted as lagoonal in origin and are divided into two discrete members, based on the age, composition, and type of lignite-bearing deposits (Koukouzas et al. 1979; Steenbrink 2001). The lower member was accumulated during the Late Miocene to possibly earliest Pliocene age. It is composed of yellow to green, fine-grained silty sandstone and sandy claystone. The upper member was deposited during the Pleistocene and is composed of repetitions of silty sandstone and sandy claystone, along with sandstone and mudstone. During the Early to Middle Pleistocene, fluvial in origin sediments consisting of loose conglomerate, sandstone, and red in color mudstone accumulate in the PTB. The fluvial deposits overlay the preexisting lagoonal deposits, and their boundary is represented by an erosional unconformity.

Vertebrate Fossils in the PTB Basin

The PTB basin hold important occurrences of vertebrate fossils, many of which have been discovered during prospecting or mining activities related to the lignite exploration and research in the basin. The most important vertebrate fossils from the PTB basin belong to two different categories. On the one hand, there are plenty and diverse localities with numerous micromammalian fossils (e.g., see Vasileiadou and Sylvestrou volume 1; Vasileiadou and Doukas this volume; and references therein). These occurrences mostly come from the lower member, as described above. On the other hand, there are several occurrences of elephants and mammoths, including some partial skeletons with exceptional preservation (see Athanassiou volume 1). Most elephant fossils come from the Pleistocene deposits of the basin.

4 The Axios-Thermaikos Basin (ATB)

Geological Setting

The Axios-Thermaikos Basin (ATB) (Figs. 1, 2, and 4) corresponds to a fault-bounded sedimentary basin (Ferentinos et al. 1981) and forms part of the North Aegean region, an area with complex geological history (Kilias et al. 2013; Maravelis et al. 2015). The ATB overlies the eastern margin of the Pelagonian Zone (the so-called Vardar-Axios Zone), a fault-related crystalline massif and part of the Internal Hellenides (Brooks and Ferentinos 1980). The Vardar-Axios Zone is NNW-SSE directed and extends through Bulgaria, Northern Crete, and Western Turkey. An important regional feature is the North Aegean Trough (NAT), a 300-km-long trough along the Tethyan Ocean suture zone (Mountrakis 2006). The western margin of the NAT is the Sporades Basin, which in turn corresponds to the offshore extension of the ATB (Brooks and Ferentinos 1980). The development of the NAT is ascribed to late collisional processes that occurred between the Apulian and Eurasian continental lithosphere during the Late Oligocene to Middle Miocene (Tranos 2009), probably combined by the westward progradation of the North Anatolia Fault (McKenzie 1972; Pavlides et al. 1990).

The study area exhibits a complex tectonic regime characterized by periods of extensional, compressional, and strike-slip tectonics since the Oligocene (Pavlides et al. 1990; Sokoutis et al. 1993; Kilias et al. 1999; Tranos 2009; Koukouvelas and Aydin 2002). Tectonic analyses suggest that the NE-SW Sporades Basin was initiated either during the (1) Early–Middle Miocene under regional contraction and strike-slip to transpressional deformation and subsequently involved into extensional basin (Tranos 2009) or (2) Middle–Late Miocene WNW-ESE extension (Tranos 2009). In both scenarios, the Sporades Basin was initiated because of late collisional processes, rather than the westward propagation of the North Anatolia Fault into the North Aegean Sea (Tranos

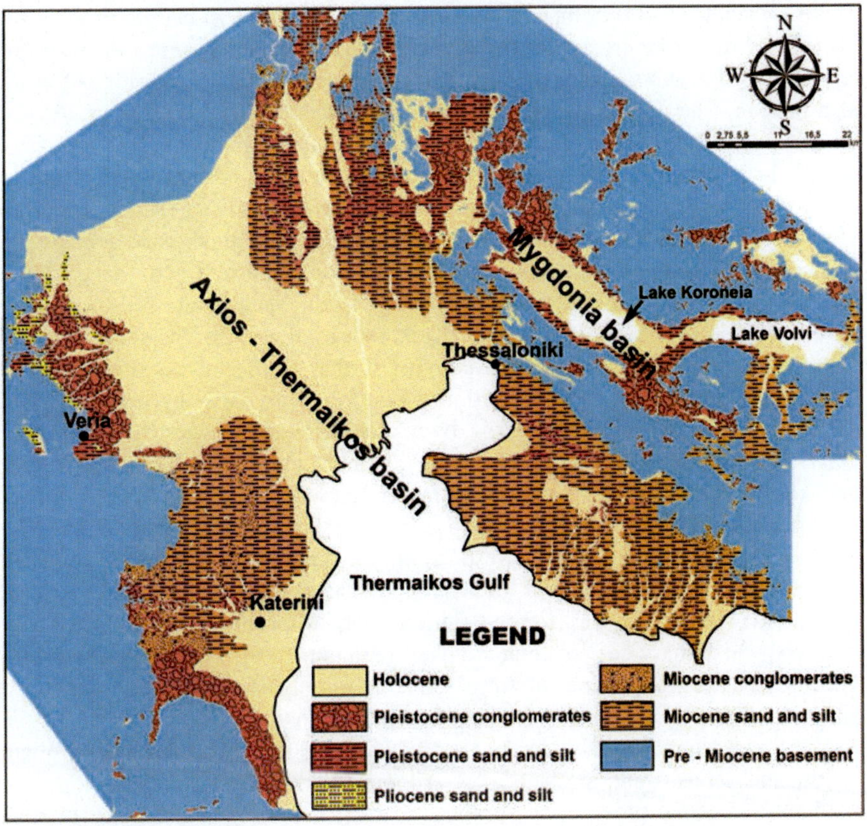

Fig. 4 Geological map of the Axios-Thermaikos basin and the accompanied Mygdonia basin

2009). Strike-slip tectonic activity played a key role in the development of the study area since the seismic reflection data suggest that basin-bounding faults display strike-slip architecture (Ferentinos et al. 1981; Roussos and Lyssimachou 1991). Normal and strike-slip interaction is also supported by fault pattern analyses, bathymetric data, and onshore and offshore seismic profiles (Koukouvelas and Aydin 2002). Since the Early Pleistocene, the deformation of the NAT is related to a NNE-SSW back-arc extension of the present-day Hellenic subduction zone (Tranos 2009).

Stratigraphic Evolution

The Neogene sedimentary succession in the ATB is ~1000 m thick (Kalkreuth et al. 1991). The ATB stratigraphy is subdivided into three major units with sediments that belong to a terrestrial to low salinity lacustrine depositional environment (Benda and Steffens 1981; Kalkreuth et al. 1991; Kotis and Papanikolaou 1999).

The lower unit is ~400 m thick and is exposed in the northern margin of the ATB because of extensive erosion of the overlying strata. It includes conglomerates of terrestrial origin and sandstone beds (up to 10 m thick). Silt and sandy clay beds are also present. The middle part is over 200 m thick and is composed of sandstone, mudstone and thick coal beds. The age of this unit, as determined by palynological data, is Lower Miocene (Benda and Steffens 1981), and it has been interpreted as being deposited in a swamp to limnic lower delta environment (Kalkreuth et al. 1991). Unconsolidated sandstones, silt and clay follow the thick-bedded coals (up to 1 m thick). The upper unit consists of sandstone and mudstone beds that accumulated during the Late Miocene age ("Pontian"). In this unit, lignite beds occur, but they are of no economic importance (Kalkreuth et al. 1991). The sediment flow direction is generally to the south and east further feeding the Sporades Basin. During the Late Quaternary, deposition is controlled primarily by prodeltaic and delta plain dynamics and, secondarily, by relative sea level changes (Lykousis et al. 2005). Three major river systems (Axios, Aliakmon, and Pinios), along with minor ones (Loudias and Gallikos), contribute freshwater and sediment to the sea. All but Pinios debouch in the north part of the Thermaikos Gulf (Lykousis et al. 2005). The excess in river-borne sediment is responsible for the development of an extensive bird-foot-type delta (Poulos et al. 1994). In the eastern margins of the ATB (western Chalkidiki), the basin was formed during early Miocene and was filled with Neogene–Quaternary deposits that are over 5 km in total thickness (Syrides 1990). The deposits are grouped into six formations: (a) Antonios Formation was deposited during the Early to Middle Miocene and Late Miocene (fluvial deposits), (b) Triglia Formation was deposited during the Late Miocene (Vallesian-Lowermost Turolian, continental deposits and red-beds), (c) Trilophos Formation was deposited during the Latest Miocene (Pontian or Turolian, brackish-lacustrine deposits), (d) Gonia Formation was deposited during the Pliocene (Ruscinian, fluvio-lacustrine deposits), and (e) Moudania Formation was deposited during from the Villafranchian onwards (continental sediments). To the west (Katerini sub-basin), the Neogene sedimentation is represented by eight Formations (Sylvestrou 2002): (1) Elatochori (alluvial fans, Early Miocene), (2) Moschopotamos (continental to lagoonal-lacustrine, Early to Middle Miocene), (3) Sykea (meandering fluvial, Middle Miocene), (4) Ryakia (meandering fluvial, Middle Miocene), (5) Lagorachi (braided fluvial, Late Miocene), (6) Sfendami (lagoonal, Late Miocene), (7) Makrygialos (braided fluvial, Late Miocene to Early Pleistocene), and (8) Lofos Formation (fluvial and lagoonal, Pleistocene to Quaternary).

Vertebrate Fossils in the ATB Basin

The ATB Basin is one of the most important regions that contain vertebrate fossils. The first collections of vertebrate fossils in the area took place in the beginning of the twentieth century by C. Arambourg (Arambourg and Piveteau 1929) and continue until present with new collections and extensive exploration. In the northern part of the basin, on each margin of the Axios river, numerous and famous localities with vertebrate fossils have concentrated enormous interest during the last 100 years. The majority of these localities are quite diverse, containing important primate (Koufos volume 1), bovid (Kostopoulos this volume), equid (Vlachou et al. this volume), carnivoran (Koufos this volume-a, -b, -c, -d), and proboscidean (Konidaris and Tsoukala volume 1) fossils. The western Chalkidiki sub-basin contains several vertebrate localities as well, mostly discovered along the coastline of the eastern border of Thermaikos gulf. These localities contain important non-mammalian fossils, including snakes (Georgalis and Delfino volume 1), turtles, and giant tortoises (Vlachos volume 1). The Katerini sub-basin is just as promising as the western Chalkidiki one, but focused fieldwork and exploration targeting vertebrate remains only recently took place (Sylvestrou 2002).

5 The Mygdonia Basin (MB)

Geological Setting

The Mygdonia basin (MB) (Fig. 4) is situated in central Macedonia, to the northeast of Thessaloniki city and is characterized by active seismic activity (Papazachos et al. 1979; Martinod et al. 1997). The MB is an E-W directed sedimentary basin (graben) that was developed during the Early-Middle Miocene because of extensional tectonic activity (Le Pichon and Angelier 1981). The MB is nowadays characterized by N-S extension (Martinod et al. 1997) that has been interpreted as the result of the spreading of the Aegean region over the subducted oceanic Mediterranean plate (Le Pichon and Angelier 1981). The extension principally influences a narrow (less than 2 km), E-W oriented zone at the southern edge of the basin (Martinod et al. 1997). The integration of seismological and neotectonic data indicates that normal faults prevail the tectonic features in the MB (Papazachos et al. 2001). Despite the variety in fault azimuths observed in some small faults, the majority of the faults and the major-magnitude seismic events are in agreement with the average N-S extension (Papazachos et al. 2001). The MB exhibits S-shape geometry, with its edges being NW-SE oriented, and the central parts being E-W oriented (Papazachos et al. 2001). The central part has drown the scientific attention because of the seismic activity of the Mw = 6.5 magnitude that affected the city of Thessaloniki (Papazachos et al. 1979; Pavlides et al. 1988). The basin is situated within the Serbomacedonian massif and is surrounded by a several sediment depocenters (Axios Basin, Strymon Basin, and North Aegean Trough). The MB receives sediments from Neogene to Quaternary that uncomformably overlay the pre-Neogene basement. Gneisses, schists, represent the basement rocks and amphibolites, metamorphic rocks that form parts of the Serbomacedonian Massif (in the central and eastern part of the MB, Kockel et al. 1977). The basement rocks in the western part of the basin correspond to phyllites, limestone, and sandstone that belong to the Circum-Rhodope Belt and have experienced low-grade metamorphism (Kockel et al. 1977).

Stratigraphic Evolution

The sedimentary pile in the MB is divided into two units: (1) the Pre-Mygdonian Group and (2) the Mygdonian Group. The Pre-Mygdonian Group encompasses Neogene and Early Pleistocene in age sediments, whereas the Mygdonian Group was deposited during the Middle Pleistocene to the Holocene and overlies uncomformably the Pre-Mygdonian Group (Koufos et al. 1995; Konidaris et al. 2015). The Pre-Mygdonian Group is further subdivided into three formations. The oldest Chrysavgi Formation (Middle Miocene, 40–50 m thick) uncomformably overlies the basement and is composed of laterally discontinuous conglomerates interbedded with sandstones. The conglomerate is unconsolidated and consists of well-rounded gravels (up to 40 cm in diameter). The gravels consist of mica-schist, gneiss, granite, quartzite, and pegmatite. Finer-grained deposits are rare and form thin-bedded lenses of silt and clay. The conglomeratic deposits mainly occupy the lower parts of the Chrysavgi Formation and exhibit an upward decrease in the gravel diameter within the formation. The upper parts of the Formation are sand-dominated and the finer-grained deposits (siltstones, silty sandstones, and silty claystones) are sporadically present. The overlying Gerakarou Formation (Early Pleistocene, over 100 m thick) is represented by red-beds consisting of repetitions of unconsolidated gravels, sand, silt, and clay. The gravelly beds are lenticular in shape and laterally discontinuous, and they have been interpreted as being accumulated in a fluvio-terrestrial environment of deposition. These red-beds erode significantly the underlying deposits and develop steep, valley-type landforms. The overlying Platanochori Formation (Early Pleistocene, 10–20 m thick) consists of sandstones and conglomerates, interbedded with silty sandstones, silty claystones, marlstones, and marly limestones. The limit between the Gerakarou and Platanochori Formations is transitional, consisting of laterally discontinuous sandstones, sandy marlstones, and red beds. The Platanochori Formation is thought as being accumulated in fluvial, fluvio-lacustrine depositional environment. This Formation represents the early stages of evolution between the continental Gerakarou Formation and the overlaying lacustrine sediments of the Mygdonian Group.

Vertebrate Fossils in the MB

Targeted exploration in the MB focusing on fossil vertebrates has contributed greatly to our knowledge of the stratigraphy of the basin; it is not a coincidence that some of the most important localities share the same name with the formations mentioned above. Vertebrate fossils are found mainly in the Pre-Mygdonian Group and are encountered in all formations. Occurrences from the great majority of mammalian families covered in volume 1 are found in the MB (e.g., Koufos 2006; Koufos et al. 1995; Tsoukala and Chatzopoulou 2005; Konidaris et al. 2015; and references therein).

6 The Strymonikos Basin (SB)

Geological Setting

The Strymonikos Basin (SB) is positioned over the boundary between the Rhodope massif and Serbo-Macedonian massif (Fig. 5). The boundary between the basement massifs is documented by the overthrusting of biotitic gneiss and marbles (Serbo-Macedonian) over the marbles of the Pangaion Mt. (Rhodope). The SB is thought as a NNW-SSE oriented graben-type basin and is arcuate in shape (Psilovikos 1994). The SB is interpreted to have been developed in response of a low-angle normal boundary fault (Dinter and Royden 1993). The fault action triggered collapse of the material and slumping toward the southwest. This tectonic event is the result of the extensional regime that controlled the basin evolution during the Neogene and extends from the Aegean Sea to Bulgaria (Dinter and Royden 1993).

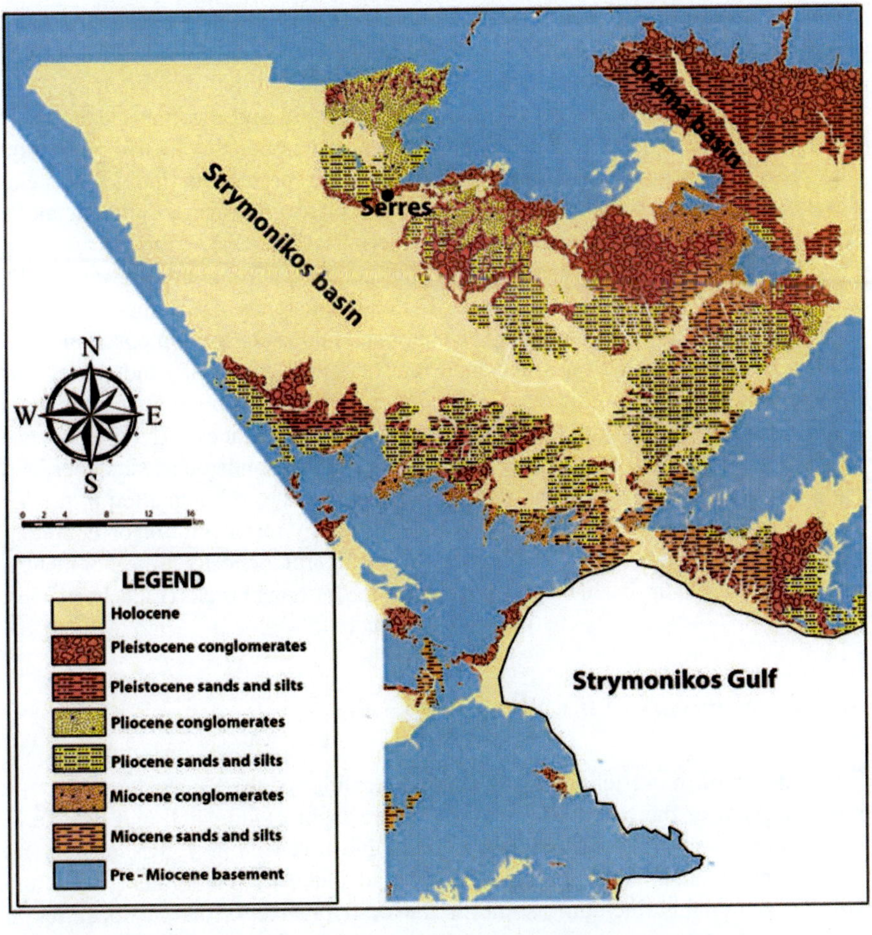

Fig. 5 Geological map of the Strymonikos basin and the accompanied Drama basin

Gravity faults affected the development of the SB, particularly the central parts, triggering post-depositional sediment deformation and anticline formation (Lalechos 1986). The uplift of the surrounding regions leads to the erosion of the basement rocks that provided the detritus for the sediment in the SB. The basin includes evidence of relative sea-level fluctuations, with stages of lagoonal-lacustrine sedimentation and stages of marine sedimentation (Lalechos 1986). At the latest stages of evolution, sedimentation in the SB is entirely continental.

Stratigraphic Evolution

The SB receives Neogene in age sediments (Miocene to Pleistocene), stemming from the erosion of the preexisting basement rocks that belong to the Rhodope massif and Serbo-Macedonian massif (Lalechos 1986). Sub-surface data indicate that the Miocene deposits correspond to both marine and lacustrine depositional environments and are subdivided into three units (Lalechos 1986). The basal unit comprises brecciated deposits, with the internal clasts being composed of fine-grained sandstones and conglomerates.

The overlying unit is interpreted as lacustrine deposits and is composed of repetitions of sandstones, mudstones, and dark brown marlstones that evolve up-section into petroliferous limestones. The younger deposits represent marine deposits and develop repetitions of sandstones, siltstones, and claystones that accumulate along with micro-brecciated deposits and lignite layers. The Pliocene sediments are interpreted to represent both brackish and continental environments of deposition (Lalechos 1986). The brackish sediments include sandstones, mudstones, siltstones, along with travertines and lignites. The continental deposits are composed of alternations of reddish sandstones and mudstones. Intercalations of sandstones, siltstones, and micro-conglomerates are common. These sediments also include travertines, lignite-bearing limestones, and chalk limestones. The Pleistocene deposits are composed of quartz gravels, coarse-grained sandstones, siltstones, and claystones. The sedimentary succession suggests that the sedimentation in the SB was dominated by lacustrine conditions and low energy sedimentation on continental deposits (Lalechos 1986). The travertinic and lignitic deposits represent materials accumulated at the basin margin, adjacent to the basement rocks (Lalechos 1986).

Vertebrate Fossils in the SB

The majority of the vertebrate fossil-bearing localities of the SB are located in the northern parts of the basin. These localities are mainly quite rich and diverse in micromammalian (e.g., see Vasileiadou and Sylvestrou volume 1; Vasileiadou and Doukas this volume; and references therein) and squamate (Georgalis and Delfino volume 1) remains, including some of the most diverse vertebrate localities in

Greece (Schmidt-Kittler 1995). These localities document the transition from the Miocene to the Pliocene in great detail. Large vertebrate remains are fewer, but include important chalicothere specimens together with associated Late Miocene remains (Tsoukala this volume and references therein). Stratigraphically younger vertebrates are known from the central and southern parts of the SB and the associated Drama basin, including mainly proboscidean remains (Athanassiou volume 1; Konidaris and Tsoukala volume 1).

References

Anastasopoulos I, Koukouzas K (1972) Geology and economy geology in the south part of coal bearing basin of Ptolemais. IGEY, Athens

Arambourg C, Piveteau J (1929) Les Vertébrés du Pontien de Salonique. Ann Paleontol 18:57–139

Argyriou T (volume 1) The fossil record of ray-finned fishes (Actinopterygii) in Greece. In: Vlachos E (ed) The fossil vertebrates of Greece Vol. 1 – basal vertebrates, amphibians, reptiles, afrotherians, glires, and primates. Springer – Nature Publishing Group, Cham

Athanassiou A (volume 1) The fossil record of non-endemic elephants and mammoths (Mammalia: Proboscidea: Elephantidae) in Greece. In: Vlachos E (ed) The fossil vertebrates of Greece Vol. 1 – basal vertebrates, amphibians, reptiles, afrotherians, glires, and primates. Springer – Nature Publishing Group, Cham

Athanassiou A (this volume) The fossil record of non-endemic hippopotamids (Mammalia: Artiodactyla: Hippopotamidae) in Greece. In: Vlachos E (ed) The fossil vertebrates of Greece Vol. 2 – laurasiatherians, artiodactyles, perissodactyles, carnivorans, and island endemics. Springer – Nature Publishing Group, Cham

Avramidis P, Zelilidis A, Kontopoulos N (2000) Thrust dissection control of deep-water clastic dispersal patterns in the Klematia-Paramythia foreland basin, western Greece. Geol Mag 137:667–685

Avramidis P, Zelilidis A, Vakalas I, Kontopoulos N (2002) Interactions between tectonic activity and sea level changes in the Pindos and Mesohellenic Basins, NE Greece: basin evolution and hydrocarbon potential. J Petrol Geol 25:53–82

Benda L, Steffens P (1981) Aufbauy and Alter des Neogens von Katerini, Griechenland. Geol Jahrb B 42:93–103

Bizon JJ, Lalechos G, Savoyat E (1968) Présence d'Eocene transgressif en Thessalie. Bull Soc Geol Fr 10:36–38

Brooks M, Ferentinos G (1980) Structure and evolution of the Sporadhes basin of the North Aegean trough, Northern Aegean Sea. Tectonophysics 68:15–30

Dinter DA, Royden LH (1993) Late Cenozoic extension in northeastern Greece: Strymon Valley detachment and Rhodope metamorphic core complex. Geology 21:45–48

Doutsos T, Koukouvelas I, Zelilidis A, Kontopoulos N (1994) Intracontinental wedging and post-orogenic collapse in the Mesohellenic trough. Geol Rundsch 83:257–275

Ferentinos G, Brooks M, Collins M (1981) Gravity-induced deformation on the north flank and floor of the Sporades Basin of the North Aegean Sea Trough. Mar Geol 44:289–302

Ferriere J, Reynaud JY, Pavlopoulos A, Bonneau M, Migiros G (2004) Geologic evolution and geodynamic controls of the Tertiary intramontane piggyback Meso-Hellenic Basin, Greece. Bull Soc Geol France 175:361–381

Georgalis GL, Delfino M (volume 1) The fossil record of lizards and snakes (Reptilia: Squamata) in Greece. In: Vlachos E (ed) The fossil vertebrates of Greece Vol. 1 – basal vertebrates, amphibians, reptiles, afrotherians, glires, and primates. Springer – Nature Publishing Group, Cham

Iliopoulos G, Svana K, Roussiakis S (volume 1) The fossil record of sea cows (Mammalia: Sirenia) in Greece. In: Vlachos E (ed) The fossil vertebrates of Greece Vol. 1 – Basal vertebrates,

amphibians, reptiles, afrotherians, glires, and primates. Springer – Nature Publishing Group, Cham
Kalkreuth W, Kotis T, Papanikolaou C, Kokkinakis P (1991) The geology and coal pe-trology of a Miocene lignite profile at Meliadi Mine, Katerini, Greece. Int J Coal Geol 17:51–67
Kilias A, Falalakis G, Mountrakis D (1999) Cretaceous–Tertiary structures and kinematics of the Serbomacedonian metamorphic rocks and their relation to the exhumation of Hellenic hinterland (Macedonia, Greece). Int J Earth Sci 88:513–531
Kilias A, Falalakis G, Sfeikos A, Papadimitriou E, Vamvaka A, Gkarlaouni C (2013) The Thrace basin in the Rhodope province of NE Greece – a tertiary supradetachment basin and its geodynamic implications. Tectonophysics 595–596:90–105
Killias A, Vamvaka A, Falalakis G, Sfeikos A, Papadimitriou E, Gkarlaouni C, Karakostas B (2015) The Mesohellenic trough and the Paleogene Thrace Basin on the Rhodope Massif, their Structural Evolution and geotectonic significance in the Hellenides. J Geol Geosci 4(2):198, 17
Kockel F, Mollat H, Walther HW (1977) Erläuterungen zur geologischen Karte der Chalkidiki und angrenzender Gebiete 1:100.000 (Nord-Griechenland). Bundesanst Geowiss Rohstoffe, Hannover, p 119
Konidaris G, Tsoukala E (volume 1) The fossil record of Neogene Proboscidea (Mammalia) in Greece. In: Vlachos E (ed) The fossil vertebrates of Greece Vol. 1 – basal vertebrates, amphibians, reptiles, afrotherians, glires, and primates. Springer – Nature Publishing Group, Cham
Konidaris GE, Tourloukis V, Kostopoulos DS, Thompson N, Giusti D, Michailidis D, Koufos GD, Harvati K (2015) Two new vertebrate localities from the Early Pleistocene of Mygdonia Basin (Macedonia, Greece): Preliminary results. C R Palevol 14:353–362
Kontopoulos N, Zelilidis A (1992) Upper Pliocene lacustrine environments in the intramontane Rio graben basin, NW Peloponnesus, Greece. N Jb Palaeontol Mh (2):102 114
Kontopoulos N, Zelilidis A, Frydas D (1996) Late Neogene sedimentary and tectonostratigraphic evolution of southwestern Crete island, Greece. N Jb Palaeontol Mh 202:287–311
Kostopoulos DS (this volume) The fossil record of bovids (Mammalia: Artiodactyla: Pecora: Ruminantia: Bovidae) in Greece. In: Vlachos E (ed) The fossil vertebrates of Greece Vol. 2 – laurasiatherians, artiodactyles, perissodactyles, carnivorans, and island endemics. Springer – Nature Publishing Group, Cham
Kotis T, Papanikolaou K (1999) Depositional environment of lignite bearing sediments in the Neogene basin of Moschopotamos Pieria, Greece. Bull Geol Soc Greece 33:17–28
Koufos GD (2006) The Neogene mammal localities of Greece: faunas, chronology and biostratigraphy. Hellenic J Geosci 41:183–214
Koufos GD (volume 1) The fossil record of the old world monkeys (Mammalia: Primates: Cercopithecidae) in Greece. In: Vlachos E (ed) The fossil vertebrates of Greece Vol. 1 – basal vertebrates, amphibians, reptiles, afrotherians, glires, and primates. Springer – Nature Publishing Group, Cham
Koufos GD (volume 1) The fossil record of the Miocene hominoids (Mammalia: Primates: Hominoidea) in Greece. In: Vlachos E (ed) The fossil vertebrates of Greece Vol. 1 – basal vertebrates, amphibians, reptiles, afrotherians, glires, and primates. Springer – Nature Publishing Group, Cham
Koufos GD (this volume-a) The fossil record of felids (Mammalia: Carnivora: Felidae) in Greece. In: Vlachos E (ed) The fossil vertebrates of Greece Vol. 2 – laurasiatherians, artiodactyles, perissodactyles, carnivorans, and island endemics. Springer – Nature Publishing Group, Cham
Koufos GD (this volume-b) The fossil record of hyaenids (Mammalia: Carnivora: Hyaenidae) in Greece. In: Vlachos E (ed) The fossil vertebrates of Greece Vol. 2 – laurasiatherians, artiodactyles, perissodactyles, carnivorans, and island endemics. Springer – Nature Publishing Group, Cham
Koufos GD (this volume-c) The fossil record of canids (Mammalia: Carnivora: Canidae) in Greece. In: Vlachos E (ed) The fossil vertebrates of Greece Vol. 2 – laurasiatherians, artiodactyles, perissodactyles, carnivorans, and island endemics. Springer – Nature Publishing Group, Cham

Koufos GD (this volume-d) The fossil record of mustelids (Mammalia: Carnivora: Mustelidae) in Greece. In: Vlachos E (ed) The fossil vertebrates of Greece Vol. 2 – laurasiatherians, artiodactyles, perissodactyles, carnivorans, and island endemics. Springer – Nature Publishing Group, Cham

Koufos GD, Syrides GE, Kostopoulos DS, Koliadimou KK (1995) Preliminary results about the stratigraphy and the palaeoenvironment of Mygdonia Basin, Macedonia, Greece. Geobios 18:243–249

Koukouvelas IK (2019) Geology of Greece. Liberal Books, Athens

Koukouvelas IK, Aydin A (2002) Fault structure and related basins of the North Aegean Sea and its surroundings. Tectonics 21(5):1046. https://doi.org/10.1029/2001TC901037

Koukouzas K, Kotis T, Ploumidis M, Metaxas A (1979) Coal exploration of Anargiri-Amynteon area, I.G.M.E. Miner Deposit Res 9:1–69. (in Greek)

Koukouzas K, Kotis T, Ploumidis M, Metaxas A (1981) Coal exploration of "Apophisis" field of Anargiri-Amynteon area, IGME. Res Energy Resour 1:1–52. (in Greek)

Koukouzas K, Kotis T, Ploumidis M, Metaxas A, Dimitriou D (1984) Coal exploration of Komnina area, Ptolemais district (W. Macedonia). IGME. Res Energy Resour 2:1–103. (in Greek)

Koukouzas K, Kotis T, Ploumidis M, Metaxas A, Dimitriou D (1985) Lignite deposits of Ptolemaida trench. Proceedings of the international meeting for the exploitation of low energy solid fuel, Ptolemaida. (in Greek)

Lalechos N (1986) Correlations and observations in molassic sediments in onshore and offshore areas of northern Greece. Miner Wealth 42:6–11

Le Pichon X, Angelier J (1981) The Aegean sea. Philos Trans R Soc Lond A 300:357–372

Lykousis V, Karageorgis AP, Chronis GTh (2005) Delta progradation and sediment fluxes since the last glacial in the Thermaikos Gulf and the Sporades Basin, NW Aegean Sea, Greece. Mar Geol 222–223:381–397

Maravelis AG, Pantopoulos G, Tserolas P, Zelilidis A (2015) Accretionary prism forearc interactions as reflected in the sedimentary fill of southern Thrace Basin (Lemnos Island, NE Greece). Int J Earth Sci 104:1039–1060

Maravelis A, Panagopoulos G, Piliotis I, Pasadakis N, Manutsoglou E, Zelilidis A (2016) Pre-Messinian (sub-salt) source-rock potential on back-stop basins of the Hellenic Trench system (Messara Basin, Central Crete, Greece). Oil Gas Sci Technol Rev IFP Energ Nouvelles 71:6

Martinod J, Hatzfeld D, Savvaidis P, Katsambalos K (1997) Rapid N-S extension in the Mygdonian graben (Northern Greece) deduced from repeated geodetic surveys. Geophys Res Lett 24:3293–3296

McKenzie DP (1972) Active tectonics of the Mediterranean region. Geophys J R Astron Soc 30:109–185

Moforis L, Kostopoulou S, Panagopoulos G, Pyliotis J, Triantaphyllou M, Manoutsoglou E, Zelilidis A (2013) Sedimentation processes and palaeogeographic evolution of Makrilia Pliocene deposits, SE Crete. Bull Geol Soc Greece 47:216–225

Moores EM (1969) Petrology and structure of the Vourinos ophiolitic complex of northern Greece. Geol Soc Am Spec Pap 118:1–74

Mountrakis D (1983) Structural Geology of the North Pelagonian Zone s.l. and Geotectonic Evolution of the Internal Hellenides (Macedonia, Greece). Habilitation Scholarship, University of Thessaloniki, Greece [in Greek with English abstract]

Mountrakis D (2006) Tertiary and quaternary tectonics of Greece. In: Dilek Y, Pavlides S (eds) Postcollisional tectonics and magmatism in the Mediterranean region and Asia. Geol Soc Am Spec Pap 409:125–136

Papazachos B, Mountrakis D, Psilovikos A, Leventakis G (1979) Surface fault traces and fault plane solutions of the May–June 1978 major shocks in the Thessaloniki area, Greece. Tectonophysics 53:171–173

Papazachos B, Vamvakaris DA, Vargemezis GM, Aidona EV (2001) A study of the active tectonics and deformation in the Mygdonia Basin (N. Greece) using seismological and neotectonic data. Bull Geol Soc Greece 34:303–309

Pasadakis N, Dagounaki V, Chamilaki E, Vafidis A, Zelilidis A, Piliotis I, Panagopoulos G, Em M (2012) Organic geochemical evaluation of Neogene formations in Messara (Heraklion, Crete) basin as source rocks of biogenetic methane. Miner Wealth 166:8–26

Pavlides S (1985) Neotectonic evolution of the Florina-Vegoritis-Ptolemais basin (W. Macedonia, Greece). Aristotle University of Thessaloniki, Thessaloniki, 256

Pavlides SB, Kondopoulou DP, Kilias AA, Westphal M (1988) Complex rotational deformations in the Serbo-Macedonian massif (north Greece): structural and palaeomagnetic evidence. Tectonophysics 145:329–335

Pavlides S, Mountrakis D (1987) Extensional tectonics of north-west Macedonia, Greece, since the Late Miocene. J Struct Geol 9:385–392

Pavlides S, Mountrakis D, Kilias A, Tranos M (1990) The role of strike-slip movements in the extensional area of Northern Aegean (Greece): a case of transtensional tectonics. Ann Tecton 4:196–211

Poulimenos G, Zelilidis A, Kontopoulos N, Doutsos T (1993) Geometry of trapezoidal fan deltas and their relationship to extensional faulting along the southwestern active margins of the Corinth rift. Basin Res 5:179–192

Poulos S, Papadopoulos A, Collins MB (1994) Deltaic progradation in Thermaikos bay, northern Greece and its socio-economic implications. Ocean Coast Manag 22:229–247

Psilovikos A (1994) Study – research of environmental impacts from protected areas constructions at upstream and downstream of Strymonas River at Lake Kerkini at torrents and at the plain area of Serres, vol 2. Research Committee Aristotle University Thessaloniki, Thessaloniki. (in Greek)

Roussos N, Lyssimachou T (1991) Structure of the Central North Aegean Trough: an active strike-slip deformation zone. Basin Res 3:39–48

Schmidt-Kittler N (1995) The vertebrate locality Maramena (Macedonia, Greece) at the Turolian–Ruscinian boundary (Neogene). Pfeil

Sokoutis D, Brun JP, Van den Driessche J, Pavlides S (1993) A major Oligocene-Miocene detachment in southern Rhodope controlling north Aegean extension. J Geol Soc Lond 150:243–246

Steenbrink J (2001) Orbital signatures in lacustrine sediments – the Late Neogene intramontane Florina-Ptolemais-Servia Basin, northwestern Greece, vol 167. Faculteit Aardwetenschappen Universiteit Utrecht, Utrecht

Sylvestrou I (2002) Stratigraphic, palaeontological and palaeogeographical study of the Neogene-quaternary deposits of Katerini basin, Northern Greece. Ph.D. thesis. Department of Geology, Aristotle University of Thessaloniki, Thessaloniki

Syrides G (1990) Lithostratigraphic, biostratigraphic and palaeogeographic study of the Neogene – quaternary sedimentary deposits of Chalkidiki Peninsula, Macedonia, Greece, Ph.D. thesis. School of Geology, Aristotle University of Thessaloniki, Thessaloniki

Tranos MD (2009) Faulting of Lemnos Island; a mirror of faulting of the North Aegean Trough (Northern Greece). Tectonophysics 467:72–88

Tsoukala E (this volume) The fossil record of chalicotheres (Mammalia: Perissodactyla: Chalicotheriidae) in Greece. In: Vlachos E (ed) The fossil vertebrates of Greece Vol. 2 – laurasiatherians, artiodactyles, perissodactyles, carnivorans, and island endemics. Springer – Nature Publishing Group, Cham

Tsoukala E, Chatzopoulou K (2005) A new Early Pleistocene (latest Villafranchian) site with mammals in Kalamotó (Mygdonia Basin, Macedonia, Greece) – preliminary report. Mitt Komm Quartärforsch Österr Akad Wiss 14:213–233

Vamvaka A, Spiegel C, Frisch W, Danisik M, Kilias A (2010) Fission track data from the Mesohellenic trough and the Pelagonian zone in NW Greece: cenozoic tectonics and exhumation of source areas. Int Geol Rev 52:223–248

Vasileiadou K, Doukas CS (this volume) The fossil record of insectivores (Mammalia: Eulipotyphla) in Greece. In: Vlachos E (ed) The fossil vertebrates of Greece Vol. 2 – laurasiatherians, artiodactyles, perissodactyles, carnivorans, and island endemics. Springer – Nature Publishing Group, Cham

Vasileiadou K, Sylvestrou I (volume 1) The fossil record of rodents (Mammalia: Rodentia) in Greece. In: Vlachos E (ed) The fossil vertebrates of Greece Vol. 1 – basal vertebrates, amphibians, reptiles, afrotherians, glires, and primates. Springer – Nature Publishing Group, Cham

Vasileiadou K, Sylvestrou I (volume 1) The fossil record of hares, rabbits and pikas (Mammalia: Lagomorpha) in Greece. In: Vlachos E (ed) The fossil vertebrates of Greece Vol. 1 – basal vertebrates, amphibians, reptiles, afrotherians, glires, and primates. Springer – Nature Publishing Group, Cham

Vlachos E (volume 1) The fossil record of turtles and tortoises (Reptilia: Testudines) in Greece. In: Vlachos E (ed) The fossil vertebrates of Greece Vol. 1 – basal vertebrates, amphibians, reptiles, afrotherians, glires, and primates. Springer – Nature Publishing Group, Cham

Vlachou T, Koufos GD, Gkeme A (this volume) In: Vlachos E (ed) The fossil record of equids (Mammalia: Perissodactyla: Equidae) in Greece, The fossil vertebrates of Greece Vol. 2 – laurasiatherians, artiodactyles, perissodactyles, carnivorans, and island endemics. Springer – Nature Publishing Group, Cham

Zelilidis A (2000) Drainage evolution in a rifted basin, Corinth graben, Greece. Geomorphology 35:69–85

Zelilidis A (2003) The geometry of fan-deltas and related turbidites in narrow linear basins. Geol J 38:31–36

Zelilidis A, Kontopoulos N (1994) Pliocene-Pleistocene fluvial/wave dominated deltaic sedimentation: the Pamisos delta in SW Peloponnesus, Greece. Geol Mag 131:653–668

Zelilidis A, Kontopoulos N (1996) Significance of fan deltas without toe-sets within rift and piggy-back basins: examples from the Corinth graben and the Mesohellenic trough, Central Greece. Sedimentology 43:253–262

Zelilidis A, Kontopoulos N (2001) Post-Miocene sedimentary evolution of south Peloponnesus, Greece. Gaia 16:1–12

Zelilidis A, Piper DJW, Kontopoulos N (2002) Sedimentation and basin evolution of the Oligocene-Miocene Mesohellenic basin, Greece. AAPG 86:161–182

Zelilidis A, Tserolas P, Chamilaki E, Pasadakis N, Kostopoulou S, Maravelis AG (2016) Hydrocarbon prospectivity in the Hellenic trench system: organic geochemistry and source rock potential of upper Miocene-lower Pliocene successions in the eastern Crete Island, Greece. Int J Earth Sci 105:1859–1878

Zidianakis G, Iliopoulos G, Zelilidis A, Kovar-Eder J (2015) Myrica from the plant assemblage of Pitsidia (Crete, Late Miocene): putting the puzzle together. Palaeontogr Abt B 293:149–171

The Fossil Record of Insectivores (Mammalia: Eulipotyphla) in Greece

Katerina Vasileiadou and Constantin S. Doukas

1 Introduction

Eulipotyphla is today the third largest mammalian order (after Rodentia and Chiroptera), comprising four extant (Talpidae, Erinaceidae, Soricidae, Solenodontidae) and several extinct families (e.g., Nesophontidae, Amphilemuridae, Dimylidae, Heterosoricidae, Plesiosoricidae). Until the end of the last century, all these families were included in the order Insectivora, which was, however, also included in the Tenrecomorpha (tenrecs and golden moles) now assigned to the order Afrotheria. They are all small (2–1 kg), mostly insectivorous animals, with a long, narrow and mobile snout, and most of them are nocturnal (Macdonald 2009). Today, the Solenodontidae inhabit only the Caribbean Islands of Hispaniola and Cuba, whereas the erinaceids (hedgehogs and moonrats), talpids (moles and desmans), and soricids (shrews) have an almost worldwide geographical distribution (except for Australia and Antarctica). They are often considered as the most primitive living placental mammals and are, thus, morphologically closer than all other living placentals to the common mammal ancestors (Macdonald 2009). In any case, the dentition is generally close to the ancestral tribosphenic condition.

K. Vasileiadou (Deceased)
Natural History Museum of the Lesvos Petrified Forest, Mytilene, Lesvos, Greece

C. S. Doukas (✉)
Department of Palaeontology and Historical Geology, University of Athens, Athens, Greece
e-mail: cdoukas@geol.uoa.gr

E. Vlachos (ed.), *Fossil Vertebrates of Greece Vol. 2*,
https://doi.org/10.1007/978-3-030-68442-6_3

Insectivore fossils have always been a by-product of large fossil rodent assemblages, and they are collected using the same methodology (sediment collection, wet screen-washing, dry sieving, picking under a stereoscopic microscope). Even though the importance of fossil rodents for biostratigraphy and paleoenvironmental reconstructions has for long been established, the use of the less common insectivores, particularly for paleoecological inferences, was only acknowledged much later. In Greece, the fossil insectivore research has essentially started in the 1970s; since then, 47 late Oligocene–late Pleistocene localities have revealed mainly isolated dental (main diagnostic elements) and a few postcranial elements. Eight species have been described from Greek Neogene localities are still recognized: three Erinaceidae (*Galerix symeonidisi*, *Parasorex kostakii*, *Schizogalerix macedonica*), two Talpidae (*Myxomygale engesseri*, *Desmanella dubia*) and one Heterosoricidae species (*Heterosorex ruemkeae*), and two Soricidae (*Amblycoptus jessiae*, *Deinsdorfia kerkhoffi*). In addition, three more species have disappeared in the synonymy (*Desmanodon meuleni*, *Galerix atticus*, *Dibolia dekkersi*).

2 Historical Overview

The first mention to fossil Eulipotyphla in a Greek locality was given by Thenius (1952), who reported the presence of *Galerix exilis* (assigned to *Schizogalerix moedlingensis* by Doukas et al. 1995) in the classic Pikermi locality. In the 1970s, studies mainly by researchers of the Geological Institute and the University of Utrecht (The Netherlands) communicated the presence of important eulipotyphlan faunas in well-studied localities such as Arnissa, Maritsa 1 and Pikermi-Chomateri (Mayhew 1978; de Bruijn et al. 1970; Rümke 1976). During the next two decades, many new eulipotyphlan faunas became known, mainly through Ph.D. theses in Greek universities (e.g., Doukas 1983; Koliadimou 1996), in Dutch universities (e.g., Rümke 1985), and also in collaborations between Greek and Dutch institutes (e.g., Reumer and Doukas 1985; Doukas et al. 1995). During the past 20 years, postgraduate studies in Greek universities resulted in the revealing of some more faunas (e.g., Sylvestrou 2002, Vasileiadou et al. 2003; Chatzopoulou 2014), and so did the ongoing collaboration of Greek and Dutch institutes (e.g., Doukas and Van den Hoek Ostende 2006; De Bruijn et al. 2012; Van den Hoek Ostende et al. 2015). In 2005, in the frame of an effort to revise the fossil record of the Eurasian Neogene insectivores known until then, Doukas presented a review of the Greek fossil record (Doukas 2005). This review made apparent that important Greek insectivore faunas had been till then only partially studied, some of which are still, after almost 15 years, not visited (e.g., the faunas of the Ptolemais Basin).

3 Phylogenetic Relationships

For over a century, the order Insectivora has been used as a "waste-basket" for extant insectivorous eutherian mammals with unspecialized dentition (Douady et al. 2002:200–2001 for short review on taxonomic concept of "Insectivora"). The monophyly of the order collapsed in the 1980s, when molecular analyses excluded from the order Afrotheria (golden moles and tenrecs), that had been included in Insectivora since the nineteenth century (Nikaido et al. 2003). Since then, Insectivora has been divided into Afrosoricida and Eulipotyphla (erinaceids, talpids, soricids and solenodons). The monophyly of Eulipotyphla has been frequently challenged (see Nikaido et al. 2003 for examples). However, studies based on nuclear genes (Douady et al. 2002), and also based on mitochondrial DNA sequences (Nikaido et al. 2003), support the monophyly of Eulipotyphla. Yet, the phylogenetic relationships within the order are controversial. As Nikaido et al. (2003:277) reported: "McKenna and Bell's (1997) morphological analysis suggested that moles and hedgehogs are more closely related to each other than either is to shrew, and Butler's (1988) morphological analysis suggested closer relationship between shrews and moles than to hedgehogs, while Murphy et al.'s (2001a, b) molecular analyses based mainly on nuclear DNA suggested a hedgehog/shrew clade excluding mole as an outgroup." Douady et al. (2002), using molecular data, indicated a sister–taxon relationship of soricids and erinaceids to the exclusion of talpids. Nikaido et al. (2003) investigated the interfamiliar relationships in eulipotyphlans using mitochondrial protein data; their results indicated that the relationships cannot be resolved. Even though this issue is still a matter of debate, most molecular studies seem to agree on a late Cretaceous origin for Eulipotyphla and that the Solenodontidae split from the other extant eulipotyphlan families soon afterwards, before the end of the period (Springer et al. 2018). For example, Bininda-Emonds et al. (2007) constructed a species-level phylogeny of extant mammals (4510 of the 4554 extant species) and estimated that Eulipotyphla originated 91.8 ± 2.6 Ma, whereas the time of their basal diversification (of Solenodontidae) was determined at 84.2 ± 2.1 Ma, whereas Springer et al. (2018) dated this basal diversification at 78.5–65.4 Ma.

4 Distribution

Fossils of Eulipotyphla have been recovered from 47 Greek localities, distributed in continental (35 localities) and insular Greece (12 localities in Lesbos, Evia, Chios, Crete, Samos, Rhodos, Kalymnos, Naxos); see Fig. 1 and Appendix for detailed information. The oldest one is correlated with the late Oligocene, six are correlated with the early Miocene, three with the middle Miocene, 11 with the late Miocene, three with the Turolian/Ruscinian boundary, eight with the Pliocene, seven with the early Pleistocene, two with the middle Pleistocene, and six with the late Pleistocene. Unfortunately, some of them have only yielded scanty and fragmentary materials,

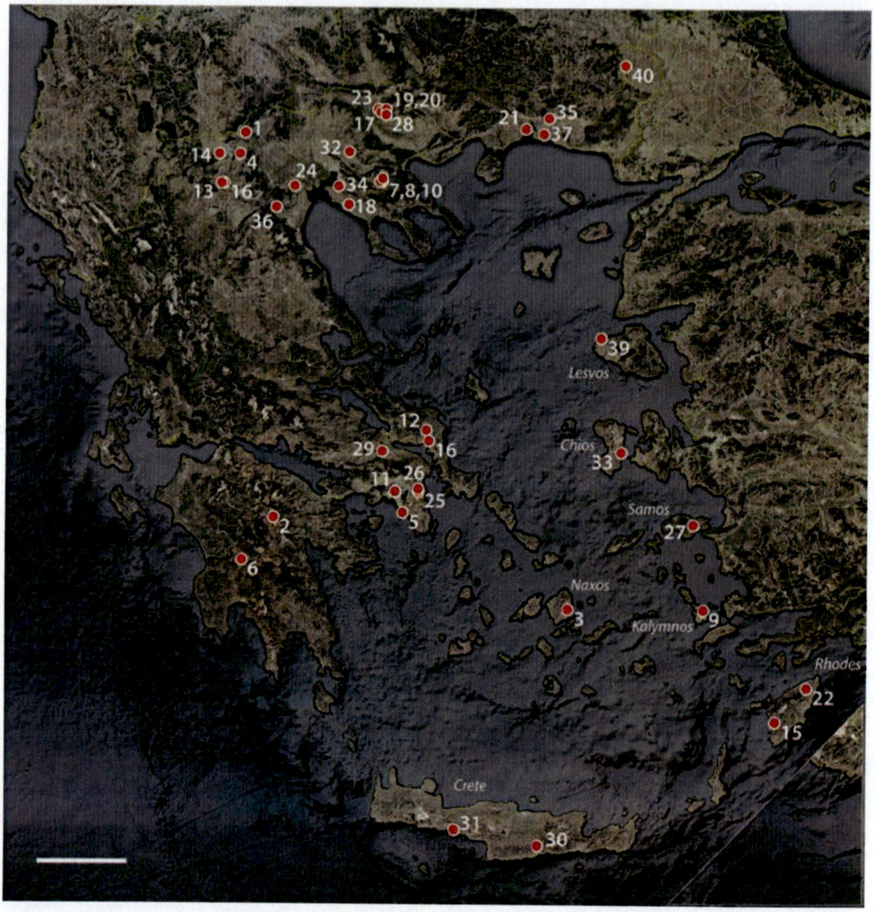

Fig. 1 Map of Greece showing the geographic distribution of the most important localities with insectivore fossils. **1**, Loutra Almopias Cave; **2**, Kitseli pothole; **3**, Naxos; **4**, Arnissa; **5**, Varkiza 2; **6**, Choremi localities; **7**, Apollonia; **8**, Ravin Voulgarakis; **9**, Kalymnos; **10**, Marathoussa; **11**, Tourkovounia localities; **12**, Limni 6; **13**, Ptolemais localities; **14**, Vevi; **15**, Apolakkia; **16**, Kardia; **17**, Spilia localities; **18**, Nea Silata; **19**, Maramena; **20**, Ano Metochi; **21**, Kessani; **22**, Maritsa 1; **23**, Monasteri; **24**, Paliambela-A; **25**, Pikermi; **26**, Chomateri; **27**, Samos-S3; **28**, Lefkon; **29**, Biodrak; **30**, Kastellios K1; **31**, Plakias; **32**, Chryssavgi 1; **33**, Thymiana localities; **34**, Antonios; **35**, Komotini; **36**, Moschopotamos; **37**, Karydia; **38**, Aliveri; **39**, Lapsarna; **40**, Kyprinos. See Appendix for more information. Image exported from Google Earth Pro © 2019, map data from US Dept. of State Geographer, SIO, NOAA, U.S. Navy, NGA, GEBCO, image from Landsat/Copernicus. Scale bar equals 80 km, North faces upward

inadequate to allow identification even to generic level (e.g., Lapsarna, Thymiana, Monasteri, Choremi), whereas materials from other localities have not been studied adequately yet (e.g., localities in the Ptolemais Basin). Some other localities, however, have offered quite diverse eulipotyphlan faunal lists: the early Miocene localities Aliveri and Karydia have revealed six and eight species, respectively (Doukas

1986; Doukas and Van den Hoek Ostende 2006), the locality Maramena correlated with the Turolian/Ruscinian boundary has revealed eleven species (Doukas et al. 1995), the Pliocene locality Apolakkia and the early Pleistocene localities Marathoussa (North Greece) and Ravin Voulgarakis six species each (Doukas 2005; Koufos et al. 2001; Koliadimou 1996), and the late Pleistocene localities Loutra Almopias Cave LAC and LAC Ia eight and six species, respectively (Chatzopoulou 2014). Nevertheless, these findings are far from enough to present a complete picture of the Neogene and Quaternary eulipotyphlan faunas in Greece and their evolution.

5 Systematic Paleontology

Eulipotyphla Waddell, Okada, and Hasegawa, 1999

Erinaceidae Fischer, 1814

Type Genus *Erinaceus* Linnaeus, 1758.

Remarks Vasileiadou and Zouros (2012) reported the presence of Erinaceidae indet. at the early Miocene locality Lapsarna: they found the metacone area of a right M1, with a size similar to that of *Schizogalerix pasalarensis* and *Schizogalerix anatolica*.

Subfamily Galericinae Pomel, 1848

Type Genus *Galerix* Pomel, 1848.

Remarks The generic assignment of many galericine species in one of the genera *Galerix*, *Schizogalerix*, and *Parasorex*, and their evolutionary relationships are still a matter of debate (Ziegler 2005; Doukas and Van den Hoek Ostende 2006; Prieto et al. 2012). Since the present work does not aim to revise the subfamily, we follow Van den Hoek Ostende (2001), Doukas and Van den Hoek Ostende (2006), Prieto et al. (2010, 2011) and Zijlstra and Flynn (2015) in the galericine taxonomy.

Parasorex von Meyer, 1865

Type Species *Parasorex socialis* von Meyer, 1865.

Other Taxa Included *P. depereti* (Crochet, 1986); *P. ibericus* (Mein and Martín-Suarez, 1993); *P. pristinus* (Ziegler, 2003); *P. kostakii* (Doukas and Van den Hoek Ostende, 2006).

Distribution From the early–late Miocene of Austria, France, Germany, Greece, Hungary, Italy, Moldova, Romania, Spain, Switzerland to the Pliocene of France, Spain.

★*Parasorex kostakii* (Doukas and Van den Hoek Ostende, 2006)

Nomenclatural and Taxonomical History *Galerix* sp. in Doukas (2005) (initial identification); *Galerix kostakii* in Doukas and van den Hoek Ostende (2006) (new species); *Parasorex kostakii* in Prieto et al. (2012) (new combination).

Type Material Left M1, KRD 3/04, Department of Paleontology and Historical Geology, University of Athens (Doukas and van den Hoek Ostende 2006: Pl. 1.5).

Type Locality Karydia, Thrace, Greece, early Miocene (MN4).

Distribution The species has until now only been found in its type locality.

Remarks Doukas and Van den Hoek Ostende (2006) described the Karydia material (one d3, one d4, six p4, eight m1, five m2, one m3, one C, 10 P3, two P4, 10 M1, 17 M2, 10 M3, including the holotype; Fig. 2) as *Galerix kostakii* (in memory of our dear colleague Constantin "Kostaki" Theocharopoulos), even though they mentioned that some of its characters are typical for *Schizogalerix* and *Parasorex*, such as the presence of a hypocone on all P3 and the bi-partitioned posterior cingulum on all M1/2. They also mentioned the absence of the protocone–metaconule connection from around 80% of the M1/2. What led them to the grouping of this species within *Galerix* is mainly the p4 morphology, which lacks a continuous paralophid connecting the tip of the protoconid to the paraconid and bears a small or even no metaconid. However, the (rare) presence of a paralophid on the p4, as well as the absence of a protocone–metaconule connection from the M1, and its rare presence on the M2 (on 3 out of 13) of the species from its type locality led Prieto et al. (2012) to classify the species within the genus *Parasorex*. The above authors explained the occasional presence of a protocone–metaconule connection on the upper molars of the species from Karydia, as well as of their *Parasorex* sp. from Sámsonháza 3 (Hungary), as the result of the relatively narrow shape of the molars of these two primitive *Parasorex* species and preferred to evaluate this character as a less important one for the generic attribution.

Galerix Pomel, 1848

Type Species *Viverra exilis* de Blainville, 1839.

Other Taxa Included *G. stehlini* (Gaillard, 1929); *G. africanus* Butler, 1956; *G. rutlandae* Munthe and West, 1980; *G. symeonidisi* Doukas, 1986; *G. aurelianensis* Ziegler, 1990; *G. saratji* Van den Hoek Ostende, 1992; *G. uenayae* Van den Hoek Ostende, 1992; *G. iliensis* (Kordikova, 2000); *G. remmerti* Van den Hoek Ostende, 2003; *G. wesselsae* Zijlstra and Flynn, 2015.

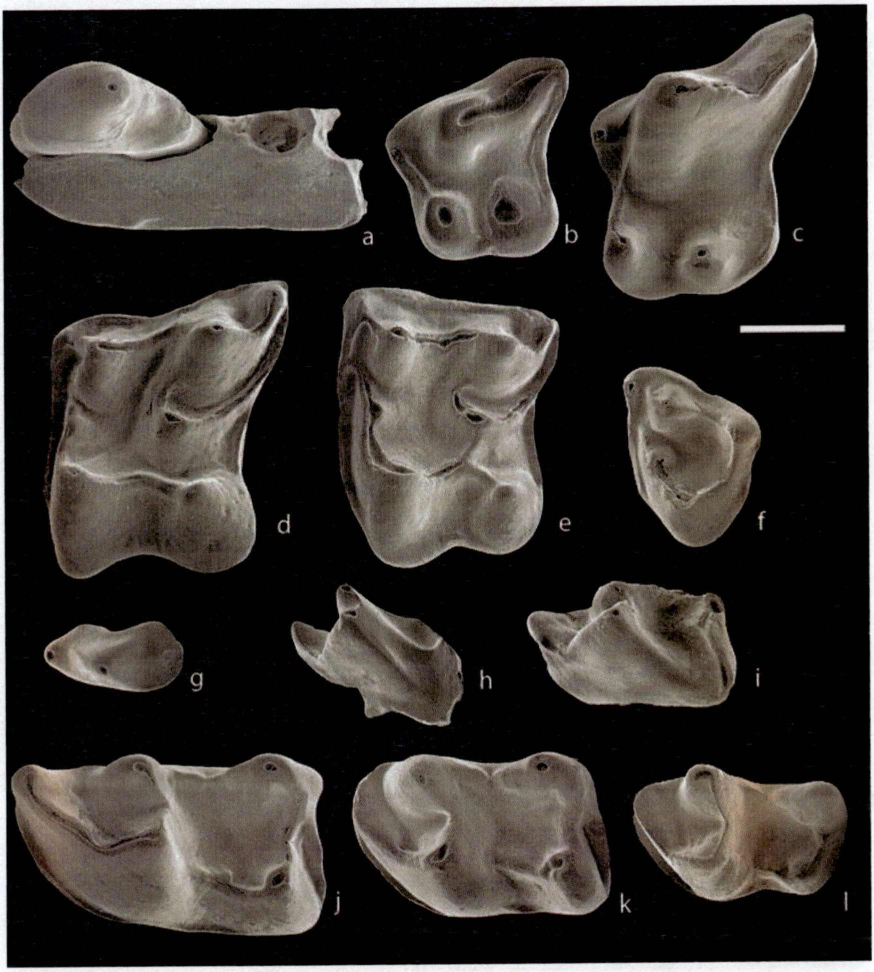

Fig. 2 *Parasorex kostakii* from Karydia (type locality). (**a**) Partial maxilla with C, (**b**) P3, (**c**) P4, (**d**) M1 holotype, (**e**) M2, (**f**) M3, (**g**) d3, (**h**) d4, (**i**) p4, (**j**) m1, (**k**) m2, (**l**) m3. Images CD, personal data. All specimens are figured as left; **f**, **h**, **i**, **k**, **l** are reversed. Scale bar equals 1 mm

Distribution Latest Oligocene of Turkey to the early–middle Miocene of Eurasia (Portugal, Spain, France, Switzerland, Austria, Germany, Czech Republic, Hungary, Moldova, Serbia, Greece, Turkey, Russia, Pakistan).

Remarks Apart from *Galerix symeonidisi* (see below), the only material from Greece attributed to this genus is a single M1 from Komotini, described as *Galerix* sp. by Doukas and van den Hoek Ostende (2006) (mentioned as *Schizogalerix* sp. in Doukas 2005), which is larger and more robust than the M1 of its contemporary galericine *Parasorex kostakii* from Karydia.

★*Galerix symeonidisi* Doukas, 1986

Nomenclatural and Taxonomical History *Galerix symeonidisi* in Doukas 1983 (nomen nudum); *Galerix symeonidisi* in Doukas 1986 (new species).

Type Material Left M2, AL 3036, Department of Paleontology and Historical Geology, University of Athens (Doukas 1986:Pl. 1.6).

Type Locality Aliveri, Evia Island, Greece, early Miocene (MN4).

Distribution Early Miocene (MN4) of Greece (Aliveri); early Miocene (MN4 to the lowermost part of MN5) of Serbia, Germany, Austria, and Spain.

Remarks The materials from Aliveri (Fig. 3) comprise 72 dental elements (including the holotype: one p2, three p3, 14 p4, two m1, nine m2, four m3, seven P3, three P4, four M1, 14 M2, 11 M3). The species was initially described in Doukas (1983; PhD thesis, in Greek). As this publication does not meet the criteria of the ICZN, it constituted nomen nudum and was later made available for the same concept by Doukas (1986). This new *Galerix* species was described on the basis of its small size, the longer p2 than p3, and the presence of two lingual cusps on the P3, of which the posterior side is noticeably emarginated.

Schizogalerix Engesser, 1980

Type Species *Schizogalerix anatolica* Engesser, 1980.

Other Taxa Included *S. zapfei* (Bachmayer and Wilson, 1970); *S. voesendorfensis* (Rabeder, 1973); *S. moedlingensis* (Rabeder, 1973); *S. pasalarensis* Engesser, 1980; *S. sarmaticum* (Lungu, 1981); *S. sinapensis* Sen, 1990; *S. macedonica* Doukas, 1995 in Doukas et al., 1995; *S. duolebulejinensis* Bi et al., 1999; *S. intermedia* Selänne, 2003; *S. evae* De Bruijn et al., 2006.

Distribution Miocene of Eurasia (France, Switzerland, Austria, Germany, Romania, Slovakia, Greece, Turkey, Moldova, Ukraine, Pakistan, Kazakhstan, China).

Remarks Dental elements attributed to *Schizogalerix* sp. have been found in several Miocene–Pliocene localities in Greece. In Antonios, *Schizogalerix* sp. is represented by one d3, three m1, one m2, one m3, two P4, two M1, six M1/2, and seven M3 and shows mixed morphology found in both *S. pasalarensis* and *S. anatolica* (Vasileiadou and Koufos 2005). De Bonis et al. (1997) reported the presence of *Schizogalerix* sp. in Thymiana A and C, but no information on the material has been published. One p2, one p4, and one M3 from Chryssavgi 1 were attributed to cf. *Schizogalerix* sp. by Koliadimou (1996). One M2 and one M3 from Kastellios K1 included in the collections of the Faculty of Geosciences, Utrecht University, the Netherlands, were attributed to *Schizogalerix* sp. by Doukas (2005), but Furió et al. (2014:166) commented that "*The material from this Greek locality fits better the*

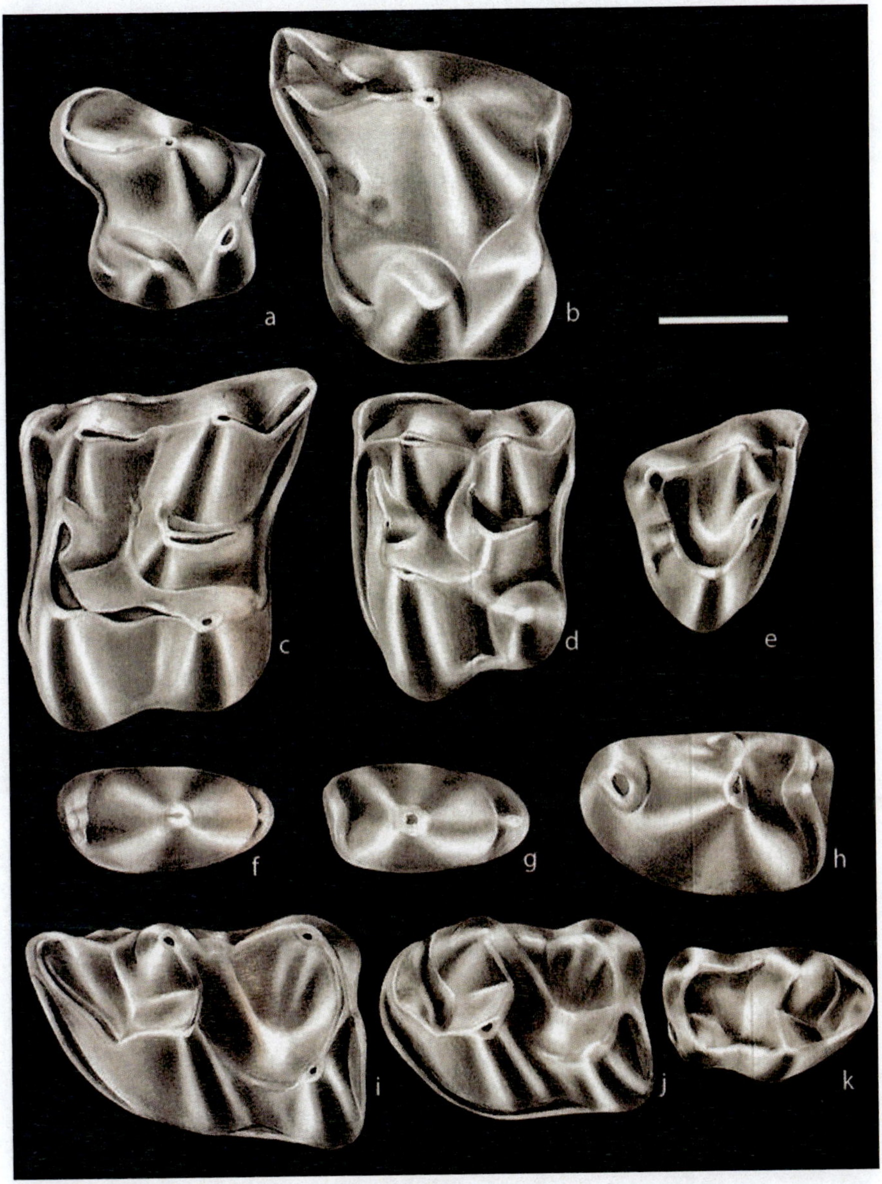

Fig. 3 *Galerix symeonidisi* from Aliveri (type locality). (**a**) P3, (**b**) P4, (**c**) M1, (**d**) M2 holotype, (**e**) M3, (**f**) p2, (**g**) p3, (**h**) p4, (**i**) m1, (**j**) m2, (**k**) m3. Images CD, personal data. All specimens are figured as left; a, b, f, g, k are reversed. Scale bar equals 1 mm

morphology of other Eastern Europe species like S. macedonica *or* S. zapfei/S. attica"). Two p4, one m1, one m2, three m3, one M1, and one M3 (most specimens fragmentary) from Biodrak were initially described as *Galerix* sp. by Rümke (1976), but transferred to *Schizogalerix* sp. by Doukas (2005). Finally, Doukas (1989) mentioned the presence of *Schizogalerix* sp. in Lefkon, but there is no additional information on this material.

★*Schizogalerix macedonica* Doukas, 1995 in Doukas et al. (1995)

Type Material Left M1, MA 3041, Department of Paleontology and Historical Geology, University of Athens (Doukas et al. 1995:Pl. 2.2).

Type Locality Maramena, Strimon basin, N Greece, Turolian/Ruscinian boundary.

Distribution The species has been reported only from Greece, from its type locality and the nearby locality Ano Metochi 2–3.

Remarks Doukas described this new species from Maramena (in Doukas et al. 1995) based on its large size, the clearly divided mesostyles, and the presence of an extra cuspule next to the hypocone on the M1–2, as well as the presence of an extra cuspid at the base of the entoconid on the ml–2 (better developed on the m2 than on the ml). This species is one of the largest and youngest *Schizogalerix* species. Maramena (Fig. 4) and Ano Metochi 2–3 have collectively yielded 40 isolated dental elements (one p3, three p4, three m1, six m2, six P3, two P4, nine M1, eight M2, two M3, including the holotype) (Doukas et al. 1995).

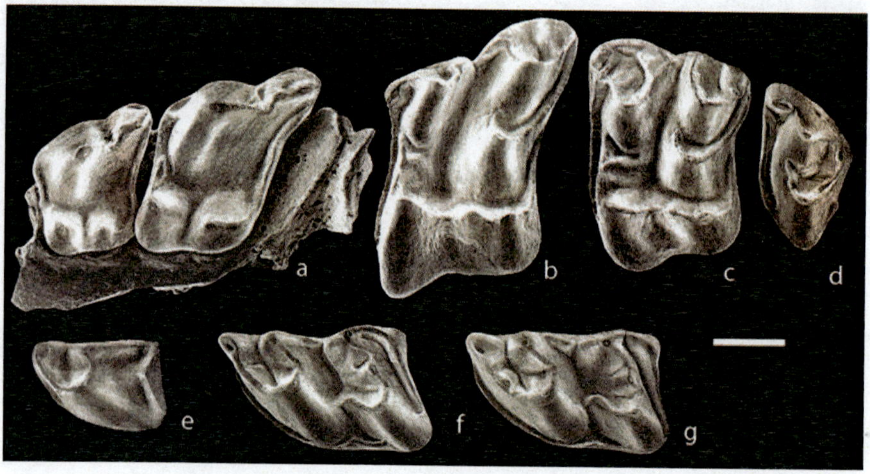

Fig. 4 *Schizogalerix macedonica* from Maramena (type locality). (**a**) Maxillary fragment with P3 and P4, (**b**) M1 holotype, (**c**) M2, (**d**) M3, (**e**) p4, (**f**) ml, (**g**) m2. Images CD, personal data. All specimens are figured as left. Scale bar equals 1 mm

Schizogalerix moedlingensis (Rabeder, 1973)

Nomenclatural and Taxonomical History *Galerix moedlingensis* in Rabeder (1973) (new species); *Schizogalerix moedlingensis* in Engesser (1980) (new combination).

Type Material Left M2, UWPI 1953/10/4, Institut für Paläontologie der Universität Wien (Rabeder 1973:Abb. 3.19).

Type Locality Eichkogel, Austria, late Miocene, early Turolian (MN11).

Distribution Late Miocene of Austria, Greece and possibly Ukraine (*S.* cf. *moedlingensis*).

Remarks The presence of a hypocone on the P3, of a posterior arm joining the metaconule with the posterior cingulum on the M1–2 and of a paralophid and a strong metaconid on the p4 led Engesser (1980) to transfer this species to *Schizogalerix*. The deeply divided mesostyle on the M1–2 is considered as an evolved character that differentiates this species from earlier *Schizogalerix*. It differs from *S. zapfei* on the presence of a hypoconid–entoconid ridge on the lower molars. Other diagnostic characteristics are the long talon on the P4 with median longitudinal ridge and the relatively large M3 with a distinct protoconule (Rabeder 1973). The species has been reported from two Greek localities: Pikermi–Chomateri and Pikermi. The Pikermi–Chomateri material consists of 59 isolated dental elements, some fragmentary (one p2, six p3, five p4, four m1, 14 m2, nine m3, two P3, one P4, seven M1, three M2, seven M3), which, according to Rümke (1976), differ slightly in morphology and size from the type material from Eichkogel. Thenius (1952) described one partial mandible with p2–m2 from the classic locality Pikermi (not to be confused with Pikermi–Chomateri) as *Galerix exilis*. The mandible, also mentioned by Rümke (1976) and Engesser (1980), was recovered from the collections of the Department of Paleontology and Historical Geology (University of Athens) and transferred to *S. moedlingensis* by Doukas et al. (1995:48) due to "*the pattern of the hypoconid–entoconid–posterior cingulum complex.*"

Schizogalerix zapfei (Bachmayer and Wilson, 1970)

Nomenclatural and Taxonomical History *Galerix zapfei* in Bachmayer and Wilson (1970) (new species); *Schizogalerix zapfei* in Engesser (1980) (new combination). ★*Galerix atticus* Rümke, 1976, is considered a junior synonym of *S. zapfei*, according to Engesser (1980).

Type Material Right lower jaw with p3–m3, No 1970/1386, Naturhistorisches Museum Wien (Bachmayer and Wilson 1970:Pl. 5. 17–17a).

Type Locality Kohfidisch, S Burgenland, Austria, late Miocene, Turolian (MN11).

Distribution Besides its type locality, this species has also been found in the late Miocene of Greece (Pikermi–Chomateri and S3 in Samos Island) (described as the species *Galerix atticus* Rümke, 1976).

Remarks The mesostyles of *S. zapfei* are slightly more developed than in *Galerix exilis*; there is a single alveolus for the p2, and the posterior cingulum of the m1–2 is continuous to the tip of the entoconid (Bachmayer and Wilson 1970).

★*Galerix atticus* Rümke, 1976

[Junior subjective synonym of *Schizogalerix zapfei* (Bachmayer and Wilson, 1970) according to Engesser (1980)].

Type Material Left M2, PK 606, Department of Palaeontology and Historical Geology, University of Athens (Rümke 1976:PI. 1.9).

Type Locality Quarry of the Pikermi Brick and Tile Factory (Pikermi–Chomateri), lignitic clay with mollusk fragments (Turolian).

Remarks In 1976, Rümke named the new species *Galerix atticus*, based on an assemblage of one p2, one p3, one p4, five m1, four m2, three m3, seven P3, one P4, two M2, one M3 (including the holotype) from the Turolian locality Pikermi–Chomateri. She mentioned that *G. atticus* differs from previously described *Galerix* species (apart from *G. zapfei*) in the isolated position of the hypoconid of the m1–2. Rümke (1976) noticed that the morphology of the talonid of her species and *G. zapfei* is very close, based on the descriptions given by Bachmayer and Wilson (1970). She also noticed the overall similarities of *G. atticus* with *G. zapfei*, with the exception of the p2, which seems to be single-rooted in the Austrian but double-rooted in the Greek species. Later, Bachmayer and Wilson (1980:358–359) noted that the type locality of *G. atticus* had only revealed one p2 with the roots broken off, and the only proof of the presence of two roots was the "*constriction of the crown enamel*." The same authors stated the probable presence of both single-rooted and double-rooted p2 in the type locality of *G. zapfei*, and the possibility that "G. atticus *may eventually prove to be a synonym of* G. zapfei" (Bachmayer and Wilson 1980:360). Engesser (1980) synonymized the two species and transferred *zapfei* to the genus *Schizogalerix*. Black et al. (1980) assigned one m1, one P4, and one M2 from the locality S3 in Samos Island to Rümke's species; these specimens can be thus safely referred to *S. zapfei*.

***Lantanotherium* Filhol, 1888**

Type Species *Erinaceus sansaniense* Lartet, 1851.

Other Taxa Included *L. robustum* Viret, 1940; *L. sanmigueli* de Villalta and Crusafont, 1944; *L. longirostre* Thenius, 1949; *L. piveteaui* Crusafont, de Villalta and Truyols, 1955; *L. sawini* James, 1963; *L. dehmi* James, 1963; *L. lactorensis* Baudelot and Crouzel, 1976; *L. sabinae* Mein and Ginsburg, 2002.

Distribution Miocene of Austria, China, Czech Republic, France, Germany, Greece, Hungary, Poland, Portugal, Serbia, Slovakia, Spain, Switzerland.

Remarks The genus name is often misspelled as *Lanthanotherium*, initially by its author Filhol in 1891 and by many authors since then (e.g., Rabeder 1973; see Furió and Alba 2011 for further examples). Furió and Alba (2011) presented a detailed discussion on the issue and concluded that the original spelling (*Lantanotherium*) is correct, and all other subsequent spellings (*Lanthanotherium*) are unjustified emendations.

Lantanotherium sanmigueli de Villalta and Crusafont, 1944

Nomenclatural and Taxonomical History *Lantanotherium sanmigueli* in de Villalta and Crusafont (1944) (new species); *Lanthanotherium sanmigueli* in Rabeder (1973) (unjustified emendation of the generic name).

Type Material Partial left mandible with m1–2, Institut de Paleontologia Miquel Crusafont de Sabadell, Barcelona (de Villalta and Crusafont 1944:Fig. 10; Pl. II.5).

Type Locality Can Purull, Vallès–Penedès Basin, Spain, late Miocene (MN10).

Distribution Late Miocene (lower Vallesian, MN9) of southern Greece (Plakias); late Miocene (Vallesian–Turolian, MN9–11) of Spain, France, Austria, Germany, Czech Republic, Hungary, China.

Remarks The species has been assigned to *Lantanotherium* based on the centrally placed and isolated metaconule on the M1–2, the presence of the "hypocone" on the M3, and the shape of the protocristid in the lower molars, characters unique, and easily recognizable for this genus. The material found in Plakias includes one p1 one p2, three p3, two p4, one m2, one P1, one P2, one M1, one M2, and one M3; De Bruijn et al. (2012) concluded that the Plakias species has a complete dental formula, unlike *L. sansaniense* and similar to *L. sanmigueli* from Austria; the small size of the dental elements also led to the allocation to the latter species.

Subfamily Erinaceinae Fischer, 1817

Type Genus *Erinaceus* Linnaeus, 1758.

Remarks The presence of the erinaceine *Palerinaceus* sp. in Ravin de la Pluie has been reported by de Bonis et al. (1986) without description of the material. During the past few years, some of the *Palerinaceus* species have been transferred to the genus *Amphechinus*, whereas the former *Palerinaceus* subgenus *Postpalerinaceus* was upgraded to generic level, leaving no species in the genus *Palerinaceus*. Since the Ravin de la Pluie material cannot be located, we cannot confirm the presence of any of the genera. As there is no information on the Ravin de la Pluie erinaceine, its presence is not mentioned in the Supplementary Information.

Erinaceus Linnaeus, 1758

Type Species *Erinaceus europaeus* Linnaeus, 1758 (extant).

Other Taxa Included Extant: *E. concolor* Martin, 1838, *E. amurensis* (Schrenk, 1858), *E. roumanicus* Barrett-Hamilton, 1900. extinct: *E. mongolicus* Schlosser, 1924, *E. praeglacialis* Brunner, 1933, *E. lechei* Kormos, 1934, *E. olgai* Young, 1934, *E. koloshanensis* Young and Liu, 1950, *E. samsonowiczi* Sulimski, 1959, *E. ostramosi* Jánossy, 1972, *E. davidi*, Jammot 1973.

Distribution From the late Miocene (MN10) to the present of Europe, Middle East, Russia, China.

Remarks *Erinaceus* sp. has been reported from the locality Nea Silata (one partial M1; Vasileiadou 2001) and from Tourkobounia 1 (one p3, two m2, and one I3; Reumer and Doukas 1985). Fossil remains of the extant *E. europaeus* have been described from the Biharian locality Apollonia (one partial mandible with i1, p4–m3; Koliadimou 1996) and of *E.* cf. *europaeus* from the late Pleistocene locality Loutra Almopias Cave LAC (one M1 and one M3; Chatzopoulou 2014).

Erinaceus praeglacialis Brunner, 1933

Type Material Brunner (1933) did not designate a holotype for his new species. The available material from Windloch based on which he described the species (thus syntypes) consists of one partial left mandible with p3–4 and m1, one partial left mandible with I1, I3 and C, one partial left mandible with m1, one partial right mandible with i1, and one isolated left m2 (Brunner 1933:Pl. 6.9–10).

Type Locality Windloch, Germany, early Pleistocene.

Distribution From the early Pleistocene of Greece to the early–middle Pleistocene of Europe (Spain, Austria, Germany, Czech Republic, Hungary, Italy).

Remarks This is a large *Erinaceus* species, with dental dimensions close to those of the extant European hedgehog *E. europaeus*. Brunner (1933) mentions that *E. praeglacialis* is characterized by a size slightly larger than that of the extant *E. europaeus*, by the elongated metastyle on the P4 and M2 and a special morphology and position of the upper P3. Rabeder (1972) noticed that the M3 of *E. praeglacialis* is larger than that of any other species. Rzebik-Kowalska (2000) considered that the longer metacrista on the M1, with a less labial and more posterior orientation, differentiates the species from *E. europaeus*. Reumer and Doukas (1985:114) differentiate their

E. praeglacialis from Tourkobounia 5 (one M1 and one M2) from *E. europaeus* based on the "considerably less trapezoidal M2" of the former species.

***Erinaceus samsonowiczi* Sulimski, 1959**

Type Material Left lower jaw with p3–m3 and partial articular processes, No. 1051, Museum of the Earth, Polish Academy of Sciences, Warsaw (Sulimski 1959:Pl. II.2).

Type Locality Węże, Poland, Pliocene, Ruscinian (MN15).

Distribution Turolian/Ruscinian boundary (MN13/14) of Greece; late Miocene (MN10) of the Czech Republic; Pliocene of Poland.

Remarks Sulimski (1959) described the species based on its small size that differentiates it from other European species (apart from *E. lechei*). He based his species in the following: "*characteristic low position of mental foramen, smaller length of the tooth-row i1–m3, strong structure of the whole lower jaw, smaller dimensions of cardinal length, differences of length in tooth-rows p3–m3 and p4–m3, shape of p3 and p4 which, as a rule, are with a slightly posteriorly curved protoconid, finally frequent strong reduction of trigonid in m3*" (Sulimski 1959:32). An assemblage of four p4, two ml, four m2, one m3, four P3, three P4, three M1, and four M2 from Maramena, initially reported as *Erinaceus* sp. in Doukas (1989), was later described and attributed to *E. samsonowiczi* by Doukas et al. (1995). It shows minor metrical and morphological differences from the material from the type locality (i.e., absence of buccal cingulum on the p4, presence of buccal cingulum on the m2 and fairly better developed metaconule on the M2; Doukas et al. 1995).

Dimylidae Schlosser, 1887

Type Genus *Dimylus* von Meyer, 1846.

***Plesiodimylus* Gaillard, 1897**

Type Species *Plesiodimylus chantrei* Gaillard, 1897.

Other Taxa Included *P. huerzeleri* Müller, 1967; *P. crassidens* Engesser, 1980; *P. bavaricus* Schötz, 1985; *P. helveticus* Bolliger, 1992; *P. johanni* Kälin and Engesser, 2001; *P. gaillardia* Mein and Ginsburg, 2002; *P. similis* Fejfar and Sabol, 2009; *P. ilercavonicus* Crespo et al., 2017.

Distribution Miocene (MN3–11) of Europe (Spain, France, Switzerland, Austria, Germany, Poland, Czech Republic, Greece) and middle Miocene (MN7/8) of Turkey.

Comments The taxonomy of *Plesiodimylus* is still mostly unresolved (see discussion in Klietmann et al. 2014:474–476), with many species considered to be invalid (e.g., *P. bavaricus* in Van den Hoek Ostende, 1995); since this work does not aim to revise the genus, we here list all the species that have been assigned to the genus, without discussing their validity.

Plesiodimylus crassidens Engesser, 1980

Type Material Left M1, Sar. 109, Naturhistorisches Museum, Basel (Engesser 1980:Abb. 27.a).

Type Locality Sari Çay, Turkey, middle Miocene (MN7/8).

Distribution From the middle Miocene of Turkey (MN7/8) and Czech Republic (MN10) to the early Miocene (MN4) of Greece.

Remarks The teeth of *P. crassidens* are large, with thick enamel, bulky cusps, and shallow valleys; the M1 bears a strong parastyle, no mesostyle and very weak or no cusp in the central valley; the m1 bears a low paraconid and a very reduced trigonid valley; the m2 has a wider trigonid valley, but the talonid is reduced (Engesser 1980). The species shows more exodaenodontous (overlapping dentary in occlusal view) and amblyodontous (inflated) teeth than all other *Plesiodimylus* species, characters that have been linked with a higher specialization in a malacophagous diet (Ziegler 1999). The few dental elements from the early Miocene locality Karydia (one p4, one m1 and two m2, two P4, two M2) have dimensions intermediate between *P. crassidens* from Sari Çay and *P. huerzeleri* from Aliveri. There are morphological similarities with the Turkish species (e.g., thick enamel, low paracristid not reaching the tip of the protoconid on the m1), but also important differences (e.g., the posterior width of the p4 and m2 is more reduced in the Turkish than in the Karydia species) (Doukas and van den Hoek Ostende 2006). As the morphology of the M1 plays a key role for the diagnosis of the *Plesiodimylus* species, the absence of this molar from the Karydia assemblage led the authors to the assignment of their material to *P.* aff. *crassidens*.

Plesiodimylus huerzeleri Müller, 1967

Nomenclatural and Taxonomical History *Plesiodimylus huerzeleri* in Müller (1967) (new species). The species has frequently been considered a junior synonym of *P. chantrei* (e.g., Engesser 1976, 1980; Doukas 1986), but other authors consider it as a valid species (e.g., Schötz 1985; Ziegler 1999; Kälin and Engesser 2001; Fejfar and Sabol 2009; Klietmann et al. 2014). We here follow the latter opinions and maintain both as valid species.

Type Material Partial left mandible with m1–2, Slg. M 1937 II 10460, Bayerische Staatssammlung für Paläontologie und Historische Geologie München (Müller 1967:Pl. 1.8a–b).

Type Locality Wintershof-West, Germany, early Miocene (MN3).

Distribution Early Miocene (MN3–4) of Central Europe (France, Germany, Czech Republic) and early Miocene (MN4) of Greece.

Remarks The diagnosis by Müller (1967) mentions that the m1–2 are equally large, the m1 is triangular, the hypoconid and entoconid are equally large, and the talonid valley is small and labially slightly open on the m1, the talonid valley opens lingually and slightly labially on the m2, the M1 is only slightly wider posteriorly than anteriorly, the parastyle is prominent, the mesostyle small and not split, and the protocone and hypocone are almost equally large. According to Klietmann et al. (2014:475), *P. huerzeleri* from its type locality includes M1 of their Morphotype I of *P.* aff. *chantrei* from Petersbuch 28: "*the mesostyle is small, the parastyle just a curved ridge; the anterior arm of the protocone ends at the parastyle or between parastyle and paracone; therefore, there are only two basins, formed by the paracone, parastyle, protocone and central cuspule or central cuspule, hypocone, metacone and posterior cingulum, respectively; the margins of all cusps are round*" (Klietmann et al. 2014:473). Doukas (1983), in his PhD thesis (in Greek) assigned two m2, one P4, and one M1 from Aliveri to *P. huerzeleri*, after comparisons with *P. chantrei* from La Grive. Later, the author agreed with Engesser (1976) in synonymizing the two species and described the Aliveri material as *P. chantrei* (Doukas 1986). Van den Hoek Ostende et al. (2015:276), in a revision of the Aliveri materials, listed the Aliveri dimylid as *P. huerzerleri*. Until the taxonomy of *Plesiodimylus* is better known, we prefer to retain both species as valid.

Heterosoricidae Viret and Zapfe, 1951

Type Genus *Heterosorex* Viret and Zapfe, 1951.

Dinosorex Engesser, 1972

Type Species *Dinosorex pachygnathus* Engesser, 1972.

Other Taxa Included *D. sansaniensis* (Lartet, 1851); *D. zapfei* Engesser, 1975; *D. anatolicus* Van den Hoek Ostende, 1995; *D. grycivensis* Rzebik-Kowalska and Topachevsky, 1997; *D. engesseri* Rabeder, 1998.

Distribution Late Oligocene–late Miocene of Europe (Spain, France, Switzerland, Austria, Germany, Poland, Slovakia, Czech Republic, Hungary, Moldova, Ukraine, Serbia, Greece) and Turkey.

Remarks Among the few representatives of Heterosoricidae in Greece, *Dinosorex* sp. is present with one lower antemolar, two m1, two m3, four A1, and three M1/2 in the late Oligocene locality Kyprinos (Doukas and Theocharopoulos 1999). In comparison with the late Oligocene *Quercysorex huerzeleri* (originally described as a

Dinosorex) and the early Miocene *D. anatolicus*—*Dinosorex* sp. from Kyprinos is larger than the former and smaller than the latter, whereas there are also some morphological differences of the Greek species with both species (Doukas and Theocharopoulos 1999). The absence of P4 from Kyprinos, an element essential for the *Dinosorex* taxonomy, hampered a specific identification. Moreover, in the absence of the diagnostic lower incisor, we even cannot exclude that the Kyprinos species represents a *Quercysorex*. Further, Van den Hoek Ostende (1995) mentioned the presence of *Dinosorex* sp. in the early Miocene locality Komotini. However, Doukas and Van den Hoek Ostende (2006), in their report on the insectivores from Karydia and Komotini, mentioned that the material seen by the latter author in 1995 could not be recovered. The occurrence of this genus in Komotini is, thus, uncertain.

Heterosorex Gaillard, 1915

Type Species *Heterosorex delphinensis* Gaillard, 1915.

Other Taxa Included *H. neumayrianus* (Schlosser, 1887); *H. neumayrianus subsequens* (Doben-Florin, 1964); *H. ruemkeae* Doukas, 1986; *H. wangi* Storch and Qiu, 1991.

Distribution From the early–middle Miocene of Europe (Spain, France, Switzerland, Austria, Germany, Poland, Slovakia, Czech Republic, Slovak Republic, Greece) to the late Miocene (MN11/12) of China; possible presence in the late Oligocene of Germany.

Remarks Apart from *H. ruemkeae* from Aliveri (see below), the only other Greek representative of the genus has been reported from Karydia, described as *Heterosorex* sp. (Doukas and Van den Hoek Ostende 2006; initially reported as *Dinosorex* sp. by Doukas, 2003). According to Doukas and Van den Hoek Ostende (2006), the Karydia species (two i1, one lower antemolar, one I1, three upper antemolars, two P4, four M1, and two M2) can be attributed to either *H. ruemkae* or *H. neumayrianus*, but due to the scarcity of the material and the absence of complete mandibles, identification at the species level is not possible.

★*Heterosorex ruemkeae* Doukas, 1986

Type Material Left M2, AL 3361, Department of Paleontology and Historical Geology, University of Athens (Doukas 1986:Pl. 4.4).

Type Locality Aliveri, Greece, early Miocene (MN4).

Distribution Early Miocene (MN4) of Greece (Aliveri) and Serbia (Snegotin; Marković 2003; Marković and Milivojević 2010).

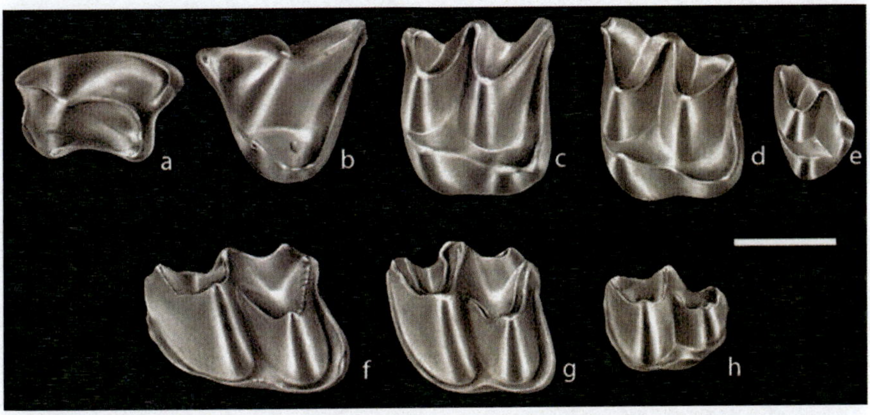

Fig. 5 *Heterosorex ruemkeae* from Aliveri (type locality). (**a**) Right A1, (**b**) right P4, (**c**) right M1, (**d**) left M2 (holotype), (**e**) left M3, (**f**) left ml, (**g**) left m2, (**h**) right m3. Images CD, personal data. All specimens are figured as left; **a**, **b**, **c**, **h** are reversed. Scale bar equals 1 mm

Remarks *Heterosorex ruemkeae* is considerably smaller than the other European *Heterosorex* species. Furthermore, its M1–2 bear an undivided mesostyle, its P4 has a strong and isolated parastyle, a narrow anterior cingulum and triangular to trapezoid outline, and its M3 bears a free-ending posterior arm on the protocone (Doukas 1986). The Aliveri material (five i1, six m1, nine m2, one m3, three I1, one A1, three P4, eight M1, four M2 and three M3; Fig. 5) was initially assigned to *Dinosorex huerzeleri* by Doukas (1983). Later, Doukas (1986) assigned the material to *Heterosorex*, based on the presence of a high entoconid crest on the m1–2 and the weak development of the lingual and posterior parts of the M1–2.

Plesiosoricidae Winge, 1917

Type Genus *Plesiosorex* Pomel, 1848.

Remarks The only plesiosoricid material found in Greece is one *Plesiosorex* sp. M2 from the early Miocene locality Karydia; although not possible to be identified to the species level, this material is important because it represents the southernmost occurrence of this family—Plesiosoricidae is present in Central Europe but absent from southern Europe (Doukas and Van den Hoek Ostende 2006).

Talpidae Fischer, 1814

Type Genus *Talpa* Linnaeus, 1758.

Remarks Talpid elements of unknown generic affinities have been reported from the late Oligocene locality Kyprinos (one m2; Doukas and Theocharopoulos 1999), the early Miocene locality Lapsarna (one m1 and one M3; Vasileiadou and Zouros 2012), and the Pliocene (MN15) locality Vevi (material under study; Doukas 2005).

Subfamily Talpinae Fischer, 1814

Type Genus *Talpa* Linnaeus, 1758.

Remarks In various cases, the finding of only a few dental elements with no mandibles and humeri hampers the precise identification of talpine assemblages, but nevertheless the attribution of these elements to the family-level is certain.

Myxomygale Filhol, 1890

Type Species *Myxomygale antiqua* Filhol, 1890.

Other Taxa Included *M. hutchisoni* (Ziegler, 1985); *M. engesseri* Doukas, 1986; *M. gracilis* Ziegler, 2003.

Distribution From the middle Oligocene to the middle Miocene of Europe (Spain, France, Austria, Germany, Czech Republic, Serbia, Greece).

★*Myxomygale engesseri* Doukas, 1986

Type Material Left M2, Al 3152, Department of Paleontology and Historical Geology, University of Athens (Doukas 1986:Pl. 2.2).

Type Locality Aliveri, Greece, early Miocene (MN4).

Distribution Early Miocene (MN4) of Greece (Aliveri) and Serbia (Snegotin; Marković and Milivojević 2010), possibly latest Oligocene of Germany (Eggingen-Mittelhart 2; Van den Hoek Ostende 1989).

Remarks The material from the type locality (Fig. 6) consists of one m2, one M1, two M2, and four M3 (including the holotype). Doukas (1986) described this species based on the small size of its dental elements, the presence of a prominent protoco-

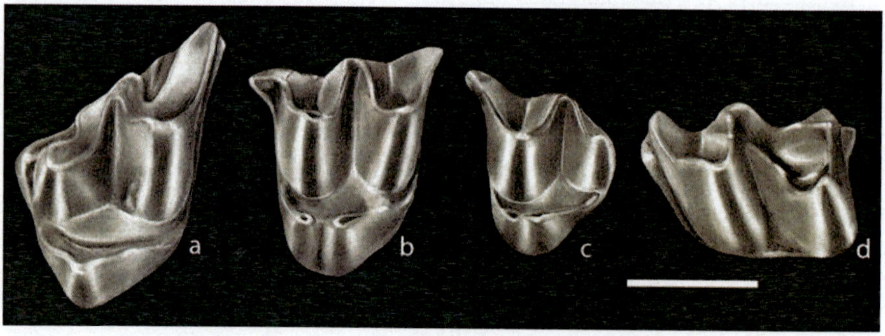

Fig. 6 *Myxomygale engesseri* from Aliveri (type locality). (**a**) M1, (**b**) M2 (holotype), (**c**) M3, (**d**) m2. Images CD, personal data. All specimens are left. Scale bar equals 1 mm

nule close to the protocone on the M2–3, the weak hypocone on all upper molars, and the absence of anterior cingulum on the Ml. Hugueney and Maridet (2017) described the new genus *Percymygale* to include species initially described as *Myxomygale* (i.e., *M. vauclusensis* Crochet, 1995 and *M. minor* Ziegler, 1990). They differentiated their genus from *Myxomygale* based on the presence of ten alveoli anterior to the p4 and the posterior mental foramen below the p4 on the mandible of the former genus versus the presence of seven alveoli anterior to the p4 and the posterior mental foramen below the m1 on the mandible of the latter genus. Since the mandible of *M. engesseri* is not known, Hugueney and Maridet (2017) mentioned that "M. engesseri *cannot be assigned to* Myxomygale *with certainty*"—however, they preferred to maintain its original combination of *M. engesseri* until more material is known.

Myxomygale hutchisoni (Ziegler, 1985)

Nomenclatural and Taxonomical History *Palurotrichus hutchisoni* in Ziegler (1985) (new species); *Myxomygale hutchisoni* in van den Hoek Ostende (1989) (new combination).

Type Material Partial left mandible with m2 and the alveoli of i1–m3, 1976 XXII 3176, Bayerische Staatssammlung für Paläontologie und Historische Geologie München (Ziegler 1985:Taf. 3.1).

Type Locality Petersbuch 2, Germany, early Miocene (MN4).

Distribution Early–middle Miocene of Europe.

Remarks Many of the diagnostic characters of the species are based on the morphol ogy of the humerus (e.g., distinctive "Scalopine ridge," with straight, mediolateral course, moderately deep fossa brachialis; moderately deep in lateral half fossa olecrani; Ziegler 1985). The dental diagnostic characters of the species include (from Ziegler 1985): the dental formula of the mandible is 3–1–3–3; the lower incisors are slightly enlarged; the p1, p2, and p3 are approximately of the same size and single-rooted; the lower molars are brachyodont; the crista obliqua on the m1 runs to approximately the middle of the trigonid posterior wall, whereas on the m2–3 it ends more lingually; mental foramens are situated under the m1 and the p1; mesostyle on the upper molars is unsplit; the metaconule is just present on the M1 and pronounced on the M2. Doukas and Van den Hoek Ostende (2006) reported the presence of one M2 and two M3 of a talpid they assigned to *M.* cf. *hutchisoni* at the Greek early Miocene locality Karydia. The molars are larger than those of *M. engesseri* and *M. gracilis* and closer in size to those of *M. hutchisoni*, but the scarcity of the material did not allow the certain assignment to the latter species (Doukas and Van den Hoek Ostende 2006).

Urotrichus Temminck, 1841

Type Species *Urotrichus talpoides* Temminck, 1841 (only extant species).

Other Taxa Included Extinct: *U. dolichochir* Gaillard, 1889; *U. giganteus* Ziegler, 2006.

Distribution From the late Miocene of Europe (France, Austria, Germany, Poland, Slovakia, Ukraine, Greece) to the present of Japan.

Remarks A few teeth of *Urotrichus* sp. have been reported from Maramena (three M1, five M2, and one M3); they were attributed to the genus based on the morphology of the M1–2, but the absence of humeri and lower dental elements hampers identification at species level (Doukas et al. 1995).

Talpa Linnaeus, 1758

Type Species *Talpa europaea* Linnaeus, 1758 (extant).

Other Taxa Included Extant: *T. caeca* Savi, 1822; *T. altaica* Nikolsky, 1883; *T. davidiana* (Milne-Edwards, 1884); *T. romana* Thomas, 1902; *T. levantis* Thomas, 1906; *T. occidentalis* Cabrera, 1907; *T. caucasica* Satunin, 1908; *T. stankovici* Martino and Martino, 1931; *T. aquitania* Nicolas et al., 2016; extinct: *T. minuta* de Blainville, 1840; *T. fossilis* Petényi, 1864; *T. minor* Freudenberg, 1914; *T. gracilis* Kormos, 1930; *T. praeglacialis* Kormos, 1930; *T. vallesensis* de Villalta and Crusafont, 1944; *T. tyrrhenica* Bate, 1945; *T. csarnotana* Kretzoi, 1959; *T. gilothi* Storch, 1978; *T. neagui* Rădulescu and Samson, 1989; *T. tenuidentata* Ziegler, 1990. This genus, which includes a large number of extant and extinct species, is in need of a revision, which is out of the scope of the present chapter; thus, all species that have been described until now are listed above but several synonymies between them have been proposed and even more are possible.

Distribution From the early Miocene to the present of Europe and Asia.

Remarks Koliadimou (1996) described six *Talpa* sp. teeth (one I, one C, two P4, one M1 and one M2) from the Pleistocene (MNQ20) locality Ravin Voulgarakis; the identification to species level is hampered by the absence of lower molars and metacranial elements. Chatzopoulou (2014) described a nearly complete right humerus from the late Pleistocene locality Loutra Almopias Cave LAC as *Talpa* sp., mentioning that its size is similar to that of the humerus of *Talpa minor*. In the same cave, in both levels LAC and LAC Ia, the extant species *T. europaea* has also been reported, represented by two M1, one M3, and the proximal part of one ulna (Chatzopoulou 2014).

Talpa fossilis Petényi, 1864

Nomenclatural and Taxonomical History *Talpa vulgaris fossilis* in Petényi (1864) (new subspecies). *T. praeglacialis* Kormos, 1930, is considered a junior synonym of *T. fossilis*, according to Kretzoi (1938).

Type Locality Beremend, Hungary, Pliocene, early Villanyian (MN16).

Distribution The earliest record of the species is from the Greek Turolian/Ruscinian locality Maramena and has a continuing presence in Europe from the Pliocene up to the middle Pleistocene (Doukas et al. 1995; Kurtén 1968).

Remarks Van Cleef-Roders and Van den Hoek Ostende (2001) provided an English translation of Petényi's (1864) original description of his subspecies, given in Old Hungarian. According to this description, the fossil bone material of this animal is similar to that of the recent common mole (*T. europaea*), in both morphology and size, the only difference being that in *T. europaea* the labial side of the mandible has only two mental foramina, one under the second premolar and one under the first molar, whereas in *T. fossilis* three foramina are present, one under the first premolar and two under the first molar. According to Kormos (1930), the presence of an anterior cingulum in m1 is an important diagnostic feature for *T. fossilis* (van den Meulen 1973). Since the size and morphology of the dental elements of *T. fossilis* are within the range for the extant *T. europaea* (a fact that has made some researchers tentatively consider the two species as synonymous, e.g., von Koenigswald 1970; Rabeder 1972; Van Cleef-Roders and Van den Hoek Ostende 2001), many of the identifications of Plio–Pleistocene assemblages might be incorrect. Van Cleef-Roders and Van den Hoek Ostende (2001:62–64) noticed that whereas the middle-sized *Talpa* is usually referred to *T. europaea* in late Pleistocene deposits, in older deposits it is often classified as *T. fossilis* in the German literature, but as *T.* cf. *europaea* in the French literature; the authors concluded that "*presumably this split in* T. fossilis *and* T. *cf.* europaea *presents a linguistic barrier only*". On the other hand, Sansalone et al. (2015) applied a geometric morphometrics analysis on a large sample of *T. fossilis* and *T. europaea* humeral remains from the Plio-Pleistocene of Hungary, and their results suggest that the two species are separate.

The material of a middle-sized *Talpa* from Maramena is quite rich, consisting of two p3, four p4, 12 m1, 24 m2, 23 m3, 17 M1, 39 M2, 24 M3, and eight humeri (some fragmented) (Doukas et al. 1995). These authors commented that the assignment of the Maramena material in *T. fossilis* should be considered as preliminary and that "*a revision of* Talpa *should particularly clarify the status of* Talpa fossilis, *and could therefore influence the identification of the Maramena species*" (Doukas et al. 1995:54). As Doukas et al. (1995:52) pointed out "*few insectivore genera are as badly in need of revision as the genus* Talpa." Here, we prefer to maintain the original identification, but with caution and in accordance with Van Cleef-Roders and Van den Hoek Ostende (2001:64), who consider *T. fossilis* a junior synonym of *T. europaea*, however, they pointed out that "*older assemblages previously assigned to* T. fossilis *(the name has been used for various Pliocene and even an Upper Miocene assemblage) could also belong to a fossil species, which is now extinct*."

Subfamily Desmaninae Thomas, 1912

Type Genus *Desmana* Güldenstädt, 1777.

Remarks The genera included in this subfamily are mainly distinguished on the basis of incisor and premolar morphological features. Koufos et al. (2001) described a small collection of desmanine dental elements (two partial mandibles, one p2, three p3, one p4, one m1, one m2, four m3, one P4, three M1, three M2, and three M3) has been recovered from the early Pleistocene locality Marathoussa (Chalkidiki, N Greece), classifying the material as Desmaninae indet.

***Archaeodesmana* Topachevski and Pashkov, 1983**

Type Species *Desmana pontica* Schreuder, 1940.

Other Taxa Included *A. verestchagini* (Topachevski, 1961), *A. vinea* (Storch, 1978), *A. bifida* (Engesser, 1980), *A. getica* (Terzea, 1980), *A. turolensis* (Rümke, 1985), *A. adroveri* (Rümke, 1985), *A. luteyni* (Rümke, 1985), *A. major* (Rümke, 1985), *A. brailloni* (Rümke, 1985), *A. acies* Dahlmann, 2001, *A. baetica* Martín-Suárez et al., 2001, *A. elvirae* Minwer-Barakat et al., 2008.

Distribution From the late Miocene to the Pliocene of Eurasia (Spain, France, Austria, Germany, Romania, Poland, Czech Republic, Moldova, Slovakia, Hungary, Bulgaria, Greece, Turkey, Ukraine).

Remarks *Archaeodesmana* was initially described by Topachevski and Pashkov (1983) as a subgenus of *Desmana* Güldenstädt, 1777, with *Desmana pontica* Schreuder, 1940 as the type species. Rümke (1985) created the genus *Dibolia* with *D. dekkersi* as a type species. This genus was found to be a junior homonym of *Dibolia* Latreille, 1829, a genus of Coleoptera, and the genus *Ruemkelia* Rzebik-Kowalska and Pawlowski, 1994, was created as a replacement name. Hutterer (1995) noticed that Rümke (1985) had included the species *Desmana pontica* in her genus *Dibolia* and thus suggested that the replacement name was not needed and included in *Archaeodesmana* all species which Rzebik-Kowalska and Pawlowski (1994) had included in *Ruemkelia* (= *Dibolia* Rümke, 1985), elevating *Archaeodesmana* to generic level. The synonymy of *Archaeodesmana* and *Ruemkelia* has also been followed by several other authors (e.g., Martín-Suárez et al. 2001, Minwer-Barakat et al. 2008). However, Rzebik-Kowalska (2005a, b) and Rzebik-Kowalska in Rzebik-Kowalska and Rekovets (2015) do not agree with the synonymy and prefer retaining the genus name *Ruemkelia* (for the species of *Dibolia* Rümke, 1985), as the two genera have different type species and there is currently no study that proves their congenerity. Despite her own opinion on the matter, Rzebik-Kowalska used the genus name *Archaeodesmana* instead of *Ruemkelia* in her reviews of the insectivores from Poland and Romania (2005a, b, respectively) "*to make uniform the generic name for this talpid in the whole volume*" (Rzebik-Kowalska 2005a:124). Since the present work cannot review the type material of *Archaeodesmana* and *Ruemkelia* species, we here tentatively follow Hutterer (1995). Dental elements of *Archaeodesmana* sp. have been reported from the Pliocene (MN15) Greek locality Vevi (Doukas 2005).

Archaeodesmana bifida (Engesser, 1980)

Nomenclatural and Taxonomical History *Desmana bifida* in Engesser (1980) (new species); *Dibolia bifida* in Rümke (1985) (new combination); *Archaeodesmana bifida* in Hutterer (1995) (new combination).

Type Material Left M1, Akc. 1, Naturhistorisches Museum Basel (Engesser 1980:Abb. 45a, 46).

Type Locality Afyon–Dinar–Akçaköy, Turkey, Pliocene, Ruscinian (MN15).

Distribution Pliocene of Turkey, Greece, Romania, and perhaps Slovakia.

Remarks According to the diagnosis by Engesser (1980), the species has a small size, similar to that of *Galemys pyrenaicus*. The I1 is straight and its scoop-shaped crown has two cusps. The P4 is significantly longer than wide, with relatively poorly developed protocone and a faint ridge anteriorly to the paracone. The P3 has three roots. The p4 lacks a lingual cuspid, but bears a clear paraconid. The canines and premolars are relatively delicate. The p2 is somewhat smaller than the p3. The roots of the p3 are straight. The p1 usually bears a non-transversally broadened root. The M1 is short lingually, as the metaconule protrudes slightly to the rear, and a relatively extended angle between the two metacone arms. The M2 is very short. The lower molars lack a lingual cingulid, and the enamel in the talonid is wrinkled. The crista obliqua of the m3 connects with the protoconid–metaconid ridge approximately in the middle of the tooth. The i2 is not much bigger than the i1 and bears a lingual cingulum. The mental foramina are very variable, but they are mostly four. Rümke (1985) gave an emended diagnosis for the species: the mean length of the p2 is about 1.32 mm, that of the p3 about 1.40 mm, and that of the p4 about 1.66 mm, the p2/p3 ratio is about 0.94, the p1 has one or two roots, the mean length of the C is about 1.40 mm, that of the P3 about 1.61 mm, and the P1 has two roots. Rümke (1985), apparently having noticed similarities of the only mandible from Ciuperceni 1 with the Turkish species, synonymized Terzea's (1980) *Desmana getica* with *D. bifida*. This opinion is not shared by most authors (e.g., Hutterer 1995, Martín-Suárez et al. 2001, Minwer-Barakat et al. 2008) who keep the two species separated, an opinion also tentatively followed here. The material from Limni 6 assigned to *A. bifida* by Rümke (1985) is rather poor, consisting of two partial mandibles and the following dental elements (including both in situ and loose teeth): one i1, one p1, four p2, one p3, one p4, two m1, two m2, two m3, one I1, three C, two P2, three P3, two P4, and two M3 (no mention on i2/3 and I2/3).

Archaeodesmana getica (Terzea, 1980)

Nomenclatural and Taxonomical History *Desmana getica* in Terzea (1980) (new species); *Ruemkelia getica* in Terzea (1997) (new combination); *Archaeodesmana getica* in Hutterer (1995) (new combination). ★*Dibolia dekkersi* Rümke, 1985, is considered a junior synonym of *A. getica*, according to Van den Hoek Ostende et al. (1989).

Type Material Lower jaw with p2–4 and m2–3, Institute of Speleology "Émile Racovitza", Bucharest (see Terzea 1980).

Type Locality Ciuperceni 1, Romania, Pliocene (MN14).

Distribution From the latest Miocene-Pliocene of Greece to the Pliocene of Europe (Spain, Romania).

Remarks The p2/p3 ratio has a middle value, the P2/P3 ratio a low value, and the C is short. The lower ml has an entostylid of medium height, whereas the lower m3 has a very small or absent entostylid. The upper P4 rarely shows a "bifid" protocone. The P1 has a single root. The lower molars have a broad posterior cingulum and lack a parastylid. The P4 lacks both a metacone and a widening of the posterocrista. The M1–2 show no metastyle. Terzea (1980) described the new species *Desmana getica* based on a mandible from the early Ruscinian Romanian locality Ciuperceni 1. Rümke (1985) included that specimen in the species *Dibolia bifida* (Engesser 1980) (see above). In the same work, she published the new species *Dibolia dekkersi*, based on material from the Greek early Ruscinian locality Kardia. Van den Hoek Ostende et al. (1989) included the Romanian mandible in the species *Dibolia dekkersi*, as its dental dimensions and the p2/p3 ratio are within the ranges of the latter species (van den Hoek Ostende pers. commun. Jan. 2019). However, as Rzebik-Kowalska (2005a) mentioned, "*the name* getica *has priority upon the name* dekkersi." Rzebik-Kowalska and Pawlowski (1994) transferred *Dibolia* species in their new genus *Ruemkelia* and Hutterer (1995) to *Archaeodesmana* (see discussion above).

Apart from Kardia (type locality of *Dibolia dekkersi*, see below), *A. getica* has also been found in Ptolemais 3 and Maramena. The material from Ptolemais 3 available to Rümke (1985) includes three i1, three c, one p1, eight p2, two p3, seven p4, 11 m1, 13 m2, 13 m3, seven I1, 10 C, four P1, 11 P2, 18 P3, 12 P4, six M1, three M2, and 14 M3 (no mention to the i2/3 and I2/3). The material from Maramena studied by Rümke (1985) included six partial mandibles and the following dental elements (including in situ and loose teeth, but with no mention to the i2/3 and I2/3): seven i1, three c, four p1, nine p2, seven p3, nine p4, 10 m1, six m2, eight m3, 10 I1, seven C, eight P1, seven P2, eight P3, six P4, seven M1, four M2, and seven M3. The material from Maramena collected after 1986 and studied by Doukas et al. (1995) is much richer, including one partial mandible with p3–4, one with p4–m2, two with p4–m1, one with m2–3, six p4, 12 m1, 13 m2, 12 m3, one partial maxilla with P1–3 and two with P4–M1, seven P2, five P3, 22 P4, 31 M1, 25 M2, 16 M3, one fragmented humerus, and one chevron bone.

★*Dibolia dekkersi* Rümke, 1985

[Junior synonym of *Archaeodesmana getica* (Terzea, 1980) according to Van den Hoek Ostende et al. (1989)].

Type Material Right p3, Ka-646, Faculty of Geosciences, Utrecht University, The Netherlands (Rümke 1985: Fig. 53.2).

Type Locality Kardia, Greece (early Ruscinian).

Remarks In the diagnosis and the differential diagnosis of her new species, Rümke (1985) mentioned that the mean length of the p2 is about 1.44 mm, of the p3 about 1.24 mm, of the p4 about 1.58 mm, the p2/p3 ratio is about 1.16 (thus average), the p1 is single-rooted, the mean length of the C is about 1.28 mm (thus short), of the P2 about 1.59 mm, of the P3 about 1.46 mm, the P2/P3 ratio is about 1.09 (thus low), and the P1 is double-rooted. The material from the type locality is rather rich, consisting of four partial mandibles and the following dental elements (including both in situ and loose ones, with no reference to the i2/3 and I2/3): 22 i1, eight c, 16 p1, 29 p2, 20 p3 (including the type specimen), 28 p4, 30 m1, 27 m2, 44 m3, 22 I1, 29 C, 18 P1, 29 P2, 31 P3, 35 P4, 34 M1, 44 M2 and 31 M3.

***Archaeodesmana verestchagini* (Topachevski, 1961)**

Nomenclatural and Taxonomical History *Desmana verestchagini* in Topachevski (1961) (new species); *Archaeodesmana verestchagini* in Martín-Suárez et al. (2001) (new combination).

Type Material Partial right mandible with p2–m3, No. 26967 (Rümke 1985:Fig. 76).

Type Locality Kosyakin, Russia, early Pliocene (MN14).

Distribution Pliocene of Bulgaria, Russia, Greece, and possibly Ukraine.

Remarks According to Rümke's (1985) emended diagnosis for the species, the mean length of the p2 is about 1.80 mm, of the p3 about 1.55 mm, of the p4 about 2.10 mm, and the p2/p3 ratio is about 1.16. She also gave some additional diagnostic features, based on material from the Greek locality Spilia 1: the mean length of the C is about 1.32 mm, of the P2 about 2.00 mm, of the P3 about 1.71 mm, the P2/P3 ratio is about 1.17, and the p1 and P1 are single-rooted. She mentioned that the species is characterized by its small size (Rümke 1985:96) and that the small size mainly of the p3 and P3 distinguishes it from all other *Desmana* species (Rümke 1985:188). Martín-Suárez et al. (2001) included the species to *Archaeodesmana*, following the diagnoses of the genus, and Minwer-Barakat et al. (2008) followed the same nomenclature. The species has been reported from the Greek localities Spilia 0 and 1, Kardia, Ptolemais 1 by Rümke (1985). The material from Spilia 1 studied by Rümke (1985) was limited, consisting of eight partial mandibles and the following dental elements (including in situ and loose teeth, no mention to i2/3 and

I2/3): six i1, five c, seven p1, eight p2, nine p3, seven p4, eight m1, six m2, eight m3, four I1, five C, five P1, four P2, 10 P3, three P4, four M1, six M2, and three M3. The scarce material from Kardia Rümke (1985) studied consisted of two i1, one c, five p2, two p3, five p4, seven m1, eight m2, eight m3, two P2, one P3, one P4, four M1, three M2, and five M3 (no mention to i2/3 and I2/3; the presence of partial mandibles is mentioned, but their quantity is not given). The material available to Rümke (1985) from Ptolemais 1 included five i1, seven c, nine p1, 15 p2, 20 p3, 24 p4, 17 m1, 12 m2, five m3, 13 C, 12 P1, 10 P2, 17 P3, 14 P4, nine M1, four M2, and five M3 (no mention to i2/3 and I2/3; the presence of partial mandibles is mentioned but their number is not given). Doukas (2005), in his review on the Greek insectivores, listed the material as *Desmana* and not *Archaeodesmana*.

Subfamily Uropsilinae Dobson, 1883

Type Genus *Uropsilus* Milne-Edwards, 1871.

***Desmanella* Engesser, 1972**

Type Species *Desmanella stehlini* Engesser, 1972.

Other Taxa Included *D. crusafonti* Rümke, 1974; *D. fejfari* Gibert, 1975; *D. dubia* Rümke, 1976; *D. sickenbergi* Engesser, 1980; *D. cingulata* Engesser, 1980; *D. amasyae* Engesser, 1980; *D. engesseri* Ziegler, 1985; *D. gardiolensis* Crochet, 1986; *D. storchi* Qiu, 1996; *D. rietscheli* Storch and Dahlmann, 2000; *D. woelfersheimensis* Dahlmann, 2001; *D. gudrunae* Van den Hoek Ostende and Fejfar, 2006.

Distribution From the late Oligocene to the late Pliocene of Eurasia (Spain, France, Switzerland, Austria, Germany, Poland, Romania, Greece, Turkey, China).

Remarks *Desmanella* sp. has been reported from the Greek localities Lefkon (no information on material; Doukas 1989) and Limni 6 (unpublished material; Doukas 2005).

★*Desmanella dubia* Rümke, 1976

Type Material Right M1, PK 565, Department of Paleontology and Historical Geology, University of Athens (Rümke 1976:Pl. 2.13).

Type Locality Pikermi (Chomateri), Greece, late Miocene, Turolian (MN12).

Distribution From the late Miocene to the Pliocene of Europe (Spain, France, Poland, Bulgaria) and late Miocene of Greece.

Remarks The diagnosis by Rümke (1976) mentions that this is a small *Desmanella* species (m1 length = 1.43 mm), with a relatively small m1, not much larger than the

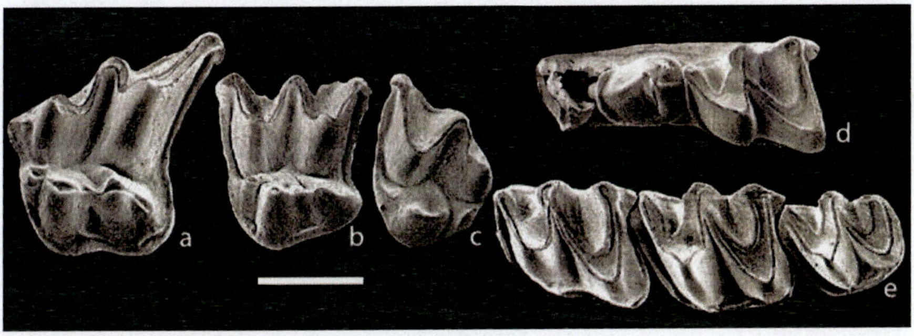

Fig. 7 *Desmanella dubia* from Maramena. (**a**) M1, (**b**) M2, (**c**) M3, (**d**) partial mandible with p4–m1, (**e**) partial mandible with m1–3. Images CD, personal data. All specimens are figured as left. Scale bar equals 1 mm

m3. The lower molars bear a continuous cingulum along their anterior, labial and posterior sides. The M1 shows a concave posterior outline, due to the strongly elongated hypocone. The lingual cusps of the M1 and M2 form slightly curved crests. The posterior cingulum of the M1 and M2 ends at the posterolabial corner of the teeth. Finally, the M3 is relatively large, with a wide valley separating the hypocone from the protocone. The material from the type locality is rather scanty, consisting of two m1 (one fragmented), two m2, two m3, one P4, one M1 (holotype), two M2 (one fragmented), and one M3 (Rümke 1976). The species name "*dubia*" reflected the author's doubt on the generic allocation, as both *dubia* and even the type species of *Desmanella* could also belong to *Asthenoscapter* Hutchison, 1974, based on the published collections back in 1976 (which did not include lower antemolars or mandibles that could reveal the number of roots, a main distinctive feature between the two genera; see Rümke (1976) for discussion). Materials published since then have confirmed the generic allocation of *dubia* to *Desmanella* (e.g., Crochet and Green 1982; Ziegler 2003; Rzebik-Kowalska 2005c). The material of *D. dubia* from Maramena (Fig. 7) is richer than that from the type locality: one partial mandible with m1–3, nine partial mandibles with p4–m1, three partial mandibles with p4–m2, 14 partial mandibles with m1–2, six partial mandibles with m2–3, 24 ml, 31 m2, 22 m3, three partial maxillae with M1–3, 23 P4, 45 M1, 51 M2, and 28 M3 (Doukas et al. 1995). The mandibles from Maramena show six alveoli for the antemolars in front of the double-rooted p4, confirming Rümke's (1976) allocation of the species to *Desmanella* instead of *Asthenoscapter* (that shows some double-rooted antemolars) (Doukas et al. 1995).

Talpidae incertae sedis

Desmanodon Engesser, 1980

Type Species *Desmanodon major* Engesser, 1980.

Other Taxa Included *D. minor* Engesser, 1980; *D. antiquus* Ziegler, 1985; *D. ziegleri* Van den Hoek Ostende, 1997; *D. burkarti* Van den Hoek Ostende, 1997; *D. daamsi* Van den Hoek Ostende, 1997; *D. crocheti* Prieto, 2009; *D. fluegeli* Prieto et al., 2010.

Distribution Early-middle Miocene of Europe (Portugal, Spain, France, Austria, Germany, Czech Republic, Slovakia, Greece) and Turkey.

Remarks Three upper molars from the early/middle Miocene Greek locality Antonios have been identified as *Paratalpa/Desmanodon* (Vasileiadou and Koufos 2005), based on the remark of Van den Hoek Ostende (2003) that the distinction between *Desmanodon* and *Paratalpa* requires information from the humeri. Since *Paratalpa* has been found only in latest Oligocene–early Miocene W and C European localities, it is probable from a paleobiogeographic and a biostratigraphic point of view that the Antonios talpid belongs to *Desmanodon*.

Desmanodon antiquus Ziegler, 1985

Taxonomical History ★*Desmanodon meuleni* Doukas, 1986, is considered a junior synonym of *D. antiquus*, according to Doukas and Van den Hoek Ostende (2006).

Type Material Right M1, 1959 XXVIII 589, Bayerische Staatssammlung für Paläontologie und Historische Geologie, München (Ziegler 1985:Taf. 6.6).

Type Locality Rembach bei Vilsbiburg, Germany, early Miocene (MN4).

Distribution Early Miocene of Europe (Austria, Germany, Czech Republic) and early Miocene of Greece.

Remarks Ziegler (1985), in his diagnosis, mentions that this is a small *Desmanodon* species. The P4 has three separate roots, and its protocone is faint. The upper molars bear a deeply split mesostyle, a clear metaconule, a short and not isolated parastyle, but lack a protoconule. The m2–3 bear a prominent precingulid, and the crista obliqua is sloping toward the middle of the trigonid wall. The p4 shows a subtle metaconid. The humerus is slender, with a long tuberculum teres, distally placed pectoral tubercle, a sharp-edged crista pectoralis and a laterally compressed caput parallel to the longitudinal axis of the humerus. The material from Aliveri, on which the species *D. meuleni* Doukas, 1986, was based, is rather scanty, consisting of three m1, six M1, ten M2, and nine M3 (Doukas 1983). It has been included in *D. antiquus* after the synonymy of the two species was established by Doukas and Van den Hoek Ostende (2006). The material from Karydia, initially mentioned as *Desmanodon* sp. in the locality faunal list given by Doukas (2005) but later described as *D. antiquus* by Doukas and Van den Hoek Ostende (2006), is even scantier: one mandibular fragment with the p4, one m2, and four M2 (Doukas and Van den Hoek Ostende 2006).

★*Desmanodon meuleni* Doukas, 1986

[Junior synonym of *Desmanodon antiquus* Ziegler, 1985, according to Doukas and van den Hoek Ostende (2006)].

Type Material Right M2, AL 3162, Faculty of Geosciences, Utrecht University, The Netherlands (Doukas 1986:Pl. 2.5).

Type Locality Aliveri, Greece, early Miocene (MN4).

Remarks Doukas first published the species *D. meuleni* in his PhD thesis in 1983, based on material from Aliveri. As his Greek thesis did not qualify as a publication, a few years later, the same name was used for the same concept, as *Desmanodon meuleni* Doukas, 1986. In the meantime, Ziegler (1985) had already published the species *D. antiquus* from Germany. As the two species seem to be conspecific, the German species predated the Greek one, and *D. meuleni* is considered as a junior synonym of *D. antiquus* (see discussion in Doukas and Van den Hoek Ostende 2006).

Desmanodon minor Engesser, 1980

Type Material Partial right mandible with p3–m3, Sar.2, Naturhistorischen Museum Basel (Engesser 1980:Abb.55b, 56b).

Type Locality Sarı Çay, Turkey, middle Miocene (MN7/8).

Distribution Besides its type locality, it is also found in the middle Miocene of Greece.

Remarks Engesser's (1980) diagnosis mentioned that the species has somewhat smaller size than the type species of *Desmanodon* and, mainly, more delicate molars. The protocone of the upper molars is less massive. The p3 has two roots. The m1 is much smaller than the m2. The M1 bears a very prominent parastyle. The lingual and posterior roots of the P4 are usually not fused. The p3–4 are relatively large. The mandible is relatively strong. The humerus is significantly smaller and less specialized than that of *D. major*, with comparatively shorter teres edges. Koliadimou (1996) reported the presence of one *D. minor* M2 from the middle Miocene locality Chryssavgi 1. Even though the material is scarce, it can be assigned to this species with certainty due to its dimensions, the absence of an anterior and lingual cingulum and the presence of a small posterior cingulum, a small cuspule on the labial area and a strong mesostyle (Koliadimou 1996), characteristics also present in the M2 from the type locality.

Soricidae Fischer, 1814

Type Genus *Sorex* Linnaeus, 1758.

Remarks Several Greek localities have yielded small collections of isolated dental elements that cannot be determined at genus level and are only reported as Soricidae indet. Vasileiadou and Zouros (2012) reported the presence of one upper incisor and one M2 of Soricidae indet. From Lapsarna De Bonis et al. (1997) listed Soricidae indet. in the faunal list from Thymiana A (unpublished material). One m2 and a partial mandible with the alveoli of the m2–3 of an unidentified soricid have been found at Paliambela-A (Sylvestrou 2002; Sylvestrou and Kostopoulos 2007). Doukas (1989, 2005) reported the presence of Soricidae indet. at Monasteri and Spilia 0,1, without giving further details on the material. Three partial upper incisors of an unidentified soricid were recovered from Kessani (Vasileiadou et al. 2012). Van de Weerd (1979) and Doukas (2005) mentioned the presence of Soricidae indet. in Ptolemais 1, with no further information. Soricidae indet. is also reported from the localities Limni 6 (Doukas 2005) and Elaiochoria (one specimen: cf. *Deinsdorfia*; Hulva et al. 2007), without further details. Undeterminable soricid dental elements were also found at Varkiza 2 (Van de Weerd 1973) and Choremi 1 and Choremi 4 (Van Vugt et al. 2000).

Subfamily Crocidosoricinae Reumer, 1987

Type Genus *Crocidosorex* Lavocat, 1951.

Remarks Two i1, one p4, 13 m1, 10 m2, two m3, three I1, one P4, four M1, and four M2 from Aliveri were initially assigned to Soricidae indet. by Doukas (1983) and later to Crocidurinae indet. by Doukas (1986), based on the crocidurine-type p4. A small number of molars (three m1, one m2, one m3, two M1, and one M2) from the early–middle Miocene locality Antonios have been assigned to Crocidosoricinae indet. (Vasileiadou and Koufos 2005), but the absence of the subfamily's diagnostic element (p4) makes this identification uncertain.

Lartetium Ziegler, 1989

Type Species *Sorex prevostianum* Lartet, 1851.

Other Taxa Included *L. dehmi* (Viret and Zapfe 1951); *L. africanum* (Lavocat 1961); *L. petersbuchense* Ziegler, 1989; *L. ziegleri* Mein and Ginsburg, 2002.

Distribution Early–middle Miocene of Europe (Spain, France, Austria, Germany, Czech Republic, Slovakia, Serbia, Greece), Turkey, and Morocco.

Lartetium dehmi (Viret and Zapfe, 1951)

Nomenclatural and Taxonomical History *Sorex dehmi* in Viret and Zapfe (1951) (new species); *Lartetium dehmi* in Ziegler (1989) (new combination).

Type Material Viret and Zapfe (1951) did not designate a holotype of their new species; lectotype, by subsequent designation by Hugueney et al. (2012): left mandible with a2–m2, MC 230, Museum of Lyon, France (Viret and Zapfe 1951:Fig. 6a–b).

Type Locality Vieux-Collonges, France, middle Miocene (MN5).

Distribution Early–middle Miocene of Europe (France, Germany, Slovakia, Serbia) and possible presence in the early Miocene of Greece.

Remarks The species is characterized by wide upper molars with a well-developed hypocone, weak entocristids and a labial termination of the oblique cristid on the lower molars, and a broad talonid and wide labial cingulum under the protoconid on the m1. Doukas and Van den Hoek Ostende (2006) assigned one m1, one m2, two P4, and four M1 from Karydia to *L.* cf. *dehmi*. The material shares metrical and morphological characteristics with the material from the type locality; however, its scantiness and the presence of only isolated dental elements instead of complete jaws did not allow the certain assignment to the species.

Subfamily Paenelimnoecinae Fejfar, Storch and Tobien, 2006

Type Genus *Paenelimnoecus* Baudelot, 1972.

Paenelimnoecus Baudelot, 1972

Type Species *Paenelimnoecus crouzeli* Baudelot, 1972.

Other Taxa Included *P. pannonicus* (Kormos, 1934); *P. micromorphus* (Doben-Florin, 1964); *P. repenningi* (Bachmayer and Wilson, 1970); *P. truyolsi* (Gibert, 1975); *P. obtusus* Storch, 1995; *P. chinensis* Jin and Kawamura, 1997.

Distribution From the early Miocene to the Pliocene of Europe (Spain, France, Austria, Germany, Poland, Moldova, Romania, Czech Republic, Slovakia, Hungary, Greece), Turkey, and China.

Remarks A single m2 from the early Miocene locality Karydia and a single m1 from the middle Miocene locality Plakias have been attributed to this genus (as *Paenelimnoecus* sp. and cf. *Paenelimnoecus* sp., respectively), based on their reduced entoconid (Doukas and Van den Hoek Ostende 2006 and De Bruijn et al. 2012, respectively). Since no additional material is available in both localities, a specific identification is impossible. However, Doukas and Van den Hoek Ostende (2006) mention that the Karydia specimen is remarkably larger than the m2 of *P. micromorphus* and *P. crouzeli*.

Subfamily Soricinae Fischer, 1814

Type Genus *Sorex* Linnaeus, 1758.

Crusafontina Gibert, 1975

Type Species *Crusafontina endemica* Gibert, 1975.

Other Taxa Included *C. kormosi* (Bachmayer and Wilson, 1970); *C. magna* (Hutchison and Bown, 1980 in Bown 1980), *C. minima* (Hutchison and Bown, 1980 in Bown 1980), *C. fastigata* Van Dam, 2004; *C. vandeweerdi* Van Dam, 2004.

Distribution From the middle Miocene to the Pliocene of Europe (Spain, France, Switzerland, Austria, Germany, Poland, Moldova, Ukraine, Czech Republic, Hungary, Greece) and Turkey.

Crusafontina endemica Gibert, 1975

Type Material Partial left mandible with p4–m2, Nr. 9002, Institut de Paleontologia Miquel Crusafont de Sabadell, Barcelona (Gibert 1975:Pl. 2.7a–b).

Type Locality Can Llobateres 1, Spain, late Miocene (MN9).

Distribution Late Miocene (MN9–MN10) of Europe (Spain, France, Austria, Moldova, Hungary) and possible presence in the late Miocene of Greece.

Remarks The species has a rather complete dental formula (1–4–3/1–2–3), as all *Crusafontina* species. The original diagnosis for the species by Gibert (1975) gives the following characteristics: two rooted P4, presence of a small postero-internal depression and presence of a labial crest but absence of a lingual one on the P4, a very wide trigonid basin and a long paralophid on the m1, presence of a hypoconid and an entoconid on the m3, a not bifid upper incisor, presence of a narrow interarticular area separating the articular facets of the mandible. Mészáros (1998:147) gave an emended diagnosis for the species: "*small* Crusafontina *species with hypoconid and entoconid on the third lower molar. The apex of the upper incisor is long, I1 L/H > 1.40. The P4 hypoconal flange is short. The second upper antemolar is relatively elongated, the A1/A2 length ratio is less than 1.62. The lower incisor may be monocuspulate or bicuspulate. The m1/m3 length ratio is less than 2.20. The mental foramen is below the m1 trigonid or the middle of the tooth.*" Van Dam (2004) discussed the differences between *C. endemica* and *C. kormosi* and proposed changes in Mészáros (1998) emended diagnosis (see van Dam 2004:746–749 for reasoning): *C. endemica* is a small *Crusafontina* species, its I1 has a long apex, and the L/H ratio is not smaller than about 1.40, the metastyle of the M1 extends more buccally than the parastyle does and the hypoconal flange does not protrude lingually, the I1 bears 2–3 cusps, the m3 shows a talonid with a small basin and the Lm3/Lm1 ratio is not smaller than about 0.40. Van Dam (2004) assigned one m1 from the locality Biodrak to *C.* cf. *endemica*, as it fits morphologically and metrically with *C. endemica*.

Crusafontina kormosi (Bachmayer and Wilson, 1970)

Nomenclatural and Taxonomical History *Anourosorex kormosi* in Bachmayer and Wilson 1970 (new species); "*Anourosorex*" *kormosi* in Storch (1978) (new combination); *Crusafontina kormosi* in Storch and Qiu (1991) (new combination). The species *Amblycoptus vicinus* Kretzoi, 1954—described based on fragmentary material from Csákvár (Hungary)—is regarded as a synonym of *C. kormosi*, even though the name *vicinus* has priority (see discussion in Mészáros 1998:149).

Type Material Right mandible with complete dentition, 1970/1389, Natural History Museum of Vienna (Bachmayer and Wilson 1970:Pl. 1.3).

Type Locality Kohfidisch, Austria, Late Vallesian (MN10).

Distribution From the late Miocene (MN10–MN13) of Europe (France, Austria, Germany, Hungary, Czech Republic, Moldova) and possible presence in the late Miocene (MN10) of Greece.

Remarks The original diagnosis by Bachmayer and Wilson (1970) mentions that this species' size is somewhat smaller than in the living *Anourosorex squamipes*, its teeth have a faint pigment, and its dental formula is 1–4–3/1–2–3. The posterior borders of the P4–M1 are emarginated, the P4 shows a somewhat pointed anterior margin, the M1 is transversally relatively wide, the m3 is reduced but shows a small basined talonid, the lower molars bear weak labial cingula and slightly crested entoconids. The upper articular condyle of the mandible has an oval rather than triangular shape, the lower articular condyle is only slightly visible in lingual view, and its anterior edge does not extend forward beyond the posterior border of the superior pterygoid fossa, and the pterygoid spicule is relatively obscure. Mészáros (1998) gave an emended diagnosis, according to which the species is relatively large among the *Crusafontina* species; the m3 bears neither a hypoconid nor an entoconid, and its trigonid is reduced to a single cusp or even without cusps or is totally missing; the I1 apex is shorter than in *C. endemica*; the P4 shows a long hypoconal flange; the second upper antemolar is long; there are always two cusps on the lower incisor; the Lm1/Lm3 ratio is higher than 2.20; the mental foramen is below the middle of the m1 or backward. As with *C. endemica*, Van Dam (2004:746–749 for reasoning) changed some points in Mészáros' (1998) emended diagnosis. According to Van Dam (2004), *C. kormosi* is a medium-sized *Crusafontina* species. The I1 shows a long apex and its L/H ration is smaller than about 1.40. On the M1, the metastyle and parastyle extend buccally about equally far, and the hypoconal flange does not protrude lingually. The i1 is bicuspulate. On the m3, the talonid might show a closed basin or a terminal heel or might be absent. The Lm3/Lm1 ratio is smaller than about 0.40. Van Dam (2004) argues that the morphology of the talonid of the m3 is not a good morphological feature to distinguish between *C. endemica* and *C. kormosi* (as Mészáros 1998 had proposed), and the Lm3/Lm1 ratio could be used instead. One m1 and one m2 from the locality Lefkon, previously mentioned as

Amblycoptus sp. by Doukas (1989), have been attributed to *C.* cf. *kormosi* by Van Dam (2004). The material is too limited to allow certain specific identification.

Amblycoptus Kormos, 1926

Type Species *Amblycoptus oligodon* Kormos, 1926.

Other Taxa Included *A. jessiae* Doukas, 1995 in Doukas et al. (1995).

Distribution From the late Miocene to the Pliocene (MN12–14) of Europe (Spain, Hungary, Russia, Greece) and Turkey.

★*Amblycoptus jessiae* Doukas, 1995 in Doukas et al. (1995)

Nomenclatural and Taxonomical History *Amblycoptus* n. sp. in Doukas (1989); *Amblycoptus jessiae* Doukas in Doukas et al. (1995) (new species).

Type Material Partial mandible with a1, p4, ml and m2, AMPG MA 3296 (Doukas et al. 1995:Pl. 6.10).

Type Locality Maramena, Greece, late Miocene, Turolian/Ruscinian boundary.

Distribution From the late Miocene to the early Pliocene (MN12–14) of Europe (Spain, Greece).

Remarks The diagnosis by Doukas et al. (1995:54) mentions that the species' size is intermediate between *A. oligodon* and *A. topali*, its A1 bears a parastyle, the P4 shows a well-developed protocone, and the ml shows "*neither well separated nor entirely merged*" entoconid and entostylid. The differential diagnosis mentions that this species "*differs from* A. oligodon *in the morphology of the talonid of ml, in having less separated entoconid and entostylid, and by having a relatively smaller m2. It also differs in the morphology (parastyle, hypoconal flange) and size of the A1. It differs from* A. topali *in the relative size and the L/W ratio of the ml and in the morphology of the A1. The entoconid/entostylid complex in* A. jessiae *is not fused as in* A. topali. *It finally differs from Amblycoptus sp. (Osztramos 7) in the less advanced morphology of the A1.*" Mészáros (1997) introduced the genus *Kordosia*, which differs from *Amblycoptus* in a number of features including the absence of an A3, and suggested that *A. jessiae* could be included in *Kordosia* with uncertainty until mandibles from the type locality Maramena are recovered. Later, Van Dam (2004:755) discussed the similarities and differences of *jessiae* with *A. oligodon* and *K. topali* and concluded that "*in short, whereas the P4 and M1 of* jessiae *are more* Amblycoptus-*like, the m1 and A1–2 are more* Kordosia-*like. Assignment to* Amblycoptus *is preferred, taking into account the probable presence of the A3.*" We here follow the opinions of Van Dam (2004) and Doukas (2005) in retaining the species in the genus *Amblycoptus*. The material from the type locality (Fig. 8) includes four lower incisors, five a1, 19 p4, 52 m1, 20 m2, a partial mandible with a1, p4–m2 (holotype), 14

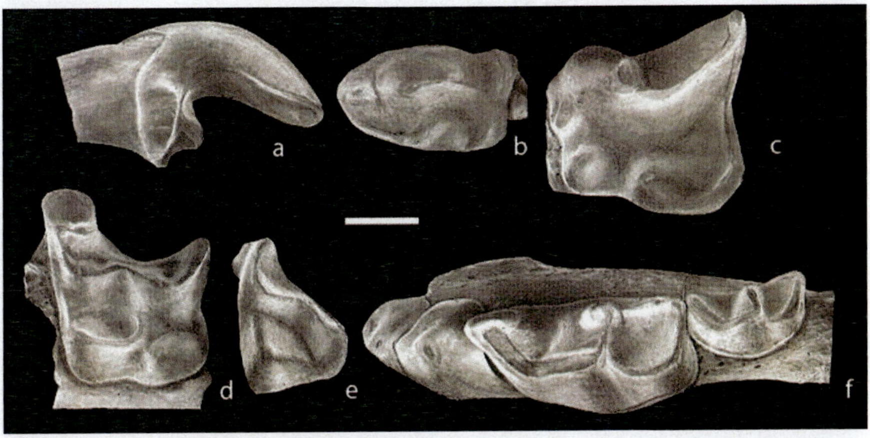

Fig. 8 *Amblycoptus jessiae* from Maramena (type locality). (**a**) I sup., (**b**) A1, (**c**) P4, (**d**) M1, (**e**) M2, (**f**) right mandibular fragment with a1, p4, ml and m2 (holotype). Images CD, personal data. All specimens are figured as left. Scale bar equals 1 mm

upper incisors, 53 A1, 12 A2, 62 P4, 59 M1, and 12 M2 (Doukas et al. 1995). Doukas (1989) mentioned the presence of *Amblycoptus* sp. in the locality Ano Metochi 3 (with no information on the material); Doukas (2005) included the insectivore material from Ano Metochi 2, 3 in that from Maramena, the type locality of the species *A. jessiae*. Apparently, the Ano Metochi 3 *Amblycoptus* also belongs to this species. Vasileiadou et al. (2003) reported the presence of one lower incisor, two m1, one m2, two upper incisors, and two M1 of *A.* cf. *jessiae* at the locality Nea Silata. The material is damaged, and no measurements are available, but the morphological features and the relative size fit well with those of the Maramena species.

Blarinella Thomas, 1911

Type Species *Sorex quadraticauda* Milne-Edwards, 1872 (extant).

Other Taxa Included Extant: *B. griselda* Thomas, 1912; *B. wardi* Thomas, 1915; Extinct: *B. dubia* (Bachmayer and Wilson, 1970); *B. europaea* Reumer, 1984.

Distribution From the late Miocene to the Pleistocene of Europe (Hungary, Greece) and Asia (China), and extant in Southeast Asia.

Remarks Storch (1995) created the genus *Alloblarinella* to separate extinct from extant species that had been included to *Blarinella* and made *B. europaea* the type species of his new genus (he included the species *dubia* in *Petenyia*, as originally allocated by Bachmayer and Wilson, 1970). Doukas (2005) mentioned the presence of cf. *Blarinella* sp. in the Pliocene locality Vevi, without additional information on the material.

Blarinella dubia (Bachmayer and Wilson, 1970)

Nomenclatural and Taxonomical History *Petenyia dubia* in Bachmayer and Wilson 1970 (new species); *Blarinella dubia* in Reumer (1984) (new combination).

Type Material Left lower jaw with m1–3, No. 1970/1387, Naturhistorisches Museum, Wien (Bachmayer and Wilson 1970:Pl. 2.6).

Type Locality Kohfidisch, Austria, late Miocene (MN11).

Distribution From the early Pliocene (Ruscinian) of Europe (Poland, Czech Republic, Hungary) and possible presence in Greece.

Remarks According to the diagnosis by Bachmayer and Wilson (1970:546), the species' size is similar to that of *Petenyia hungarica*, its lower incisor is relatively smooth in dorsal profile, with some pigment, "*but not as intense as now preserved*," and its coronoid spicule is not particularly prominent. Later, Reumer (1984) transferred the species to *Blarinella* and gave an emended diagnosis: the P4 shows a normally developed protocone, the M1–2 bear a protocone and a talon but lack an endoloph and a hypocone, the lower molars show a very strong cingulum on both lingual and labial sides, as well as high entoconid crests, and there is a horizontal bar in the internal temporal fossa. De Bruijn et al. (1970) reported the presence of one partial mandible with i, p4–m2 and one partial maxilla with P4–M2 of *Blarinella* sp. at the locality Maritsa 1 (Rhodos Island). Later, Reumer (1984) assigned these two specimens to *B. dubia* based on the absence of a continuous endoloph on the upper molars.

Blarinella europaea Reumer, 1984

Type Material Partial maxilla with M1–2, no. V.12705, Vt.99, Hungarian State Geological Survey, Budapest (Reumer 1984:Pl. 21.6).

Type Locality Csarnóta 2, Hungary, early Pliocene, Ruscinian (MN14–15).

Distribution Early Pliocene (Ruscinian) of Europe.

Remarks The diagnosis, as given by Reumer (1984), mentions that this is a large *Blarinella* species with heavily pigmented teeth, both upper and lower incisors bear a strongly spatulate apex, the P4 shows a faint protocone, the upper molars bear a well-developed paracone and metacone as well as a continuous endoloph, the lower molars are quadrate, with sharp buccal edges of the protoconids and the hypoconids due to sharp angles of the lophids and with extremely high entoconid crests. Two lower incisors from the late Ruscinian Greek locality Apolakkia were assigned to *Blarinella* sp. by Van de Weerd et al. (1982), based on their similarity with the lower incisor on the partial mandible from Maritsa 1. Reumer (1984) assigned the specimens to *B.* cf. *europaea*.

Petenyia Kormos, 1934

Type Species *Petenyia hungarica* Kormos, 1934.

Other Taxa Included *P. robusta* Rzebik-Kowalska, 1989; *P. katrinae* Qiu and Storch, 2000 (see Rzebik-Kowalska 1989 for synonymies among *Petenyia* species and transfer of former *Petenyia* species to other genera).

Distribution From the early Pliocene to the early Pleistocene of Europe (Spain, France, Switzerland, Germany, The Netherlands, Poland, Italy, Slovakia, Romania, Russia, Hungary, Bulgaria, Greece), and Asia (Turkey and China).

Petenyia hungarica Kormos, 1934

Type Material Skull with nearly complete dentition, no. Ob.3684, Hungarian State Geological Survey, Budapest (Reumer 1984:Pl. 15.1).

Type Locality Villány Kalkberg (= Villány 3), Hungary, early Pleistocene, late Villanyian (MNQ17).

Distribution From the early Pliocene to the early Pleistocene (early Ruscinian to early Biharian) of Europe (Spain, France, Switzerland, Germany, The Netherlands, Poland, Italy, Slovakia, Romania, Russia, Hungary, Bulgaria, Greece) and Turkey.

Remarks Kormos (1934) gave only a description for his species and not a formal diagnosis. His description includes the following morphological features: the rostral part of the skull above the antemolars is excavated; the upper incisor has a strong talon; there are four upper antemolars, which reduce in size backward; the P4 and M1–2 show nearly straight posterior margins; the mandible is short and robust; the mental foramen is situated between the paraconid and the protoconid of the m1; the coronoid process shows concave anterior edge and an undulate tip; the coronoid spicule is strongly pronounced; the lower incisor is bicuspulate and lacks a buccal cingulum; the m3 bears a single-cusped talonid (Reumer 1984). Reumer (1984) commented that Kormos' (1934) description of the position of the mental foramen is probably wrong, as all the specimens he studied bear the mental foramen between the protoconid and the hypoconid of the m1. Reumer (1984) also proposed some characters as diagnostic: the upper incisor is not fissident; the upper molars lack a hypocone, but show a small metaloph; the lower molars bear entoconids close to the metaconids and usually high entoconid crests; the lower incisor reaches buccally the end of the p4; the anterior edge of the coronoid process shows a small spicule that results in a slight "3″-shape in side view; the external temporal fossa is divided into two roughly equal parts by the coronoid spicule; the lower condylar facet is high; the interarticular area is broad and low; all teeth are pigmented dark red. This species has been found in Maramena, where it is represented by one p4, five m1, four m2, three m3, one upper incisor, four P4, nine M1, and four M2 (Doukas et al. 1995). The same material had been reported as *Petenyia* sp. by Doukas (1989).

Sulimskia Reumer, 1984

Type Species *Sorex kretzoii* Sulimski, 1962.

Other Taxa Included *S. ziegleri* Qiu and Storch, 2000.

Distribution Pliocene of Europe (Poland, Slovakia, Hungary, Greece) and Asia (China).

Sulimskia kretzoii (Sulimski, 1962)

Nomenclatural and Taxonomical History *Sorex kretzoii* in Sulimski (1962) (new species); *Sulimskia kretzoii* in Reumer (1984) (new combination).

Type Material Rostral fragment of a skull with the mandibles (left mandible nearly complete, only lacking the angular process; right mandible with a damaged coronoid process; both mandibles with complete dentition), No. M.Z. VIII/ Vm–312/1, Museum of the Earth, Polish Academy of Sciences, Warsaw (Sulimski 1962:Text-Plate 2, Fig. 10).

Type Locality Weze 1, Poland, Pliocene, Ruscinian (MN15).

Distribution From the Pliocene to the early Pleistocene (Ruscinian to Villanyian) of Europe (Germany, Poland, Slovakia, Hungary, Greece).

Remarks The diagnostic features of *S. kretzoii* given by Sulimski (1962) are the following: the dental formula is 3–1–3–3/2–0–1–3; the i2 is small, compressed, with one cusp and a strong cingulum; the p4 is high, with a robust cingulum, strongly posteriorly directed; all cingulums of the lower molars are thick and uniform; the entoconid is separated on all molars; the m3 bears a small entoconid; the mental foramen is situated in a shallow depression between the proto- and hypoconid; the coronoid process is wide and spade-like; the masseteric crest bears a spine; the pterygoid fossa is triangular, wide and deep; the upper articular facet of the condyloid process is oblique, expanded lingually; the lower articular facet of the condyloid process is long and wide, not so oblique, bears a slight notch on its upper edge and its lingual end distinctly curves downward; the I1 shows a strong talon and a cingulum, but lacks a cusp; the arrangement ratio of the upper unicuspid antemolars and premolars is: 2:2:1:1:0.5; the P3 is small, pushed below the P4; the M3 is small and reduced; the rostrum of the skull is short. Two almost complete m1 and three partial lower molars (two trigonids and one talonid) found in Apolakkia (Rhodos Island) were attributed to Blarinini indet. by Van de Weerd et al. (1982), based on the absence of an entoconid crest. The authors mentioned that two partial lower incisors, one partial P4 and one partial M1 from the locality might also belong to the same species (Van de Weerd et al. 1982:93). Reumer (1984) attributed the fragmentary lower molars from Apolakkia to *S. kretzoii*, based on their size, the absence of an entoconid crest, the anterior position of the entoconid, the deep trigonid valley and the narrow and pointed metaconid; however, he did not mention the lower incisors and upper teeth from Apolakkia.

Asoriculus Kretzoi, 1959

Type Species *Crocidura gibberodon* Petényi, 1864.

Other Taxa Included *A. similis* (Hensel, 1855); *A. corsicanus* (Bate, 1945); *A. castellarini* (Pasa, 1947); *A. tornensis* (Jánossy, 1973); *A. borsodensis* (Jánossy, 1973); *A. adroveri* (Jammot, 1977); *A. thenii* (Malez and Rabeder, 1984); *A. maghrebiensis* Rzebik-Kowalska, 1988; *A. burgioi* Masini and Sarà, 1998. Some of the species (*similis, corsicanus, burgioi*) are Plio–Pleistocene endemic insular species from Mediterranean islands (Sardinia, Corsica, Sicily) and the generic name *Nesiotites* Bate, 1945, also used for the endemic shrew from the Balearic Islands, has been proposed for them (e.g., Reumer 1984:132), an opinion not widely accepted in the literature (e.g., Masini and Sarà 1998). Furthermore, some synonymies have been proposed among the above species [e.g., *A. tornensis* and *A. borsodensis* are thought to be synonyms of *A. gibberodon* by Reumer (1984:89)]. Since this work is not a review of the genus *Asoriculus* and original material has not been studied, we give here the complete list of species that have occasionally been thought to belong to the genus.

Distribution From the late Miocene to the Pleistocene of Europe (Portugal, Spain, France, Austria, Germany, Poland, Slovakia, Czech Republic, Romania, Italy, Croatia, Bulgaria, Hungary, Greece), Asia (Turkey),and North Africa (Morocco).

Remarks *Asoriculus* was initially described as a subgenus of *Soriculus* by Kretzoi (1959) and was later raised to generic rank by the same author (Kretzoi 1962). It is a widely distributed genus in the Plio–Pleistocene of Europe, where it is represented mainly by the species *A. gibberodon*, a species also present in many Greek localities. *Asoriculus* sp. is mentioned by Doukas (2005) in Vevi, with no additional information on the available material.

Asoriculus castellarini (Pasa, 1947)

Nomenclatural and Taxonomical History *Neomys castellarini* in Pasa (1947) (new species); *Asoriculus castellarini* in Hutterer (1994) (new combination).

Type Material See Pasa 1947.

Type Locality Cava Sud, North Italy, Pleistocene, Biharian.

Distribution Pleistocene of Europe (Italy, Slovakia, Greece).

Remarks The species has a large number of features in common with *A. gibberodon*, including the size of the dental elements. There seem to be some differences in the area of the mandibular articulation, but since *A. castellarini* is not known from many localities and *A. gibberodon* shows a large variation in the mandibular articulation morphology, the differences cannot be fully evaluated. Both Reumer (1984) and Rzebik-Kowalska (2013) mentioned that *A. castellarini* might be a junior synonym of *A. gibberodon*. We here prefer to include the species as valid without, however, implying that Reumer's and Rzebik-Kowalska's opinions are rejected; the material from the type locality of the species needs to be revised before definite conclusions are drawn. Koliadimou (1996) attributed two lower incisors, three m1, five m2, five m3, one partial mandible with m1–3, one partial mandible with m2–3, one partial mandible with m1–2, one upper incisor, two P4, and two M1 from the Biharian locality Ravin Voulgarakis to *A.* cf. *castellarini*. The author based the identification on the morphology of the vertical blade of the mandible, which is closer to that of *A. castellarini* than to that of *A. gibberodon*. However, the limited number of specimens that retain this part of the mandible attributed to *A. castellarini* does not allow a good assessment of its morphological variation. Finally, as stated above, detailed study of the *A. castellarini* type material might prove that the species is synonymous with *A. gibberodon*, and thus, the Ravin Voulgarakis material should be attributed to the latter species.

Asoriculus gibberodon (Petényi, 1864)

Nomenclatural and Taxonomical History *Crocidura gibberodon* in Petényi (1864) (new species); *Soriculus gibberodon* in Kretzoi (1956) (new combination); *Soriculus (Asoriculus) gibberodon* in Kretzoi (1959) (new subgenus); *Asoriculus gibberodon* in Kretzoi (1962) (new combination); *Episoriculus gibberodon* in Repenning (1967) (new combination). *Soriculus kubinyii* Kormos, 1934, is considered a junior synonym of *A. gibberodon*, according to Reumer (1984).

Type Material The type material of *Crocidura gibberodon* Petényi, 1864, from Beremend has been lost and the type material of *Soriculus kubinyii* Kormos, 1934, from Villány 3 (Hungary) has been proposed as a neotype (Reumer 1984): skull with nearly complete dentition, Ob.3685, Hungarian State Geological Survey, Budapest (Reumer 1984:Pl. 27.3).

Type Locality Villány Kalkberg (Villány 3), Hungary, Pliocene, Villanyian (late MN17) (original locality of the neotype, type locality of *Soriculus kubinyii* Kormos, 1934).

Distribution From the late Miocene (Turolian, MN13) to the Pleistocene of Europe (Spain, France, Austria, Germany, Poland, Romania, Slovakia, Hungary, Italy, Bulgaria, Greece) and Turkey.

Remarks Reumer (1984) mentioned that some parts of Kormos' (1934) original diagnosis for his species *S. kubinyii* can be used as diagnostic features for *A. gibberodon*. These include the following: the upper incisor is fissident, and the lower one bearing weak cuspules; pigmentation is only present at some tips. Furthermore,

Reumer (1984) gave an emended diagnosis, according to which the species is rather small and bears weakly pigmented teeth; it bears four upper antemolars, of which the A4 is variable in its development; the upper molars show a variable morphology, with a moderate posterior emargination; the lower incisor is bicuspulate and bears a well-developed buccal cingulum; the a1 is elongate; the lower molars have high entoconid crests; the m3 shows a talonid basin; the mental foramen is situated below the posterior half of the m1. A large number of Greek localities have yielded dental elements of this species. De Bruijn et al. (1970) reported the presence of one complete mandible, three partial mandibles with m1–3, three partial maxillas and numerous isolated lower and upper teeth of *A. gibberodon* in Maritsa 1 (Rhodos Island). The locality Maramena has yielded five lower incisors, 17 m1, 20 m2, 10 m3, six upper incisors, five P4, nine M1, and 12 M2 of this species (Doukas et al., 1995; material previously mentioned as *Episoriculus* sp. by Doukas, 1989). One lower incisor, one m1, one m2, and one M2 from Kessani and one lower incisor, three m1, two upper incisors, nine P4, four M1, seven M2, and one M3 from Nea Silata have also been assigned to this species (Vasileiadou et al. 2012 and Vasileiadou 2001, respectively). Van de Weerd et al. (1982) have assigned one fragmentary lower incisor, one m1, and one m2 from Apolakkia (Rhodos Island), and Reumer and Doukas (1985) have assigned six lower incisors, one p4, 11 m1, six m2, one m3, 11 upper incisors, ten P4, 17 M1, 16 M2, two M3, and nine condyles from Tourkobounia 1 to *A. gibberodon*. Finally, one lower incisor, four m1, two m2, three upper incisors, eight M1, and three M2 from the Biharian locality Marathoussa (Chalkidiki, N Greece) also belong to this species (Koufos et al. 2001).

Deinsdorfia Heller, 1963

Type Species *Sorex hibbardi* Sulimski, 1962.

Other Taxa Included *D. fallax* (Heller, 1936); *D. kordosi* Reumer, 1984; *D. janossyi* Reumer, 1984; *D. insperata* Rzebik-Kowalska, 1990; *D. reumeri* Rzebik-Kowalska, 1990; *D. kerkhoffi* Reumer, 1995 in Doukas et al. (1995); *D. doukasi* Furió and Mein, 2008.

Distribution From the late Miocene to the early Pleistocene of Europe (Spain, France, Switzerland, Germany, Poland, Slovakia, Romania, Bulgaria, Hungary, Greece).

⋆*Deinsdorfia kerkhoffi* Reumer in Doukas et al. (1995)

Type Material Partial ramus with m1–3, MA 6092, Department of Paleontology and Historical Geology, University of Athens (Doukas et al. 1995:Pl. 8.6).

Type Locality Maramena, Strimon basin, Greece, Miocene/Pliocene boundary.

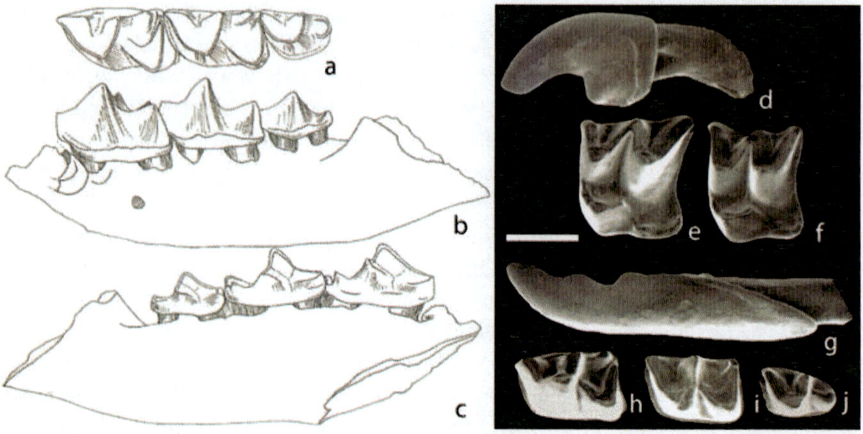

Fig. 9 *Deinsdorfia kerkhoffi* from Maramena (type locality; **a–c**; drawings CD, personal data.) and Silata (**d–j**; images KV, personal data). (**a**) Partial right mandible with m1–3 holotype—occlusal view, (**b**) same—buccal view, (**c**) same—lingual view, (**d**) I sup., (**e**) M1, (**f**) M2, (**g**) i inf., (**h**) m1, (**i**) m2, (**j**) m3. All specimens are figured as left; **a–c**, **f**, **h** are reversed. Scale bar equals 1 mm

Distribution From the late Miocene/early Pliocene (Turolian/Ruscinian boundary) of Greece and Bulgaria (*D.* cf. *kerkhoffi* in Dorkovo).

Remarks The diagnosis of the species in Doukas et al. (1995) includes the following features: *Deinsdorfia* species of a large size; the lower incisor bears three weak cuspules; the lower molars bear short talonid and moderately high entoconid crest; the m3 has a reduced, single-cusped talonid; the upper incisor is not fissident; the upper molars bear a metaloph and a low and small hypoconal flange, surrounded by a weak cingular ridge; the M3 is relatively long and has a small pointed parastyle; the coronoid process has a forward-pointing apex and its lower part leans backward; the condyle is large; the coronoid spicule is well-developed. Apart from the holotype, the type locality has also yielded (Fig. 9) three lower incisors, one p4, 10 m1, 14 m2, four m3, eight upper incisors, two upper antemolars, four P4, 17 M1, 11 M2, one M3, and three rami with condyles (Doukas et al. 1995). This species has also been reported from two localities in N Greece of similar age with Maramena (Turolian/Ruscinian boundary): 11 lower incisors, nine m1, 12 m2, eight m3, 12 upper incisors, two P4, seven M1, and 12 M2 have been found in Nea Silata (Vasileiadou 2001) and one m2, one m1, and one M1 in Kessani (Vasileiadou et al. 2012).

Beremendia **Kormos, 1934**

Type Species *Beremendia fissidens* (Petényi, 1864).

Other Taxa Included *B. pohaiensis* (Kowalski and Li, 1963); *B. minor* Rzebik-Kowalska, 1976.

Distribution From the early Pliocene to the middle Pleistocene of Europe (Spain, France, Switzerland, Austria, Germany, The Netherlands, United Kingdom, Poland, Italy, Czech Republic, Slovakia, Romania, Russia, Hungary, Bulgaria, Greece) and Asia (Siberia and China).

Beremendia fissidens **(Petényi, 1864)**

Nomenclatural and Taxonomical History *Crossopus fissidens* in Petényi (1864) (new species); *Neomys fissidens* in Kormos (1911) (new combination); *Beremendia fissidens* in Kormos (1930) (new combination). *Beremendia ucrainica* (Pidoplichko, 1956) is a junior synonym of *B. fissidens*, according to Rzebik-Kowalska (1976).

Type Material (syntypes) Six maxillary and 37 mandibular fragments, Hungarian Natural History Museum, Department of Paleontology and Geology (Petényi 1864:Pl. 1.5a–o).

Type Locality Beremend limestone quarry (locality 2), Hungary, Pliocene, early Villanyian (MN16).

Distribution Early Pliocene to middle Pleistocene of Europe.

Remarks According to the original diagnosis (in Old Hungarian) by Petényi (1864), this species has a bifid upper incisor and an acuspulate lower one (Botka and Mészáros 2014). Botka and Mészáros (2014) gave an emended diagnosis that includes the following: the teeth are intensively pigmented dark red; the upper incisor is strongly fissident and shows a bifid apex; there are four upper antemolars; the A1–2 have almost the same size, the A3 is smaller, and the A4 is even smaller, hidden and not visible in buccal view or might be completely absent; the P4 and M1–2 show a moderate posterior emargination; the M1 has a moderately developed parastyle; the M3 is relatively small; the lower incisor is grooved and acuspulate, with an upward curved apex; the a2 is bicuspid and has a posterolingual basin; the m3 is reduced; the ramus is strong and massive; the coronoid process is robust and anterolaterally curved with a variable angle; the angular process is short and flat; the external pterygoid fossa is more or less deeply pocketed, with a narrow upper facet of which the shape varies from elliptic to oval; the interarticular area is broad and shows a lingual margin; the lower facet leans strongly anteriorly and is invisible in buccal view; the internal temporal fossa is small, deep, and pocketed. One upper antemolar from Tourkobounia 1, one lower incisor, two lower a1, two m1, two m3, two partial mandibles with m1, two upper incisors, three M1, and one M2 from Ravin Voulgarakis and one m1, one partial mandible with m1, two P4, two M1, one M2, and one M3 from Marathoussa (N Greece) have been attributed to this species (Reumer and Doukas 1985; Koliadimou 1996; Koufos et al. 2001).

Sorex **Linnaeus, 1758**

Type Species *Sorex araneus* Linnaeus, 1758.

Other Taxa Included As reported by Reumer (1984:22): "*the genus* Sorex *has long been the waste-basket for soricid remains. Large numbers of* 'Sorex' *species have been described, many of which belong to other genera, tribes or even to other subfamilies*." Although many of these issues have been resolved at the end of the last century and at the beginning of the present, a revision of the fossil material attributed to this genus is certainly needed. For the list of extant *Sorex* species see Hutterer (2005). Many of the extant species can be found in fossil assemblages correlated back to the Pliocene (e.g., *S. minutus* MN15, today). According to Rzebik-Kowalska (2008), at least 16 extinct forms have been described from Asia, 34 from North America and 28 from Europe. One of the extinct subgenera of *Sorex* is *Drepanosorex* Kretzoi, 1941, originally described as a separate genus, but later included as a subgenus in *Sorex* by Reumer (1985), an opinion shared by many authors since then and also followed herein. The subgenus includes five species from the Pleistocene of Europe: *S.* (*D.*) *savini* Hinton, 1911, *S.* (*D.*) *margaritodon* Kormos, 1930, *S.* (*D.*) *praearaneus* Kormos, 1934, *S.* (*D.*) *austriacus* Kormos, 1937, *S.* (*D.*) *rupestris* Zaitsev and Baryshnikov, 2002.

Distribution From the early Pliocene (Ruscinian) to the present of Eurasia and North America.

Remarks The oldest occurrence of *Sorex* in Greece is *Sorex* sp. from the Turolian/Ruscinian boundary locality Maramena, where the genus is represented by one broken m1, one toothless ramus, two upper incisors, two P4, seven M1, and four M2 (Doukas et al. 1995; material previously mentioned as "*Sorex*" sp. in Doukas 1989). The authors discussed that this *Sorex* is of medium size, larger than both *S. minutus* and the early Ruscinian *S. bor*, but similar in size with the late Ruscinian *S. bor* from Osztramos 7, with which both morphological similarities and differences can be observed. *Sorex* is a common element of Greek Pleistocene localities, where it is present with the extant species *S. minutus* Linnaeus, 1766 (Eurasian pygmy shrew) and *S. araneus* Linnaeus, 1758 (common Eurasian shrew). The former species is represented by six lower incisors, four m1, nine m2, one m3, one partial mandible, ten upper incisors, two P4, six M1, and one M2 from Marathoussa (N Greece; Koufos et al. 2001), one lower incisor, two m1 and five partial mandibles from Arnissa (Mayhew 1978), two partial mandibles with m2, one upper incisor, and two P4 from Loutra Almopias Cave LAC and one partial mandible with i–m1, one partial mandible with m1–3, one upper incisor, one M2, and one partial maxilla with P4–M1 from Loutra Almopias Cave LAC Ia (Chatzopoulou 2014), whereas Koliadimou (1996) assigned one lower incisor, one m1, two partial mandibles with m1–2, one partial mandible with m1–3, one upper incisor, three M1, and one partial maxilla with P4–M1 from Ravin Voulgarakis to *S.* cf. *minutus*; the dental elements show similarities with *S. minutus*, but some metrical parameters are slightly higher than those of the corresponding elements from Plio–Pleistocene Hungarian and Dutch localities, hence the uncertain identification. *Sorex araneus* is represented in Arnissa by one lower incisor, one partial mandible, one P4, and one M2 (Mayhew 1978), in Loutra Almopias Cave LAC by one m1, one M1 and one M2 and in Loutra Almopias Cave LAC Ia by one p4, one partial mandible with i–p4, one partial mandible with m2, one upper incisor, one P4, one M1, one partial maxilla with A2–3, one partial maxilla with P4, and one partial maxilla with M1–2 (Chatzopoulou 2014).

Sorex (Drepanosorex) praearaneus Kormos, 1934

Nomenclatural and Taxonomical History *Sorex praearaneus* in Kormos (1934) (new species); *Drepanosorex* n. g. *praearaneus* in Kretzoi (1941) (new combination); *Sorex (Drepanosorex) praearaneus* in Reumer (1985). *Sorex pachyodon* Pasa, 1947, is a junior synonym of *S. praearaneus*, according to Reumer (1985).

Type Material Skull with partial dentition, no. Ob.3681, Hungarian State Geological Survey, Budapest (Kormos 1934:Fig. 32; Reumer 1984:Pl. 9.1).

Type Locality Villány Kalkberg (= Villány 3), Hungary, early Pleistocene, late Villanyian (MNQ17).

Distribution From the early to the middle Pleistocene of Europe (Spain, The Netherlands, Poland, Italy, Czech Republic, Slovakia, Romania, Hungary, Greece).

Remarks Reumer (1984) used the following characters given by Kormos (1934) as diagnostic features for the species: the A5 is smaller than the A4; the mental foramen is situated below the posterior edge of the p4; the external temporal fossa reaches the upper condylar facet; the internal temporal fossa reaches half-way up the coronoid process. Reumer (1984) gave an emended diagnosis for the species, according to which *praearaneus* is a small *Drepanosorex* species with fissident upper incisor, slightly exoedaenodont dentition, five upper antemolars of which the A1–2 are the largest, and the A4 is slightly smaller than the A3 and A5, poorly developed hypocones on the P4 and the upper molars, a tricuspulate lower incisor with bulbous cuspules and a well-developed talonid on the m3. Koufos et al. (2001) described six lower incisors, one a1, six p4, six m1, nine m2, six m3, ten partial mandibles, 12 upper incisors, 14 upper antemolars, six P4, three M1, eight M2, and four M3 of *S. (D.) praearaneus* from Marathoussa (N Greece); the small degree of exoedaenodonty in the antemolars and the small size of the dental elements assigned them to the most primitive *Drepanosorex* species known. Furthermore, Koliadimou (1996) described three lower incisors, two p2, one partial mandible with p4–m2, three upper incisors, and two P4 of *S. (D.)* cf. *praearaneus* from Ravin Voulgarakis; the reason for the uncertain determination is the slightly larger width of the lower premolar and molars from Ravin Voulgarakis in comparison with that of the corresponding teeth from the Hungarian and Dutch materials described by Reumer (1984).

Neomys Kaup, 1829

Type Species *Sorex fodiens* Pennant, 1771 (extant).

Other Included Taxa At present, there are three extant species of *Neomys* widespread in Eurasia: *N. fodiens* (Pennant, 1771) (Eurasian water shrew), *N. anomalus* Cabrera, 1907 (Mediterranean water shrew) and *N. teres* Miller, 1908 (Transcaucasian water shrew). Furthermore, four extinct species have been described up to now from

European fossil localities: *N. newtoni* Hinton, 1911, *N. browni* Hinton, 1911, *N. hintoni* Zaitsev and Baryshnikov, 2002, and *N. intermedius* Brunner, 1952.

Distribution From the early Pliocene (Ruscinian) to the present in Eurasia and North America.

Remarks The presence of *Neomys* sp. in the Biharian locality Tourkobounia 2 (one upper incisor) has been reported by Reumer and Doukas (1985), as well as in the late Pleistocene localities Loutra Almopias Cave LAC (one partial mandible with i–m1, one partial mandible with m1, two upper incisors, one P4) and Loutra Almopias Cave LAC Ia (three lower incisors, two m1, one m3, one partial mandible with p4–m3, one partial mandible with m1–2, one partial mandible with m2, three upper incisors, one partial maxilla with I–A2, one partial maxilla with I–A1, one partial maxilla with P4–M1, one partial maxilla with P4–M2, one partial maxilla with M1–2, one partial maxilla with M2) has been reported by Chatzopoulou (2014). The species from the Loutra Almopias Cave could belong to either *N. fodiens* or *N. anomalus*, as it shows metrical and morphological similarities with both; the difference between the two species is located mainly in the height of the coronoid process, an area which is not preserved in the Loutra Almopias Cave material (Chatzopoulou 2014).

Subfamily Crocidurinae Milne-Edwards, 1872

Type Genus *Crocidura* Wagler, 1832.

Crocidura Wagler, 1832

Type Species *Sorex leucodon* Hermann, 1780.

Other Included Taxa Today there are at least 183 valid *Crocidura* species living in South Eurasia and Africa (for the list of extant species see Hutterer 2005). Some of them are present in Greek Pleistocene fossil assemblages (see below).

Distribution From the Pliocene (Ruscinian) to the present of Eurasia and Africa.

Remarks The oldest occurrence of *Crocidura* in Europe seems to be the single m3 described as *Crocidura* sp. from the Pliocene locality Apolakkia, showing the characteristic reduction of the talonid to a single cuspid, i.e., the hypoconid (Van de Weerd et al. 1982; Reumer 1984). Furthermore, a number of fossil assemblages have been described as *Crocidura* sp. from Greek Pliocene–Pleistocene localities: one lower incisor, two m1, four m2, two m3, one upper incisor, one M1, and one fragmentary M2 from Tourkobounia 1 were described by Reumer and Doukas (1985) and are considered as the oldest *Crocidura* of the European mainland; Kotsakis et al. (1980) listed two species of *Crocidura* sp. for Kalymnos without giving details on the material; Van der Geer et al. (2014) figured a partial mandible with m1–3 and a partial maxilla with P4–M2 found in the Faculty of Geosciences,

Utrecht University, The Netherlands, together with material from Naxos that had never been reported before; Tsoukala et al. (2006) assigned to *Crocidura* sp. a partial mandible with m2, one M1 and two M2 from Kitseli pothole. Additionally, three extant species have been found in late Pleistocene localities: the greater white-toothed shrew *C. russula* is represented by two partial mandibles at Arnissa (Mayhew 1978); the bicoloured shrew *C. leucodon* is known with two partial mandibles, two upper incisors, two P4, two M1, and two M2 from Arnissa (Mayhew 1978), four lower incisors, one a1, six m1, three m2, one m3, one partial mandible with i–a2, one partial mandible with m1–2, seven upper incisors, one A1, one M1, and one M2 from Loutra Almopias Cave LAC and one partial mandible with i–m2, one upper incisor, and one partial maxilla with A2–M2 from Loutra Almopias Cave LAC Ia (Chatzopoulou 2014); finally, the lesser white-toothed shrew *C. suaveolens* is known with two m2, one partial mandible with m1–2, one upper incisor, and two M1 from Loutra Almopias Cave LAC and one partial mandible with i–m1, three partial mandibles with m1–2 and one upper incisor from Loutra Almopias Cave LAC Ia (Chatzopoulou 2014). The endemic Cretan shrew *C. zimmermanni* has been reported in late Pleistocene localities in Crete (Lyras et al. this volume).

Crocidura kornfeldi Kormos, 1934

Type Material Skull with partial dentition (the left A4 got lost after 1934), no. Ob. 3686, Hungarian State Geological Survey, Budapest (Reumer 1984:Pl. 1.1).

Type Locality Villány Kalkberg (= Villány 3), Hungary, early Pleistocene, late Villanyian (MNQ17).

Distribution From the early to the middle Pleistocene of Europe (Spain, Austria, Czech Republic, Slovakia, Romania, Hungary, Greece).

Remarks Since Kormos (1934) did not give a formal diagnosis but only a description for his new species, Reumer (1984) separated some of the features given by Kormos as diagnostic: the P4, M1 and M2 are relatively short and broad and the molars show a short emargination; the lower incisor is acuspulate; the al–2 are broad; a thick bar is present between the anterior base of the coronoid process and the condyle, under the internal temporal fossa, below which there is a deep groove. Furthermore, Reumer (1984) gave an emended diagnosis, according to which *kornfeldi* is a fairly small *Crocidura* species with relatively short and broad P4 and molars. Furthermore, the A1 is much larger than the A2–3, while the A2 is slightly smaller than the A3, which is nearly as high as the parastyle of the P4 in buccal view. The P4 has a buccally situated protocone far from the antero-lingual corner of the premolar. The lower molars bear a narrow, undulating cingulum. The ascending ramus is robust, the internal temporal fossa is bordered ventrally by a thick bar, and the condyle is relatively large. There are several Greek localities that have revealed elements of this extinct *Crocidura*. Three lower incisors, two p4, 16 m1, 11 m2, two m3, two condyles, five upper incisors, one A1, one A2/3, two P4, five M1, and two M2 have been reported from Tourkobounia 2, one m1, four m2, one m3, and two condyles from Tourkobounia 3, and two m1, one m2, one m3, one condyle and one P4 from Tourkobounia 5 (Reumer and Doukas 1985). Koliadimou (1996) described one mandibular fragment with m1–m3, one with m1–m2, one upper incisor, and five M1 from the late Biharian locality Rema Voulgarakis. Finally, the richest *C. kornfeldi* material has been reported from Marathoussa (N Greece) by Koufos et al. (2001): five lower incisors, four a1, six p4, 12 m1, 10 m2, five m3, 13 upper incisors, six P4, eight A1, five A2/3, five M1, 11 M2, and one M3. The materials from Ravin Voulgarakis and Marathoussa are morphologically identical and metrically very close to *C. kornfeldi* from Osztramos 3/2 described by Reumer (1984) (Koufos et al. 2001).

6 Concluding Remarks

The Greek insectivore fossil record, even though represented by fragmentary or inadequately studied assemblages, can give some valuable information on the evolution of the faunas in the southern part of the Balkans. The highest peak in the Greek eulipotyphlan diversity occurred during the early Miocene, when all six families were present (with at least three erinaceid, two dimylid, one heterosoricid, one plesiosoricid, three talpid, and two soricid species), whereas by the late Miocene, only the three extant ones remained. As the extinct families disappeared, soricids started showing a larger diversity (with at least five species in the late Miocene of Greece and afterwards, versus two species in the early Miocene). Extant genera started emerging in the late Miocene in Greece (*Erinaceus*, *Talpa*, *Blarinella*, *Sorex*), and by the end of the early Pleistocene, all extinct genera had disappeared from the Greek faunas (last representatives being *Asoriculus* and *Beremendia*). All extant genera present in Greece since the early Pleistocene are also represented today in the area, albeit some of them by different species (e.g., *Erinaceus europaeus* in the Pleistocene but *Erinaceus roumanicus* today). The importance of the Greek insectivore fossil record is also reflected in the description of eight valid new species from Greek Neogene faunas (*Galerix symeonidisi*, *Parasorex kostakii*, *Schizogalerix macedonica*, *Myxomygale engesseri*, *Desmanella dubia*, *Heterosorex ruemkeae*, *Amblycoptus jessiae*, *Deinsdorfia kerkhoffi*).

Acknowledgments We are indebted to Dr. Wilma Wessels for access to important literature and to original images and drawings used for most of the figures. We also want to express our gratitude to Frans Verdaasdonk and Dr. Wilma Wessels for all the hard and tedious work of scanning original images and drawings. The help of Panagiotis Skandalos, Dr. Marc Furió, and Dr. Lars van den Hoek Ostende in collecting necessary literature is also very much appreciated. The latter colleague, in addition, and together with Johannes Klietmann, also made some remarks in his review of our chapter.

Editorial Note The works in this book (lagomorphs, rodents, insectivores) are the last of Katerina, who left us recently after a long battle with cancer. Katerina deeply believed in this collective project and worked with passion and dedication for its implementation, giving even her very last strength. Unfortunately, she could not deal with the constructive comments and suggestions of the reviewers, a task undertaken by the co-authors of the articles and her friends in the same scientific field.

Appendix

Occurrences of insectivores in Greece. Locality numbers refer to the collection numbers of the PaleoBiology Database (PBDB). Type localities are marked with boldface

Localities[PBDB no.]	Age (MN, GPTS)	Taxon
Loutra Almopias Cave LAC Ia[204954]	Younger Dryas (13,000–14,500 BP)	*Talpa europaea*[1] *Sorex araneus*[1] *Sorex minutus*[1] *Neomys* sp.[1] *Crocidura leucodon*[1] *Crocidura suaveolens*[1]
Kitseli pothole[204955]	Late Pleistocene	*Crocidura* sp.[2]
Naxos[203848]	Late Pleistocene	*Crocidura* sp.[3]
Arnissa[204957]	Late Pleistocene [penultimate interglacial (Eemian) age]	*Sorex araneus*[4] *Sorex minutus*[4] *Crocidura russula*[4] *Crocidura leucodon*[4]
Loutra Almopias Cave LAC[203847]	Middle Würm (42,500 BP)	*Erinaceus* cf. *europaeus*[1] *Talpa europaea*[1] *Talpa* sp.[1] *Sorex araneus*[1] *Sorex minutus*[1] *Neomys* sp.[1] *Crocidura leucodon*[1] *Crocidura suaveolens*[1]
Varkiza 2[204167]	Middle or late Pleistocene	Soricidae indet.[5]
Choremi 4[200082]	Middle Pleistocene (late Biharian, ~580–620 ky)	Soricidae indet.[6]
Choremi 1[200079]	Middle Pleistocene, (late Biharian, ~840–880 ky)	Soricidae indet.[6]
Apollonia[34784]	Pleistocene (MNQ20)	*Erinaceus europaeus*[7]
Ravin Voulgarakis[34783]	Pleistocene (MNQ20)	*Talpa* sp.[7] *Asoriculus* cf. *castellarini*[7] *Beremendia fissidens*[7] *Sorex* cf. *minutus*[7] *Sorex* cf. *praearaneus*[7] *Crocidura kornfeldi*[7]
Kalymnos[34770]	Early Pleistocene (Biharian, MNQ19)	*Crocidura* sp. (two species)[8]
Marathoussa[34765]	Early Pleistocene (Biharian, MNQ19)	Desmaninae indet.[9] *Asoriculus gibberodon*[9] *Beremendia fissidens*[9] *Sorex minutus*[9] *Sorex praearaneus*[9] *Crocidura kornfeldi*[9]
Tourkovounia 5[34592]	Early Pleistocene (Biharian)	*Erinaceus praeglacialis*[10] *Crocidura kornfeldi*[10]
Tourkovounia 3[34592]	Early Pleistocene (Biharian)	*Crocidura kornfeldi*[10]

(continued)

Localities[PBDB no.]	Age (MN, GPTS)	Taxon
Tourkovounia 2[34767]	Early Pleistocene (Biharian, MNQ19)	*Crocidura kornfeldi*[10] *Neomys* sp.[10]
Limni 6[34590]	Pliocene (Villanyian, MN16)	*Archaeodesmana bifida*[11] *Desmanella* sp.[12] Soricidae indet.[12]
Tourkovounia 1[34589]	Pliocene (Villanyian, MN16)	*Erinaceus* sp.[10] *Asoriculus gibberodon*[10] *Beremendia fissidens*[10] *Crocidura* sp.[10]
Ptolemais 3[35088]	Pliocene (late Ruscinian, MN15, ~4.95–4.77 Ma)	*Archaeodesmana getica*[11–12]
Vevi[214354]	Pliocene (late Ruscinian, MN15)	*Archaeodesmana* sp.[12] Talpidae indet.[12] *Asoriculus* sp.[12] cf. *Blarinella* sp.[12]
Apolakkia[202499]	Pliocene (late Ruscinian, MN15)	Galericinae indet.[12] *Asoriculus gibberodon*[14] *Blarinella* cf. *europaea*[14, 15] *Sulimskia kretzoii*[14, 15] *Crocidura* sp.[14, 15] Echinosoricinae indet.[12]
Ptolemais 1[35086]	Pliocene (uppermost part of the early Ruscinian, MN14, ~5.04–4.95 Ma)	*Archaeodesmana verestchagini*[11] Soricidae indet.[12]
Kardia[35087]	Pliocene (early Ruscinian, MN14, ~5.23–5.22 Ma)	*Archaeodesmana verestchagini*[11] *Archaeodesmana getica*[11]
Spilia 0, 1[31931,31,299]	Pliocene (early Ruscinian, MN14, ~5.23–4.95 Ma)	*Archaeodesmana verestchagini*[11] Soricidae indet.[16]
Nea Silata[191612]	Miocene/Pliocene, (Turolian/ Ruscinian, MN13/14, ~5.4–5.23 Ma)	*Erinaceus* sp.[18, 19] *Amblycoptus* cf. *jessiae*[18, 19] *Asoriculus gibberodon*[18, 19] *Deinsdorfia kerkhoffi*[18, 19]
Maramena[32189a]	Miocene/Pliocene, (Turolian/ Ruscinian, MN13/14, ~5.4–5.23 Ma)	*Erinaceus samsonowiczi*[16, 20] ***Schizogalerix macedonica***[20] *Archaeodesmana getica*[11, 20] *Desmanella dubia* [20] *Urotrichus* sp.[20] *Talpa fossilis*[20] ***Amblycoptus jessiae***[20] *Asoriculus gibberodon*[16, 20] *Sorex* sp.[16, 20] *Petenyia hungarica*[16, 20] ***Deinsdorfia kerkhoffi***[20]
Kessani[205296–7]	Miocene/Pliocene, (Turolian/ Ruscinian, MN13/14, ~5.4–5.23 Ma)	*Asoriculus gibberodon*[21] *Deinsdorfia kerkhoffi*[21] Soricidae indet.[21]
Maritsa 1[68046]	Late Miocene (Turolian, MN13)	*Asoriculus gibberodon*[22] *Blarinella dubia*[22, 15]

(continued)

Localities[PBDB no.]	Age (MN, GPTS)	Taxon
Monasteri[31940]	Late Miocene (Turolian, uppermost MN13, ~5.96–5.33 Ma)	Soricidae indet.[16]
Paliambela-A[205335]	Late Miocene (Turolian, uppermost MN13, ~5.96–5.33 Ma)	Soricidae indet.[23, 24]
Pikermi[68214]	Late Miocene (end of middle Turolian, MN 12, ~7.1 Ma)	*Schizogalerix moedlingensis*[26, 20]
Chomateri[195562]	Late Miocene (Turolian, MN12)	*Schizogalerix moedlingensis*[27] *Schizogalerix zapfei* = ***Galerix atticus*[27]** ***Desmanella dubia*[27]**
Samos-S3[205351]	Late Miocene (middle Turolian, MN12, near the fossil level MYT, ~7.3 Ma, Chron C3Br.2r)	*Schizogalerix zapfei*[28]
Lefkon[31223]	Late Miocene (latest Vallesian, MN10)	*Schizogalerix* sp.[16] *Desmanella* sp.[16] *Crusafontina* cf. *kormosi*[30]
Biodrak[205342]	Late Miocene (latest Vallesian, MN10)	*Schizogalerix* sp.[27, 12] *Crusafontina* cf. *endemica*[30]
Kastellios K1[205360]	Late Miocene (late Vallesian, MN10, 9.3–9.1 Ma, C4Ar.1r)	*Schizogalerix* sp.[32]
Elaiochoria[219100]	Late Miocene (late Vallesian to early Turolian, MN10–11)	Soricidae indet.[33]
Plakias[183151]	Late Miocene (early Vallesian, lower part of MN9, ~9.9 Ma or 9.311–9.6 Ma)	*Lantanotherium sanmigueli*[34] Galericinae indet.[34] cf. *Paenelimnoecus* sp.[34]
Chryssavgi 1[205366]	Middle Miocene (late Astaracian, MN7 and 8)	cf. *Schizogalerix* sp.[7] *Desmanodon minor*[7]
Thymiana C[205390]	Middle Miocene (Orleanian, MN5, ~15.5 Ma, interval of Chrons C5Bn.1n—C5Cr)	*Schizogalerix* sp.[36]
Thymiana A[205389]	Middle Miocene (Orleanian, MN5, ~15.5 Ma, interval of Chrons C5Bn.1n, C5Cr)	*Schizogalerix* sp.[36] Soricidae indet.[36]
Antonios[73861]	Early Miocene (Orleanian, MN4/5)	*Schizogalerix* sp.[37] *Desmanodon* sp.[37] Crocidosoricinae indet.[37]
Komotini[68085]	Early Miocene (Orleanian? MN4)	*Galerix* sp.[38]
Moschopotamos[208601]	Early Miocene (Orleanian, MN4)	Eulipotyphla indet.[39]
Karydia[69435]	Early Miocene (Orleanian, MN4)	***Parasorex kostakii*[38]** *Plesiodimylus* aff. *crassidens*[38] *Heterosorex* sp.[38] *Plesiosorex* sp.[38] *Myxomygale* cf. *hutchisoni*[38] *Desmanodon antiquus*[38] *Lartetium* cf. *dehmi*[38] *Paenelimnoecus* sp.[38]

(continued)

Localities[PBDB no.]	Age (MN, GPTS)	Taxon
Aliveri[68032]	Early Miocene (Orleanian, MN4, 18–17.5 Ma)	***Galerix symeonidisi*[40]** *Plesiodimylus huerzeleri*[40, 41] ***Heterosorex ruemkeae*[40]** ***Myxomygale engesseri*[40]** *Desmanodon antiquus*[40, 38] Soricidae indet.[40]
Lapsarna[186558]	Early Miocene (Orleanian, MN4, older than 18.4 ± 0.5 Ma)	Erinaceidae indet.[42] Talpidae indet.[42] Soricidae indet.[42]
Kyprinos[196839]	late Oligocene	*Dinosorex* sp.[43] Talpidae indet.[43]

MN Mammal Neogene Zone, *MNQ* Mammal Quaternary Zone

[1]Chatzopoulou (2014), [2]Tsoukala et al. (2006), [3]van der Geer et al. (2014), [4]Mayhew (1978), [5]van de Weerd (1973), [6]van Vugt et al. (2000), [7]Koliadimou (1996), [8]Kotsakis et al. (1980), [9]Koufos et al. (2001), [10]Reumer and Doukas (1985), [11]Rümke (1985), [12]Doukas (2005), [13]Hordijk and De Bruijn (2009), [14]van de Weerd et al. (1982), [15]Reumer (1984), [16]Doukas (1989), [17]Koufos and Vasileiadou (2015), [18]Vasileiadou (2001), [19]Vasileiadou et al. (2003), [20]Doukas et al. (1995), [21]Vasileiadou et al. (2012), [22]de Bruijn et al. (1970), [23]Sylvestrou (2002), [24]Sylvestrou and Kostopoulos (2007), [25]Koufos (2013), [26]Thenius (1952), [27]Rümke (1976), [28]Black et al. (1980), [29]Koufos et al. (2009), [30]van Dam (2004), [31]Zachariasse et al. (2011), [32]Furió et al. (2014), [33]Hulva et al. (2007), [34]de Bruijn et al. (2012), [35]Koufos (2006), [36]de Bonis et al. (1997), [37]Vasileiadou and Koufos (2005), [38]Doukas and van den Hoek Ostende (2006), [39]Benda and Steffens (1981), [40]Doukas (1986), [41]van den Hoek Ostende et al. (2015), [42]Vasileiadou and Zouros (2012), [43]Doukas and Theocharopoulos (1999)

[a]The insectivore fauna from the locality Maramena includes also the material from Ano Metochi 2–3, as the two localities are very close by and the faunas are thought to be identical (Doukas 2005)

References

Bachmayer F, Wilson RW (1970) Die Fauna der altpliozänen Höhlen– und Spaltenfüllungen bei Kohfidisch, Burgenland (Österreich). Ann Naturhist Mus Wien 74:533–587

Bachmayer F, Wilson RW (1980) A third contribution to the fossil small mammal fauna of Kohfidisch (Burgenland). Ann Naturhist Mus Wien 83:351–386

Benda L, Steffens P (1981) Aufbauy and Alter des Neogens von Katerini, Griechenland. Geol Jahrbuch B42:93–103

Bininda-Emonds OR, Cardillo M, Jones KE, MacPhee RD, Beck RM, Grenyer R, Price SA, Vos RA, Gittleman JL, Purvis A (2007) The delayed rise of present–day mammals. Nature 446(7135):507–512

Black C, Krishtalka L, Solounias N (1980) Mammalian fossils of Samos and Pikermi. Part I. The Turolian rodents and insectivores of Samos. Ann Carnegie Mus 49:359–378

Botka D, Mészáros LG (2014) *Beremendia* (Mammalia, Soricidae) remains from the late Early Pleistocene Somssich Hill 2 locality (Southern Hungary) and their taxonomic, biostratigraphic, palaeoecological and palaeobiogeographical relations. Fragm Palaeontol Hung 31:79–111

Brunner G (1933) Eine präglaziale Fauna aus dem Windloch bei Sackdilling (Oberpfalz). N Jahrb Miner Geol Paläontol B 71:303–328

Butler PM (1988) Phylogeny of the insectivores. In Benton MJ (ed) The Phylogeny and Classification of the Tetrapods. Vol.2 Mammals. The Systematics Association Special Volume 35B:117–141. Clarendon Press

Chatzopoulou A (2014) The micromammals of the Quaternary deposits from Cave A, Loutra Almopias (Pella, Northern Greece) Stratigraphy–Taphonomy–Paleoenvironment. PhD thesis. Department of Geology, School of Geology, Aristotle University of Thessaloniki, Thessaloniki

Crochet J–Y, Green M (1982) Contributions à l'étude des micromammifères du gisement miocène supérieur de Montredon (Hérault). 3 Les insectivores. Palaeovertebrata 12:119–131

de Bonis L, Bouvrain GD, Koufos G, Melentis J (1986) Succession and dating of the late Miocene primates of Macedonia. In: Else JG, Lee PC (eds) Primate evolution. Cambridge University Press, Cambridge, pp 107–114

de Bonis L, Koufos GD, Sen S (1997) The Sanithera (Mammalia, Suoidea) from the Middle Miocene of Chios Island, Aegean Sea, Greece. Rev Paléobiol Genève 16(1):259–270

de Bruijn H, Dawson M, Mein P (1970) Upper Pliocene Rodentia, Lagomorpha and Insectivora (Mammalia) from the isle of Rhodes (Greece). Proc Kon Neder Akad Weten 73(5):535–584

de Bruijn H, Doukas CS, van den Hoek Ostende LW, Zachariasse WJ (2012) New finds of rodents and insectivores from the Upper Miocene at Plakias (Crete, Greece). Swiss J Palaeontol 131:61–75

de Villalta JF, Crusafont M (1944) Nuevos insectívoros del Mioceno continental del Vallés–Panadés. Notas Com Inst Geol Miner España 12:1–26

Douady CJ, Chatelier PI, Madsen O, de Jong WW, Catzeflis F, Springer MS, Stanhope MJ (2002) Molecular phylogenetic evidence confirming the Eulipotyphla concept and in support of hedgehogs as the sister group to shrews. Mol Phylogenet Evol 25(1):200–209

Doukas CS (1983) Insectivores (Insectivora, Mammalia) from the Lower Miocene of Aliveri (Island of Evia, Greece). PhD Thesis. University of Athens (in Greek), Athens

Doukas CS (1986) The mammals from the Lower Miocene of Aliveri (Island of Evia, Greece). Part 5. The insectivores. Kon Neder Akad Weten 89:15–38

Doukas CS (1989) Smaller mammals from the Upper Pliocene of Strimon basin, Greece. Part 2: Insectivora. Boll Soc Paleontol It 28:213–216

Doukas CS (2003) The MN4 faunas of Aliveri and Karydia, Greece. In López–Martínez N, Peláez–Campomanes P, Hernández Fernández M (eds) En torno a fósiles de mamíferos: datación, evolución y paleoambiente. Coloquios Paleontol 1:127–132

Doukas CS (2005) Greece. In van den Hoek Ostende, LW, Doukas CS, Reumer JWF (eds) The fossil record of the Eurasian Neogene Insectivores (Erinaceomorpha, Soricomorpha, Mammalia), Part I. Scripta Geol 5(Spec Issue):99–112

Doukas CS, Theocharopoulos CD (1999) Smaller mammals from the Oligocene of Kyprinos (Thrace, N. Greece). In Reumer JWF, de Vos J (eds) Elephants have a Snorkel! Papers in honour of Paul Y. Sondaar. Deinsea 7:133–146

Doukas CS, van den Hoek Ostende LW (2006) Insectivores (Erinaceomorpha, Soricomorpha; Mammalia) from Karydia and Komotini (Thrace, Greece; MN4/5). Beitr Paläontol 30:109–131

Doukas CS, van den Hoek Ostende LW, Theocharopoulos C, Reumer JWF (1995) Insectivora (Erinaceidae, Talpidae, Soricidae, Mammalia). In Schmidt-Kittler N (ed) The vertebrate locality Maramena (Macedonia, Greece) at the Turolian–Ruscinian boundary. Münchner Geowissen Abh A28:43–64

Engesser B (1976) Zum Milchgebiss der Dimyliden (Insectivora, Mammalia). Eclogae Geol Helvetiae 69:795–808

Engesser B (1980) Insectivora und Chiroptera (Mammalia) aus dem Neogen der Türkei. Schweiz Paläontol Abh 102:47–149

Fejfar O, Sabol M (2009) Middle Miocene *Plesiodimylus* from the Devínska Nová Ves–Fissures site (western Slovakia). Bull Geosci 84:611–624

Filhol H (1891) Études sur les Mammifères fossiles de Sansan. Ann Sci Géol 21:1–319

Furió M, Alba D (2011) Aspectos problemáticos del género *Lantanotherium* (Galericinae, Erinaceidae, Mammalia). In: Pérez-García A, Gascó F, Gasulla JM, Escaso F (eds) Viajando a Mundos Pretéritos. Ayuntamiento de Morella, Castellón, pp 123–130

Furió M, van Dam J, Kaya F (2014) New insectivores (Lipotyphla, Mammalia) from the Late Miocene of the Sivas Basin, Central Anatolia. Bull Geosci 89:163–181

Gibert J (1975) New insectivores from the Miocene of Spain. Proc Kon Neder Akad Weten B 78:108–133

Hordijk K, De Bruijn H (2009) The succession of rodent faunas from the Mioc/Pliocene lacustrine deposits of the Florina-Ptolemais-Servia Basin (Greece) Hellenic. Journal of Geosciences 44:21–103

Hugueney M, Maridet O (2017) Evolution of Oligo-Miocene talpids (Mammalia, Talpidae) in Europe: focus on the genera *Myxomygale* and *Percymygale* n. gen. Hist Biol 30:1–9

Hugueney M, Mein P, Maridet O (2012) Revision and new data on the Early and Middle Miocene soricids (Soricomorpha, Mammalia) from central and south–eastern France. Swiss J Palaeontol 131:23–49

Hulva P, Horáček I, Benda P (2007) Molecules, morphometrics and new fossils provide an integrated view of the evolutionary history of Rhinopomatidae (Mammalia: Chiroptera). BMC Evol Biol 7:165. https://doi.org/10.1186/1471-2148-7-165

Hutterer R (1994) Generic limits among neomyine and soriculine shrews (Mammalia: Soricidae). In: Neogene and quaternary mammals of the Palaearctic, conference in honour of Professor Kazimierz Kowalski, Abstracts. Kraków, p 32

Hutterer R (1995) *Archaeodesmana* Topachevski & Pashkov, the correct name for *Dibolia* Rümke, a genus of fossil water moles (Mammalia: Talpidae). Bonner zool Beit 45:171–172

Hutterer R (2005) In: Wilson DE, Reeder DM (eds) Mammal species of the world: A taxonomic and geographic reference, 3rd edn. Johns Hopkins University Press, Baltimore, MD

Kälin D, Engesser B (2001) Die jungmiozäne Säugetierfauna vom Nebelbergweg bei Nunningen (kanton Solothurn, Schweiz). Schweiz Paläontol Abh 121:1–61

Klietmann J, Nagel D, Rummel M, van den Hoek Ostende LW (2014) Enlightening complexity. The Dimylidae of Petersbuch 28. Palaeobiodiv Palaeoenviron 94:463–479

Koliadimou K (1996) Palaeontological and biostratigraphic study of the fossil small mammals of the Mygdonia Basin (N. Greece) Ph.D. thesis. Aristotle University Thessaloniki, Thessaloniki

Kormos T (1911) Canis (Cerdocyon) Petényii n. sp. és egyéb érdekes leletek Baranya megyéből. Magyar Királyi Földtani Intézet Évkönyve 19(4):153–178

Kormos T (1930) Új adatok a püspökfürdői Somlyóhegy preglaciális faunájához. Állattani Közlemények 27:40–62

Kormos T (1934) Neue Insektenfresser, Fledermäuse und Nager aus dem Oberpliozän der Villanyer Gegend. Földtani Közlöny 64:296–321

Kotsakis T, Petronio C, Sirna G (1980) The Quaternary vertebrates of the Aegean islands: Palaeogeographical implications. Ann Géol Pays Helléniques 30:31–64

Koufos GD (2006) The Neogene mammal localities of Greece: faunas, chronology and biostratigraphy. Hellenic J Geosci 41:183–214

Koufos GD, Kostopoulos DS, Vlachou Th D (2009) The Late Miocene Mammal Faunas of the Mytilinii Basin, Samos Island Greece: New Collection. 16 Biochronology-Beitr. Palaont Wien 31:397–408

Koufos GD (2013) Neogene mammal biostratigraphy and chronology of Greece. In: Wang X, Flynn LJ, Fortelius M (eds) Fossil mammals of Asia. Neogene biostratigraphy and chronology. Columbia University Press, New York, NY, pp 595–621, 732

Koufos GD, Vasileiadou K (2015) Miocene/Pliocene mammal faunas of southern Balkans: implications for biostratigraphy and palaeoecology. Palaeobiodivers Palaeoenviron 95(3):285–303

Koufos GD, Vassiliadou A, Koliadimou K, Syrides G (2001) Early Pleistocene small mammals from Marathousa, a new locality in the Mygdonia basin, Macedonia, Greece. Deinsea 8:49–102

Kretzoi M (1938) Die Raubtiere von Gombaszög nebst einer Übersicht der Gesamtfauna. Ann Hist Nat Mus Nat Hung 31:88–157

Kretzoi M (1941) Die unterpleistozäne Säugetierfauna von Betfia bei Nagyvárad. Földtani Közlöny 71:308–335

Kretzoi M (1956) Die altpleistozänen Wirbeltierfaunen des Villányer Gebirges. Geol Hungarica Seri Paleontol 27:1–264

Kretzoi M (1959) Insectivoren, Nagetiere und Lagomorphen der jüngstpliozänen Fauna von Csarnóta im Villányer Gebirge (Südungarn). Vertebr Hung 1:237–246
Kretzoi M (1962) Fauna und faunenhorizont von Csarnota. Jahres Ungar Geol Anstalt 1959:297–395
Kurtén B (1968) Pleistocene mammals of Europe. Weidenfeld and Nicholson, London
Macdonald DW (ed) (2009) The encyclopedia of mammals. The Brown Reference Group Ltd, Oxford University Press, Oxford
Marković Z (2003) The Miocene small mammals of Serbia, a review. Deinsea 10:393–398
Marković Z, Milivojević M (2010) The Neogene small mammals from Serbia – collection methods and results. Bull Nat Hist Mus 3:105–114
Martín-Suárez E, Bendala N, Freudenthal M (2001) *Archaeodesmana baetica*, sp. nov. (Mammalia, Insectivora, Talpidae) from the Mio–Pliocene transition of the Granada Basin, southern Spain. J Vertebr Paleontol 21:547–554
Masini F, Sarà M (1998) *Asoriculus burgioi* sp. nov.[Soricidae, Mammalia] from the Monte Pellegrino faunal complex [Sicily]. Acta Zool Cracov 41:111–124
Mayhew DF (1978) Late Pleistocene small mammals from Arnissa (Macedonia, Greece). Proc Kon Neder Akad Weten 81:302–321
McKenna CM, Bell SK (1997) Classification of Mammals above the species level. Columbia University Press, New York, 136 pp 3
Mészáros LG (1997) *Kordosia*, a new genus for some Late Miocene Amblycoptini shrews (Mammalia, Insectivora). N Jahr Geol Paläontol Monats 1997:65–78
Mészáros LG (1998) *Crusafontina* (Mammalia, Soricidae) from Late Miocene localities in Hungary. Senckenb Lethaea 77:145–159
Minwer-Barakat R, García-Alix A, Freudenthal M (2008) Desmaninae (Talpidae, Mammalia) from the Pliocene of Tollo de Chiclana (Guadix Basin, Southern Spain). considerations on the phylogeny of the genus *Archaeodesmana*. Geobios 41:381–398
Müller A (1967) Die Geschichte der Familie Dimylidae (Insectivora, Mamm.) auf Grund der Funde aus tertiären Spaltenfüllungen Süddeutschlands. Thesis. Bayerische Akad Wissen, München
Murphy WJ (2001a) Resolution of the Early Placental Mammal Radiation Using Bayesian Phylogenetics. Science 294(5550):2348–2351
Murphy WJ, Eizirik E, Johnson WE, Zhang YP, Ryder OA, O'Brien SJ (2001b) Molecular phylogenetics and the origins of placental mammals. Nature 409(6820):614–618
Nikaido M, Cao Y, Harada M, Okada N, Hasegawa M (2003) Mitochondrial phylogeny of hedgehogs and monophyly of Eulipotyphla. Mol Phylogenet Evol 28(2):276–284
Pasa A (1947) I mammiferi di alcune antiche brecce Veronesi. Memo Mus Civico Storia Nat di Verona 1:1–111
Petényi SJ (1864) Hátrahagyott munkái. AMagyar Tudományos, Akad Kiadasá
Prieto J, Gross M, Böhmer C, Böhme M (2010) Insectivores and bat (Mammalia) from the late Middle Miocene of Gratkorn (Austria): biostratigraphic and ecologic implications. N Jahr Geol Paläontol Abh 258:107–119
Prieto J, van den Hoek Ostende LW, Böhme M (2011) Reappearance of *Galerix* (Erinaceomorpha, Mammalia) at the Middle to Late Miocene transition in South Germany biostratigraphic and palaeoecologic implications. Contr Zool 80:179–189
Prieto J, van den Hoek Ostende LW, Hír J (2012) The Middle Miocene insectivores from Sámsonháza 3 (Hungary, Nógrád County): biostratigraphical and palaeoenvironmental notes near to the Middle Miocene Cooling. Bull Geosci 87:227–240
Rabeder G (1972) Die Insectivoren und Chiropteren (Mammalia) aus dem Altpleistozän von Hundsheim (Niederösterreich). Ann Naturhist Mus Wien 76:375–474
Rabeder G (1973) *Galerix* und *Lanthanotherium* (Erinaceidae, Insectivora) aus dem Pannon des Wiener Beckens. N Jahr Geol Paläontol Monats 1973:429–446
Repenning CA (1967) Subfamilies and genera of the Soricidae. Geol Survey Professional Paper, Washington
Reumer JWF (1984) Ruscinian and early Pleistocene Soricidae (Insectivora, Mammalia) from Tegelen (The Netherlands) and Hungary. Scripta Geol 73:1–173

Reumer JWF (1985) The generic status and species of *Drepanosorex* reconsidered (Mammalia, Soricidae). Rev Paléobiol 4:53–58

Reumer JWF, Doukas CS (1985) Early Pleistocene Insectivora (Mammalia) from Tourkobounia (Athens, Greece). Proc Kon Neder Akad Weten 88:111–121

Rümke CG (1976) Insectivora from Pikermi and Biodrak (Greece). Proc Kon Neder Akad Weten 79:256–270

Rümke CG (1985) A review of fossil and recent Desmaninae (Talpidae, Insectivora). Utrecht Micropaleontol Bull 4(Spec Pub):1–241

Rzebik-Kowalska B (1976) The Neogene and Pleistocene insectivores (Mammalia) of Poland. III. Soricidae: *Beremendia* and *Blarinoides*. Acta Zool Cracov 21:359–385

Rzebik-Kowalska B (1989) Pliocene and Pleistocene Insectivora (Mammalia) of Poland. V. Soricidae: *Petenyia* Kormos, 1934 and *Blarinella* Thomas, 1911. Acta Zool Cracov 32:521–546

Rzebik-Kowalska B (2000) Insectivora (Mammalia) from the Early and early Middle Pleistocene of Betfia in Romania. II. Erinaceidae Bonaparte, 1838 and Talpidae Gray, 1825. Acta Zool Cracov 43:55–77

Rzebik-Kowalska B (2005a) Poland. In: van den Hoek Ostende LW, Doukas CS, Reumer JWF (eds) The fossil record of the Eurasian Neogene Insectivores (Erinaceomorpha, Soricomorpha, Mammalia), Part I. Script Geol 5(Spec Issue):119–134

Rzebik-Kowalska B (2005b) Romania. In van den Hoek Ostende LW, Doukas CS, Reumer JWF (eds) The fossil record of the Eurasian Neogene Insectivores (Erinaceomorpha, Soricomorpha, Mammalia), part I. Scripta Geol 5(Spec Issue):135–147

Rzebik-Kowalska B (2005c) Erinaceomorpha and Soricomorpha (Mammalia) from the Miocene of Bełchatów, Poland. IV. Erinaceidae FISCHER VONWALDHEIM, 1817 and Talpidae FISCHER VON WALDHEIM, 1817. Acta Zool Cracov 48A:71–91

Rzebik-Kowalska B (2008) Insectivores (Soricomorpha, Mammalia) from the Pliocene and Pleistocene of Transbaikalia and Irkutsk region (Russia). Quaternary Int 179:96–100

Rzebik-Kowalska B (2013) *Sorex bifidus* n. sp. and the rich insectivore mammal fauna (Erinaceomorpha, Soricomorpha, Mammalia) from the Early Pleistocene of Żabia Cave in Poland. Palaeontol Electron 16(2/12A):1–35

Rzebik-Kowalska B, Pawlowski B (1994) *Ruemkelia* [Mammalia, Insectivora, Talpidae] nom. nov. for *Dibolia* Rümke, 1985 [nec Latreille, 1829]. Acta Zool Cracov 1:75–76

Rzebik-Kowalska B, Rekovets LI (2015) Recapitulation of data on Ukrainian fossil insectivore mammals (Eulipotyphla, Insectivora, Mammalia). Acta Zool Cracov 58:137–171

Sansalone G, Kotsakis T, Piras P (2015) *Talpa fossilis* or *Talpa europaea*? Using geometric morphometrics and allometric trajectories of humeral moles remains from Hungary to answer a taxonomic debate. Palaeontol Electron 18:1–17

Schötz M (1985) Die Dimyliden (Mammalia, Insectivora) aus dem Kiesgrube Niederaichbach und Maßendorf (Obere Süßwassermolasse Niederbayerns). Mitt Bayer Staatss Paläontol Hist Geol 25:95–130

Springer MS, Murphy WJ, Roca AL (2018) Appropriate fossil calibrations and tree constraints uphold the Mesozoic divergence of solenodons from other extant mammals. Mol Phylogenet Evol 121:158–165

Storch G (1978) Die turolische Wirbeltierfauna von Dorn–Dürkheim, Rheinhessen (SW–Deutschland). 2. Mammalia: Insectivora. Senckenb Lethaea 58:421–449

Storch G (1995) The Neogene mammalian faunas of Ertemte and Harr Obo in Inner Mongolia (Nei Mongol), China.–11. Soricidae (Insectivora). Senckenb Lethaea 75:221–251

Storch G, Qiu Z (1991) Insectivores (Mammalia: Erinaceidae, Soricidae, Talpidae) from the Lufeng Hominoid locality, Late Miocene of China. Geobios 24:601–621

Sulimski A (1959) Pliocene Insectivores from Węże. Study on the Tertiary bone breccia fauna from Węże near Dialoszyn in Poland. Acta Palaeontol Pol 4:119–163

Sulimski A (1962) Supplementary studies on the insectivores from Weze 1 (Poland). Acta Palaeontol Pol 7:441–503

Sylvestrou I (2002) Stratigraphic, palaeontological and palaeogeographical study of the neogene–quaternary deposits of Katerini Basin, Northern Greece. PhD thesis, Aristotle Univ Thessaloniki

Sylvestrou I, Kostopoulos DS (2007) Pseudomeriones megistos nov. sp. (Gerbillinae, Mammalia) from the Latest Miocene of Northern Greece and its phylogenetic relationships. Geobios 40:833–848

Terzea E (1980) Deux micromammifères du pliocène de Roumanie. Travaux l'Institut Spéol Émile Racov 19:191–201

Terzea E (1997) Biochronologie du pliocène du bord méridional du Bassin Dacique (Roumanie). In Aguilar J-P, Legendre S, Michaux J (eds) Actes du Congrès BiochroM'97. Mém Travaux l'Inst Montpellier 21:649–660

Thenius E (1952) *Galerix exilis* aus den pontischen Schichten von Pikermi (Attika). Ann Géol Pays Helléniques 4:195–201

Topachevski VA (1961) Novyi Pliotsenovykh vid vykhukholi iz predkavkazya. Paleontol Z 4:131–137

Topachevski VA, Pashkov AV (1983) Nadvidovaya sistemática vykhukholei roda Desmana (Insectivora, Talpidae). Vestnik Zool 3:39–45

Tsoukala E, Bartsiokas A, Chatzopoulou K, Lazaridis G (2006) Quaternary mammalian remains from the Kitseli pothole (Alea, Nemea, Peloponnese). Sci Ann 98:273–284

van Dam JA (2004) Anourosoricini (Mammalia, Soricidae) from the Mediterranean region: a pre-quaternary example of recurrent climate–controled North–South range shifting. J Paleontol 78:741–764

van de Weerd A (1973) Rodentia from two Pleistocene fissure fillings near Athens. Proc Kon Neder Akad Weten 76:148–166

van de Weerd A (1979) Early Ruscinian rodents and lagomorphs (Mammalia) from the lignites near Ptolemais (Macedonia, Greece). Proc Kon Neder Akad Weten 82:127–170

van de Weerd A, Reumer JWF, de Vos J (1982) Pliocene mammals from Apolakkia Formation (Rhodes, Greece). Proc Kon Neder Akad Weten 85:89–112

van den Hoek Ostende LW (1989) The Talpidae (Insectivora, Mammalia) of Eggingen–Mittelhart (Baden–Württemberg, (F.R.G.)) with special reference to the *Paratalpa–Desmanodon* lineage. Stuttgarter Beit Natur B 152:1–29

van den Hoek Ostende LW (1995) Insectivore faunas from the Lower Miocene of Anatolia. part 2: Heterosoricidae. Proc Kon Neder Akad Weten 98:1–18

van den Hoek Ostende LW (2001) A revised generic classification of the Galericini (Insectivora, Mammalia) with some remarks on their palaeobiogeography and phylogeny. Geobios 34:681–695

van den Hoek Ostende LW (2003) Insectivores (Erinaceomorpha, Soricomorpha, Mammalia) from the Ramblian of the Daroca–Calamocha area. Coloq Paleontol Spec 1:281–310

van den Hoek Ostende LW, Rümke CG, Hogeweg P (1989) The use of time–constrained minimal spanning subtrees in the reconstruction of the phylogeny of the European desmanine moles (Desmaninae, Talpidae, Insectivora). Proc Kon Neder Akad Weten 92:47–60

van den Hoek Ostende LW, Mayda S, Oliver A, Madern A, Hernández-Ballarín V, Peláez-Campomanes P (2015) Aliveri revisited, a biogeographical appraisal of the early Miocene mammals from the eastern Mediterranean. Palaeobiodivers Palaeoenviron 95:271–284

van den Meulen AJ (1973) Middle Pleistocene smaller mammals from the Monte Peglia (Orvieto, Italy) with special reference to the phylogeny of Microtus (Arvicolidae, Rodentia). Quaternaria 17:144

van der Geer AA, Lyras GA, van den Hoek Ostende LW, de Vos J, Drinia H (2014) A dwarf elephant and a rock mouse on Naxos (Cyclades, Greece) with a revision of the palaeozoogeography of the Cycladic Islands (Greece) during the Pleistocene. Palaeogeogr Palaeoclimatol Palaeoecol 404:133–144 van C-RJT, van den Hoek Ostende LW (2001) Dental morphology of Talpa europaea and Talpa occidentalis (Mammalia: Insectivora) with a discussion of fossil Talpa in the Pleistocene of Europe. Zool Med 75:51–68

van Vugt N, de Bruijn H, van Kolfschoten T, Langereis CG (2000) Magneto- and cyclostratigraphy and mammal-fauna's of the Pleistocene lacustrine Megalopolis Basin, Peloponnesos, Greece. Geol Ultrajectina 189:69–92

Vasileiadou K (2001) Study of the Late Miocene/Early Pliocene micromammalian fossil assemblage from Nea Silata, Chalkidiki, North Greece. MSc thesis. Aristotle University of Thessaloniki Greece, Thessaloniki. (in Greek)

Vasileiadou K, Koufos GD (2005) The micromammals from the Early/Middle Miocene locality of Antonios, Chalkidiki, Greece. Ann Paléontol 91:197–225

Vasileiadou K, Zouros N (2012) Early Miocene micromammals from the Lesvos Petrified Forest (Greece): preliminary results. Palaeobiodivers Palaeoenviron 92:249–264

Vasileiadou KV, Koufos GD, Syrides GE (2003) Silata, a new locality with micromammals from the Miocene/Pliocene boundary of the Chalkidiki peninsula, Macedonia, Greece. In: Reumer JWF, Wessels W (eds) Distribution and migration of tertiary mammals in Eurasia. A volume in honour of Hans de Bruijn. Deinsea 10:549–562

Vasileiadou K, Konidaris GE, Koufos GD (2012) New data on the micromammalian locality of Kessani (Thrace, Greece) at the Mio–Pliocene boundary. Palaeobiodivers Palaeoenviron 92:211–237

Viret J, Zapfe H (1951) Sur quelques soricidés miocènes. Ecol Geol Helvetiae 44:411–426

von Koenigswald W (1970) Mittelpleistozane Kleinsäugerfauna aus der Spaltenfullung Petersburch bei Eichstatt. Mitt Bayer Staatsamml Paläontol Hist Geol 10:407–432

Zachariasse WJ, van Hinsbergen DJ, Fortuin AR (2011) Formation and fragmentation of a late Miocene supradetachment basin in central Crete: implications for exhumation mechanisms of high-pressure rocks in the Aegean forearc. Basin Res 23(6):678–701

Ziegler R (1985) Talpiden (Mammalia, Insectivora) aus dem Orleanium und Astaracium Bayerns. Mitt Bayer Staatsamml Paläontolo hist Geol 25:131–175

Ziegler R (1989) Heterosoricidae und Soricidae (Insectivora, Mammalia) aus dem Oberoligozän und Untermiozän Süddeutschlands. Stuttgarter Beit Natur Ser B (Geol Palaeontol) 1989:1–73

Ziegler R (1999) Order Insectivora. In: Rössner GE, Heissig K (eds) The Miocene land mammals of Europe. Verlag Dr. Friedrich Pfeil, München, pp 53–74

Ziegler R (2003) Moles (Talpidae) from the late Middle Miocene of South Germany. Acta Palaeontol Pol 48:617–648

Ziegler R (2005) Erinaceidae and Dimylidae (Lipotyphla) from the Upper Middle Miocene of South Germany. Senckenb Lethaea 85:131–152

Zijlstra JS, Flynn LJ (2015) Hedgehogs (Erinaceidae, Lipotyphla) from the Miocene of Pakistan, with description of a new species of *Galerix*. In: Van den Hoek Ostende LW, Pelaez-Campomanes P, Wessels W (eds) Old worlds, new ideas. A tribute to Albert van der Meulen. Palaeobiodivers Palaeoenviron 95:477–495

The Fossil Record of Bats (Mammalia: Chiroptera) in Greece

Pavlos Piskoulis and Katerina Chatzopoulou

1 Introduction

At present, bats (Order Chiroptera) are the second most diverse mammalian order after the rodents, with more than 1100 living species (>20% of the total extant species) distributed worldwide (apart from the poles and some isolated islands) (Simmons 2005a, b). Despite the abundant information on modern bats, their fossils are rare mostly because of their small and delicate skeleton (Gunnell and Simmons 2005). However, their distribution is universal with the oldest chiropteran fossils coming from the early Eocene of Europe (Coimbra District, Portugal), North America (Wyoming, USA), South America (Chubut, Argentina), Africa (Chambi, Tunisia or El Kohol, Algeria), Oceania (Queensland, Australia), and Asia (Gujarat, India) (Tabuce et al. 2009; Simmons et al. 2008; Tejedor et al. 2005; Sigé 1991; Ravel et al. 2011; Hand et al. 1994; Smith et al. 2007). It is evident that the appearance of Chiroptera in early Eocene is sudden, almost global, and it seems to be an isochronous event (Smith et al. 2007). Nevertheless, some of the most important questions that have yet to be answered are the temporal, geographical, and phylogenetic origins of Chiroptera.

Regarding the Greek region, recent zoological studies recorded 32 chiropteran species (Hanák et al. 2001). In the fossil record, the earliest documented chiropteran specimens come from the early Miocene locality of Lapsarna, Lesvos (Vasileiadou and Zouros 2012; Vasileiadou et al. 2017).

P. Piskoulis (✉) · K. Chatzopoulou
School of Geology, Faculty of Sciences, Aristotle University of Thessaloniki, Thessaloniki, Greece
e-mail: pavlospi@geo.auth.gr

E. Vlachos (ed.), *Fossil Vertebrates of Greece Vol. 2*,
https://doi.org/10.1007/978-3-030-68442-6_4

2 Historical Overview

The knowledge concerning the fossilized chiropteran faunas in Greece is still limited, and the Greek archive of fossilized Chiroptera is sparse. The studies so far have mainly conducted basic systematics, usually on a limited amount of specimens, and there are several factors that did not lead researchers to work with fossil Chiroptera. First, the process for the collection of microfossils requires the use of specialized equipment (sieves), unlimited water supply, etc. (for more details on the collection of microfossils and relevant information about the study of micromammals see van den Hoek Ostende 2003), which was rarely used at paleontological excavations, as researchers that worked on fossil material from the Greek region, organized excavations mainly for large mammals. Moreover, even when researchers were looking for microfossils, the chances of retrieving fossil Chiroptera were low, due to the fact that most excavations have taken place in open sites and not caves, the natural habitat for a plethora of bats. If we take all the above into account, in addition to the difficulty of the fossilization of chiropteran remains, this sparsity can be partly explained. However, the Greek archive of fossilized Chiroptera spans from early Miocene until Early/Middle Holocene. The localities where Chiroptera are documented in their faunal lists are given in Box 1.

Box 1

Early Miocene
Lapsarna, Lesvos (Vasileiadou and Zouros 2012; Vasileiadou et al. 2017)

Early/Middle Miocene
Antonios, Chalkidiki (Vasileiadou and Koufos 2005; Koufos 2006)

Late Miocene
Elaiochoria 2, Chalkidiki (Horáček 1991; Hulva et al. 2007; Sigé et al. 2014), Mytilinii, Samos (Revilliod 1922)

Miocene/Pliocene
Nea Silata, Chalkidiki (Vasileiadou et al. 2003; Koufos 2006), Maramena, Serres (Schmidt-Kittler et al. 1995; Koufos 2006)

Late Pliocene
Tourkobounia 1, Athens (Reumer and Doukas 1985)

Early Pleistocene
Tourkobounia 2, 3, 5, Athens (Reumer and Doukas 1985), Vathy, Kalymnos (Kuss 1973; Kotsakis et al. 1979)

Middle Pleistocene
Latomi 1, Chios (Storch 1975; Kotsakis et al. 1979), Petralona Cave, Chalkidiki (Sickenberg 1964, 1971; Kretzoi 1977; Kretzoi and Poulianos 1981; Horacek and Poulianos 1988; Tsoukala 1989)

(continued)

Late/Middle Pleistocene
Varkiza 1, 2, Athens (van de Weerd 1973)

Late Pleistocene
Kalamakia Cave, Lakonia (Harvati et al. 2013; Darlas and Psathi 2016; Kolendrianou et al. 2020), Loutra Almopias Cave, Pella (Chatzopoulou et al. 2001; Chatzopoulou 2003, 2005; Tsoukala et al. 2006; Piskoulis 2018, 2019, 2020), Naxos, Cyclades (van der Geer et al. 2014), Kharoumes 5, Lasithi (Kuss 1970; Mayhew 1977; Dermitzakis 1977; Doukas and Papayianni 2016), Liko, Chania (Mayhew 1977; Dermitzakis 1977; Doukas and Papayianni 2016), Bate Cave, Rethymno (Kotsakis et al. 1976)

Late Pleistocene/Early Holocene
Franchthi Cave, Argolis (Stiner and Munro 2011), Anonymous Cave of Schisto at Keratsini, Piraeus (Mavridis et al. 2013)

Holocene
Charkadio Cave, Tilos (Symeonidis et al. 1973; Kotsakis et al. 1979), Vraona Cave, Attica (Symeonidis et al. 1980)

Early/Middle Holocene
Sarakenos Cave, Boeotia (Pereswiet-Soltan 2016)

Box 2

The Loutra Almopias Cave (LAC) is located in Pella, Macedonia (Northern Greece) at the eastern slopes of Mount Voras (2524 m), about 120 km northwest of Thessaloniki. The cave is developed on the northern slope of the V-shaped gorge of Thermopotamos Stream, on the altitude of 540 m. The coordinates of the site are: N 40°58.267′, E 021°54.850′. The accumulation of sediments in the cave was in cyclic intervals (clastic and chemical sediments). The fossiliferous bed is mainly brown gravelly sand of fluvial origin. Thousands of Late Pleistocene faunal remains were collected during the excavation and through sieving procedure. The cave-site has yielded abundant faunal remains comprised of large and small mammals, birds, fish, and reptiles. The major part of the large mammalian material belongs to the cave bear (Tsoukala et al. 2006). Thirty species of rodents, insectivores, and lagomorphs have been recognized within the floor sediments providing a remarkable diversity to the LAC microfauna (Chatzopoulou 2014).

Regarding the chiropteran fauna of Loutra Almopias Cave (for more information about it, see Box 2), it is worth to mention that it is currently under study within the framework of the doctorate thesis of the first author. Here are presented the first preliminary results (up to date) of the chiropteran fauna, based on fossil specimens that have been retrieved from the sediments of the cave's floor (Fig. 1).

Fig. 1 Finding chiropteran fossils from Loutra Almopias Cave: (**a**) chiropteran tooth (in circle) found in the sieved sediments, (**b**) screening the sieved sediments for microremains, (**c**) box containing cataloged chiropteran fossils, (**d**) over 3000 of cataloged specimens found in the sieved sediments, (**e**) overview of collected microremains, and (**f**) drawing of specimens with the use of camera lucida

3 Phylogenetic Relationships

Chiroptera is clearly distinguished from other mammalian taxonomic groups because of their ability for powered flight and, for most of the bats, to echolocate (Gunnell and Simmons 2005; Simmons 2005b; Teeling et al. 2005). Despite the fact of being the only known mammals with the aforementioned characteristics, the monophyly of Chiroptera was largely debated since the late 1980s (Gunnell and Simmons 2005 and references therein); however, the intensification of research related to this topic and, thus, the acquirement of plethora of data (morphological, DNA hybridization and nucleotide sequence data), eventually further supported the hypothesis that all chiropteran taxa have a common (flying) ancestor (Gunnell and Simmons 2005; Teeling et al. 2005).

Table 1 All chiropteran families, extinct and extant. As classification above family level is still a matter of dispute, only the most widely accepted distinct families are presented (after Gunnell and Simmons 2005; Miller-Butterworth et al. 2007; Simmons et al. 2008; Lack et al. 2010). Chiropteran families present in the fossil record of Greece are indicated with boldface

Extinct families	Extant families		
Icaronycteridae	Pteropodidae	Emballonuridae	Furipteridae
Archaeonycteridae	**Rhinolophidae**	Myzopodidae	Natalidae
Palaeochiropterygidae	Hipposideridae	Mystacinidae	Molossidae
Hassianycteridae	**Megadermatidae**	Phyllostomidae	**Vespertilionidae**
Tanzanycteridae	**Rhinopomatidae**	Mormoopidae	**Miniopteridae**[a]
Philisidae	Craseonycteridae	Noctilionidae	Cistugidae[a]
Onychonycteridae	Nycteridae	Thyropteridae	

[a]Miniopteridae and Cistugidae are distinct families according to Miller-Butterworth et al. (2007) and Lack et al. (2010), respectively

The aforementioned debate, resulted in the abandonment of the traditional division of Chiroptera into two sub-orders—Megachiroptera (one extant family of non-echolocating Old World fruit bats) and Microchiroptera (19 extant families of echolocating bats) (Gunnell and Simmons 2005; Simmons 2005b)—and led to the establishment of the sub-orders of Yinpterochiroptera, which includes Pteropodidae, Rhinolophidae, Hipposideridae, Megadermatidae, Craseonycteridae and Rhinopomatidae (Springer et al. 2001), and Yangochiroptera, which includes the remaining chiropteran families (Gunnell and Simmons 2005). Accordingly, echolocation in bats was evolved separately for either the two sub-orders or once with Pteropodidae (Old World fruit bats) losing it at a later stage (Teeling et al. 2005; Simmons 2005b).

As stated in the Introduction, Chiroptera appear in the fossil record during the early Eocene. New data on the phylogeny of fossil bats indicate that Onychonycteridae is the basal family for the order Chiroptera, with the rest of the extinct and extant families being sister groups (Simmons et al. 2008). The chiropteran families (20 extant and seven extinct) up to date are given in Table 1. In respect to the Greek chiropteran faunas, no phylogenetic analyses have been carried out until now. As a result, no comments can be made on the phylogeny of Greek bats.

4 Distribution

Even though the fossil record of Chiroptera is a not well-studied subject in Greece, there are quite a few localities across different regions of the country. All localities are East of Pindus Mountain Range (the absence of fossiliferous localities West of Pindus Mountain Range is because of its uplift and the absence of sedimentary basins). Out of 23 localities (Fig. 2), six are in the geographic region of Macedonia (Loutra Almopias Cave, Maramena, Antonios, Petralona Cave, Nea Silata,

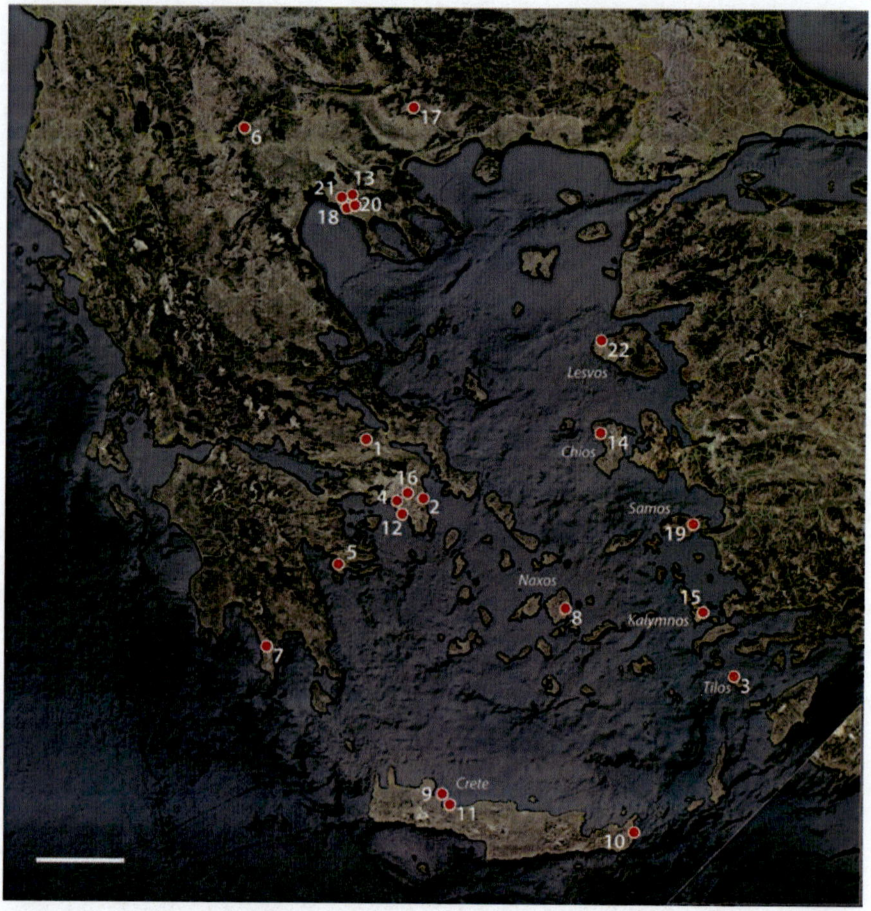

Fig. 2 Map with all the localities, where Chiroptera have been described/recorded: **1** Sarakenos Cave, **2** Vraona Cave, **3** Charkadio Cave, **4** Anonymous Cave of Schisto at Keratsini, **5** Franchthi Cave, **6** Loutra Almopias Cave, **7** Kalamakia Cave, **8** Naxos, **9** Liko Cave, **10** Kharoumes 5, **11** Bate Cave, **12** Varkiza 1 and 2, **13** Petralona Cave, **14** Latomi 1, **15** Vathy village, **16** Tourkobounia 1, 2, 3 and 5, **17** Maramena, **18** Nea Silata, **19** Mytilinii, **20** Elaiochoria 2, **21** Antonios, **22** Lapsarna. See Appendix for more information. Image exported from Google Earth Pro © 2019, map data from US Dept. of State Geographer, SIO, NOAA, U.S. Navy, NGA, GEBCO, image from Landsat/Copernicus. Scale bar equals 80 km, North faces upward

Elaiochoria 2), three of North Aegean Islands (Lapsarna, Latomi 1, Mytilinii), six of Central Greece (Sarakenos Cave, Anonymous Cave of Schisto at Keratsini, Tourkobounia 1, Tourkobounia 2, 3, 5, Vraona Cave, Varkiza 1, 2), two of Peloponnese (Franchthi Cave, Kalamakia Cave), one of Cyclades (Naxos) two of Dodecanese Islands (Vathy village, Charkadio Cave), and three of Crete (Liko, Bate Cave, Kharoumes 5). Further details on all localities are given in the Appendix.

5 Systematic Paleontology

As Chiroptera is the only known mammal group capable of self-powered flight, there are several morphological characteristics that can easily distinguish them from other extant and extinct mammals. Most importantly, their forelimbs and hind limbs are highly specialized for flight (wings) and upside-down roosting respectfully (Gunnell and Simmons 2005 and references therein). In addition to that, the chiropteran families that evolved echolocation have multiple specializations of their auditory system (Gunnell and Simmons 2005 and references therein). As described in the Introduction, complete chiropteran fossils are extremely rare, limiting their study mostly from their dental elements (including mandibles and maxillae) and humeri, which can be easily fossilized due to their harder nature (Sevilla 2016). All the above have morphological characteristics that allow paleontologists to identify fossil specimens up to species level. The identifications that follow were based on the aforementioned characteristics. Moreover, because of the limited information on most of the localities containing chiropteran fossils in Greece, this section contains information only up to the generic level. For further details on the mentioned species, see Simmons (2005a), Dietz et al. (2009), etc. *Samonycteris* is the only extinct chiropteran genus/species named from Greece (Revilliod 1922). It is also worthy to note that one extant chiropteran species, *Myotis alcathoe*, has been named from Greece (von Helversen et al. 2001).

Chiroptera Blumenbach, 1779[1]

Rhinolophidae Gray, 1825

Type Genus *Rhinolophus* Lacépède, 1799.

***Rhinolophus* Lacépède, 1799**

Type Species *Vespertilio ferrum-equinum* Schreber, 1774.

Included Taxa *Rhinolophus* gr. *delphinensis* Gaillard, 1899, *Rhinolophus ferrumequinum* Schreber, 1774 (Greater Horseshoe Bat) (Fig. 3a, b), *Rhinolophus ferrumequinum topali* Kretzoi 1977, *Rhinolophus* (cf.) *hipposideros* Bechstein, 1800 (Lesser Horseshoe Bat) (Fig. 3c, d), *Rhinolophus euryale* Blasius, 1853 (Mediterranean Horseshoe Bat), *Rhinolophus* (cf.) *mehelyi* Matschie, 1901 (Mehely's Horseshoe Bat) (Fig. 3f), *Rhinolophus blasii* Peters, 1866 (Blasius's

[1] For the purpose of this chapter, we follow Simmons (2005a). We are aware that new species have been described since then, however, as plenty of them are still under discussion and because the purpose of this chapter is not to provide an updated taxonomy of extant Chiroptera, we follow the last widely accepted valid work of this kind.

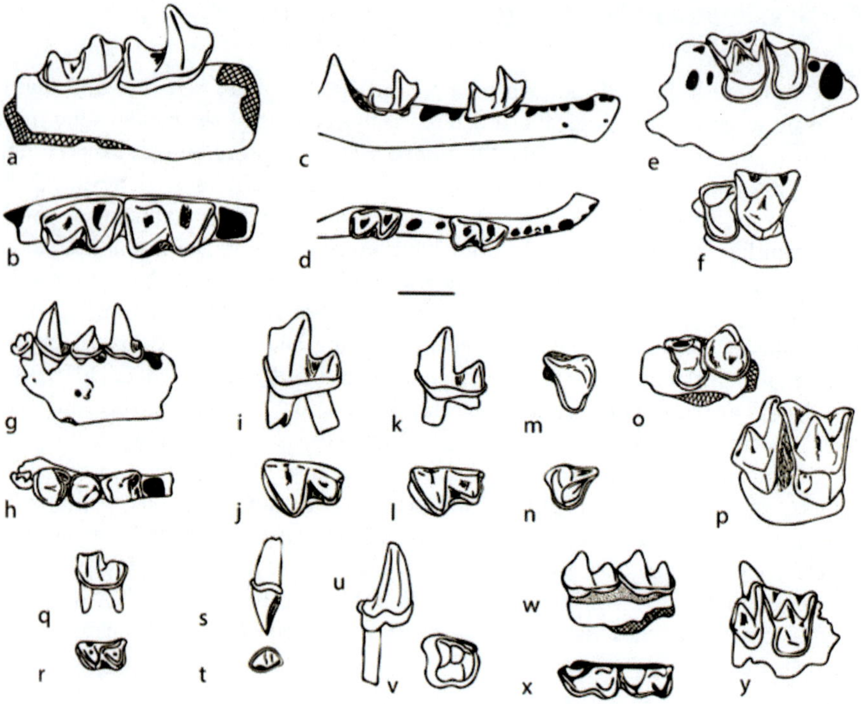

Fig. 3 (**a**, **b**) *Rhinolophus ferrumequinum* (LAC 20600) right partial hemimandible in labial (**a**) and occlusal (**b**) view, (**c**, **d**) *Rhinolophus hipposideros* (LAC 16660) right partial hemimandible in labial (**c**) and occlusal (**d**) view, (**e**) *Rhinolophus* sp. (LAC 17496) right partial hemimaxilla in occlusal (**e**) view, (**f**) *Rhinolophus mehelyi* (LAC 16667) left partial hemimaxilla in occlusal (**f**) view, (**g**, **h**) *Rhinolophus mehelyi/euryale/blasii* (LAC 16744) left partial hemimandible in labial (**g**) and occlusal (**h**) view, (**i**, **j**) *Myotis myotis* (LAC 22236) left m3 in labial (**i**) and occlusal (**j**) view, (**k**, **l**) *Myotis blythii* (LAC 22232) left m3 in labial (**k**) and occlusal (**l**) view, (**m**) *Myotis blythii* (LAC 20468) right P4 in occlusal (**m**) view, (**n**) *Myotis emarginatus* (LAC 20473) left P4 in occlusal (**n**) view, (**o**) *Vespertilio murinus* (LAC 16379) right partial hemimaxilla in occlusal (**o**) view, (**p**) *Nyctalus noctula* (LAC 19325) right partial hemimaxilla in occlusal (**p**) view, (**q**, **r**) *Barbastella barbastellus* (LAC 17977) left m1 in labial (**q**) and occlusal (**r**) view, (**s**, **t**) *Pipistrellus pipistrellus* (LAC 18052) left C in labial (**s**) and occlusal (**t**) view, (**u**, **v**) *Eptesicus serotinus* (LAC 21077) right p4 in labial (**u**) and occlusal (**v**) view, (**w**, **x**) *Barbastella barbastellus* (LAC 16371) left partial hemimandible in labial (**w**) and occlusal (**w**) view, and (**y**) *Miniopterus schreibersii* (LAC 16383) right partial hemimaxilla in occlusal (**y**) view

Horseshoe Bat), *Rhinolophus mehelyi/euryale/blasii* (Fig. 3g, h), *Rhinolophus* sp. (Fig. 3e), *Rhinolophus* sp. I (euryale group), *Rhinolophus* sp. II, *Rhinolophus* sp. III.

Comments *Rhinolophus* appears in the fossil record during late Eocene–Oligocene at Quercy, France (Sigé 1978). Five Horseshoe Bat species can be found nowadays in Greece (Hanák et al. 2001; Simmons 2005a). All of those species are also present in the Greek fossil record, including *R.* gr. *delphinensis*, a fossil species that appears in the European fossil record, between MN 6 and MN 15 (Ziegler 2003). *R.* gr. *del-*

phinensis is present in Elaiochoria 2, *R. ferrumequinum* and *R. hipposideros* are present in Sarakenos Cave, *R. hipposideros* and *R. blasii* are present in Charkadio Cave, *R. ferrumequinum*, *R. hipposideros*, *R. euryale*, *R. mehelyi*, *R. blasii*, *R. mehelyi/euryale/blasii*, and *Rhinolophus* sp. are present in Loutra Almopias Cave, *Rhinolophus* sp. is present in Kalamakia Cave, *R. ferrumequinum topali*, *R.* cf. *hipposideros*, *R.* cf. *mehelyi*, *Rhinolophus* sp. I (euryale group), II and III are present in Petralona Cave.

Vespertilionidae Gray, 1821

Type Genus *Myotis* Kaup, 1829.

Comments Vesper Bats appear in the fossil record during early Eocene (Miller-Butterworth et al. 2007). Several species have been identified in the Greek fossil record at different taxonomic levels. Vespertilionidae sp.1 and sp.2 are present at Nea Silata.

★*Samonycteris* Revilliod, 1922

Type Species ★*Samonycteris majori* Revilliod, 1922.

Included Taxa *Samonycteris majori* Revilliod, 1922.

Comments Major (1891) assigned an incomplete chiropteran skull retrieved from the late Miocene Mytilinii basin (Samos) as Chiroptera indet. Later on, Revilliod (1919) refers to this specimen for the first time as *S. majori*, while its description was published in 1922 by the same author. *Samonycteris* has a single robust upper incisor and its skull has a morphological resemblance to that of *Otonycteris* and/or *Eptesicus*, which, however, differs in several morphological features (e.g., the cochlea) and is also a little smaller than the latter (Revilliod 1922; Horáček 1991). *S. majori* is a monotypic taxon.

Myotis Kaup, 1829

Type Species *Vespertilio myotis* Borkhausen, 1797.

Included Taxa *Myotis myotis* Borkhausen, 1797 (Greater Mouse-eared Bat) (Fig. 3i, j), *Myotis blythii* Tomes, 1857 (Lesser Mouse-eared Bat) (Fig. 3k–m), *Myotis blythii oxygnathus* Monticelli, 1885, *Myotis bechsteinii* Kuhl, 1817 (Bechstein's Bat), *Myotis emarginatus* Geoffroy, 1806 (Geoffroy's Bat) (Fig. 3n), *Myotis capaccinii* Bonaparte, 1837 (Long-fingered Bat), *Myotis* cf. *daubentonii* Kuhl, 1817 (Daubenton's Bat), *Myotis mystacinus*? Kuhl, 1817 (Whiskered Bat), *Myotis myotis/blythii*, *Myotis* sp., *Myotis* sp. I, *Myotis* sp. II, cf. *Myotis* (small sized).

Comments *Myotis* appears in the fossil record during early Oligocene (Gunnell et al. 2017). Ten to eleven Mouse-eared Bat species currently inhabit the Greek region (Hanák et al. 2001), and some of them are also present in the fossil record. *M. blythii* and *M. myotis/blythii* are present in Sarakenos Cave, *Myotis* sp. is present in Vraona Cave, *M. blythii* and *Myotis* sp. are present in Charkadio Cave, *M.* cf. *blythii* is present in Kalamakia Cave, *M. myotis*, *M. blythii*, *M. bechsteinii*, *M. emarginatus*, *M. capaccinii*, *M. mystacinus*?, *M. myotis/blythii* and *Myotis* sp. are present in Loutra Almopias Cave, *M. myotis*, *M. blythii* ssp., *M. blythii oxygnathus*, *M. emarginatus*, *M.* cf. *daubentonii*, *Myotis* sp. I and II are present in Petralona Cave,[2] cf. *Myotis* (small sized) is present in Elaiochoria 2.

Nyctalus Bowdich, 1825

Type Species *Nyctalus verrucosus* Bowdich, 1825 (= *Vespertilio leisleri* Kuhl, 1817).

Included Taxa *Nyctalus lasiopterus*? Schreber, 1780 (Greater Noctule Bat), *Nyctalus leisleri* Kuhl, 1817 (Leisler's Bat), *Nyctalus noctula* Schreber, 1774 (Noctule Bat) (Fig. 3p).

Comments *Nyctalus* appears in the fossil record during the Oligocene (Horáček 2001). Three Noctule Bat species can be found nowadays in the Greek region (Hanák et al. 2001; Simmons 2005a). In the fossil record of Greece, *N. noctula* is present in Sarakenos and Loutra Almopias Caves, *N.* cf. *noctula* is present in Petralona Cave, *N. lasiopterus*? and *N. leisleri* are present only in Loutra Almopias Cave.

Pipistrellus Kaup, 1829

Type Species *Vespertilio pipistrellus* Schreber, 1774.

Included Taxa *Pipistrellus pipistrellus* Schreber, 1774 (Common Pipistrelle Bat) (Fig. 3s, t), *Pipistrellus*(?) sp.

Comments *Pipistrellus* appears in the fossil record during Middle Pleistocene (Horáček and Jahelková 2005). Four Pipistrelle Bat species can be found nowadays in the Greek region (Hanák et al. 2001; Simmons 2005a). In the Greek fossil record,

[2] *M. blythii oxygnathus* is described for the first time by Kretzoi (1977). He states in his description of this taxon that "*Though there are some differences in comparison with the living European form, the material at our disposal is not enough to define the subspecific distinction.*" Later on, Horacek and Poulianos (1988) add *M. blythii* in the chiropteran faunal list of Petralona Cave; however, they state that their "*sample is too small to enable a detailed comparison as to its subspecific relationship that undoubtedly is quite a complicated (cf.* oxygnathus, punnicus, omari *that may come in consideration in the Mediterranean region).*" Consequently, *M. blythii* should be considered a different subspecies from *M. blythii oxygnathus* (most possibly its nominal subspecies *M. blythii blythii*).

P. pipistrellus is present in Loutra Almopias Cave, *Pipistrellus*(?) sp. is present in Petralona Cave, and *Pipistrellus* sp. is present in Kalamakia Cave.

Hypsugo Kolenati, 1856

Type Species *Vespertilio savii* Bonaparte, 1837.

Included Taxa *Hypsugo savii* Bonaparte, 1837 (Savi's Pipistrelle Bat).

Comments *Hypsugo* was included in *Pipistrellus* until mid-1980s, however, since then it gained separate generic status (Horáček and Hanák 1985). One Asian Pipistrelle Bat species can be found nowadays in the Greek region (Hanák et al. 2001), which is also present in the Greek fossil record. *H. savii* is present only in Petralona Cave.

Vespertilio Linnaeus, 1758

Type Species *Vespertilio murinus* Linnaeus, 1758.

Included Taxa *Vespertilio murinus* Linnaeus, 1758 (Parti-colored Bat) (Fig. 3o).

Comments *Vespertilio* appears in the fossil record during the Pliocene (Gunnell and Simmons 2005). One Parti-colored Bat species can be found nowadays in the Greek region (Hanák et al. 2001), which is also present in the Greek fossil record. *V. murinus* is present in Loutra Almopias and Petralona Caves.

Eptesicus Rafinesque, 1820

Type Species *Eptesicus melanops* Rafinesque, 1820 (= *Vespertilio fuscus* Beauvois, 1796).

Included Taxa *Eptesicus serotinus* Schreber, 1774 (Serotine Bat) (Fig. 3u, v), *Eptesicus* sp.

Comments *Eptesicus* appears in the fossil record during the early Miocene (Gunnell and Simmons 2005). Three Serotine Bat species can be found nowadays in the Greek region, with *E. bottae* only present in Rhodes (Hanák et al. 2001; Simmons 2005a), and at least one of them is also present in the Greek fossil record. *E. serotinus* is present in Loutra Almopias and Sarakenos Caves, *Eptesicus* sp. is present in Petralona Cave.

Barbastella Gray, 1821

Type Species *Vespertilio barbastellus* Schreber, 1774.

Included Taxa *Barbastella barbastellus* Schreber, 1774 (Western Barbastelle Bat) (Fig. 3q, r, w, x).

Comments *Barbastella* appears in the fossil record during the Early Pleistocene (Rydell and Bogdanowicz 1997). One Barbastelle Bat species can be found nowadays in the Greek region (Hanák et al. 2001), which is also present in the Greek fossil record. *B. barbastellus* is present only in Loutra Almopias Cave.

Plecotus Geoffroy, 1818

Type Species *Vespertilio auritus* Linnaeus, 1758.

Included Taxa *Plecotus austriacus* Fischer, 1829 (Gray Long-eared Bat), *P. auritus/austriacus*.

Comments *Plecotus* appears in the fossil record during the middle Miocene (Gunnell and Simmons 2005). Three Long-eared Bat species are currently present in the Greek region (Hanák et al. 2001). *P. austriacus* is present in the fossil record of Charkadio Cave and *P. auritus/austriacus* is present in the fossil record of Loutra Almopias Cave.

Miniopteridae Mein and Tupinier, 1977[3]

Type Genus *Miniopterus* Bonaparte, 1837.

Miniopterus Bonaparte, 1837

Type Species *Vespertilio ursinii* Bonaparte, 1837 (= *Vespertilio schreibersii* Kuhl, 1817).

Included Taxa *Miniopterus schreibersii* Kuhl, 1817 (Schreibers' Bent-winged Bat) (Fig. 3y).

Comments *Miniopterus* appears in the fossil record during middle Miocene (Horáček et al. 2013). One Bent-winged Bat species can be found nowadays in Greece (Hanák et al. 2001), which is also present in the Greek fossil record. *M. schreibersii* is present in Loutra Almopias Cave, Petralona Cave, and Vathy village.

Rhinopomatidae Bonaparte, 1838

Type Genus *Rhinopoma* Geoffroy, 1818.

[3] For details on the classification of this family, see Sect. 3 and references therein.

***Rhinopoma* Geoffroy, 1818**

Type Species *Vespertilio microphyllus* Brünnich, 1782.

Included Taxa *Rhinopoma* aff. *hardwickii* Gray, 1831 (Lesser Mouse-tailed Bat).

Comments The Mouse-tailed Bats (*R.* aff. *hardwickii*) appear in the fossil record during Late Miocene at Elaiochoria 2, Greece, which is the earliest and sole occurrence of this family in Europe (Hulva et al. 2007).

Megadermatidae Allen, 1864

Type Genus *Megaderma* Geoffroy, 1810.

Included Taxa cf. Megadermatidae gen. and sp. indet. (False Vampire Bats).

Comments The False Vampire Bats appear in the fossil record during late Eocene (Agnarsson et al. 2011). Fossils of Megadermatidae have been retrieved from several European sites (Gunnell and Simmons 2005), including Greece. cf. Megadermatidae gen. and sp. indet. is present in Antonios.

Chiroptera indet.

Comments Undetermined species of Chiroptera (also mentioned as "Chiroptera," "Bats," etc.) have been found at Anonymous Cave of Schisto at Keratsini, Franchthi Cave, Loutra Almopias Cave, Naxos, Liko, Kharoumes 5, Bate Cave, Varkiza 1, 2, Latomi 1, Tourkobounia 1, 2, 3, 5, Vathy village, Maramena, and Lapsarna.

6 Concluding Remarks

It is evident that the fossil record of Chiroptera is not well-studied in Greece, with sparse finds and most of the works reaching up to generic level. Nevertheless, the following taxa have been identified in Greece: *Rhinolophus* gr. *delphinensis*, *Rhinolophus ferrumequinum*, *Rhinolophus ferrumequinum topali*, *Rhinolophus* (cf.) *hipposideros*, *Rhinolophus euryale*, *Rhinolophus* (cf.) *mehelyi*, *Rhinolophus blasii*, *Rhinolophus mehelyi/euryale/blasii*, *Rhinolophus* sp., *Rhinolophus* sp. I (euryale group), *Rhinolophus* sp. II, *Rhinolophus* sp. III, *Samonycteris majori*, *Myotis myotis*, *Myotis* (cf.) *blythii* (ssp.), *Myotis blythii oxygnathus*, *Myotis bechsteinii*, *Myotis emarginatus*, *Myotis capaccinii*, *Myotis* cf. *daubentonii*, *Myotis mystacinus*?, *Myotis myotis/blythii*, *Myotis* sp., *Myotis* sp. I, *Myotis* sp. II, cf. *Myotis* (small sized), *Nyctalus lasiopterus*?, *Nyctalus leisleri*, *Nyctalus* (cf.) *noctula*, *Pipistrellus pipistrellus*, *Pipistrellus*(?) sp., *Hypsugo savii*, *Vespertilio murinus*, *Eptesicus serotinus*, *Eptesicus* sp., *Barbastella barbastellus*, *Plecotus austriacus*, *Plecotus auritus/austriacus*, *Miniopterus schreibersii*, *Rhinopoma* aff. *hardwickii*, cf.

Megadermatidae gen. and sp. indet., Vespertilionidae sp.1, Vespertilionidae sp.2, and Chiroptera indet.

Rhinopoma aff. *hardwickii* from Late Miocene site Elaichoria 2 is the earliest and sole occurrence of Rhinopomatidae in Europe and consequently Greece. Furthermore, the study of Chiroptera from Loutra Almopias Cave, which is presented here for the first time as an updated preliminary list of taxa, will give us the first complete chiropteran record in the Greek region. In addition, the revision of the existing specimens retrieved from the localities mentioned herein might allow us to reach to a more detailed identification. In any case, the acquisition of new material and basic systematic study will give us the opportunity to carry out more targeted research on this relatively unexplored topic (i.e., phylogenetic analysis) that will allow us to extent our knowledge on the chiropteran evolution in the Balkan Peninsula.

Acknowledgments This research (P.P.) is co-financed by Greece and the European Union (European Social Fund; ESF) through the Operational Programme "Human Resources Development, Education and Lifelong Learning" in the context of the project "Strengthening Human Resources Research Potential via Doctorate Research" (MIS-5000432), implemented by the State Scholarships Foundation (IKY). We would also like to thank Prof. E. Tsoukala (supervisor), Prof. P. Sevilla and Prof. D. Kostopoulos (members of the Ph.D. committee of the first author) for their continuous help and support. We would also like to thank Prof. V. Popov for his critical review that helped on the improvement of the manuscript.

Appendix

List of fossil localities with occurrences of Chiroptera in Greece. Locality numbers refer to the collection numbers of the PaleoBiology Database (PBDB). Type localities are marked with boldface

Localities[PBDB no]	Age (MN)	Taxon
Sarakenos Cave[203845]	Early/Middle Holocene	*Rhinolophus ferrumequinum*[1] *Rhinolophus hipposideros*[1] *Myotis blythii*[1] *Myotis myotis/blythii*[1] *Nyctalus noctula*[1] *Eptesicus serotinus*[1]
Vraona Cave[183139]	Holocene	*Myotis* sp.[2]
Charkadio Cave[182809]	Holocene	*Rhinolophus hipposideros*[3] *Rhinolophus blasii*[3] *Myotis blythii*[3] *Myotis* sp.[3] *Plecotus austriacus*[3]
Anonymous Cave of Schisto at Keratsini[203846]	Late Pleistocene/Early Holocene	Chiroptera indet.[4]
Franchthi Cave[182709]	Late Pleistocene/Early Holocene	Chiroptera indet.[5]

(continued)

Localities[PBDB no]	Age (MN)	Taxon
Loutra Almopias Cave[203847]	Late Pleistocene, Torringian (MNQ 26)	*Rhinolophus ferrumequinum*[a] *Rhinolophus hipposideros*[a] *Rhinolophus euryale*[a] *Rhinolophus mehelyi*[a] *Rhinolophus blasii*[a] *Rhinolophus mehelyi/euryale/blasii*[a] *Rhinolophus* sp.[a] *Myotis myotis*[a] *Myotis blythii*[a] *Myotis bechsteinii*[a] *Myotis emarginatus*[a] *Myotis capaccinii*[a] *Myotis mystacinus*?[a] *Myotis myotis/blythii*[a] *Myotis* sp.[a] *Nyctalus lasiopterus*?[a] *Nyctalus leisleri*[a] *Nyctalus noctula*[a] *Pipistrellus pipistrellus*[a] *Plecotus auritus/austriacus*[a] *Vespertilio murinus*[a] *Eptesicus serotinus*[a] *Barbastella barbastellus*[a] *Miniopterus schreibersii*[a] Chiroptera indet.[a]
Kalamakia Cave[184245]	Late Pleistocene	*Myotis* cf. *blythii*[6] *Pipistrellus* sp.[6] *Rhinolophus* sp.[6]
Naxos[203848]	Late Pleistocene	Chiroptera indet.[7]
Liko Cave[92902]	Late Pleistocene	Chiroptera indet.[8]
Kharoumes 5[204088]	Late Pleistocene	Chiroptera indet.[9]
Bate Cave[183122]	Late Pleistocene	Chiroptera indet.[10]
Varkiza 2[204167]	Late/Middle Pleistocene	Chiroptera indet.[11]
Varkiza 1[204166]	Late/Middle Pleistocene	Chiroptera indet.[11]
Petralona Cave[183123]	Middle Pleistocene	*Rhinolophus ferrumequinum topali*[12] *Rhinolophus* cf. *hipposideros*[12] *Rhinolophus* cf. *mehelyi*[12] *Rhinolophus* sp. I (*euryale* group)[12] *Rhinolophus* sp. II[12] *Rhinolophus* sp. III[12] *Myotis myotis*[12] *Myotis blythii* ssp.[12] *Myotis blythii oxygnathus*[12] *Myotis emarginatus*[12] *Myotis* cf. *daubentonii*[12] *Myotis* sp. I[12] *Myotis* sp. II[12] *Nyctalus* cf. *noctula*[12] *Pipistrellus*? sp.[12] *Hypsugo savii*[12] *Vespertilio murinus*[12] *Eptesicus* sp.[12] *Miniopterus schreibersii*[12]

(continued)

Localities[PBDB no]	Age (MN)	Taxon
Latomi 1[204077]	Middle Pleistocene	Chiroptera indet.[13]
Vathy village	Early Pleistocene	*Miniopterus schreibersii*[14] Chiroptera indet.[14]
Tourkobounia 3–5[34592]	Late Pliocene, Early Biharian (MNQ 19)	Chiroptera indet.[15]
Tourkobounia 2[34767]	Late Pliocene, Early Biharian (MNQ 19)	Chiroptera indet.[15]
Tourkobounia 1[34589]	Late Pliocene, Early Villanyian, MN 16	Chiroptera indet.[15]
Maramena[32189]	Miocene/Pliocene, Late Turolian/Early Ruscinian (MN 13/14)	Chiroptera indet.[16]
Nea Silata[191612]	Miocene/Pliocene, Late Turolian/Early Ruscinian (MN 13/14)	Vespertilionidae sp.1[17] Vespertilionidae sp.2[17]
Mytilinii[202120]	Late Miocene (MN 12)	***Samonycteris majori***[18]
Elaiochoria 2[219100]	Late Miocene, Early Turolian (MN 10/11)	*Rhinolophus* gr. *delphinensis*[19], *Rhinopoma* aff. *hardwickii*[19] cf. *Myotis* (small sized)[19]
Antonios[73861]	Early/Middle Miocene, Orleanian (MN 4/5)	cf. Megadermatidae gen. and sp. indet.[20]
Lapsarna[186558]	Early Miocene, Orleanian, MN 4	Chiroptera indet.[21]

[1]Pereswiet-Soltan (2016), [2]Symeonidis et al. (1980), [3]Symeonidis et al. (1973), [4]Mavridis et al. (2013), [5]Stiner and Munro (2011), [6]Kolendrianou et al. (2020), [7]van der Geer et al. (2014), [8]Mayhew (1977), [9]Kuss (1970), [10]Kotsakis et al. (1976), [11]van de Weerd (1973), [12]Horacek and Poulianos (1988), [13]Storch (1975), [14]Kuss (1973), [15]Reumer and Doukas (1985), [16]Schmidt-Kittler et al. (1995), [17]Vasileiadou et al. (2003), [18]Revilliod (1922), [19]Hulva et al. (2007), [20]Vasileiadou and Koufos (2005), [21]Vasileiadou and Zouros (2012)
[a]This study

References

Agnarsson I, Zambrana-Torrelio CM, Flores-Saldana NP, May-Collado LJ (2011) A time-calibrated species-level phylogeny of bats (Chiroptera, Mammalia). PLoS Curr 3:RRN1212

Chatzopoulou K (2003) The Late Pleistocene small mammal fauna from Loutra Aridea Bear-cave (Pella, Macedonia, Greece). Atti Mus Civ Stor Nat Trieste 49:35–45

Chatzopoulou K (2005) The small mammal fauna from the Loutra Aridea bear-Cave (Pella, Macedonia, Greece) with emphasis on the third chamber. Abh Nathist Ges Nürnberg 45:57–64

Chatzopoulou K (2014) The micromammals of the Quaternary deposits from cave A, Loutra Almopias (Pella, Northern Greece), Stratigraphy-Taphonomy-Paleoenvironment. Dissertation. Aristotle University of Thessaloniki, Thessaloniki

Chatzopoulou K, Vasileiadou A, Koliadimou K, Tsoukala E, Rabeder G, Nagel D (2001) Preliminary report on the Late Pleistocene small mammal fauna from the Loutraki Bear-cave (Pella, Macedonia, Greece). Cad Lab Xeol Laxe 26:485–495

Darlas A, Psathi E (2016) The middle and upper Paleolithic on the western coast of the Mani Peninsula (southern Greece). In: Harvati K, Roksandic M (eds) Paleoanthropology of the Balkans and Anatolia: Human evolution and its context. Springer, Dordrecht, pp 95–117

Dermitzakis M (1977) The quarternary fossil mammals in the caves and karstic-holes of Crete island and their significance. Bull Speleol Soc Grece 14:152–190

Dietz C, Nill D, von Helversen O (2009) Bats of Britain, Europe and Northwest Africa. A & C Black Publishers, London, pp 1–400

Doukas CS, Papayianni K (2016) Small mammals in the Plio/Pleistocene sediments of Greece. In: Harvati K, Roksandic M (eds) Paleoanthropology of the Balkans and Anatolia: Human evolution and its context. Springer, Dordrecht, pp 291–302

Gunnell GF, Simmons NB (2005) Fossil evidence and the origin of bats. J Mamm Evol 12:209–246

Gunnell GF, Smith R, Smith T (2017) 33 million year old *Myotis* (Chiroptera, Vespertilionidae) and the rapid global radiation of modern bats. PLoS One 12:e0172621

Hanák V, Benda P, Ruedi M, Horáček I, Sofianidou TS (2001) Bats (Mammalia: Chiroptera) of the Eastern Mediterranean. Part 2. New records and review of distribution of bats in Greece. Acta Soc Zool Bohem 65:279–346

Hand S, Novacek M, Godthelp H, Archer M (1994) First Eocene bat from Australia. J Vert Paleontol 14:375–381

Harvati K, Darlas A, Bailey SE, Rein TR, El Zaatari S, Fiorenza L, Kullmer O, Psathi E (2013) New Neanderthal remains from Mani peninsula, southern Greece: the Kalamakia middle Paleolithic cave site. J Hum Evol 64:486–499

Horáček I (1991) Enigma of *Otonycteris*: ecology, relationship, classification. Myotis 29:17–30

Horáček I (2001) On the early history of vespertilionid bats in Europe: the Lower Miocene record from the Bohemian Massif. Lynx NS 32:123–154

Horáček I, Hanák V (1985) Generic status of *Pipistrellus savii* (Bonaparte, 1837) and remarks on systematics of the genus *Pipistrellus*. Bat Res News 26:62

Horáček I, Jahelková H (2005) History of the *Pipistrellus pipistrellus* group in Central Europe in light of its fossil record. Acta Chiropterol 7:189–204

Horacek I, Poulianos N (1988) Further data on bats of the Early Pleistocene site, Petralona, Greece. Abstracts of the 2nd Panellenic congress of anthropology, Athens, 27–29 May 1988

Horáček I, Maul LC, Smith KT, Barkai R, Gopher A (2013) Bat remains (Mammalia, Chiroptera) from the Middle Pleistocene site of Qesem Cave, Israel, with the first Pleistocene record of fruit bats in the Mediterranean region. Palaeontol Electron 16:1–23

Hulva P, Horáček I, Benda P (2007) Molecules, morphometrics and new fossils provide an integrated view of the evolutionary history of Rhinopomatidae (Mammalia: Chiroptera). BMC Evol Biol 7:165

Kolendrianou M, Ligkovanlis S, Maniakas I, Tzortzi M, Iliopoulos G (2020) The Palaeolithic cave of Kalamakia (Mani Peninsula), Greece: new insights on the palaeoenvironment using microvertebrates and mesowear analysis of ruminant teeth. Heliyon 6:e03958

Kotsakis T, Melentis J, Sirna G (1976) Seconda spedizione paleontologica Lincea nell'isola di Creta (1975). Rend Accad Naz Lincei 223:3–10

Kotsakis T, Petronio C, Sirna G (1979) The quaternary vertebrates of the Aegean islands: palaeogeographical implications. Ann Geol Pays Helleniques 30:31–64

Koufos GD (2006) The Neogene mammal localities of Greece: faunas, chronology and biostratigraphy. Hellenic J Geosci 41:183–214

Kretzoi M (1977) The fauna of small vertebrates of the Middle Pleistocene at Petralona. Anthropos 4:131–143

Kretzoi M, Poulianos NA (1981) Remarks on the middle and lower Pleistocene vertebrate fauna in the Petralona Cave. Anthropos 8:57–72

Kuss SE (1970) Abfolge und Alter der Pleistozänen Säugetierfaunen der Insel Kreta. Ber Nat forsch Ges Freiburgi Br 60:35–83

Kuss SE (1973) Die pleistozänen Säugetierfaunen der ostmediterranen Inseln. Ihr Alter und ihre Herkunft. Ber Naturf Ges Freiburg Br 63:49–71

Lack JB, Roehrs ZP, Stanley Jr CE, Ruedi M, Van Den Bussche RA (2010) Molecular phylogenetics of *Myotis* indicate familial-level divergence for the genus *Cistugo* (Chiroptera). J Mammal 91:976–992

Major CF (1891) Considérations nouvelles sur la faune des Vertébrés du Miocène supérieur dans l'île de Samos. C R Hebd Séances Acad Sci 113:608–610

Mavridis F, Kormazopoulou L, Papadea A, Apostolikas O, Yamaguchi D, Tankosic Z, Kotzamani G, Trantalidou K, Karkanas P, Maniatis Y, Papagianni K, Lambropoulos D (2013) Anonymous cave of Schisto at Keratsini, Attika: a preliminary report on a diachronic cave occupation from the Pleistocene/Holocene transition to the Byzantine times. In: Mavridis F, Jensen JT (eds) Stable places and changing perceptions: cave archaeology in Greece. Archaeopress, Oxford, pp 248–284

Mayhew DF (1977) The endemic Pleistocene murids of Crete I-II. P K Ned Akad B Phys 80:182–214

Miller-Butterworth CM, Murphy WJ, O'brien SJ, Jacobs DS, Springer MS, Teeling EC (2007) A family matter: conclusive resolution of the taxonomic position of the long-fingered bats, *Miniopterus*. Mol Biol Evol 24:1553–1561

Pereswiet-Soltan A (2016) First record of serotine bat *Eptesicus serotinus* (Chiroptera: Vespertilionidae) from Early-Mid Holocene in Southern Europe (Beotia–Greece). Hystrix It J Mamm 27(2)

Piskoulis P (2018) *Rhinolophus ferrumequinum* (Chiroptera): tooth size variation from late to latest Pleistocene of Loutra Almopias Cave (Macedonia, Greece) XXXIV Jornadas de Paleontologia & IV Congresso Ibérico de Paleontologia, Vila Real (Portugal), 26–29 Sep 2018

Piskoulis P (2019) Late Pleistocene bats (Mammalia: Chiroptera) from Loutra Almopias Cave (Pella, Macedonia, Greece): preliminary results. 4th International meeting of early-stage researchers in palaeontology, Cuenca (Spain), 11–14 Jun 2019

Piskoulis P (2020) Potential precipitation-driven body size differentiation of *Rhinolophus ferrumequinum* from the Late to latest Pleistocene of Loutra Almopias Cave A (Pella, Macedonia, Greece). N Jb Geol Paläont Abh 297:311–323

Ravel A, Marivaux L, Tabuce R, Adaci M, Mahboubi M, Mebrouk F, Bensalah M (2011) The oldest African bat from the early Eocene of El Kohol (Algeria). Naturwissenschaften 98:397–405

Reumer JWF, Doukas CS (1985) Early Pleistocene Insectivora (Mammalia) from Tourkobounia (Athens, Greece). P K Ned Akad B Phys 88:111–121

Revilliod P (1919) L'état actuel de nos connaissances sur les chiroptères fossiles (note préliminaire). C R Soc Sci Phys Nat Genève 36:93–96

Revilliod P (1922) Contribution a l'étude des chiroptères des terrains tertiaires. Troisième partie et fin. Mém Soc Paléont Suisse 45:133–195

Rydell J, Bogdanowicz W (1997) *Barbastella barbastellus*. Mamm Species 557:1–8

Schmidt-Kittler N, de Bruijn H, Doukas K (1995) The vertebrate locality Maramena (Macedonia, Greece) at the Turolian-Ruscinian boundary (Neogene). 1. General introduction. Münchner Geowiss Abh A 2:9–18

Sevilla P (2016) Bats from Azokh caves. In: Fernández-Jalvo Y, King T, Yepiskoposyan L, Andrews P (eds) Azokh cave and the transcaucasian corridor. Springer, Dodrecht, pp 177–189

Sickenberg O (1964) Die Säugetierfauna der Höhle Petralona bei Thessaloniki (Vor. Mit.). Geol Geophys Res 9:1–16

Sickenberg O (1971) Revision der Wirbeltierfauna der Höhle petralona (Griech. Mazedonien). Ann Geol Pays Helleniques 23:230–264

Sigé B (1978) La poche à phosphate de Ste-Néboule (Lot) et sa faune de vertébrés du Ludien supérieur. 8. Insectivores et Chiroptères. Palaeovertebrata 8:249–268

Sigé B (1991) Rhinolophoidea et Vespertilionoidea (Chiroptera) du Chambi (Eocèneinférieur de Tunisie). Aspects biostratigraphique, biogéographique et paléoécologique de l'origine des chiroptèresmodernes. N Jb Geol Paläont Abh 182:355–376

Sigé B, Mein P, Jousse H, Aguilar JP (2014) Un nouveau Rhinopomatidae (Chiroptera) du Paléokarst miocène de Baixas (Pyrénées-Orientales, France); apport zoogéographique. Geodiversitas 36:257–282

Simmons NB (2005a) Order Chiroptera. In: Wilson DE, Reeder DM (eds) Mammal species of the world: a taxonomic and geographic reference, vol 1. JHU Press, Baltimore, MD, pp 312–529

Simmons NB (2005b) An Eocene big bang for bats. Science 307:527–528

Simmons NB, Seymour KL, Habersetzer J, Gunnell GF (2008) Primitive Early Eocene bat from Wyoming and the evolution of flight and echolocation. Nature 451:818–821

Smith T, Rana RS, Missiaen P, Rose KD, Sahni A, Singh H, Singh L (2007) High bat (Chiroptera) diversity in the Early Eocene of India. Naturwiss 94:1003–1009

Springer MS, Teeling EC, Madsen O, Stanhope MJ, de Jong WW (2001) Integrated fossil and molecular data reconstruct bat echolocation. Proc Natl Acad Sci U S A 98:6241–6246

Stiner MC, Munro ND (2011) On the evolution of diet and landscape during the Upper Paleolithic through Mesolithic at Franchthi Cave (Peloponnese, Greece). J Hum Evol 60:618–636

Storch G (1975) Eine Mittelpleistozäne Nager-Fauna von der Insel Chios, Ägais (Mammalia, Rodentia). Senckenberg Biol 56:165–189

Symeonidis N, Bachmayer F, Zapfe H (1973) Grabungen in der Zwergelefanten-Höhle "Charkadio" auf der Insel Tilos (Dodekanes, Griechenland). Ann Naturhistor Mus Wien 77:133–139

Symeonidis N, Bachmayer F, Zapfe H (1980) Ergebnisseweiterer Grabungen in der Höhle von Vraona (Attika, Griechenland). Ann Geol Pays Helleniques 30:291–299

Tabuce R, Antunes MT, Sigé B (2009) A new primitive bat from the earliest Eocene of Europe. J Vert Paleontol 29:627–630

Teeling EC, Springer MS, Madsen O, Bates P, O'brien SJ, Murphy WJ (2005) A molecular phylogeny for bats illuminates biogeography and the fossil record. Science 307:580–584

Tejedor MF, Czaplewski NJ, Goin FJ, Aragón E (2005) The oldest record of South American bats. J Vert Paleontol 25:990–993

Tsoukala E (1989) Contribution to the study of the Pleistocene fauna of large mammals (Carnivora, Perissodactyla, Artiodactyla) from Petralona Cave (Chalkidiki, N. Greece). Dissertation, Aristotle University of Thessaloniki. Sci Ann School Geol 1(8):1–360

Tsoukala E, Chatzopoulou K, Rabeder G, Pappa S, Nagel D, Withalm G (2006) Paleontological and stratigraphical research in Loutra Arideas Bear Cave (Almopia Speleopark, Pella, Macedonia, Greece). Sci Ann School Geol 98:41–67

van de Weerd A (1973) Rodentia from two Pleistocene fissure fillings near Athens. P K Ned Akad B Phys 76:148–166

van den Hoek Ostende LW (2003) Fossils explained 42. Micromammals. Geol Today 19:70–76

van der Geer AA, Lyras GA, van den Hoek Ostende LW, de Vos J, Drinia H (2014) A dwarf elephant and a rock mouse on Naxos (Cyclades, Greece) with a revision of the palaeozoogeography of the Cycladic Islands (Greece) during the Pleistocene. Palaeogeogr Palaeoclimatol Palaeoecol 404:133–144

Vasileiadou K, Koufos GD (2005) The micromammals from the Early/Middle Miocene locality of Antonios, Chalkidiki, Greece. Ann Paleontol 91:197–225

Vasileiadou K, Zouros N (2012) Early Miocene micromammals from the Lesvos Petrified Forest (Greece): preliminary results. Palaeobiodiv Palaeoenviron 92:249–264

Vasileiadou KV, Koufos GD, Syrides GE (2003) Silata, a new locality with micromammals from the Miocene/Pliocene boundary of the Chalkidiki peninsula, Macedonia, Greece. Deinsea 10:549–562

Vasileiadou K, Böhme M, Neubauer TA, Georgalis GL, Syrides GE, Papadopoulou L, Zouros N (2017) Early Miocene gastropod and ectothermic vertebrate remains from the Lesvos Petrified Forest (Greece). Pal Z 91:541–564

von Helversen O, Heller KG, Mayer F, Nemeth A, Volleth M, Gombkötö P (2001) Cryptic mammalian species: a new species of whiskered bat (*Myotis alcathoe* n. sp.) in Europe. Naturwissenschaften 88:217–223

Ziegler R (2003) Bats (Chiroptera, Mammalia) from Middle Miocene karstic fissure fillings of Petersbuch near Eichstätt, Southern Franconian Alb (Bavaria). Geobios 36:447–490

The Fossil Record of Bovids (Mammalia: Artiodactyla: Ruminantia: Pecora: Bovidae) in Greece

Dimitris S. Kostopoulos

1 Introduction

Often simplified in the common sense as oxen, goats, and antelopes, bovids are certainly more than just that. With over 50 living genera and 130 species, and more than double fossil ones, Bovidae Gray, 1821, is possibly one of the most fascinating mammal families with an intriguing, complicated, and still little understood evolutionary history, traced back to the middle Miocene (McKenna and Bell 1998; Bibi et al. 2009; Groves and Grubb 2011). The archetypal image of a bovid is defined by the presence of sheathed unbranched horncores, a character that has been proved to be the single undisputed apomorphy of the family in scientific terms (Janis and Scott 1987). Living and fossil taxa distribute almost globally (except Oceania and South America), seen in a great variety of natural environments from alpine tundra and steppes, to tropical forests, desert and arctic barrens (McDonald 2009). Nevertheless, there is a sharp contradiction between the scarcity of the recent "out of Africa" bovid record compared to the past; during late Neogene to early Quaternary times, Eurasian bovids exhibited a tremendous diversification. The link between bovids and humans is probably as old as the man himself, and bovids have always been a basic source of food and other raw materials. This relationship is visualized in the cave paintings of prehistoric humans and is completed almost 10,000 years ago in the Near East by the domestication of the cattle, goat and sheep, among others.

Definitely, Bovidae systematics is an intricate and quite delicate process, especially in supra-generic level. Eleven extant tribes, at least, are coupled with an unknown number of extinct ones in a multifarious taxonomic scheme, still far from being fully resolved (e.g., Gentry 1992; Groves and Grubb 2011). The implication

D. S. Kostopoulos (✉)
School of Geology, Aristotle University of Thessaloniki, Thessaloniki, Greece
e-mail: dkostop@geo.auth.gr

E. Vlachos (ed.), *Fossil Vertebrates of Greece Vol. 2*,
https://doi.org/10.1007/978-3-030-68442-6_5

of extinct taxa in the restoration of the Bovidae phylogeny, though essential and desirable, can increase the vulnerability of the produced reconstructions and evolutionary scenarios (Reif 2003). Short-time events of important radiation, multiple intracontinental dispersals, parallelism and convergences, introgression phenomena, temporally and spatially discontinuous fossil records, conflicting molecular and morphological evidence, all conspire to keep discussion alive. Evidently, the study of such a multidimensional problem—applying different theories and methodologies on fully distinct or partially comparable data sources (i.e., paleontological, zoological, molecular)—could provide various taxonomic solutions.

Apart from introduced (*Ovis ammon*, *Bubalus*, *Bison*) and domesticated bovids, representatives of the family are extremely rare today in the Greek landscape, including only two taxa: the almost threatened (category NT of IUCN) *Rupicapra rupicapra balkanica* Bolcay, 1925, that mostly lives in Pindus and Rhodope Mountains, and the endangered (category EN of IUCN) *Capra aegagrus cretica* Schinz, 1838, from Crete Island (Papageorgiou 1990; Sfouggaris 2009). This image does not, however, reflect to a minimum the paleontological record of the Greek bovids, one of the richest in Eurasia. The aim of this work is therefore to expose as far as possible the great diversity of the Greek fossil Bovidae through time, providing taxonomical, phylogenetic, biogeographic, and chronological information. The extremely large bibliography on the topic has been condensed to the minimum possible, and the taxonomic opinion is restricted in several cases to the author's point of view. Taxa originally described from Greece are marked with a star.

Museum Abbreviations AMNH: American Museum of Natural History, New York; AeMNH.Samos: Aegean Museum of Natural History, Samos island, Greece; AMPG: Athens Museum of Paleontology and Geology of the National and Kapodistrian University of Athens; BSPM: Bayerische Staatssammlung für Paläontologie und Geologie, Paläontologisches Museum, Munich; IGPM: Institute für Geologie und Paläontologie Universität Münster; LGPUT: Laboratory of Geology and Paleontology-Museum of Geology-Paleontology-Paleoanthropology, Aristotle University of Thessaloniki; MNB: Museum für Naturkunde Berlin; MNH.MG: Museum of Natural History Milias, Grevena, Greece; MNHN.P: Museum national d'Histoire naturele, Paris; NHMUK: Natural History Museum, London; NHMW: Naturhistorisches Museum Wien; SMNS: Staatliches Museum für Naturkunde, Stuttgard.

2 Historical Overview

A spiral horned antelope and a gazelle (i.e., "*Antilope*" *lindermayeri* and "*Antilope*" *capricornis*; nowadays referred to the genera *Palaeoreas* and *Gazella,* respectively) were likely the first fossil bovids ever described from Greece by the German paleontologist, zoologist, and archeologist Johann Andreas Wagner (1848). The late nineteenth to early twentieth century study of the Greek fossil Bovidae is firmly linked with the classical Upper Miocene fossiliferous sites of Pikermi (near Athens) and Samos Island, with pioneering and important works of Roth and Wagner (1854), Wagner (1857), Gaudry (1861a, b, 1862–1867), Dames (1883), Weithofer (1888), Forsyth Major (1888, 1891a, b, 1894), Schlosser (1904), Andrée (1926), and Sickenberg (1932, 1933, 1936). In the wake of the First World War, Pilgrim & Hopwood published the "*Catalogue of the Pontian Bovidae of Europe*" (Pilgrim and Hopwood 1928), where they summarized and reviewed all data available at that time, largely based on the Greek findings. One year later, Arambourg and Piveteau (1929) announced a third Greek area of high paleontological impact, Axios Valley (near Thessaloniki), from where they described several bovid taxa. The second period of the study of Greek fossil bovids was marked by John Melentis, professor of Paleontology at the Aristotle University of Thessaloniki, who in the late 1960s continued previous researches on the Miocene taxa, and introduced to the international scientific community the important, but unknown till that time, Pleistocene artiodactyl record of Greece (Melentis 1966, 1967, 1969). One of the most emblematic figures in the study of the fossil bovids during the second half of the twentieth century, Alan Gentry marked the entrance to the modern times, by his 1971 work "*The earliest goats and other antelopes from Samos* Hipparion *fauna*" (Gentry 1971). In clear distinction from previous periods, Gentry introduced the study of fossil bovids into an evolutionary context, influencing all upcoming scholars. Nikos Solounias (1981) provided a new important review of the Samos bovids, whereas during the 1980s and 1990s, Genevieve Bouvrain—taking the advantage of the full exploration of the wealth Axios Valley fossil sites—registered several new bovid taxa in the Greek and European fossil record (Bouvrain 1978–2001). Most importantly, however, both latter scholars put the study of the Greek fossil bovids into a wider biogeographic and paleoecological frame. During the same decades, the research carried out by George Koufos and his team from the Aristotle University of Thessaloniki provided important new data on the Greek Plio–Pleistocene bovids, greatly expanding the influence of the local record on a pan-European scale.

Over one and a half century after Wagner, the study of the Greek fossil bovid record remains astonishing by means of wealth, diversity, and importance for the understanding of the evolutionary history of this mammal family. As a whole, 52 genera and 83 species are currently known from Greece; 23 of these genera and 58 species are originally based on Greek type material. It is important to note that many more extinct bovid taxa have been named from Greece, but they are currently considered invalid or unavailable, and therefore, they are not covered in detail herein. The phylogenetic relationships of many taxa are, however, barely resolved, and further studies are required on this topic. In addition, the contribution of bovids to the study of paleoenvironmental changes, as demonstrated by a series of modern approaches (e.g., Merceron et al. 2007; Rey et al. 2013), entails the full exploitation of the Greek record in this direction.

3 Phylogenetic Relationships

Members of the even-toed hoofed ruminants—Mammalia: Artiodactyla Owen, 1848 (clade Cetartiodactyla Montgelard, Catzeflis, Douzery, 1997): Ruminantia Scopoli, 1777 (clade Ruminantiomorpha Spaulding O'Leary, Gatesy, 2009)—Bovidae are part of the monophyletic Pecora Flower, 1883 infraorder, which separated from other ruminants likely at Eocene-Oligocene times. Among extant pecora, Moschidae stands out today as the most possible sister group of Bovidae on the basis of molecular data (Hassanin and Douzery 2003; Hassanin et al. 2012) but alternatively Antilocapridae, Cervidae, and "Cervidae+Giraffidae" clades have also been proposed in the past (Janis and Scott 1987; Gentry and Hooker 1988; Beintema et al. 2003). Furthermore, Miocene taxa clustering with *Hypsodontus* Sokolov, 1949, share with Bovidae unbranched horncores associated, however, by precociously specialized dentition compared to the rest of contemporaneous bovids; hence, "hypsodontines" have been interpreted as the possible sister group of all other bovids (Gentry et al. 1999; Bibi et al. 2009; Gentry 2010), though other scholars suggest them as basal members of Caprini or nested within Antilopinae (Dmitrieva 2007; Kostopoulos 2014).

The family of Bovidae is widely accepted as monophyletic (e.g., Hassanin and Douzery 1999; Marcot 2007; Hassanin et al. 2012; Bibi et al. 2009) defined by probably a single apomorphic feature, i.e., the presence of sheathed permanent and unbranched horncores (Janis and Scott 1987). Other skeletal morphological features that are thought to characterize the family include a single lachrymal orifice inside the orbital rim and the distally open dorsal sagittal gulley on the metatarsals, though their full apomorphic status remains disputable (e.g., Janis and Scott 1987; Groves and Grubb 2011). A wide consensus on that Bovidae are divided deeply in time into two subgroups of subfamily taxonomic rank is installed based on both morphological and molecular evidence (Vrba and Schaller 2000; Matthee and Davis 2001; Marcot 2007; Groves and Grubb 2011; Hassanin et al. 2012; Bibi 2013; Yang et al. 2013). The first subfamily, Bovinae, includes boselaphines, bovines, and tragelaphines, all characterized by ox-like dentition (or böodont teeth), whereas Antilopinae include all the remaining bovids, commonly characterized as aegodonts, though bovids without goat-like dentition are also included (Vrba and Schaller 2000; Hassanin and Douzery 1999; Bibi et al. 2009 and references therein).

The emergence of the family is placed in Eurasia at about 19 Ma, a period of major faunal interchanges with the Afro-Arabian continent (Gentry 1992, 2000a; Vrba and Schaller 2000 and literature herein). During Early Miocene (19–17 Ma), Bovinae appear to emerge in Eurasia and Antilopinae (or Antilopini) in Africa documenting the first major split into the family (Gentry 1992, 2000a; Hassanin and Douzery 1999; Vrba and Schaller 2000; Ropiquet and Hassanin 2005). The middle Miocene global warming (16–13.5 Ma) constrained a subsequent explosive radiation of tribes associated by a second important phase of faunal interchanges between Eurasia and Africa (Vrba and Schaller 2000; Bibi et al. 2009). Several modern groups (i.e., tragelaphines, bovines, caprines, reduncines, aepycerotines, hippotragines, alcelaphines) arise at the end of Miocene (10–5 Ma) with the expansion of the grassland environments. During the Pliocene and Pleistocene bovid evolution took place mostly at a continental scale, leading gradually to the Middle to Late Pleistocene "modernization" of the bovid associations.

4 Distribution

The Greek bovid record covers almost the entire time range of the family, with the earliest known taxa dated at 16.5 Ma. Late Middle Miocene bovids are barely known from Greece but an explosive diversity of bovid taxa is documented during the late Miocene (10–5 Ma), accounting more than 25 genera and half of the local diversity through time. The Pliocene Greek bovid record is rather poor, due to both inadequate sampling and restricted suitable lithostratigraphic formations. Pleistocene bovids are abundant and fairly well known from Greece, represented by more than 15 genera of both subfamilies. The relationships of all these genera and related species with each other as well as with the living representatives of the family are poorly documented, especially as we go back in time, and rarely tested on a sound cladistics context. An important number of the recorded genera appears monotypic, and the monophyly of some others remains disputable, while several genera show no clear affiliation with any living tribe, leaving them as incertae sedis. Most of the Greek bovids through time seem to represent the westernmost populations of taxa with a West Asian distribution, though taxa with a more restricted South European/European distribution are also present especially during the Late Pliocene and Early Pleistocene. See Fig. 1 for a map with the most important bovid occurrences and the Appendix for a detailed list.

5 Systematic Paleontology

Bovidae Gray, 1821

Bovinae Gray, 1821 (*sensu* Kingdon 1982)

Tragoportacini Bibi, Bukhsianidze, Gentry, Geraads, Kostopoulos, Vrba, 2009

***Miotragocerus* Stromer, 1928**

Type Species *Miotragocerus monacensis* Stromer, 1928, by original designation.

Distribution Late Miocene (Vallesian–Turolian) of Europe, Asia Minor, India/Pakistan, China, and Africa.

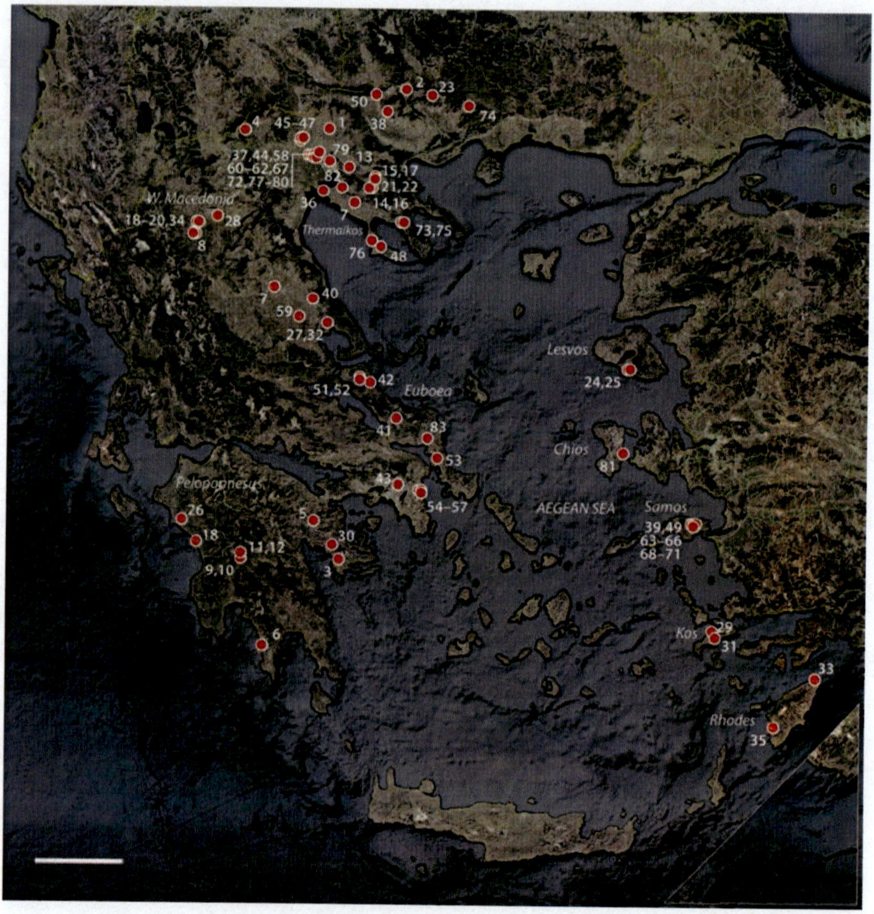

Fig. 1 Map of the occurrences of fossil bovids from Greece. See Appendix for further information. **1**, Agios Georgios Cave; **2**, Kataphyto Cave; **3**, Franchthi Cave; **4**, Loutra Almopias Cave; **5**, Klisoura Cave; **6**, Kalamakia Cave; **7**, Petralona Cave; **8**, Ambelia; **9**, Marathousa 1; **10**, Megalopolis; **11**, Marathousa; **12**, Kyparissia 1; **13**, Apollonia; **14**, Kalamoto-1; **15**, Platanochori-1; **16**, Kalamoto-2; **17**, Tsiotra Vryssi; **18**, Kaiafas; **19**, Livakos; **20**, Polylakkos; **21**, Vassiloudi 1; **22**, Gerakarou; **23**, Volax; **24**, Vatera DS; **25**, Vatera F; **26**, Pyrgos; **27**, Sesklon; **28**, Dafnero; **29**, Kos; **30**, Karnezeika; **31**, Kardamena; **32**, Sesklon lower level; **33**, Damatria; **34**, Milia; **35**, Apolakkia-2; **36**, Megalo Emvolon; **37**, Gefira-1; **38**, Maramena; **39**, Samos, unknown level; **40**, Alifakas; **41**, Triada; **42**, Achladion; **43**, Pyrgos Vassilissis; **44**, Ravin R; **45**, Dytiko-3; **46**, Dytiko-2; **47**, Dytiko-1; **48**, Kryopigi; **49**, Samos Q5; **50**, Thermopigi; **51**, Kerassia-4; **52**, Kerassia-1; **53**, Halmyropotamos; **54**, Chomateri; **55**, Pikermi Valley-3; **56**, Pikermi Valley-1; **57**, Pikermi; **58**, Ravin X; **59**, Perivolaki; **60**, Prochoma-1; **61**, Vathylakkos-3; **62**, Vathylakkos-2; **63**, Samos Mytilinii 1A; **64**, Samos Mytilinii 1B; **65**, Samos Q; **66**, Samos Q4; **67**, Ravin C; **68**, Samos Mytilinii-3; **69**, Samos Mytilinii-4; **70**, Samos Q2; **71**, Samos Q6; **72**, Ravin de Zouaves-5; **73**, Nikiti-2; **74**, Platania; **75**, Nikiti-1; **76**, Fourka; **77**, Ravin de la Pluie; **78**, Ravin de Zouaves-1; **79**, Xirochori; **80**, Pentalophos-1; **81**, Thymiana; **82**, Antonios; **83**, Aliveri. See Appendix for more information. Image exported from Google Earth Pro © 2019, map data from US Dept. of State Geographer, SIO, NOAA, US Navy, NGA, GEBCO, image from Landsat/Copernicus. Scale bar equals 80 km, North faces upward

Other Taxa Included *M. valensiennesi* (Gaudry, 1861b); *M. gradiens* (Pilgrim, 1937); *M. pannoniae* (Kretzoi, 1941); *M. maius* (Meladze, 1967); *M. bonus* Kototkevich, 1971; *M. citus* Korotkevich, 1982; *M. abyssinicus* (Haile-Selassie et al., 2009).

Remarks Bibi et al. (2009) suggested that Late Miocene boselaphine-like bovids belong to a distinct tribe, Tragoportacini, challenging the monophyly of Boselaphini Knottnerus-Meyer, 1907. Solounias (1981), Spassov and Geraads (2004), and Kostopoulos (2005), among others discussed the validity of the genus and included species, whereas Fuss et al. (2015) revised the type species *M. monacensis* from South Germany and Lower Austria. Bouvrain (2001) suggested re-installing *Dystychoceras* Kretzoi, 1941, for both the Pikermi and Dytiko species, and more lately Spassov et al. (2018) re-activated *Pikermicerus gaudryi* Kretzoi, 1941, for the Pikermian taxon, here suggested as a synonym of *M. valenciennesi* (Kostopoulos 2005). *Miotragocerus* appears in Greece as early as at the very beginning of late Vallesian in Pentalophos site (Axios Valley) and extends well within the Turolian. Several late Vallesian samples from Greece (Nikiti-1, Platania) are referred to this genus, but a species level taxonomy is still pending (Kostopoulos 2016; Vasileiadis et al. 2019). Turolian samples referred to as *Miotragocerus* sp. are coming from the faunas of Kryopigi (MN12/13) and Samos Mytilinii 4 (MN11).

★*Miotragocerus (Pikermicerus) valenciennesi* (Gaudry, 1861b)

Nomenclatural and Taxonomic History *Tragocerus valenciennesi* in Gaudry 1861b (new species); *Graecoryx valenciennesi* in Pilgrim and Hopwood 1928 (new combination); *Pikermicerus gaudryi* in Kretzoi 1941 (new combination); *Miotragoceus valenciennesi* in Solounias 1981 (new combination); *Tragoportax gaudryi* in Moyà-Solà 1983; *Dystychocerus gaudryi* in Bouvrain 2001.

Type Material MNHN.P PIK 2367 (lectotype selected by Pilgrim and Hopwood 1928), female frontlet illustrated by Gaudry (1862–1867: Pl. XLVIII, fig. 2) (Fig. 2).

Type Locality Pikermi, Greece, Upper Miocene.

Age Late Miocene, Turolian (MN12), ca. 7.3 Ma.

Distribution Late Miocene (Vallesian–Turolian) of Greece, Northern Macedonia, Bulgaria, France, Spain, and Turkey. Other Greek occurrences include those from Samos Qx, Q2, Q6 (MN11), Chomateri (MN12), Thermopigi (MN12), Samos Mytilinii 1A, 1B, 1C (MN12), Samos Q5 (MN12/13).

Remarks Although starting from a different taxonomic point of view, Spassov and Geraads (2004) and Kostopoulos (2005, 2009a) provide similar distinctive features such as: narrow and high braincase; grooved basioccipital; postcornual area not depressed or raised as a low plateau and weakly rugose; weak temporal crests; narrow intercornual area; moderately long horncores with tear-shaped cross-section

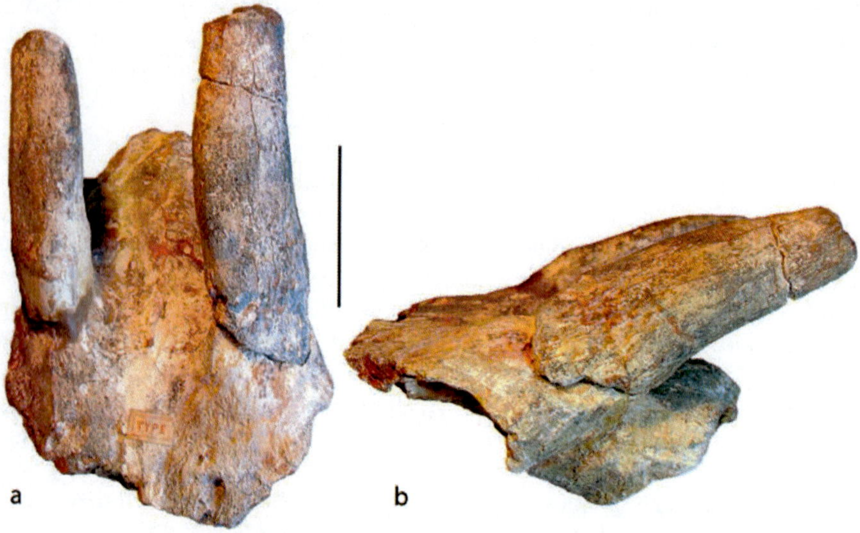

Fig. 2 Female frontlet (MNHN.P PIK 2367), lectotype of *Miotragocerus (Pikermicerus) valenciennesi* (Gaudry, 1861b) from Pikermi, in (**a**) frontodorsal, and (**b**) left-lateral view. Scale bar equals 5 cm. Photographs courtesy of MNHN.P

marked anteriorly by a strong keel; horncores showing distally one demarcation (step) at least; and primitive, long premolars compared to the molars.

★*Miotragocerus* (*Pontoportax*) *parvidens* (Schlosser, 1904)

Nomenclatural and Taxonomic History *Tragocerus amaltheus parvidens* in Schlosser 1904 (new subspecies); *Pontoportax parvidens* in Kretzoi 1941 (new combination and elevation to the species level). It has been considered as a junior synonym of *Tragoportax rugosifrons* (part) by Bouvrain (1994) and Spassov and Geraads (2004). It is here considered as the senior synonym of *Tragocerus punjabicus* Pilgrim, 1910, *Tragocerus frolovi* Pavlow, 1913, *Tragocerus curvicornis* Andrée, 1926, and *Tragocerus browni* Pilgrim, 1937.

Type Material BSPM #1899 VII 522 (lectotype selected by Pilgrim and Hopwood 1928), partial cranium illustrated by Schlosser (1904: Pl. XII, fig. 5) (Fig. 3).

Type Locality Samos (unknown level), Greece, Upper Miocene.

Age Late Miocene, Turolian (MN11–13).

Distribution Late Miocene (likely early–middle Turolian) of Greece, Bulgaria, India/Pakistan, possibly also in Ukraine and Afghanistan. In particular, other Greek occurrences include those from Nikiti-2 (MN11), Ravin des Zouaves 5 (MN11),

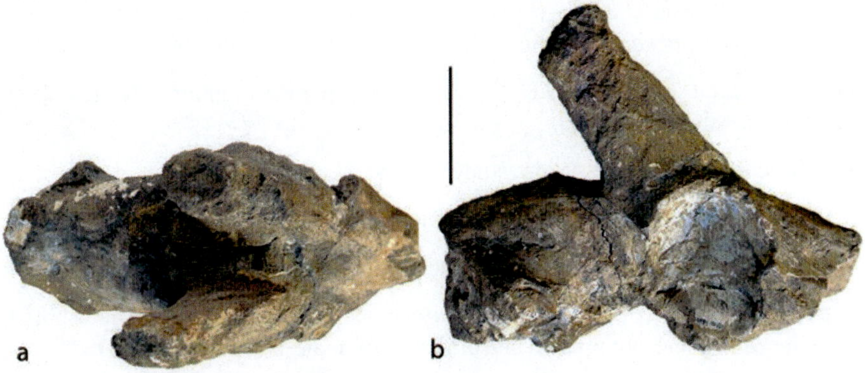

Fig. 3 Partial cranium (BSPM #1899 VII 522), lectotype of *Miotragocerus (Pontoportax) parvidens* (Schlosser, 1904) from Samos, in (**a**) dorsal, and (**b**) right-lateral view. Scale bar equals 5 cm. Photographs courtesy of BSPM and A Athanassiou

Ravin C (MN11/12), Prochoma (MN12), Perivolaki (MN12), and Samos (unknown level).

Remarks Schlosser (1904) described under "*Tragocerus amaltheus* var. *parvidens*" a partially preserved cranium from Samos. The taxon was long considered as a junior synonym of *Tragoportax rugosifrons* (e.g., Bouvrain 1994; Spassov and Geraads 2004), though Kretzoi (1941) used it as the lectotype specimen of *Pontoportax parvidens*, genotype of his new genus *Pontoportax*. Solounias (1981), Moyà-Solà (1983), Spassov and Geraads (2004), and Kostopoulos (2009a, 2016) suggested as likely conspecific several Eurasian crania attributed to different species originally described by Pilgrim (1910, 1937), and Andrée (1926). I propose to unite all of them under the subgenus *Miotragocerus* (*Pontoportax*) Kretzoi, 1941. This taxon appears intermediate in cranial morphology between *Miogragocerus* and *Tragoportax* in having long narrow and high braincase with well-expressed rugose area on the dorsal surface behind the horncores, marked laterally by moderately strong temporal crests and caudally by a blunt hump. Horncores are long, strongly curved posteriorly in their distal part, tear-shaped in cross-section, without or only a smooth distal demarcation. Dentition is advanced and more "*Tragoportax*"-like. Bouvrain (1994), Kostopoulos (2006a), and Spassov and Geraads (2004) described as *T. rugosifrons* several crania from N. Greece and S. Bulgaria that I would now include in this taxon, which seems to have been expanded significantly eastwards.

★*Miotragocerus* (*Pontoportax?*) *macedoniensis* (Bouvrain, 1988)

Nomenclatural and Taxonomic History *Tragoportax gaudryi macedoniensis* in Bouvrain 1988 (new subspecies); *Dystychoceras macedoniensis* in Bouvrain 2001 (new combination, referral to the species level); *Tragoportax macedoniensis* in

Spassov and Geraads 2004 (original combination, referral to the species level); *Miotragocerus macedoniensis* in Kostopoulos 2005 (new combination).

Type Material LGPUT DTK 53 (holotype), partial cranium, illustrated by Bouvrain (1988: Pl. 1, fig. 1).

Type Locality Dytiko-1, Axios Valley, Greece, Upper Miocene.

Age Late Miocene, late Turolian (MN13).

Distribution Late Miocene (middle–late Turolian) of Greece.

Remarks Bouvrain (1988) distinguished the taxon by its smaller size and shorter, straighter and more compressed medio-laterally male horncores compared to the Pikermi *Miotragocerus*. Spassov and Geraads (2004) included the species in *Tragoportax*, but I think it represents a *Miotragocerus*, likely of the subgenus *Pontoportax*. Apart from Dytiko, the species in reported from Pyrgos Vassilissis near Athens (MN12–13) (Boehme et al. 2017).

Tragoportax Pilgrim, 1937

Type Species *Tragoportax salmontanus* Pilgrim, 1937, by original designation.

Distribution Late Miocene (Turolian) of Europe, Asia Minor, India/Pakistan, and Africa.

Other Taxa Included *T. amalthea* (Roth and Wagner, 1854); *T. rugosifrons* (Schlosser, 1904); *T. cyrenaicus* Thomas, 1979; *T. acrae* (Gentry, 1974); *T. ventiensis* Morales, 1984.

Remarks Moyà-Solà (1983), Bouvrain (1988), Spassov and Geraads (2004), and Kostopoulos (2009a), among others, commented on the genus and its contents, but a general consensus is still missing. In distinction from *Miotragocerus*—with which it usually coexists in the Late Miocene faunas of the Greco-Irano-Afghan and Indo-Pakistani provinces—*Tragoportax* is characterized by well-keeled anteriorly, twisted heteronymously, and more inclined backward horncores, without or only weekly expressed steps, and a flattened posterior surface, commonly marked by a posterolateral keel. The braincase is wider and lower than in *Miotragocerus*, and the dorsal surface behind the horncores forms a well-marked rugose area, defined laterally by strong temporal crests and caudally by a step leading to a slightly raised plateau. The intercornual area is wider and shorter than in *Miotragocerus* and the premolars shorter than in *M. (Pikermicerus)*.

★*Tragoportax amalthea* (Roth and Wagner, 1854)

Nomenclatural and Taxonomic History *Capra amalthea* in Roth and Wagner 1854 (new species); *Tragocerus amaltheus* in Gaudry 1865 (new combination and mandatory change of spelling); *Tragoportax amalthea* in Moyà-Solà 1983 (new combination).

Type Material BSPM #ASII 648 (holotype), right horncore illustrated by Roth and Wagner (1854: Pl. XII, fig. 2) (Fig. 4a).

Type Locality Pikermi, Greece, Upper Miocene.

Age Late Miocene, middle Turolian (MN12), ca. 7.3 Ma.

Distribution Late Miocene (Turolian) of Greece, N. Macedonia, Bulgaria, Turkey, Iran. In particular, other Greek occurrences include those from Nikiti-2 (MN11), Ravin des Zouaves 5 (MN11), Vathylakkos 2,3 (VTK, VAT), Ravin X (MN12), Perivolaki (MN12), Halmyropotamos (MN12), Kerassia 1 (MN12), Thermopigi (MN12), Kryopigi (MN12/13), Achladion (MN11–13), Alifakas (MN11–13), Ravin R (MN11–13), Triada (MN11–13), ?Maramena (MN13/14).

Fig. 4 Type specimens of *Tragoportax* spp. from Greece. **(a)** Right horncore (BSPM #ASII 648), holotype of *Tragoportax amalthea* (Roth and Wagner, 1854) from Pikermi, in medial view. **(b)** Partial cranium (BSPM #1899 VII 18), lectotype of *Tragoportax rugosifrons* (Schlosser, 1904) from Samos, in left-lateral view. Scale bars equal 5 cm. Photographs courtesy of BSPM and A Athanassiou

Remarks *Tragoportax amalthea* is characterized by moderately long horncores, weakly to moderately twisted heteronymously, regularly diverging from the base up, and quite often with anterior steps. Kostopoulos and Bernor (2011) and Kostopoulos (2016) provided recently some additional hints on this taxon.

★*Tragoportax rugosifrons* (Schlosser, 1904)
(= ★*Tragocerus recticornis* Andrée, 1926)

Nomenclatural and Taxonomic History *Tragocerus rugosifrons* in Schlosser 1904 (new species); *Tragoportax rugosifrons* in Moyà-Solà 1983 (new combination). This species is widely considered as the senior synonym of *Tragocerus recticornis* Andrée, 1926, which was named from a similarly unknown level from the Turolian of Samos Island (Pilgrim and Hopwood 1928); the type specimen of the latter species is SMNS 13269 (figured by Andrée 1926).

Type Material BSPM #1899 VII 18 (lectotype selected by Pilgrim and Hopwood 1928), partial cranium, illustrated by Schlosser (1904: Pl. XII, fig. 6) (Fig. 4b).

Type Locality Samos (unknown level), Greece, Upper Miocene.

Age Late Miocene, Turolian (MN11–13).

Distribution Late Miocene (likely early–middle Turolian) of Greece and Asia Minor but maybe had a much wider distribution, reaching Afghanistan. In particular, other Greek occurrences include those from Samos Q6 (MN11), Samos Q1 (MN12), and Samos Mytilinii 1A (MN12).

Remarks Typical *Tragoportax rugosifrons* from Samos differs from *T. amalthea* in the longer and untwisted horncores, parallel braincase sides, less posteriorly expanded rugose area, shorter basioccipital, shallower and probably larger lacrimal fossa and trapezoidal-shaped occiput. Bouvrain (1994), Spassov and Geraads (2004), and Spassov et al. (2018) ascribed to *T. rugosifrons* several specimens from the Upper Miocene sites of Axios Valley (Greece), southern Bulgaria and N. Macedonia but I think that at least a part of them may represent *T. amalthea* while others likely belong to *Miotragocerus (Pontoportax)*.

Bovini Gray, 1821

***Alephis* Gromolard, 1980**

Type Species *Alephis lyrix* Gromolard, 1980, by original designation.

Distribution Late Miocene–middle Pliocene of southern Europe (MN13–16).

Other Included Taxa *Alephis tigneresi* Michaux et al., 1991; according to some authors, *Parabos*? *boodon* (Gervais, 1853) from Alcoy (Spain) has also to be ascribed into *Alephis* (e.g., Montoya et al. 2006).

Remarks During the in-depth review of the west European latest Miocene–early Pliocene primitive bovines and advanced boselaphines, Gromolard (1980, 1981) recognized the Roussillon large bovid as a new taxon, *Alephis lyrix*, fully distinct from species of *Parabos* or *Palaeoryx* with which was previously confused. The taxon was exclusively known from SW Europe (Spain and S. France; Italian records still pending confirmation), but Crégut-Bonnoure and Tsoukala (2017b) refer a horncore from Milia, Grevena basin (MN16) to this genus.

★*Grevenobos* Crégut-Bonnoure and Tsoukala, 2017b

Type Species *Grevenobos antiquus* Crégut-Bonnoure and Tsoukala, 2017b, by original designation.

Distribution middle Pliocene of Greece (MN16a).

Remarks Monotypic genus. According to Crégut-Bonnoure and Tsoukala (2017b), *Grevenobos* may be related to *Alephis*.

★*Grevenobos antiquus* Crégut-Bonnoure and Tsoukala, 2017b

Type Material MNH.MG MIL 401a (holotype), partial cranium illustrated by Crégut-Bonnoure and Tsoukala (2017b: fig 2).

Type Locality Milia, Greece, middle Pliocene.

Age middle Pliocene, early Villafranchian (MN16a).

Distribution Known only from the type locality.

Remarks According to the original diagnosis and descriptions by Crégut-Bonnoure and Tsoukala (2017b), *Grevenobos antiquus* is a large Bovini with a mixture of primitive and advanced features within the tribe. The braincase is relatively long and high, with trapezoidal shaped occipital. The long, massive, and weakly heteronymously twisted horncores with triangular cross-section, keeled both anteriorly and posteriorly, are inserted above the back half of the orbits and rather laterally, strongly diverging from each other. Frontals and horncores are extensively pneumatized. Teeth moderately hypsodont without cement; rugose enamel and strong entostyles/ectostylids. Metapodials are long and slender for a bovine.

Leptobos Rütimeyer, 1877–1878

Type Species *Leptobos falconeri* Rütimeyer, 1877–1878, by original designation.

Fig. 5 Right horncore and partial cranium (LGPUT AQP-1) of *Leptobos etruscus* from Haliakmon-Q profile site, in frontoventral view. Scale bar equals 5 cm

Distribution middle Pliocene to Early Pleistocene of Eurasia.

Other Included Taxa *L. stenometopon* (Sismonda, 1846); *L. etruscus* (Falconer, 1859); *L. vallisarni* Merla, 1949; *L. brevicornis* (Hu and Qi, 1978); *L. crassus* (Jia and Wang, 1978); *L. merlai* DeGiuli, 1987; *L. furtivus* Duvernois and Guerin, 1989.

Remarks Duvernois (1989, 1990), Masini (1989), Masini et al. (2013), Mead et al. (2014), and Cherin et al. (2019), among others, exhaustively discuss the taxonomic content of *Leptobos* and the problem of possible synonymy among different species. Most authors accept two main lineages within the genus: *L. stenometopon–L. merlai–L. furtivus*, and *L. etruscus–L. vallisarni*. Although *Leptobos* is very common and wealthy represented in the Late Pliocene to Early Pleistocene SW European fossil record (Spain, Italy, France), it is rather rare in the Balkans and Greece. *Leptobos* sp. is referred from the early Villafranchian (MN16) fauna of Damatria in Rhodes Island (but the age of this site needs confirmation), and the late Villafranchian (MN17) faunas of Pyrgos and Vatera, whereas *L. etruscus* or closely related forms are documented in the Early Pleistocene faunas of Kos, Gerakarou, Vassiloudi, Libakos, Haliakmon-Q profile (Fig. 5), and Tsiotra Vryssi, though represented by poor material (Koufos 2006).

Bison Hamilton-Smith, 1827

Type Species *Bos bison* Linnaeus, 1758, by Linnean tautonomy.

Distribution Early Pleistocene to Present of Eurasia and N. America.

Other Included Eurasian Taxa *B. bonasus* (Linnaeus, 1758); *B. priscus* Bojanus, 1827; *B. sivalensis* Lydekker, 1878; *B. mutus* Przewalski, 1883; *B. schoetensaki* Freudenberg, 1910; *B. palaeosinensis* Teilhard de Chardin and Piveteau, 1930; *B. baicalensis* Veretchgagin, 1954; *B. tamanensis* Vereshchagin, 1959; *B. georgicus* (Burchak-Abramovich and Vekua, 1994), *B. menneri* Sher, 1997; *B. degiulii* (Masini et al., 2013).

Remarks *Bison* is one of the most widespread Pleistocene mammal taxa, distributed from SW Europe to C. America. Its taxonomic history is quite complicated, and several authors propose the inclusion of all extant Bovina in the single genus *Bos* Linaeus, 1758 (e.g., Groves and Grubb 2011; Hassanin 2015; but see also Kostopoulos et al. 2018). Early small-sized bisons invaded S. Europe and the Balkans at the first part of the post-Olduvai Pleistocene. In the Greek fossil record, early bisons ascribed to *Bison* cf. *degiulii* are known from Mygdonia Basin (Tsiotra Vryssi, Krimni, Kalamoto 1, 2, Platanochori 1, Apollonia), and likely Grevena area (Libakos), while heavy priscoid forms (*B. priscus*) certainly occur in the late Middle Pleistocene faunas of Petralona cave and Megalopolis basin (sites Marathousa 1, Kyparissia-1, Kyparissia-4) (Tsoukala 1992; Kostopoulos et al. 2018; Athanassiou et al. 2018a).

Bos Linnaeus, 1758

Type Species *Bos taurus* Linnaeus, 1758, by Linnean tautonomy.

Distribution Emerged during the Early to Middle Pleistocene transition, the genus originally had a Eurasian late Middle Pleistocene to Present distribution; domesticated forms have now a worldwide distribution.

Other Included Taxa *Bos primigenius* Bojanus, 1827; *Bos planifrons* Lydekker, 1878; *Bos acutifrons* Lydekker, 1898; *Bos palaeosondaicus* Dubois, 1908; *Bos buiaensis* Martinez-Navarro et al., 2010; for extant species see Groves and Grubb (2011).

Remarks Martínez-Navarro et al. (2007) suggested that the African *Pelorovis* should be included in *Bos*, transmitting the origin of the genus to the Early Pleistocene of Africa. The Aurochs (*B. primigenius*) rapidly spread across Eurasia from Middle Pleistocene onwards. It invaded Southern Balkans and Greece likely between 0.6 and 0.4 Ma, as evidenced by its presence in Haliakmon Valley, Petralona Cave, and Megalopolis basin (Tsoukala 1992). Late Pleistocene Greek records are known from Larissa area (Thessaly), Agios Georgios Cave (Kilkis), Loutra Almopias, Ampelia (Grevena), Kalamakia, Franchthi, and Klisoura Cave (Peloponnese), as well as from Kataphyto cave (Drama) (Samartzidou et al. in press, pers. com. 2020), but by means of abundance, the species did not reach the optimum seen in isochronous western and northern European sites. The last aurochs survived in Northern Europe until the early seventeenth century, but the domestic cattle already appears in Southern Europe as early as 8000–6000 BC (Ajmone Marsan et al. 2010; Pandolfi et al. 2011).

Bubalus Hamilton-Smith, 1827

Type Species *Bubalus bubalis* Linnaeus, 1758, by absolute tautonomy.

Distribution Pleistocene to Present of Eurasia.

Other Included Taxa *B. palaeindicus* (Falconer, 1868); *B. platyceros* (Lydekker, 1877); *B. sivalensis* (Rütimeyer, 1878); *B. palaeokerabau* Dubois, 1908; *B. mephistopheles* Hopwood, 1925; *B. murrensis* Berckhemer, 1927; *B. wansjocki* (Boule and Teilhard de Chardin, 1928); *B. teilhardi* Young, 1932; *B. brevicornis* Young, 1936; *B. tingi* Bohlin, 1938; *B. youngi* Chow and Hsu, 1957; *B. guzhensis* Liu and Zhen, 1981; *B. triangulatus* Liu, 1981; *B. cebuensis* Croft et al., 2006; *B. grovesi* Rozzi, 2017; for extant taxa, see Croft et al. (2006) and Groves and Grubb (2011).

Remarks Though well distributed and diversified in Asia, *Bubalus* is a rare taxon in Pleistocene European faunas. *B. murrensis* invaded Europe during the interglacial periods of the Middle Pleistocene and disappeared soon after (von Koenigswald et al. 2019). The Greek Pleistocene record is extremely poor in *Bubalus* remains. Boessneck (1965) described as *Bubalus* cf. *arnee* a broken horncore from Larissa area (Thessaly), but von Koenigswald et al. (2019) challenged this attribution. Sickenberg (1976) also described from Marathousa-Megalopolis some fossil remains of *Bubalus* proposing a new species (see Species inquirenda section).

Bovinae Incertae Sedis

Eotragus Pilgrim, 1939

Type Species *Antilope clavata* Gervais, 1850 (= *Antilope sansaniensis* Lartet, 1851), by original designation.

Distribution Early to early middle Miocene of India, Middle East, and Europe; north African and Chinese records are debatable.

Other Taxa Included *E. cristatus* (Biedermann, 1873); *E. artenensis* Ginsburg and Heintz, 1968; *Eotragus noyei* Solounias et al., 1995; *E. minus* Ginsburg et al., 2001 (= *E. noyei* according to some authors).

Remarks Considered as the (or one of the) earliest bovids known from horncores, *Eotragus* appears 18–19 Ma ago in India/Pakistan and it is characterized by short, spike-like horncores with concave to asymmetric sigmoidal anterior profile and convex posterior one. Several authors tentatively proposed it as an early or stem Bovidae (Solounias and Moelleken 1992; Azanza and Morales 1994; Solounias et al. 1995), while others (e.g., van der Made 2012) indicate it as a primitive boselaphine. The Greek record of *Eotragus* is extremely limited, reported but not described so far only from the early Orleanian (MN3) fauna of Aliveri (van den Hoek Ostende et al. 2015) and the late Orleanian (MN5) faunas of Antonios and Thymiana, Chios Island (Vasileiadou and Koufos 2005 and ref. therein).

★*Samokeros* Solounias, 1981

Type Species *Samokeros minotaurus* Solounias, 1981, by original designation.

Distribution Late Miocene (Turolian) of Greece and Iran.

Remarks *Samokeros* is a monotypic genus. Solounias (1981) sees *Samokeros* as more closely allied to *Tragoportax* than to other contemporaneous genera or to the late boselaphine *Parabos* Arambourg and Piveteau, 1929, with which it shares massive horncores and a rather low braincase. Gentry et al. (1999), instead, propose *Samokeros* as a possible bovine.

★*Samokeros minotaurus* Solounias, 1981

Nomenclatural and Taxonomic History *Tragelaphus*? in Andree 1926 (initial identification); *Samokeros minotaurus* in Solounias 1981 (new genus and species).

Type Material IGPM (PIM) 99 (holotype), partial cranium, illustrated by Andrée (1926: Pl. 13, fig. 1, and Pl. 15, fig. 9) (Fig. 6).

Type Locality Samos (unknown level), Greece, Upper Miocene.

Age Late Miocene, Turolian (MN11–13).

Distribution Late Miocene (Turolian) of Greece and Iran.

Remarks The species is originally diagnosed by Solounias (1981) with later additions and modifications by Kostopoulos and Bernor (2011). It represents a medium- to large-sized bovid with massive, keeless and moderately diverging horncores inserted above the orbits, and lacking distal demarcations. The postcornual frontal area is less sculptured than in *Tragoportax* and the premolars shorter. In Greece, it has been recognized only in Samos.

Parabos Arambourg and Piveteau, 1929

Type Species *Parabos cordieri* (de Christol, 1832), by the subsequent designation of Gromolard and Guérin (1980).

Distribution Early Pliocene (Ruscinian–early Villafranchian) of S. Europe and Turkey?

Other Taxa Included *P. soriae* Morales, 1984; *P. savelisi* Crégut-Bonnoure and Tsoukala, 2017a. *Parabos*? *boodon* is currently considered as belonging to *Alephis* (see below).

Remarks *Parabos* Barbour and Schultz, 1941 is a junior homonym of *Parabos* Arambourg and Piveteau, 1929. According to Gromolard and Guérin (1980) and Gromolard (1981), the genus *Parabos* represents one of the last European representatives of Boselaphini, but phylogenetic relationships have been never tested cladistically. Bouvrain and de Bonis (2007) discussed several postcranials from Dytiko 2 fauna (MN13, Axios Valley) whose proportions are compatible with *Parabos*, which may indicate one of the earliest European evidences of this genus.

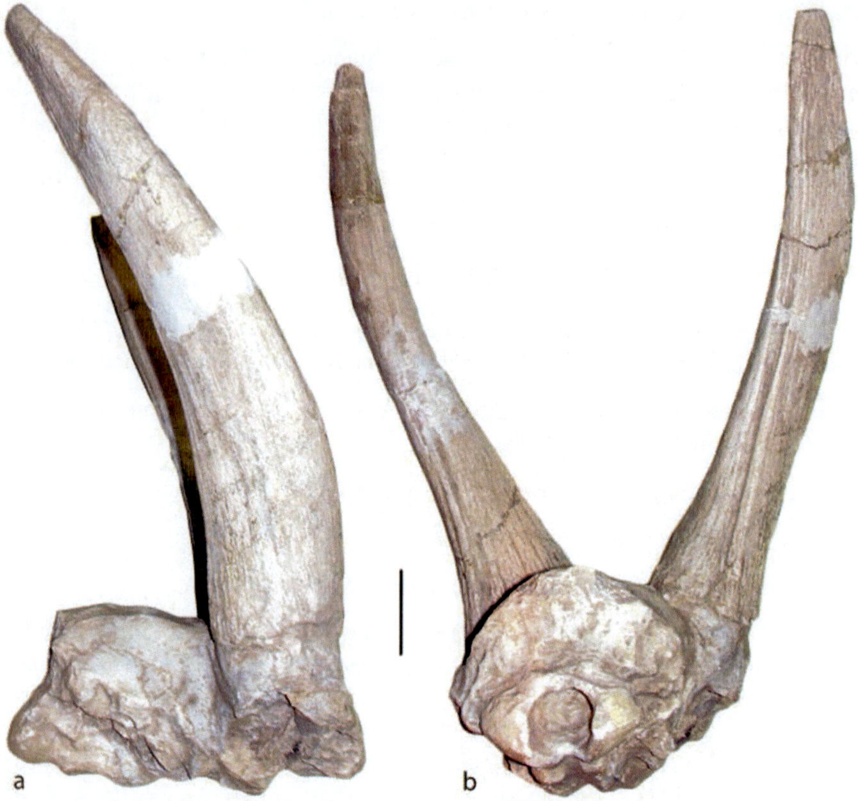

Fig. 6 Partial cranium IGPM (PIM) 99, holotype of *Samokeros minotaurus* Solounias, 1981, from the late Miocene of Samos, in (**a**) right-lateral and (**b**) caudal view. Scale bar equals 5 cm. Photographs courtesy of IGPM

★*Parabos savelisi* Crégut-Bonnoure and Tsoukala, 2017a

Type Material LGPUT GAS 9 (holotype), partial cranium illustrated by Crégut-Bonnoure and Tsoukala (2017a: fig. 2) (Fig. 7a, b).

Type Locality Gephyra, Greece, Lower Pliocene.

Age Early Pliocene, Ruscinian (MN15).

Distribution Early Pliocene (late Ruscinian) of Greece.

Remarks According to the diagnosis provided by Crégut-Bonnoure and Tsoukala (2017a), *P. savelisi* differs from the type species of the genus, *P. cordieri*, in the less divergent and stronger inclined backward horncores, having more rounded cross-section and a posterior groove. Trifonov et al. (2018) provisionally referred a cranium from Erikdere terrace V (Turkey) to this taxon, but I think it is more closely allied to *Tragoportax* than to *Parabos*.

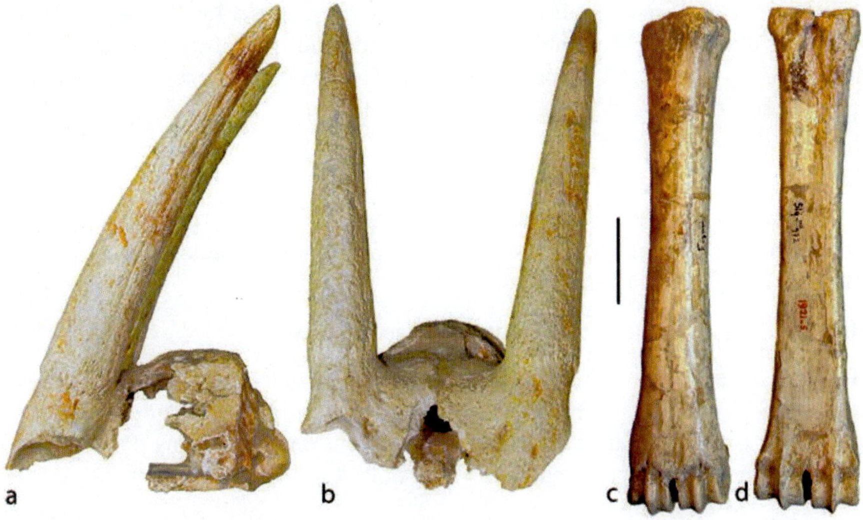

Fig. 7 Type specimens of *Parabos* spp. from Greece. (**a, b**) Partial cranium (LGPUT GAS 9), holotype of *Parabos savelisi* Crégut-Bonnoure and Tsoukala, 2017a, from Gephyra, (**a**) in left-lateral and (**b**) frontal view. (**c, d**) right metacarpal (MNHN.P. Slq-972), lectotype of *Parabos*? *macedoniae* Arambourg and Piveteau, 1929, from Megalo Emvolo in **c** dorsal and **d** palmar view. Photographs courtesy of MNHN.P. Scale bar equals 5 cm

★*Parabos*? *macedoniae* Arambourg and Piveteau, 1929

Nomenclatural and Taxonomic History *Parabos macedoniae* in Arambourg and Piveteau 1929 (new species); *Parabos*? *macedoniae* in Gromolard 1981 (doubtful generic combination).

Type Material Arambourg and Piveteau (1929) did not indicate any type. The syntype MNHN.P Slq 972, right metacarpal (Arambourg and Piveteau 1929: Pl. IX, fig. 4), here chosen as a lectotype (Fig. 7c, d).

Type Locality "Falaise de Karabouroun" (= Megalo Emvolon), Greece, Lower Pliocene.

Age Early Pliocene, Ruscinian (MN15).

Distribution Early Pliocene (late Ruscinian) of Greece. In particular, other Greek occurrences include those from Ptolemais basin (unknown level) (according to Gromolard 1981).

Remarks The species is based on very fragmentary material (i.e., some juvenile maxillae and postcranials; Arambourg and Piveteau 1929: 117, Gromolard 1981) from the Lower Pliocene deposits exposed along the shoreline of Megalo Emvolo near Thessaloniki (see Koufos et al. 1991). Arambourg and Piveteau (1929) did not provide any diagnostic feature, and Gromolard (1981) in her revision remains skeptical about the genus identification. Morphometric analysis by Gromolard (1981) indicates closer dental resemblance of *P.*? *macedoniae* with *Parabos cordieri* than with *Alephis lyrix* but on the other hand most postcranials show morphological and proportional compatibility to *Alephis*.

★*Pheraios* Kostopoulos and Koufos, 2006

Type Species *Pheraios chryssomallos* Kostopoulos and Koufos, 2006, by original designation.

Distribution Late Miocene (middle Turolian) of Greece.

Remarks *Pheraios* is a monotypic genus. Kostopoulos and Koufos (2006) proposed *Pheraios* as a possible stem tragelaphine, but a later cladistic analysis conducted by Bibi (2009 and pers. com.) failed to confirm it, indicating this taxon as a member of the Antilopinae subfamily. Pending a new analysis, the taxon is here referred to as incertae sedis.

★*Pheraios chryssomallos* Kostopoulos and Koufos, 2006

Type Material LGPUT PER-810 (holotype), cranium of a male individual, illustrated by Kostopoulos and Koufos (2006: figs 2–4).

Type Locality Perivolaki, Thessaly, Greece, Upper Miocene.

Age Late Miocene, middle Turolian (MN 12).

Distribution Known only from the type locality.

Remarks As originally diagnosed by Kostopoulos and Koufos (2006), *Pheraios chryssomallos* represents a small- to medium-sized bovid with horncores that are long, compressed, loosely anticlockwise spiraled and smoothly keeled caudally, strongly curved caudo-laterally, and set above the back of the orbits on very short pedicles lacking sinuses. The braincase is moderately shallow and wide and the frontals are narrowly raised between the hon-core bases.

Antilopinae Gray, 1821 (*sensu* Kingdon 1982**)**

Antilopini Gray, 1821

***Gazella* Blainville, 1816**

Type Species *Capra dorcas* Linnaeus, 1758, by original designation.

Distribution Late Miocene to Present of Eurasia and Africa.

Other Taxa Included More than 30 Neogene–Quaternary fossil species are referred to the genus *Gazella* s.l. (e.g., NOW database: http://www.helsinki.fi/science/now; https://www.biolib.cz/en/taxon); for extant taxa see Groves and Grubb (2011).

Remarks Under the name *Gazella* s.l. are grouped small- to medium-sized antilopines with simple conical to saber-like untwisted (or marginaly twisted) horncores weakly to moderately curved and inclined backward. Neontologists split living gazelle species in four distinct genera: *Gazella, Eudorcas, Nanger*, and *Procapra* (Groves and Grubb, 2011). Fossil gazelle-like bovids appear in the Eurasian record as early as at the end of early Vallesian (MN9) and though generally attributed to the genus *Gazella* may not belong to *Gazella* s.s. Authors, especially from the Russian school of Paleontology, provided several alternatives like *Vetagazella* Dmitrieva, 1970, *Gazella (Miogazella)* Korotkevich, 1976, *Procapra (Vetaprocapra)* Korotkevich, 1976, in order to describe the Eurasian Miocene gazelle record, but there is no current consensus and most of these (sub-)generic names are not in common use today. *Gazella*-like bovids occur in the Greek fossil record in the latest early Vallesian (MN9) fauna of Pentalophos fauna (Kostopoulos et al. 2020a). During late Vallesian, one more unnamed species is recorded in the fauna of Xirochori (Kostopoulos et al. 2020a), whereas a taxon close to *Gazella ancyrensis* Tekkaya, 1973 is present in the latest Vallesian/earliest Turolian of Drama basin and the early Turolian of Samos (Kostopoulos 2009a; Vasileiadis et al. 2019). From the early Turolian onwards, gazelles become common faunal elements of the Greek Miocene mammal associations, usually represented by two co-existing species, some of them originally described from Greece (see below). *G. schlosseri* Pavlow, 1913 (Dytiko 3, MN13) and *G. deperdita* (Gervais, 1847) (Kerasia, MN12; Dytiko 1, 2, MN13) are also known from the mid–late Turolian of Greece (Bouvrain 1996; Theodorou et al. 2003). During the Early Pliocene (Ruscinian, MN15), the European *G. borbonica* Depéret, 1884 makes its appearance in Greece at the Megalo Emvolon site and survives until the middle Villafranchian (localities Sesklo, Vatera, Kos; MN17) (Kostopoulos and Athanassiou 1997; Athanassiou 2018b; Kokotini et al. 2019). Gazelles appear to vanish at around 2.0 Ma in Western Europe. At the same period, some more gazelle-like taxa originated from the East shortly invaded southern Balkans and Greece.

★*Gazella capricornis* (Wagner, 1848) (= ★*Antilope brevicornis* Roth and Wagner, 1854)

Nomenclatural and Taxonomic History *Antilope capricornis* in Wagner 1848 (new species); *Gazella capricornis* in Pilgrim and Hopwood 1928 (new combination). This species is also generally considered the senior synonym of *Antilope brevicornis* Roth and Wagner, 1854 (also from Pikermi), a species that appears under the combination *Gazella brevicornis* in Gaudry (1862–1867).

Type Material Isolated horncore (holotype) illustrated by Wagner (1848: Pl. 4, fig. 6); according to Pilgrim and Hopwood (1928), the specimen belongs to BSPM collection, but it was not located during our 2014 survey.

Type Locality Pikermi Greece, Upper Miocene.

Age Late Miocene, middle Turolian MN12, ca. 7.3 Ma.

Distribution Late Miocene (Turolian) of Greece, Northern Macedonia, Romania, Hungary, Ukraine, Turkey, and Iran. In particular, other Greek occurrences include those from Nikiti-2 (MN11), Vathylakkos (MN12), Kerasia (MN12), Samos Mytilinii 1A, 1B (MN12), Halmyropotamos (MN12), Samos Q1 (MN12), ?Thermopigi (MN12), Kryopigi (MN12/13), and Samos Q5 (MN12/13).

Remarks *Gazella capricornis* is likely the most common *Gazella* species in the Turolian faunas from the Balkans and eastwards, sometimes confused with either *G. deperdita* or *G. pilgrimi*. According to the emended diagnosis proposed by Kostopoulos (2009a, 2016), the species is characterized by moderately long, thinly grooved horncores, moderately spaced on the frontals and strongly divergent in their distal part, gently inclined and curved to the rear with an oval at the base to rounded toward the tips cross-section; pedicles short anteriorly; postcornual grooves rather large and deep; supraorbital foramina sunken into large, triangular-shaped pits; nasals short and domed in lateral view.

★*Gazella pilgrimi* Bohlin, 1935
(= ★*Gazella longicornis* Andrée, 1926)

Nomenclatural and Taxonomic History *Gazella gaudryi* in Schlosser, 1904 (new species). This species name is a secondary homonym of *Protetraceros gaudryi* Schlosser, 1903, from Mongolia. However, soon after the establishment of both nomina, both were considered as members of *Gazella*. As such, a new replacement was proposed: *Gazella pilgrimi* Bohlin, 1935 (new replacement name). The species is considered as a synonym of *Gazella longicornis* Andree, 1926. Some authors also merge it with *Gazella capricornis* (e.g., in Solounias 1981).

Type Material A frontlet illustrated by Schlosser (1904: Pl. XIII, fig. 1) has been selected by Pilgrim and Hopwood (1928) as the lectotype of the species *Gazella gaudryi* Schlosser, 1904; according to the same authors, the specimen belongs to BSPM collection, but it was not located during our 2014 survey.

Type Locality Samos (unknown level), Greece, Upper Miocene.

Age Late Miocene, Turolian (MN11–13).

Distribution Late Miocene (Turolian) of Greece and Turkey. In particular, other Greek occurrences include those from Nikiti-2 (MN11), Ravin des Zouaves 5 (MN11), Samos Qx (MN11), Prochoma (MN12), Vathylakkos 2, 3 (MN12), Samos

Mytilinii 4, 3, 1A, 1B (MN11-12), Perivolaki (MN12), Kryopigi (MN12/13), and Samos Q5 (MN12/13).

Remarks A gazelle species rather characteristic of the early to middle Turolian faunas of Greece. It has moderately tilted backward, relatively long, well grooved, and subparallel horncores with elliptical cross-section, set moderately apart on the frontals, and slightly curved to the rear; the pedicles are long anteriorly and the supraorbital foramina sunken into large, triangular-shaped pits (Bouvrain, 1996; Kostopoulos 2006a, 2009a, 2016).

★*Gazella mytilinii* Pilgrim, 1926

Nomenclatural and Taxonomic History *Gazella schlosseri* in Andrée, 1926 [new species; name preoccupied by *Gazella schlosseri* Pavlow, 1913]; *Gazella mytilinii* in Pilgrim, 1926 (new replacement name).

Type Material NHMW A4777 (lectotype selected by Pilgrim and Hopwood 1928), cranium illustrated by Andrée (1926: Pl. XVI, figs 1, 6, 7) (Fig. 8).

Type Locality Samos (unknown level), Greece, Upper Miocene.

Age Late Miocene, Turolian (MN11–13).

Distribution Late Miocene (Turolian) of Greece and probably Moldavia, Turkey, and Iran. In particular, other Greek occurrences include those from Samos Mytilinii 1A, 1B (MN 12), and Samos Q1, Q5 (MN12/13).

Remarks Pilgrim and Hopwood (1928), Solounias (1981), and Kostopoulos (2009a) discuss the morphological features of this species, characterized by uprightly inserted, almost parallel, and moderately long and robust horncores, distinctly grooved, closely settled on the frontals and strongly curved backward, sometimes showing a weak torsion. The cross-sections are elliptical at the base to oval at the apexes. The postcornual grooves are small.

★*Gazella bouvrainae* Kostopoulos and Athanassiou, 1997

Type Material LGPUT GER-46 (holotype), frontlet (Kostopoulos and Athanassiou 1997: fig. 2a, b) (Fig. 9).

Type Locality Gerakarou, Greece, Lower Pleistocene.

Age Early Pleistocene, late Villafranchian (MNQ 18).

Distribution Late early Villafranchian (MNQ16b–18) of Greece. In particular, other Greek occurrences include those from Sesklo (lower horizon; ?MN16b), Dafnero 1 (MNQ17), Sesklo (MNQ17), Vatera (MNQ17), and Vassiloudi (MNQ18).

Fig. 8 Cranium (NHMW A4777), lectotype of *Gazella mytilinii* Pilgrim, 1926, from Samos in right-lateral view. Not scaled (specimen in exhibition). Photograph courtesy of NHMW

Remarks According to the descriptions by Kostopoulos and Athanassiou (1997), this *Gazella* species is characterized by short weakly inclined and diverging horn-cores, well but finely striated, widely settled on the frontals and gently curved backward, with an oval to rounded cross-section; the opisthocranium is long and slopes gently down caudally; the orbits are large. The species probably entered southern Balkans during MN16 (Athanassiou 2018b) and lasted until the Olduvai subchron; no gazelle record after this date exists in Greece. Kostopoulos and Athanassiou (1997) suggested *G. bouvrainae* as closely linked with some contemporaneous Asian *Gazella*-like taxa.

★*Gazella aegea* Athanassiou, 2002a

Type Material AMPG Σ-350 (holotype), frontlet illustrated by Athanassiou (2002a: Pl. I, fig. 1).

Type Locality Sesklo, Greece, Lower Pleistocene.

Age Early Pleistocene, middle Villafranchian (MN17).

Distribution Middle Villafranchian of Greece.

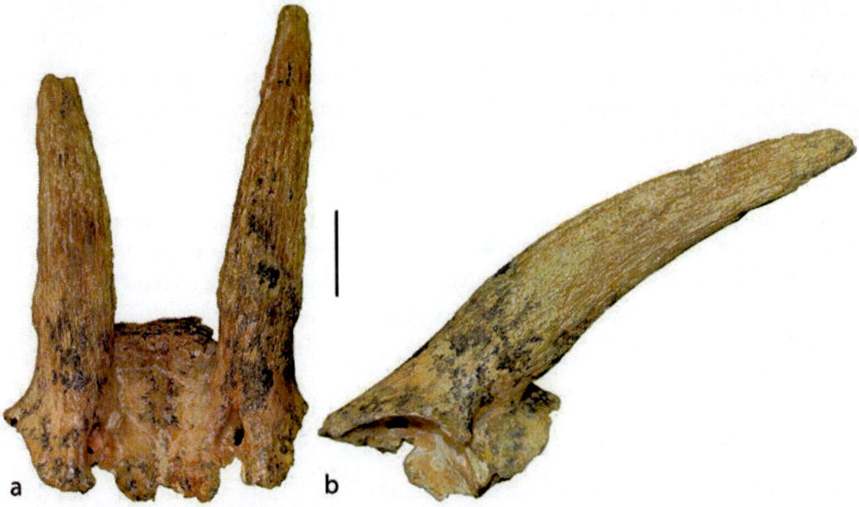

Fig. 9 Frontlet (LGPUT GER-46), holotype of *Gazella bouvrainae* Kostopoulos and Athanassiou, 1997, from Gerakarou, in (**a**) frontal and (**b**) left-lateral view. Scale bar equals 5 cm

Remarks This large gazelle species, known from Sesklo (Thessaly, MNQ17) and likely Vatera (Lesvos Island, MNQ17), has slightly curved, long, and divergent horncores of elliptical cross-section; deep postcornual fossae and large supraorbital foramina sunken in triangular-shaped pits (Athanassiou 2002a).

Prostrepsiceros Forsyth Major, 1891

Type Species *Tragelaphus*? *houtumschindleri* Rodler et Weitfofer, 1890, by original designation.

Distribution Late Miocene of Balkans and central-western Asia; African occurrences are questioned.

Other Taxa Included *P. rotundicornis* (Weithofer, 1888), *P. zitteli* (Schlosser, 1904), *P. fraasi* (Andrée, 1926), *P. vinayaki* (Pilgrim, 1939), *P. vallesiensis* Bouvrain, 1982; *P. syridisi* Kostopoulos and Koufos, 1996, and *P. axiosi* Kostopoulos, 2004.

Remarks Gentry (1971) and Bouvrain (1982) revised the taxonomy of this widely distributed genus and provided emended diagnoses of several of the included species. Bouvrain (1982) distinguished two subgenera: *P. (Prostrepsiceros)* with mediolaterally compressed horncores and *P. (Helicotragus)* with rostrocaudally compressed horncores, but later studies failed to confirm such a strict division. Instead, there is a growing impression among scholars that *Prostrepsiceros*, as currently stated, may be polyphyletic (e.g., Kostopoulos 2009b). Earliest evidence of the genus is coming from the late Vallesian (MN10) of Turkey and Greece, whereas the last occurrences do not exceed the Mio-Pliocene boundary.

★*Prostrepsiceros rotundicornis* (Weithofer, 1888)

Nomenclatural and Taxonomic History *Helicoceras rotundicorne* in Weithofer 1888 (new genus, new species); *Helicotragus rotundicornis* in Palmer, 1903 (new combination and justified emendation of the specific name); *Prostrepsiceros rotundicornis* in Gentry, 1971 (new combination).

Type Material A frontlet (lectotype selected by Pilgrim and Hopwood 1928) figured by Weithofer (1888: Pl. 18, figs 1–2); according to Pilgrim and Hopwood (1928), the specimen belongs to the Palaeontological Museum of the University of Vienna, but S. Roussiakis (pers. com.) located only the paralectotype of the species, figured by Weithofer (1888: Pl. 18, figs 3–4) (Fig. 10).

Type Locality Pikermi, Greece, Upper Miocene.

Age Late Miocene, middle Turolian (MN12), ca. 7.3 Ma.

Distribution Late Miocene (early–middle Turolian) of Greece, Northern Macedonia, Bulgaria, Turkey, and Iran. In particular, other Greek occurrences include those from Ravin des Zouaves 5 (MN11), Perivolaki (MN12), Halmyropotamos (MN12), Chomateri (MN12), and Thermopigi (MN12).

Remarks *P. rotundicornis* is one of the most widely distributed and commonly reported species of the genus in the Turolian faunas of the Balkano-Irano-Afgan province. Horncores are openly spiraled with clear double flexion (backward and then upward/medially), barely keeled, and/or distinctly grooved.

★*Prostrepsiceros zitteli* (Schlosser, 1904)

Nomenclatural and Taxonomic History *Protragelaphus zitteli* in Schlosser, 1904 (new species); *Hemistrepsiceros zitteli* in Pilgrim and Hopwood, 1928 (new combination); *Prostrepsiceros zitteli* in Bouvrain, 1982 (new combination). Gentry (1971; and others, see below) considered this species as a junior synonym of *Prostrepsiceros houtumschindleri* Rodler and Weithofer, 1890.

Type Material BSPM #1899 VII 552 (lectotype selected by Pilgrim and Hopwood 1928), frontlet figured by Schlosser (1904: Pl. VI, fig. 5) (Fig. 11).

Type Locality Samos (unknown level), Greece, Upper Miocene.

Age Late Miocene, Turolian (MN11–13).

Distribution Late Miocene (Turolian) of Greece (Samos Q5, MN12/13), Turkey; reported Ukrainian occurrences need confirmation.

Remarks *Prostrepsiceros zitteli* was considered as a junior synonym of *P. houtumschindleri* (Rodler and Weithofer, 1890) by Gentry (1971) and Solounias (1981), followed lately by Bibi and Güleç (2008). However, Bouvrain (1982), Kostopoulos

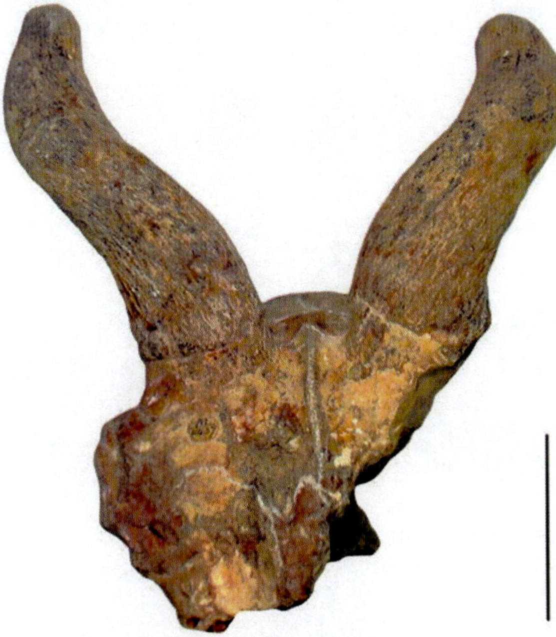

Fig. 10 Frontlet NHMUK M11437 of *Prostrepsiceros rotundicornis* (Weithofer, 1888) from Pikermi in frontal view. Scale bar equals 5 cm. Photograph courtesy of NHMUK

(2004), Gentry and Heizmann (1996), and Roussiakis (2009), among others, recognized the species as valid, fully distinguishable from *P. houtumschindleri* from Maragheh, Iran by the strong anterior keel. In the more recent review, Kostopoulos (2004) revised the species content by excluding samples from Axios Valley previously incorporated into *P. zitteli* by Bouvrain (1982). Horncores of *P. zitteli* are strongly bent backward, moderately spaced on frontals, rostrocaudaly compressed and closely spiraled, with a strong anterior keel descending anteromedially and describing 1½ coil.

★*Prostrepsiceros fraasi* (Andrée, 1926)

Nomenclatural and Taxonomic History *Helicoceras fraasi* in Andrée 1926 (new species); *Helicotragus fraasi* in Pilgrim and Hopwood 1928 (new combination); *Prostrepsiceros fraasi* in Bouvrain 1982 (new combination). *Prostrepsiceros fraasi* was considered as a synonym of *P. rotundicornis* (Weithofer, 1888) by Gentry (1971) and Solounias (1981), but Bouvrain (1982) recognized it as a valid species. Her opinion has been followed by most later scholars.

Type Material SMNS 13278 (holotype), cranium figured by Andrée (1926: Pl. XI, fig. 4, and Pl. XV, fig. 1).

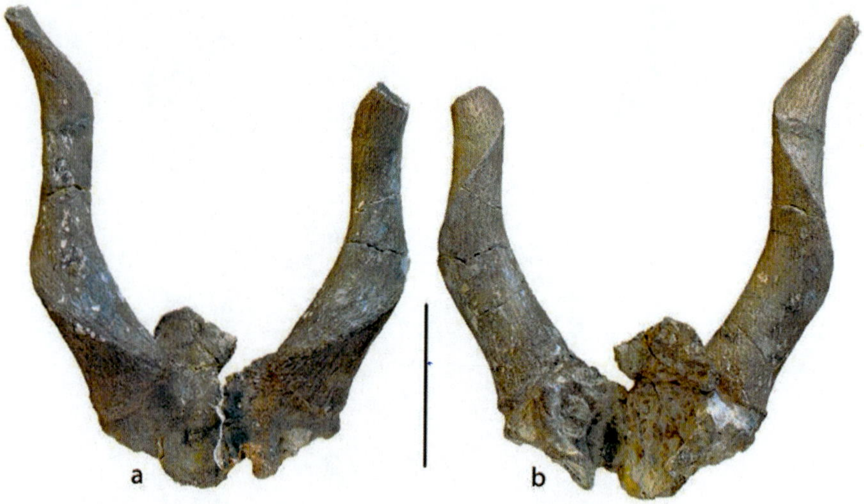

Fig. 11 Frontlet (BSPM #1899 VII 552), lectotype of *Prostrepsiceros zitteli* (Schlosser, 1904) from Samos, in (**a**) frontal and (**b**) caudal view. Scale bar equals 5 cm. Photographs courtesy of BSPM and A Athanassiou

Type Locality Samos (unknown level), Upper Miocene.

Age Late Miocene, likely early–middle Turolian, MN11–12.

Distribution Late Miocene of Greece and Iran.

Remarks According to the emended diagnosis provided by de Bouvrain (1982) and the revision of the Maragheh material by Kostopoulos and Bernor (2011), *P. fraasi* represents a medium-sized *Prostrepsiceros* with inclined face compared to the braincase and horncores inserted vertically above the orbits, set well apart on frontals, strongly divergent, and openly spiraled bearing a trace of an anterior keel.

★***Prostrepsiceros vallesiensis*** **Bouvrain,** 1982

Type Material LGPUT RPl-234 (holotype), cranium illustrated by Bouvrain (1982: figs 2–3) and Kostopoulos (2009b: fig. 1).

Type Locality Ravin de la Pluie, Axios Valley, Greece, Upper Miocene.

Age Late Miocene, late Vallesian (MN10), ca. 9.3 Ma.

Distribution Late Miocene (late Vallesian) of Greece and possibly in Turkey.

Remarks As originally diagnosed by Bouvrain (1982), *P. vallesiensis* represents a small *Prostrepsiceros* with strongly compressed mediolaterally horncores, short nasal bones ending anteriorly at a single point, weak cranial flexion, and small supraorbital foramens not sunken into pits. Based on new material from the type locality, Kostopoulos (2009b) showed that the holotype cranium likely represents a female individual and provided an emended diagnosis of the species. Apart from Greece, this species is reported from the contemporaneous Middle Sinap sites, in Turkey (Gentry 2003).

★*Prostrepsiceros syridisi* Kostopoulos and Koufos, 1996

Nomenclatural and Taxonomic History *Prostrepsiceros houtumschindleri syridisi* in Kostopoulos and Koufos 1996 (new subspecies); *Prostrepsiceros syridisi* in Kostopoulos 2006b, Bibi and Güleç 2008, Koufos et al. 2016 (elevation to the species level under its original combination).

Type Material LGPUT NKT-119 (holotype), cranium illustrated by Kostopoulos and Koufos (1996: Pl. 2, figs A–C) and Koufos et al. (2016: fig. 8b–c).

Type Locality Nikiti-1, Greece, Upper Miocene.

Age Late Miocene, latest Vallesian-early Turolian MN10/11.

Distribution Late Miocene (latest Vallesian-early Turolian) of Greece and Turkey. Also, possibly known from Platania (MN10/11).

Remarks *Prostrepsiceros syridisi* is characterized by long, slender, and loosely spiraled horncores, widely spaced on the frontals, and openly diverging with a blunt posterior and a weak to absent anterior keel. Horncore compression varies from mediolaterally to barely anteroposterior and frontal flexion is stronger than in *P. rotundicornis* and *P. axiosi*. Premolars are long but p4 morphology quite advanced. Apart from Greece, the species is reported from Sivas, Turkey (Bibi and Güleç 2008).

★ *Prostrepsiceros axiosi* Kostopoulos, 2004

Nomenclatural and Taxonomic History *Prostrepsiceros zitteli* in Bouvrain 1982 (initial identification); *Prostrepsiceros axiosi* in Kostopoulos 2004 (new species).

Type Material LGPUT RZO-251 (holotype), frontlet illustrated by Kostopoulos (2004: fig. 3) (Fig. 12)

Type Locality Ravin dez Zouaves 5, Axios Valley, Greece, Upper Miocene.

Age Late Miocene, early Turolian (MN11), ca. 8.2 Ma.

Distribution Late Miocene (early–middle Turolian) of Greece, Northern Macedonia, and possibly Ukraine. In particular, other Greek occurrences include those from Vathylakkos 3 (MN12), Prochoma (MN12), and Kryopigi (MN12/13).

Remarks According to the diagnosis provided by Kostopoulos (2004), *P. axiosi* represents a small to medium-sized *Prostrepsiceros*. The anteroposteriorly com-

Fig. 12 Frontlet (LGPUT RZO-251), holotype of *Prostrepsiceros axiosi* Kostopoulos, 2004, from Ravin des Zouaves-5 (Axios Valley) in dorsofrontal view. Scale bar equals 5 cm

pressed, slender, and openly spiraled horncores are inserted almost vertically above the orbits and bear a strong anterior keel, descending medially and a posterior keel marked only distally.

Nisidorcas Bouvrain, 1979

Type Species (cf. *Antilope*) *planicornis* Pilgrim, 1939, by original designation.

Distribution Late Miocene of Greece, Turkey, Iran, and N. India.

Remarks *Nisidorcas* is a monotypic genus. Its relationships with contemporaneous and/or extinct antilopines is still pending (Bouvrain 1992).

Nisidorcas planicornis (Pilgrim, 1939)

Nomenclatural and Taxonomic History (cf. *Antilope*) *planicornis* in Pilgrim 1939 (new species); *Nisidorcas planicornis* in Bouvrain 1979 (new combination; new genus). It has been considered a senior synonym of *Antilospira incarinatus* Tekkaya, 1969.

Type Material NHMUK M37264 (holotype), right horncore illustrated by Pilgrim (1939: Pl. I, fig. 7).

Type Locality Perim Island, India, Upper Miocene.

Age Late Miocene.

Distribution Late Miocene (early–middle Turolian) of Greece, Turkey, Iran, and India. In particular, other Greek occurrences include those from Nikiti-2 (MN11), Ravin des Zouaves 5 (MN11), Vathylakkos 1, 2, 3 (MN12), Prochoma (MN12), Thermopigi (MN12), Perivolaki (MN12), Ravin X (MN12), and Kryopigi (MN12/13).

Remarks A small-sized antelope with straight, strongly inclined horncores inserted above the orbits, moderately twisted, and having a strong posterior keel. Although the type comes from India, the bulk of the known material of this species is currently from Greece. The species was included in several recent revisions (Bouvrain 1992; Kostopoulos 2006a, 2016).

★*Majoreas* Kostopoulos, 2004

Type Species *Prostrepsiceros woodwardi* Pilgrim and Hopwood, 1928, by original designation.

Distribution Late Miocene of Greece and Turkey.

Other Included Taxa *Majoreas elegans* (Ozansoy, 1965).

Remarks This genus of heteronymously spiral horned antelopes erected in 2004 by Kostopoulos on the basis of Greek and Turkish samples previously attributed to either *Palaeoreas* or *Prostrepsiceros*. Gentry (2003) and Kostopoulos (2004) assumed close phylogenetic relationships with *Nisidorcas*.

★*Majoreas woodwardi* (Pilgrim and Hopwwod, 1928)

Nomenclatural and Taxonomic History *Prostrepsiceros woodwardi* in Major 1891 (nomen nudum); *Prostrepsiceros woodwardi* in Pilgrim and Hopwood 1928 (new species); *Majoreas woodwardi* in Kostopoulos 2004 (new combination, new genus). This species has been considered as a junior synonym of *Prostrepsiceros zitteli* by Bouvrain (1982).

Type Material NHMUK M4192 (holotype), partial cranium illustrated by Pilgrim and Hopwood (1928: Pl. VII, fig. 1), Gentry et al. (1999: fig. 23.17), and Kostopoulos (2004: fig. 4).

Type Locality Samos (unknown level), Upper Miocene.

Age Late Miocene, Turolian (MN11–13).

Distribution Late Miocene (Turolian) of Greece and Turkey. In particular, other Greek occurrences include those from Samos Qx, Q6, Mytilinii 3 (MN 11/12), Dytiko 1 (MN13), Ano Metochi 2,3 (MN13), and ?Tanagra (MN12–13).

Remarks According to the diagnosis provided by Kostopoulos (2004) and additional diagnostic features given by Kostopoulos and Soubise (2018), the species is characterized by weakly raised frontals and moderate craniofacial flexion, supraorbital foramens not directly open into the orbits and placed in large triangular pits, wider than high braincase and large mastoid exposure facing mostly caudally, grooved basioccipital, and moderately long, closely spiraled horncores bearing a weak, straight posterior keel and a strong anterior one descending rostro-medially.

Antilope Pallas, 1766

Type Species *Capra cervicapra* Linnaeus, 1758, by original designation.

Distribution Pleistocene to Present of Eurasia.

Other Included Taxa *A. subtorta* Pilgrim, 1937; *A. koufosi* (Kostopoulos, 1998); *A. intermedia* Khan and Akhtar, 2014.

Remarks The living blackbuck (*Antilope cervicapra*) is native to the Indian subcontinent, from where Pilgrim (1939) reports *Antilope subtorta* and Kahn and Akhtar (2014) *A. intermedia* from the Pliocene deposits of Siwaliks, India. Outside India, the genus is recorded only in the Early Pleistocene of Greece but relations of all three described fossil species with the extant one are based on a very limited number of characters and thus friable.

★*Antilope koufosi* (Kostopoulos, 1998)

Nomenclatural and Taxonomic History *Parastrepsiceros koufosi* in Kostopoulos 1998 (new species); *Antilope koufosi* in Kostopoulos and Koufos 2006 (new combination).

Type Material LGPUT GER-273 (holotype), right horncore illustrated by Kostopoulos (1998: fig. 4b) (Fig. 13).

Type Locality Gerakarou, Greece, Lower Pleistocene.

Age Early Pleistocene (MNQ 18).

Fig. 13 Right horncore (LGPUT GER-273), holotype of *Antilope koufosi* (Kostopoulos, 1998) from Gerakarou site in (**a**) frontal and (**b**) lateral view. Scale bar equals 5 cm

Distribution Known only from the type locality.

Remarks Following Kostopoulos (1998), *Antilope koufosi* is a gazelle-sized heternonymous spiral horned antelope. The horncores are weakly compressed mediolaterally, inserted above the orbits and very openly spiraled, bearing a vestigial posterior keel and fine discontinuous longitudinal grooves. Supraorbital foramina are sunken in large and deep pits. The species is based on restricted material and, although different from any other contemporaneous bovid taxon of the wider area, its phylogenetic relationships with the living taxon is hard at the moment to be tested under parsimony.

Gazellospira Pilgrim and Schaub, 1939

Type Species *Antilope torticornis* Aymard in Dorlhac, 1854, by original designation.

Distribution Pliocene to Early Pleistocene of Europe, Middle East, Turkey, Tajikistan, southern Russia, and probably China.

Other Included Taxa *G. zdansky* (Teilhard de Chardin and Trassaert, 1938) (= *Antilospira zdanskyi* Teilhard de Chardin and Trassaert, 1938 = *G. gromovae* Dmitrieva, 1975); *G. sokolovi* (Vekua, 1968).

Remarks The genus *Gazellospira* was subject to a thorough revision by Pilgrim and Schaub (1939). Heintz (1966), Duvernois and Guérin (1989), Garrido (2008), Rodrigo (2011), and Hermier et al. (2020) provide valuable additional information and opinions on the type species, *Gazellospira torticornis*. Bai et al. (2019) suggest close phylogenetic relationships with the Chinese genus *Antilospira,* whereas Hermier et al. (2020) discuss possible synonymies with Plio-Pleistocene heteronymoys Eurasian spiral horned antelopes. Position within the extant tribes is still debatable as several authors suggest *Gazellospira* either as a member of Antilopini or Tragelaphini (Bai et al. 2019; Rodrigo 2011), though the former hypothesis seems more reasonable. In Greece, *Gazellospira torticornis* has a rich local record, restricted however to the first part of the Early Pleistocene.

Distribution In Greece, *Gazellospira torticornis* is known from Dafnero 1, 2, 3 (MNQ17), Sesklo (MNQ17), Volax (MNQ17), Vatera (MNQ17), Karnezeika (?MN17), Pyrgos (?MNQ18), and possibly Gerakarou (MNQ18).

Oiocerina Pilgrim, 1934

★*Oioceros* Gaillard, 1902

Type Species *Antilope rothii* Wagner, 1857, by original designation.

Distribution Late Miocene of Greece, Turkey, Italy, Georgia, Iran, and Mongolia.

Other Included Taxa *Oioceros atropatenes* (Rodler and Weithofer, 1890); *Oioceros occidentalis* (Masini and Thomas, 1989).

Remarks *Oioceros* is central to a group of small- to medium-sized, Late Miocene bovids, characterized by homonymously torsioned horncores with a clear lateral sulcus. Taxa clustered around *Oioceros* are usually referred to an extinct Antilopinae family-group taxon of tribal or subtribal level: Oiocerina Pilgrim, 1934 (e.g., Gentry 2010; Kostopoulos 2014).

★*Oioceros rothii* (Wagner, 1857)

Nomenclatural and Taxonomic History *Antilope rothii* in Wagner 1857 (new species); *Oioceros rothii* in Gaillard 1902 (new genus and new combination).

Type Material BSPM #AS II 601 (holotype), frontlet illustrated by Wagner (1857: Pl. VIII, fig. 20) (Fig. 14a).

Type Locality Pikermi, Greece, Upper Miocene.

Age Late Miocene, middle Turolian (MN12) ca. 7.3 Ma.

Distribution Late Miocene (middle Turolian) of Greece, Turkey, Iran, Georgia, and Mongolia. In particular, other Greek occurrences include those from Axios Valley (unknown site; Arambourg and Piveteau 1929) and Ravin X (MN12).

Remarks Roussiakis (2003) described the first complete cranium (Fig. 14b) of this species from the type locality and provided an emended diagnosis. *Oioceros rothii* is a small- to medium-sized bovid, characterized by homonymously twisted horncores of one whorl, inserted above the orbits and lyrate in frontal view, moderately compressed throughout their length with a blunt anterior keel and a strong posterolateral one usually associated by a deep furrow.

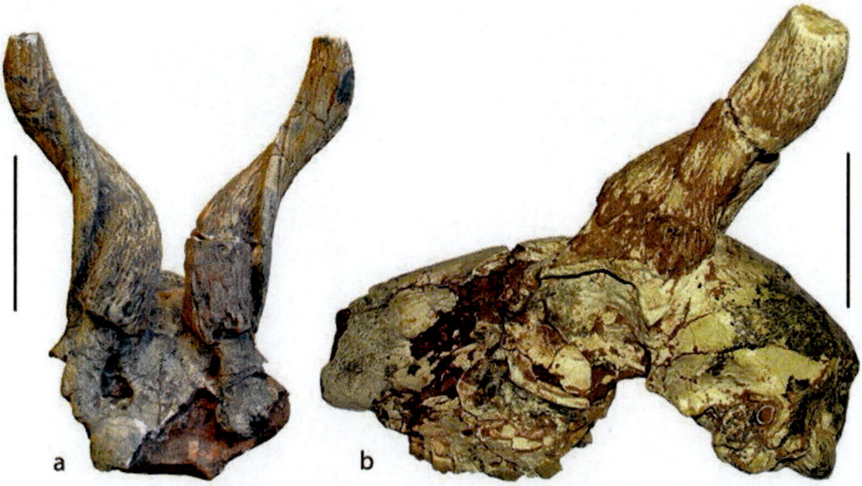

Fig. 14 Fossil specimens of *Oioceros rothii* (Wagner, 1857) from Pikermi. (**a**) The holotype frontlet (BSPM #AS II 601) in frontal view, and (**b**) the referred cranium (APGM PG 95/1502a) in left-lateral view. Scale bars equal 5 cm. Photographs courtesy of BSPM, APGM, A Athanassiou and S Roussiakis

★ *Samotragus* Sickenberg, 1936

Type Species *Samotragus crassicornis* Sickenberg, 1936, by original designation.

Distribution Late Miocene of Greece.

Other Included Taxa *S. praecursor* Bouvrain and de Bonis, 1985.

Remarks *Samotragus* includes small- to medium-sized bovids with basally robust, homonymously spiraled horncores, set closely each other on the frontals and tapering fast upward. Solounias (1981) suggested synonymizing *Samotragus* with *Sinotragus* Bohlin, 1935, but Bouvrain and de Bonis (1985: 285) and Geraads et al. (2002: 485) have subjectively rejected such a possibility. Solounias (1981: 167) included the specimen NHMW A4778 from Samos in *Prosinotragus* Bohlin, 1935, and suggested it as a non-named new species. *Prosinotragus* is a barely known Chinese taxon and likely a synonym of *Sinotragus* (e.g., Geraads et al. 2002). MNHW A4778 (Fig. 15) substantially differs from both *Sinotragus* and *Prosinotragus* in the relatively longer, non-flattened medially and loosely spiraled horncores of subtriangular at the base to elliptical at the top cross-section and may represent a variety or distinct species of *Samotragus*. Kostopoulos (2014) provides a recent generic revision of *Samotragus*, currently known from Greece and possibly N. Macedonia, and spanning from the late Vallesian to the middle Turolian.

Fig. 15 *Samotragus* sp. frontlet (NHMW A4778) from Samos in frontal view. Scale bar equals 5 cm. Photograph courtesy of NHMW

★*Samotragus crassicornis* Sickenberg, 1936

Nomenclatural and Taxonomic History *Samotragus crassicornis* in Sickenberg 1936 (new species); *Sinotragus crassicornis* in Solounias 1981 (new combination).

Type Material NHMW A4787 (holotype), frontlet illustrated by Sickenberg (1936: Pl. III, figs 1, 2).

Type Locality Samos (unknown fossil horizon), Upper Miocene.

Age Late Miocene, Turolian (MN11–13); likely middle Turolian, MN12 (Kostopoulos 2014).

Distribution Known only from the type locality.

Remarks Large-sized *Samotragus* species with moderately long and narrow braincase, protruding orbits, and moderately elevated and strongly angled frontals pneumatized rostrally. Proximally heavy horncores, uprightly inserted above the orbits, very close to each other at the base, strongly curved caudally, abruptly tapering, and strongly diverging distally; cross-section sub-square at the base with flattened posterior surface to roughly triangular at mid-height to round toward the tips. A deep lateral sulcus on the proximal part is exaggerated caudally by a variably developed "exostosis"-like swelling (Solounias 1981; Kostopoulos 2014).

★*Samotragus praecursor* Bouvrain and de Bonis, 1985

Type Material LGPUT RPl-480 (holotype), cranium illustrated by Bouvrain and de Bonis (1985: figs 1, 3) and Kostopoulos (2014: fig. 5B).

Type Locality Ravin de la Pluie, Axios Valley, Greece, Upper Miocene.

Age Late Miocene, late Vallesian (MN10) ca. 9.3 Ma.

Distribution Late Miocene (late Vallesian) of Greece.

Remarks A small *Samotragus* species, with hornless females diagnosed by relatively short and box-like braincase with a rough dorsal surface around the frontoparietal suture, shallow and rather short face moderately inclined compared to the braincase, rather short nasals, and moderately short premolars compared to the molars. The horncores are more inclined posteriorly and less spiraled than in *S. crassicornis* (Bouvrain and de Bonis 1985). Kostopoulos (2014) provisionally referred to the same taxon a frontlet, as well as some isolated horncores and postcranials, from the somewhat older locality Ravin des Zouaves 1 in Axios Valley.

Paraoioceros Meladze, 1985

Type Species *Oioceros wegneri* Andrée, 1926, by subsequent designation by Kostopoulos (2014).

Distribution Late Miocene of Greece, Turkey, and Georgia.

Remarks Gentry and Heinzmann (1996) firstly commented on the possible synonymy between *Paraoioceros improvisus* Meladze, 1985, and *Oioceros wegneri* Andrée, 1926, whereas the referral of *wegneri* to *Oioceros* was further challenged by Roussiakis (2003). Kostopoulos (2014) formally revised the genus and associated it with Oiocerina Pilgrim, 1934.

★*Paraoioceros wegneri* (Andrée, 1926)

Nomenclatural and Taxonomic History *Oioceros wegneri* in Andrée 1926 (new species); *Paraoioceros wegneri* in Kostopoulos 2014 (new combination). This species has been considered also as the senior synonym of *Paraoioceros improvisus* Meladze, 1985, from the late Miocene of Georgia.

Type Material IGPM (PIM) 141 (holotype), cranium illustrated by Andrée (1926: Pl. XV, figs 3, 6).

Type Locality Samos (unknown fossil horizon), Upper Miocene.

Age Late Miocene, Turolian (MN11–13).

Distribution Late Miocene (Turolian) of Greece, Turkey, and Georgia.

Remarks This medium-sized bovid is characterized by moderately thick and long horncores strongly homonymously twisted and closely spiraled, significantly diverging distally, and bearing multiple deep, longitudinal furrows with keel-like edges. The face is long and deep, the opisthocranium is short, the basicranial flexion is strong, the basioccipital is short and grooved, and the frontals are thick, pneumatized, and strongly elevated between the horncores with pinched interfrontal suture. The premolars are short compared to the molars which bear clear goat folds. Apart from Samos Island, the species is recorded in several Turkish faunas of early–middle Turolian age, as well as at the latest Vallesian/earliest Turolian fauna of Rustavi, Georgia, altogether suggesting a basically early Turolian distribution.

Hispanodorcas Thomas et al., 1982

Type Species *Hispanodorcas torrubiae* Thomas et al., 1982, by original designation.

Distribution Late Miocene to Pliocene of Spain and Greece.

Other Taxa Included *H. orientalis* Bouvrain and de Bonis, 1988; *H. heintzi* Alcalá and Morales, 2006; *H.? pilgrimi* (Azanza et al., 1998).

Remarks Members of *Hispanodorcas* are small-sized bovids with short, thin, gradually tapering, and weakly homonymously twisted horncores bearing a variably wide and shallow lateral depression. The taxon has been recently revised by Kostopoulos (2014), who assigned it to Oiocerina Pilgrim, 1934. Although originally known from the middle–late Turolian, new data by Alcalá and Morales (2006) from Spain and Kostopoulos (2014) from Greece expand the genus distribution from at least the latest Vallesian to the early Ruscinian of Europe.

★*Hispanodorcas orientalis* Bouvrain and de Bonis, 1988

Type Material LGPUT DKO–4 (holotype), partial skull with horncores and associated parts of maxillae and mandible illustrated by Bouvrain and de Bonis (1988: figs 1, 2) (Fig. 16).

Type Locality Dytiko 3, Axios Valley, Greece, Upper Miocene.

Age Late Miocene, Turolian (MN13).

Distribution Late Miocene (latest Vallesian to late Turolian) of Greece.

Remarks Originally diagnosed by Bouvrain and de Bonis (1988), *H. orientalis* differs from other species of *Hispanodorcas* in its slightly smaller size and horncores that are more transversely compressed, weakly keeled anteriorly, and stronger diverging distally. *Hispanodorcas orientalis* was until recently known exclusively from its type locality (Fig. 16, but new data from the latest Vallesian Nikiti-1 fauna in Chalkidiki peninsula (Greece) provide evidence for a much earlier presence in the southern Balkans (Kostopoulos 2014).

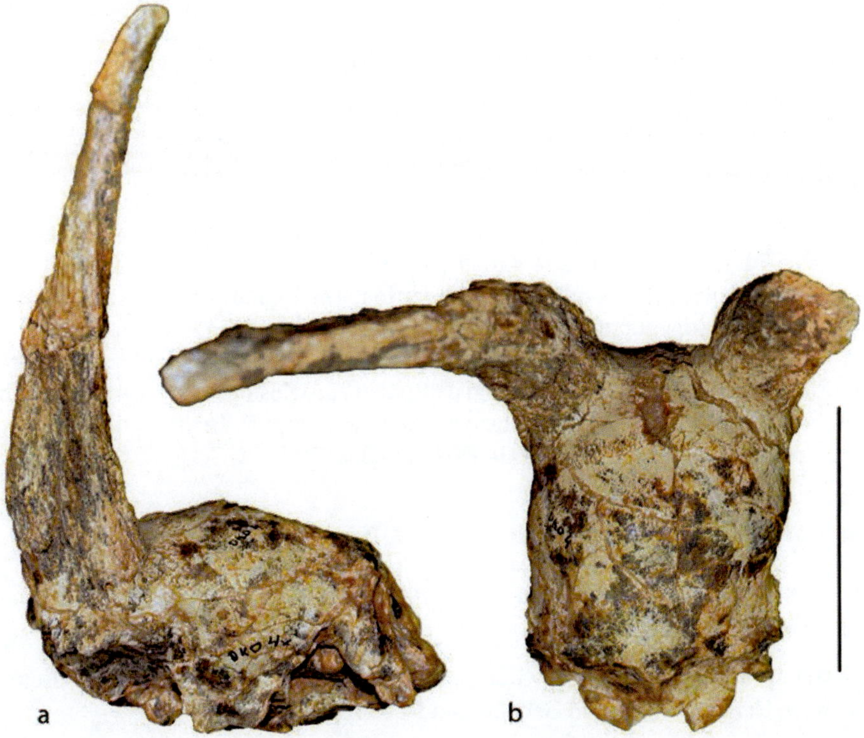

Fig. 16 Partial cranium (LGPUT DKO–4), holotype of *Hispanodorcas orientalis* Bouvrain and de Bonis, 1988, from Dytiko, in (**a**) left-lateral and (**b**) dorsal view. Scale bar equals 5 cm

★*Samodorcas* Bouvrain and de Bonis, 1985

Type Species *Ovis kuhlmani* Andrée, 1926, by original designation.

Distribution Late Miocene of Greece.

Remarks Monotypic genus, discussed by Solounias (1981), Bouvrain and de Bonis (1985), and Kostopoulos (2014).

★*Samodorcas kuhlmani* (Andrée, 1926)

Nomenclatural and Taxonomic History *Ovis kuhlmani* in Andrée 1926 (new species); *Prosinotragus kuhlmani* in Solounias 1981 (new combination); *Samodorcas kuhlmani* in Bouvrain and de Bonis 1985 (new genus and new combination).

Type Material IGPM (PIM) 142 (holotype), cranium (mistakenly referred to as PIM 143 in Bouvrain and de Bonis 1985), illustrated by Andrée (1926: Pl. XIII, fig. 5 and Pl. XV, fig. 11) and Solounias (1981: fig. 53).

Type Locality Samos (unknown fossil horizon), Greece, Upper Miocene.

Age Late Miocene, Turolian (MN11–13).

Distribution Known only from the type locality.

Remarks *Samodorcas kuhlmani* is an extremely rare small- to medium-sized bovid, known only from its holotype and some uncertainly assigned dental material (Solounias 1981), all from Samos—but from an unknown stratigraphic location. The species is diagnosed by its uniquely long, homonymously twisted well-striated and anteroposteriorly compressed horncores, inserted just behind the orbits and directed backward; they bear a strong anteromedial keel proximally, as well as a strong posterolateral keel on their distal portion. The frontals are elevated between the horncores, the supraorbital foramina are sunken in deep, large, and closely spaced depressions, the face is importantly inclined compared to the braincase, the opisthocranium is short (Bouvrain and de Bonis 1985; Kostopoulos 2014). Although originally linked to the sheep lineage (Andrée 1926), *S. kuhlmani* is currently included in Oiocerina Pilgrim, 1934 (Kostopoulos 2014).

★*Dytikodorcas* Bouvrain and de Bonis, 2007

Type Species *Dytikodorcas longicornis* Bouvrain and de Bonis, 2007, by original designation.

Distribution Late Miocene of Greece; for a possible African expansion see discussions in Geraads (2019) and Geraads and Bobe (2020).

Remarks Bouvrain and de Bonis (2007) formally referred to their new genus *Dytikodorcas* only the type-species *D. longicornis* from the Upper Miocene site of

Dytiko, Axios Valley, Greece, but they also suggest as possible congeneric *Prostrepsiceros libycus* Lehmann and Thomas, 1987 from Sahabi, Libya. Geraads (2019) proposed, instead, closer relationships between *Pr. lybicus* and *Afrotragus* Geraads, 2019. Geraads et al. (2013) and Geraads and Bobe (2020) re-defined *Aepyceros* sp. of Harris et al. (2003) from Kanapoi, Kenya as Genus indet., aff. *Dytikodorcas* sp., leaving open the possibility that the Kanapoi taxon may be related to either *Dytikodorcas* or *Prostrepsiceros*.

★*Dyticodorcas longicornis* Bouvrain and de Bonis, 2007

Type Material LGPUT DKO 107 (holotype), cranium illustrated by Bouvrain and de Bonis (2007: fig. 7).

Type Locality Dytiko 3, Axios Valley, Greece, Upper Miocene.

Age Late Miocene, late Turolian (MN13).

Distribution Known only from the type locality.

Remarks At the size of a gazelle, *Dytikodorcas longicornis* is diagnosed by its long, mediolaterally compressed, and weakly heteronymously twisted horncores set on high pedicles, curved gently backward and bearing a predominant wide and shallow anterior longitudinal furrow with sharp rostral rim descending medially-anteromedially. The supraorbital foramina are large, the basioccipital narrow and grooved, and the teeth are mesodont with simple premolars and a weak goat fold on molars (Bouvrain and de Bonis 2007).

Urmiatherium Rodler, 1889

Type Species *Urmiatherium polaki* Rodler, 1889, by original designation.

Distribution Late Miocene of China, Iran, Iraq, Turkey, and Greece.

Remarks *Parurmiatherium* Sickenberg, 1932 is regarded—provisionally by Gentry et al. (1999) and conclusively by Kostopoulos (2009a)—as a junior synonym of *Urmiatherium*. Kostopoulos (2009a) also provided an emended generic diagnosis.

★*Urmiatherium rugosifrons* (Sickenberg, 1932)

Nomenclatural and Taxonomic History *Parurmiatherium rugosifrons* in Sickenberg 1932 (new species); *Urmiatherium rugosifrons* in Kostopoulos 2009a (new combination).

Type Material NHMW A4758 (lectotype selected by Solounias 1981), partially preserved cranium illustrated by Sickenberg (1933: Pl. V) and Kostopoulos (2009a: fig. 13B).

Type Locality Samos, Greece (unknown level), Upper Miocene.

Age Late Miocene, Turolian (MN11–13).

Distribution Late Miocene (likely middle–late Turolian) of eastern Greece and western Turkey (Asia Minor). In particular, other Greek occurrences include those from Samos Mytilinii 1A (MN12) and Samos Q5 (MN12/13).

Remarks This small-sized species of *Urmiatherium* has very short and robust, medially unfused, strongly homonymously twisted and grooved horncores, prolonged anteriorly on the frontals by low buttresses. It is originally known from Samos Island, Greece, and certainly present at the neighboring and roughly isochronous localities of Serefkoy-2 and Sahalipasalar, Turkey (Kaya et al. 2012). Bouvrain et al. (1995) also referred as *Parurmiatherium rugosifrons*, a partial cranium and toothrows from the late Vallesian to early Turolian fauna of Djebel Hamrin, Iraq, but the genus and species assignment of these specimens remain questionable (e.g., Kostopoulos and Bernor 2011).

★*Urmiatherium kassandriensis* Lazaridis et al., 2017

Nomenclatural and Taxonomic History *Urmiatherium* sp. in Lazaridis 2015 (initial identification); *Urmiatherium kassandriensis* in Lazaridis et al. 2017 (new species).

Type Material LGPUT FRK19 (holotype), partial cranium illustrated by Lazaridis et al. (2017: figs 2–3).

Type Locality Fourka sand deposits, Chalkidiki Peninsula, Upper Miocene.

Age Late Miocene, likely late Vallesian (MN10).

Distribution Known only from the type locality.

Remarks A medium-sized *Urmiatherium* with short, conical horncores without torsion and buttresses and with voluminous ellipsoidal-shaped posterior tuberosities of the basioccipital that converge rostrally and bear caudally two large sub-triangular facets for the atlas (Lazaridis et al. 2017). This species, along with the Djebel Hamrin, Iraq taxon represent the earliest occurrences of the *Urmiatherium*-like cranial morphotype in SE Mediterranean.

Caprini Gray, 1821

★*Protoryx* Forsyth Major, 1891

Type Species *Protoryx carolinae* Forsyth Major, 1891, by original designation.

Distribution Late Miocene of Greece and possibly in Turkey.

Other Taxa Included *P. capricornis* (Schlosser, 1904).

Remarks Several Miocene to Pliocene species have been clustered around the genus *Protoryx*, but recent revisions moved most of them to other genera. Pilgrim and Hopwood (1928) and Solounias (1981) suggested synonymizing *Pachytragus* Schlosser, 1904 with *Protoryx*, but Gentry (1971) provided full argumentation for their generic distinction. Chen and Zhang (2007) revised the majority of Chinese taxa referred to *Protoryx*, suggesting they belong to the new genera *Huabeitragus* Chen and Zhang, 2007, and *Macrotragus* Chen and Zhang, 2007. *Protoryx tadzhikistanica* Dmitrieva, 1977, definitely belongs to *Sporadotragus*. The generic attribution of *Pachytragus solignaci* Robinson, 1971 (commonly also referred to as *Protoryx solignaci*) and *Protoryx enanus* Köhler, 1987, is still debatable. The Pliocene taxa *Protoryx paralaticeps* Dmitrieva, 1977, from Kuruksay, Tadjikistan, and *Pr. heinrichi* Vekua, 1972, from Kvabebi, Georgia need further investigation, though the former appears closer to *Capra*, and the latter was recently regarded as close to *Gazella borbonica* (Bukshianidze and Koiava 2018). In Greece, "protoryxoid" bovids appear as early as the Vallesian, in the faunas of Pentalophos (MN9/10) and Xirochori (MN10), but their identification even at the generic level remains uncertain, and they may not represent *Protoryx* s.s. at all (Bouvrain 1997; Kostopoulos et al. 2020a).

★*Protoryx carolinae* Forsyth Major, 1891

Type Material MNHN.P PIK 2455 (holotype), partial cranium illustrated by Gaudry (1862–1867; Pl. LII, fig. 1) (Fig. 17).

Type Locality Pikermi, Greece, Upper Miocene.

Age Late Miocene, middle Turolian (MN12), ca. 7.3 Ma.

Distribution Late Miocene (Turolian) of Greece, including occurrences in Axios Valley (likely Vathylakkos according to Arambourg and Piveteau 1929).

Remarks The species has been extensively discussed by Pilgrim and Hopwood (1928), Gentry (1971), Solounias (1981) and Kostopoulos (2009a). An emended diagnosis is provided by Gentry (1971).

★*Protoryx capricornis* (Schlosser, 1904)
(= ★*Pachytragus crassicornis* Schlosser, 1904 = ★*Pseudotragus longicornis* Andrée, 1926)

Nomenclatural and Taxonomic History *Pseudotragus capricornis* in Schlosser 1904 (new species); *Protoryx capricornis* in Kostopoulos 2009a (new combination). This species is also considered as a senior synonym of *Pachytragus crassicornis* Schlosser, 1904 and *Pseudotragus longicornis* Andrée, 1926 (Kostopoulos 2009a).

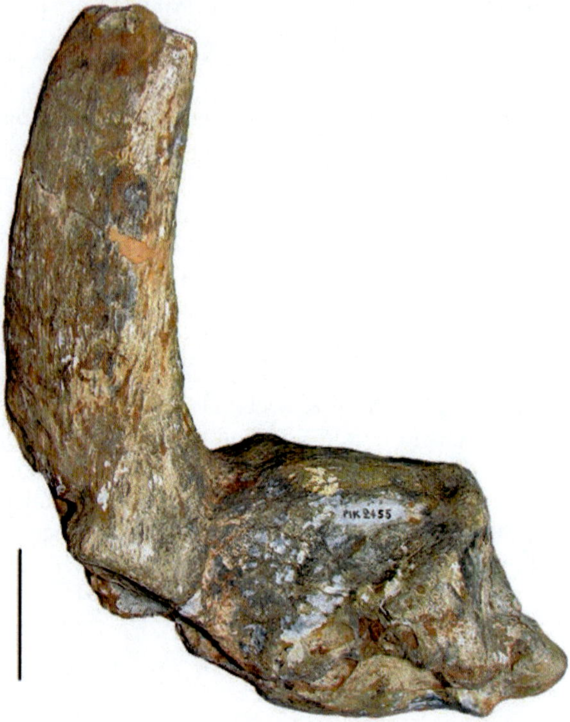

Fig. 17 Partial cranium (MNHN.P PIK 2455), holotype of *Protoryx carolinae* Forsyth Major, 1891 from Pikermi, in left-lateral view. Scale bar equals 5 cm. Photograph courtesy of MNHN.P

Type Material The partially preserved skull (lectotype selected by Pilgrim and Hopwood 1928) illustrated by Schlosser (1904: Pl. X, fig. 7). Pilgrim and Hopwood (1928) indicate that the holotype is part of the BSPM collection, but it was not located during our 2014 survey.

Type Locality Samos (unknown level), Greece, Upper Miocene.

Age Late Miocene, Turolian (MN11–13).

Distribution Known only from the type locality.

Remarks Among several crania recorded in Samos Upper Miocene deposits, Kostopoulos (2009a) suggested synonymizing *Pseudotragus capricornis* Schlosser, 1904, *Pachytragus crassicornis* Schlosser, 1904, and *Pseudotragus longicornis* Andrée, 1926 under *Protoryx capricornis* (Schlosser, 1904), indicating that these specimens differ from other similar Samos taxa in the smaller crania widened at the orbits, having closely settled and increasingly divergent, long horncores that are openly curved, flattened laterally, and anteriorly keeled with a rather subtriangular basal cross-section.

★*Palaeoryx* Gaudry 1861b

Type Species *Antilope pallasi* Wagner, 1857, by original designation.

Distribution Late Miocene (late Vallesian–Turolian) of Greece, Turkey, Ukraine, Moldova, Iran, and China.

Other Taxa Included *P. majori* Schlosser, 1904; *P. sinensis* Bohlin, 1935; *P. minor* Vasileiadis et al., 2019.

Remarks The genus *Palaeoryx* takes part of a group of medium- to large-sized Late Miocene antelopes with caprine-like appearance, recorded from Eastern Mediterranean to China. Gentry (1971) and Kostopoulos (2009a) deduced possible affiliations between *Palaeoryx* and the Plio-Pleistocene *Megalovis* Schaub, 1923, and *Gallogoral* Guérin, 1965. Earliest reports of *Palaeoryx* in Greece are coming from the site of Ravin de la Pluie, Axios Valley (MN10), but the material is insufficient for certain identification (Koufos 2006). Last Greek occurrences are from Dytiko fauna (Bouvrain and de Bonis 2007) of late Turolian (MN13) age.

★*Palaeoryx pallasi* (Wagner, 1857)

Nomenclatural and Taxonomic History *Antilope pallasi* in Wagner 1857 (new species); *Palaeoryx pallasi* in Gaudry (1861b, 1862–1867) (new genus and new combination). This species is widely considered as the senior synonym of *Palaeoryx woodwardi* Pilgrim and Hopwood, 1928.

Type Material BSPM #AS II 642 (lectotype selected by Pilgrim and Hopwood 1928), partial cranium with horncores illustrated by Wagner (1857: Pl. VII, fig. 21) (Fig. 18).

Type Locality Pikermi, Greece, Upper Miocene.

Age Late Miocene, middle Turolian (MN12) ca. 7.3 Ma.

Distribution Late Miocene (early–middle Turolian) of Greece, Turkey, Iran, and possibly Ukraine and Moldova. In particular, other Greek occurrences include those from Samos Q6, Q4 (MN11), Samos Mytilinii 4 (MN11), Samos Mytilinii 1A (MN12), Perivolaki (MN12), Halmyropotamos (MN12), Ravin X (MN12), Achladion (MN11–13), Triada (MN11–13), Ravin Ar. (MN11-13), and Rhodes (MN11–13).

Remarks Following the emended diagnosis by Kostopoulos (2009a), *P. pallasi* is a large antelope with laterally inserted, moderately divergent, strongly tilted backward and long, almost straight horncores, placed well behind the orbits and weakly recurving inwards at their tips. The opisthocranium is relatively short with more open frontoparietal than occipitoparietal angle. The premolars are large with premolar row shorter compared to the molars than in other species of the genus.

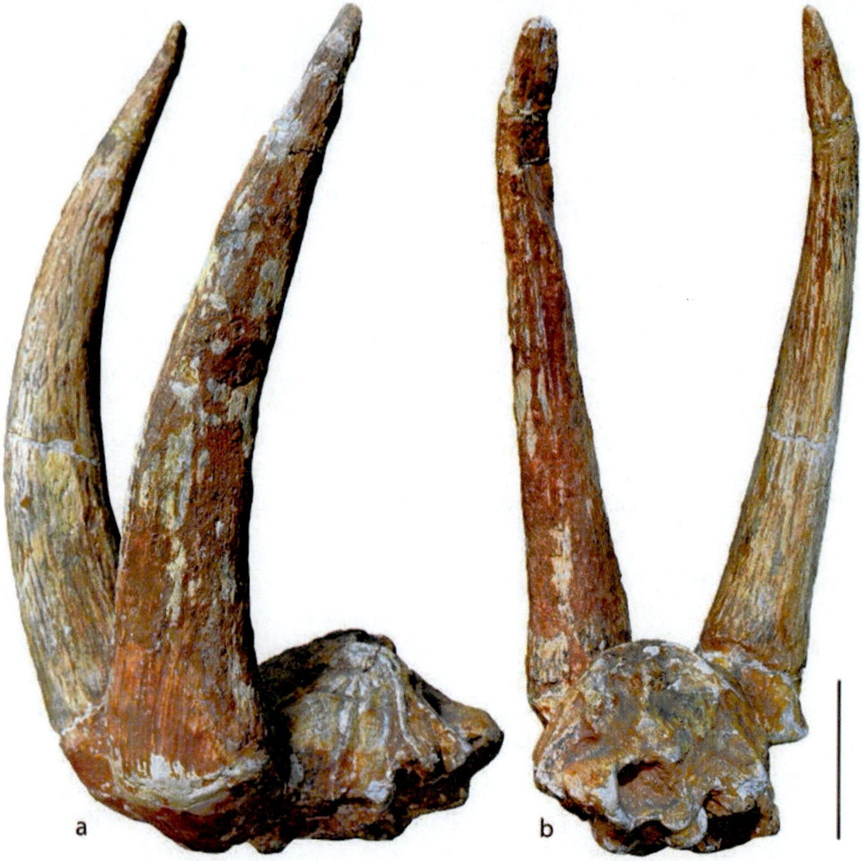

Fig. 18 Partial cranium (BSPM #AS II 642), lectotype of *Palaeoryx pallasi* (Wagner, 1857) from Pikermi, in (**a**) frontolateral and (**b**) caudal view. Scale bar equals 5 cm. Photographs courtesy of BSPM and A Athanassiou

★*Palaeoryx majori* Schlosser, 1904

Nomenclatural and Taxonomic History *Palaeoryx majori* in Schlosser, 1904 (new species). This species has been considered as a senior synonym of *Palaeoryx laticeps* Andrée, 1926, and as a junior synonym of *Palaeoryx pallasi* in Gentry (1971) and Solounias (1981). Kostopoulos (2005) re-validated the species.

Type Material BSPM #1899 VII 21 (lectotype selected by Pilgrim and Hopwood 1928), partial cranium illustrated by Schlosser (1904: Pl. VII, fig. 5) (Fig. 19).

Type Locality Samos (unknown level), Greece, Upper Miocene.

Age Late Miocene, Turolian (MN11–13). In particular, other Greek occurrences include those from Samos Mytilinii 1A, 1B (MN12), and Halmyropotamos (MN12).

Fig. 19 Partial cranium (BSPM #1899 VII 21), lectotype of *Palaeoryx majori* Schlosser, 1904 from Samos in frontodorsal view. Scale bar equals 5 cm. Photograph courtesy of BSPM and A Athanassiou

Distribution Late Miocene (likely middle Turolian) of Greece and Turkey.

Remarks The species has long been considered as a synonym of *P. pallasi* (e.g., Gentry 1971, Solounias 1981), but based on new material from Samos and Turkey Kostopoulos (2005, 2009a) recently re-validated it. Following the emended diagnosis provided by the latter author, *P. majori* represents a large antelope with strongly laterally inserted and significantly divergent horncores placed behind the orbits, but upright compared to the cranial roof. The horncores curve backward and significantly bent inwards at their tips with weak mediolateral compression and feeble distal twisting. The opisthocranium is relatively short with more open occipitoparietal than frontoparietal angle. The premolar row is longer compared to the molars than in *P. pallasi*. *Palaeoryx sinensis* from China looks like a larger version of *P. majori* and close relationships are suspected (Kostopoulos 2005, 2009a).

★*Palaeoryx minor* Vasileiadis et al., 2019

Type Material LGPUT PLD 150 (holotype), incomplete cranium illustrated by Vasileiadis et al. (2019: fig. 6).

Type Locality Platania, Drama basin, Greece, Upper Miocene.

Age Late Miocene, late Vallesian to early Turolian (MN10/11).

Distribution Known only from the type locality.

Remarks *Palaeoryx minor* represents the smallest taxon of its genus by means of cranial proportions, although dentition is metrically compatible to other species. The rather straight and keeless horncores are inserted above the back half of the orbits, tilted backward and weakly divergent. Although quite close in morphology to *P. pallasi*, *P. minor* differs in the shorter facial area, the shallower and less-defined lacrimal depression, the more closely set supraorbital foramina, and the less well-grooved basioccipital. Close phylogenetic relationships between the two taxa are very likely, though they have to be confirmed by cladistic analysis.

★*Skoufotragus* Kostopoulos, 2009a

Type Species *Pachytragus schlosseri* Andrée, 1926, by original designation.

Distribution Late Miocene (Turolian) of Greece, Turkey, and Iran.

Other Taxa Included *Skoufotragus laticeps* (Andrée, 1926); *Skoufotragus zemalisorum* Kostopoulos, 2009a.

Remarks The nomen novum *Skoufotragus* Kostopoulos, 2009a, has been introduced to incorporate several taxa and samples of Late Miocene large-sized antelopes previously merged mainly within the unavailable *Pachytragus* Schlosser, 1904 (e.g., Gentry 1971). A primer phylogenetic analysis by Gentry (2000b) indicates ten synapomorphies between *Skoufotragus schlosseri* (= *Pachytragus crassicornis* therein) and *Capra*, suggesting closer phylogenetic relationships of *Skoufotragus* with Caprini.

★*Skoufotragus schlosseri* (Andrée, 1926)

Nomenclatural and Taxonomic History *Pachytragus crassicornis* (in part) in Schlosser 1904 (initial identification); *Pachytragus schlosseri* in Andrée 1926 (new species); *Skoufotragus schlosseri* in Kostopoulos 2009a (new genus, new combination). This species has been considered as a junior synonym of *crassicornis*, under the combinations *Pachytragus crassicornis* in Gentry 1971 and *Protoryx crassicornis* in Solounias 1981.

Type Material NHMW A4780 (lectotype selected by Pilgrim and Hopwood 1928), cranium figured by Andrée (1926:148; Pl. XIII: fig. 3; Pl. XIV, fig. 4) and Kostopoulos (2009a: fig. 9A, B).

Type Locality Samos (unknown levels), Greece, Upper Miocene.

Age Late Miocene, middle-late Turolian (MN12–13).

Distribution Late Miocene (middle–late Turolian) of Greece and Turkey. In Greece, it is also known from Samos Q5 (MN12/13) (Fig. 20).

Remarks The species is characterized by its strongly goat-like appearance, the smaller size, shortened braincase and the shorter, less curved, stronger keeled anteriorly, and more divergent horncores than *Skoufotragus laticeps*. A full, emended, diagnosis is provided by Kostopoulos (2009a).

★*Skoufotragus laticeps* (Andrée, 1926)

Nomenclatural and Taxonomic History *Pachytragus laticeps* in Andrée 1926 (new species); *Protoryx laticeps* in Solounias 1981 (new combination); *Skoufotragus laticeps* in Kostopoulos 2009a (new combination). Full synonymy list in Gentry (1971) and Solounias (1981).

Type Material IGPM (PIM) 8 (holotype), partially preserved cranium illustrated by Andrée (1926: Pl. XII, figs. 5, 9).

Type Locality Samos (unknown level), Greece, Upper Miocene.

Age Late Miocene, Turolian (MN11–13).

Fig. 20 Lateral view of the cranium AMNH 20568 of *Skoufotragus schlosseri* (Andrée, 1926) from Q5, Samos in the AMNH, New York. Scale bar equals 5 cm. Photograph courtesy of AMNH

Distribution Late Miocene (likely early–middle Turolian) of Greece, Turkey, and Iran. In particular, other Greek occurrences include those from Samos Q4 (MN11), Samos Q1 (MN12), Samos Mytilinii 1A, 1B (MN12), Perivolaki (MN12), and possibly in Kryopigi (MN12/13). The generic attribution of protoryxoid taxa from Vathylakkos 3 (MN12) and Prochoma (MN12) is still open.

Remarks An emended diagnosis is provided by Gentry (1971). Horncores moderately long, strongly curved backward especially in their distal part, mediolaterally compressed, and lacking keels. Orbital rims are narrow to moderately wide. The braincase appears with two distinct morphotypes: the "long-brained" and the "short brained" variety (Gentry 1971; Kostopoulos 2009a). The species is likely the most widespread and common of its genus recorded in several Late Miocene faunas ranging from Southern Balkans to Iran.

★*Skoufotragus zemalisorum* Kostopoulos, 2009a

Type Material AMNH.Samos MYT (holotype), partially preserved skull illustrated by Kostopoulos (2009a: fig. 11).

Type Locality Mytilinii 3 (MYT), Samos Island, Greece, Upper Miocene.

Age Late Miocene, early middle Turolian (MN12).

Distribution Known only from the type locality.

Remarks Smaller than other species of this genus, *S. zemalisorum* is characterized by keeless and less uprightly inserted horncores, weakly curved backward, and relatively large toothrow compared to the cranium size. Kostopoulos (2009a) suggested that the three *Skoufotragus* species recorded in Samos may represent a chronocline.

★*Sporadotragus* Kretzoi, 1968

Type Species *Microtragus schafferi* Andrée, 1926, by original designation; currently considered as a junior synonym of *Sporadotragus parvidens* (Gaudry 1861b).

Distribution Late Miocene (Turolian) of Greece, Bulgaria, Northern Macedonia, Turkey, Tajikistan, and Afghanistan.

Other Taxa Included *Sp. tadzhikistanicus* (Dmitrieva, 1977); *Sp. vasili* Geraads et al., 2006.

Remarks Kretzoi (1968) indicated that *Microtragus* Andrée, 1926, was a name preoccupied by the coleoptere *Microtragus* White, 1846, and instead proposed the nomen novum *Sporadotragus*. Detailed synonymy lists are provided by Solounias (1981). Extended discussions on the taxonomic status of the genus are given by Gentry (1971, 2000b), Solounias (1981), Köhler (1987), Bosscha-Erdbrink (1988), Geraads et al. (2006), and Kostopoulos (2009a). *Pseudotragus* Schlosser, 1904, is herein considered as representing a different taxon, following Solounias (1981), Geraads et al. (2006), and Kostopoulos (2009a). Kostopoulos (2009a) also suggests that *Sporadotragus vasili* from Bulgaria may be closer to *Pseudotragus* than to *Sporadotragus*. Gentry (2000b) suggests possible relations of *Sporadotragus* with Caprini.

★*Sporadotragus parvidens* (Gaudry, 1861b) (= ★*Microtragus schafferi* Andrée, 1926)

Nomenclatural and Taxonomic History *Palaeoryx parvidens* in Gaudry 1861b (new species); *Microtragus parvidens* in Pilgrim and Hopwood 1929 (new combination); *Sporadotragus parvidens* in Solounias 1981 (new combination); *Pseudotragus parvidens* in Köhler 1987 (new combination). This species is considered as a senior synonym of *Microtragus schafferi* Andrée, 1926.

Type Material MNHN.P PIK 2453 (holotype), incomplete cranium (Gaudry 1861b: Pl. IX, fig. 4; 1862–67: Pl. XLVII, figs. 6, 7); note that both illustrations show a "reverse" (left as right) view of the same specimen.

Type Locality Pikermi, Greece, Upper Miocene.

Age Late Miocene, Turolian (MN12) ca. 7.3 Ma.

Distribution Late Miocene (Turolian) of Greece and Turkey. In particular, other Greek occurrences include those from Samos Qx (MN11), Samos Q1 (MN12), Samos Mytilinii 3 (MN11/12), Samos Mytilinii 1A (MN12), and Samos Q5 (MN12/13).

Remarks Geraads et al. (2006) emended diagnosis of this species indicates a short face strongly angled on the basicranium, complex sutures with pinched mid-frontal one, low and broad occipital making an obtuse angle with the dorsal braincase surface, and large horncores relatively to the skull size, rather long and well curved posteriorly, and weakly compressed medio-laterally with smooth surface lacking keels.

★*Tragoreas* Schlosser, 1904

Type Species *Tragoreas oryxoides* Schlosser, 1904, by original designation

Distribution Late Miocene (Turolian) Greece, China, and Mongolia.

Other Taxa Included *T. palaeosinensis* Schlosser, 1904; *T.? largelii* Bohlin, 1935.

Remarks *Tragoreas* is a poorly known genus, known originally from Samos (Greece). According to Dmitrieva and Serdyuk (2011), it is distinguished from other related genera by the small size, almost horizontal dorsal part of the braincase, widely open frontal angle and strongly inclined backward, rather short, almost straight, and sub-parallel horncores.

★*Tragoreas oryxoides* Schlosser, 1904

Type Material BSPM #1899 VII 11 (lectotype selected by Pilgrim and Hopwood 1928), partial cranium illustrated by Schlosser (1904: Pl. VI, figs 1, 9) (Fig. 21).

Type Locality Samos Island (unknown level), Upper Miocene.

Age Late Miocene, Turolian (MN11–13). In particular, other Greek occurrences include those from Samos Qx and Q6 (MN11).

Distribution Known only from Samos.

Remarks Discussed by Gentry (1971) and Solounias (1981), *Tragoreas oryxoides* has been suggested as a synonym of *Pseudotragus capricornis* Schlosser, 1904, by Gentry (2000a, b) but as a valid taxon by Kostopoulos (2009a) and Dmitrieva and Serdyuk (2011). The species is characterized by small size compared to other related taxa, weakly divergent and moderately to strongly tilted backward, rather long, straight, and keelless horncores of elliptical cross-section.

Qurliqnoria Bohlin, 1938

Type Species *Qurliqnoria cheni* Bohlin, 1938, by original designation.

Distribution Late Miocene–Pliocene of China, late Miocene of Anatolia (Vallesian–early Turolian) and possibly in Greece.

Fig. 21 Partial cranium (BSPM #1899 VII 11), lectotype of *Tragoreas oryxoides* Schlosser, 1904, from Samos in left-lateral view. Scale bar equals 5 cm. Photograph courtesy of BSPM and A Athanassiou

Other Taxa Included *Q. bohlini* (Ozansoy, 1965), *Q. chorakensis* Kostopoulos et al., 2020b.

Remarks Originally known from the peri-Tibetan area exclusively (Bohlin 1937; Wang et al. 2019), the genus was lately recognized in the Vallesian of Anatolia (Kostopoulos et al. 2020b). It represents medium-sized bovids with well-divergent, gently heteronymoysly twisted, moderately to strongly keeled anteriorly, strongly grooved, weakly curved backward and gently to strongly flaring outwards horn-cores inserted obliquely on short pedicles. In Greece, the genus has been recently reported from the late Vallesian to early Turolian (MN10/11) fauna of Platania in Drama basin (Vasileiadis et al. 2019; Kostopoulos et al. 2020b).

★*Norbertia* Köhler et al., 1995

Type Species *Norbertia hellenica* Köhler et al., 1995, by original designation.

Distribution Latest Miocene to earliest Pliocene of Greece.

Remarks *Norbetia* is a monotypic genus. Köhler et al. (1995: Fig. 1) suggested *Norbertia* as a stem Caprini, an opinion adopted by Gentry et al. (1999) and Gentry (2000b).

★*Norbertia hellenica* Köhler et al., 1995

Type Material AMPG MA 723 (holotype), part of frontals with the complete right and the basis of the left horncore illustrated by Köhler et al. (1995: Pl. 1, fig. 1).

Type Locality Maramena, Greece, Upper Miocene/Lower Pliocene.

Age Latest Miocene, latest Turolian/earliest Ruscinian (MN 13/14), ca. 5.3 Ma.

Distribution Known only from the type locality.

Remarks As originally diagnosed, *N. hellenica* had the size of the extant goat (*Capra* ssp.) with very thick frontals between the horncores, strongly pneumatized pedicles and frontals, laterally protruding orbits, simple but proximally large horn-cores closely set each other on the frontals and gently curved backward, smoothly keeled anteriorly, inserted above the orbits and running in parallel each other. The dental morphology is caprine, with distolabially projected metastyle on M3, and well-developed goat fold on lower molars (Köhler et al. 1995).

★*Koufotragus* de Bonis and Bouvrain, 1996

Type Species *Gazella bailloudi* Arambourg and Pivetau, 1929, by original designation.

Distribution Early Pliocene of Greece.

Remarks *Koufotragus* is a monotypic genus. Gentry (1970) excludes relations with gazelles and suggests affinities with rupicaprins.

★*Koufotragus bailloudi* (Arambourg and Piveteau, 1929)

Nomenclatural and Taxonomic History *Gazella bailloudi* in Arambourg and Piveteau 1929 (new species); *Koufotragus bailloudi* in de Bonis and Bouvrain 1996 (new genus, new combination).

Type Material MNHN.P Slq-984 (holotype), cranium illustrated by Arambourg and Piveteau (1929: Pl. 6, fig. 7) and de Bonis and Bouvrain (1996: figs 1–3) (Fig. 22).

Type Locality Megalo Emvolon (Karabournou, Karaburun and other variant spellings), Thessaloniki, Greece, Lower Pliocene.

Age Early Pliocene, late Ruscinian (MN15).

Distribution Known only from the type locality.

Remarks According to the emended diagnosis provided by de Bonis and Bouvrain (1996), *Koufotragus bailloudi* is a small- to medium-sized bovid with simple horncores lacking keels, spiraling and longitudinal grooves, inserted perpendicu-

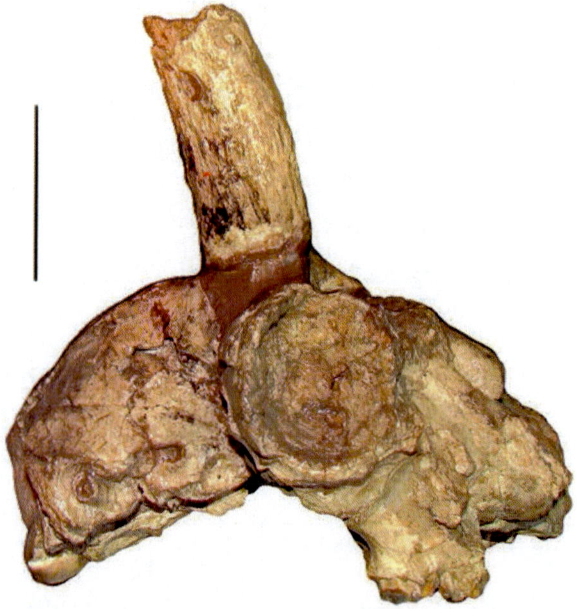

Fig. 22 Cranium (MNHN.P Slq-984), holotype of *Koufotragus bailloudi* (Arambourg and Piveteau, 1929) from Megalo Emvolon, in right-lateral view. Scale bar equals 5 cm. Photograph courtesy of MNHN.P

larly on the caudal parts of the prominent orbits. Their cross-section is oval to round. There are no sinuses in the horncore bases. The postcornual groove is large and shallow. The occipital face is low and the mastoids face mostly caudally. The basioccipital is short with weak tuberosities.

★*Euthyceros* Athanassiou, 2002b

Type Species *Euthyceros thessalicus* Athanassiou, 2002b, by original designation.

Distribution Early Pleistocene of Greece.

Remarks *Euthyceros* is a monotypic genus. Athanassiou (2002b) suggested possible relations with either caprins or ovibovins but as the author states, the material is insufficient for certain conclusions.

★*Euthyceros thessalicus* Athanassiou, 2002b

Type Material AMPG Σ-408 (holotype), left horncore illustrated by Athanassiou (2002b: fig. 2).

Type Locality Sesklo, Greece, Lower Pleistocene.

Age Early Pleistocene, middle Villafranchian (MN17).

Distribution Known only from the type locality.

Remarks As originally diagnosed (Athanassiou 2002b), *Euthycerus thessalicus* represents a medium to large bovid with relatively short and straight horncores of semicircular cross section, flatten laterally. Horncores appear strongly pneumatized internally as in Caprini, but the lower dentition indicates rather wide and brachyodont teeth and long premolars compared to the molars.

Procamptoceras Schaub, 1923

Type Species *Procamptoceras brivatense* Schaub, 1923, by original designation.

Distribution Early Pleistocene (middle–late Villafranchian) of S. Europe.

Remarks Viret (1961) and Guérin (1965) discussed *Procamtoceras* from the Villafranchian of Europe and Duvernois and Guérin (1989) provided an emended diagnosis of this monotypic and rare genus, possibly affiliated to rupicaprins (Crégut-Bonnoure and Guérin 1996). The face is very short and strongly inclined to the braincase; the horncores are closely set on the frontals, long, well-striated, gently curved frontwards, and trending parallel each other. In Greece, *Procamptoceras* is only provisionally reported from the middle–late Villafranchian faunas of Volax (MNQ17) and Vassiloudi (MNQ18, Mygdonia basin) (Kostopoulos 1997a, 2006b).

Gallogoral Guérin, 1965

Type Species *Palaeoryx meneghinii* Rütimeyer, 1878, by original designation.

Distribution Early Pleistocene (middle–late Villafranchian) of S. Europe. Greek occurrences of *Gallogoral meneghinii* include those from Karnezeika in Peloponnesus (MN17?), Volax (MN17), Dafnero-1, Dafnero-3 (MN17), and Sesklo (MN17).

Remarks Guérin (1965) and Duvernois and Guérin (1989) suggested *Gallogoral* as a large-sized rupicaprine with long, fairly straight, divergent, and keeless horncores inserted behind the orbits, and short moderately robust metapodials. Rodrigo (2011) prefers keeping this taxon insertae sedis within Caprini. The single known species, *Gallogoral meneghinii*, is known from the middle–late Villafranchian faunas of Georgia, France, Spain, and Italy.

★*Gallogoral meneghinii sickenbergi* Kostopoulos, 1997a

Type Material LGPUT VOL-1, cranium (Kostopoulos 1997a: Pl. I, Fig. 23).

Type Locality Volax, Drama basin.

Age Early Pleistocene, middle Villafranchian (MN17).

Distribution Known only from the type locality.

Remarks Kostopoulos (1997a) recognized the Volax taxon as distinct from the typical one at the subspecies level, based on a wider cranium, larger toothrow, stouter horncores and longer metapodials. Rodrigo (2011) recognized another subspecies (*G. m. heintzi*) from La Puebla de Valverde, Spain.

Pliotragus Kretzoi, 1941

Type Species *Antilope ardea* Depéret, 1883, by original designation.

Distribution Late Pliocene–Early Pleistocene (Villafranchian) of S. Europe.

Other Included Taxa *Pl. macedonicus* Crégut-Bonnoure and Tsoukala, 2005.

Remarks Gentry (1970) suggested synonymizing *Megalovis* Schaub, 1923, *Pliotragus* Kretzoi, 1941, *Soergelia* Schaub, 1951, and *Hesperidoceras* Crusafont, 1965, though no consensus exists. Duvernois and Guérin (1989) provided an extended review of *Pliotragus*, suggesting synonymy with *Hesperidoceras*, but Rodrigo (2011) provided much more data distinguishing these two extremely rarely reported genera. Though a deep phylogenetic analysis is still missing, most authors include *Pliotragus* in Caprini. In Greece, *Pliotragus sp.* is reported from the fauna of Karnezeika in Peloponnesus (?MN17) (Kokotini et al. 2019).

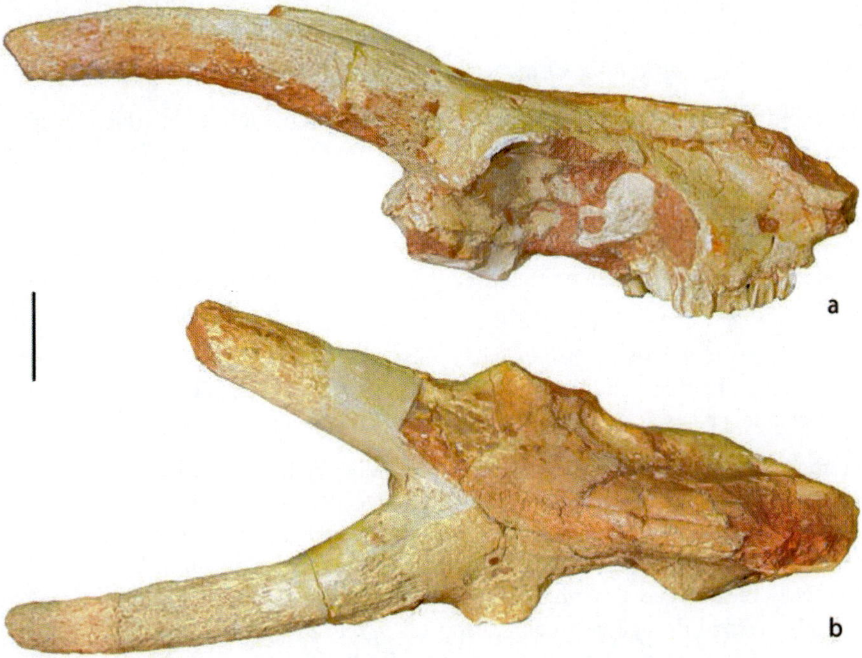

Fig. 23 The holotype cranium (LGPUT VOL-1) of *Gallogoral meneghinii sickenbergi* Kostopoulos, 1997a, from Volakas site in (**a**) right-lateral, and (**b**) dorsal view. Scale bar equals 5 cm

★*Pliotragus macedonicus* Crégut-Bonnoure and Tsoukala, 2005

Type Material LGPUT PEC 234 (holotype), upper second molar illustrated by Crégut-Bonnoure and Tsoukala (2005: fig. 4).

Type Locality Petralona cave, Greece, Middle–Upper Pleistocene.

Age Middle? Pleistocene.

Distribution Known only from the type locality.

Remarks Crégut-Bonnoure and Tsoukala (2005) recognized within the rich bovid sample from the hominin-bearing cave of Petralona an M2 and two M3s, which are identified as representing a distinct species of *Pliotragus,* namely *P. macedonicus*. The poverty of the material makes any assumption doubtful, though the authors suggest the survival of the genus in Greece until much later than in the rest of Europe due to refugia phenomena.

Capra Linnaeus, 1758

Type Species *Capra hircus* Linnaeus, 1758, by Linnean tautonomy.

Distribution Pleistocene to present; although *Capra* has today a worldwide distribution, it is originally a Eurasian genus, maximally expanded in North and NE Africa. In Greece, apart from Petralona Cave, *Capra ibex* is also recorded in the late Middle to latest Pleistocene cave faunas of Apidima, Kalamakia, Perama, Loutra Almopias, Franchiti, and Klisoura, as well as in the Peneios river plains. All known evidences of *C. aegagrus* are from Holocene times (subfossils and archeological sites) (Geskos 2013).

Other Included Taxa *C. primaeva* Arambourg, 1979; *C.*? *alba* Moyà-Solà, 1987; *C. dalii* Bukshianidze and Vekua, 2006; *C. baetica* Arribas and Garrido, 2008; likely nine extant species as well (e.g., Groves and Grubb 2011).

Remarks Molecular data suggest that *Capra* diverged 3 Ma ago in Asia (Manceau et al. 1999); "*Protoryx*" *paralaticeps* Dmitrieva, 1977 from Kuruksay, Tadjikistan, dated at about 2.5–2.0 Ma may indicate such emergence together with *Capra primaeva* from Aïn Brimba, Tynisia. *Capra* is practically unknown in Europe before the Early Pleistocene, though sporadic evidence from Spain and Georgia suggest it already invaded Europe between 2.0 and 1.5 Ma. *Capra* becomes a common faunal element of European faunas from the late Middle Pleistocene onwards; Petralona cave in N. Greece may represent one of the oldest known European records of *Capra ibex*, dated at around 300 ka (Crégut-Bonnoure and Tsoukala 2005). Sickenberg (1971), followed by Tsoukala (1989), credited subspecies value on the Petralona goat (★***Capra ibex macedonica*** Sickenberg, 1971).

Ovis Linnaeus, 1758

Type Species *Ovis aries* Linnaeus, 1758, by Linnean tautonomy.

Distribution Pleistocene to present; originally Eurasian to N. American, but domestic forms widely distributed around the world.

Other Included Taxa *O. antiqua* Pomerol, 1879; *O. shantungensis* Matsumoto, 1926; *O. zdanskyi* Bohlin, 1938; the number of extant species varies from 7 to 20 depending on the authors (e.g., Groves and Grubb 2011).

Remarks Wang et al. (2016) place the origin of *Ovis* in the peri-Tibetan area between 5 and 3 Ma, linked with *Protovis* Wang et al., 2016. Mead and Taylor (2005) suggest however that N. American *Sinocapra* Chen, 1991, dated at about 6 Ma may be more closely related to the origin of *Ovis*. In China, *Ovis* appears as early as at the beginning of the Pleistocene. The first European record of the taxon comes from Senéze in France (but the age of the findings is disputable) and Slivnitsa in Bulgaria (Crégut-Bonnoure 1992; Spassov and Crégut-Bonnoure 1999). *Ovis* becomes a common faunal element in European faunas only after 400 ka. The Greek fossil record of *Ovis* is extremely poor. The taxon is reported, but not described, from the late Early Pleistocene fauna of Apollonia (Mygdonia Basin, MNQ20) (Koufos 2001), and it is recorded in Charkadio Holocene cave fauna of Tilos Island (Symeonidis 1973).

Soergelia Schaub, 1951

Type Species *Soergelia elisabethae* Schaub, 1951, by original designation.

Distribution Early–Middle Pleistocene of Europe.

Other Taxa Included *S. minor* Moyà-Solà, 1987; *S. brigittae* Kostopoulos, 1997b; *S. intermedia* Crégut-Bonnoure and Dimitrijevic, 2006.

Remarks Crégut-Bonnoure (2002) provided an emended diagnosis of this ovibovin genus, known exclusively from the Early–Middle Pleistocene of Europe. There is no agreement among scholars not only about the number of the valid species but even about the genus-level identification of several local samples (e.g., Crégut-Bonnoure 2002; Moullé et al. 2004; Crégut-Bonnoure and Dimitrijevic 2006). Martínez-Navarro et al. (2012) proposed grouping most *Soergelia* samples recorded around the European Mediterranean region during the earliest part of the late Villafranchian under *S. minor* Moyà-Solà, 1987.

★*Soergelia brigittae* Kostopoulos, 1997b

Type Material LGPUT APL-383 (holotype), right mandible illustrated by Kostopoulos (1997b: fig. 9c).

Type Locality Apollonia 1, Greece, Lower Pleistocene.

Age Early Pleistocene, ca. 1.4–1.0 Ma.

Distribution Early Pleistocene (latest Villafranchian/earliest Gallerian) of Greece. Crégut-Bonnoure (2002) provisionally refers the Le Vallonet (France) taxon to the same species.

Remarks As originally diagnosed (Kostopoulos 1997b), *Soergelia brigittae* is morphometrically in between *S. elisabethae* and *S. minor* and distinguished from both taxa mostly by its particular premolar features (Crégut-Bonnoure 2002).

Praeovibos Staudinger, 1908

Type Species *Praeovibos priscus* Staudinger, 1908, by original designation.

Distribution Pleistocene of Eurasia and North America.

Other Taxa Included *Pr. recticornis* (Ryziewicz, 1933); *Pr. beringiensis* Sher, 1971; *Pr. mediterraneus* Crégut-Bonnoure, 2002.

Remarks Ovibovin genus that is considered as a close relative of *Ovibos* de Blainville, 1816 and a possible ancestor of the North American *Bootherium* Leidy, 1852. McDonald et al. (1991) and Crégut-Bonnoure (2002, 2006) revised its world-wide geographic and stratigraphic occurrence. A poorly documented early representative of *Praeovibos* known by several South European post-Olduvai faunas is recognized by Crégut-Bonnoure (2002) as a distinct species, *Praeovibos mediterraneus*, diagnosed by slender metapodials and horncores lacking a posterior keel and with their bases rising less above the skull than in the type species. In Greece, *Praeovibos* is only reported from the Early Pleistocene fauna of Apollonia (Mygdonia basin, MNQ20) (Kostopoulos 1997b), ascribed later to *P. mediterraneus* by Crégut-Bonnoure (2002).

Antilopinae Incertae Sedis

Kubanotragus Gabunia, 1973

Type Species *Kubanotragus sokolovi* Gabunia, 1973, by original designation.

Distribution Middle Miocene of Eurasia.

Other Taxa Included *K. gaopoensis* Chen, 1990.

Remarks Originally erected based on a single partial horncore from Belometchetskaye site in North Caucasus (e.g., Köhler 1987; Pickford et al. 2000), the genus and related species are part of a disputed group of ante-Vallesian bovids referred to as hypsodontines (Hypsodontinae sensu Köhler 1987) due to their precociously hypsodont dentition (e.g., Köhler 1987; Gentry et al. 1999; Geraads 2003; Dmitrieva 2007). Suggested synonymy between *Kubanotragus* Gabunia, 1973 and *Hypsodontus* Sokolov, 1949 (e.g., de Bonis et al. 1998; Pickford et al. 2000) is not widely accepted (e.g., Geraads 2003; Dmitrieva 2007).

Hypsodontinae or Hypsodontini are assumed to be diphyletic in respect to the rest of Bovidae (Gentry et al. 1999) or interpreted as either basal members of Caprini sensu lato (Dmitrieva 2007) or nested within Antilopini (Kostopoulos 2014). A frontlet and several toothrows of a hypsodontine bovid are known from the MN5 Thymiana fauna in Chios Island (Fig. 24). de Bonis et al. (1998) described the Thymiana taxon as "*Hypsodontus* cf. *gaopense*" unnecessarily altering the species name from "*gaopoensis*" to "*gaopense.*" As the synonymy between *Hypsodontus* and *Kubanotragus* is debated (e.g., Geraads 2003; Dmitrieva 2007), I suggest referring the Chios material to *Kubanotragus* cf. *K. gaopoensis* Chen, 1990, pending a full revision of the genus and included species. *Kubanotragus gaopoensis* is a small- to medium-sized bovid characterized by slender, solid, moderately long and slightly homonymously twisted horncores anteroposteriorly compressed and lacking keels.

Fig. 24 The frontlet (LGPUT THB-34) of *Kubanotragus* cf. *K. gaopoensis* from Thymiana site (Chios island), in frontal view. Scale bar equals 2 cm. Photograph courtesy of GD Koufos

Tethytragus Azanza and Morales, 1994

Type Species *Tethytragus langai* Azanza and Morales, 1994, by original designation.

Distribution Middle Miocene of Europe and Turkey.

Other Taxa Included *Te. koehlerae* Azanza and Morales, 1994; *Te. stehlini* (Thenius, 1951).

Remarks Still of uncertain phylogenetic relationships, *Tethytragus* together with some other ante-Vallesian bovids (i.e., *Gentrytragus* and *Caprotragoides*) is suggested as close or into the stem group leading to the large crown group incorporating Caprini, Hippotragini, and Alcelaphini (Azanza and Morales 1994; Gentry 2000b; Bibi et al. 2009). *Te. koehlerae*, originally identified from Çandir, Turkey, is diagnosed by little pneumatized frontals, long pedicles, horncores of weakly homonymous torsion and long, narrow teeth (Azanza and Morales 1994). A single hemimandible from the MN5 fauna (late Orleanian) of Thymiana in Chios is the only specimen from Greece provisionally referred to this species (de Bonis et al. 1998).

★*Mesembriacerus* Bouvrain, 1975

Type Species *Mesembriacerus melentisi* Bouvrain, 1975, by original designation.

Distribution Late Miocene (Vallesian) of Greece and Northern Macedonia.

Remarks *Mesembriacus* is a monotypic genus known from the Vallesian of Southern Balkans. A preliminary phylogenetic analysis by Bouvrain and de Bonis (1984) based on 25 cranial features allowed the authors to refer this taxon to as Ovibovini, though later studies challenged the monophyly of this tribe. At present time, *M. melentisi* stands as one of the oldest bovids with specialized cranial features analogous (?) to *Ovibos* and its Pleistocene allies.

★*Mesembriacerus melentisi* Bouvrain, 1975

Type Material LGPUT RPl-032 (holotype), cranium illustrated by Bouvrain (1975: Fig. 1) and Bouvrain and de Bonis (1984: Figs 1–5).

Type Locality Ravin de la Pluie, Axios Valley, Greece, Upper Miocene.

Age Late Miocene, late Vallesian (MN10) ca. 9.3 Ma.

Distribution Late Miocene of Greece and Northern Macedonia.

Remarks *Mesembriacus melentisi* definitely represents the predominant bovid of the *Ouranopithecus*-bearing Ravin de la Pluie locality, known by numerous crania, dentitions and postcranials. At the size of a goat, *M. melentisi* cranium shows several particular features such as the thin and long horncores inserted behind the orbits and strongly inclined and curved caudolaterally, the high and long face with well-developed ante-orbital fossa, the presence of parietal rugosities, the presence of additional stop-facets for the atlas on the posterior tuberosities of the basioccipital and the short premolars compared to the molars (Bouvrain and de Bonis 1984).

★*Criotherium* Forsyth Major 1891a

Type Species *Criotherium argalioides* Forsyth Major, 1891a, by original designation.

Distribution Late Miocene (Turolian) of Greece, Bulgaria, and Turkey.

Other Taxa Included *Cr. nikolovi* Geraads and Spassov, 2008.

Remarks Initially allocated to Bubalidinae (a synonym of Alcelaphinae) by Schlosser (1904) and Pilgrim and Hopwood (1928), the genus was later linked to Ovibovini (Bouvrain and de Bonis 1984) or to Antilopini (Gentry 2003), but it is considered as an incertae sedis by most scholars.

★*Criotherium argalioides* Forsyth Major, 1891a

Type Material NHMUK M-4199 (lectotype selected by Pilgrim and Hopwood 1928) cranium, NHMUK M-4201 (paralectotype) crania, designated and illustrated by Pilgrim and Hopwood (1928: Figs 2–3; note however that the authors erroneously listed M-4200 as the paratype in their Fig. 1) (Fig. 25).

Type Locality Samos (unknown fossil horizon), Greece, Upper Miocene.

Age Late Miocene, likely early Turolian (MN 11).

Distribution Late Miocene (likely early Turolian) of Greece and Turkey. In particular, other Greek occurrences include those from Samos Q2, Q6, and Q4 (MN11).

Remarks Recently revised by Geraads and Spassov (2008), *Criotherium argalioides* is a large bovid with tightly twisted heteronymously horncores, uprightly inserted behind the orbits and close to the occipital, bearing strong keels and furrows and gently curved forwards in their distal parts. The braincase is very short, the occipital lies nearly at the same plane with the braincase roof, the muzzle is high and narrow, the premaxillae do not contact the nasals, the teeth are hypsodont and the premolar row not particularly reduced compared to the molars.

★*Stryfnotherium* Kostopoulos and Soubise, 2018

Type Species *Stryfnotherium exophthalmon* Kostopoulos and Soubise, 2018, by original designation.

Distribution Late Miocene (early Turolian) of Greece.

Remarks *Stryfnotherium* is a monotypic genus. Kostopoulos and Soubise (2018) discussed possible relations of *Stryfnotherium* with *Palaeoreas* and *Criotherium* but thorough phylogenetic analysis is pending.

★*Stryfnotherium exophthalmon* Kostopoulos and Soubise, 2018

Nomenclatural and Taxonomic History *Palaeoreas zouavei* (part) in Bouvrain 1980 (initial identification); *Stryfnotherium exophthalmon* in Kostopoulos and Soubise 2018 (new genus and species).

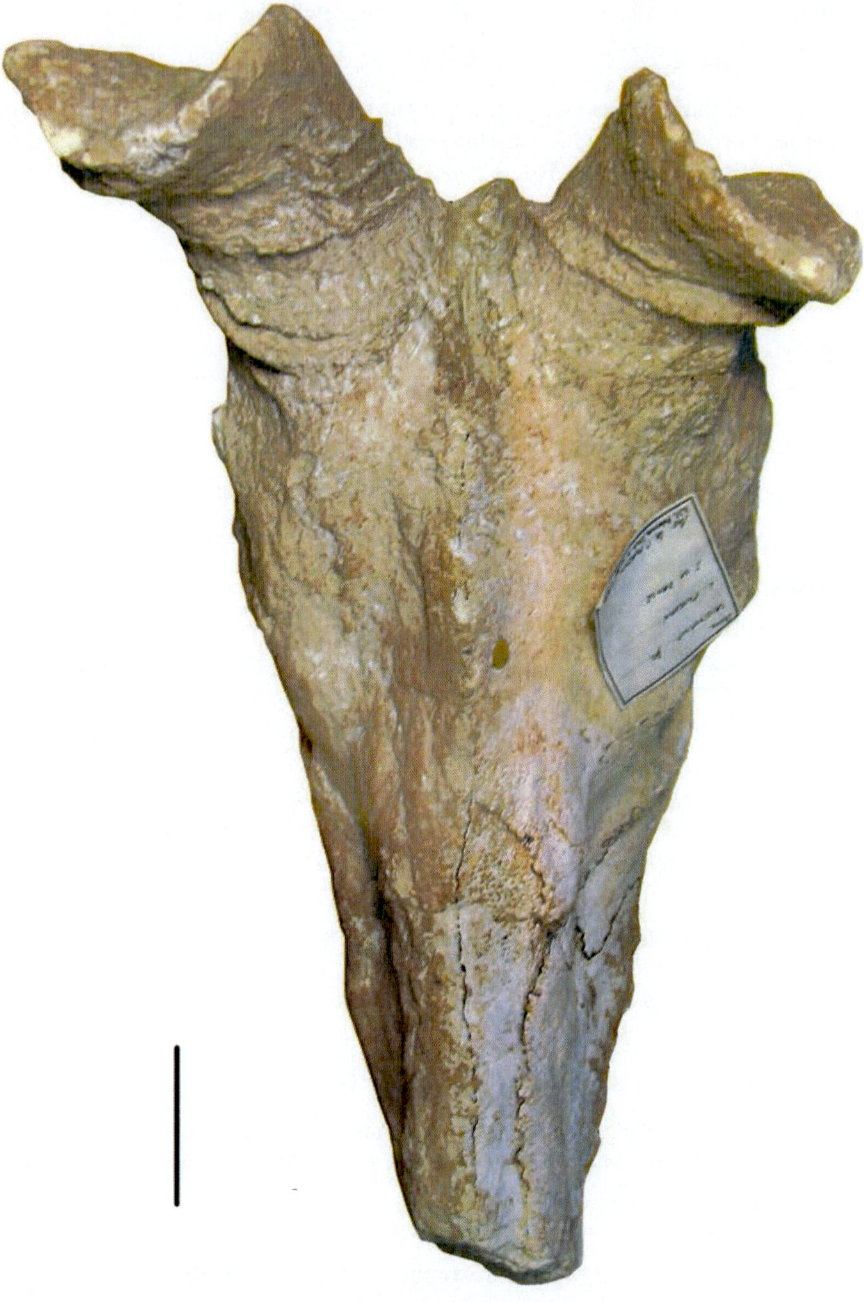

Fig. 25 Cranium (NHMUK M-4199), lectotype of *Criotherium argalioides* Forsyth Major, 1891a from Samos in dorsal view. Scale bar equals 5 cm. Photograph courtesy of NHMUK

Type Material LGPUT RZO-068 (holotype), male cranium illustrated by Kostopoulos and Soubise (2018: Fig. 10).

Type Locality Ravin de Zouaves-5, Axios Valley, Greece, Upper Miocene.

Age Late Miocene, Early Turolian (MN 11) ca. 8.2 Ma.

Distribution Known only from the type locality.

Remarks LGPUT RZO-068 was originally included by Bouvrain (1980) in her new species *Palaeoreas zouavei* from Axios Valley, the validity of which was recently challenged by Kostopoulos and Soubise (2018). The same authors recognized LGPUT RZO-068 cranium as distinctly different from *Palaeoreas* and proposed a new binominal name for it. *Stryfnotherium exophthalmon* is diagnosed as a medium-sized heteronymous spiral horned antelope with straight, moderately long, twisted, and divergent horncores set widely apart on the frontals, and bearing a fairly strong posterior keel describing a full whorl from the base to the tip. The frontals are strongly elevated between the horncores, the orbits are salient in their caudal part, the face is short and narrows quickly in dorsal view, and the braincase is wide and flattened dorsally with the broad occipital set almost in the same plane as the parietal. The basioccipital is long and triangular, lacking median groove and having strong, laterally shifted posterior tuberosities forming additional stop facets for the atlas. *Stryfnotherium exophthalmon* resembles *Palaeoreas* in the horncore pattern but *Criotherium* in the cranial one, though it differs from both of them in the unique combination of features including apomorphies such as the broadening of the cranium at the orbital region, the obliterated interfrontal suture and the particular basioccipital morphology. Its phylogenetic relationships remain unknown pending a detailed analysis.

★*Helladorcas* Bouvrain, 1997

Nomenclatural History Gentry (2003: 356) suggested the spelling *Helladodorcas* instead of *Helladorcas* as the genitive of *Hellas* [Greece] is "*Hellado-*". Kostopoulos et al. (2018) suggest however, this act is unnecessary as Bouvrain was already aware of this and her "*Helladorcas*" is in fact a combination of "*Hellado-*" and "*Dorcas*" by simplifying the compound word (i.e., omitting one "*do*"), a common practice in the Greek grammar.

Type Species *Helladorcas geraadsi* Bouvrain, 1997, by original designation.

Distribution Late Miocene (Vallesian) of Greece.

Remarks *Helladorcas* is a monotypic genus. Although the single representative of the genus, *H. geraadsi*, is exclusively known from the Vallesian of Greece (Pentalophos, Axios Valley), Kostopoulos (2006a), and Lazaridis (2015)—based on limited material from the middle Turolian faunas of Perivolaki (Thessally, Greece, MN12) and Kryopigi (Chalkidiki peninsula, Greece, MN12/13) respectively—refer some antelopes with larger but morphologically compatible horncores and similar dental features to *Helladorcas*. These records may expand the time frame of the genus into the Turolian.

★*Helladorcas geraadsi* Bouvrain, 1997

Type Material LGPUT PNT-090 (holotype), male partial cranium illustrated by Bouvrain (1997: Fig. 1; Pl. II, fig. 4).

Type Locality Pentalophos, Axios Valley, Greece, Upper Miocene.

Age Late Miocene, late early Vallesian (end of MN9), ca. 9.7 Ma.

Distribution Known from the type locality.

Remarks As originally diagnosed by Bouvrain (1997), this small- to medium-sized heteronymous spiral-horned antelope with hornless females is characterized by mediolaterally compressed horncores, strongly twisted along their longitudinal axis, and bearing a sharp posterior keel describing a full whorl from the base to the tip. The narrow ethmoidal fissure is principally outlined by the lachrymal and frontal bones, the long nasals lead rostrally to two points and the premolar row is short compared to the molars. By its horncore morphology, *Helladorcas geraadsi* recalls the Turolian and larger *Protragelaphus*, but an analysis of the possible phylogenetic relationships of these taxa is still pending.

★*Ouzocerus* Bouvrain and de Bonis, 1986

Type Species *Ouzocerus gracilis* Bouvrain and de Bonis, 1986, by original designation.

Distribution Late Miocene (Vallesian) of Greece and possibly Turkey.

Other Taxa Included *O. pentalophosi* Bouvrain, 1997.

Remarks *Ouzocerus* may stand as a possible forerunner of some Turolian antelopes with horncores twisted along their axis but a thorough analysis of their phylogenetic relationships is not yet available. Apart from the Vallesian of Axios Valley, the genus is possibly present in Sinap Turkey (Kostopoulos et al. 2020a). Köhler et al. (1995) also recorded *Ouzocerus* in the Turolian/Ruscinian (MN13/14) fauna of Maramena, Greece, and although the extremely badly preserved cranial material does not allow confirmation of this identification, the dental material indicates significant differences.

★*Ouzocerus gracilis* Bouvrain and de Bonis, 1986

Type Material LGPUT RZ1-018 (holotype), partial cranium illustrated by Bouvrain and de Bonis (1986: Pl. I, figs 1–2).

Type Locality Ravin de Zouaves-1, Axios Valley, Greece, Upper Miocene.

Age Late Miocene, late Vallesian (MN10).

Distribution Known from the type locality.

Remarks This small-sized antelope is characterized by horncores implanted above the orbits, weakly compressed mediolaterally, moderately divergent distally and heteronymously twisted along their longitudinal axis, bearing a strong posterior keel but only hints of an anterior one. The supraorbital foramens are large, the opisthocranium is short, and the trapezoidal shaped and grooved basioccipital bears strong anterior tuberosities.

★*Ouzoceros pentalophosi* Bouvrain, 1997

Type Material LGPUT PNT-259 (holotype), left horncore illustrated by Bouvrain (1997: Pl. I, fig. 3); LGPUT PNT-332 (allotype), female cranium illustrated by Bouvrain (1997: Pl. I, fig. 1).

Type Locality Pentalophos, Axios Valley, Greece, Upper Miocene.

Age Late Miocene, late early Vallesian, (end of MN-9) ca. 9.7 Ma.

Distribution Known only from the type locality. This species could be possibly known from Xirochori-1 (MN10) as well.

Remarks This species differs from the type species of the genus in the slightly longer horncores, bearing a strong anterior keel and a weaker posterior one and the smaller supraorbital pits. The face is long, the ante-orbital fossa large but shallow and the long premolars long compared to the molars. Females appear hornless (Bouvrain 1997). Kostopoulos et al. (2020a) extend the occurrence of the species across late MN9 to early MN10.

★*Palaeoreas* Gaudry, 1861b

Type Species *Antilope lindermayeri* Wagner, 1848.

Distribution Late Miocene (Turolian) of Greece, Bulgaria and Turkey, and possibly in Iran as well.

Remarks Kostopoulos and Soubise (2018) lately revised the genus, considering *Palaeoreas zouavei* as a junior synonym of *Palaeoreas lindermayeri* and recognizing the genus as monotypic. Gaudry (1861b, 1862–1867) originally placed the taxon in Bovinae as sister to *Oreas* (i.e., a synonym at that time of the African Tragelaphini *Taurotragus*). Gentry suggested *P. lindermayeri* as more closely related to Antilopinae and place it initially in Antilopini and later in Caprini as a probable ovibovine (Gentry 1971 and references therein). Bouvrain (1992), in turn, highlights some common characters with Reduncini, whereas more recently

Geraads et al. (2003) conclude that it represents a distinct extinct taxon of unknown phylogenetic relationships with the extant ones.

★*Palaeoreas lindermayeri* (Wagner, 1848) (= ★*Palaeoreas zouavei* Bouvrain, 1980, part)

Nomenclatural and Taxonomic History *Antilope lindermayeri* in Wagner 1848 (new species); *Palaeoreas lindermayeri* in Gaudry 1861b (new combination).

Type Material BSPM #530 (holotype), right horncore illustrated by Wagner (1848: Pl. 12, fig. 5).

Type Locality Pikermi, Greece, Upper Miocene.

Age Late Miocene, middle Turolian (MN12) ca. 7.3 Ma.

Distribution Late Miocene (Turolian) of Greece, Bulgaria, Turkey, and perhaps in Iran. In particular, other Greek occurrences include those from Nikiti 2 (MN11), Ravin des Zouaves 5 (MN11), Vathylakkos 1, 3 (MN12), Thermopigi (MN12), Halmyropotamos (MN12), Chomateri (MN12), Kerasia (MN12), and Dytiko 1 (MN13).

Remarks *Palaeoreas lindermayeri* is a small to moderate bovid with moderately long and massive horncores heteronymously twisted on a straight axis, closely spaced on frontals and moderately diverging toward the tips with a sharp posterior keel descending caudo-laterally and a variable but always less strong anterior keel. The teeth are mesodont and the premolars relatively long compared to the molars. The species shows a mainly peri-Aegean distribution.

★*Protragelaphus* Dames, 1883

Type Species *Protragelaphus skouzesi* Dames, 1883, by original designation.

Distribution Late Miocene (Turolian) of Greece, Bulgaria, Northern Macedonia, Turkey, and Iran.

Other Taxa Included *Pr. theodori* Bouvrain, 1978.

Remarks Originally suggested as Tragelaphini (Dames 1883), the genus was later attributed to Antilopinae (Pilgrim and Schaub 1939; Gentry 1971). Bouvrain (1978, 1992) mentioned, however, that *Protragelaphus* differs significantly from Antilopini and approaches better *Aepyceros* and especially Alcelaphini and Caprini, with which it appears to share several apomorphic features.

★***Protragelaphus skouzesi*** **Dames, 1883**

Type Material MNB Ma.42647 (holotype), frontlet (Fig. 26).

Type Locality Pikermi, Greece, Upper Miocene.

Age Late Miocene, middle Turolian (MN 12) ca. 7.3 Ma.

Distribution Late Miocene (middle Turolian) of Greece, Turkey, Iran, ?Ukraine. In particular, other Greek occurrences include those from Chomateri (MN12), Kerasia (MN12), Halmyropotamos (MN12), and Samos (unknown level).

Remarks Dames (1883) originally described but not figured a frontlet from Pikermi stored in the Museum für Naturkunde, Berlin as *Protragelaphus skouzesi*.

Fig. 26 The recovered holotype (MNB Ma.42647) of *Protragelaphus skouzesi* Dames, 1883, in frontal view. Scale bar equals 5 cm. Photograph courtesy of MNB, F Bibi and Th Schossleitner

Pilgrim and Hopwood (1928: 88) suggested taking Wagner's figure of *Antilope lindermayeri* (Wagner 1857: Pl. VII, fig. 18) as the paratype of the species, though the illustration itself shows only the palate in occlusal view. A recent search in the collections of MNB (F. Bibi and Th. Schossleitner pers. com. 2018) revealed among the restricted Pikermi material stored in the Museum, a single frontlet of *Protragelaphus skouzesi* that perfectly fits Dames' descriptions. I suggest, therefore, taking the specimen MB Ma.42647 as the recovered holotype of the species. Emended diagnoses of *Protragelaphus skouzesi* were provided by Gentry (1971) and Bouvrain (1978, 1992). The species is medium sized, characterized by relatively long, weakly compressed mediolaterally and constantly divergent from the base up horncores, heteronymously twisted along a straight to closely spiraled axis, and having a strong posterior keel descending caudolaterally.

★*Protragelaphus theodori* Bouvrain, 1978

Type Material LGPUT DTK-059X (holotype), cranium illustrated by Bouvrain (1978: Pl. I).

Type Locality Dytiko-1, Axios Valley, Greece, Upper Miocene.

Age Late Miocene, late Turolian (MN13).

Distribution Known only from the type locality.

Remarks Bouvrain (1978) diagnosed *Protragelaphus theodori* as a taxon larger than the type species, with less twisted horncores set closely to each other at their bases, longer face, grooved basioccipital with strong anterior tuberosities extended caudally by crests, more raised interfrontal suture, larger supraorbital foramina, weaker to absent postcornual fossae, and shorter premolars compared to the molars. The species is known from the Dytiko-1 site by three complete and six partial crania. A re-examination of the differential features of this species suggest that neither the insertion of the horncores on the frontals nor the reduction of the premolar row separate it satisfactorily from *Protragelaphus skouzesi*. Instead, it shows less elevated frontals and a weaker torsion with the closely spiraled horncores running in parallel in their distal part. The duct from the supraorbital foramina ends to the lachrymal fossa instead of the orbit itself. Bouvrain (1978) also referred to *Protragelaphus theodori* a single partial cranium LGPUT DKO-217X from Dytiko-3. This specimen shows, however, significantly more massive horncores and a higher braincase that *Pr. theodori* from Dytiko-1, the occipital plane slopes more abruptly on the dorsal parietal surface, and the basioccipital has no medial groove, differences that may deserve a distinction at the species level.

***Procobus* Khomenko, 1913**

Type Species *Procobus melania* Khomenko, 1913 by original designation.

Distribution Late Miocene (?Turolian) of Moldavia, Ukraine, and Greece.

Remarks As the name of the genus itself implies, the taxon was originally taken as a renduncine, due to the concave front edge of the horncores, but Gentry et al. (1999) and Gentry (2000b) already questioned such an affiliation. Khomenko (1913) described from the same locality (Taraklia, Moldova) two distinct species: *Procobus melania,* based on a well-preserved frontlet, and *Procobus brauneri*, based on a seriously damaged and partially preserved cranium. Original comparison of the two specimens (ONU#1974 and ONU#2391, respectively, Odessa National University and Museum of Paleontology) suggest close resemblance and I would include both specimens into the same taxon: *Procobus melania*. ONU#2391 shows rather thinner horncores compared to the teeth size, more rounded, and more widely apart on the frontals that ONU#1974 features that may be related to sexual dimorphism. Gentry et al. (1999) and Gentry (2000b) suggest possibly including into the same taxon *Moldoredunca amalthea* Lungu, 1979, but I think that horncores related to this species represent females of *Miotragocerus*. Basic cranial features of *Procobus* are summarized in Gentry (2000b). *Procobus* was unknown from Greece, but a recently discovered cranium and a partially preserved frontlet from the late Turolian fauna of Dytiko-1 (DTK, Axios Valley, Greece) can be definitely attributed to this genus (pers. data).

Pontoceros Vereshchagin et al., 1971

Type Species *Pontoceros ambiguus* Vereshchagin et al., 1971, by original designation.

Distribution Early–Middle Pleistocene of Ukraine, Moldova, Georgia, and Greece.

Other Taxa Included *Po. surprine* Vekua, 2012.

Remarks *Pontoceros* represents one of the very few Pleistocene Eurasian bovids with homonymously twisted horncores. Crégut-Bonnoure and Guérin (1996) rely the sparsely documented *Pontoceros* to Tragelaphini, but later studies suggest closer relations to Caprini and Oiocerina (e.g., Kostopoulos 2006b; Vekua 2012). Although Vekua (2012) refers the homonymously spiral horned antelope from Dmanisi to a new species of *Pontoceros*, he clearly states (idem: 142–143) that this decision is rather unlikely. This taxon is known in Greece from Libakos (?MNQ19), Tsiotra Vryssi (MNQ19), Platanochori (MNQ20), and Apollonia (MNQ20).

★*Pontoceros ambiguus mediterraneus* Kostopoulos, 1997b

Type Material LGPUT APL-39 (holotype), frontlet (Kostopoulos 1997b: fig. 14; Fig. 27).

Type Locality Apollonia-1, Mygdonia basin.

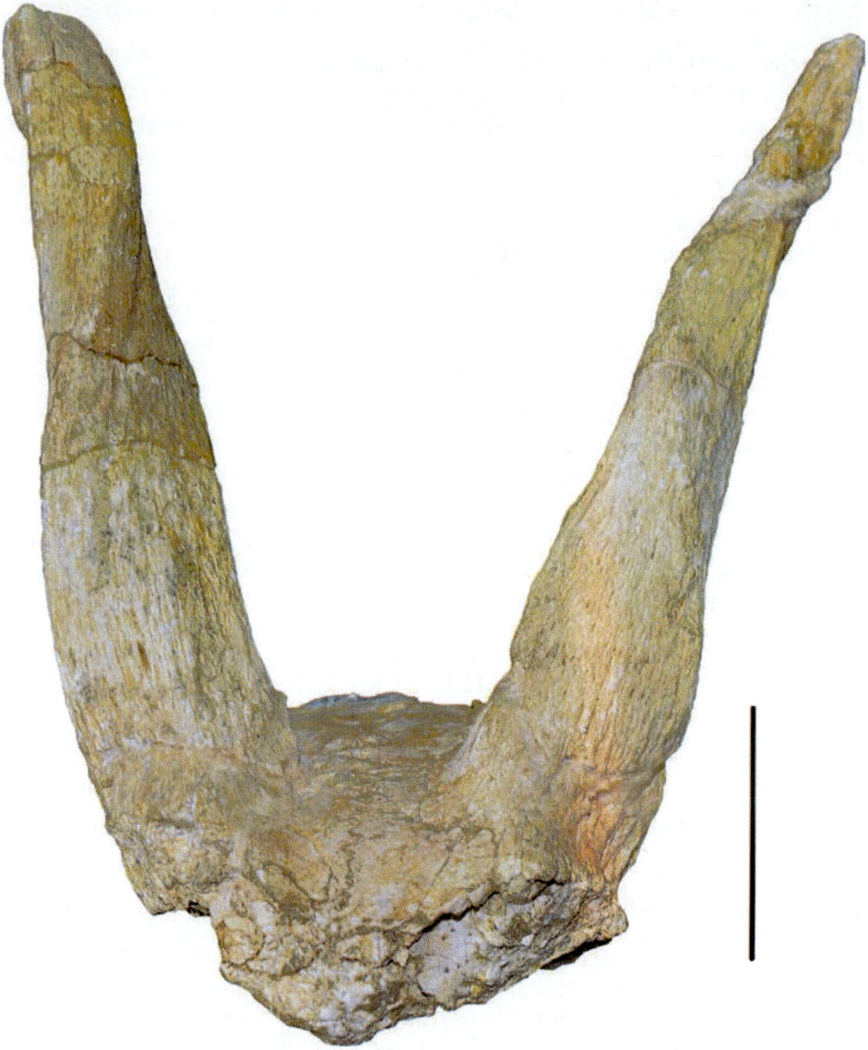

Fig. 27 The holotype frontlet (LGPUT APL-39) of *Pontoceros ambiguus mediterraneus* Kostopoulos, 1997b, from Apollonia, in frontal view. Scale bar equals 5 cm

Age Late Early Pleistocene (MNQ20).

Distribution Known only from the type locality.

Remarks *Pontoceros ambiguus* is known from four Greek fossil faunas spanning the entire Early Pleistocene (Steensma 1988; Kostopoulos 1997b; Konidaris et al. 2015; pers. data). Kostopoulos (1997b) distinguishes the Apollonia population from the north Black Sea one at the subspecies level, based on the more massive and looser twisted horncores with less sharp keels and less triangular basal cross-section.

Species inquirenda

★*Bubalus marathousae* Sickenberg, 1976

Type Material Unknown hosting institution; the holotype is a left horncore illustrated by Sickenberg (1976: Pl. V, figs 1–4), associated by parts of a cranium.

Type Locality Marathousa–Megalopolis.

Age Middle Pleistocene.

Distribution Known only from the type locality.

Remarks A partially preserved cranium, a mandibular fragment and a few postcranials are the basis on which Sickenberg (1976) established his new species of water buffalo *Bubalus marathousae* from the Middle Pleistocene lake deposits of Megalopolis in Peloponnesus. The taxon has received little attention so far. According to the original diagnosis by Sickenberg (1976), the species is moderately large with comparatively short, robust horncones that do not curve distally; basal part of the horncores on the dorsal frontal level; horncore cross-section at mid-length equilateral triangular; dorsal horncore surface slightly convex; anterior proximal surface marked by a blunt keel; no posterior keel; rounded apexes. von Koenigswald et al. (2019) comment that although the described cranial specimen may well-represent *Bubalus* is insufficient for species identification. Hence, I propose treating *B. marathousae* as a *species inquirenda*.

★*Oioceros*? *proaries* Schlosser, 1904

Type Material A hornless partial cranium described and figured by Schlosser (1904: 73, Pl. XIII, figs 10, 11, 13). The specimen is preserved in BSPM without catalog number.

Type Locality Samos (unknown level).

Remarks The specimen represents most likely a female of another Samos taxon. Pilgrim and Hopwood (1928) believed it may be a female of the *Oioceros* group taxa recorded at Samos, whereas Solounias (1981) regarded it as a possible female of *Sporadotragus parvidens*.

★*Prodamaliscus gracilidens* Schlosser, 1904

Type Material A female cranium described and figured by Schlosser (1904: 73, Pl. XIII, figs 10, 11, 13). According to Solounias (1981), this cranium destroyed during the Second World War.

Type Locality Samos (unknown level).

Remarks Schlosser (1904) suggested affinities to Alcelaphini and African *Damaliscus*. Solounias (1981), on the other hand, hypothesized synonymy with *Samokeros*.

6 Conclusions

Past bovid biodiversity of Greece was enormously higher than today, including 83 species and 52 genera at least spanning from Orleanian (Middle Miocene) to latest Pleistocene. The Greek fossil bovid record is, however, unevenly distributed through time. Not much is known for the Middle Miocene bovids of Greece and the same is rather true for the Early–Middle Pliocene ones as well. The majority of the 80 fossil localities that contain Greek fossil bovids are concentrated in two main chronological windows: the Late Miocene and the Early Pleistocene (Appendix). Furthermore, most of pre-Mid-Pleistocene findings come from open sites usually related to fluviolacustrine depositional environments, whereas later records are mostly from cave deposits.

The peak of the Greek bovid biodiversity took place certainly during the Turolian, reaching a maximum of 8–10 species per locality between 8 and 7 Ma. Tragoportacins, a group of bovids closely related to extant boselaphins are represented in the Greek record by five species, whereas bovines are known by seven species. Antilopinae are represented by 19 antilopin species, 10 oiocerins, and 23 caprins, whereas the tribal affinities of a significant number of other taxa (15 genera; 20 species) remains debatable. Fifty eight (58) of the recorded species are named from Greek type localities mostly from Pikermi, Samos, and Axios Valley, and twenty-three species (23) represent genotypes. Three more species named from Greek type localities are extremely poorly known and need further investigation. The relatively high number of monotypic genera (14) is rather indicative of the low phylogenetic resolution attained until now, than real generic distinction.

Among bovid genera recorded in Greece, *Gazella* s.l. shows the widest temporal distribution with a fairly continuous record from the late Vallesian to the late Villafranchian (ca. 7 My). From a geographic point of view, most Miocene taxa recorded in Greece show a principally west-Eurasian distribution controlled to the North by the Alpine Orogene. During Pliocene and Pleistocene times, however, Greek bovids seem to belong to a more restricted European/south European zoogeographic unit, though Asiatic influences were always present as may be expected from a territory at the junction of East and West. Bovids of undisputed African affinities are not recorded in the Greek fossil record; on the contrary, several scholars suggest that some African late Neogene/Quaternary lineages may in fact originate from the wealth Late Miocene bovid stock of Greece and surroundings.

Acknowledgments I am grateful to all colleagues and staff of Natural History Museums and other Institutions that I have had the opportunity to visit over the years. Special thanks are due to Dr. F. Bibi and Th. Schossleitner for helping me locate the holotype of *Protragelaphus skouzesi* in

NMB during this work. I would also like to thank S. Roussiakis, A. Athanassiou, G. Iliopoulos for fruitful discussions, and D. Geraads for carefully reading the manuscript and providing valuable suggestions.

Appendix

List of the Greek localities containing fossils bovids. Locality numbers refer to the collection numbers of the PaleoBiology Database (PBDB). Type localities and species/subspecies based on Greek fossil material are marked in bold

Localities[PBDB No]	Age (ELMA; MNQ)	Taxa	Refs.
Agios Georgios Cave[207192]	latest Pleistocene	*Bos primigenius*	1
Kataphyto Cave, Drama	latest Pleistocene	*Bos primigenius*	2
Franchthi Cave[182709]	latest Pleistocene	*Bos primigenius*, *Capra* sp.	3
Loutra Almopias Cave[203847]	Late Pleistocene	*Bos primigenius*, *Capra ibex*,	4
Klisoura Cave[184243]	latest Pleistocene	*Bos primigenius*, *Rupicapra rupicapra*, *Capra* cf. *ibex*	5
Kalamakia Cave[184245]	latest Pleistocene	*Bos primigenius*, *Capra ibex*	6
Petralona Cave[183123]	Late Pleistocene	*Bos primigenius*, *Bison priscus*, ***Capra ibex macedonica***, ***Pliotragus macedonicus***	7, 8
Penios riverbank[204397]	Late Pleistocene	*Bos primigenius*, *Bubalus* sp., *Capra ibex*	9, 10
Ambelia[204557]	Middle Pleistocene	*Bos primigenius*	11
Marathousa 1, Megalopolis [187637]	Middle Pleistocene	*Bison* sp.	12
Marathousa, Megalopolis	Middle Pleistocene	***Bubalus "marathousae"***	13,a
Kyparissia 1, 4[194472, 194475]	Middle Pleistocene	*Bison* sp.	14
Apollonia 1[34784]	MNQ20	*Bison* cf. *degiulii*, *Praeovibos mediterraneus*, ***Soergelia brigittae***, *Ovis* sp., ***Pontoceros ambiguus mediterraneus***	15–17
Kalamoto-1[200083]	MNQ20	*Bison* cf. *degiulii*	17
Platanochori-1[204658]	MNQ20	*Bison* sp.	18

(continued)

Localities[PBDB No]	Age (ELMA; MNQ)	Taxa	Refs.
Kalamoto-2[200084]	MNQ19–20	cf. *Leptobos etruscus*, *Bison* cf. *degiulii*	17, 19
Tsiotra Vryssi[197943]	MNQ19–20	*Leptobos* cf. *etruscus*, *Bison* cf. *degiulii*	17, 18
Kaiafas[34766]	MNQ19	*Gazella* sp.	20
Livakos[34764]	MN18–19	*Leptobos* sp., *Pontoceros ambiguus*	15, 21
Polylakkos[34763]	MNQ18–19	*Leptobos* sp.	20
Krimni[34762]	MNQ18–19	*Bison* cf. *degiulii*, *Gazellospira* cf. *torticornis*	17,20
Vassiloudi 1[34651]	MNQ18	*Leptobos etruscus*, *Procamptoceras* sp., *Gazella* cf. *bouvrainae*	20
Gerakarou 1[34617]	MNQ18	*Leptobos* cf. *etruscus*, *Gazellospira* sp., ***Gazella bouvrainae***, ***Antilope koufosi***	22, 23
Haliakmon Q[34812]	MNQ17–20	*Leptobos* cf. *etruscus*	20
Volax[34593]	MN17	cf. *Leptobos* sp., ? *Procamptoceras* sp., *Gazellospira* cf. *torticornis*, ***Gallogoral meneghinii sickenbergi***, *Gazella* sp.	24
Vatera DS[183344]	MN17	cf. *Leptobos* sp., aff. *Gazella borbonica*	25
Vatera F[183341]	MN17	cf. *Leptobos* sp., cf. *Gazellospira torticornis*, aff. *Gazella borbonica*, cf. *Gazella bouvrainae*	25
Pyrgos[34591]	MN17	*Leptobos* sp., *Gazellospira* cf. *torticornis*	20
Sesklon[34614]	MN17	*Gazellospira torticornis*, ***Euthyceros thessalicus***, *Gallogoral meneghinii*, *Gazella borbonica*, *Gazella bouvrainae*, ***Gazella aegaea***	10, 26, 27
Dafnero[345594]	MN17	*Gazellospira torticornis*, *Gallogoral* cf. *meneghinii*, *Gazella bouvrainae*	20
Kos	MN17	*Leptobos* cf. *etruscus*, *Gazella borbonica*	20
Karnezeika[202122]	MN17	*Pliotragus* sp., *Gazellospira torticornis*, *Gallogoral meneghinii*, *Gazella* sp.	28, a
Kardamena[204662]	MN16–17	*Leptobos elatus*	29, a

(continued)

Localities[PBDB No]	Age (ELMA; MNQ)	Taxa	Refs.
Sesklon lower level[205043]	MN16	*Gazellospira torticornis*, *Gazella* cf. *bouvrainae*	27
Damatria[34591]	MN16	*Leptobos* sp.	20
Milia[185859]	MN16	*Alephis* sp., ***Grevenobos antiquus***, *Gazella borbonica*	30
Apolakkia-2[34607]	MN15	*Gazella* sp.	31
Megalo Emvolon[36579]	MN15	***Parabos macedoniae***, ***Koufotragus bailloudi***, *Gazella borbonica*	32, 33
Gefira-1[182685]	MN15	***Parabos savelisi***	34
Maramena[32189]	MN13/14	Tragoportacini indet., *Gazella* sp., ?*Protragelaphus* sp., ***Norbertia hellenica***	35, a
Samos, unknown level: **NHMW collection[182751]** **NHML collection[202120]** **BSPM collection[207137]** **IGPM collection[207193]**	MN11–13	***Miotragocerus parvidens***, ***Tragoportax rugosifrons***, ***Samokeros minotarus***, ***Criotherium argalioides***, ***Urmiatherium rugosifrons***, ***Samotragus crassicornis***, ***Majoreas woodwardi***, ***Samodorcas kuhlmani***, ***Paraoioceros wegneri***, ***Gazella pilgrimi***, *Gazella mytilinii*, *Gazella ancyrensis*, ***Prostrepsiceros zitteli***, ***Prostrepsiceros fraasi***, ***Protoryx capricornis***, *Protragelaphus skouzesi*, ***Skoufotragus schlosseri***, ***Skoufotragus laticeps***, ***Palaeoryx majori***, ***Tragoreas oryxoides***, **"*Oioceros*? *proaries*"**, **"*Prodamaliscus gracilidens*"**	36–41, 64, a
Alifakas[207134]	MN11–13	*Tragoportax amalthea*? , *Palaeoryx* sp.	42, 43
Triada[207135]	MN11–13	*Tragoportax amalthea*, *Palaeoryx pallasi*	43, 44
Achladion[207133]	MN11–13	*Tragoportax amalthea*, *Palaeoryx* cf. *pallasi*	43, 44
Pyrgos Vassilissis[195555]	MN11–13	*Miotragocerus macedoniensis*, cf. *Tragoportax amalthea*, *Gazella* sp.	45
Ravin R	MN11–13	*Tragoportax amalthea*	43

(continued)

Localities[PBDB No]	Age (ELMA; MNQ)	Taxa	Refs.
Dytiko-3[32376]	MN13	*Protragelaphus* sp., ***Hispanodorcas orientalis***, *Gazella schlosseri*, ***Dytikodorcas longicornis***	46, 47
Dytiko-2[32375]	MN13	*Tragoportax* sp., *Miotragocerus macedoniensis*, *Palaeoreas lindermayeri*, *Palaeoryx* sp., *Procobus* sp., *Gazella deperdita*	46–48, a
Dytiko-1[32374]	MN13	***Miotragocerus macedoniensis***, *Tragoportax* sp., *Palaeoreas lindermayeri*, ***Protragelaphus theodori***, *Gazella deperdita*, *Majoreas woodwardi*, ? *Palaeoryx* sp.	47–49
Kryopigi[157582]	MN12–13	*Tragoportax amalthea*, *Miotragocerus* sp., *cf. Helladorcas* sp., *Nisidorcas planicornis*, *Prostrepsiceros axiosi*, *Gazella capricornis*, *Gazella pilgrimi*, *Skoufotragus* cf. *laticeps*	51
Samos Q5[95690]	MN12/13	*Tragoportax rugosifrons*, *Miotragocerus parvidens*, *Gazella pilgrimi*, *Gazella capricornis*, *Gazella mytilinii*, *Prostrepsiceros zitteli*, *Sporadotragus parvidens*, *Skoufotragus schlosseri*, *Palaeoryx* sp.	39, 40
Thermopigi[73553]	MN12	*Tragoportax* cf. *amalthea*, *Miotragocerus valenciennesi*, *Palaeoreas lindermayeri*, ? *Criotherium* sp., *Prostrepsiceros* cf. *rotundicornis*, *Nisidorcas planicornis*, *Gazella* sp., *Sporadotragus*? sp.	52, a
Kerassia-4[195435]	MN12	*Gazella* cf. *capricornis*	53, a
Kerassia-1[195432]	MN12	*Tragoportax* cf. *amalthea*, *Gazella* cf. *deperdita*	53

(continued)

Localities[PBDB No]	Age (ELMA; MNQ)	Taxa	Refs.
Halmyropotamos[202213]	MN12	*Tragoportax amalthea*, *Palaeoreas lindermayeri*, *Protragelaphus skouzesi*, *Palaeoryx pallasi*, *Palaeoryx majori*, *Majoreas woodwardi*, *Prostrepsiceros rotundicornis*, *Gazella pilgrimi*	54
Chomateri[195562]	MN12	*Miotragocerus valenciennesi*, *Prostrepsiceros rotundicornis*, *Palaeoreas lindermayeri*, *Protragelaphus skouzesi*	55
Pikermi Valley-3[202631]	MN12	*Palaeoreas lindermayeri*, *Tragoportax amalthea*, *Gazella capricornis*	56
Pikermi Valley-1[202630]	MN12	*Palaeoreas lindermayeri*, *Protragelaphus skouzesi*, *Palaeoryx pallasi*, *Tragoportax amalthea*, *Gazella capricornis*	56
Pikermi[182754]	MN12	***Miotragocerus valenciennesi***, ***Palaeoreas lindermayeri***, ***Palaeoryx pallasi***, ***Tragoportax amalthea***, ***Sporadotragus parvidens***, ***Oioceros rothii***, ***Prostrepsiceros rotundicornis***, ***Gazella capricornis***, ***Protragelaphus skouzesi***, ***Protoryx carolinae***	37, 58–65
Ravin X[182745]	MN12	*Tragoportax amalthea*, *Palaeoryx pallasi*, *Prostrepsiceros rotundicornis*, *Oioceros rothi*, *Gazella pilgrimi*	32, 65
Perivolaki[194879]	MN12	*Tragoportax amalthea*, *Miotragocerus parvidens*, *Gazella* cf. *pilgrimi*, *Gazella* sp., *Nisidorcas planicornis*, *Prostrepsiceros* cf. *rotundicornis*, cf. *Helladorcas* sp., *Skoufotragus* aff. *laticeps*, *Palaeoryx* aff. *pallasi* ***Pheraios chryssomallos***	66, a

(continued)

Localities[PBDB No]	Age (ELMA; MNQ)	Taxa	Refs.
Prochoma-1[202222]	MN12	*Miotragocerus parvidens*, *Nisidorcas planicornis*, *Prostrepsiceros axiosi*, *Gazella pilgrimi*	67, a
Vathylakkos-3[182750]	MN12	*Tragoportax amalthea*, *Miotragocerus* sp., *Palaeoreas lindermayeri*, *Nisidorcas planicornis*, *Prostrepsiceros axiosi*, *Helicotragus rotundicornis*, *Gazella capricornis*, *Gazella pilgrimi*, *Protoryx carolinae*	32, 65, 67
Vathylakkos-2[202703]	MN12	*Tragoportax amalthea*, *Nisidorcas planicornis*, *Gazella* sp.	65, 67
Samos Mytilinii 1A[202215] (MTLA) Samos Mytilinii 1B[202216] (MTLB)	MN12	*Miotragocerus vallenciennesi*, *Tragoportax rugosifrons*, *Gazella* cf. *capricrnis*, *Gazella pilgrimi*, *Gazella mytilinii*, *Sporadotragus parvidens*, *Palaeoryx pallasi*, *Palaeoryx majori*, *Skoufotragus laticeps*, *Urmiatherium rugosifrons*	40
Samos Q1[95691]	MN12	*Miotragocerus vallenciennesi*, *Miotragocerus parvidens*, *Tragoportax rugosifrons*, *Gazella mytilinii*, *Sporadotragus parvidens*, *Skoufotragus laticeps*, *Criotherium argalioides*	39, 40
Samos Q4[95689]	MN12	*Tragoportax rugosifrons*, *Sporadotragus parvidens*, *Skoufotragus laticeps*	39, 40
Samos **Mytilinii-3**[202219] **(MYT)**	MN12	*Gazella pilgrimi*, *Majoreas*? sp., *Sporadotragus parvidens*, *Palaeoryx*? sp., ***Skoufotragus zemalisorum***	40
Ravin C[182749]	MN11/12	*Miotragocerus parvidens*, *Prostrepsiceros rotundicornis*, *Gazella pilgrimi*	32, 65, a
Samos Mytilinii-4 (MLN)	MN11/12	*Miotragocerus* sp., *Tragoportax* sp., *Gazella pilgrimi*, *Palaeoryx*? sp.	40

(continued)

Localities[PBDB No]	Age (ELMA; MNQ)	Taxa	Refs.
Samos Q2	MN11/12	*Tragoportax* sp., *Miotragocerus vallenciennesi*	39
Samos Q6[206462]	MN11/12	*Tragoportax* sp., *Gazella* sp., *Majoreas* sp., *Criotherium argalioides*, *Palaeoryx* sp., *Protoryx capricornis*	39, 40
Samos Qx	MN11	*Sporadotragus parvidens*, *Gazella pilgrimi*	39, 40
Ravin de Zouaves-5[195489]	MN11	*Tragoportax amalthea*, *Miotragocerus parvidens*, *Palaeoreas lindermayeri*, *Nisidorcas planicornis*, *Prostrepsiceros rotundicornis*, ***Prostrepsiceros axiosi***, *Gazella pilgrimi*, ***Stryfnotherium exophthalmon***	48, 65–68, a
Nikiti-2[73869]	MN11	*Tragoportax amalthea*, *Miotragocerus parvidens*, *Palaeoreas lindermayeri*, *Nisidorcas planicornis*, *Gazella* cf. *capricornis*, *Gazella pilgrimi*, *Palaeoryx* cf. *pallasi*	69, a
Platania[182682]	MN10/11	cf. *Palaeoreas* sp., *Qurliqnoria* sp., ***Palaeoryx minor***, *Prostrepsiceros* aff. *syridisi*, *Miotragocerus* sp., *Gazella* cf. *ancyrensis*, *Gazella* sp.	70, 71
Nikiti-1[202729]	MN10	*Miotragoceros* sp., ***Prostrepsiceros syridisi***, *Hispanodorcas* cf. *orientalis*	69, 72, 73
Fourka[202330]	MN10	***Urmiatherium kassandriensis***	74
Ravin de la Pluie[191070]	MN10	*Palaeoryx* sp., ***Prostrepsiceros vallesiensis***, ***Mesembriacerus melentisi***, ***Samotragus praecursor***	76–77
Ravin de Zouaves-1[182746]	MN10	*Mesembriacerus melentisi*, *Samotragus praecursor*, ***Ouzocerus gracilis***	78, 79
Xirochori	MN10	"*Protoryx*" sp., *Gazella* sp., *Ouzocerus* cf. *pentalophosi*	80

(continued)

Localities[PBDB No]	Age (ELMA; MNQ)	Taxa	Refs.
Pentalophos-1 202119	MN9	*Miotragocerus* sp., "*Protoryx*" sp., *Gazella* sp., ***Helladorcas geraadsi***, ***Ouzoceros pentalophosi***	[81, a]
Thymiana[182752]	MN5	*Eotragus* sp., *Kubanotragus* cf. *gaopoensis*, *Tethytragus* cf. *koehlere*	[79]
Antonios[73861]	MN5	*Eotragus* sp.	[82]
Aliveri[68032]	MN3	*Eotragus* cf. *artenensis*	[83]

ELMA European Land Mammal Age, *MN* Mammal Neogene Zone, *MNQ* Mammal Neogene–Quaternary Zone; *Refs* Main references

[1]Tsoukala (1992), [2]Samartzidou et al. (in press), [3]Stiner and Munro (2011), [4]Tsoukala et al. (2006), [5]Starkovich (2012), [6]Harvati et al. (2013), [7]Tsoukala (1989), [8]Crégut-Bonnoure and Tsoukala (2005), [9]Boessneck (1965), [10]Athanassiou (2001), [11]Tsoukala and Lister (1998), [12]Konidaris et al. (2018), [13]Sickenberg (1976), [14]Athanassiou (2018a), [15]Kostopoulos (1997b), [16]Crégut-Bonnoure (2002), [17]Kostopoulos et al. (2018), [18]Konidaris et al. (2015), [19]Tsoukala and Chatzopoulou (2005), [20]Koufos (2001), [21]Steensma (1988), [22]Kostopoulos and Athanassiou (1997), [23]Kostopoulos (1998), [24]Kostopoulos (1997a), [25]de Vos et al. (2002), [26]Athanassiou (2002a, b), [27]Athanassiou (2018b), [28]Kokotini et al. (2019), [29]Airaghi (1928), [30]Crégut-Bonnoure and Tsoukala (2017b), [31]van de Weerd et al. (1982), [32]Arambourg and Piveteau (1929), [33]de Bonis and Bouvrain (1996), [34]Crégut-Bonnoure and Tsoukala (2017a), [35]Köhler et al. (1995), [36]Schlosser (1904), [37]Andrée (1926), [38]Pilgrim and Hopwood (1928), [39]Solounias (1981), [40]Kostopoulos (2009a), [41]Sickenberg (1936), [42]Melentis and Schneider (1966), [43]Koufos (2006), [44]Mitzopoulos (1947), [45]Böhme et al. (2017), [46]Bouvrain and de Bonis (1988), [47]Bouvrain and de Bonis (2007), [48]Kostopoulos and Soubise (2018), [49]Bouvrain (1988), [50]Bouvrain (1978), [51]Lazaridis (2015), [52]Geraads et al. (2007), [53]Theodorou et al. (2003), [54]Melentis (1969), [55]Roussiakis (2009), [56]Roussiakis et al. (2014), [57]Wagner (1848), [58]Gaudry (1861a, b), [59]Wagner (1848), [60]Roth and Wagner (1854), [61]Wagner (1857), [62]Dames (1883), [63]Weithofer (1888), [64]Forsyth Major (1891a, b), [65]Koufos (2006), [66]Kostopoulos (2006a), [67]de Bonis et al. (1986), [68]Kostopoulos (2004), [69]Kostopoulos (2016), [70]Vasileiadis et al. (2019), [71]Kostopoulos et al. (2020b), [72]Kostopoulos and Koufos (1996), [73]Kostopoulos (2014), [74]Lazaridis et al. (2017), [75]Bouvrain (1975), [76]Bouvrain (1982), [77]Bouvrain and de Bonis (1985), [78]Bouvrain and de Bonis (1986), [79]de Bonis et al. (1988), [80]Kostopoulos et al. (2020a), [81]Bouvrain (1997), [82]Koufos and Syrides (1997), [83]van den Hoek Ostende et al. (2015)
[a]This study

References

Ajmone Marsan P, Garcia JF, Lenstra JA, the Global Division Consortium (2010) On the origin of cattle: how aurochs became cattle and colonized the world. Evol Anthropol 19:148–157

Alcalá L, Morales J (2006) Antilopinae (Bovidae, Mammalia) from the Lower Pliocene of Teruel Basin (Spain). Estud Geol-Madrid 62:559–570

Andrée J (1926) Neue Cavicornier aus dem Pliocän von Samos. Palaeontographica 67:135–175

Arambourg C, Piveteau J (1929) Les vertébrés du Pontien de Salonique. Ann Paleont 18:59–138

Athanassiou A (2002a) A new gazelle species (Artiodactyla, Bovidae) from the Late Pliocene of Greece. Ann Geol Pays Helleniques 39:299–310

Athanassiou A (2002b) *Euthyceros thessalicus*, a new bovid from the Late Pliocene of Sesklo (Thessaly, Greece). Neues Jahrb Geol P M 2:113–128

Athanassiou A (2018a) Pleistocene vertebrates from the Kyparíssia lignite mine, Megalopolis Basin, S. Greece: Rodentia, Carnivora, Proboscidea, Perissodactyla, Ruminantia. Quat Int 497:198–221

Athanassiou A (2018b) A Villafranchian *Hipparion*-bearing mammal fauna from Sésklo (E. Thessaly, Greece): implications for the question of *Hipparion–Equus* sympatry in Europe. Quaternary 1:12

Azanza B, Morales J (1994) *Tethytragus* nov. gen. et *Gentrytagus* nov. gen. deux nouveaux Bovidés (Artiodactyla, Mammalia) du Miocène moyen. Relations phylogénétiques des Bovidés anté-Vallésiens. Proc K Ned Akad Wet 97:249–282

Bai W, Dong W, Zhang L, Wei Q, Liu W, Chen Z (2019) New material of the Early Pleistocene spiral horned antelope Spirocerus (Artiodactyla, Mammalia) from North China and discussion on its evolution. Quat Int 522:94–102

Beintema JJ, Breukelman H, Dubois J-Y, Warmels H (2003) Phylogeny of ruminants secretory ribonuclease gene sequences of pronghorn (*Antilocapra americana*). Mol Phylogenet Evol 26:18–25

Bibi F (2013) A multi-calibrated mitochondrial phylogeny of extant Bovidae (Artiodactyla, Ruminantia) and the importance of the fossil record to systematics. BMC Evol Biol 13:166

Bibi F, Güleç ES (2008) Bovidae (Mammalia: Artiodactyla) from the Late Miocene of Sivas, Turkey. J Vertebr Paleontol 28:501–519

Bibi F, Bukhsianidze M, Gentry AW, Geraads G, Kostopoulos DS, Vrba ES (2009) The fossil record and evolution of Bovidae: state of the field. Paleontol Electronica 1(3):10A, 10p

Boehme M, Spassov N, Ebner M, Geraads D, Hristova L, Kirscher U, Koetter S, Linnemann U, Prieto J, Roussiakis S, Theodorou G, Uhlig G, Winklhofer M (2017) Messinian age and savannah environment of the possible hominin Graecopithecus from Europe. PLoS ONE 12(5): e0177347. https://doi.org/10.1371/journal. pone.0177347

Boessneck J (1965) Die jungpleistozänen Tierknochenfunde aus dem Peneiostal bei Larissa in Thessalien. In: Milojcic V, Boessneck J, Jung D, Schneider HL (eds) Paläolithikum um Larissa in Thessalien, Beiträge zur ur- und frühgeschichtlichen Archäologie des Mittelmeerkulturraumesl, vol 1, pp 42–69. Rudolf Habelt Verlag, Bonn.

Bohlin B (1937) Eine tertiäre Saugetier-Fauna aus Tsaidam. Palaeontol Sinica C 14:1–111

Bosscha-Erdbrink DP (1988) *Protoryx* from three localities East of Maragha, NW Iran. Proc K Ned Akad Wet 91:101–159

Bouvrain G (1975) Un nouveau bovidé du Vallésien de Macédoine, Grèce. C R Acad Sci Paris 280D:1357–1359

Bouvrain G (1978) *Protragelaphus theodori* n. sp. (Mammalia, Artiodactyla, Bovidae) du Miocène de Macédoine (Grèce). Geol Mediter 5:229–236

Bouvrain G (1979) Un nouveau genre de bovidé de la fin du Miocène. B Soc Geol Fr 21:507–511

Bouvrain G (1980) Le genre *Palaeoreas* (Mammalia, Artiodactyla, Bovidae), systématique et éxtension géographique. Palaeontolog Z 54:55–65

Bouvrain G (1982) Révision du genre *Prostrepsiceros* Major, 1891 (Mammalia, Bovidae). Palaeontolog Z 56:113–124

Bouvrain G (1988) Les *Tragoportax* (Bovidae, Mammalia) des gisements du Miocène supérieur de Dityko (Macédoine, Grèce). Ann Paleontol 74:43–63

Bouvrain G (1992) Antilopes à chevilles spiralées du Miocène supérieur de la province Gréco-Iranienne: nouvelles diagnoses. Ann Paleontol 78:49–65

Bouvrain G (1994) Un bovide du Turolien inferieur d'Europe orientale *Tragoportax rugosifrons*. Ann Paleontol 80:61–87

Bouvrain G (1996) Les gazelles du Miocène supérieur de Macédoine, Grèce. Neues Jahrb Geol Palaeontol Abh 199:111–132

Bouvrain G (1997) Les bovidés du Miocène supérieur de Pentalophos (Macèdoine, Grèce). Münchner Geowiss Abh 34:5–22

Bouvrain G (2001) Les Bovidés (Mammalia, Artiodactyla) des gisements du Miocène supérieur de Vathylakkos (Grèce du Nord). Neues Jahrb Geol Palaeontol Abh 220:225–244

Bouvrain G, de Bonis L (1984) Le genre *Mesembriacerus* (Bovidae, Artiodactyla, Mammalia): une oviboviné primitif du Vallésien (Miocène supèrieur) de Macédoine (Grèce). Palaeovertebrata 14:201–223

Bouvrain G, de Bonis L (1985) Le genre *Samotragus* (Artiodactyla, Bovidae), une antilope du Miocène supérieur de Grèce. Ann Paleontol 71:257–299

Bouvrain G, de Bonis L (1986) *Ouzocerus gracilis* ng, n. sp., bovidae (Artiodactyla, Mammalia) du Vallésien (Miocène supérieur) de Macédoine (Grèce). Geobios 19(5):661–667

Bouvrain G, de Bonis L (1988) Découverte du genre *Hispanodorcas* (Bovidae, Artiodactyla) dans le Turolien de Grèce septentrionale. Ann Paleontol 74:97–112

Bouvrain G, de Bonis L (2007) Ruminants (Mammalia, Artiodactyla : Tragulidae, Cervidae, Bovidae) des gisements du Miocène supérieur (Turolien) de Dytiko (Grèce). Ann Paleontol 93:121–147

Bouvrain G, Sen S, Thomas H (1995) *Parurmiatherium rugosifrons* Sickenberg, 1932, un Ovibovinae (Bovidae) du Miocène supérieur d'Injana (Djebel Hamrin, Irak). Geobios 28:719–726

Bukshianidze M, Koiava K (2018) Synopsis of the terrestrial vertebrate faunas from the Middle Kura Basin (Eastern Georgia and Western Azerbaijan, South Caucasus). Acta Palaeontol Polonica 63:441–461

Chen C-F, Zhang Z-Q (2007) Restudy of Chinese fossils referred to *Protoryx*. Vertebrat Palasiatic 4:98–109. (in Chinese with English summary)

Cherin M, D'Allestro V, Masini F (2019) New bovid remains from the early Pleistocene of Umbria (Italy) and a reappraisal of *Leptobos merlai*. J Mamm Evol 26:201–224

Crégut-Bonnoure E (1992) Intérêt biostratigraphique de la morphologie dentaire de *Capra* (Mammalia, Bovidae). Ann Zool Fenn 28:273–290

Crégut-Bonnoure E (2002) Les Ovibovini, Caprini et Ovini (Mammalia, Artiodactyla, Bovidae, Caprinae) du Plio-Pléistocène d'Europe : systématique, évolution et biochronologie. PhD Thesis Univ Lyon I, 1-2002, 430p

Crégut-Bonnoure E (2006) European Ovibovini, Ovini and Caprini (Caprinae, Mammalia) from the Plio-Pleistocene: new interpretations. In: Kahlke RD, Maul L, Mazza P (eds) Proceedings 18th SNG conference. Cour Forsch Senck, vol 256, pp 139–158

Crégut-Bonnoure E, Dimitrijevic V (2006) *Megalovis balcanicus* sp. nov. and *Soergelia intermedia* sp. nov. (Mammalia, Bovidae, Caprinae), new Ovibovini from the Early Pleistocene of Europe. Rev Paleobiol 25:723–773

Crégut-Bonnoure E, Guérin C (1996) V. Famille Bovidae. In: Guérin C, Patou-Mathis M (eds) Les grands mammifères Plio-Pléistocènes d'Europe. Masson, Paris, pp 62–106

Crégut-Bonnoure E, Tsoukala E (2005) The Pleistocene bovids from the Petralona cave (Macedonia, Greece): new interpretations and biogeographical implications. Quaternaire hs 2:161–178

Crégut-Bonnoure E, Tsoukala E (2017a) The Pliocene Artiodactyla and Proboscidea (Mammalia) from Gephyra (lower Axios Valley, Macedonia, Greece). Discovery of a new Boselaphine. Quat Int 445:200–214

Crégut-Bonnoure E, Tsoukala E (2017b) The late Pliocene Bovidae and Cervidae (Mammalia) of Milia (Grevena, Macedonia, Greece). Quat Int 445:215–249

Croft DA, Heaney LR, Flynn JJ, Bautista AP (2006) Fossil remains of a new, diminutive *Bubalus* (Artiodactyla: Bovidae: Bovini) from Cebu island. Philipp J Mammal 87:1037–1051

Dames WB (1883) Eine neue Antilope aus dem Pliocän von Pikermi in Attica. Sber Ges naturf Freunde Berlin 6:95–97

de Bonis L, Bouvrain G (1996) Une nouveau genre d'antilope dans le Ruscinien de Macédoine (Pliocène, Grèce). Neues Jahrb Geol Palaeontol M 11:692–700

de Bonis L, Koufos GD, Sen S (1998) Ruminants (Bovidae and Tragulidae) from the Middle Miocene (MN5) of the island of Chios, Aegean Sea (Greece). Neues Jahrb Geol Palaeontol Abh 210:399–420

Dmitrieva EL (2007) Caprinae (Bovidae, Artiodactyla, Mammalia) from the Neogene of Mongolia. Paleontolog J 41:671–682

Dmitrieva EL, Serdyuk NV (2011) Hippotraginae (Bovidae, Artiodactyla, Mammalia) from the late Miocene of Tuva. Paleontolog J 45:665–673

Duvernois M-P (1989) Les *Leptobos* (Mammalia, Artiodactyla) du Villafranchien. C R Acad Sci Paris 309:769–775

Duvernois M-P (1990) Les *Leptobos* (Mammalia, Artiodactyla) du Villafranchien d'Europe Occidentale. Doc Lab Geol Lyon 113:1–213

Duvernois M-P, Guérin C (1989) Les Bovidae (Mammalia, Artiodactyla) du Villafranchien supérieur d'Europe occidentale. Geobios 22:339–379

Forsyth Major CIF (1888) Sur un gisement d'ossements fossiles dans l'ile de Samos, contemporains de l' age de Pikermi. C R hebd Acad Sci Paris 107:1178–1181

Forsyth Major CIF (1891a) Considerations nouvelles sur la faune des vertebres du Miocene superieur dans l'ile de Samos. C R hebd Acad Sci Paris 113:608–610

Forsyth Major CIF (1891b) Sur l' age de la faune de Samos. C R hebd Acad Sci Paris 113:708–710

Forsyth Major CIF (1894) Le gisement ossifère de Mytilini et catalogue d' osséments fossiles. George Bridel & Cie (eds), Lausanne, pp 1–51

Fuss J, Prieto J, Böhme M (2015) Revision of the boselaphine bovid *Miotragocerus monacensis* Stromer, 1928 (Mammalia, Bovidae) at the Middle to Late Miocene transition in Central Europe. Neues Jahrb Geol Palaeontol Abh 276(3):229–265

Gaillard C (1902) Le belier de Mendes ou le mouton domestique de l'ancienne Egypte. B Soc Anthropol Biol Lyon 20:70–103

Garrido G (2008) La muestra más moderna y completa conocida de *Gazellospira torticornis* (Bovidae, Artiodactyla, Mammalia) en el Plioceno superior terminal de Europa occidental (Fonelas P-1, Cuenca de Guadix, Granada). In: Arribas A (ed) Vertebrados del Plioceno terminal en el suroeste de Europa: Fonelas P-1 y el proyecto Fonelas. Publ Inst Geol Min Espana, Cuadernos del Museo Geominero 10:413–460

Gaudry A (1861a) Résultats des fouilles enterprises en Grèce sous les auspices de l' Académie. C R hebd Acad Sci Paris 52:238–241. & 297–299

Gaudry A (1861b) Notes sur les antilopes trouvées à Pikermi (Grèce). Bull Soc Geol Fr 18:388–400

Gaudry A (1862–1867) Animaux fossiles et géologie de l'Attique. Savy, Paris

Gentry AW (1970) The Bovidae (Mammalia) of the Fort Ternan fossil fauna. In: Leakey LSB, Savage RJG (eds) Fossil vertebrates of Africa, vol 2. Academic Press, London, pp 243–323

Gentry AW (1971) The earliest goats and other antelopes from the Samos *Hipparion* Fauna. B Brit Mus (Nat Hist) Geol 20:229–296

Gentry AW (1992) The subfamilies and tribes of the family Bovidae. Mammal Rev 22:1–32

Gentry AW (2000a) The ruminant radiation. In: Vrba E, Schaller G (eds) Antelopes, deer and relatives. Yale University Press, New Haven, pp 11–25

Gentry AW (2000b) Caprinae and Hippotragini (Bovidae, Mammalia) in the Upper Miocene. In: Vrba E, Schaller G (eds) Antelopes, deer and relatives. Yale University Press, New Haven, pp 65–83

Gentry AW (2003) Ruminantia (Artiodactyla). In: Fortelius M, Kappelmann J, Sen S, Bernor R (eds) Geology and paleontology of the Miocene Sinap formation, Turkey. Columbia University Press, New York, pp 332–379

Gentry AW (2010) Bovidae. In: Werdelin L, Sanders WJ (eds) Cenozoic mammals of Africa. University California Press, Berkeley, pp 741–796

Gentry AW, Heinzmann P (1996) Miocene ruminants of the Central and Eastern Tethys and Paratethys. In: Bernor R, Fahlbusch V, Mittmann H-W (eds) The evolution of Western Eurasian Neogene mammal faunas. Columbia University Press, New York, pp 378–391

Gentry AW, Hooker P (1988) The phylogeny of the Artiodactyla. In: Benton MJ (ed) The phylogeny and classification of the Tetrapods, vol 2, Mammals. Systematics Association sp vol 35B, pp 235–272

Gentry AW, Rössner GE, Heizmann PJ (1999) Suborder ruminantia. In: Rössner G, Heissig K (eds) The Miocene land mammals of Europe. Verlag Dr. F Pfeil, Munich, pp 225–258
Geraads D (2003) Ruminants, other than Giraffidae from the middle Miocene hominoid locality of Çandir (Turkey). Cour Forsch Senck 240:181–199
Geraads D (2019) A reassessment of the Bovidae (Mammalia) from the Nawata Formation of Lothagam, Kenya, and the Late Miocene diversification of the family in Africa. J Syst Palaeontol 17:169–182
Geraads D, Bobe R (2020) Ruminants (Giraffidae and Bovidae) from Kanapoi. J Hum Evol 140:102383. https://doi.org/10.1016/j.jhevol.2017.08.006
Geraads D, Spassov N (2008) A new species of *Criotherium* (Bovidae, Mammalia) from the Late Miocene of Bulgaria. Hell J Geosci 43:21–27
Geraads D, Güleç E, Kaya T (2002) *Sinotragus* (Bovidae, Mammalia) from Turkey and the Late Miocene Asiatic Province. Neues Jahrb Geol Palaeontol M 8:477–489
Geraads D, Spassov N, Kovachev D (2003) *Palaeoreas lindermayeri* (Wagner, 1848) (Mammalia, Bovidae) from the upper Miocene of Bulgaria, and a revision of the species. Geodiversitas 25:405–415
Geraads D, Spassov N, Kovachev D (2006) A new *Sporadotragus* (Bovidae, Mammalia) from the late Miocene of Bulgaria. Riv Ital Paleontol S 112:473–479
Geraads D, Bobe R, Manthi FK (2013) New ruminants (Mammalia) from the Pliocene of Kanapoi, Kenya, and a revision of previous collections, with a note on the Suidae. J African Earth Sci 85:53–61
Geskos A (2013) Past and present distribution of the genus *Capra* in Greece. Acta Theriol 58:1–11
Gromolard C (1980) Une nouvelle interprétation des grands Bovidae (Artiodactyla, Mammalia) du Pliocène d'Europe occidentale classés jusqu'à présent dans le genre *Parabos*: *Parabos cordieri* (de Christol) emend.,?*Parabos boodon* (Gervais) et *Alephis lyrix* n. gen., n. sp. Geobios 13:767–775
Gromolard C (1981) Les grands Bovidae (Artiodactyla, Mammalia) du Pliocène d'Europe Occidentale. Biostratigraphie, Paleoneurologie, Systématique, Evolution, Paléoécologie. PhD thesis, Université Claude-Bernard Lyon
Gromolard C, Guérin C (1980) Mise au point sur *Parabos cordieri* (de Christol) un Bovidé (Mammalia, Artiodactyla) du Pliocène d'Europe occidentale. Geobios 13:741–755
Groves C, Grubb P (2011) Ungulate taxonomy. The John Hopkins University Press, Baltimore
Guérin C (1965) *Gallogoral* (nov. gen.) *meneghinii* (Rutimeyer, 1878) un rupicapriné du Villafranchien d'Europe occidentale. Doc Lab Geol Lyon 11:1–353
Harris JM, Leakey MG, Cerling TE, Winkler AJ (2003) Early Pliocene tetrapod remains from Kanapoi, Lake Turkana Basin, Kenya. In: Harris JM, Leakey MG (eds) Geology and vertebrate paleontology of the Early Pliocene site of Kanapoi, Northern Kenya (contributions in Science 498). Natural History Museum of Los Angeles County, Los Angeles, pp 39–113
Hassanin A (2015) Systematics and phylogeny of cattle. In: Garrick DJ, Ruvinsky A (eds) The genetics of cattle, 2nd edn. CAB International, Wallingford, pp 1–19
Hassanin A, Douzery EJP (1999) The tribal radiation of the family Bovidae (Artiodactyla) and the evolution of the mitochondrial cytochrome b gene. Mol Phylogenet Evol 13:227–243
Hassanin A, Douzery EJP (2003) Molecular and morphological phylogenies and the alternative position of the Moschidae. Syst Biol 52:206–228
Hassanin A, Delsuc F, Ropiquet A, Hammere C, van Vuuren BJ, Matthee C, Ruiz-Garcia M, Catzeflis F, Areskoug V, Nguyen TT, Couloux A (2012) Pattern and timing of diversification of Cetartiodactyla (Mammalia, Laurasiatheria), as revealed by a comprehensive analysis of mitochondrial genomes. C R Palevol 335:32–50
Heintz E (1966) La présence de *Gazellospira torticornis* Aymard (Ruminant) dans le gisement villafranchien de Saint-Vallier (Drôme). Essai de répartition géographique et stratigraphique. Bull Soc Geol Fr 8:25–30
Hermier R, Merceron G, Kostopoulos DS (2020) The emblematic Eurasian Villafranchian antelope *Gazellospira* (Mammalia: Bovidae): new insights from the Lower Pleistocene Dafnero fossil sites (Northern Greece). Geobios 61:11–29

Janis C, Scott K (1987) The interrelationships of higher ruminant families with special emphasis on the members of the Cervoidea. Am Mus Novit 2893:1–85

Kahn MA, Akhtar M (2014) Antelopes (Mammalia, Ruminantia, Bovidae) from the Upper Siwaliks of Tatrot, Pakistan, with description of a new species. Paleontolog J 48:79–89

Kaya TT, Mayda S, Kostopoulos DS, Alcicek MC, Merceron G, Tan A, Karakütük S, Giesler AK, Scott RS (2012) Şerefköy-2, a new late Miocene mammal locality from the Yatağan Formation, Muğla, SW Turkey. C R Palevol 11:5–12

Khomenko J (1913) La faune méotique du village Taraklia du distinct de Bendery. Ann Geol Minéral Russie 15:107–143

Kingdon J (1982) East African mammals: an atlas of evolution in Africa. Academic Press, London

Köhler M (1987) Boviden des türkischen Miozäns (Känozoikum und Braunkohlen der Türkei). Paleontol i Evol 21:133–246

Köhler M, Moyà-Solà S, Morales J (1995) The vertebrate locality Maramena (Macedonia, Greece) at the Turolian-Ruscinian boundary (Neogene). 15. Bovidae and Giraffidae. Münchner Geowiss Abh 28:167–180

Kokotini M, Kargopoulos N, Iliopoulos G, Roussiakis S, Skandalos P, Michailidis D, Svorligkou G, Kampouridis P, Theodorou G (2019) Preliminary fauna of the new Villafranchian locality of Karnezeika (Argolis, Peloponnese). B Geol Soc Gr, sp publ 7A: ext abstract GSG2019-104

Konidaris GE, Tourloukis E, Kostopoulos DS, Thompson N, Giusti D, Michailidis D, Koufos GD, Harvati K (2015) Two new vertebrate localities from the Early Pleistocene of Mygdonia Basin (Macedonia, Greece): preliminary results. C R Palevol 14:353–362

Kostopoulos DS (1997a) The Plio-Pleistocene artiodactyls from Macedonia, Greece: 2. The fossiliferous locality of Volakas, VOL (Volakas Basin, Drama, NE Greece). Paleontol i Evol 30–31:83–92

Kostopoulos DS (1997b) The Plio-Pleistocene artiodactyls (Vertebrata, Mammalia) of Macedonia 1. The fossiliferous site “Apollonia-1”, Mygdonia basin of Greece. Geodiversitas 19:845–875

Kostopoulos DS (1998) *Parastrepsiceros koufosi* n.sp.; note on the possible presence of a *Prostrepsiceros* descendant in the latest Pliocene of Northern Greece. Acta Zool Cracov 41:101–109

Kostopoulos DS (2004) Revision of some late Miocene spiral horned antelopes (Bovidae, Mammalia). Neues Jahrb Geol Palaeontol Abh 231:167–190

Kostopoulos DS (2005) The Bovidae (Artiodactyla, Mammalia) from the Late Miocene mammal locality of Akkaşdağı (Central Anatolia, Turkey). Geodiversitas 27:747–791

Kostopoulos DS (2006a) 9. Cervidae and Bovidae. In: Koufos GD (ed) The Late Miocene vertebrate locality of Perivolaki, Thessaly, Greece. Palaeontogr Abt A 276:151–183

Kostopoulos DS (2006b) Greek bovids through time. Hell J Geosci 41:141–152

Kostopoulos DS (2009a) 13. Bovidae. In: Koufos GD, Nagel D (eds) The Late Miocene mammal faunas of the Mytilinii Basin, Samos Island, Greece: new collection. Beit Paleontol 31:345–389

Kostopoulos DS (2009b) Contribution to the systematics and phylogeny of *Prostrepsiceros vallesiensis* Bouvrain, 1982 (Mammalia, Bovidae). Geodiversitas 31:879–891

Kostopoulos DS (2014) Taxonomic re-assessment and phylogenetic relationships of Miocene homonymously spiral-horned antelopes. Acta Palaeontol Polonica 59:9–29

Kostopoulos DS (2016) Artiodactyla. In: Koufos GD, Kostopoulos DS (eds) Palaeontologie of the Upper Miocene vertebrate localities of Nikiti (Chalkidiki Peninsula, Macedonia, Greece). Geobios 49:119–134

Kostopoulos DS, Athanassiou A (1997) Les gazelles du Pliocène moyen-terminal de la Grèce continentale (Macédoine, Thessalie). Neues Jahrb Geol Palaeontol Abh 205:413–430

Kostopoulos DS, Bernor RL (2011) The Maragheh bovids (Mammalia, Artiodactyla): systematic revision and biostratigraphic-zoogeographic interpretation. Geodiversitas 33:649–708

Kostopoulos DS, Koufos GD (1996) Late Miocene bovids (Mammalia, Artiodactyla) from the locality “Nikiti–1” (NKT), Macedonia, Greece. Ann Paleontol 81:251–300

Kostopoulos DS, Koufos GD (2006) *Pheraios chryssomallos*, gen. et sp. nov. (Mammalia, Bovidae, Tragelaphini), from the Late Miocene of Thessaly (Greece): implications for tragelaphin biogeography. J Vertebr Paleontol 26:436–445

Kostopoulos DS, Soubise J (2018) *Palaeoreas, Majoreas*, and *Stryfnotherium* gen. nov. (Mammalia: Artiodactyla: Bovidae) from the Late Miocene of Greece. Ann Paleontol 104:231–247

Kostopoulos DS, Maniakas I, Tsoukala E (2018) Early bison remains from Mygdonia Basin (Northern Greece). Geodiversitas 40:283–319

Kostopoulos DS, de Bonis L, Koufos GD (2020a) The Vallesian bovid association from the *Ouranopithecus*-bearing site of Xirochori-1 (Axios Valley, Greece). Palaeontolog Z 94:179–194

Kostopoulos DS, Sevim Erol A, Mayda S, Yavuz AY, Tarhan E (2020b) *Qurliqnoria* (Bovidae, Mammalia) from the upper Miocene of Çorakyerler (Central Anatolia, Turkey) and its biogeographic implications. Palaeoworld 29(3):629–635. https://doi.org/10.1016/j.palwor.2019.10.003

Koufos GD (2001) The Villafranchian mammalian faunas and biochronology of Greece. Boll Soc Paleontol Ital 40:217–223

Koufos GD (2006) The Neogene mammal localities of Greece: faunas, chronology and biostratigraphy. Hell J Geosc 41:183–214

Koufos GD, Syrides GE (1997) A new mammalian locality from the Early-Middle Miocene of Macedonia, Greece. C R Acad Sc Paris 325:511–516

Koufos GD, Syrides G, Koliadimou K (1991) A Pliocene primate from Macedenia (Greece). J Hum Evol 21:283–294

Koufos GD, Kostopoulos DS, Vlachou DT (2016) Revision of the Nikiti 1 (NKT) fauna with description of new material. In: Koufos GD, Kostopoulos DS, (eds) Palaeontology of the Upper Miocene vertebrate localities of Nikiti (Chalkidiki Peninsula, Macedonia, Greece). Geobios 49:11–22

Kretzoi M (1941) Neue antilopen-form aus dem Soproner Sarmat. Földtani Közlöny Mayer 71:336–343

Kretzoi M (1968) New generic names for homonyms. Vertebr Hungarica 10(1–2):163–166

Lazaridis G (2015) Study of the Late Miocene vertebrate locality of Kryopigi and other localities of Kassandra Peninsula, Chalkidiki (Greece). Systematics, taphonomy, paleoecology, biochronology. PhD thesis, Aristotle Univ of Thessaloniki, Sc Annals School Geol, 174

Lazaridis G, Kostopoulos DS, Lyras G, Rousiakis S (2017) A new Late Miocene ovibovine-like bovid (Bovidae, Mammalia) from the Kassandra Peninsula (Chalkidiki, Northern Greece) and implications to the phylogeography of the group. Paleontolog Z 91:427–437

Manceau V, Despres L, Bouvet J, Taberlet P (1999) Systematics of the Genus *Capra* inferred from mitochondrial DNA sequence data. Mol Phylogenet Evol 13:504–510

Marcot JD (2007) Molecular phylogeny of terrestrial artiodactyls. In: Prothero D, Foss SE (eds) The evolution of Artiodactyls. John Hopkins University press, Baltimore, pp 4–18

Martínez-Navarro B, Pérez-Claros JA, Palombo MR, Rook L, Palmqvist P (2007) The Olduvai buffalo *Pelorovis* and the origin of *Bos*. Quatern Res 68:220–226

Martínez-Navarro B, Sardella R, Rook L, Bellucci L, Ros-Montoya S (2012) First occurrence of *Soergelia* (Ovibovini, Bovidae, Mammalia) in the Early Pleistocene of Italy. Quat Int 267:98–102

Masini F (1989) I bovini villafranchiani dell'Italia. PhD thesis, Univ Modena-Bologna-Firenze-Roma

Masini F, Palombo MR, Rozzi R (2013) A reappraisal of the Early to Middle Pleistocene Italian Bovidae. Quat Int 288:45–62

Matthee CA, Davis SK (2001) Molecular insights into the evolution of the family Bovidae: a nuclear DNA perspective. Mol Biol Evol 18:1220–1230

McDonald DW (ed) (2009) The encyclopedia of mammals. Oxford University Press, Oxford

McDonald JN, Ray CE, Harington CR (1991) Taxonomy and zoogeography of the musk ox genus *Praeovibos* Staudinger, 1908. In: Purdue JR, Klippel WE, Styles BW (eds) Beamers, Bobwhites and blue-points: tribute to the career of Paul W. Parmalle. Illinois State Mus Sci Pap 23:285–314

McKenna MC, Bell SK (1998) Classification of mammals: above the species level. Columbia University Press, New York

Mead JI, Taylor LH (2005) New species of *Sinocapra* (Bovidae, Caprinae) from the Lower Pliocene Panaca Formation, Nevada, USA. Palaeontol Electronica 8:11A:20p

Mead JI, Jin C, Wei G, Sun C, Wang Y, Swift SL, Zheng L (2014) New data on *Leptobos crassus* (Artiodactyla, Bovidae) from Renzidong Cave, Early Pleistocene (Nihewanian) of Anhui, China, and an overview of the genus. Quat Int 354:139–146

Melentis JK (1966) Studien über fossile Vertebraten Griechenlands. 7. Die Boviden des Jungpleistozäns des Beckens von Megalopolis im Peloponnes (Griechenland). Ann Geol Pays Hell 16:446–472

Melentis JK (1967) Studien über fossile Vertebraten Griechenlands: 19. Die Pikermifauna von Halmyropotamos (Euboea, Griechenland). I Teil: Odontologie und Kraniologie. Ann Geol Pays Hell 19:283–411

Melentis JK (1969) Paläontologische Ausgrabungen auf der Insel Samos (eine vorläufige Mitteilung). Proc Acad Athens 43:344–349

Merceron G, Blondel C, Viriot L, Koufos GD, de Bonis L (2007) Dental microwear analysis of bovids from the Vallesian (late Miocene) of Axios Valley in Greece: reconstruction of the habitat of *Ouranopithecus macedoniensis* (Primates, Hominoidea). Geodiversitas 29:421–433

Montoya P, Ginsbourg L, Alberti M-T, van der Made J, Morales J, Soria MD (2006) Fossil large mammals from the Early Pliocene locality of Alcoy (Spain) and their importance in biostratigraphy. Geodiversitas 28:137–173

Moullé P-E, Échassoux A, Martinez-Navarro B (2004) *Ammotragus europaeus*: une nouvelle espèce de Caprini (Bovidae, Mammalia) du Pléistocène inférieur de la grotte du Vallonnet (France). C R Palevol 3:663–673

Moyà-Solà S (1983) Los Boselaphini (Bovidae, Mammalia) del Neogeno de la Peninsula Iberica. Publicaciones de Geologia 18:1–236

Palmer TS (1903) Some new generic names of mammals. Science NY 17:873

Pandolfi L, Petronio C, Salari L (2011) *Bos primigenius* Bojanus, 1827 from the early Late Pleistocene deposit of Avetrana (Southern Italy) and the variation in size of the species in Southern Europe: preliminary report. J Geol Res 2011:245408

Papageorgiou N (1990) Biology of wild fauna. University Studio Press, Thessaloniki

Pickford M, Gabunia L, Mein P, Morales J, Azanza B (2000) The Middle Miocene mammalian site of Belometchetskaya, North Caucasus: an important biostratigraphic link between Europe and China. Geobios 33:257–267

Pilgrim EG (1910) Notices of new mammalian genera and species from the tertiaries of India. Rec Geol Survey India 40:63–71

Pilgrim EG (1937) Siwalik antelopes and oxen in the American Museum of Natural History. Bull Am Mus Nat Hist 72:729–847

Pilgrim EG (1939) The fossil Bovidae of India. Paleontolol Indica n s 1:1–365

Pilgrim EG, Hopwood A (1928) Catalogue of the Pontian Bovidae of Europe. British Museum (Natural History), London, p 106

Pilgrim EG, Schaub S (1939) Die schraubenhörnige Antilope des europäischen Oberpliocaens und ihre systematische Stellung. Abh Schweizer Paleontolog Gesellsch 62:1–30

Reif WE (2003) Problematic issues of cladistics: 1.ancestor recognition and phylogenetic classification. Neues Jahrb Geol Palaeontol Abh 230:97–143

Rey K, Amiot R, Lécuyer C, Koufos GD, Martineaua F, Fourel F, Kostopoulos DS, Merceron G (2013) Late Miocene climatic and environmental variations in northern Greece inferred from stable isotope compositions (δ18O, δ13C) of equid teeth apatite. Palaeogeogr Palaeocl 388:48–57

Rodrigo MA (2011) Los Bóvidos Villafranquienses de La Puebla de Valverde y Villarroya: sistemática, filogenia y paleobiología. PhD thesis, Univ de Ciencias de Zaragoza, 317 p

Ropiquet A, Hassanin A (2005) Molecular phylogeny of caprines (Bovidae, Antilopinae): the question of their origin and diversification during the Miocene. J Zool Syst Evol Res 43:49–60

Roth J, Wagner A (1854) Die fossilen knochenuberreste von Pikermi in Griechenland. Abh bayer Akad Wissensch Mathem naturwiss Klasse Munchen 7:371–464

Roussiakis S (2003) *Oioceros rothii* (Wagner, 1857) from the Late Miocene of Pikermi (Greece): cranial and dental morphology, comparison with related forms. Geobios 25:717–735

Roussiakis S (2009) *Prostrepsiceros* and *Protragelaphus* (Artiodactyla, Mammalia) from the Late Miocene locality of Chomateri (Attica, Greece). Ann Paleontol 95:181–195

Samartzidou E, Tsoukala E, Maniatis Y, Stoulos S (in press) *Bos primigenius* Bojanus, 1827 from the Late Pleistocene of Greece. Chronology, Taphonomy, Paleobiology. A revision of the species within the Pleistocene and Palaeolithic faunas of Greece

Schlosser M (1904) Die fossilen Cavicornier von Samos. Beitr Palaeont Geol Österreich-Ungarn 17:21–118

Sfouggaris T (2009) Mammals. In: Legakis A, Maragos P (eds) The Red Book of endangered animals of Greece. Greek Zoological Society, Athens, pp 356–425

Sickenberg O (1932) Eine neue Antilope, *Parurmiatherium rugosifrons* nov. gen., nov. sp. aus dem Unterpliozän von Samos. Anzr Osterreich Akad Wissensch Mathem naturwissen Klasse Wien 1:1–8

Sickenberg O (1933) *Parurmiatherium rugosifrons* ein neuer Bovide aus dem Unterpliozän von Samos. Palaeobiologica 5:81–102

Sickenberg O (1936) Über *Samotragus crassicornis* nov. gen. et spec. aus dem Unterpliozän von Samos. Paleontlog Z 18:90–94

Sickenberg O (1971) Revision der wirbeltier Fauna der Höhe Petralona (Griech. Mazedonien). Ann Geol Pays Hell 23:230–264

Sickenberg O (1976) Eine Säugetierfauna des tieferen Bihariums aus dem Becken von Megalopolis (Peloponnes, Griechenland). Ann Geol Pays Hell 27:25–73

Solounias N (1981) The Turolian fauna from the island of Samos, Greece. Contr Vertebr Evol 6:1–232

Solounias N, Moelleken SMC (1992) Cranial restoration of *Eotragus sansaniensis* (Mammalia: Ruminantia) one of the oldest known bovids. J Vertebr Palaeontol 12:250–255

Solounias N, Barry C, Bernor RL, Lindsay EH, Raza SM (1995) The oldest bovid from the Siwaliks, Pakistan. J Vertebr Palaeontol 15:806–814

Spassov N, Crégut-Bonnoure E (1999) Premières données sur les Antilopinae et les Caprinae villafranchiens de Bulgarie. Richesse faunique et signification biozoogéographique. C R Acad Sci Paris 328:493–498

Spassov N, Geraads D (2004) *Tragoportax* Pilgrim, 1937 and *Miotragocerus* Stromer, 1928 (Mammalia, Bovidae) from the Turolian of Hadjidimovo, Bulgaria, and a revision of the late Miocene Mediterranean Boselaphini. Geodiversitas 26:339–370

Spassov N, Geraads D, Hristova L, Markov GN, Garevska B, Garevski R (2018) The Late Miocene mammal faunas of the Republic of Macedonia (FYROM). Palaeontogr Abt A 311:1–85

Steensma KJ (1988) Plio-Pleistozäne Grossäugetiere (Mammalia) aus dem Beckem von Kastoria/ Grevena, südlich von Neopolis-NW Griechenland. PhD thesis, Technical University of Clausthal

Symeonidis N (1973) Die Entdeckung von Zwergelephanten in der Hohle 'Charkadio' auf der Insel Tilos (Dodekanes, Griechenland). Ann Geol Pays Hell 24:445–461

Theodorou G, Athanassiou A, Roussiakis S, Iliopoulos G (2003) Preliminary remarks on the Late Miocene herbivores of Kerassiá (Northern Euboea, Greece). DEINSEA 10:519–530

Trifonov VG, Çelik H, Simakova AN, Bachmanov DM, Frolov PD, Trikhunkov YI, Tesakov AS, Titov VM, Lebedev VA, Ozherelyev DV, Latyshev AV, Sychevskaya EK (2018) Pliocene – Early Pleistocene history of the Euphrates valley applied to Late Cenozoic environment of the northern Arabian Plate and its surrounding, eastern Turkey. Quat Int 493:137–165

Tsoukala E (1989) Contribution to the study of the Pleistocene fauna of large mammals (Carnivora, Perissodactyla, Artiodactyla) from Petralona cave Chalkidiki (N. Greece). PhD thesis, University of Thessaloniki

Tsoukala E (1992) Quaternary faunas of Greece. Cour Forsch Senck 153:79–92

van den Hoek Ostende L, Mayda S, Oliver A, Madern A, Hernández-Ballarín V, Peláez-Campomanes P (2015) Aliveri revisited, a biogeographical appraisal of the early Miocene mammals from the eastern Mediterranean. Palaeobio Palaeoenv 95:271–284 van der Made J (2012) *Eotragus clavatus* (Artiodactyla, Bovidae, Boselaphini) from Sansan. In: Peigné S, Sen S (eds) Mammiferes de Sansan, Mem Mus natl hist nat 203:145–199

van de Weerd A, Reumer JWF, de Vos J (1982) Pliocene mammals from Apolakkia Formation (Rhodes, Greece) Proc K Ned Akad Wet B 85(1):89–112

Vasileiadis N, Tsoukala E, Kostopoulos DS (2019) The Late Miocene bovids from Platania (Drama Basin, Greece), with description of a new species of *Palaeoryx*. Geobios 55:57–76

Vasileiadou K, Koufos GD (2005) The micromammals from the Early/Middle Miocene locality of Antonios, Chalkidiki, Greece. Ann Paleontol 91:197–225

Vekua A (2012) New spiral-horned antelope in Dmanisi fauna. Bull Georgian Nat Acad Sci 6:139–144

Viret J (1961) Artiodactyla. In: Piveteau J (ed) Traité de Paleontol. Masson, Paris, pp 887–1084

von Koenigswald W, Schwermann AH, Keiter M, Menger F (2019) First evidence of Pleistocene *Bubalus murrensis* in France and the stratigraphic occurrences of *Bubalus* in Europe. Quat Int 522:85–93. https://doi.org/10.1016/j.quaint.2019.06.019

Vrba ES, Schaller GB (2000) Phylogeny of Bovidae based on behavior, glands, skulls, and post-crania. In: Vbra ES, Schaller GB (eds) Antelopes, deer and relatives. Yale University Press, New Haven, pp 203–222

Wagner A (1848) Urweltliche Saugethier-Überreste aus Griechenland. Abh bayer Akad Wissensch Mathem naturwiss Klasse Munchen 5:333–378

Wagner A (1857) Neue Beitraege zur Kenntniss der fossilen Säugethier-Überreste von Pikermi. Abh bayer Akad Wissensch Mathem naturwiss Klasse Munchen 8:111–158

Wang X, Li Q, Takeuchi GT (2016) Out of Tibet: an early sheep from the Pliocene of Tibet, *Protovis himalayensis*, gen. et sp. nov. (Bovidae, Caprini), and origin of ice age mountain sheep. J Vertebr Paleontol 36(5):e1169190

Wang SQ, Yang Q, Zhao Y, Li CX, Shi QQ, Zong LY, Ye J (2019) New *Olonbulukia* material and its related assemblage reveal an early radiation of stem Caprini along the north of the Tibetan Plateau. J Paleo 93:385–397

Weithofer KA (1888) Beiträge zur Kenntnis der Fauna von Pikermi bei Athen. Beitr Paleont Geol Österreich-Ungarn 6:225–292

Yang C, Xiang C, Qi W, Sh X, Tu F, Zhang X, Moermond T, Yue B (2013) Phylogenetic analyses and improved resolution of the family Bovidae based on complete mitochondrial genomes. Biochem Syst Ecol 48:136–143

The Fossil Record of Continental Fossil Deer (Mammalia: Artiodactyla: Cervidae) in Greece

Athanassios Athanassiou

1 Introduction

Although less diversified than the bovids, among the pecoran ruminants, the Cervidae Gray, 1821, still remain a very rich family with regard to the number of included genera and species. According to recent taxonomical accounts, the family today comprises 16–19 genera and 40–52 species (McKenna and Bell 1997; Hutchins et al. 2003; Grubb 2005; Groves and Grubb 2011; Groves 2014), although Groves and Grubb (2011) split many of these species in distinct allopatric populations, raising their total number to more than 100. The species currently included in Cervidae range in body size from very small (such as the montane *Pudu mephistophiles*, with a body mass of about 3–6 kg) to very large (the boreal species *Alces alces*, which can reach a body mass of 600–700 kg) (Geist 1998, pp. 120, 230). Throughout their history, the deer dispersed to most continents, except for sub-Saharan Africa, Australia, and Antarctica. The recent populations are adapted to a great variety of environments, from tropical forests to the tundra and arid steppes. Most typically, though, they inhabit temperate regions with a forest or woodland character (particularly concerning the larger-sized species).

Morphologically, the Cervidae are defined on an array of characters, most conspicuous of which is the development of antlers, annually deciduous osseous cranial appendages that lack a keratin sheath. These are typically found only in male individuals. Other diagnostic morphological features include the presence of a double lacrimal orifice, a lacrimal fossa, as well as of a praeorbital vacuity in the splachno-

A. Athanassiou (✉)
Ephorate of Palaeoanthropology–Speleology, Hellenic Ministry of Culture and Sports, Athens, Greece

Museum of Paleontology and Geology, National and Kapodistrian University of Athens, Athens, Greece
e-mail: aathanas@geol.uoa.gr

E. Vlachos (ed.), *Fossil Vertebrates of Greece Vol. 2*,
https://doi.org/10.1007/978-3-030-68442-6_6

cranium, and the generally brachyodont proportions of the dentition (Brooke 1878; Janis and Scott 1987). Otherwise, the dentition is selenodont (i.e., showing crescent-shaped lophs/lophids on the occlusal surfaces) and lacks upper incisors, quite like the other ruminant families. However, there are certain peculiarities observed in the family's dental morphology, particularly in comparison to the bovids: the premolar section is proportionally rather long; the crescent-shaped cusps of the cheek teeth remain unfused to each other (in occlusal view), except in cases of advanced wear; the lingual and buccal walls of the cheek teeth are markedly convergent (Heintz 1970). Postcranially, the members of the family exhibit metatarsals III–IV with a dorsal median gully which is covered distally, unlike the condition observed in other ruminant families (Heintz 1970; Bouvrain et al. 1989; Janis and Theodor 2014, fig. 9), while a soft tissue defining character is the absence of the gallbladder (Brooke 1878; Groves and Grubb 2011).

The antlers are positioned postorbitally, each borne on a cylindrical pedicle, and are formed as outgrowths of the frontal bones. They consist of an outer layer of compact bone that surrounds a cancellous inner part. A layer of fur-covered skin (velvet) shrouds them during their growth, which dies and is shed when they are fully grown. The antler morphology varies greatly, depending mainly on the species and the ontogenetic stage (Heintz 1970; Goss 1983; Lister 1987a; Geist 1998). Adult individuals of different species typically exhibit different morphological patterns in their antlers, which, when fully grown, can be simple spikes, bifurcate, or branched, with tines branching out of the main beam. In certain taxa the beam and tines are interconnected with additional bone tissue, forming a palmation. Within a species, younger individuals have smaller and simpler antlers, which become larger and more branched each subsequent year. Senile individuals, however, may have again simpler or malformed antlers. Additionally, ecological and physiological factors, such as poor nutrition and endocrinal problems, can result in abnormal antler development, since their very rapid yearly growth is hormonally driven and intensely demanding in food resources. As a result, the antlers exhibit very high intraspecific variation, even including bilateral asymmetries in the same individual. Among different species, the antlers' size and complexity are also correlatable to the body size, with larger species usually exhibiting increasingly larger and more complex antlers (Gould 1973; Goss 1983, p. 90).

Methods Due to the considerable polymorphy of the antlers among different genera, it is not always possible to trace the homologies of their tines. Thus, they are usually referred to as ordinal numbered, starting from the most proximal one. However, already established naming systems in certain groups are also used in parallel (e.g., basal, middle, posterior tine in Megacerini, and brow, bez, trez tine in *Cervus*). Upper and lower molars are abbreviated as "M" and "m," premolars as "P" and "p," respectively. The geochronologic and stratigraphic framework follows Gradstein et al. (2012). Geochronologic ages are given, were available, in thousand or million years before present, abbreviated as "ka" or "Ma" respectively. "MN" and "MNQ" stand for Mammal Neogene and Quaternary Zones (Guérin 1990; Mein 1999), "ICZN" for "International Code of Zoological Nomenclature." The taxonomy generally follows Croitor (2018b). Transliterated local geographic names are typed accented when necessary, to help with their correct pronunciation.

2 Phylogenetic Relationships and Taxonomy

The family Cervidae is of Eurasian origin, descending from a cervoid, apparently antlerless, ancestor during the early Miocene (Janis 1982; Gentry 2000). For almost three quarters of their evolutionary history, the members of the family remained in their native continent as least-specialized herbivores, expanding their biogeographic range in new habitats in a rather opportunistic way (Geist 1998, p. 1). During the latest Miocene certain East Asian taxa colonized North America (Webb 2000) and subsequently spread to South America during a major dispersal event of the Great American Biotic Interchange, which took place in the Early Pleistocene (Cione et al. 2015). More recently, at the beginning of the Late Pleistocene, they entered North Africa, where they exist as patchy populations until today (Geraads 1982).

Biochronologically, the origin of the family is traced back to the early Miocene, more specifically to the biozone MN3, when the basal taxa, like *Procervulus* Gaudry, 1877, appear in the fossil record (Gentry 1994, 2000; Groves 2007). Other nearly contemporary, closely related genera are *Lagomeryx*, *Dicrocerus*, and *Stephanocemas*, all of which were often placed in a separate family, Lagomerycidae Pilgrim, 1941, but are currently included within the Cervidae (see, e.g., Janis and Scott 1987 and Groves 2007 for relevant discussions). Close sister groups of the Cervidae are considered the extinct family Palaeomerycidae Lydekker, 1883, and the extant Moschidae Gray, 1821, all three forming the superfamily Cervoidea Goldfuß, 1820 (Janis and Scott 1987; Gentry 1994). The North American Antilocapridae Gray, 1821, are often also included in the same superfamily (McKenna and Bell 1997; Gentry 2000).

Taxonomically, the family Cervidae is typically divided into two monophyletic clades of subfamily rank, the Cervinae Gray, 1821, and the Capreolinae Brookes, 1828 (= Odocoileinae Pocock, 1923), a division originally recognized by Brooke (1878). Both are principally characterized by the morphology of their reduced lateral (II and V) metacarpals. The former—also informally known as "Plesiometacarpalia" or "Old World deer"—retain only the proximal parts of the lateral metacarpals, while in the latter—also informally known as "Telemetacarpalia" or "New World deer"—only the distal parts of the same bones are present. Most later taxonomic studies have accepted this dichotomy, also supporting it with additional morphological characters, such as the morphology of the otic region (Bouvrain et al. 1989; Mennecart et al. 2017), and ethologic observations (Cap et al. 2002, 2008). Moreover, the subdivision into these two clades is also supported by molecular phylogenetic studies (Douzery and Randi 1997; Hassanin and Douzery 2003; Pitra et al. 2004; Kuznetsova et al. 2005; Gilbert et al. 2006; Zurano et al. 2019). Nevertheless, older taxonomic subdivisions of the family included very often a separate clade for the Chinese water deer *Hydropotes* (subfamily Hydropotinae

Trouessart, 1898), due to the antlerlessness of this genus, and/or a clade for the muntjac-like deer (subfamily Muntiacinae Pocock, 1923) (e.g., among the most recent studies, McKenna and Bell 1997; Grubb 2005; Hernández Fernández and Vrba 2005). Oversplitting schemes have also been published, such as that of Bubenik (1990), who recognized the families Muntiacidae, Odocoileidae, and Capreolidae, as distinct from Cervidae, while overlumping ones include the Moschidae as a subfamily of Cervidae (e.g., Hutchins et al. 2003, p. 341, among recent classifications). Other important and influential taxonomic schemes proposed for the Cervidae include those of Simpson (1945), Azzaroli (1953), Vislobokova (1990), McKenna and Bell (1997), Grubb (2000), and Di Stefano and Petronio (2002). Recent overviews of the cervid taxonomy are included in Croitor (2014, 2018b).

This considerable degree of variation in the proposed taxonomic schemes for the family Cervidae is probably caused partially by an extensive character homoplasy among different lineages, which obscure their true phylogenetic relationships, as well as by the uncommon morphologies of some taxa (such as the genera *Muntiacus*, *Hydropotes*, and *Alces*), whose evolutionary pathways through the fossil record are largely unresolved. Genus and species-level taxonomic studies have to cope with the rather high uniformity in the cranial and dental morphology observed among living and fossil cervids. Postcranial characters are even more problematic, because they most probably reflect a population's ecologic adaptations, not its phylogenetic relationships (Croitor 2018b). Thus, the main morphological character used in genus and species-level taxonomy—particularly in fossil forms—is the shape of the antlers, because of their special significance in reproductive ethology, which is unique for each species (Goss 1983; Geist 1998).

3 Historical Overview and Distribution

Fossils of the family Cervidae are not very common in Greece, because of the predominantly open-landscape faunas represented in most localities. The number of deer-bearing sites is high (particularly the Late Pleistocene ones), but the number of taxa and individuals represented in each one is generally low. The currently known fossil localities cover a temporal distribution from the early middle Miocene (MN5) until the latest Pleistocene (see Appendix), while the family continues to be present in Greece until today. Geographically, the deer-bearing localities are spread throughout continental Greece (including close-to-shore islands), from Corfu to Rhodes, and from Vólakas in E. Macedonia to S. Peloponnese and Kýthera Island (see Fig. 1). Below I present some of the most important localities with fossil Cervidae, in terms of wealth of material or taxonomic distinctiveness. A comprehensive list of the localities and the recovered taxa are given in Fig. 1 and the Appendix.

The oldest known scientific description of a cervid in the fossil record of Greece was published in 1865 in the monumental monograph by Albert Gaudry on the fauna of Pikermi (Gaudry 1862–1867). The author coined the species name

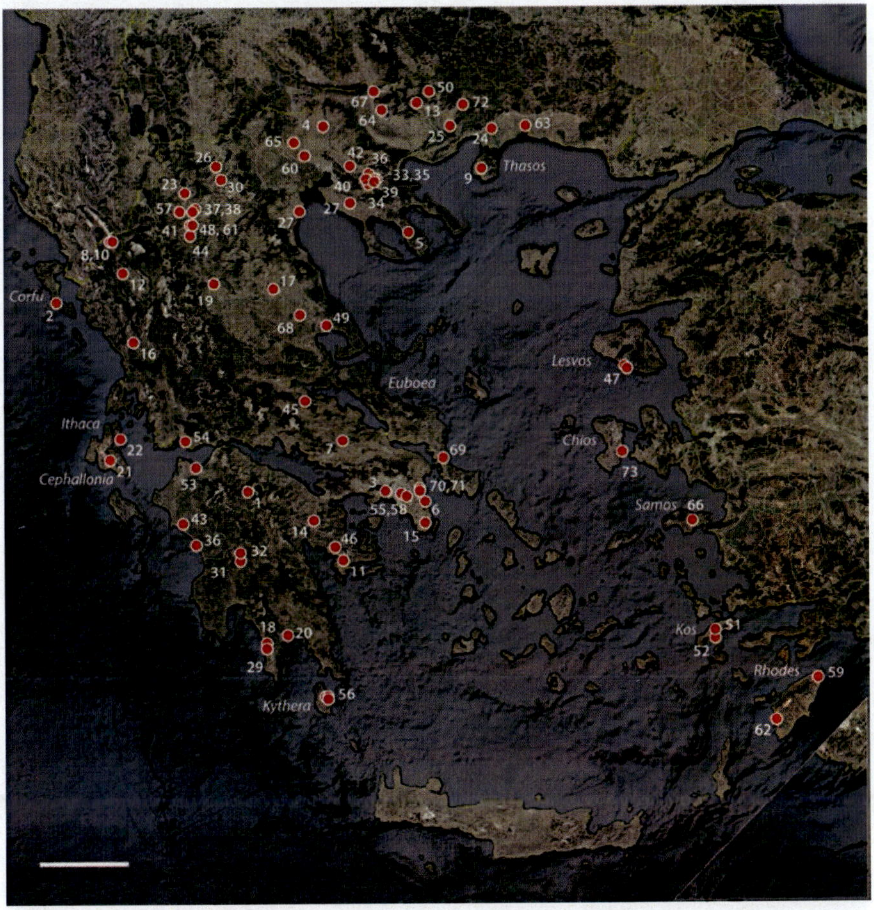

Fig. 1 Map of Greece showing the geographic distribution of the most important localities with non-endemic Cervidae. **1**, Limnón Cave; **2**, Gráva Rockshelter; **3**, Schistó Cave; **4**, Agios Geórgios Cave; **5**, Dragoudélis; **6**, Vraóna Doline; **7**, Seidí Rockshelter; **8**, Boíla Rockshelter; **9**, Tzínes; **10**, Klithí Rockshelter; **11**, Franchthi Cave; **12**, Kastrítsa Rockshelter; **13**, Angítis Cave; **14**, Klissoúra Cave; **15**, Kítsos Cave; **16**, Asprocháliko Rochshelter; **17**, Peniós Valley; **18**, Kalamákia Cave; **19**, Theópetra Cave; **20**, Lakonís; **21**, Phytídi Cave; **22**, Nymphs Cave; **23**, Lágoura; **24**, Xeriás; **25**, Phílippi; **26**, Sotíras; **27**, Petralona Cave; **28**, Aeginio; **29**, Apídima Caves; **30**, Perdíkkas; **31**, Marathousa 1; **32**, Kyparíssia localities; **33**, Kalamotó localities; **34**, Apollonía; **35**, Platanochóri; **36**, Kaiáfas; **37**, Kapetánios; **38**, Libákos; **39**, Krímni; **40**, Tsiótra Vrýssi; **41**, Polýlakkos; **42**, Gerakaroú; **43**, Pyrgos; **44**, Q-Profil; **45**, Reghínio; **46**, Karnezéika; **47**, Vaterá localities; **48**, Dafneró; **49**, Sésklo; **50**, Vólakas; **51**, Antimáchia; **52**, Kardámaena; **53**, Kastrítsi; **54**, Makýnia; **55**, Tourkoboúnia; **56**, Kýthera localities; **57**, Tsotýli; **58**, Korydallós; **59**, Rhodes airport; **60**, Géphyra; **61**, Miliá localities; **62**, Apolakkiá; **63**, Kessáni 2; **64**, Maraména; **65**, Dytikó 1; **66**, Samos localities; **67**, Thermopigí; **68**, Perivoláki; **69**, Halmyropótamos; **70**, Chomaterí; **71**, Pikermi; **72**; Plataniá; **73**, Thymianá. See Appendix for more information. Image exported from Google Earth Pro © 2019, map data from US Department of State Geographer, SIO, NOAA, U.S. Navy, NGA, GEBCO, image from Landsat/Copernicus. Scale bar equals 80 km. North faces upward

Fig. 2 Cranial and antler specimens of Turolian age, referred to *Procapreolus pentelici* (Dames, 1883). (**a**) Antler pair, the holotype of the species, in rostral view. The specimen comes from the excavations carried out in 1853 by H. Mitzopoulos at Pikermi (Melentis 1969a). Note that the left antler is wrongly prepared, as its basal part looks incorrectly oriented. (**b**) Frontlet in rostral view, bearing both antlers (the distal part of the right one is reconstructed), excavated in 1972 at Chomaterí, close to Pikermi (Symeonidis 1973, 1974). Both specimens, as well as their original photographs used for this figure, belong to the collections of the Museum of Palaeontology and Geology, University of Athens. Scale bar equals 10 cm

Dremotherium? *pentelici*, a very small-sized cervid, while he recognized the presence of an additional somewhat larger species in Pikermi, which remained unnamed, though referred to the genus *Dremotherium*? as well. Dames (1883) described a pair of antlers from Pikermi (Fig. 2a), based on a draft sketch of the specimens (they were later redescribed in detail by Melentis 1969a). Other early publications on members of the family were mostly simple reports, lacking descriptions or pictorial documentation (see Appendix for a more complete locality listing). Systematic studies on fossil Cervidae were initiated during the 1960s, with the work of Melentis (1963, 1966b) on the rich sites in the Megalopolis Basin (Peloponnese). The cervid sample includes some nice craniodental and postcranial specimens, which the author attributed to three species, *Cervus (Dama) somonensis*, *Cervus elaphus*, and *Capreolus* sp., the former being the most abundantly represented in the studied material. Shortly later, Melentis described an exceptional find from the Haliákmon River basin, a frontal fragment bearing both antlers, which he identified as *Orthogonoceros verticornis* (Dawkins, 1872) (Melentis 1967, 1968). Here, this specimen (Fig. 7) is referred to *Praemegaceros pliotarantoides* (De Alessandri, 1903), a species less derived than *P. verticornis*. The Haliákmon specimen is currently the best preserved *Praemegaceros* fossil in Greece. The locality seems, though, not to be well documented. Melentis (1967) mentioned that the skull had been found shortly before the publication near the town of Véria (Central Macedonia). However, according to documentation provided in Zografou and Bonda-Doumanaki (2014), this specimen was excavated in 1905 in the area of the village Polýlakkos (W. Macedonia) and was subsequently sent to Prof. Th. Skuphos at the University of Athens, soon after its discovery. Unfortunately, no relevant information was found in the document archives of the University of Athens (Lyras pers. comm. 2020).

In the late 1960s, Melentis worked on the late Miocene (Turolian) fauna of Halmyropótamos (S. Euboea), excavated in 1916 by Prof. Th. Skuphos (Melentis 1970). The entire cervid material, consisting of an antlered cranial fragment and several antler parts, was referred to *Pliocervus pentelici* (Gaudry, 1865).

Sickenberg (1971) revisited the Pleistocene fauna from the Petralona Cave in Chalkidikí, based on the whole fossil material available until that time. The author referred the deer samples to three species, *Cervus elaphus*, *Dama dama*, and *Praemegaceros verticornis*, similarly to previous preliminary reports. A few years later, Symeonidis (1973, 1974) reported on a new find of the incompletely known late Miocene species *Pliocervus pentelici*, excavated in 1972 in the Turolian locality of Chomaterí, near Pikermi (Attica). The specimen (Fig. 2b) is a dorsal cranial part preserving the complete left antler and the proximal part of the right one. Sickenberg (1976) examined a fossil sample from the Megalopolis Basin, which was collected in the early 1960s during geological fieldwork in the area. The available material, though fragmentary, seems to include four distinct cervid taxa, one of which was named as a new species: *Cervus peloponnesiacus*. A few finds belong to a somewhat smaller form and were referred by the author to ?*Capreolus* sp. Cervid finds of middle–large and very large size were assigned to *Cervus s.l.* sp. and *Praemegaceros verticornis*, respectively. The former are probably identifiable with *Cervus elaphus* Linnaeus, 1758, and/or *Dama* sp. *Praemegaceros verticornis* was better represented by several cranial and antler fragments, as well as dental and postcranial specimens.

Four years later, Symeonidis et al. (1980) published the results of the excavation in the doline of Vraóna, east of Athens, a site dominated by leopard fossils. The fauna presented in the paper, which dates from the latest Pleistocene, includes remains of *Cervus elaphus* and an undeterminable species about the size of fallow deer. Updated faunal lists of the early 1990s (Symeonidis and Theodorou 1994; Rabeder 1995) include three deer species, namely *C. elaphus*, *Dama dama*, and *Capreolus capreolus*.

During a geological mapping fieldwork carried out in 1969–1970 by the University of Utrecht in the area of Apolakkiá (SW part of Rhodes Island), several surface finds of Pliocene fossil vertebrates were collected (van de Weerd et al. 1982). The identifiable cervid specimens were referred to *Cervus* aff. *philisi*, including an almost complete, three-tined antler (Fig. 3; see also van de Weerd et al. 1982, pl. 5, fig. 3, text-fig. 5).

The rich fossiliferous region of Neápolis (W. Macedonia), known previously from occasional reports on isolated fossils, was revisited during a 1988 PhD study (Steensma 1988), based on material collected during the previous years by a team

Fig. 3 Unshed antler from the Ruscinian of Apolakkiá, Rhodes Island, referred to the species aff. *Metacervocerus rhenanus*. The specimen, as well as the original photograph used for this figure, belongs to the collections of the Museum of Palaeontology and Geology, University of Athens. Scale bar equals 10 cm

from the Technical University of Clausthal (Germany). Steensma described craniodental and postcranial deer specimens from four localities (Libákos, Kapetánios, Q-Profil, and Polýlakkos), most important of which are an *Eucladoceros* antler part from Kapetánios, referred to *Eucladoceros* aff. *boulei* Teilhard de Chardin and Piveteau, 1930, as well as a dorsal skull fragment with the basal parts of both antlers, and a shed antler, both from Libákos, and referred to Cervidae indet. The cranial and antler parts from Libákos belong to a large-size species, with strongly divergent, dorsolaterally directed, and weakly ventrally curving antler beams. There are two palmate tines, one close to the burr, and another about 20 cm further up. Although Steensma (1988) did not refer these finds to a genus, later authors identified them as *Megaloceros* aff. *savini* (Dawkins 1887) (van der Made and Tong 2008; Vislobokova 2012). Due to its morphologic peculiarities, this species is often classified in the separate genus *Praedama* Portis, 1920 (e.g., Vislobokova 2012; Croitor 2014, 2018b). This taxonomy is followed here. It is noted that a morphologically similar partly preserved cranial specimen was also described from the locality Kapetánios (Steensma 1988, pp. 174–175). The locality of Libákos has also yielded medium-sized deer remains, referred by Steensma (1988) to *Cervus* cf. *nestii eurygonos* (Azzaroli, 1947), here assigned to the species *Dama vallonnetensis* (de Lumley et al., 1988).

The very rich mammal fauna from the Petralona Cave, a site well known for its anthropological content, was revised by Tsoukala (1989), based on the material stored in the collections of the Aristotle University of Thessaloniki. Tsoukala identified three cervid taxa, just like Sickenberg (1971), but refrained to refer the giant deer finds to a species. Her taxonomic list includes *Praemegaceros* sp., *Cervus elaphus*, and *Dama dama*.

Azanza (1995) studied the cervid sample from Maraména, which includes mainly dental specimens and a fairly complete shed antler, and erected a new species, *Pliocervus graecus*. The stratigraphic position of the locality close to the Miocene/Pliocene boundary let the author to revise the taxonomy of the late Miocene deer from Greece, based on the Maraména sample: Most of the material attributed until then to *Pl. pentelici* (e.g., Dames 1883; Melentis 1969a, 1970) was transferred to the new species, while *Pl. pentelici* was confined to the pair of very small-sized mandibles described by Gaudry (1862–1867).

In a PhD study on the Plio-Pleistocene artiodactyls of Macedonia, Kostopoulos (1996) described numerous deer samples from five localities: Dafneró, Vólakas, Gerakaroú, Krímni, and Apollonía. Two taxa were found at Dafneró, "*Cervus*" *philisi* cf. *valliensis* (= *Metacervocerus rhenanus*) and *Eucladoceros senezensis* (= *E. ctenoides*). The biochronologically similar (MN17) fauna of Vólakas yielded the same species, as well as the small-sized *Croizetoceros ramosus*, documented on dental and osteological material. The more recent fauna of Gerakaroú presents the same three-species cervid association, typical of Villafranchian faunas. However, the medium-sized sample was not referred to a species, due to the poor available material ("*Cervus*" sp.), while the much better represented small sized was described as a new subspecies, *Croizetoceros ramosus gerakarensis*. The geochronologically younger faunas of Krímni and Apollonía yielded only one large-sized cervid each, *E. senezensis* (= *E. ctenoides*) and *Megaloceros* sp. respectively. Van der Made (2001) assigned the Apollonía deer to *Eucladoceros giulii* Kahlke, 1997, while a later systematic study split this sample to two taxa, *Praemegaceros pliotarandoides* (De Alessandri, 1903) and *Arvernoceros* cf. *verestchagini* David, 1992 (Croitor and Kostopoulos 2004). Quite recently, Croitor (2018a) revised again the Apollonía deer and created the new species *Rucervus gigans* for the sample, which had been previously identified as *A.* cf. *verestchagini* (Fig. 5). In the same publication, Croitor suggested that all *Eucladoceros* samples from Greece, particularly that from Gerakaroú, be better referred to *Rucervus radulescui* Croitor, 2018; however, this is rather a premature proposal, in view of the persisting lack of antler specimens from the Greek fossil record. The samples studied by Kostopoulos (1996) have been also presented preliminarily, reviewed, or revised in other publications (Koufos et al. 1992; Kostopoulos and Koufos 1994; Kostopoulos 1997a, b; Kostopoulos and Athanassiou 2005). Geochronologically close to Dafneró and Vólakas is also the diverse fauna of Sésklo, which includes the *Eucladoceros–Metacervocerus–Croizetoceros* association as well (Athanassiou 1998; Kostopoulos and Athanassiou 2005).

The locality Kalamotó in Central Macedonia (Tsoukala and Chatzopoulou 2005) includes two sites of late Early Pleistocene age (MNQ19) with a rich mammalian fauna, which includes an impressive *Praemegaceros* skull, found at Kalamotó 1. It is partially preserved, bearing both antlers to a great length, lacking only the facial part. The antlers exhibit a beam that bends abruptly at the level of the posterior tine, while they lack a middle tine and a distal palmation. These characters let the authors to refer the find to *P. pliotarantoides*. A similar partial skull from Kalamotó 2 preserves the neurocranium and the basal parts of both antlers, and it was referred to the same species. Both Kalamotó sites yielded *Cervus* and *Dama*-sized cervids as well.

The late Miocene exceptionally diverse fauna of Perivoláki includes also some scanty dental elements of small size (Kostopoulos 2006). The lack of antlers or any other cranial remains makes their identification rather ambiguous. The author referred them, as well as those from Samos Q6, provisionally to the Spanish genus *Lucentia* Azanza and Montoya, 1995, until more material becomes available.

A small deer was also present in the late Miocene fauna from Dytikó, Axiós Valley, Central Macedonia (Bouvrain and de Bonis 2007), but, again, it is known only from dental material, dimensionally similar to that from Perivoláki. The authors referred the finds to ?*Procapreolus* sp., although they also find dental similarities with the Asian genus *Pavlodaria* Vislobokova, 1980.

Crégut-Bonnoure and Tsoukala (2017a) have recently described cervids from the lower Axiós Valley locality of Géphyra (Central Macedonia). The locality is one of the few in Greece dated in the Pliocene (early Villafranchian, MN16a). Among the fossils recovered there is a fine cranial part, bearing the left three-pointed antler, which the authors referred to *Procapreolus cusanus* (Croizet and Jobert, 1928). Despite its later age, and based on the antlers' slenderness, Croitor (2018b) considered the Géphyra deer as closer to *Procapeolus pentelici*, than to the initially reported species. The same authors also studied the cervids from another Pliocene locality, Miliá, W. Macedonia (Crégut-Bonnoure and Tsoukala 2017b), which is roughly contemporary to Géphyra, dated in MN16a as well (Vlachos et al. 2018). The fauna contains the cervid taxa *Croizetoceros ramosus*, Cervidae indet. (medium sized), and *Praeelaphus* cf. *lyra* (Azzaroli, 1992). The former species is the best documented by a wealth of antler parts, dental, and postcranial material. The latter was identified only on a part of the antler beam and is in need of more material to better support it.

The discovery of very large-sized antler fragments in the collections of the Palaeontological–Historical Museum of Ptolemaís was announced in a conference in 2017 (Kevrekidis and Kostopoulos 2017). The specimens have been found in the area of the village Pentávryssos, in deposits of Middle Pleistocene age, and based on their morphology, they were referred to *Cervalces latifrons*. This was the first documented occurrence of an Alceini in Greece, as other such allegations are either unsupported (Melentis 1966a; Stratigopoulos 1986) or referring to undescribed material (van der Meulen and van Kolfschoten 1988).

In 2018, two studies focused on recently excavated samples from localities within the Megalopolis Basin (Peloponnese), an area known since the nineteenth century for its paleontological wealth (see above in this section). Marathoúsa 1, an elephant butchering site (Konidaris et al. 2018), has also yielded deer fossils, often with cut marks. Konidaris et al. (2018) identified two taxa, *Cervus elaphus* and *Dama* sp., based on dental and postcranial material. The second locality, Kyparíssia, is much richer in deer diversity, but the available specimens are often surface finds and derive from two stratigraphic levels (Athanassiou 2018). The presence of *Praemegaceros verticornis* is documented by cranial and postcranial elements, including antler parts, which are characterized by a bent of the beam at the level of the posterior tine, and a well-developed distal palmation. Other identified taxa include *Cervus elaphus*, *Dama* sp., and "*Cervus*" *peloponnesiacus*, based on antlers and postcranial elements. A single antler specimen with a quite unusual, *Muntiacus*-like morphology was referred to Cervidae indet.

Deer are also common faunal elements in Paleolithic archaeological contexts (typically in caves or rockshelters), where the deposits, although partly anthropogenic, are reflecting to some extent the local faunal composition of their time. Examples of such localities are Gráva on Corfu, Kastrítsa, Asprocháliko, Klithí, Megálakkos and Boíla in Epirus, Theópetra in Thessaly, Kítsos and Schistó in Attica, Franchthi, Klissoúra, Lakonís, Kalamákia, Kolomínitsa, and Melitziá in Peloponnese. The localities and the identified taxa are given in Fig. 1 and the Appendix. As a general pattern, three species occur in Paleolithic sites: *Capreolus capreolus*, *Dama dama*, and *Cervus elaphus*. The former is always rare, while *C. elaphus* is usually most frequent in northern localities, and *D. dama* usually prevails in southern ones. More recent, post-Paleolithic archaeozoological samples are reasonably considered as representing rather the local economic practices, than the surrounding natural environment, and as such are excluded from the present review.

4 Systematic Paleontology

This section includes the continental, non-endemic deer taxa, whose presence is documented in the fossil faunas of Greece, not considering the numerous insular forms, which evolved on palaeoislands during the Pleistocene (see Lyras et al. this volume). It has to be pointed out that, due to the paucity of cervid specimens, particularly antlers, in most localities, the attributions to taxa may be tentative and a result of divergent taxonomic interpretations by different authors. In general, the most current and better supported taxonomic opinions are presented below, but, again, some taxa still remain not well established and in need of further study and more complete samples. As far as we presently know, the taxa present in the fossil record of Greece are of advanced antler morphology, with rather short pedicels, and two or more bifurcations, unlike the ancestral forms with high, supraorbital pedicels and two-tined antlers. This is due to the fact that pre-Turolian cervids are essentially as yet hardly known.

Valid Taxa

Cervidae Goldfuß, 1820

Type Genus *Cervus* Linnaeus, 1758.

Remarks The family is presently represented in the fossil record of Greece by sixteen species, classified in at least 12 genera. Some rare and incompletely preserved specimens quite probably represent additional species, which remain unidentified yet, pending the recovery of better samples. An example is the small-sized Cervidae indet. from Kyparíssia, represented by a basal antler fragment (Athanassiou 2018). On the contrary, there are several cases in the literature, where the reported specimens were referred to a genus, or even to a species, although they were in fact not identifiable further than the family level. In a similar way, old authors (mainly of the nineteenth century) used to refer their cervid finds sweepingly to the genus *Cervus*, meant in a much broader sense, essentially equivalent to Cervidae indet. in the current taxonomy. For a list of these samples, the reader is referred to the Appendix.

Capreolinae Brookes, 1828

Type Genus *Capreolus* Frisch, 1775.

Procapreolus Schlosser, 1924

Type Species *Procapreolus latifrons* Schlosser, 1924, selected as such in 1963 by Korotkevich (cited in Croitor 2018b).

Remarks *Procapreolus* is a Eurasian genus that comprises small- to medium-sized late Miocene and Pliocene forms with simple, three-pointed antlers, inserted on long, divergent pedicels.

★*Procapreolus pentelici* (Dames, 1883)

Nomenclatural and Taxonomical History All late Miocene deer fossils from Greece, like those from Pikermi, Chomaterí, and Halmyropótamos (Melentis 1969a, 1970; Symeonidis 1974), have been traditionally referred to *Pliocervus pentelici* (Gaudry, 1865), a species based on very small-sized dental material from Pikermi. Initially classified in *Dremotherium*?, its validity was challenged by Dames (1882,

1883), who considered it, together with Gaudry's *Dremotherium*? sp., as potential synonyms of *Cervus matheroni* Gervais, 1859, from the French locality of Mont Léberon, and in any case as belonging to a single, sexually dimorphic species, referred to the genus *Cervus*. Dames (1883) named this species as new, defined on antlers (Fig. 2a), but confusingly used the very same species name coined by Gaudry: *Cervus pentelici* Dames, 1883. Later authors accepted this species name, either with Gaudry's or Dames' authorship, but referred it to different genera, most frequently to *Pliocervus* Hilzheimer, 1922. Azanza (1995), while studying the cervid material from Maraména, observed that the antlers, which had been referred to this species so far, are too large with regard to the type material published by Gaudry (1862–1867). Thus, she coined a new name, *Pliocervus graecus*, for the Maraména sample, to which she referred most of the late Miocene antler specimens from Greece, as well as the larger mandible from Pikermi, referred by Gaudry to *Dremotherium*? sp. Later, Gentry (2005) suggested that *Pliocervus* should better be restricted to its type species, *P. matheroni*, and included the other species, previously referred to it, to *Procapreolus*. Croitor and Stefaniak (2009, p. 31) supported Gentry's suggestion by noting that *P. matheroni* may actually have had four-tined antlers. They also added that the antlers and dental samples published so far under the species name *graecus* are hardly distinguishable from those of *Procapreolus moldavicus* (Janovskaya, 1954) (see also Croitor 2014). In a well-supported reasoning, Croitor (2018b, pp. 31–32) considered *Procapreolus pentelici* (Dames, 1883) as the only valid name for the roe deer-sized Turolian deer from Greece, a senior synonym of *Pl. graecus*, and different from Gaudry's smaller-sized and little known *Dremotherium*? *pentelici*, whose antlers are unknown. It should be also noted, however, that a comparison of the available dental material from Greek localities indicated affinities with the Spanish genus *Lucentia* (see Kostopoulos 2006, p. 154), suggesting a potentially more diverse taxonomic scheme.

Type Material Pair of antlers (AMPG V 1919), collected at Pikermi in 1853 and described by Dames (1883) and Melentis (1969a). The specimens (Fig. 2a) are kept in the Museum of Palaeontology and Geology of the University of Athens.

Type Locality Upper Miocene of Pikermi (MN12).

Distribution The species is rare in Greece. Apart from its type locality, it has been documented so far at Chomaterí, Halmyropótamos, and Maraména. It is also possibly present in the faunas of Samos, Dytikó, and Perivoláki, but the lack of antler finds there does not allow a reasonably safe taxonomic attribution.

Taxonomic Remarks *Procapreolus pentelici* is a small-sized capreoline species, characterized by three-pointed, slender antlers borne on distantly situated, long and divergent pedicels (Fig. 2). The beam is circular in cross section, weakly curved, and is often ornamented by parallel, proximo-distally directed furrows. The first bifurcation occurs very high, closer to the second bifurcation than to the burr. The antlers from Chomaterí (Fig. 2b) are somewhat more robust, inserted more obliquely to the burr, and much more curved laterally. Those from Halmyropótamos are somewhat slenderer and less ornamented (rugose) than the pair from Pikermi. The taxonomic significance of these differences is currently unknown.

Procapreolus cusanus (Croizet and Jobert, 1828)

Nomenclatural and Taxonomical History The species was originally referred to the genus *Cervus*. Several junior synonyms have been coined for the same taxon during the nineteenth century, all based on samples from French localities (Heintz 1970).

Type Material The holotype is a partial skull with the right antler, figured by Croizet and Jobert (1828, pl. VIII, figs 1, 2). Because of the figure not corresponding closely to any existing museum specimen, Heintz (1970) designated as a neotype a frontal fragment bearing the right antler. The author notes, though, that the holotype and the neotype may actually be the same specimen.

Type Locality Upper Pliocene of Les Étouaires, Auvergne, France (MN16b).

Distribution In Greece, this species is only known from the Pliocene of Géphyra.

Taxonomic Remarks *Procapreolus cusanus* is a small, roe deer-sized species, characterized by rather short, three-pointed antlers borne on distantly situated, long and divergent pedicels. Certain mature individuals may develop as many as five tines arranged sagittally, which is a peculiarity of this species among others in the same genus (Heintz 1970; Croitor 2018b). The beam is oval in cross section, exhibiting a lateral compression, especially between the first and the second bifurcation. This latter character is not observed in the Géphyra skull (Crégut-Bonnoure and Tsoukala 2017a, fig. 7A–D), which prompted Croitor (2018b) to question the attribution of the Géphyra sample to *Pr. cusanus*, considering it as closer to the Turolian *Pr. pentelici*. This observation is certainly well founded; however, the Géphyra deer is retained here provisionally in *Pr. cusanus* for biostratigraphical reasons.

⊙*Capreolus* Frisch, 1775

Type Species *Cervus capreolus* Linnaeus, 1758.

Remarks It is a small-sized and slender-built telemetacarpal cervid, bearing short, simple, three-pointed antlers, with nearly straight beam. It is extant with two species, one in Europe, *C. capreolus*, and one in Asia, the somewhat larger-sized *C. pygargus* (Pallas, 1771), living in Siberia and China. The genus is known in the European fossil record since the late Early Pleistocene with the species *C. cusanoides* Kahlke, 2001, from Untermaßfeld, Germany, a roe deer similar in body size to the extant European one, but closer to the supposedly ancestral genus *Procapreolus* in antler and dental morphology (Kahlke 2001, p. 465). Another species defined on a sample from Germany, *C. suessenbornensis* Kahlke, 1956, from the lower Middle Pleistocene of Süßenborn, is a large-sized roe deer, but otherwise morphologically similar to *C. capreolus*. Fossils of late Middle Pleistocene age, or more recent, are referred to the extant species.

⊙*Capreolus capreolus* (Linnaeus, 1758)

Nomenclatural and Taxonomical History Classified by Linnaeus in the genus *Cervus* in its original very broad meaning, *C. capreolus* was soon separated in the genus or subgenus level from the other cervids. It has a long list of synonyms based on local samples (see, e.g., Sempéré et al. 1996).

Type Material Unknown.

Type Locality Undefined region in Sweden.

Distribution The species is frequently found in Middle and Upper Pleistocene sites of Greece, but less often than the contemporary *Cervus* and *Dama*. It is recorded so far in at least 13 localities, being the rarest cervid species in each fossil sample.

Taxonomic Remarks *Capreolus capreolus* is characterized by its small-sized and slender bones and its short, three-pointed antlers. These are borne on closely positioned, caudally inclined pedicels. The beam is weakly curved and bears pearling in its proximal part. The first bifurcation is found very high above the burr, closer to the second bifurcation.

⊙*Alces* Gray, 1821

Type Species *Cervus alces* Linnaeus, 1758.

Remarks *Alces* is the largest-sized extant cervid genus and one of the most aberrant morphologically, featuring a long head with laterally developing antlers, a short neck, a massive trunk, and very long legs. The antlers are usually palmate, although this character varies intraspecifically, even within the same population. Fossil samples are often classified in different genera, such as *Cervalces* Scott, 1885, and *Libralces* Azzaroli, 1952 (the latter currently regarded as a junior synonym of the former genus), which are distinguished from the living species by different body sizes, the less retracted nasals and the larger and more laterally expanded antlers. However, these differences seem to occur gradually among chronologically successive populations, likely indicating an anagenetic evolutionary procedure. If this is true, then both the extant and the fossil genera would be paraphyletic. Moreover, morphologically and dimensionally marginal samples are often impossible to refer to a genus or species. For these reasons, all the fossil species are often classified in the extant genus (e.g., Lister 1993; Croitor 2014, 2018b). This classification is followed here.

The extant species, *A. alces*, has been reported from an undefined locality in the Haliákmon River valley, based on a partial skull (Melentis 1966a). The specimen does not exhibit, though, the widely divergent, essentially horizontal pedicels of this species. Given its very large size, the find is probably better attributable to a megacerine genus, such as *Praemegaceros* Portis, 1920.

Alces latifrons (Johnson, 1874)

Nomenclatural and Taxonomical History It was originally classified in the genus *Cervus* as a very large-sized species, larger than the extant *A. alces*. It is often indistinguishable from the geochronologically older and smaller-sized *A. carnutorum* (Laugel, 1862), a name which has priority over *latifrons*, in case a synonymy between the two be accepted.

Type Material Frontal bone fragment bearing the left antler, M.6227, curated in the Natural History Museum in London.

Type Locality Middle Pleistocene of Hasbro, a coastal site between Cromer and Yarmouth, Norfolk, Great Britain.

Distribution The species has been reported from Pentávryssos, Ptolemaís Basin. Another reported occurrence at Kaiáfas, Peloponnese, is unconfirmed.

Taxonomic Remarks It is an elk species of very large size. The skull has less retracted nasals with regard to the extant elk, so that they are in contact with the premaxillaries. The antlers are massive, with laterally expanded beam, which is intermediate in length between those of *A. gallicus* and *A. alces*.

Cervinae Goldfuß, 1820

Type Genus *Cervus* Linnaeus, 1758.

Metacervocerus Dietrich, 1938

Type Species *Cervus pardinensis* Croizet and Jobert, 1828.

Remarks This genus includes early Pliocene–Early Pleistocene (Villafranchian) European taxa of *Axis*-like deer, characterized by simple antlers with two bifurcations (three-pointed antlers). *Metacervocerus* is of medium body size, similar to that of the fallow deer. Several named species belong to this genus-rank group, which was usually referred to as "*Cervus*" or *Cervus sensu lato*, acknowledging its difference from the true *Cervus*. Nevertheless, some authors classified this group to the genus *Cervus*, under its current strict meaning (e.g., Spaan 1992; de Vos et al. 1995). Other proposed names of genus rank include *Pseudodama* Azzaroli, 1992, *Rusa* Smith, 1827, and *Dama* Frisch, 1775, as these Villafranchian populations have been considered ancestral to the extant fallow deer (Azzaroli 1992; Pfeiffer 1999). Croitor and Bonifay (2001) and Croitor (2006a) recognized the independent evolutionary identity of this group and resurrected the genus name *Metacervocerus* (originally coined in subgenus rank), as the oldest available and appropriate name.

Metacervocerus rhenanus (Dubois, 1904)

Nomenclatural and Taxonomical History The species has a fairly complicated nomenclatural history, because each major sample from European localities was considered taxonomically distinct and was given a different name. Species names that have been coined so far include *Cervus (Axis) rhenanus* Dubois, 1904, *C. philisi* Schaub, 1941, *C. perolensis* Azzaroli, 1952, and *C. ischnoceros* Bœuf, Geraads and Guth, 1992. *Cervus philisi* is the most common name in the literature, because it was used for the very rich medium-sized deer samples from the French localities of Senèze, Saint-Vallier, and Chilhac (Viret 1954; Heintz 1970; Boeuf 1983; Valli 2004), which have since been used as reference material. Viret (1954) had already recognized the similarities between the samples from Tegelen (*C. rhenanus*) and Saint-Vallier, but it was since the work of Spaan (1992) and the subsequent publications of de Vos et al. (1995) and Kahlke (1997) that these four species names were considered as true synonyms, with *C. rhenanus* having priority. *Metacervocerus rhenanus* has been split to various local subspecies, such as *valliensis* Heintz, 1970 (from Saint-Vallier), *philisi* Schaub, 1941 (from Senèze), *rhenanus* Dubois, 1904 (from Tegelen), and *perolensis* Azzaroli, 1952 (from Peyrolles) (arranged in chronological succession, from older to more recent).

Type Material The species was defined on the left antler Ha-15-777 (holotype), figured and described by Dubois in 1904 and 1905 (Spaan 1992, pp. 23–24). The holotype is kept in the Teylers Museum in Haarlem, The Netherlands.

Type Locality Clays of Tegelen, The Netherlands, Lower Pleistocene (MNQ18).

Distribution This species is documented in faunas of Villafranchian type across the territory of Greece—Vólakas, Dafneró, Sésklo, Gerakaroú, and Tsiótra Vrýssi—while a similar form occurs at Tourkoboúnia. The species has been also reported, without any information about the available finds, from an undefined site on the island of Kos, as well as from Apolakkiá on Rhodes (Fig. 3). Note, however, that the latter sample is morphologically aberrant and considered to be of Pliocene (Ruscinian) age, significantly older than the other localities, which are placed biostratigraphically in the Early Pleistocene (MNQ17–18, possibly also MNQ19).

Taxonomic Remarks *Metacervocerus rhenanus* is the most commonly occurring deer in the faunas of Villafranchian type. It features an underived cranial morphology with long neurocranium that is little flexed with regard to the face. The antlers are simple, three-pointed, developing on rather long and caudally inclined pedicels. The upper teeth have well-developed cingulum. The p4 is usually weakly molarized in variable degree. There is no *Palaeomeryx* fold in the lower molars. Detailed morphological descriptions were given by Heintz (1970) and more recently by Croitor (2006a, 2018b).

The sample from Apolakkiá, referred to *Cervus* aff. *philisi* by van de Weerd et al. (1982), is dimensionally close to *M. rhenanus*, although it presents a fairly high variation. The antler morphology diverges from the typical one of *M. rhenanus* in having a disproportionally long beam between the first and the second bifurcation

(Fig. 3; compare, e.g., Viret 1954, p. 106; Heintz 1970, p. 152; Spaan 1992, pp. 25–26). Moreover, the beam becomes mediolaterally flattened, even concave, below the second bifurcation, which seems to form a quite acute angle. According to Croitor (pers. com. 2020), this morphology may indicate an affinity to *Praeelaphus*. The whole sample certainly needs a revision.

Eucladoceros Falconer, 1868

Type Species *Cervus (Eucladoceros) sedgwickii* Falconer, 1868.

Remarks *Eucladoceros* is a large-sized Eurasian cervid genus, similar in size with the extant red deer, which lived during the late Pliocene and Early Pleistocene. It is characterized by long laterodorsally directed antlers with four to six tines. The tines developed dorsal to the beam, subparallel to each other in a comb-like form, and were often branched. A certain diversity in the antler patterns has caused the erection of several species, based on specimens of different populations or ontogenetic stages. Currently, two species are accepted in Europe, *E. dicranios* (Nesti, 1841) and *E. ctenoides* (Nesti, 1841) (de Vos et al. 1995; Croitor 2018b). The former is the most advanced, characterized by bi- or trifurcated tines that give a bush-like appearance to the skull, obscuring the underlying comb-like structure. It occurs scarcely in the fossil record, being known from Italy, the Azov region (Russia) and Great Britain (originally described as *Cervus sedgwickii* Falconer, 1868). *Eucladoceros ctenoides* had simpler antlers with conspicuous comb-like morphology and is commonly found in European Villafranchian faunas. It is also a frequently reported species from Greek localities. The genus is known in Asia as well, particularly in China. One of the Chinese species, *E. boulei* Teilhard de Chardin and Piveteau, 1930, is morphologically close to an antler find from Kapetánios, N. Greece, according to Steensma (1988).

This partially preserved antler (Fig. 4) has some morphological peculiarities, such as the wide and mediolaterally flattened beam, and the closely positioned second to fourth tines. The first tine is positioned directly above the burr, unlike the Chinese species *E. boulei*. Compared to the European *E. ctenoides*, it has a more robust beam and an additional tine, though weakly developed with regard to the others. Supposing that the presence of this tine might be an aberrant feature of this individual, an attribution to *E. ctenoides* is not unlikely. Croitor (2018b, p. 81) hypothesized that the *Eucladoceros* from Kapetánios might have been the ancestral form of all European populations of this genus, which is, however, not consistent with the site's late age (MNQ19). Due to its unusual morphology, the Kapetánios antler is better referred for the moment to *Eucladoceros* sp., in the hope that more relevant material from this locality will become available in the future.

Eucladoceros cf. *ctenoides* (Nesti, 1841)

Nomenclatural and Taxonomical History The species was originally classified in the genus *Cervus* in its broader sense, based on a part of an antler from Upper

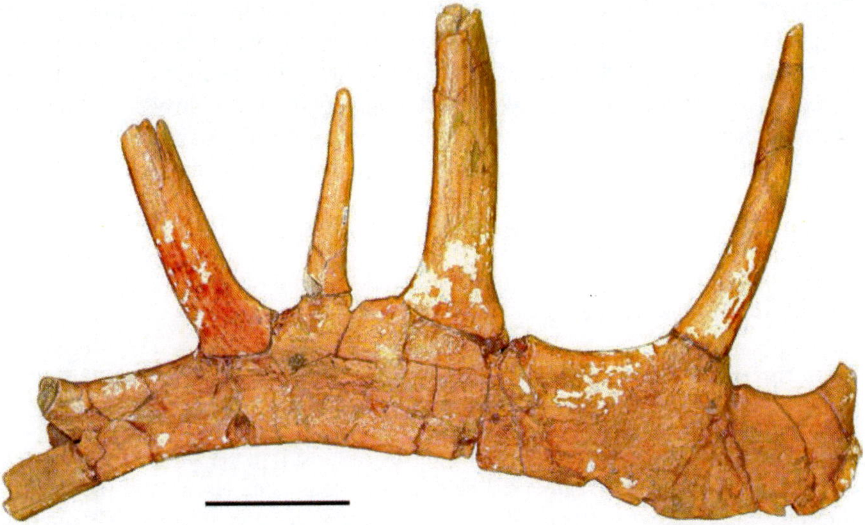

Fig. 4 Left antler part, in lateral view (proximal end is on the left), from the Early Pleistocene locality of Kapetánios, W. Macedonia, referred to *Eucladoceros* sp. The specimen was described by Steensma (1988) as *Eucladoceros* aff. *boulei*. It belongs to the collections of the Geology–Palaeontology Museum of the Aristotle University of Thessaloniki. The original photograph used for this figure was provided by Prof. D. Kostopoulos. Scale bar equals 10 cm

Valdarno, N. Italy. In the following years several similar specimens from many European localities were described as separate species, such as *Cervus orticeros* Nesti, 1841, *C. falconeri* Dawkins, 1868, *C. tetraceros* Dawkins, 1878, *C. tegulensis* Dubois, 1904, *C. senezensis* Depéret, 1910, etc. (see de Vos et al. 1995 for a complete list). However, as shown by de Vos et al. (1995), all these species share the same pattern in their antler shape, and they are also biometrically similar. Consequently, they are considered as synonyms, with *C. ctenoides* having priority. Some of them are occasionally used in subspecific rank to characterize local samples.

Type Material The species was defined on the right partially preserved antler IGF-377 from Upper Valdarno in N. Italy, kept in the Museum of Natural History of the University of Florence.

Type Locality Upper Valdarno, N. Italy, Lower Pleistocene.

Distribution *Eucladoceros* cf. *ctenoides* is present in four faunas in Greece: Vólakas, Dafneró, Krímni, and Gerakaroú. Another occurrence, at undefined locality/ies on Kos Island is unconfirmed, because the species was just mentioned in a faunal list. The large cervids from Sésklo and Reghínio could belong to the same species, but they are not identifiable further than the genus level. The *E.* cf. *ctenoides*-bearing faunas of Greece are placed biostratigraphically in the Early Pleistocene (MNQ17–19).

Taxonomic Remarks *Eucladoceros ctenoides* is diagnostically identified by the peculiar comb-like pattern of its antlers, consisting of a long beam with four to five tines developing in parallel on the dorsorostral side of the beam and perpendicular to it. The skull remains least derived, exhibiting very obtuse angle between the neurocranium and the splachnocranium, large ethmoidal openings and small tympanic bullae. The presence of this species in Greece is based on dental similarities with samples from W. Europe. However, the lack of antlers in all four localities makes this specific attribution just tentative.

⊙*Rucervus* Hodgson, 1838

Type Species *Cervus elaphoides* Hodgson, 1835 (= *Cervus duvaucelii* Cuvier, 1823).

Remarks *Rucervus* is an extant genus of medium to large-sized deer, geographically confined today to fragmented habitats of South Asia (India and Indochina). Recently, the genus was expanded to include European samples previously referred to the very large-sized deer *Arvernoceros* Heintz, 1970, on the basis of similarities in the cranial and antler morphology (Croitor 2018a). *Rucervus* is characterized by long antlers, bearing a basal tine (often associated with a small accessory prong) and a multipointed crown ramification. There is no middle tine. The crown tines develop caudally and are often branched or palmate. The genus is known in Greece during the late Early Pleistocene.

★*Rucervus gigans* Croitor, 2018

Nomenclatural and Taxonomical History The sample from Apollonía 1 (Central Macedonia), currently assigned to *R. gigans*, was initially described by Kostopoulos (1996, 1997a) as *Megaloceros* sp. A few years later, van der Made (2001) included the Apollonía deer in *Eucladoceros giulii* Kahlke, 1997. A first revision of the available material in 2004 resulted in referring a part of the initial "*Megaloceros*" sample to *Arvernoceros* cf. *verestchagini* David, 1992 (Croitor and Kostopoulos 2004). Recently, Croitor (2018a) coined this new species name for the sample previously referred to *A.* cf. *verestchagini*, considering *Arvernoceros* as a junior synonym of *Rucervus*, or at least a subgenus within it. The species, however, is placed into the nominal subgenus *Rucervus (Rucervus)*, due to the low position of the basal tine (Croitor 2018a).

Type Material The species was defined on a series of syntypes that include cranial and postcranial specimens. The type material is kept in the Geology–Palaeontology Museum of the Aristotle University of Thessaloniki.

Type Locality Apollonía 1, Greece, upper Lower Pleistocene, (MNQ19).

Taxonomic Remarks This is a very large-sized species, apparently the largest classified in the genus *Rucervus*. Its antlers are characterized by a low-positioned

Fig. 5 Dorsal cranial fragment with the proximal parts of both antlers, in rostral view, from the late Early Pleistocene locality of Apollonía 1, Central Macedonia, referred to *Rucervus gigans*. The specimen belongs to the collections of the Geology–Palaeontology Museum of the Aristotle University of Thessaloniki and has a complex nomenclatural history (see text). The original photograph used for this figure was provided by Prof. D. Kostopoulos. Scale bar equals 10 cm

basal tine, which is flattened mediolaterally, and forms an obtuse angle with the beam (Fig. 5). The dentition is of extremely large size, with wide upper teeth, bilobed upper premolars, and semi-molarized p4. The limb bones are exceptionally long, suggesting a high-level-browser palaeoecological niche for this species (Croitor and Kostopoulos 2004; Croitor 2018a).

⊙*Cervus* Linnaeus, 1758

Type Species *Cervus elaphus* Linnaeus, 1758.

Remarks *Cervus* is the most typical and recognizable extant genus of the family Cervidae, as it is the most broadly dispersed. It is found from tropical to cold, arid to wet, and lowland to highland environments, being very flexible ecologically. Biochronologically, it is known in the European fossil record since the late Early Pleistocene. *Cervus* is a morphologically conservative genus, with little derived cranial and dental morphology. The antlers are large, dorsolaterally directed, with one or two tines close to the burr, a middle tine, and the crown, which is formed by two or multiple tines.

⊙*Cervus elaphus* Linnaeus, 1758

Distribution The species has been ubiquitous in Greece, being present in most Middle and Late Pleistocene localities (almost 40).

Taxonomic Remarks *Cervus elaphus* is a large-sized deer. Morphologically, it is characterized by the presence of large, multipointed antlers. The first tine (brow tine) is situated rather close to the burr, but not in contact with it, and it is accompanied by the bez tine, almost in contact with the former, but developing in a more lateral direction. The crown is a simple, sagittally directed bifurcation, in early forms (often named *C. elaphus acoronatus* Beninde, 1937), or a true crown of three or more tines, in later forms. In between the basal tines and the crown there exists a middle tine (trez tine). The dentition is characterized by a molarized p4. Due to the extended geographic range and the rather long evolutionary history of *C. elaphus*, several subspecies of local or temporal character have been coined.

⊙*Dama* Frisch, 1775

Type Species *Cervus dama* Linnaeus, 1758.

Remarks It is a medium-sized extant cervid genus, which is known since the late Early Pleistocene. It is characterized by particularly large antlers with regard to its body size, with strong first (brow) tine, situated immediately above the burr and forming with the beam a well obtuse angle. The distal tines are palmate in the advanced species, or form a simple bifurcation directed sagittally. The cranial morphology is derived, with ventrally flexed neurocranium with regard to the face, short frontal region, large orbits, large ethmoidal vacuities, and inflated tympanic bullae.

The genus is very common in Greece, particularly during the Late Pleistocene, but the antler finds are rather scarce, so that the taxonomic attributions are mainly based on size. Geologically older samples, when more species were present, are, thus, taxonomically problematic. Middle Pleistocene antler samples from Megalopolis, which have been published by Melentis (1966b, pl. III, figs. 1–3) as *Cervus (Dama) somonensis* Desmarest, 1820, do not exhibit the typical morphology of *Dama* in the relative development and orientation of the brow tine, being rather *Cervus* like. On the contrary, a basal antler fragment from Megalopolis referred by Sickenberg (1976, pl. IX, fig. 1) to *Cervus* sp., has the morphology of *Dama*. A peculiar find, which has been assigned to *Dama*, is a natural cast of a brain from the fissure-filling site Halmyrós, near Kalamáta, SW Peloponnese (Psarianos and Thenius 1954).

⊙*Dama dama* (Linnaeus, 1758)

Distribution *Dama dama* has been very common in the fossil record of Greece, being present in most deer-bearing localities (at least 18) of Middle–Late Pleistocene age.

Taxonomic Remarks This species is characterized by impressive, distally palmate antlers. The brow tine is strong, positioned very close to the burr, and forms with the beam a well obtuse angle. Older, Middle Pleistocene forms had, however, less elaborate palmations and are often classified in distinct subspecies.

Dama vallonnetensis (de Lumley, Kahlke, Moigne, and Moullé, 1988)

Nomenclatural and Taxonomical History This species was originally described as a subspecies, *"Cervus" (s.l.) nestii vallonnetensis*. In more recent publications, it was raised to the species rank. Its classification in the genus *Dama* is based on the morphology of the proximal antler, featuring a very low-positioned and robust first tine, which forms an obtuse angle with the beam.

Type Material Partially preserved right shed antler Val 10170, designated as the holotype. It is kept in the Institut de Paléontologie Humaine in Paris.

Type Locality Lower Pleistocene of the cave Vallonnet, Alpes-Maritimes, France. The type layer was dated at 980–910 ka.

Distribution Fossils exhibiting the morphology of *D. vallonnetensis* were found at Libákos.

Taxonomic Remarks The antler fragments from Libákos (Steensma 1988, pp. 194–198) belong to a *Dama* with a caudally inclined beam and a robust and strongly curved brow tine, which is in contact with the burr and forms an extremely obtuse angle with the beam (Fig. 6). Although the finds have been tentatively referred by Steensma (1988) to *C. nestii eurygonos* (= *D. eurygonos*), they better conform to the morphology of *D. vallonnetensis* (brow tine in contact to the burr; de Lumley et al. 1988; Croitor 2006a). The state of the middle tine (lacks in *D. vallonnetensis* but is present in *D. eurygonos*) is not well known in the Libákos sample, because there is no complete antler. Nevertheless, one specimen described by Steensma (1988, p. 197) preserves a sufficiently long part of the beam, which is tineless above the brow tine, in full accordance with a specific attribution to *D. vallonnetensis*.

Praemegaceros Portis, 1920

Type Species *Cervus dawkinsi* Newton, 1882.

Remarks This is a very large-sized cervid genus, present in upper Lower Pleistocene and Middle Pleistocene deposits of Europe. Insular endemic forms persist in the Mediterranean during the Late Pleistocene. Possibly related Asian forms of Early Pleistocene age, referred by Vislobokova (2012) to *Praemegaceros* sp., extend the chronological and geographic range of the genus. It is characterized by wide skull with divergent pedicels and large antlers that develop almost laterally in their proximal part and then expand dorsally, forming an abrupt bent of the beam. There is a tine close to the burr that is inserted in the dorsal side of the beam ("dorsal

Fig. 6 Cranial fragment with proximal part of the right antler, in medial view, from the Lower Pleistocene locality of Libákos, W. Macedonia, referred to *Dama vallonnetensis*. It is currently kept in the collections of the Geology–Palaeontology Museum of the Aristotle University of Thessaloniki. The original photograph used for this figure was provided by Prof. D. Kostopoulos. Scale bar equals 10 cm

tine"), while a vestigial one may exist between the latter and the burr. A tine extending caudally ("posterior tine") exists in the area of the beam bent. The crown part can form a palmation. The mandible is not pachyostotic. The species of this group were initially classified in a broadly meant genus *Cervus*, and later in *Megaceros* or *Megaloceros*. The currently used name, *Praemegaceros*, is based on a derived species, *P. dawkinsi*, endemic in Britain; thus its validity for other European species has been questioned in the past. Other proposed names of genus rank include *Megaceroides* Joleaud, 1914, and *Orthogonoceros* Kahlke, 1956. Although *Megaceroides* has priority over *Praemegaceros* and has been used extensively in the past, particularly by Italian authors, it is not considered valid, because it is based on an aberrant and phylogenetically distinct species, *M. algericus*, endemic in N. Africa. According to Croitor (2006b), the genus *Praemegaceros* in its current meaning includes three distinct clades of subgenus rank, which are named using existing, previously coined names: *P. (Praemegaceros)* Portis, 1920, *P. (Orthogonoceros)* Kahlke, 1956 (which comprises the two species present in the Greek fossil record), and *P. (Nesoleipoceros)* Radulesco and Samson, 1967. This division is, however, not universally accepted (see, e.g., Vislobokova 2013).

Praemegaceros pliotarantoides (De Alessandri, 1903)

Nomenclatural and Taxonomical History This species was erected for an antler from the Italian locality of Cortiglione Monferrato (Piedmont). It was initially referred to the genus *Cervus* (in its broad sense). A very similar specimen from Psekups Valley (Northern Caucasus, Russia) was described by Radulesco and Samson (1967) as the new genus and species *Psekupsoceros orientalis*, but the two species were readily considered as synonyms.

Type Material Left shed antler, stored in the Geological Museum of Turin, Italy.

Type Locality Cortiglione Monferrato (Piedmont, Italy), upper Lower Pleistocene.

Distribution *Praemegaceros pliotarantoides* is known from four sites in Greece: Haliákmon Valley (Véria or Polýlakkos; see relevant comment in Sect 3; Fig. 7), Apollonía, and Kalamotó.

Taxonomic Remarks The antlers of this species are characterized by the absence of a middle tine, as well as of a distal palmation (Fig. 7). The dorsal tine is rather short and may be accompanied by a vestigial one. The beam is cylindrical and bends sharply at the level of the posterior tine. The crown has two branched tines. Due to

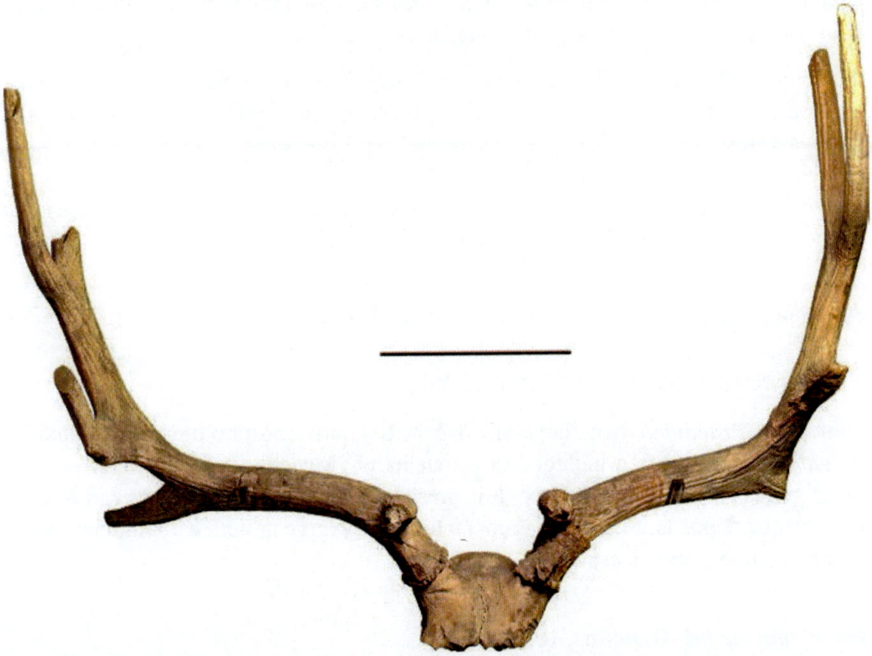

Fig. 7 Cranial part with both antlers of *Praemegaceros pliotarantoides* (De Alessandri, 1903) from the Haliákmon Valley; rostral view. The specimen was published by Melentis (1967, 1968) and belongs to the collections of the Museum of Palaeontology and Geology, University of Athens, which provided the original photograph. Scale bar equals 30 cm

the overall similarity with *P. verticornis*, the validity of *P. pliotarantoides* has been challenged, being considered a junior synonym of the former by most authors (e.g., Azzaroli 1979; Pfeiffer 2002; Abbazzi 2004). Croitor (2006b) observed that the defining characters of the species are constant within each population, so that they cannot be attributed to intraspecific variation and resurrected the species name. This taxonomic view is provisionally followed here, although it is still not generally accepted (e.g., Vislobokova 2013).

Praemegaceros verticornis (Dawkins, 1872)

Nomenclatural and Taxonomical History Dawkins (1872) based this species on two antlers from the Pakefield (Great Britain), one of which is adult and the other juvenile. He referred his new species to *Cervus*, in its broad sense of that time.

Type Material Dawkins (1872) did not designate a type for this species. In 1882, Newton chose the adult specimen described by Dawkins—kept in the Norwich Museum—as the lectotype of the species (Radulesco and Samson 1967).

Type Locality Middle Pleistocene of Pakefield, Suffolk, Great Britain.

Distribution In Greece, the species was documented at Kyparíssia, while its presence in other sites within the Megalopolis Basin is also probable. Its occurrence at Kaiáfas is reported in a faunal list and is not confirmed.

Taxonomic Remarks This is the most commonly found species of *Praemegaceros* in the European faunas. It has also been regarded as a senior synonym of most of the species referred to the genus, considering the differences among them as mere expression of intraspecific variation (e.g., Pfeiffer 2002). Its differences with *P. pliotarantoides* consist of the presence of a middle tine, developing rostrally at the level of the beam bent, as well as of a crown palmation.

Praedama Portis, 1920

Type Species *Cervus savini* Dawkins, 1887.

Remarks *Praedama* is a large-sized cervid genus morphologically related to *Megaloceros*, and often included in it, but its phylogenetic relationships with the latter genus remain yet undefined. It is present in European upper Lower to Middle Pleistocene deposits. *Praedama* is characterized by large antlers with flattened basal tine and a non-palmate crown.

Praedama savini (Dawkins, 1887)

Nomenclatural and Taxonomical History The species was originally referred to the genus *Cervus*. In more recent publications, it was grouped with other large-sized species in *Megaceros*, or *Megaloceros*. Its classification in the separate genus *Praedama* is based on differences in antler morphology. *Dolichodoryceros sues-*

senbornensis Kahlke, 1956, is a junior synonym (Croitor 2014; van der Made 2015). The recently described species *Megaloceros novocarthaginiensis* van der Made, 2015, has also *Praedama* morphology (van der Made 2015; Croitor 2018b).

Type Material Partially preserved left shed antler M. 6093, designated by Dawkins as the holotype. It is kept in the Natural History Museum in London.

Type Locality Middle Pleistocene of Trimingham, Norfolk, Great Britain.

Distribution Antler and cranial specimens morphologically similar to *P. savini* are present in the faunas of Libákos and Kapetánios (Steensma 1988, pp. 168–175; van der Made and Tong 2008).

Taxonomic Remarks *Praedama savini* is a rare species and it is incompletely known. Its antlers bear a low and wide basal tine, at least in its basal part. The crown is formed by three long tines, lacking a palmation.

Megaloceros Brookes, 1828

Type Species *Alce gigantea* Blumenbach, 1799.

Remarks *Megaloceros* is an emblematic genus, as it comprises the impressive true giant deer of the European Pleistocene. It is perhaps the largest deer ever lived. It is known in the fossil record from the late Middle Pleistocene to the early Holocene, but most of the available samples date from the very last part of its biochronologic range. In the current taxonomic practice, the genus is essentially monotypic (van der Made and Tong 2008; Vislobokova 2012, 2013; Croitor 2014), although Croitor (2018b) and van der Made (2019) added two additional species from the Middle East and Spain respectively, while *Praedama savini* is often included in *Megaloceros* (e.g., van der Made 2019). Despite its antiquity, the name *Megaloceros* (emended spelling of the original *Megalocerus*) has only recently started to be used systematically in the literature, since the recognition of its priority over the previously commonly used *Megaceros* Owen, 1844 (Lister 1987b; see also Simpson 1945, p. 154). Moreover, *Megaceros* and *Megaloceros* were used for many decades with a much broader meaning, grouping all the Pleistocene very large-sized species (most importantly those currently classified in *Praemegaceros*), despite their different morphologies. *Megaloceros* developed immense, laterodorsally expanded antlers with very large distal palmation, and characteristically pachyostotic cranial skeleton.

Megaloceros giganteus (Blumenbach, 1799)

Nomenclatural and Taxonomical History Initially classified in the genus *Alce* as *Alce gigantea*, the species was later transferred to the genera *Cervus* (in a meaning roughly equivalent to Cervini), *Megaloceros*, and *Megaceros*. Several species names have been coined in the past for local samples of *M. giganteus*, all of which are unequivocally considered as junior synonyms of it, or are used occasionally in the subspecies level.

Type Material The material originally described by Blumenbach is unknown and its whereabouts cannot be traced in any collection.

Type Locality Upper Pleistocene of Ireland, unknown local provenance.

Distribution The species' distribution in Greece is very patchy. It is recorded so far at Angítis, based on the very large size of the recovered dental material and the age of the locality; its presence in Agios Geórgios and Apídima Caves is also probable for the same reasons.

Taxonomic Remarks *Megaloceros giganteus* is readily characterized by its remarkable antlers, with dorsoventrally flattened basal tine and an immense distal palmation. The skull roof is very thick, with sutures fused in an ontogenetically early stage, long nasals, and very small ethmoidal vacuities. The mandible is extremely thick mediolaterally.

Cervidae incertae sedis

Croizetoceros Heintz, 1970

Type Species *Cervus ramosus* Croizet and Jobert, 1828.

Remarks This is a monotypic late Pliocene–Early Pleistocene genus that lived in Western and Southern Europe. It was never reported outside Europe. Despite its small body size, *Croizetoceros* had long and impressive antlers with relatively high-positioned first tine and several crown tines, emerging closely to each other on the rostral side of the beam. This morphology is unique and separates this genus clearly from its contemporary genera. Apart from its genotype, other species, such as *Cervus pyrenaicus* Depéret, 1890, are often classified in this genus (but also as subspecies of *C. ramosus*). The medium-sized *Croizetoceros proramosus* Dong, 1996, from the Pliocene (MN15) of Serrat d'en Vacquer, France, is known only from dental specimens, not antlers (Dong 1996); thus its classification in this genus is yet uncertain.

Croizetoceros ramosus (Croizet and Jobert, 1828)

Nomenclatural and Taxonomical History The species was originally described as belonging to the genus *Cervus* in a broader sense, a practice that was followed until 1970, when the genus *Croizetoceros* was erected for it (Heintz 1970). Samples belonging to this species have also been described under different names. The synonym list in Heintz (1970) includes, among others, *Cervus cladocerus* Pomel, 1853, *Cervus polycladus* Gervais, 1859, and *Cervus cylindrocerus* Dawkins, 1878. Moreover, the species *ramosus* has been classified under *Anoglochis* Croizet and Jobert, 1826, *Polycladus* Pomel, 1853, and *Cylindroceros* Teilhard de Chardin and Piveteau, 1930, in genus or subgenus rank. Local samples of *Cr. ramosus* have been assigned to subspecies to underline subtle differences from the type sample in size or morphology. Among them is *Croizetoceros ramosus gerakarensis* Kostopoulos and Athanassiou, 2005, erected initially in 1996 for the sample for Gerakaroú, Central Macedonia (Kostopoulos 1996). The authorship of the subspecies refers to its first published description, because the original one does not fulfill the requirements of the ICZN for the publication of scientific names.

Type Material The species was defined on an antler figured by Croizet and Jobert (1828). As none of the available museum specimens corresponds to the figured holotype, Heintz (1970) designated as a neotype another antler figured in the same publication, which is curated in the Natural History Museum in Paris.

Type Locality Les Etouaires, Auvergne, France, upper Pliocene (MN16b).

Distribution This species is reported from several localities of Villafranchian type across the territory of Greece: Miliá, Vólakas, Sésklo, Kos Island (undefined localities), Kastrítsi, and Gerakaroú. Its biostratigraphic range in the Greek fossil record is MN16–MNQ18, the same as in W. Europe.

Taxonomic Remarks This is a rather small-sized deer species, characterized by the presence of large and complex antlers, featuring a long and curved tine situated at some distance from the burr, and a cluster of distal subparallel tines that develop toward a dorsorostral direction. The frontal area is wide and almost flat between the pedicels, but becomes narrow caudally to them. The dentition is derived, exhibiting constantly molarized p4 (Heintz 1970). The single subspecies defined on a sample from Greece, *Cr. ramosus gerakarensis*, is metrically smaller than the W. European samples, while it also differs morphologically in its shorter and lower-positioned first tine, and its proportionally shorter premolar row (Kostopoulos 1996; Kostopoulos and Athanassiou 2005).

Genus incertae sedis

★*"Cervus" peloponnesiacus* Sickenberg, 1976

Nomenclatural and Taxonomical History The species was founded for a collection of small-sized antler fragments, as well as dental and postcranial material from several findspots within the Megalopolis Basin, Peloponnese.

Type Material Fragment of frontal bone bearing the basal part of the antler (9-I-2), kept in the Museum of Palaeontology and Geology, University of Athens.

Type Locality Middle Pleistocene Marathousa Member, 600 m NNW of the village Karvounáris, Megalopolis Basin, Peloponnese.

Distribution Known only in sites and findspots within the Megalopolis Basin.

Taxonomic Remarks This is a very incompletely known species of small size. Its antlers bear a first tine, situated close but clearly above the burr, and forming an obtuse angle with the beam. The beam is slightly oval in cross section. Except for the proximal (basal) part, the morphology of the rest of the antler remains unknown. Sickenberg (1976) referred also dental and postcranial specimens to his new species, based on their size. However, the presence of an additional small cervid in the Megalopolis Basin, documented from a muntjak-like antler fragment of obscure taxonomic affinities (Cervidae indet. in Athanassiou 2018), excludes all non-antler material from the hypodigm of "*C.*" *peloponnesiacus*.

Invalid Taxa

Several poor samples, lacking diagnostic characters, have been identified in the past to the species level, without any proper documentation. These include *Libralces gallicus* from Sotíras, W. Macedonia (Stratigopoulos 1986), *Eucladoceros falconeri* and *Cervus philisi* from Perdíkkas, W. Macedonia (Stratigopoulos 1988, 1989), *Alces alces* from an undefined site along the Haliákmon Valley (Melentis 1966a), and *Cervus dicranius?* from Kardámaena on Kos Island (Airaghi 1928). The latter belongs to a significantly smaller-sized deer than *Eucladoceros*, as is evident from the dimensions of the sample's most diagnostic specimen (a mandibular part with d4–m2) (compare, e.g., Heintz 1970, tables 84, 87).

Paraskevaidis (1940) described a middle Miocene fauna from Thymianá, Chios Island, which, according to him, includes two fragmentary specimens referable to Cervidae. The author erected the new genus and species *Georgiomeryx georgalasi* based on a mandibular fragment, while he referred the second specimen, a part of a maxilla, to *Cervus* aff. *lunatus*. However, more recent research in this locality, which produced more and better preserved material, has shown that *Georgiomeryx* actually belongs to the family Giraffidae (Lehmann and Tobien 1995; de Bonis et al. 1997; see also Iliopoulos and Roussiakis this volume), while *C.* aff. *lunatus* is a bovid (Lehmann and Tobien 1995). Nevertheless, Lehmann and Tobien (1995) also reported new cervid material from the same locality, which they referred to aff. *Euprox furcatus* (Hensel, 1859). The available sample (4 isolated teeth) is, though, insufficient for any taxonomic identification further than the family level.

5 Conclusions

Although deriving from a hundred localities, the deer fossil record of Greece is comparatively not so rich in number of taxa and recovered specimens, particularly with regard to the record of another ruminant family, the Bovidae (see Kostopoulos this volume). Nevertheless, it exhibits a considerable taxonomic diversity, particularly in post-Pliocene faunas. Five species names are based on samples from Greece: *Dremotherium*? *pentelici*, *Cervus pentelici*, "*Cervus*" *peloponnesiacus*, *Pliocervus graecus*, and *Rucervus gigans*. *Pl. graecus* is considered as a junior synonym of *C. pentelici*, while *Dr.*? *pentelici* remains of uncertain status. Moreover, a subspecies, *Croizetoceros ramosus gerakarensis*, is based on a local sample. Most of the cervid-bearing faunas are dated to the Middle and Late Pleistocene, and comprise most abundantly the extant genera *Cervus* and *Dama*. The increased representation of deer in these faunas may be related to the climate cooling toward the end of the Pleistocene. Similarly increased is the presence of deer in localities with rich vegetation cover, like the Middle Pleistocene lacustrine basin of Megalopolis.

Acknowledgments The author wishes to thank S. Roussiakis and G. Lyras (National and Kapodistrian University of Athens) as well as D. Kostopoulos (Aristotle University of Thessaloniki) for providing specimen photographs and for useful discussions and valuable help during the writing of this chapter. A review of the manuscript by Roman Croitor greatly improved its quality.

Appendix

List of fossil Cervidae-bearing localities in Greece, arranged in approximate geochronological order (chronological data taken from the corresponding original publications cited herein and from Facorellis 2013). The taxonomic identifications follow the present revision. Locality numbers refer to the collection numbers of the PaleoBiology Database (PBDB)

Localities[PBDB No]	Age (biozone)	Taxon
Limnón Cave	latest Pleistocene?	*Cervus elaphus*[1]
Gráva Rockshelter	Late Pleistocene (MNQ26)	*Cervus elaphus*[2] *Dama* sp.[2] *Capreolus* sp.[2]
Schistó Cave	Late Pleistocene (MNQ26)	*Cervus* cf. *elaphus*[3]
Agios Geórgios Cave	Late Pleistocene (MNQ26)	*Cervus elaphus*[4] *Megaloceros* sp.[4]
Dragoudélis	Late Pleistocene (MNQ26)	*Capreolus capreolus*[5]
Vraóna Doline	Late Pleistocene (MNQ26)	*Cervus elaphus*[6] *Dama dama*[6] *Capreolus capreolus*[6]
Seidí Rockshelter	Late Pleistocene (MNQ26)	*Cervus elaphus*[7]
Boíla Rockshelter	Late Pleistocene (MNQ26)	*Cervus elaphus*[8] *Dama dama*[8] *Capreolus capreolus*[8]
Tzínes, Thassos	Late Pleistocene (MNQ26)	*Cervus elaphus*[9] *Dama dama*[9] *Capreolus capreolus*[9]
Megálakkos Rockshelter	Late Pleistocene (MNQ26)	*Cervus elaphus*[10]
Klithí Rockshelter	Late Pleistocene (MNQ26)	*Cervus elaphus*[11]
Franchthi Cave	Late Pleistocene (MNQ26)	*Cervus elaphus*[12]
Kastrítsa Rockshelter	Late Pleistocene (MNQ26)	*Cervus elaphus*[13] *Dama dama*[13] *Capreolus capreolus*[13]
Tripsána Cave	Late Pleistocene (MNQ26)	*Cervus elaphus*[14] *Dama dama*[14]
Melitziá Cave	Late Pleistocene (MNQ26)	*Cervus elaphus*[14]

(continued)

Localities[PBDB No]	Age (biozone)	Taxon
Kastánis Cave	Late Pleistocene (MNQ26)	*Cervus elaphus*[14] *Dama dama*[14]
Glypháda Cave, Dyrós	Late Pleistocene (MNQ26)	*Dama dama*[15]
Skiní 4 Cave	Late Pleistocene (MNQ26)	*Cervus elaphus*[14] *Dama dama*[14]
Kolomínitsa Cave	Late Pleistocene (MNQ26)	*Cervus elaphus*[14] *Dama dama*[14] *Capreolus capreolus*[14]
Angítis[204391]	Late Pleistocene (MNQ26)	*Megaloceros giganteus*[16] *Cervus elaphus*[16]
Klissoúra Cave[184243]	Late Pleistocene (MNQ26)	*Cervus elaphus*[17] *Dama dama*[17] *Capreolus capreolus*[17]
Kephalári Cave	Late Pleistocene (MNQ26)	*Cervus elaphus*[17] *Dama dama*[17] *Capreolus capreolus*[17]
Kítsos Cave	Late Pleistocene (MNQ26)	*Cervus elaphus*[18] *Dama dama*[18]
Asprocháliko Rochshelter	Late Pleistocene (MNQ26)	*Cervus elaphus*[13] *Dama dama*[13] *Capreolus capreolus*[13]
Peniós Valley[204397]	Late Pleistocene (MNQ26)	*Megaceros* sp.[19] *Cervus elaphus*[19] *Dama* sp.[19] *Capreolus capreolus*[19]
Kalamákia Cave[184245]	Late Pleistocene (MNQ26)	*Cervus elaphus*[14] *Dama dama*[14] *Capreolus capreolus*[14]
Theópetra Cave	Late Pleistocene (MNQ26)	*Cervus elaphus*[20] *Dama dama*[20] *Capreolus* sp.[20]
Lakonís	Late Pleistocene (MNQ26)	*Cervus elaphus*[21] *Dama dama*[21] *Capreolus capreolus*[21]
Phytídi Cave	Late Pleistocene?	*Cervus elaphus*[22]
Nymphs Cave, Ithaca	Late Pleistocene?	*Cervus elaphus*[a]
Lágoura[204558]	Late Pleistocene?	?*Megaloceros* sp.[23]
Xeriás[182679]	Middle–Late Pleistocene	*Megaloceros* sp.[24]
Phílippi[204747]	Middle–Late Pleistocene	*Cervus elaphus*[25]
Sotíras[182678]	Middle–Late Pleistocene	Cervidae indet.[26,a] *Cervus elaphus*[25]
Petralona Cave[183123]	Middle–Late Pleistocene	*Praemegaceros* sp.[27] *Cervus elaphus*[27] *Dama dama*[27]
Aeginio	Middle–Late Pleistocene	*Cervus elaphus*[28]
Apídima Cave B	Middle–Late Pleistocene	*Cervus elaphus*[29] *Dama dama*[29]

(continued)

Localities[PBDB No]	Age (biozone)	Taxon
Apídima Cave C[32082]	Middle–Late Pleistocene	*Megaloceros* sp.[29] *Cervus elaphus*[29] *Dama dama*[29]
Ptolemaís Basin (unspecified locality)	Middle Pleistocene	?*Praemegaceros* sp.[26,a]
Pentávryssos	Middle Pleistocene	*Alces latifrons*[30]
Perdíkkas[204753]	Middle Pleistocene	Cervidae indet.[31,a]
Haliákmon Valley (unspecified locality)	Middle Pleistocene?	?*Praemegaceros* sp.[28,a]
Marathousa 1[187637]	Middle Pleistocene (MNQ23)	*Cervus elaphus*[32] *Dama* sp.[32]
Kyparíssia 1[194472]	Middle Pleistocene (MNQ22)	*Praemegaceros verticornis*[33] *Cervus elaphus*[33] *Dama* sp.[33]
Kyparíssia 2	Middle Pleistocene (MNQ22)	*Praemegaceros verticornis*[33]
Kyparíssia 3[194474]	Middle Pleistocene (MNQ22)	*Praemegaceros verticornis*[33] *Cervus elaphus*[33] *Dama* sp.[33]
Kyparíssia 4[194475]	Middle Pleistocene (MNQ22)	*Praemegaceros verticornis*[33] *Cervus elaphus*[33] *Dama* sp.[33]
Kyparíssia KYP	Middle Pleistocene (MNQ22)	*Praemegaceros verticornis*[33] *Cervus elaphus*[33] *Dama* sp.[33] *"Cervus" peloponnesiacus*[33] Cervidae indet.[33]
Megalopolis Basin[182721]	Middle Pleistocene	*Praemegaceros verticornis*[34] *Cervus elaphus*[34] *Dama* sp.[34,a] *"Capreolus"* sp.[34] *"Cervus" peloponnesiacus*[34]
Apollonía[34784]	late Early Pleistocene (MNQ19)	*Rucervus gigans*[36] *Praemegaceros pliotarandoides*[37]
Kalamotó 1[200083]	late Early Pleistocene (MNQ19)	*Praemegaceros pliotarandoides*[35] *Cervus* sp.[35] *Dama* sp.[35]
Kalamotó 2[200084]	late Early Pleistocene (MNQ19)	*Praemegaceros pliotarandoides*[35] *Cervus* sp.[35] *Dama* sp.[35]
Platanochóri[204658]	late Early Pleistocene (MNQ19)	Cervidae indet.[38]
Kaiáfas[34766]	late Early Pleistocene (MNQ19)	*Praemegaceros verticornis*[39] *Alces* cf. *latifrons*[39] Cervidae indet.[39]

(continued)

Localities[PBDB No]	Age (biozone)	Taxon
Kapetánios[34781]	Early Pleistocene	*Eucladoceros* sp.[40,a] *Praedama* aff. *savini*[40,a]
Libákos[34764]	late Early Pleistocene (MNQ19)	*Praedama* aff. *savini*[40,41] ?*Eucladoceros* sp.[40] *Dama vallonnetensis*[40,a]
Krímni	Early Pleistocene (MNQ 18–19)	*Eucladoceros* cf. *ctenoides*[42,43,a]
Tsiótra Vrýssi[197943]	Early Pleistocene (MNQ18–19)	*Metacervocerus rhenanus*[38] Cervidae indet.[38]
Haliákmon Valley (Véria or Polýlakkos)	late Early Pleistocene?	*Praemegaceros pliotarandoides*[44]
Polýlakkos[34763]	Early Pleistocene (MNQ18–19)	Cervidae indet.[40]
Gerakaroú[34617]	Early Pleistocene (MNQ18)	*Eucladoceros* cf. *ctenoides*[42,43,a] *Metacervocerus rhenanus*[42,43] *Croizetoceros ramosus gerakarensis*[42,43]
Pyrgos	Early Pleistocene	*Cervus* sp.[39] cf. *Metacervocerus pardinensis*[39] Cervidae indet.[39]
Q-Profil[34812]	Early Pleistocene	Cervidae indet.[40]
Reghínio[204661]	Early Pleistocene	*Eucladoceros* sp.[45]
Karnezéika[202122]	Early Pleistocene (MNQ17?)	*Eucladoceros* sp.[46]
Vaterá F[183341]	Early Pleistocene (MNQ17)	cf. *Metacervocerus rhenanus*[47]
Vaterá DS[183344]	Early Pleistocene (MNQ17)	Cervidae indet.[47]
Dafneró[34594]	Early Pleistocene (MNQ17)	*Eucladoceros* cf. *ctenoides*[42,43,a] *Metacervocerus rhenanus*[42,43]
Sésklo[34614]	Early Pleistocene (MNQ17)	*Eucladoceros* sp.[43,48,a] *Metacervocerus rhenanus*[43,48] *Croizetoceros ramosus*[43,48]
Vólakas[34593]	Early Pleistocene (MNQ17)	*Eucladoceros* cf. *ctenoides*[42,43,a] *Metacervocerus rhenanus*[42,43] *Croizetoceros ramosus*[42,43]
Antimáchia[207130]	Early Pleistocene	Cervidae indet.[49,a]
Kardámaena[204662]	Early Pleistocene	Cervidae indet.[50,a] ?*Capreolus* sp.[51]
Kos Island (unspecified localities)	Early Pleistocene	*Eucladoceros ctenoides*[39] cf. *Metacervocerus rhenanus*[39] *Croizetoceros ramosus*[39]
Kastrítsi	Early Pleistocene	cf. *Croizetoceros ramosus*[52]
Makýnia	Early Pleistocene	Cervidae indet.[52]

(continued)

Localities[PBDB No]	Age (biozone)	Taxon
Tourkoboúnia[34589]	Early Pleistocene	cf. *Metacervocerus rhenanus*[53]
Vrouléa, Kýthera	Pleistocene	Cervidae indet.[54]
Tourkoboúni, Kýthera	Pleistocene	Cervidae indet.[54]
Mylopótamos, Kýthera	Pleistocene	Cervidae indet.[54]
Tsotýli	Pleistocene	Cervidae indet.[25]
Korydallós	Pleistocene	Cervidae indet.[55]
Halmyrós	Pleistocene?	*Dama* sp.[56]
Dafni, Rhodes	Pleistocene?	Cervidae indet.[57]
Kritinía, Rhodes	Pleistocene?	*Cervus* sp.[58]
Rhodes airport	Early Pleistocene	Cervidae indet.[59]
Géphyra[182685]	late Pliocene (MN16a)	*Procapreolus cusanus*[60]
Miliá[185859]	late Pliocene (MN16a)	cf. *Praeelaphus lyra*[61] *Croizetoceros ramosus*[61] Cervidae indet.[61]
Apolakkiá[184242]	early Pliocene (MN15)	aff. *Metacervocerus rhenanus*[62]
Kessáni 2[205297]	latest Miocene/earliest Pliocene (MN13/14)	cf. *Procapreolus* sp.[63]
Maraména[32189]	latest Miocene/earliest Pliocene (MN13/14)	*Procapreolus pentelici*[64]
Dytikó 1[32374]	late Miocene (MN13)	?*Procapreolus* sp.[65]
Samos Q5[95690]	late Miocene (MN12/13)	*Procapreolus* sp.[66]
Thermopigí[73553]	late Miocene (MN12)	Cervidae indet.[67]
Perivoláki[194879]	late Miocene (MN12)	cf. *Lucentia* sp.[68]
Halmyropótamos[202213]	late Miocene (MN12)	*Procapreolus pentelici*[69]
Chomaterí[195562]	late Miocene (MN12)	*Procapreolus pentelici*[70]
Pikermi[182754]	late Miocene (MN12)	*Procapreolus pentelici*[71] "*Dremotherium*?" *pentelici*[72]
Andrianós (Samos)	late Miocene (MN12)	"*Dremotherium*?" *pentelici*[73]
Samos Q6[206462]	late Miocene (MN11)	cf. *Lucentia* sp.[68]
Plataniá[182682]	late Miocene (MN10–11)	Cervidae indet.[74]
Thymianá[182752]	early middle Miocene (MN5)	Cervidae indet.[75,a]

MN/MNQ Mammal Neogene–Quaternary Zone

[1]Melentis (1969b) (but Giannopoulos 2000 considered the material heterogeneous and of dubious provenance, at least partly), [2]Sordinas (1969), [3]Trantalidou in Mavridis et al. (2013), [4]Tsoukala (1992, 1999), [5]Tsoukala (1981), [6]Symeonidis and Theodorou (1994), Rabeder (1995), [7]Schmid (1965), [8]Kotjabopoulou (2001), [9]Koukouli-Chrysanthaki and Weisgerber (1997), [10]Sinclair (1997), [11]Bailey (1992), [12]Payne in Jacobsen (1973), [13]Bailey et al. (1983), [14]Darlas and Psathi (2016), [15]Giannopoulos (2000), [16]Koufos (1981) and Trantalidou (2013), [17]Starkovich and Ntinou (2017), [18]Jullien (1981), [19]Boessneck in Milójčić et al. (1965), [20]Newton (2003), [21]Panagopoulou et al. (2004), [22]Petrochilou (1969), [23]Panagopoulou et al. (2006), [24]Tsoukala et al. (2011), [25]Marinos (1964) and Soulios (1972), [26]Stratigopoulos (1986), [27]Sickenberg (1971) and Tsoukala (1989), [28]Melentis (1966a), [29]Tsoukala (1999), [30]Kevrekidis and Kostopoulos (2017), [31]Stratigopoulos (1988, 1989), [32]Konidaris et al. (2018), [33]Athanassiou (2018), [34]Melentis (1966b) and Sickenberg (1976), [35]Tsoukala and Chatzopoulou (2005), [36]Croitor (2018a), [37]Croitor and Kostopoulos (2004),

(continued)

[38]Konidaris et al. (2015), [39]van der Meulen and van Kolfschoten (1988), [40]Steensma (1988), [41]van der Made and Tong (2008), [42]Kostopoulos (1996), [43]Kostopoulos and Athanassiou (2005), [44]Melentis (1967) and Croitor (2005, 2006b), [45]Athanassiou (2006), [46]Kokotini et al. (2019), [47]de Vos et al. (2002), [48]Athanassiou (1998), [49]Forsyth Major (1887), [50]Airaghi (1928), [51]Charrier and Giglio (1969), [52]Symeonidis et al. (1986), [53]Symeonidis and de Vos (1977), [54]Jameson (1836) and Leonhard (1899), [55]Paraskevaidis (1961), [56]Psarianos and Thenius (1954), [57]Merla (1949), [58]Kuss (1975), [59]Sondaar (1971), [60]Crégut-Bonnoure and Tsoukala (2017a), [61]Crégut-Bonnoure and Tsoukala (2017b), [62]van de Weerd et al. (1982), [63]Syrides et al. (1997) and Vasileiadou et al. (2012), [64]Azanza (1995) and Croitor (2018b), [65]Bouvrain and de Bonis (2007), [66]Kostopoulos et al. (2003) and Kostopoulos (pers. com.), [67]Tsoukala (2018), [68]Kostopoulos (2006), [69]Melentis (1970), [70]Symeonidis (1973, 1974), [71]Dames (1883), Melentis (1969a), and Croitor (2018b), [72]Gaudry (1862–1867), [73]Forsyth Major (1894), [74]Vasileiadis et al. (2019), [75]Lehmann and Tobien (1995)
[a]This study

References

Abbazzi L (2004) Remarks on the validity of the generic name *Praemegaceros* Portis, 1920, and an overview on *Praemegaceros* species in Italy. Rend Fis Acc Lincei 15:115–132

Airaghi C (1928) Mammiferi pliocenici dell'isola di Coo (Dodecaneso). Atti Soc It Sci Nat Mus Civ St Nat Milano 67:125–135

Athanassiou A (1998) [Contribution to the study of the fossil mammals of Thessaly]. PhD thesis (1996), National and Kapodistrian University of Athens (in Greek)

Athanassiou A (2006) Reghínio, a new mammal locality from the Plio-Pleistocene of Central Greece. N Jb Geol Pal Mh 2006:116–128

Athanassiou A (2018) Pleistocene vertebrates from the Kyparíssia lignite mine, Megalopolis Basin, S. Greece: Rodentia, Carnivora, Proboscidea, Perissodactyla, Ruminantia. Quat Int 497:198–221

Azanza B (1995) The vertebrate locality Maramena (Macedonia, Greece) at the Turolian–Ruscinian boundary (Neogene). 14. Cervidae (Artiodactyla, Mammalia). Münchner Geowiss Abh 28:157–166

Azzaroli A (1953) The deer of the Weybourn Crag and Forest Bed of Norfolk. Bull Br Mus (Nat Hist) 2:1–96

Azzaroli A (1979) Critical remarks on some giant deer (genus *Megaceros* Owen) from the Pleistocene of Europe. Palaeont It 71:5–16

Azzaroli A (1992) The cervid genus *Pseudodama* n.g. in the Villafranchian of Tuscany. Palaeont It 79:1–41

Bailey G (1992) The Palaeolithic of Klithi in its wider context. Ann Br Sch Athens 87:1–28

Bailey GN, Carter PL, Gamble CS, Higgs HP (1983) Asprochaliko and Kastritsa: further investigations of Palaeolithic settlement and economy in Epirus (North-West Greece). Proc Prehist Soc 49:15–42

Boeuf O (1983) Le site villafranchien de Chilhac (Haute-Loire, France). Etude paléontologique et biochronologique. Thèse, Université de Paris VII

Bouvrain G, de Bonis L (2007) Ruminants (Mammalia, Artiodactyla: Tragulidae, Cervidae, Bovidae) des gisements du Miocène supérieur (Turolien) de Dytiko (Grèce). Ann Paléont 93:121–147

Bouvrain G, Geraads D, Jehenne Y (1989) Nouvelles données relatives à la classification des Cervidae (Artiodactyla, Mammalia). Zool Anz 223:82–90

Brooke V (1878) On the classification of the Cervidae with a synopsis of the existing species. Proc Zool Soc London 1878:883–928

Bubenik GA (1990) Epigenetical, morphological, physiological, and behavioral aspects of evolution of horns, pronghorns, and antlers. In: Bubenik GA, Bubenik AB (eds) Horns, pronghorns, and antlers: evolution, morphology, physiology, and social significance. Springer, New York, pp 3–113

Cap H, Aulagnier S, Deleporte P (2002) The phylogeny and behaviour of Cervidae (Ruminantia Pecora). Ethol Ecol Evol 14:199–216

Cap H, Deleporte P, Joachim J, Reby D (2008) Male vocal behavior and phylogeny in deer. Cladistics 24:917–931

Charrier G, Giglio A (1969) Primi risultati di una campagna di rilevamento geologico nell'isola di Coo (Sporadi meridionali – Mare Egeo). Boll Assoc Min Subalp 6:482–516

Cione AL, Gasparini GM, Soibelzon E, Soibelzon LH, Tonni EP (2015) The Great American Biotic Interchange: a South American perspective. Springer Briefs in Earth System Sciences. Springer, Dordrecht

Crégut-Bonnoure E, Tsoukala E (2017a) The Pliocene Artiodactyla and Proboscidea (Mammalia) from Gephyra (lower Axios valley, Macedonia, Greece). Discovery of a new boselaphine. Quat Int 445:200–214

Crégut-Bonnoure E, Tsoukala E (2017b) The Late Pliocene Bovidae and Cervidae (Mammalia) of Milia (Grevena, Macedonia, Greece). Quat Int 445:215–249

Croitor R (2005) Large-sized deer from the early Pleistocene of South-East Europe. Acta Palaeont Rom 4:97–104

Croitor R (2006a) Early Pleistocene small-sized deer of Europe. Hell J Geosci 41:89–117

Croitor R (2006b) Taxonomy and systematics of large-sized deer of the genus *Praemegaceros* Portis, 1920 (Cervidae, Mammalia). Cour Forsch-Inst Senck 256:91–116

Croitor R (2014) Deer from Late Miocene to Pleistocene of Western Palearctic: matching fossil record and molecular phylogeny data. Zitteliana (B) 32:115–153

Croitor R (2018a) A description of two new species of the genus *Rucervus* (Cervidae, Mammalia) from the Early Pleistocene of Southeast Europe, with comments on hominin and South Asian ruminants dispersals. Quaternary 1(17):1–32

Croitor R (2018b) Plio-Pleistocene deer of Western Palearctic: taxonomy, systematics, phylogeny. Institute of Zoology of the Academy of Sciences of Moldova, Chişinău

Croitor R, Bonifay MF (2001) Etude préliminaire des cerfs du gisement pléistocène inférieur de Ceyssaguet (Haute-Loire). Paléo 13:129–144

Croitor R, Kostopoulos DS (2004) On the systematic position of the large-sized deer from Apollonia, Early Pleistocene, Greece. Paläont Z 78:137–159

Croitor R, Stefaniak K (2009) Early Pliocene deer of Central and Eastern European regions and inferred phylogenetic relationships. Palaeontographica 287:1–39

Croizet JB, Jobert A (1828) Recherches sur les ossements fossiles du département du Puy-de-Dôme, Paris

Dames W (1882) Über das Vorkommen fossiler Hirsche in den Pliocänablagerungen von Pikermi in Attika. Sitz-Ber Gesell Naturforsch Fr Berlin 1882:71–72

Dames W (1883) Hirsche und Mäuse von Pikermi in Attica. Zeitschr D Geol Gesell 35:92–101

Darlas A, Psathi E (2016) The middle and upper Paleolithic on the western coast of the Mani Peninsula (Southern Greece). In: Roksandic KHM (ed) Paleoanthropology of the Balkans and Anatolia: human evolution and its context. Vertebrate paleobiology and paleoanthropology series. Springer, Dordrecht, pp 95–117

Dawkins WB (1872) On the Cervidæ of the Forest-bed of Norfolk and Suffolk. Quart J Geol Soc 28:405–410

de Bonis L, Koufos GD, Sen S (1997) A giraffid from the middle Miocene of the island of Chios, Greece. Palaeontology 40:121–133

de Lumley H, Kahlke H-D, Moigne A-M, Moullé P-E (1988) Les faunes de grands mammifères de la grotte du Vallonnet, Roquebrune-Cap-Martin, Alpes-Maritimes. L'Anthropol 92:465–495

de Vos J, Mol D, Reumer JWF (1995) Early Pleistocene Cervidae (Mammalia, Artiodactyla) from the Oosterschelde (the Netherlands), with a revision of the cervid genus *Eucladoceros* Falconer, 1868. Deinsea 2:95–121

de Vos J, van der Made J, Athanassiou A, Lyras G, Sondaar PY, Dermitzakis MD (2002) Preliminary note on the Late Pliocene fauna from Vatera (Lesvos, Greece). Ann Géol Pays Hellén 39:37–70
Di Stefano G, Petronio C (2002) Systematics and evolution of the Eurasian Plio-Pleistocene tribe Cervini (Artiodactyla, Mammalia). Geol Rom 36:311–334
Dong W (1996) Les Cervidae (Artiodactyla) rusciniens (Pliocene) du Languedoc et du Roussillon (France). Bull Mus natl Hist nat 18:133–163
Douzery E, Randi E (1997) The mitochondrial control region of Cervidae: evolutionary patterns and phylogenetic content. Mol Biol Evol 14:1154–1166
Facorellis Y (2013) Radiocarbon dates from archaeological sites in caves and rockshelters in Greece. In: Mavridis F, Jensen JT (eds) Stable places and changing perceptions: cave archaeology in Greece, BAR International Series, vol 2558. Archaeopress, Oxford, pp 19–72
Forsyth Major CJ (1887) Faune mammalogiche delle isole di Kos e di Samos. Atti Soc Tosc Sc Nat, Proc Verb 5:272–275
Forsyth Major CJ (1894) Le gisement ossifère de Mytilini et catalogue d'ossements fossiles recueillis à Mytilini, île de Samos, et déposés au Collège Galliard, à Lausanne. Georges Bridel & Cie, Lausanne
Gaudry A (1862–1867) Animaux fossiles et géologie de l'Attique. E. Savy, Paris
Geist V (1998) Deer of the world: their evolution, behavior, and ecology. Stackpole, Mechanicsburg
Gentry AW (1994) The Miocene differentiation of Old World Pecora (Mammalia). Hist Biol 7:115–158
Gentry AW (2000) The ruminant radiation. In: Vrba ES, Schaller GB (eds) Antelopes, deer, and relatives. Yale University Press, New Haven, pp 11–25
Gentry AW (2005) Ruminants of Rudabánya. Palaeont It 90:283–302
Geraads D (1982) Paléobiogéographie de l'Afrique du nord depuis le Miocène terminal, d'après les grands mammifères. Geobios mém spéc 6:473–481
Giannopoulos VI (2000) [Contribution to the study of recent and old environments of the most important Greek caves]. PhD thesis, National and Kapodistrian University of Athens (in Greek)
Gilbert C, Ropiquet A, Hassanin A (2006) Mitochondrial and nuclear phylogenies of Cervidae (Mammalia, Ruminantia): systematics, morphology, and biogeography. Mol Phylogenet Evol 40:101–117
Goss RJ (ed) (1983) Deer antlers: regeneration, function, and evolution. Academic Press, New York
Gould SJ (1973) Positive allometry of antlers in the "Irish elk", *Megaloceros giganteus*. Nature 244:375–376
Gradstein FM, Ogg JG, Schmitz MD, Ogg GM (eds) (2012) The geologic time scale 2012. Elsevier, Oxford
Groves CP (2007) Family Cervidae. In: Prothero DR, Foss SE (eds) The evolution of Artiodactyls. John Hopkins University Press, Baltimore, pp 249–256
Groves C (2014) Current taxonomy and diversity of crown ruminants above the species level. Zitteliana (B) 32:5–14
Groves C, Grubb P (2011) Ungulate taxonomy. Johns Hopkins University Press, Baltimore
Grubb P (2000) Valid and invalid nomenclature of living and fossil deer, Cervidae. Acta Theriol 45:289–307
Grubb P (2005) Order Artiodactyla. In: Wilson DE, Reeder DM (eds) Mammal species of the world: a taxonomic and geographic reference, vol 1. Johns Hopkins University Press, Baltimore, pp 637–722
Guérin C (1990) Biozones or Mammal Units? Methods and limits in biochronology. In: Lindsay EH, Fahlbusch V, Mein P (eds) European Neogene mammal chronology. NATO ASI Series (A: Life Sciences). Plenum, New York, pp 119–130
Hassanin A, Douzery EJP (2003) Molecular and morphological phylogenies of Ruminantia and the alternative position of the Moschidae. Syst Biol 52:206–228
Heintz E (1970) Les Cervidés villafranchiens de France et d'Espagne. Mém Mus Nat Hist Nat 22:1–303,1–206

Hernández Fernández M, Vrba ES (2005) A complete estimate of the phylogenetic relationships in Ruminantia: a dated species-level supertree of the extant ruminants. Biol Rev 80:269–302

Hutchins M, Kleiman DG, Geist V, McDade MC (eds) (2003) Grzimek's animal life encyclopedia, vol 15, 2nd edn. Gale, Farmington Hills

Iliopoulos G, Roussiakis S (this volume) The fossil record of giraffes (Mammalia: Giraffidae) in Greece. In: Vlachos E (ed) The fossil vertebrates of Greece Vol. 2 – laurasiatherians, artiodactyles, perissodactyles, carnivorans, and island endemics. Springer – Nature Publishing Group, Cham

Jacobsen TW (1973) Excavation in the Franchthi Cave, 1969–1971, Part I. Hesperia 42:45–88

Jameson R (1836) Notes on the natural history and statistics of the island of Cerigo and its dependencies. Edinburgh New Philos J 21:263–279

Janis CM (1982) Evolution of horns in ungulates: ecology and paleoecology. Biol Rev 57:261–318

Janis CM, Scott KM (1987) The interrelationships of higher ruminant families with special emphasis on the members of the Cervoidea. Am Mus Nov 2893:1–85

Janis CM, Theodor JM (2014) Cranial and postcranial morphological data in ruminant phylogenetics. Zitteliana (B) 32:15–31

Jullien R (1981) La faune des vertébrés, à l'exclusion de l'homme, des oiseaux, des rongeurs et des poissons. In: Lambert N (ed) La grotte préhistorique de Kitsos (Attique): missions 1968–1978, vol II. A.D.P.F. – École Française d'Athènes, Athènes, pp 569–590

Kahlke HD (1997) Die Cerviden-Reste aus dem Unterpleistozän von Untermaßfeld. In: Kahlke RD (ed) Das Pleistozän von Untermaßfeld bei Meiningen (Thüringen), Teil 1. Monographien des Römisch-Germanischen Zentralmuseums Mainz, vol 40 (1). Habelt, Mainz, pp 181–275

Kahlke HD (2001) Neufunde von Cerviden-Resten aus dem Unterpleistozän von Untermaßfeld. In: Kahlke RD (ed) Das Pleistozän von Untermaßfeld bei Meiningen (Thüringen), Teil 2. Monographien des Römisch-Germanischen Zentralmuseums Mainz, vol 40 (2). Habelt, Mainz, pp 461–481

Kevrekidis C, Kostopoulos DS (2017) The southernmost occurrence of *Cervalces latifrons* (Johnson, 1874) (Artiodactyla: Cervidae) in Europe. In: 44. Treffen des Arbeitskreises Wirbeltierpaläontologie, Münster, 2017

Kokotini M, Kargopoulos N, Iliopoulos G, Roussiakis S, Skandalos P, Michailidis D, Svorligkou G, Kampouridis P, Theodorou G (2019) Karnezeika (Argolis, Peloponnese): preliminary data concerning a new Villafranchian locality of Southern Greece. In: Proceedings of the 15th international congress of the Geological Society of Greece, Athens, 2019. Bull Geol Soc Greece spec publ 7:GSG2019-104

Konidaris GE, Tourloukis V, Kostopoulos DS, Thompson N, Giusti D, Michailidis D, Koufos GD, Harvati K (2015) Two new vertebrate localities from the Early Pleistocene of Mygdonia Basin (Macedonia, Greece): preliminary results. CR Palevol 14:353–362

Konidaris GE, Athanassiou A, Tourloukis V, Thompson N, Giusti D, Panagopoulou E, Harvati K (2018) The skeleton of a straight-tusked elephant (*Palaeoloxodon antiquus*) and other large mammals from the Middle Pleistocene butchering locality Marathousa 1 (Megalopolis Basin, Greece): preliminary results. Quat Int 497:65–84

Kostopoulos DS (1996) [The Plio-Pleistocene Artiodactyls of Macedonia (Greece): systematics, palaeoecology, biochronology, biostratigraphy]. PhD thesis, Aristotle University of Thessaloniki (in Greek)

Kostopoulos DS (1997a) The Plio-Pleistocene Artiodactyls (Vertebrata, Mammalia) of Macedonia (N. Greece): 1. The fossiliferous site "Apollonia-1", Mygdonia basin of Greece. Geodiversitas 19:845–875

Kostopoulos DS (1997b) The Plio-Pleistocene Artiodactyls from Macedonia, Greece: 2. The fossiliferous site of Volakas, VOL (Volakas basin, Drama, NE Greece). Paleont i Evol 30–31:83–92

Kostopoulos DS (2006) The late Miocene vertebrate locality of Perivolaki, Thessaly, Greece: 9. Cervidae, and Bovidae. Palaeontographica 276:151–183

Kostopoulos DS (this volume) The fossil record of bovids (Mammalia: Artiodactyla: Ruminantia: Pecora: Bovidae) in Greece. In: Vlachos E (ed) The fossil vertebrates of Greece Vol. 2 – laurasiatherians, artiodactyles, perissodactyles, carnivorans, and island endemics. Springer – Nature Publishing Group, Cham

Kostopoulos DS, Athanassiou A (2005) In the shadow of bovids: suids, cervids and giraffids from the Plio-Pleistocene of Greece. Quaternaire hors-série 2:179–190

Kostopoulos DS, Koufos GD (1994) The Plio-Pleistocene artiodactyls of Macedonia (Northern Greece) and their biostratigraphic significance; preliminary report. CR Acad Sci Paris 318:1267–1272

Kostopoulos DS, Sen S, Koufos GD (2003) Magnetostratigraphy and revised chronology of the late Miocene mammal localities of Samos, Greece. Intern J Earth Sci 92:779–794

Kotjabopoulou E (2001) Patterned fragments and fragments of patterns: Upper Palaeolithic rock-shelter faunas from Epirus, northwestern Greece. PhD thesis, University of Cambridge

Koufos GD (1981) A new Late Pleistocene (Würmian) mammal locality from the basin of Drama (Northern Greece). Sci Ann Fac Phys Mathem Univ Thessaloniki 21:129–148

Koufos GD, Syrides GE, Kostopoulos DS, Koliadimou KK (1992) Apollonia, a new vertebrate site in the Pleistocene of the Mygdonia basin (Macedonia, Greece); the first fossil freshwater mollusks in the area. CR Acad Sci Paris 315:1041–1046

Koukouli-Chrysanthaki Ch, Weisgerber G (1997) [Prehistoric ochre mines on Thasos]. Arch Res Macedonia Thrace 7:541–558 (in Greek)

Kuss SE (1975) Die pleistozänen Hirsche der ostmediterranen Inseln Kreta, Kasos, Karpathos und Rhodos (Griechenland). Ber Naturf Ges Freiburg i Br 65:25–79

Kuznetsova MV, Kholodova MV, Danilkin AA (2005) Molecular phylogeny of deer (Cervidae: Artiodactyla). Russian J Gen 41:742–749

Lehmann U, Tobien H (1995) Artiodactyle Fossilien (Mammalia) aus dem Miozän von Thymianá, Chios. Ann Géol Pays Helléн 36:403–414

Leonhard R (1899) Die Insel Kythera. Eine geographische Monographie. Justus Perthes, Gotha

Lister AM (1987a) Diversity and evolution of antler form in Quaternary deer. In: Wemmer CM (ed) Biology and management of the Cervidae. Smithsonian Institution Press, Washington, DC, pp 81–98

Lister AM (1987b) *Megaloceros* Brookes, 1828 (Mammalia, Artiodactyla): proposed emendation of the original spelling. Bull Zool Nomencl 44:255–256

Lister AM (1993) Evolution of mammoths and moose: the Holarctic perspective. In: Martin RA, Barnosky AD (eds) Morphological change in Quaternary mammals of North America. Cambridge University Press, New York, pp 178–204

Lyras GA, Athanassiou A, van der Geer AAE (this volume) The fossil record of insular endemic mammals from Greece. In: Vlachos E (ed) The fossil vertebrates of greece Vol. 2 – laurasiatherians, artiodactyles, perissodactyles, carnivorans, and island endemics. Springer – Nature Publishing Group, Cham

Marinos G (1964) [Beiträge zur Kenntnis der Verbreitung des Pleistozäns in Mazedonien]. Sci Annals, Fac Phys Mathem, Univ Thessaloniki 9:95–111 (in Greek)

Mavridis F, Kormazopoulou L, Papadea A, Apostolikas O, Yamaguchi D, Tankosic Z, Kotzamani G, Trantalidou K, Karkanas P, Maniatis Y, Papagianni K, Lambropoulos D (2013) Anonymous cave of Schisto at Keratsini, Attika: a preliminary report on a diachronic cave occupation from the Pleistocene/Holocene transition to the Byzantine times. In: Mavridis F, Jensen JT (eds) Stable places and changing perceptions: cave archaeology in Greece. BAR International Series, vol 2558. Archaeopress, Oxford, pp 248–284

McKenna MC, Bell SK (1997) Classification of mammals above the species level. Columbia University Press, New York

Mein P (1999) European Miocene mammal biochronology. In: Rössner GE, Heissig K (eds) The Miocene land mammals of Europe. Dr. Friedrich Pfeil, München, pp 25–38

Melentis JK (1963) [Die pleistozänen Cerviden des Beckens von Megalopolis im Peloponnes]. Prakt Akad Ath 38:555–564 (in Greek)

Melentis JK (1966a) Die pleistozäne Säugetierfauna des Beckens von Haliakmon (Griechenland). Ann Géol Pays Hellén 17:247–265

Melentis JK (1966b) Die pleistozänen Cerviden des Beckens von Megalopolis im Peloponnes (Griechenland). Ann Géol Pays Hellén 16:1–92
Melentis JK (1967) *Orthogonoceros verticornis* aus dem Altpleistozän des Beckens von Haliakmon (Griechenland). Ann Géol Pays Hellén 18:447–455
Melentis JK (1968) *Orthogonoceros verticornis* aus dem Altpleistozän des Beckens von Haliakmon (Griechenland). Prakt Akad Ath 42:79–88
Melentis JK (1969a) Zur Morphologie und systematischen Stellung von *Pliocervus pentelici* (Gaudry) aus dem Pont von Attika. Prakt Akad Ath 43:5–18
Melentis JK (1969b) [Die quartären Vertebraten der Höhle der Seen von Kleitoria (im Gebiet der Aroania-Gebirge)]. Prakt Akad Ath 43:350–363 (in Greek)
Melentis JK (1970) Die Pikermifauna von Halmyropotamos (Euböa, Griechenland) — 1. Teil: Odontologie und Kraniologie. Ann Géol Pays Hellén 19:283–411
Mennecart B, DeMiguel D, Bibi F, Rössner GE, Métais G, Neenan JM, Wang S, Schulz G, Müller B, Costeur L (2017) Bony labyrinth morphology clarifies the origin and evolution of deer. Sci Rep 7:13176
Merla G (1949) I *Leptobos* Rütim. italiani. Palaeont It 46:41–155
Milójčić V, Boessneck J, Jung D, Schneider H (1965) Paläolithikum um Larissa in Thessalien. Beiträge zur Ur- und Frühgeschichtlichen Archäologie des Mittelmeerkulturräumes, vol 1. Rudolf Habelt, Bonn
Newton S (2003) Theopetra Cave and the Palaeolithic–Mesolithic transition in southern Europe. In: Kotjabopoulou E, Hamilakis Y, Halstead P, Gamble C, Elefanti P (eds) Zooarchaeology in Greece: recent advances. British School at Athens, London, pp 115–122
Panagopoulou E, Karkanas P, Tsartsidou G, Kotjabopoulou E, Harvati K, Ntinou M (2004) Late Pleistocene archaeological and fossil human evidence from Lakonis Cave, Southern Greece. J Field Archaeol 29:323–349
Panagopoulou E, Harvati K, Karkanas P, Athanassiou A, Elefanti P, Frost SR (2006) [The West Macedonia Palaeolithic survey]. Arch Res Macedonia Thrace 18:631–640 (in Greek)
Paraskevaidis I (1940) Eine obermiocäne Fauna von Chios. N Jb Min Geol Paläont 83:363–442
Paraskevaidis I (1961) [Neue Funde quartärer Wirbeltierreste in Attika]. Ann Géol Pays Hellén 12:149–152 (in Greek)
Petrochilou A (1969) [The cave Phytídi at Sami, Kephallenía]. Bull Soc Spél Gr 10:26–30 (in Greek)
Pfeiffer T (1999) Die Stellung von *Dama* (Cervidae, Mammalia) im System plesiometacarpaler Hirsche des Pleistozäns: phylogenetische Rekonstruktion, metrische Analyse. Cour Forsch-Inst Senck 211:1–218
Pfeiffer T (2002) The first complete skeleton of *Megaloceros verticornis* (Dawkins, 1868), Cervidae, Mammalia, from Bilshausen (Lower Saxony, Germany): description and phylogenetic implications. Mitt Mus Natkd Berlin, Geowiss 5:289–308
Pitra C, Fickel J, Meijaard E, Groves C (2004) Evolution and phylogeny of Old World deer. Mol Phylogenet Evol 33:880–895
Psarianos P, Thenius E (1954) Ein fossiles Cerviden "Gehirn" aus dem Quartär des Peloponnes (Griechenland). Ann Géol Pays Hellén 6:13–32
Rabeder G (1995) Jungpleistozäne und Frühholozäne Säugetierreste aus der Höhle von Vraona auf Attika, Griechenland. Ann Géol Pays Hellén 36:47–56
Radulesco C, Samson P (1967) Sur un nouveau cerf mégacérin du Pléistocène moyen de la dépression de Brasov (Roumanie). Geol Rom 6:317–344
Schmid E (1965) Die Seïdi-Höhle eine jungpaläolitische Station in Griechenland. In: IVème Colloque International de Speléologie, Athens, 1963, pp 163–174
Sempéré AJ, Sokolov VE, Danilkin AA (1996) *Capreolus capreolus*. Mamm Spec 538:1–9
Sickenberg O (1971) Revision der Wirbeltierfauna der Höhle Petralona (Griech. Mazedonien). Ann Géol Pays Hellén 23:230–264
Sickenberg O (1976) Eine Säugetierfauna des tieferen Bihariums aus dem Becken von Megalopolis (Peloponnes, Griechenland). Ann Géol Pays Hellén 27:25–73

Simpson GG (1945) The principles of classification and a classification of mammals. Bull Am Mus Nat Hist 85:i–xvi, 1–350

Sinclair A (1997) Lithic and faunal assemblages from Megalakkos: some problems in the interpretation of small sites. In: Bailey GN (ed) Klithi: Palaeolithic settlement and Quaternary landscapes in northwestern Greece, vol 2. McDonald Institute for Archaeological Research, Cambridge, pp 415–426

Sondaar PY (1971) Paleozoogeography of the Pleistocene mammals from the Aegean. In: Strind A (ed) Evolution in the Aegean. Opera Botan 30:65–70

Sordinas A (1969) Investigations of the prehistory of Corfu during 1964–1966. Balkan Studies 10:393–424

Soulios G (1972) *Cervus elaphus* de la region Amyntaion (Macédoine). Folia Bioch et Biol Graeca 9:37–46

Spaan A (1992) A revision of the deer from Tegelen (Province of Limburg, The Netherlands). Scripta Geol 98:1–85

Starkovich BM, Ntinou M (2017) Climate change, human population growth, or both? Upper Paleolithic subsistence shifts in southern Greece. Quat Int 428:17–32

Steensma KJ (1988) Plio-/Pleistozäne Großsäugetiere (Mammalia) aus dem Becken von Kastoria/Grevena, südlich von Neapolis — NW Griechenland. Inaugural-Dissertation, Technische Universität Clausthal

Stratigopoulos S (1986) [Antlers, remains of the extinct cervids *Libralces gallicus* Azz. and *Megaceros verticornis* D. from Ptolemais basin]. Bull Geol Soc Gr 18:173–182 (in Greek)

Stratigopoulos SK (1988) [The Villafranchian stage in Ptolemais basin — New species of mammalian paleofauna and correlations]. Bull Geol Soc Gr 20:167–180 (in Greek)

Stratigopoulos S (1989) [Sur la présence de *Cervus philisi* Schaub, 1941 dans le Villafranchien du bassin de Ptolemais, en Macédoine]. Bull Geol Soc Gr 23:101–111 (in Greek)

Symeonidis N (1973) Ein bemerkenswertes Geweih von *Pliocervus pentelici* (Gaudry) aus dem Altpliozän von Pikermi (Griechenland). Anz Math-Naturwiss Kl Österr Akad Wiss 110:134–137

Symeonidis NK (1974) Ein vollständiges Geweih von *Pliocervus pentelici* (Gaudry) aus Pikermi (Griechenland). Ann Géol Pays Hellén 25:308–316

Symeonidis NK, Theodorou GE (1994) [Palaeontological excavations in Vraona in Attica]. Bull Soc Spél Gr 21:269–279 (in Greek)

Symeonidis N, de Vos J (1977) Großsäuger-Funde aus den plistozänen Spaltenfüllungen von Turkovunia in Athen. Ann Géol Pays Hellén 28:135–144

Symeonidis NK, Bachmayer F, Zapfe H (1980) Ergebnisse weiterer Grabungen in der Höhle von Vraona (Attica, Griechenland). Ann Géol Pays Hellén 30:291–299

Symeonidis NK, Theodorou GE, Schütt H, Velitzelos E (1986) [Palaeontological and stratigraphical observations in the area of Achaia and Etoloakarnania (W. Greece)]. Ann Géol Pays Hellén 33:329–365 (in Greek)

Syrides GE, Koliadimou KK, Koufos GD (1997) New Neogene molluscan and mammalian sites from Thrace, Greece. CR Acad Sci Paris 324:427–433

Trantalidou K (2013) [On the banks of the river Angitis: Environmental data for the basin of Drama in Prehistoric times]. In: Proceedings of the 5th Scientific Meeting, Drama, 2006. pp 41–156 (in Greek)

Tsoukala E (1981) [Fossil and extant samples of *Capreolus capreolus* Linné from the area of Sithonia, Chalkidiki]. Sci Annals, Fac Phys Mathem, Univ Thessaloniki 21:237–254 (in Greek)

Tsoukala E (1989) [Contribution to the study of the Pleistocene fauna of large mammals (Carnivora, Perissodactyla, Artiodactyla) from Petralona Cave, Chalkidiki (N. Greece)]. PhD thesis, Aristotle University of Thessaloniki (in Greek)

Tsoukala ES (1992) The Pleistocene large mammals from the Agios Georgios Cave, Kilkis (Macedonia, N. Greece). Geobios 25:415–433

Tsoukala E (1999) Quaternary large mammals from the Apidima Caves (Lakonia, S. Peloponnese, Greece). Beitr Paläontol 24:207–229

Tsoukala E (2018) Rhinocerotidae from the Late Miocene and Late Pliocene of Macedonia, Greece. A revision of the Neogene–Quaternary Rhinocerotidae of Greece. Rev Paléobiol 37:609–630

Tsoukala E, Chatzopoulou K (2005) A new Early Pleistocene (latest Villafranchian) site with mammals in Kalamotó (Mygdonia Basin, Macedonia, Greece)—preliminary report. Mitt Komm Quartärforsch Österr Akad Wiss 14:213–233

Tsoukala E, Mol D, Pappa S, Vlachos E, van Logchem W, Vaxevanopoulos M, Reumer J (2011) *Elephas antiquus* in Greece: new finds and a reappraisal of older material (Mammalia, Proboscidea, Elephantidae). Quat Int 245:339–349

Valli AMF (2004) Les Cervidae du gisement Pliocène supérieur (Villafranchien moyen) de Saint-Vallier (Drôme, France). Geobios 37:S191–S232

van de Weerd A, Reumer JWF, de Vos J (1982) Pliocene mammals from the Apolakkia Formation (Rhodes, Greece). Proc Konin Nederl Akad Wet 85:89–112

van der Made J (2001) Les ongulés d'Atapuerca. Stratigraphie et biogeographie. L'Anthropol 105:95–113

van der Made J (2015) The latest early Pleistocene giant deer *Megaloceros novocarthaginiensis* n. sp. and the fallow deer *Dama* cf. *vallonnetensis* from Cueva Victoria (Murcia, Spain). Mastia 11–13:269–323

van der Made J (2019) The dwarfed "giant deer" *Megaloceros matritensis* n.sp. from the Middle Pleistocene of Madrid—a descendant of *M. savini* and contemporary to *M. giganteus*. Quat Int 520:110–139

van der Made J, Tong H (2008) Phylogeny of the giant deer with palmate brow tines *Megaloceros* from west and *Sinomegaceros* from east Eurasia. Quat Int 179:135–162

van der Meulen AJ, van Kolfschoten T (1988) Review of the late Turolian to early Biharian mammal faunas from Greece and Turkey. Mem Soc Geol It 31:201–211

Vasileiadis N, Tsoukala E, Kostopoulos DS (2019) The late Miocene bovids from Platania (Drama Basin, Greece), with description of a new species of *Paleoryx*. Geobios 55:57–76

Vasileiadou K, Konidaris G, Koufos GD (2012) New data on the micromammalian locality of Kessani (Thrace, Greece) at the Mio-Pliocene boundary. Palaeobiodiv Palaeoenv 92:211–237

Viret J (1954) Le loess à bancs durcis de Saint-Vallier (Drôme) et sa faune de mammifères villafranchiens. Nouv Arch Mus Hist Nat Lyon 4:1–200

Vislobokova IA (1990) [The fossil deer of Eurasia]. Trudy Paleontologicheskogo Instituta, Akademija Nauk SSSR, vol 240. Nauka, Moskva (in Russian)

Vislobokova IA (2012) Giant deer: origin, evolution, role in the biosphere. Paleont J 46:643–775

Vislobokova IA (2013) Morphology, taxonomy, and phylogeny of megacerines (Megacerini, Cervidae, Artiodactyla). Paleont J 47:833–950

Vlachos E, Tsoukala E, Crégut-Bonnoure E, Guérin C, Mol D (2018) The paradise lost of Milia (Grevena, Greece; Late Pliocene, early Villafranchian, MN15/MN16a): faunal composition and diversity. Quaternary 1:13. (1–16)

Webb SD (2000) Evolutionary history of New World Cervidae. In: Vrba ES, Schaller GB (eds) Antelopes, deer, and relatives. Yale University Press, New Haven, pp 38–64

Zografou E, Bonda-Doumanaki C (2014) [From the cultural heritage of Siatista]. Published by the authors, Siatista (in Greek)

Zurano JP, Magalhães FM, Asato AE, Silva G, Bidau CJ, Mesquita DO, Costa GC (2019) Cetartiodactyla: updating a time-calibrated molecular phylogeny. Mol Phylogenet Evol 133:256–262

The Fossil Record of Suoids (Mammalia: Artiodactyla: Suoidae) in Greece

Dimitris S. Kostopoulos and Ioanna Sylvestrou

1 Introduction

Suoids (Superfamily Suoidea) include today 7 genera and 16 (or 19) species distributed in 2 families of distinct geographic ranges, the Old World suids or pigs (Family Suidae; 5 genera and 13–16 species) and the New World tayssuids or peccaries (Family Tayssuidae; 2–3 genera and 3 species) (McDonald 2009; Groves and Grubb 2011). Members of both families represent primarily omnivorous artiodactyls with a simple nonruminating stomach, a fairly characteristic cuneate-shaped cranium, absence of frontal appendages, complete tooth formula with low (apart from warthogs) crowned bunodont (or neobunodont sensu Stehlin 1899/1900) to bunolophodont molars, and as a rule, short tetradactyle acropodials (Hünnermann 1999; Harris and Liu 2007). Apomorphic features uniting Suoidea are the presence of ossified tympanic bullae, the dorsally open external auditory meatus, the elongated tympanic process of the squamosal, the rootless lower canine of males, the paraconid-metaconid fusion (or paraconid reduction), the lack of cingulum on the lingual wall of upper molars and the similarly shaped first and second lower incisors (Liu 2003).

Earliest known members of suoids are known from the Eocene of China and Thailand (Liu 2001), which makes Asia the most likely center of their origin. The early history of the group is still little understood and opinions among authors differ substantially. Primitive, Oligocene suoids, commonly referred to as "Old World peccaries" or Palaeochoeridae Matthew, 1924, are strongly debatable (e.g., McKenna and Bell 1998); some of them are currently seen as members of Suidae family while others as stem Suidae (Liu 2003; but see also van der Made 2010, 2020).

The first unequivocal occurrences of the Old World family Suidae are dated at the early Miocene of Eurasia and Africa, suggesting that suids already radiated

D. S. Kostopoulos (✉) · I. Sylvestrou
School of Geology, Aristotle University of Thessaloniki, Thessaloniki, Greece
e-mail: dkostop@geo.auth.gr; sylvest@geo.auth.gr

E. Vlachos (ed.), *Fossil Vertebrates of Greece Vol. 2*,
https://doi.org/10.1007/978-3-030-68442-6_7

during the Oligocene (Pickford 1993). During the Miocene Suidae show a remarkable morphological diversity. More than 20 suid genera are recorded, distributed in 6–9 subfamilies. Listriodontin, cainochoerin, and hyotheriin suids went extinct before the Pliocene and tetraconodontins during the Early Pleistocene, while at the same time suins diversified strongly (Franz et al. 2016; Harris and Liu 2007). Today suids may be found along a strickingly broad environmental spectrum, from tropical islands and rainforests to semi-desert grasslands and the high Himalayan plateaus (Meijaard et al. 2015).

As for other members of Artiodactyla, suids have been a basic source of food and other raw materials for prehistoric men. Cave paintings in Leang Bulu' Sipong 4 Cave in Sulawesi, Indonesia, dated at about 44,000 years nicely portray such relationship (Aubert et al. 2019). Zooarchaeological records from China, Near East, and Greece suggest that pigs were already domesticated by 8500–8000 years ago (Payne 1985; Franz et al. 2016).

Extant suoids are represented in Greece exclusively by the wild boar, *Sus scrofa* L., naturally recorded in Central and Northern Greece, as well as in Samos Island. Suoids were never exceptionally abundant or diversified according to the available Greek fossil record; still they are represented by 10 species at least during the last 17 Myr. The aim of this work is to present the Greek fossil Suoidea through time, providing taxonomical, phylogenetic, biogeographic, and chronological information. The bibliography on the topic has been condensed to the minimum possible and in cases of debatable taxonomic issues we follow the most recent reviews or author's point of view.

2 Historical Overview

Professors J. Roth and J.A. Wagner of the University of Munich were likely the first who reported on fossil suids from Greece. They described and illustrated (Roth and Wagner 1854: Pl. 11, fig. 4) a partial mandible from Pikermi, near Athens, introducing a new species, *Sus erymanthius*, inspired by the legend of the Erymanthian boar, from the feats of Hercules. In the following years more material of this taxon became known through the classic works of Gaudry (1862–1865), Forsyth-Major (1888, 1894), and Arambourg and Piveteau (1929). At the dawn of World War II, Professor I. Paraskevaidis of the Agricultural University of Athens extended the Greek suoid record to early Miocene by describing sanithere remains from "Michalos clay-pit" in Chios Island. During the 1950s Professor E. Thenius from the University of Vienna (Thenius 1950, 1955) reported two more taxa from Greece, *Sus antiquus* Kaup, 1833, and *Potamochoerus (Postpropotamochoerus) hyotheriodes* (Schlosser, 1903) from Sophades, Thessaly and Samos Island respectively. All of the aforementioned taxa participated in a series of taxonomic revisions in the coming years, while new paleontological excavations constantly enriched the Greek suid record by means of both specimens and species. Over the next decades new material from several sites partially completed the Greek suid record up to the Pleistocene (Steffens et al. 1979; Koufos 1986, 2007; Tsoukala

1989; de Bonis et al. 1997; Sylvestrou and Kostopoulos 2006, 2009). At the same time Bonis and Bouvrain (1996) provided the first revision of Greek Late Miocene suids.

3 Phylogenetic Relationships

Both morphological and molecular evidence suggest that the sister extant Families Suidae Gray, 1821, and Tayssuidae Palmer, 1897, included into the suborder Suina Gray, 1868 (or Suiformes Jaeckel, 1911), represent together a distinct monophyletic clade of the even-toed hoofed mammals—Mammalia: Artiodactyla Owen, 1848 (e.g., Boiserie et al. 2005; Marcot 2007). Relationships of Suina with other groups of artiodactyls are however less clear. Boiserie et al. (2005) and Boisserie and Lihoreau (2006) recognized Suina as the sister group of a clade uniting Hippopotamidae, Anthracotheriidae, and Ceracea, whereas Spaulding et al. (2009) based on morphological data suggest sister group relationships between Hippopotamidae and Suidae. Pickford (1993) also proposed the ancestry of hippos among "Palaeochoerinae."

Most authors agree that Suidae, Tayssuidae, and the extinct Sanitheriidae Simpson, 1945 (or Saniitheriinae according to other scholars), constitute a taxon of superfamily level, Suoidea. Suidae appears to be the most diversified among suoid families; Harris and Liu (2007) recognize seven subfamilies: Listriodontinae Lydekker, 1884, Kubanochoerinae Gabunia, 1958, Tetraconodontinae Lydekker, 1876, Namachoerinae Pickford, 1995, Cainochoerinae Pickford, 1988, Taucanaminae van der Made, 1997 (= Schizochoerinae Golpe-Posse, 1972), and Suinae Gray, 1821 but opinions among scholars vary considerably (e.g., Orliac et al. 2010: tab. 1; van der Made 2010).

4 Distribution

The Greek fossil record of Suoidea ranges from the middle Miocene to the Late Pleistocene. Earliest known occurrences are from the middle Aragonian (MN5–6; Orleanian/Astaracian) faunas of Antonios, and Thymiana (Chios Island) representing mostly sanithere and listriodontine taxa (Appendix). During the Late Miocene the Greek record is monopolized by the genus *Hippopotamodon* (=*Microstonyx*) though sparse evidences of *Propotamochoerus* are also present. From the early Ruscinian onwards, suids are represented in Greece exclusively by the genus *Sus*. Suid diversity remains always low in the Greek fossil record reaching a maximum of only two species per site (Fig. 1 and Appendix). As a whole, 5 suoid genera and 10 species at least are known from Greece during the last 17 Myr. Geographic affinities of Greek suoids are exclusively Eurasian and most recorded taxa show a wide pan-Eurasian distribution. A list of localities with the most important suoid occurrences is given in the Appendix.

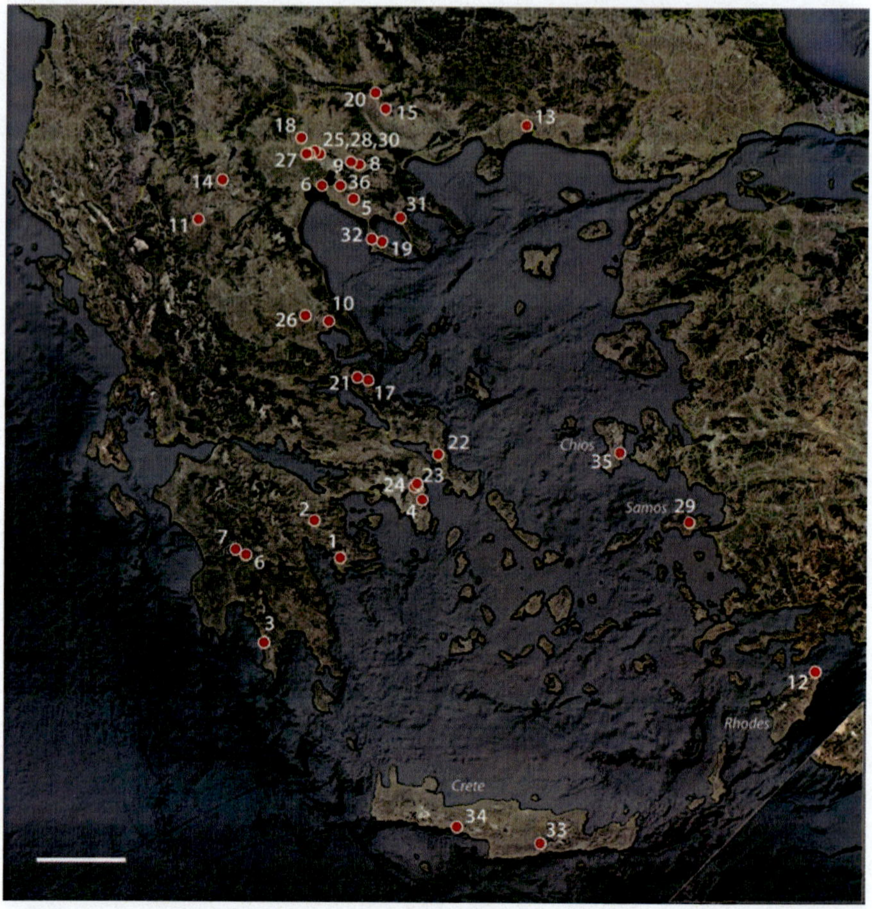

Fig. 1 Map of Greece showing the geographic distribution of the most important localities with suoid fossils. **1**, Franchthi Cave; **2**, Klisoura Cave; **3**, Kalamakia Cave; **4**, Vraona Cave; **5**, Petralona Cave; **6**, Megalopolis; **7**, Kyparíssia localities; **8**, Vassiloudi 1; **9**, Gerakarou 1; **10**, Sesklon lower level; **11**, Milia localities; **12**, Damatria; **13**, Kessani localities; **14**, Kardia; **15**, Maramena; **16**, Samos, various localities; **17**, Achladion; **18**, Dytiko 1; **19**, Kryopigi; **20**, Thermopigi; **21**, Kerassia localities; **22**, Halmyropotamos; **23**, Chomateri; **24**, Pikermi localities; **25**, Ravin X; **26**, Perivolaki; **27**, Prochoma-1; **28**, Vathylakkos localities; **29**, Samos localities; **30**, Ravin de Zouaves-5; **31**, Nikiti-1; **32**, Fourka; **33**, Kastellios K2; **34**, Plakias; **35**, Thymiana; **36**, Antonios. See Appendix for more information. Image exported from Google Earth Pro © 2019, map data from US Dept. of State Geographer, SIO, NOAA, U.S. Navy, NGA, GEBCO, image from Landsat/Copernicus. Scale bar equals 80 km, North faces upward

5 Systematic Paleontology

Suoidea Gray, 1821

Sanitheriidae Simpson, 1945
***Sanitherium* von Meyer, 1866**
(=*Xenochoerus* Zdarsky, 1909)

Type Species *Sanitherium schlaginweiti* von Meyer, 1866, by original designation.

Distribution Early-middle Miocene of Africa, Europe, Turkey, Pakistan, and India. *Sanitherium* sp. is also mentioned from the Oligocene of Pakistan.

Other Taxa Included *S. leobense* (Zdarsky, 1909); *S. jeffreysi* (Forster-Cooper, 1913); *S. africanus* (Stromer, 1926); *S. nadirum* Wilkinson, 1976.

Remarks Pickford (1984) recognized within the family Sanitheriidae both *Diamantohyus* Stromer, 1926, and *Sanitherium* von Meyer, 1866, as distinct genera but van der Made and Hussain (1992) drastically revised the systematics of the group by recognizing a single genus *Sanitherium* with only three valid species (*S. schlaginweiti, S. africanus,* and *S. jeffreysi*). Harris and Liu (2007) follow Pickford in retaining *Diamantohyus* for *D. africanus* and *D. jeffreysi* and recognized *S. leobense* as a valid species. According to Pickford (1984) *Sanitherium* is characterized by a P4 with six subequally developed cusps and wider lower molars than in *Diamantohyus* but van der Made and Hussain (1992) doubt most morphometric differences between the two genera.

***Sanitherium schlaginweiti* von Meyer, 1866**
(= ★*Sanitherium masticum* Paraskevaidis, 1940)

Nomenclatural and Taxonomic History See Pickford (1984), van der Made and Hussain (1992), van der Made (1998).

Type Material Left m2–3 (von Meyer 1866: Pl. 2, figs 9–12).

Type Locality Koshialgarh Punjab, India, middle Miocene.

Age Early-middle Miocene.

Distribution Early-middle Miocene of Austria, Greece, Indo-Pakistan, and possibly East Africa.

Fig. 2 Left lower toothrow AMPG(v)47 of *Sanitherium schlaginweiti* from Thymiana Chios (Orleanian) in lateral (buccal) view. The specimen is a syntype of *Sanitherium masticum*. Scale bar equals 1 cm. Photograph courtesy of S. Roussiakis

Remarks According to van der Made (1998) *S. leobense* from Seegraben, Austria cannot distinguish from *S. schlaginweiti* from Koshialgarh, India, although van der Made and Hussain (1992) stated that premolars of *S. leobense* may have a less complicated structure, which would re-validate the species. Paraskevaidis (1940) described from "Michalos clay-pit" in Chios Island, a site later known as Thymiana, rich sanithere material attributed to the new species *S. masticum* (Fig. 2). *S. masticum* is considered as a junior synonym of *S. leobense* by Pickford (1984), whereas van der Made (1998) suggested both taxa as synonymous of *S. schlaginweiti*. De Bonis et al. (1997) described more sanithere material from the MN5 Thymiana B fauna in Chios. They confirmed close similarity between Chios and the Leoben taxa, and they accepted both samples as better attributable to *S. schlaginweiti*. Koufos (2007) provided new material of *S. schlaginweiti* from the MN4/5 fauna of Antonios in Chalkidiki Peninsula.

Suidae Gray, 1821

Listriodontinae [Authorship debated; see van der Made 1996; Pickford and Morales 2003].

Listriodon von Meyer, 1846 (=*Bunolistriodon* Arambourg, 1963)

Type Species *Listriodon splendens* von Meyer, 1846, by original designation.

Distribution Early to late Miocene of Europe, Africa, and Asia.

Other Taxa Included *L. lockharti* (Pomel, 1848); *L. pentapotamiae* (Falconer, 1868); *L. latidens* (Biedermann, 1873); *L. affinis* (Pilgrim, 1908); *L. guptai* (Pilgrim, 1926); *L. jeanneli* (Arambourg, 1933); *L. intermedius* Liu and Li, 1963; *L. akatikubas* (Wilkinson, 1976); *L. anchidens* van der Made, 1996; *L. meidamon* Fortelius, van der Made and Bernor, 1996; *L. bartulensis* Pickford, 2001; *L. retamaensis* Pickford and Morales, 2003.

Remarks Pickford and Morales (2003) suggested *Bunolitriodon* Arambourg, 1963 (sensu van der Made 1996a), as a junior synonym of *Listriodon* von Meyer, 1846. Both taxa are thoroughly discussed in van der Made (1996a). They represent bunodont to sublophodont suids with strong, laterally expanded processes above the canines, decreasing bunodonty and increasing postcranial robusticity in later forms, some of them additionally characterized by high crowned canines curved outward and upward (van der Made 1996a; Harris and Liu 2007; Pickford and Morales 2003; Orliac 2009).

Listriodon lockharti (Pomel, 1848)

Nomenclatural and Taxonomic History *Sus lockharti* in Pomel 1848 (new species); *Bunolistriodon lockharti* in Arambourg 1963 (new combination); *Listriodon lockhardi* in Stehlin 1899/1900 (new combination); see also Pickford and Morales (2003).

Type Material Museum National d'Histoire Naturelle, Paris MNHN CHE 30, mandible (lectotype selected by van der Made 1996a).

Type Locality Chevilly, France, lower Miocene MN4.

Age Early Miocene, MN4.

Distribution Early-middle Miocene of Europe and Saudi Arabia.

Remarks According to van der Made (1996a) and Pickford and Morales (2003) the species is characterized by large size, increased bunodonty, mesiodistally long incisors with the first one having one or three lobes, and strongly curved upper canines. The species has a rather good record in Western Europe (especially in France) but in Greece is so far known with certainty only from the Orleanian (MN4/5) fauna of Antonios, Chalkidiki peninsula (Koufos 2007). An upper third molar from Thymiana in Chios Island is also referred to as *Listriodon* (n. sp.?) *lockharti* var. *michali* by Paraskevaidis (1940). Van der Made (1996a) suggested however that the molar size is too small for this species and, although the absence of incisors and canines makes any assignment difficult, van der Made (2020) implies that the Chios species may turn out to be a senior synonym of *Eurolistriodon adelli* Pickford and Moyà Solà, 1995.

Suinae Gray, 1821

Propotamochoerus Pilgrim, 1925 (= *Korynochoerus* Schmidt-Kittler, 1971)

Type Species *Sus hysudricus* Stehlin, 1899/1900, by original designation.

Distribution Late middle Miocene to late Pliocene of Europe, Turkey, India, Myanmar China; Late Miocene of Africa.

Other Included Taxa *P. palaeochoerus* (Kaup, 1833); *P. provincialis* (Gervais, 1859); *P. hyotheroides* (Schlosser, 1903); *P. wui* van der Made and Han, 1994; *Propotamochoerus* sp. nov.

Remarks According to van der Made et al. (1999) *Propotamochoerus* represents small dicoryphochoerin suids with relatively large canines. Their cranium shows a relatively long face with square snout in cross section, strongly expanded laterally zygomatic arches and large supra-canine flanges (Pickford 1988; Harris and Liu 2007). The dentition is characterized by a lower third molar with simple talonid and a predominantly tall protoconid on p4. *Propotamochoerus* possibly arose from an Asian hyotherin suid at about 11 Ma ago and dispersed toward SW Asia and Europe during the Late Miocene (Pickford 1988; Fortelius et al. 1996).

Propotamochoerus palaeochoerus (Kaup, 1833)

Nomenclatural and Taxonomic History *Sus palaeochoerus* in Kaup 1833 (new species); *Korynochoerus palaeochoerus* in Schmidt-Kittler (1971) (new combination); *Propotamochoerus palaeochoerus* in Pickford (1988) and van der Made et al. (1992) (new combination).

Type Material Hessisches Landesmuseum, Darmstadt, HLD Din 3, right mandible with p3–m3 (lectotype selected by Hünermann 1968: Pl. 1, fig. 1).

Type Locality Eppelsheim, Germany, Upper Miocene.

Age Early late Miocene, early Vallesian (MN9).

Distribution Late middle-late Miocene (latest Aragonian–Vallesian) of Europe.

Remarks *P. palaeochoerus* is the oldest representative of its genus known from several west European faunas ranging from the latest Aragonian (MN8; late Astaracian) to the late Vallesian (MN10). The species is extensively discussed in Schmidt-Kittler (1971) and later commented by Pickford (1988), van der Made and Moyà-Solà (1989), and Fortelius et al. (1996). Its eastern most known record comes from Ukraine (van der Made et al. 1999). It is about the size of a wild boar, characterized by a concave cranial profile and elevated occiput, inflated zugomatic arches, large upper third incisor, proportionally small third molar, and strong metaconid on the lower fourth premolar (Schmidt-Kittler 1971; van der Made and Moyà-Solà 1989; van der Made et al. 1999). van der Made (1996b) refers a lower second milk molar from the likely early Vallesian (MN9) fauna of Plakias, Crete (de Bruijn et al. 2012), to as cf. *Propotamochoerus palaeochoerus*, whereas Lazaridis (2015) reports a lower m2–m3 and an astragalus from the late Vallesian mammal assemblage of Fourka sand-pits (Chalkidiki Peninsula) to *P.* cf. *palaeochoerus*. Hellmund (1995) also refers a partial cranium and some isolated teeth and toothrows from the latest Miocene (end of MN13) fauna of Maramena, Serres basin, to *P. palaeochoerus* but Geraads et al. (2008) doubt this attribution and van der Made et al. (1999) suggest it represents a different SE European taxon.

★*Propotamochoerus* sp. nov.

Nomenclatural and Taxonomic History The new species has been originally described by Lazaridis in his PhD thesis (2015) but is still pending for a formal publication (under ICZN rules) and, therefore, it is referred here to as *Propotamochoerus* sp. nov. (Lazaridis et al. 2021 under review)

Age Late Miocene, Turolian (MN11–13).

Distribution Late Miocene (Vallesian–Turolian) of Greece, Northern Macedonia, Bulgaria, Turkey.

Remarks Turolian propotamochoerid samples recorded in SE Europe, especially around the Aegean Sea, and attributed to several species (e.g., *P. hysudricus* in Bonis and Bouvrain 1996; *P. palaeochoerus* in Hellmund 1995; *P. hyotherioides* in Thenius 1955, and Fortelius et al. 1996) are part of a long lasting discussion (e.g., van der Made and Moyà-Solà 1989; Fortelius et al. 1996; van der Made et al. 1999; Geraads et al. 2008). Geraads et al. (2008) suggested that all these records likely represent a new *Propotamochoerus* species which has to be formally established on the basis of adequate cranial material. Lazaridis (2015) based on a fairly complete skull from Kryopigi, Chalkidiki Peninsula, revised previous data and proposed a new, though still unpublished, species, characterized by medium size, converging but not merging temporal lines, presence of diastemata between the first two premolars and the canine, wide third premolars, and comparatively large lower third molars with a hexaconid. The earliest occurrence of *Propotamochoerus* sp. nov. in Greece is in Ravin des Zouaves 5 (MN11; Fig. 3) and the latest likely in Maramena (MN13/14) (van der Made et al. 1999; Pickford 2013; Lazaridis 2015; Lazaridis et al. 2021 under review).

Hippopotamodon Lydekker, 1877
(= *Microstonyx* Pilgrim, 1926 = *Dicoryphochoerus* Pilgrim, 1926 = *Eumaiochoerus* Hürzeler, 1982)

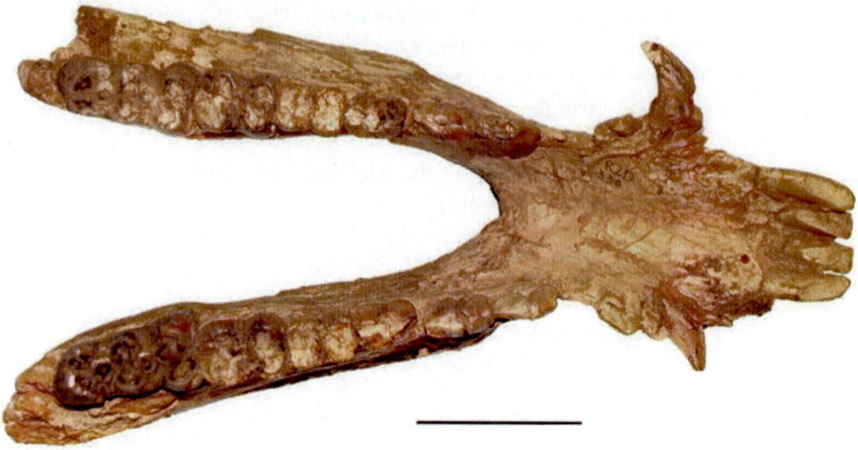

Fig. 3 Lower mandible LGPUT RZO-330 of *Propotamochoerus* sp. nov. from Ravin des Zouaves 5, Axios valley (early Turolian) in occlusal view. The specimen has been illustrated as *Propotamochoerus* cf. *hysudricus* in Bonis and Bouvrain (1996: fig. 9). Scale bar equals 5 cm

Type Species *Hippopotamodon sivalense* Lydekker, 1877, by original designation.

Distribution Late Miocene of Europe; late Miocene to Early Pleistocene of Asia.

Other Included Taxa *H. antiquus* (Kaup, 1833); *H. major* (Gervais, 1850); *H. ultimus* (Han, 1987); *H. pilgrimi* (Pickford, 1988); *H. etruscus* (Michelotti, 1861).

Remarks Giant European Late Miocene dicoryphochoerin suins referred to the genus *Microstonyx* Pilgrim, 1926, and their Asian counterparts, referred to the genus *Hippopotamodon* Lydekker, 1877, build a still unresolved taxonomic group extensively discussed over the years (e.g., Pickford 1988, 2015; van der Made et al. 1992, 2013; van der Made and Hussain 1989; Fortelius et al. 1996; Liu et al. 2004). Here we follow the most recent review by Pickford (2015) in recognizing synonymy between these two genera. *Hippopotamodon* is characterized by large to very large size, the male canines vary from relatively large and flaring (in *H. sivalense*) to reduced, usually associated by prominent supra-canine flanges, the snout is elongated, the braincase is broad and flat with elevated occiput, and the zygomatic arches are strongly inflated.

Hippopotamodon major (Gervais, 1848–1852) (= ★*Sus erymanthius* Roth and Wagner, 1854)

Nomenclatural and Taxonomic History See Trofimov (1954). There is no current consensus among scholars about the species context. Here we follow Liu et al. (2004) in recognizing *Sus erymanthius* Roth and Wagner, 1854 (type locality

Fig. 4 Male cranium of *Hippopotamodon major* AMNH. Samos MTLA-537 from Mytilinii 1A, Samos (middle Turolian) in dorsal view. The specimen is illustrated as *Microstonyx major* in Sylvestrou and Kostopoulos (2009: Pl. 1). Scale bar equals 5 cm

Pikermi, Greece), as a junior synonym of *Sus major* Gervais, 1848 (but see also van der Made et al. 2013; Pickford 2015).

Type Material Upper left third molar from Cucuron, France, illustrated by Gervais (1850: Pl. 12, fig. 2; *fide* Pickford 2015).

Type Locality Cucuron, France, upper Miocene.

Age Late Miocene, Turolian.

Distribution Late Miocene (latest Vallesian to latest Turolian; end MN10–endMN13) of Eurasia.

Remarks Liu et al. (2004) analysis of several *Microstonyx* samples across Eurasia revealed no temporal or spatial pattern on observed morphometrical variability, and thus the species *M. major* is proposed as polymorphic, absorbing past distinctions at species or subspecies level. *M. major* is characterized by smaller size, longer snout, deep infraorbital incisure, weaker canines, tendency of losing the first premolars, and longer diastemata, compared to the isochronous Indian taxon (Fig. 4). *M. major* is by far the most common among Neogene suins in Greece, recorded in almost every Turolian mammal assemblage. Greek material is discussed in detail by Bonis and Bouvrain (1996), Liu et al. (2004), and Sylvestrou and Kostopoulos (2006, 2009). Its earliest known local record comes from the latest Vallesian fauna of Nikiti-1, Chalkidiki Peninsula, and the latest from Dytiko-1, Axios Valley. According to Fortelius et al. (1996) and Kostopoulos et al. (2001), *H. antiquus* record from Sophades (Thenius 1950) needs further documentation.

Sus Linnaeus, 1758

Type Species *Sus scrofa* Linnaeus, 1758, by original designation.

Distribution Pliocene to Pleistocene of Eurasia.

Other Included Taxa *S. arvernensis* Croizet and Jobert, 1828; *S. strozzii* Forsyth-Major, 1881; *S. falconeri* Lydekker, 1884; *S. macrognathus* Dubois, 1908; *S. brachygnathus* Dubois, 1908 (=? *S. sangirensis* von Koenigswald, 1963); *S. lydekkeri* Zdansky, 1928; *S. bijiashanensis* Han et al., 1975 (=? *S. liuchengensis* Han, 1987); *S. australis* Han, 1987 (=?*S. peii* Han, 1987); *S. xiaozhu* Han, 1987, *Sus sondaari* van der Made, 1999; for extant taxa see Groves and Grubb (2011).

Remarks *Sus* likely emerged at the end of Miocene in Asia and by the Late Pliocene, had colonized most of Eurasia (see Franz et al. 2016 for a brief evolutionary history). Depending mostly on the type of male lower canine, *Sus* species are divided into two groups: the "verrucosic" and the "scrofic" one, already separated each other during the Pliocene. Pickford (2012) ascribes Eurasian taxa of the "verrucosic" group to the genus *Dasychoerus* Gray, 1873, but this taxonomic opinion has been challenged by Cherin et al. (2018) and it is not widely adopted (e.g., Liu et al. 2017; Iannucci et al. 2020a). During the Middle Pleistocene *S. scrofa* replaced most other Eurasian *Sus* species, becoming the only suin of European Mid-Late Pleistocene continental faunas (Franz et al. 2016).

Sus arvernensis Croizet and Jobert, 1828
(= *Sus minor* Gervais, 1848)

Nomenclatural and Taxonomic History *Aper arvernensis* in Croizet and Jobert 1828 (new species); *Sus provincialis* var. *minor* in Depéret 1890 (new subspecies); *Sus minor* in Azzaroli 1954 (new combination); *Sus arvernensis* in Gervais, 1859 (new combination).

Type Material MNHN-no cat. No, juvenile mandible and maxilla (holotype) illustrated by Croizet and Jobert (1828: Pl. 13, figs. 3–5).

Type Locality Les Étouaires, Perrier, France, Pliocene.

Age Early Villafranchian (MN16).

Distribution Ruscinian to middle Villafranchian (MN14–MN17) of Europe.

Remarks The smaller among Plio-Pleistocene suins of continental Europe characterized by proportionally long premolar row, simple third molars, and low anterior and posterior ends on the lower third premolar. The Greek record of the species is rather poor restricted in a few isolated teeth and partial toothrows from the Ruscinian faunas of Kessani 1,2, and Megalo Emvolo, as well as from the early Villafranchian faunas of Milia and the lower levels of Sesklo (Koufos et al. 1991; Syrides et al. 1997; Guérin and Tsoukala 2013; Athanassiou 2018).

Sus strozzii Forsyth-Major, 1881

Nomenclatural and Taxonomic History See Azzaroli (1954).

Type Material Museum of Geology and Paleontology of Florence, IGF 424, adult male skull, mandible, and partial skeleton (lectotype selected by Azzaroli, 1954).

Type Locality Upper Val d'Arno, Italy (unknown fossil horizon), Plio-Pleistocene.

Age Late Pliocene–Early Pleistocene (MNQ17–19).

Distribution Late Pliocene (late early Villafranchian, MN16b) to Middle Pleistocene of Southern Europe, Azov Sea region, Azerbaijan, and Israel.

Remarks Cherin et al. (2018) recently rediscussed the diagnostic features of this large sized suin, characterized basically by a narrow cranium on the parietal region, and smoothly undulating dorsally, inflated and broad in the middle zygomatic arches, long and rugose male supra-canine flanges, verrucosic lower canines, and elongated lower third molars with a single cuspid between the first and second lobe. *Sus strozzii* may appear at the SE of Greece (Damatria, Rhodes Island) as early as late early Villafranchian (van der Made 1988). It possibly occurs at Sesklo during MN17, though it is characteristically missing from other contemporaneous Greek assemblages, and it certainly occurs during the late Villafranchian in Mygdonia Basin (sites Gerakarou, Fig. 5a and Vassiloudi) (Koufos 1986; Kostopoulos and Athanassiou 2005). The taxon seems to disappear temporarily from Europe during late Villafranchian and re-dispersed at the beginning of Epivillafranchian, ca. 1.2 Ma ago (Cherin et al. 2020).

Sus scrofa Linnaeus, 1758

Nomenclatural and Taxonomic History Extant; see Groves and Grubb (2011).

Distribution Middle Pleistocene to Recent; originally Eurasian and North African.

Remarks *S. scrofa* distinguishes by a fairly long snout, rather shallow preorbital fossa, discontinuous postorbital bar, short premolar row, relatively long third upper molars, elevated mesial and distal end of the second and third upper premolars, and lower male canines of "scrofic" type (van der Made and Moyà-Solà 1989; Groves and Grubb 2011). Already present in Atapuerca TD6, Spain (~0.9 Ma; van der Made 1988), early members of *Sus scrofa,* commonly referred to as *S. s. priscus*, expanded across Europe at the first part of Middle Pleistocene, as evidenced by data from Gombaszög (Slovakia), Dorn Dürkheim 2, Voigtstedt and Mosbach (Germany), Pakefield and West Runton (Britain), and likely Kyparissia 1, 4 in Megalopolis

Fig. 5 Selected specimens of *Sus* from Greece. (**a**) Partial cranium LGPUT GER-51 of *Sus strozzii* from Gerakarou, Mygdonia Basin (late Villafranchian) in lateral view, originally described by Koufos (1986). (**b**) Left mandible LGPUT PEC of *Sus scrofa* from Petralona Cave, Chalkidiki Peninsula (Middle Pleistocene) in occlusal view, originally described by Tsoukala (1989). Scale bars equal 5 cm

Basin, Greece (Athanassiou et al. 2018; Cherin et al. 2020). Based on some teeth and postcranials, again from the Middle Pleistocene of Megalopolis Basin, Melentis (1965) was likely the first to report the species in Greece. Tsoukala and Guérin (2016) also described primitive *Sus scrofa* from the Middle Pleistocene of Petralona Cave (Fig. 5b), whereas the species appears to be quite common in Late Pleistocene faunal assemblages discovered in several cave deposits (Tsoukala 1992 and Appendix).

Suinae Insertae Sedis

"Sus" provincialis (de Blainville, 1847)

Nomenclatural and Taxonomic History *Sus provincialis* de Blainville, 1847 (new species); for a full history of the species see Pickford (2013).

Type Material University of Montpellier II UM SM 460, right M3 (lectotype selected by Pickford 2013: fig.1).

Type Locality Montpellier France, Lower Pliocene.

Age Early Pliocene, Ruscinian.

Distribution Latest Miocene to early Pliocene of Europe.

Remarks Pickford (2013) recently revised the suin species originally known from the Pliocene marine sands at Montpellier, suggesting that authorship has also to be transferred from Gervais (1859) to Blainville (1847), who first illustrated and named the taxon. The species is commonly referred to as *Propotamochoerus* or to as "*Propotamochoerus*," is part of a long discussion (e.g., Fortelius et al. 1996; van der Made et al. 1999; Geraads et al. 2008) but Pickford (2013) avoids a direct reference to a particular genus as he considers that the type material does not have enough diagnostic characters. Dental size places "*Sus*" *provincialis* between *S. arvernensis* and *S. strozzii*. A female cranium from the Lower Pliocene coal deposits of Kardia, Ptolemais, is so far the only evidence of this species in Greece (van der Made and Moyà-Solà 1989), although Iannucci et al. (2020b) recently questioned this affiliation.

6 Conclusions

Two suoid families, 5 genera and 10 species have been so far reported from the Greek fossil record, suggesting a diachronously much lower diversity than other artiodactyls or ungulates, and a rather even temporal distribution of one to two species per time interval. Among the recorded taxa, two species have been originally named based on Greek material (*Sanitherium masticum* Paraskevaidis, 1940 and *Sus erymanthius* Roth and Wagner, 1854); both however have been later proved or suggested as junior synonyms of other taxa. The longer lasting taxon appears to be *Hippopotamodon major*, known from the latest Vallesian to the latest Turolian, hence for about 3.5 Myr.

Known major faunal turnovers are fairly well evidenced on the Greek suoid record in accordance with other European data (e.g., Fortelius et al. 1996; Cherin et al. 2020), though the Greek Astaracian-early Vallesian data are extremely poor prohibiting certain conclusions about the middle-late Miocene transition. Listriodontines and sanitheres predominate during the early Aragonian (Orleanian). They will be replaced toward the Vallesian by dicoryphochoerin suids of the genus *Propotamochoerus*. *Hippopotamodon antiquus* is so far unknown from Greece but it is present nearby (e.g., Geraads et al. 2005; Kostopoulos and Sen 2016) suggesting its absence from Greece as rather accidental. *Hippopotamodon major* (= *Microstonyx major*) enters in the local record at the end of Vallesian, at about 8.7–9.0 ma ago, followed by a new SE European species of *Propotamochoerus*, and marking quite sharply the rise of the Pikermian faunas in the area. They both went extinct during or soon after the Messinian Salinity Crisis, replaced by early members of the genus *Sus*. The Mid-Pleistocene transition will finally allow *Sus scrofa* entering the local record, leading to the extinction of verrucosic taxa such as *Sus strozzii*.

Acknowledgments We would like to thank S. Roussiakis for providing us with photo material of the Paraskevaidis sample from Chios, as well as J. van der Made and E. Tsoukala for fruitful comments and suggestions.

Appendix

List of fossil localities with occurrences of suoids in Greece. Locality numbers refer to the collection numbers of the PaleoBiology Database (PBDB)

Localities[PBDB No]	Age (MN, MNQ)	Taxa	Refs
Franchthi Cave[182709]	latest Pleistocene	*Sus scrofa*	1
Klisoura Cave[184243]	latest Pleistocene	*Sus scrofa*	2
Kalamakia Cave[184245]	latest Pleistocene	*Sus scrofa*	3
Vraona Cave[183130]	latest Pleistocene	*Sus scrofa*	4
Petralona Cave[183123]	Middle Pleistocene	*Sus scrofa priscus*	5
Megalopolis	Middle Pleistocene	*Sus scrofa*	6
Kyparissia 1, 4[194472, 194,475]	Middle Pleistocene	*Sus scrofa*	7
Vassiloudi 1[34651]	Early Pleistocene (MNQ18)	*Sus strozzii*	8
Gerakarou 1[34617]	Early Pleistocene (MNQ18)	*Sus strozzii*	9
Sesklon lower level[205043]	Late Pliocene (MN16)	*Sus arvernensis*	10
Milia[185859]	Late Pliocene (MN16)	*Sus arvernensis arvernensis*	11
Damatria[34591]	Late Pliocene (MN16)	*Sus strozzii*	12
Megalo Emvolon[36579]	Early Pliocene (MN15)	*Sus arvernensis*	13
Kessani 1[205296], 2[205297]	Early Pliocene (MN14)	*Sus arvernensis*	14
Kardia[204752]	Early Pliocene (MN14–15)	*"Sus" provincialis*	15
Maramena[32189]	latest Miocene/earliest Pliocene (MN13/14)	*Propotamochoerus* sp. nov.	16, a
Samos, unknown level: NHMW collection[182751] NHML collection[202120] BSPM collection[207137] IGPM collection[207193]	Late Miocene (MN11–13)	*Hippopotamodon major* *Propotamochoerus* sp. nov.	17, a
Achladion[207133]	Late Miocene (MN11–13)	*Propotamochoerus?* sp. nov.	18, a
Dytiko-1[32374]	Late Miocene (MN13)	*Hippopotamodon major*	19
Kryopigi[157582]	Late Miocene (MN12–13)	*Propotamochoerus* sp. nov. *Hippopotamodon major*	20, a
Samos Q5[95690]	Late Miocene (MN12/13)	*Hippopotamodon major*	17
Thermopigi[73553]	Late Miocene (MN12)	*Hippopotamodon major* *Propotamochoerus* sp. nov.	21, a
Kerassia-1[195432]	Late Miocene (MN12)	*Hippopotamodon major*	22

(continued)

Localities[PBDB No]	Age (MN, MNQ)	Taxa	Refs
Halmyropotamos[202213]	Late Miocene (MN12)	*Hippopotamodon major*	23
Chomateri[195562]	Late Miocene (MN12)	*Hippopotamodon major*	18
Pikermi Valley-3[202631]	Late Miocene (MN12)	*Hippopotamodon major*	24
Pikermi Valley-1[202630]	Late Miocene (MN12)	*Hippopotamodon major*	24
Pikermi[182754]	Late Miocene (MN12)	*Hippopotamodon major*	25
Ravin X[182745]	Late Miocene (MN12)	*Hippopotamodon major*	19, 26
Perivolaki[194879]	Late Miocene (MN12)	*Hippopotamodon major*	27
Prochoma-1[202222]	Late Miocene (MN12)	*Hippopotamodon major*	19
Vathylakkos-3[182750]	Late Miocene (MN12)	*Hippopotamodon major*	19
Vathylakkos-2[202703]	Late Miocene (MN12)	*Hippopotamodon major*	19
Samos Mytilinii 1A[202215] (MTLA)	Late Miocene (MN12)	*Hippopotamodon major*	17
Samos Q1[95691]	Late Miocene (MN12)	*Hippopotamodon major*	17
Samos Q4[95689]	Late Miocene (MN12)	*Hippopotamodon major*	17
Mytilinii-4	Late Miocene (MN11/12)	*Hippopotamodon* sp.	17
Ravin de Zouaves-5[195489]	Late Miocene (MN11)	*Hippopotamodon major* *Propotamochoerus* sp. nov.	19, 20, a
Nikiti-1[202729]	Late Miocene (MN10)	*Hippopotamodon major*	28
Fourka sand pits	Late Miocene (?MN10)	*Propotamochoerus* cf. *palaeochoerus*	20
Kastellios K2	Late Miocene (MN10)	*Taucanamo?*	29
Plakias	Late Miocene (MN9)	*Propotamochoerus* cf. *palaeochoerus*	29
Thymiana[182752]	early Middle miocene (MN5)	*Listriodon* sp. *Sanitherium schlaginweiti*	30 31
Antonios[73861]	early Middle miocene (MN5)	*Listriodon lockharti* *Sanitherium schlaginweiti*	32

MN, Mammal Neogene

[1]Stiner and Munro (2011), [2]Starkovich (2012), [3]Harvati et al. (2013), [4]Symeonidis et al. (1980), [5]Tsoukala and Guérin (2016), [6]Melentis (1965), [7]Athanassiou et al. (2018), [8]Kostopoulos and Athanassiou (2005), [9]Koufos (1986), [10]Athanassiou (2018), [11]Guérin and Tsoukala (2013), [12]Van der Made (1988), [13]Steffens et al. (1979), [14]Syrides et al. (1997), [15]Van der Made and Moyà-Solà (1989), [16]Hellmund (1995), [17]Sylvestrou and Kostopoulos (2009), [18]Koufos (2006), [19]Bonis and Bouvrain (1996), [20]Lazaridis (2015), [21]Geraads et al. (2007), [22]Theodorou et al. (2003), [23]Melentis (1969), [24]Roussiakis et al. (2014), [25]Gaudry (1862-65), [26]Arambourg and Piveteau (1929), [27]Sylvestrou and Kostopoulos (2006), [28]Kostopoulos (1994), [29]Van der Made (1996b), [30]Paraskevaidis (1940), [31]de Bonis et al. (1997), [32]Koufos (2007)

[a]Lazaridis et al. (2021 under review)

References

Arambougr C, Piveteau J (1929) Les vertébrés du Pontien de Salonique. Ann Paleont 18:59–138

Athanassiou A (2018) A Villafranchian *Hipparion*-bearing mammal fauna from Sésklo (E. Thessaly, Greece): implications for the question of *Hipparion–Equus* sympatry in Europe. Quaternary 1:12

Athanassiou A, Michailidis D, Vlachos E, Tourloukis V, Thompson N, Harvati K (2018) Pleistocene vertebrates from the Kyparíssia lignite mine, Megalopolis Basin, S. Greece: Testudines, Aves, Suiformes. Quat Int 497:178–197

Aubert M, Lebe R, Oktaviana AA, Tang M, Burhan B, Hamrullah, Jusdi A, Abdullah, Hakim B, Zhao J, Geria IM, Hadi Sulistyarto P, Sardi R, Brumm A (2019) Earliest hunting scene in prehistoric art. Nature 576:442–445

Azzaroli A (1954) Revisione della fauna dei terreni fluvio-lacustri del Valdarno Superiore, V. Filogenesi e biologia di *Sus strozzii* e di *Sus minor*. Palaeontogr Ital 48:41–76

Boiserie JR, Lihoreau F, Brunet M (2005) The position of Hippopotamidae within Cetartiodactyla. PNAS 102:1537–1541

Boisserie J-R, Lihoreau CR (2006) Emergence of Hippopotamidae: new scenarios. C R Palevol 5:749–756

Bonis L de, Bouvrain G (1996) Suidae du Miocène supérieur de Grèce. Bull Mus Nat Hist Nat C 18(1):107–132

Bonis L de, Koufos GD, Sen S (1997) The sanitheres (Mammalia, Suoidea) from the middle Miocene of Chios Island, Aegean Sea, Greece. Rev Palaeobiol 16:259–270

Bruijn H, Doukas CS, van den Hoek Ostende LW, Zachariasse WJ (2012) New finds of rodents and insectivores from the upper Miocene at Plakias (Crete, Greece). Swiss J Paleontol 131:61–75

Cherin M, Sorbelli L, Crotti M, Iurino DA, Sardella R, Souron A (2018) New material of *Sus strozzii* (Suidae, Mammalia) from the early Pleistocene of Italy and a phylogenetic analysis of suines. Quat Sci Rev 194:94–115

Cherin M, Alba DM, Crotti M, Menconero S, Moullée P-E, Sorbelli L, Madurell-Malapeira J (2020) The post-Jaramillo persistence of *Sus strozzii* (Suidae, Mammalia) in Europe: new evidence from the Vallparadís section (NE Iberian Peninsula) and other coeval sites. Quat Int 233:106234

Croizet JB, Jobert A (1828) Recherches sur les ossemens fossiles du département du Puy-de-Dôme. Adolphe Delahays, Paris, pp 1–226

Forshyth Major CIF (1888) Sur un gisement d'ossements fossiles dans l'ile de Samos, contemporains de l' age de Pikermi. C R Hebd Acad Sci Paris 107:1178–1181

Forsyth Major CIF (1894) Le gisement ossifère de Mytilini et catalogue d' osséments fossiles. Lausanne, pp. 1–51

Fortelius M, van der Made J, Bernor R (1996) Middle and late Miocene Suoidea of Central Europe and the eastern Mediterranean: evolution, biogeography and paleoecology. In: Bernor RL, Fahlbusch V, Mittmann HW (eds) Evolution of Western Eurasian Neogene mammal faunas. Columbia University Press, New York, pp 348–377

Franz L, Meijaard E, Gongora J, Haile J, Groenen MAM, Larson G (2016) The evolution of Suidae. Annu Rev Anim Biosci 4:3.1–3.25

Gaudry A (1862-1865) Animaux fossiles et géologie de l'Attique. Savy, Paris

Geraads D, Kaya T, Mayda S (2005) Late Miocene large mammals from Yulaflı, Thrace region, Turkey, and their biogeographic implications. Acta Palaeont Pol 50:523–544

Geraads D, Tsoukala E, Spassov N (2007) A skull of *Ancylotherium* (Chalicotheriidae, Mammalia) from the late Miocene of Thermopigi (Serres, N. Greece) and the relationships of the genus. J Vert Paleont 27:461–466

Geraads D, Spassov N, Garevski R (2008) New specimens of *Propotamochoerus* (Suidae, Mammalia) from the late Miocene of the Balkans. N Jb Geol Pal Abh 248:103–113

Groves C, Grubb P (2011) Ungulate taxonomy. The John Hopkins University Press, Baltimore

Guérin C, Tsoukala E (2013) The Tapiridae, Rhinocerotidae and Suidae (Mammalia) of the early Villafranchian site of Milia (Grevena, Macedonia, Greece). Geodiversitas 35:447–489

Harris JM, Liu LP (2007) Superfamily Suoidea. In: Prothero D, Foss SE (eds) The evolution of artiodactyls. John Hopkins University Press, Baltimore, pp 130–150

Harvati K, Darlas A, Bailey SE, Rein TR, El Zaatari S, Fiorenza L, Kullmer O, Psathi E (2013) New Neanderthal remains from Mani peninsula, southern Greece: the Kalamakia middle Paleolithic cave site. J Hum Evol 64:486–499

Hellmund M (1995) The vertebrate locality Maramena (Macedonia, Greece) at the Turolian-Ruscinian boundary (Neogene) 13. Suidae (Artiodactyla, Mammalia). Münchner Geowiss Abh A 28:143–156

Hünermann KA (1968) Die Suidae (Artiodactyla, Mammalia) aus den Dinotheriensanden (Unterpliozän: Pont) Rheinhessens (Sudwestdeutschland). Schweiz Paläont Abh 86:1–96

Hünnermann KA (1999) Superfamily Suoidea. In: Rössner GE, Heissig GE (eds) The Miocene land mammals of Europe. Dr Pfeil, München, pp 209–216

Iannucci A, Gasparik M, Sardella R (2020a) First report of *Sus strozzii* (Suidae, Mammalia) from the early Pleistocene of Hungary (Dunaalmas) and species distinction based on deciduous teeth. Sci Nat 107:5

Iannucci A, Cherin M, Sorbelli L, Sardella R (2020b) Suidae transition at the Miocene-Pliocene boundary: a reassessment of the taxonomy and chronology of *Propotamochoerus provincialis*. J Mam Evol https://doi.org/10.1007/s10914-020-09521-w

Kostopoulos DS (1994) *Microstonyx major* (Suidae, Artiodactyla) from the late Miocene locality of Nikiti-1 (Macedonia, Greece); some remarks about the species. Bull Geol Soc Greece 30(1):341–355

Kostopoulos DS, Athanassiou A (2005) In the shadow of bovids: suids, cervids and giraffids from the Plio-Pleistocene of Greece. Quaternaire H.S. 2:179–190

Kostopoulos DS, Sen S (2016) Suidae, Tragulidae, Giraffidae, and Bovidae. In: Sen S (ed) Late Miocene mammal locality of Küçükçekmece, European Turkey. Geodiversitas, vol 38, pp 273–298

Kostopoulos DS, Spassov N, Kovachev D (2001) Contribution to the study of *Microstonyx*: evidence from Bulgaria and the SE European populations. Geodiversitas 23:411–437

Koufos GD (1986) The presence of *Sus strozzii* in the Villafranchian (Villanyan) of Macedonia (Greece). Palaeont Z 60:341–351

Koufos GD (2006) The Neogene mammal localities of Greece: faunas, chronology and biostratigraphy. Hell J Geosci 41:183–214

Koufos GD (2007) Suoids (Mammalia, Artiodactyla) from the early/middle Miocene locality of Antonios, Macedonia, Greece. Senckenb Lethaea 87:171–186

Koufos GD, Syrides G, Koliadimou K (1991) A Pliocene primate from Macedonia (Greece). J Hum Evol 21:283–294

Lazaridis G (2015) Study of the late Miocene vertebrate locality of Kryopigi and other localities of Kassandra Peninsula, Chalkidiki (Greece). Systematics, Taphonomy, Paleoecology, Biochronology. PhD thesis, Aristotle University of Thessaloniki, Sc Annals School Geol, p. 174

Lazaridis G, Tsoukala E, Kostopoulos DS (2021) Validation of a prematurely abolished new *Propotamochoerus* species (Mammalia, Suidae) from SE Mediterranean. C R Palevol (under review)

Liu LP (2001) Eocene suoids (Artiodactyla, Mammalia) from Bose and Yongle basins, China and the classification and evolution of the Paleogene suoids. Vertebrata PalAsiatica 39:115–128

Liu LP (2003) Chinese fossil Suoidea-systematics, evolution, and paleoecology. PhD thesis, University of Helsinki. ISBN 902-10-1197-1

Liu LP, Kostopoulos DS, Fortelius M (2004) Late Miocene *Microstonyx* remains (Suidae, Artiodactyla) from Hezheng areas. Geobios 37:49–64

Liu WH, Dong W, Zhang LM, Zhao WJ, Li KQ (2017) New material of early Pleistocene *Sus* (Artiodactyla, Mammalia) from Yangshuizhan in Nihewan Basin, North China. Quat Int 434:32–47

Marcot JD (2007) Molecular phylogeny of terrestrial artiodactyls. In: Prothero D, Foss SE (eds) The evolution of artiodactyls. John Hopkins University Press, Baltimore, pp 4–18

McDonald DW (ed) (2009) The encyclopedia of mammals. Oxford University Press, Oxford

McKenna MC, Bell SK (1998) Classification of mammals: above the species level. Columbia University Press, New York

Meijaard E, d'Huart J-P, Oliver W (2015) Family Suidae (pigs). In: Wilson DE, Mittermeier RA (eds) Handbook of the mammals of the world, vol 2. Lynx Edicions, Barcelona, pp 2–19

Melentis J (1965) Studien über fossile Vertebraten Griechenlands 6. *Sus scrofa* L. aus dem Jungpleistozän des Beckens von Megalopolis in Peloponnes. Ann Geol Pays Hell 16:346–445

Melentis J (1969) Studien über fossile Vertebraten Griechenlands 28. Die Pikermifauna von Halmyropotamos (Euböa, Griechenland) teil 2: Osteologie. Ann Geol Pays Hell 21:217–307

Orliac M (2009) The differentiation of bunodont Listriodontinae (Mammalia, Suidae) of Africa: new data from Kalodirr and Moruorot, Kenya. Zool J Linnean Soc 15:653–678

Orliac M, Antoine P-O, Ducrocq S (2010) Phylogenetic relationships of the Suidae (Mammalia, Cetartiodactyla): new insights on the relationships within Suoidea. Zool Scr 39:315–330

Paraskevaidis I (1940) Eine obermiocaene Fauna von Chios. N Jb Mineral Geol Palaeont 83:363–442

Payne S (1985) Zooarchaeology in Greece: a reader's guide. In: Wilkie NC, Coulson WDE (eds) Contribution to Aegean Archaeology, studies in honor of WA McDonald Kendall Hunt, Dobuque, pp 211–244

Pickford M (1984) A revision of the Sanitheriidae, a new family of Suiformes (Mammalia). Geobios 17:133–154

Pickford M (1988) Revision of the Miocene Suidae of the Indian subcontinent. Münchner Geowiss Abh A 12:1–91

Pickford M (1993) Old world suid systematics, phylogeny, biogeography, and biostratigraphy. Paleontol Evol 26–27:237–269

Pickford M (2012) Ancestors of Broom's pigs. Trans R Soc S Afr 67:17–35

Pickford M (2013) Re-assessment of the suids from the sables marins de Montpellier and selection of a lectotype for *Sus provincialis* Blainville, 1847. Geodiversitas 35:655–668

Pickford M (2015) Late Miocene Suidae from Eurasia: the *Hippopotamodon* and *Microstonyx* problem revisited. Münchner Geowiss Abh A 42:1–126

Pickford M, Morales J (2003) New Listriodontinae (Mammalia, Suidae) from Europe and a review of listriodont evolution, biostratigraphy and biogeography. Geodiversitas 25:347–404

Roth J, Wagner A (1854) Die fossilen knochenuberreste von Pikermi in Griechenland. Abh bayer Akad Wissensch Mathem naturwiss Klasse Munchen 7:371–464

Roussiakis S, Athanassiou A, Michailidis D, Mitsopoulou V, Solomos Ch, Theodorou G (2014) Remarks on new Proboscidean remains from the classical Late Miocene locality of Pikermi and their associated fauna. Sci Ann School Geol Aristotle University of Thessaloniki, spec. vol 102, pp 171–172

Schmidt-Kittler N (1971) Die Obermiozäne Fossillagerstätte Sandelzhausen 3. Suidae, Artiodactyla, Mammalia. Mitt Bayer Staats Pal Hist Geol 11:129–170

Spaulding M, O'Leary MA, Gatesy J (2009) Relationships of Cetacea (Artiodactyla) among mammals: increased taxon sampling alters interpretations of key fossils and character evolution. PLoS One 4(9):e7062

Starkovich BM (2012) Intensification of small game resources at Klissoura cave 1 (Peloponnese, Greece) from the middle Paleolithic to Mesolithic. Quat Int 264:17–31

Steffens P, de Bruijn H, Meulenkamp JE, Benda L (1979) Field guide to the Neogene of northern Greece (Thessaloniki area and Strimon basin). Publ Dept Geol Paleont Univ Athens Ser A 35:1–14

Stiner MC, Munro ND (2011) On the evolution of diet and landscape during the upper Paleolithic through Mesolithic at Franchthi cave (Peloponnese, Greece). J Hum Evol 60:618–636

Sylvestrou I, Kostopoulos DS (2006) The late Miocene vertebrate locality of Perivolaki, Thessaly, Greece. 7. Suidae. Palaeontograph Abt A 276:121–133

Sylvestrou I, Kostopoulos DS (2009). 12. Suidae. In: Koufos GD, Nagel D (eds) The late Miocene mammal faunas of the Mytilinii Basin, Samos Island, Greece: new collection. Beitr Paläontol 31:283–297

Symeonidis N, Bachmayer F, Zapfe H (1980) Ergebnisse weiterer Grabungen in der Hohle von Vraona (Attika, Griechenland). Ann Geol Pays Hell 30:291–299

Syrides G, Koliadimou K, Koufos GD (1997) New Neogene molluscan and mammalian sites from Thrace, Greece. C R Acad Sci Paris 324:427–433

Thenius E (1950) *Postpropotamochoerus* nov. subgen., *hyotheriodes* aus dem Unterpliozän von Samos (Griechenland) und die Herkunft der Potamochoeren. Sitzungsberichte Akad Wissensch mathemat-naturwiss Kl 159(1–5):25–36

Thenius E (1955) Studien über fossile Vertebraten Griechenlands IV: *Sus antiquus* aus Ligniten von Sophades (Thessalien) und die Alterstellung der Fundschichten. Ann Geol Pays Hell 6:199–205

Theodorou G, Athanassiou A, Roussiakis S, Iliopoulos G (2003) Preliminary remarks on the late Miocene herbivores of Kerassiá (northern Euboea, Greece). DEINSEA 10:519–530

Trofimov BA (1954) The fossil suids of the genus *Microstonyx*. [in:] tertiary mammals, part 2: on the Mammalia of the southern SSSR and Mongolia. Dokl Akad Nauk SSSR 47:61–99

Tsoukala E (1989) Contribution to the study of the Pleistocene fauna of large mammals (Carnivora, Perissodactyla, Artiodactyla) from Petralona cave Chalkidiki (N. Greece). PhD Thesis, University of Thessaloniki

Tsoukala E (1992) Quaternary faunas of Greece. Cour Forsch Senck 153:79–92

Tsoukala E, Guérin C (2016) The Rhinocerotidae and Suidae of the middle Pleistocene from Petralona cave (Macedonia, Greece). Acta Zool Bulg 68:243–264

van der Made J (1988) *Sus nanus* nov. sp., a Pliocene dwarf pig from capo Figari (northeastern Sardinia). Boll Soc Paleont Ital 27:367–378

van der Made J (1996a) Listriodontinae (Suidae, Mammalia), their evolution, systematics and distribution in time and space. Contr Tert Quatern Geol 33:3–254

van der Made J (1996b) Pre-Pleistocene land mammals from Crete. In: Reese DS (ed) Pleistocene and Holocene fauna of Crete and its first settlers, Monographs in world Archaeology, vol 28. Prehistory Press, Madison, pp 69–79

van der Made J (1998) *Aureliachoerus* from Oberdorf and other Aragonian pigs from Styria. Ann Naturhist Mus Wien 99A:225–277

van der Made J (2010) The pigs and "Old World peccaries" (Suidae and Palaeochoeridae, Suoidea, Artiodactyla) from the Miocene of Sandelzhausen (southern Germany): phylogeny and an updated classification of the Hyotheriinae and Palaeochoeridae. Paläont Z 84:43–121

van der Made J (2020) The Suoidea from the middle Miocene of Gracanica (Bugojno Basin, Bosnia and Herzegovina)-evolution, taxononomy, and biostratigraphy. Palaeobiodivers Palaeoenviron 100:321–349

van der Made J, Hussain T (1989) *"Microstonyx" major* (Suidae, Artiodactyla) from the type area of the Nagri formation, Siwalik group, Pakistan. Estudios geol 45:409–416

van der Made J, Hussain T (1992) Sanitheres from the Miocene Manchar formation of Sind, Pakistan and remarks on sanithere taxonomy and stratigraphy. Proc Koninkl Nederl Akad Wetensch 95:81–95

van der Made J, Moyà-Solà S (1989) European Suinae (Artiodactyla) from the late Miocene onwards. Boll Soc Pal Ital 28:329–339

van der Made J, Montoya P, Alcalá L (1992) *Microstonyx* (Suidae, Mammalia) from the upper Miocene of Spain. Geobios 25:395–413

van der Made J, Krakhmalnaya T, Kubiak H (1999) The pig *Propotamochoerus palaeochoerus* from the upper Miocene of Grytsiv, Ukraine. Estudios geol 55:283–292

van der Made J, Savaş Güleç E, Erkman AC (2013) *Microstonyx* (Suidae, Artiodactyla) from the upper Miocene of Hayranlı-Haliminhanı, Turkey. Turk J Zool 37:1–17

The Fossil Record of Anthracotheres (Mammalia: Artiodactyla: Hippopotamoidea) in Greece

Dimitris S. Kostopoulos

1 Introduction

Anthracotheres derive their legendary name from the Greek words *άνθραξ* (*anthrax* = coal; genitive: *άνθρακος* = anthrakos) and *θηρίο* (*therium* = beast, animal) as their earliest findings come from Paleogene European coal deposits (Cuvier 1822). Paradoxically, they are extremely rare in the Greek fossil record. The family comprises bunodont to bunoselenodont suid-like artiodactyls ranging considerably in size from that of a mouse-deer to a hippo. Earliest anthracotheres occur at the end of middle Eocene in SE Asia and subsequently radiated and dispersed to the rest of the Old World and North–Central America (e.g., Rincon et al. 2013). The family declined dramatically during the middle Miocene, but last representatives still survived into the late Miocene of Africa and the late Pliocene of Asia (van der Made 1999; Holroyd et al. 2010; Lihoreau and Ducrocq 2007).

Anthracotheres have been pivotal in discussions of both paleogeographic interpretations (e.g., Grandi and Bona 2017; Scherler et al. 2019) and mammalian phylogeny (e.g., Boiserie et al. 2005; Gatesy et al. 2013; Lihoreau et al. 2015), and although they were believed to represent mainly semiaquatic animals, recent studies demonstrated a wide array of preferred environments and diets (e.g., Colbert 1935; Lihoreau and Ducrocq 2007; Alloing-Séguier et al. 2014; Cooper et al. 2016).

D. S. Kostopoulos (✉)
School of Geology, Aristotle University of Thessaloniki, Thessaloniki, Greece
e-mail: dkostop@geo.auth.gr

E. Vlachos (ed.), *Fossil Vertebrates of Greece Vol. 2*,
https://doi.org/10.1007/978-3-030-68442-6_8

2 Historical Overview

In the first decade of the twentieth century, Professor Th. Skoufos of the University of Athens, undertook a series of paleontological excavations at various fossil sites in the Greek territory of that time. Among them in the basin(s) of Aliveri and Kymi, Euboea where since the mid-nineteenth century extensive exploitation of early Neogene coal deposits has been taking place. Skoufos (1905) was probably the first to record the presence of anthracotheres in Greece, based on a badly preserved mandible from the Enzi coal mine near Kalimeriani village. In this first report, Skoufos does not provide taxonomic identifications, descriptions, or illustrations. It seems, however, that in the newspapers and magazines of the time, the finding was widely discussed as "Kymi's Anthracothere." Several years later, Professor I. Melentis rediscovered the same specimen in the drawers of the Museum of Paleontology at the National and Kapodistrian University of Athens accompanied by the label "1911, *Anthracotherium kalimerianum*," a name apparently given by Skoufos himself (Melentis 1965). Melentis (1965) appropriately published the specimen as *Brachyodus onoideus*. Just a few years before, Lüttig and Thenius (1961) described from the Greek part of Thrace another small anthracothere, *Elomeryx*, already known from the surroundings of Southern Bulgaria and Eastern (Turkish) Thrace (Kostopoulos et al. 2012). The third and so far the last record of Greek anthracotheres came years later and, again, accidentally. The discovery of some dental remains into a lignite sample forgotten in the drawers of the Museum of Geology–Paleontology–Palaeoanthropology of the Aristotle University of Thessaloniki allowed Kostopoulos et al. (2012) to recognize the second *Elomeryx* occurrence.

3 Phylogenetic Relationships

Members of the even-toed hoofed mammals—Mammalia: Artiodactyla Owen, 1848 (clade Cetartiodactyla Montgelard, Catzeflis, Douzery 1997): Hippopotamoidea Gray, 1821 (sensu Gentry and Hooker 1988), Anthracotheriidae Leidy, 1869—are nowadays accepted as paraphyletic since multiple lines of evidence suggest the deep nesting of hippopotamids within African anthracotheres (e.g., Boiserie et al. 2005, 2010, 2017; Boisserie and Lihoreau 2006; Geisler et al. 2007; Orliac et al. 2010; Alloing-Séguier et al. 2014; Lihoreau et al. 2015, 2017 but see also Pickford 2008 for an opposite view; Boisserie et al. 2011 and Alloing-Séguier et al. 2014 summarize main phylogenetic alternatives). Spaulding et al.'s (2009) total evidence analysis including phenotypic and molecular features proposes Anthracotheriidae as polyphyletic, whereas an analysis restricted only on skeletal and dental characters unites anthracotheres with some other primitive artiodactyls (entelodonts, raoelliids, hyotheriins) in a clade that appears as the sister group of hippopotamids + suids (but see discussion in Boisserie et al. 2011). Three anthracothere subfamilies are currently recognized: Bothriodontinae Scott, 1940 and Anthracotheriinae Leidy, 1869, distinguished each other mainly by divergent degrees of molar selenodonty, and Microbunodontinae Lihoreau and Ducrocq, 2007 characterized by small size, slender limbs, and long flattened upper canines (e.g., Kron and Manning 1998; Lihoreau and Ducrocq 2007; Scherler et al. 2019). Among those, archaic bothriodontines clustered around *Bothriogenys* are proposed as stem group of the hippos clade (e.g., Boisserie et al. 2010; Orliac et al. 2010; Alloing-Séguier et al. 2014; Lihoreau et al. 2015).

4 Distribution

The extremely poor anthracothere record of Greece covers a very restricted part of their otherwise wide Old World distribution (Fig. 1). All Greek evidence comes from coal deposits dated from the Oligocene to the end of Lower Miocene (Table 1).

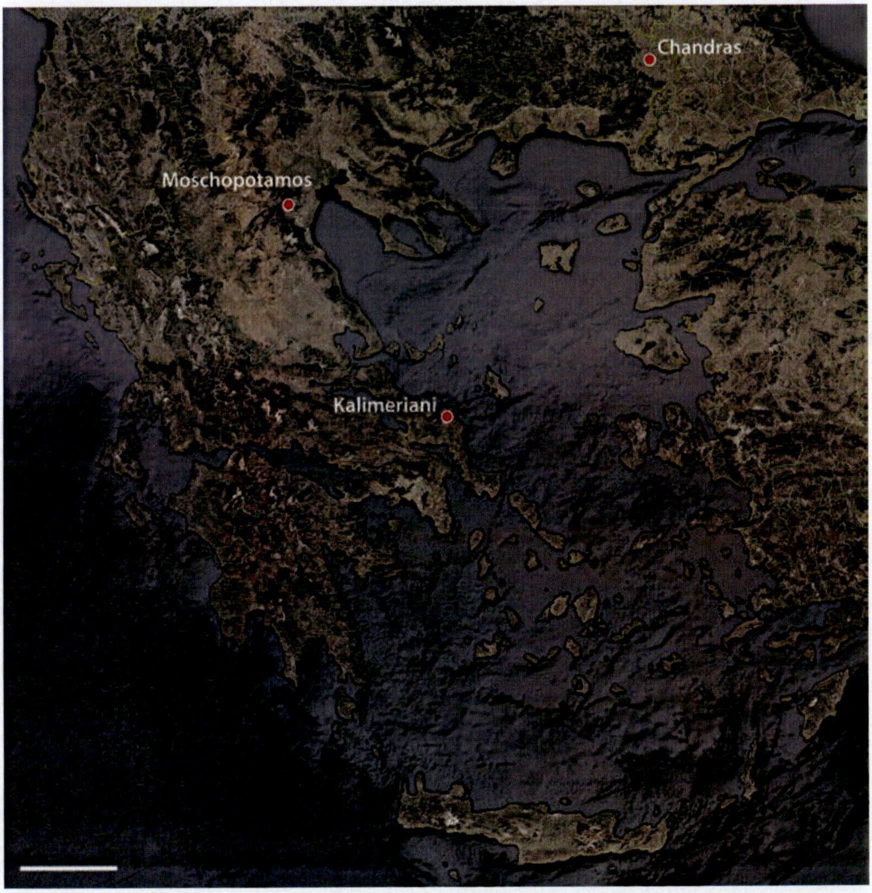

Fig. 1 Map of the occurrences of fossil anthracotheriids from Greece. See Table 1 for further information and text for details. Image exported from Google Earth Pro © 2019, map data from US Dept. of State Geographer, SIO, NOAA, U.S. Navy, NGA, GEBCO, image from Landsat/ Copernicus. Scale bar equals 100 km, North faces upward

Table 1 List of the Greek localities containing anthracothere fossils. Locality numbers refer to the collection numbers of the PaleoBiology Database (PBDB)

Localities[PBDB No]	Age	Taxon
Moschopotamos[208601]	Early Miocene	*Elomeryx* sp.[1]
Kalimeriani[202961]	Early Miocene	*Brachyodus onoideus*[2]
Chandras[208602]	Late Oligocene	*Elomeryx crispus*[3]

[1]Kostopoulos et al. (2012), [2]Melentis (1965), [3]Lüttig and Thenius (1961)

The fragmentary Greek record is a result of both the proportionally limited exposure of Oligocene–Lower Miocene terrestrial sedimentary formations and the absence of systematic research in these areas.

5 Systematic Paleontology

Hippopotamoidea Gray, 1821 (sensu Gentry and Hooker 1988)

Bothriodontinae Scott, 1940

Brachyodus Depéret, 1895

Type Species *Anthracotherium onoideum* Gervais, 1859, by original designation.

Distribution Early Miocene to early middle Miocene of Europe, SE Asia, and Africa.

Other Taxa Included *B. intermedius* Mayer, 1908; *B. depereti* (Fourteau, 1918); *B. mogharensis* Pickford, 1991; *B. aequatorialis* McInnes, 1951.

Remarks *Brachyodus* is one of the earliest recognized anthracothere genera. It represents a large bothriodontine with shallow mandible and long, wide snout. It is suggested as possible relative of the NE African-SE Asian Eocene–Oligocene *Bothriogenys* Schmidt, 1913 (Hellmund 1991; van der Made 1999; Lihoreau and Ducrocq 2007; Alloing-Séguier et al. 2014; Lihoreau et al. 2015).

Brachyodus onoideus (Gervais, 1859)

Nomenclatural and Taxonomic History *Anthracotherium onoideum* Gervais, 1859 (new species); *Brachyodus onoideus* in Depéret 1895 (new combination and mandatory change of species epithet). See Dineur (1982), Pickford (1991), Hellmund (1991) for further information.

Type Material Musée des Sciences Naturelles, Orleans, MHNO 314 (holotype), part of right mandible with p2–m3 discussed by Gervais (1859).

Type Locality Neuville-aux-bois, Loiret, France, Lower Miocene.

Age Early Miocene.

Distribution Early Miocene (MN3–4) of France, Portugal, Switzerland, Austria, and Germany. Material from Ban NA Sai Thailand is also reported to this species (Ducrocq et al. 2003). In Greece, it is known only from the Lower Miocene locality of Kalimeriani.

Remarks The species is revised by Dineur (1982; unpublished thesis) and Hellmund (1991) who, following Depéret (1895), include it in *Brachyodus*. Melentis (1965) fully described the single known mandible (AMPG 1968/8sin and 1968/9dex) from Kalimeriani (Euboea) and originally compared it with the material of this genus in Vienna. The teeth are in advanced wear stage, and limited dental details are therefore accessible. According to Melentis (1965), the dentition is characterized by large dimensions (the length p2–m3 is 184.3 mm; the length m3–m2 is 84.4 mm, m3 measures 50 × 29 mm), a small p1 in line with p2, the enamel is thinning toward the basis of the molars, and the talonid of m3 is single cuspid (Fig. 2). A fresh look on the Kalimeriani specimen is likely needed as its m3 appears significantly shorter than the material from the type area near Orléans, France (F. Lihoreau, pers. com. 2020).

Fig. 2 Left mandible (AMPG 1968/8) of *Brachyodus onoideus* from Kalimeriani, Euboia in (**a**) occlusal and (**b**) lingual views. Scale bar equals 5 cm. Photograph courtesy of S. Rousiakis, National and Kapodistrian University of Athens, Museum of Paleontology and Geology

Elomeryx Marsh, 1894

Type Species *Elomeryx armatus* Marsh, 1894, by original designation.

Distribution Middle Eocene to early Miocene of Eurasia.

Other Taxa Included *E. cluai* (Depéret, 1906); *E. crispus* (Gervais, 1849); *E. borbonicus* (Geais, 1934); *E. japonicus* (Matsumoto in Tokunaga, 1925).

Remarks Hellmund (1991), Ducrocq and Lihoreau (2006), Lihoreau and Ducrocq (2007), Lihoreau et al. (2009), and Tsubamoto and Khono (2011) revised the genus context and its main features. According to these authors, *Elomeryx* represents small- to medium-sized bothriodontines with sexually dimorphic upper canines characterized by serrated posterior edge on males, short rostrum, and absence of diastemata between premolars. The genus is seen as paraphyletic by Lihoreau and Ducrocq (2007) and stem of several anthracothere lineages, though more recent studies (Lihoreau et al. 2017) doubt previous suggested relations between *Elomeryx* and advanced bothriodontines.

Elomeryx crispus (Gervais, 1849)

Nomenclatural and Taxonomic History *Hyopotamus crispus* Gervais, 1849 (new species); *Elomeryx crispus* in de Bonis (1964) (new combination); for a detailed taxonomic history see Hellmund (1991).

Type Material Musée national d'Histoire naturelle, Paris, MNHN. Paris LDB 134, part of left upper toothrow with dP4–M1 illustrated by Hellmund (1991: Pl. 7, fig. 1).

Type Locality La Débruge, France, Upper Eocene (Priabonian).

Age Late Eocene (MP18).

Distribution Late Eocene to early late Oligocene (MP18–26) of France, Germany, Switzerland, UK, Czech Republic, Turkish Thrace. In Greece, it is known only from the Lower Miocene locality of Chandras.

Remarks The Chandras palate (IPUW 16397, Department of Paleontology, University of Vienna) from Western Thrace (Fig. 3) was originally described as *Elomeryx woodi* by Lüttig and Thenius (1961), a name later recognized as a junior synonym of *Elomeryx crispus* (Hellmund 1991). The overall size of the Chandras dentition and the strong ectometacristule on its molars suggest, indeed, close affiliation to *E. crispus* (Lüttig and Thenius 1961), though some tooth proportions and morphological features look more advanced than in that species (see Kostopoulos et al. 2012).

Fig. 3 Palate (cast IPUW 16397) of *Elomeryx* from Chandras, Western Thrace, in occlusal view. Scale bar equals 5 cm

Elomeryx sp.

Kostopoulos et al. (2012) described some more dental material of *Elomeryx* from the Lower Miocene Moschopotamos coal pits (Katerini Basin), near Thessaloniki and suggested it represents a taxon metrically close to *E. crispus*, but with some dental features shared with *E. borbonicus* and *E. japonicus*, such as the strongly compressed mediolaterally upper canine, the presence of an accessory cusp on the mesial crest of the lower premolars, and the strongly developed parastyle–metastyle on p4. More material is definitely needed for certain taxonomic attribution.

6 Conclusions

Current knowledge of Greek anthracotheres is extremely limited based only on very few and scattered evidences of disputable stratigraphic/chronologic context. Three bothriodontine taxa have been so far recognized in Greece: *Elomeryx crispus*, *Elomeryx* sp., and *Brachyodus onoideus*. In all cases, however, anthracothere findings lack associating faunas. The earliest Greek evidence comes from Western (Greek) Thrace. Lüttig and Thenius (1961) supposed the Chandras *Elomeryx* is of early Oligocene age, but lithostratigraphic and mollusk data likely date the fossil-bearing deposits to the Upper Oligocene (Kostopoulos et al. 2012 and references therein). Coal deposits of Kalimeriani, Euboea from where *Brachyodus* mandible is unearthed, are suggested as of Lower Miocene (Burdigalian) age and fairly isochronous with the coal deposits of Aliveri, Euboea (Melentis 1965 and references therein). Aliveri fossil site provided a much richer, MN4 (middle Orleanian) mammal association, including among others *Anchitherium*, *Lagomeryx*, and *Eotragus* (van den Hoek Ostende et al. 2015). If indeed Kalimeriani and Aliveri highlight the same early Miocene mammal paleocenosis, then they point together to an early MN4 fauna, dated at around 17.5–17.0 Ma. Of similar or slightly younger age is *Elomeryx* sp. from Moschopotamos, Katerini (Kostopoulos et al. 2012 and references therein), which marks the last known occurrence of anthracotheres in Greece. If and when these age estimates are confirmed, it would mean that *Elomeryx* may lasted longer in the SE than in the rest of Europe and that its local biostratigraphic range possibly overlaps for a short time with that of the forthcoming *Brachyodus*.

Acknowledgments Thanks are due to S. Roussiakis for providing photographs of the Kalimeriani mandible and to F. Lihoreau for fruitful comments and suggestions.

References

Alloing-Séguier L, Lihoreau F, Boisserie J-R, Charruault A-L, Orliac M, Tabuce R (2014) Enamel microstructure evolution in anthracotheres (Mammalia, Cetartiodactyla) and new insights on hippopotamoid phylogeny. Zool J Linnean Soc 171:668–695

Boiserie J-R, Lihoreau F, Brunet M (2005) The position of Hippopotamidae within Cetartiodactyla. PNAS 102:1537–1541

Boisserie J-R, Lihoreau F (2006) Emergence of Hippopotamidae: new scenarios. C R Palevol 5:749–756

Boisserie J-R, Lihoreau F, Orliac M, Fisher RE, Weston EM, Ducrocq S (2010) Morphology and phylogenetic relationships of the earliest known hippopotamids (Cetartiodactyla, Hippopotamidae, Kenyapotaminae). Zool J Linnean Soc 158:325–366

Boisserie J-R, Fisher RE, Lihoreau F, Weston EM (2011) Evolving between land and water: key questions on the emergence and history of the Hippopotamidae (Hippopotamoidea, Cetancodonta, Cetartiodactyla). Biol Rev 86:601–625

Boisserie J-R, Kiarie C, Lihoreau F, Nengo I (2017) Middle Miocene *Kenyapotamus* (Cetartiodactyla, Hippopotamidae) from Napudet, Turkana Basin, Kenya. J Vertebr Paleontol 37(1):e1272055

Colbert EH (1935) The phylogeny of the Indian Suidae and the origin of the Hippopotamidae. Am Mus Nov 799:1–24

Cooper LN, Clementz MT, Usip S, Bajpai S, Hussain ST, Hieronymus TL (2016) Aquatic habits of cetacean ancestors: integrating bone microanatomy and stable isotopes. Integr Comp Biol 56:1370–1384

Cuvier G (1822) Recherches sur les ossemens fossiles, où l'on rétablit les caractères de plusieurs animaux, dont les révolutions du globe ont détruit les espèces Tome V. E. d'Ocagne, Paris

Depéret C (1895) Über die Fauna von miocänen Wirbelthieren aus der ersten Mediterranstufe von Eggenburg. Sitzungsber Akad Wiss math naturw Cl. Wien C4 Abth 1:395–416

Dineur H (1982) Le genre Brachyodus, Anthracotheriidae (Artiodactyla, Mammalia) du Miocène inférieur d'Europe et d'Afrique [Unpublished PhD thesis]. Université Paris, Paris

Ducrocq S, Lihoreau F (2006) The occurrence of bothriodontines (Artiodactyla, Mammalia) in the Palaeogene of Asia with special reference to *Elomeryx*: palaeobiogeographical implications. J Asian Earth Sci 27:885–891

Ducrocq S, Chaimanee Y, Suteethorn V, Jaeger J-J (2003) Occurrence of the anthracotheriid *Brachyodus* (Artiodactyla, Mammalia) in the early middle Miocene of Thailand. C R Palevol 2:261–268

Gatesy J, Geisler JH, Chang J, Buell C, Berta A, Meredith RW, Springer MS, McGowen MR (2013) A phylogenetic blueprint for a modern whale. Mol Phylogenet Evol 66:479–506

Geisler JH, Theodor JM, Uhen MD, Foss SE (2007) Phylogenetic relationships of cetaceans to terrestrial artiodactyls. In: Prothero DR, Foss SE (eds) The evolution of artiodactyls. Johns Hopkins University Press, Baltimore, pp 19–31

Gentry AW, Hooker P (1988) The phylogeny of the Artiodactyla. In: Benton MJ (ed) The phylogeny and classification of the Tetrapods, vol 2, mammals. Syst Assoc 35B:235–272

Gervais P (1859) Zoologie et paleontologie françaises (animaux vertebres), 2nd ed. Paris

Grandi F, Bona F (2017) *Prominatherium dalmatinum* from the late Eocene of Grancona (Vicenza, NE Italy). The oldest terrestrial mammal of the Italian peninsula. C R Palevol 16:738–745

Hellmund M (1991) Revision der europäischen species der Gattung *Elomeryx* Marsh 1894 (Anthracotheriidae, Artiodactyla, Mammalia) - odontologische Untersuchungen. Palaeontographica Abt A 220:1–101

Holroyd PA, Lihoreau F, Gunnel GF, Miller ER (2010) Anthracotheriidae. In: Werdelin L, Sanders WJ (eds) The Cenozoic mammals of Africa. University of California Press, Berkeley, pp 843–851

Kostopoulos DS, Koufos GD, Christanis K (2012) On some anthracotheriid (Artiodactyla, Mammalia) remains from northern Greece: comments on the palaeozoogeography and phylogeny of *Elomeryx*. Swiss J Palaeontol 131:303–315

Kron DG, Manning E (1998) Anthracotheriidae. In: Janis CM, Scott KM, Jacobs LL (eds) Evolution of tertiary mammals of North America. Cambridge University Press, Cambridge, pp 381–388

Leidy J (1869) The extinct mammalian fauna of Dakota and Nebraska. J Philos Acad Nat Sci 7:1–472

Lihoreau F, Ducrocq S (2007) Family Anthracotheriidae. In: Prothero DR, Foss SE (eds) The evolution of artiodactyls. Johns Hopkins University Press, Baltimore, pp 89–105

Lihoreau F, Ducrocq S, Antoine PO, Vianey-Liaud M, Rafay S, Garcia G, Valentin X (2009) First complete skulls of *Elomeryx crispus* (Gervais, 1849) and of *Protaceratherium albigense* (Roman, 1912) from a new Oligocene locality near Moissac (SW France). J Vertebr Paleontol 29:242–253

Lihoreau F, Boisserie J-R, Manthi FK, Ducrocq S (2015) Hippos stem from the longest sequence of terrestrial cetartiodactyl evolution in Africa. Nat Commun 6:6264. https://doi.org/10.1038/ncomms7264

Lihoreau F, Alloing-Séguier L, Antoine P-O, Boisserie J-R, Marivaux L, Métais G, Welcomme J-L (2017) Enamel microstructure defines a major Paleogene hippopotamoid clade: the Merycopotamini (Cetartiodactyla, Hippopotamoidea). Hist Biol 29:947–957

Lüttig G, Thenius E (1961) Über einen Anthracotheriiden aus dem Alttertiär von Thrazien (Griechenland). Paläont Z 35:179–186

Marsh OC (1894) A new Miocene mammal. Am J Sci 48:409

Melentis J (1965) Der erste Nachweiss von *Brachyodus onoideus* (Mammalia, Anthracotheriidae) aus Griechenland und die Datierung der Fundschichten. Ann Acad Athens 40:406–423

Orliac M, Boisserie J-R, MacLatchyd L, Lihoreau F (2010) Early Miocene hippopotamids (Cetartiodactyla) constrain the phylogenetic and spatiotemporal settings of hippopotamid origin. PNAS 107:11871–11876

Pickford M (1991) Revision of the Neogene Anthracotheriidae of Africa. Geol Libya 4:1491–1525

Pickford M (2008) The myth of the hippo-like anthracothere: the eternal problem of homology and convergence. Rev Esp Paleont 23:31–90

Rincon AF, Bloch JI, Macfadden BJ, Jaramillo CA (2013) First central American record of Anthracotheriidae (Mammalia, Bothriodontinae) from the early Miocene of Panama. J Vertebr Paleontol 33:421–433

Scherler L, Lihoreau F, Becker D (2019) To split or not to split Anthracotherium? A phylogeny of Anthracotheriinae (Cetartiodactyla: Hippopotamoidea) and its palaeobiogeographical implications. Zool J Linnean Soc 185:487–510

Scott WB (1940) The mammalian fauna of the White River Oligocene part IV, Artiodactyla. Trans Am Philos Soc (N S) 28(4):361–746

Skoufos T(1905) Űber die paläontologischen Ausgrabungen in Grieclienland in Beziehung auf das Vorhandensein des Menschen. In: C R 1st Congr Intern Achaeol, pp 231–236

Spaulding M, O'Leary MA, Gatesy J (2009) Relationships of Cetacea (Artiodactyla) among mammals: increased taxon sampling alters interpretations of key fossils and character evolution. PLoS One 4(9):e7062

Tsubamoto T, Kohno N (2011) Reappraisal of "*Brachyodus" japonicus*, an Oligocene anthracotheriid cetartiodactyl from Japan. Palaeontol Res 15:115–124

van den Hoek Ostende LW, Mayda S, Oliver A, Madern A, Hernández-Ballarín V, Peláez-Campomanes P (2015) Aliveri revisited, a biogeographical appraisal of the early Miocene mammals from the eastern Mediterranean. Palaeobio Palaeoenv 95:271–284

van der Made J (1999) Superfamily Hippopotamidae. In: Rossner G, Heissig K (eds) The Miocene land mammals of Europe. Verlag Dr Friedrich Pfeil, Munich, pp 203–208

The Fossil Record of Continental Hippopotamids (Mammalia: Artiodactyla: Hippopotamidae) in Greece

Athanassios Athanassiou

1 Introduction

The family Hippopotamidae Gray, 1821, comprises primarily middle- to large-sized semiaquatic taxa, with graviportal body shapes, well-adapted to fluvial and lacustrine environments. They are of African origin. Their first representatives are usually believed to appear in the eastern part of the African continent during the middle Miocene, rapidly replacing ecologically similar anthracotheriid taxa (Pickford 1983; Boisserie 2007; Boisserie et al. 2010; Weston and Boisserie 2010). However, Orliac et al. (2010) moved the family's first appearance back to the lowest Miocene (about 21 Ma) by including the East African genera *Morotochoerus* Pickford, 1998, and *Kulutherium* Pickford, 2007, within the Hippopotamidae (a taxonomic decision not universally accepted—see, e.g., Pickford 2011). Members of the family migrated repeatedly out of their continent of origin, starting from the late Miocene, as evidenced by scarce occurrences in Southern Europe and a better representation in Southern Asian localities (Coryndon 1978; Kahlke 1990; Boisserie 2007). During the late Pliocene and the Pleistocene, they were widely dispersed to Africa and South Eurasia (occasionally reaching also northern latitudes), but after the Late Pleistocene, they are, again, geographically confined to Africa.

The Hippopotamidae have a distinct and unique ecology among other mammalian families: they are characteristically amphibious, leading a life between fresh water and land, where they usually feed. This mode of life is mostly typical of the continental species of the genus *Hippopotamus*, but other genera of the family, like,

A. Athanassiou (✉)
Ephorate of Palaeoanthropology–Speleology, Hellenic Ministry of Culture and Sports, Athens, Greece

Museum of Paleontology and Geology, National and Kapodistrian University of Athens, Athens, Greece
e-mail: aathanas@geol.uoa.gr

E. Vlachos (ed.), *Fossil Vertebrates of Greece Vol. 2*,
https://doi.org/10.1007/978-3-030-68442-6_9

e.g., *Choeropsis* and *Hexaprotodon*, do not diverge significantly from it (Boisserie 2005). Morphologically, the more advanced genera of the family that constitute the subfamily Hippopotaminae (the less derived Kenyapotaminae are still incompletely known) are united as a monophyletic group by several synapomorphies, which, according to Boisserie (2007), include: prolonged to permanently growing incisors and canines; procumbent lower incisors; tetracuspidate cheek teeth; triangular to trefoliate molar wear pattern; wide and tall muzzle; intercanine position of main palatal foramina; large temporal fossae; poorly compressed postglenoid area of the braincase; wide and subvertical occipital face; thick and wide mandibular symphysis; shallow mandibular constriction; ventrally extended angular process; digits II and V slightly reduced. Members of the family also share evolutionary trends, such as canine process expansion, angular process expansion, muzzle elongation, orbit and sagittal crest elevation, incisor differentiation and losses, premolar row shortening and simplification, and increase of molar crown height.

Because of their semiaquatic habits and their swimming abilities, *Hippopotamus* have colonized several Mediterranean islands, including Greek ones (van der Geer et al. 2010). The insular forms, whenever became genetically isolated from the continental populations, evolved to endemic dwarf species, adapted to the usually more rugged and much drier island environments. These taxa are reviewed by Lyras et al. (this volume) and are not considered here.

Methods The geochronologic and stratigraphic framework follows Pillans and Gibbard (2012). Geochronologic ages are given in thousand or million years before present, abbreviated as “ka” or “Ma,” respectively. “MN” and “MNQ” stand for Mammal Neogene and Quaternary Zones (Guérin 1990; Mein 1999). Transliterated local geographic names are typed accented when necessary, to help with their correct pronunciation.

2 Phylogenetic Relationships and Taxonomy

Unlike most artiodactyl families, the Hippopotamidae evolved for the major part of their history in the African continent, where they also exhibited most of their diversity. Relatively soon after their first appearance in the Middle Miocene (seemingly as early as in the Late Miocene), they dispersed out of the continent several times, colonizing the southern regions of Eurasia, and briefly also North Europe (Coryndon 1978; Kahlke 1990; Boisserie 2005). About 40 valid species, fossil and recent, have been described (Coryndon 1978; Boisserie 2005). The origin of the family is debated. Traditionally, based on their morphological characters, they are considered to have branched off from an anthracotheriid stock (e.g., Colbert 1935; see also Boisserie 2005; Boisserie et al. 2010), a suggestion based on a supposed evolutionary pathway of the dentition, but also influenced by the morphological and ecological (semiaquatic way of life) resemblance of certain anthracotheriids (notably *Merycopotamus*) to the hippos. This concept has survived until today, also based on recently discovered fossils (Boisserie 2005; Boisserie et al. 2005a, b, 2010, 2011; Lihoreau et al. 2015). Another early hypothesis, put forth by Pearson (1927), relates the hippos to the Eocene Cebochoeridae. A third suggestion (Pickford 1983) links the Hippopotamidae with the Tayassuidae on the basis of numerous shared derived characters (see also Pickford 1989, 2008). A related assertion had been already made earlier, though very preliminarily, by Matthew (1929, p. 452) (see also Colbert 1935; Pickford 1983).

Modern phylogenetic studies considering molecular data have shown close relationships between the Hippopotamidae and the Cetacea, which form a clade excluding the rest of the extant artiodactyl groups (Gatesy et al. 1996; Gatesy 1997; Nikaido et al. 1999; Zurano et al. 2019). This renders the traditional artiodactyls (Artiodactyla Owen, 1848) paraphyletic, as they do not include the Cetacea, and the Suiformes polyphyletic, as they include the hippopotamids together with the Suoidea. Further, it implies a vast temporal gap between the pre-Eocene Cetacea+Hippopotamidae common ancestor and the emergence of the Hippopotamidae (Miocene). Other suiforms, such as the aforementioned anthracotheriids and tayassuids, as well as paleochoerids and cebochoerids, have been proposed as potential links to fill this gap, at least partially (Pickford 1983, 2008; Boisserie et al. 2005b, 2010, 2011; Lihoreau et al. 2015); however, the phylogenetic pathways that led to the emergence of the hippopotamids are still debated, remaining largely unresolved.

The oldest genus, which is unambiguously referred to the family Hippopotamidae, *Kenyapotamus* Pickford, 1983, is known with two comparatively small-sized species from the middle and late Miocene of Kenya (Pickford 1983), and it is taxonomically placed in a separate subfamily (Kenyapotaminae Pickford, 1983), because of its less derived, brachyodont, dentition. Its morphology is, however, incompletely known. The more advanced genera (*Archaeopotamus*, *Hexaprotodon*, *Saotherium*, *Hippopotamus*, and *Choeropsis*) are grouped under the subfamilial taxon Hippopotaminae Gray, 1821 (McKenna and Bell 1997, p. 399; Boisserie 2007; Weston and Boisserie 2010). Despite the family's past diversity, there are only two extant genera, *Hippopotamus Linnaeus,* 1758, and *Choeropsis* Leidy, 1853, each one represented by a single species (*Hip. amphibius* and *Ch. liberiensis*, respectively)—both geographically confined to Africa. *Hippopotamus*, which is known in the fossil record since the latest Miocene, has today a quite patchy geographic distribution along the freshwater bodies in the sub-Saharan part of the continent, while *Choeropsis* only occurs in the tropical forests of West Africa. *Hippopotamus amphibius* is a highly specialized semiaquatic mega-herbivore mammal with derived anterior dentition and short and robust limbs adapted for locomotion on muddy substrates. Less-derived fossil forms have comparatively less-diversified hexaprotodont (i.e., with six upper and six lower incisors, instead of four) anterior dentition, deeply grooved upper canines, lower canines with smooth enamel, longer premolar rows, less hypsodont molars, shorter snout, lower-positioned orbits, and less robust limbs.

The first hippopotamid migration out of Africa occurred in latest Miocene times. The migrant populations, probably belonging to the genus *Hexaprotodon* Falconer and Cautley, 1836, reached Southern Europe and became well established in Southern Asia. The European fossil record is poor, and the available finds are generally very scanty, thus their taxonomy is uncertain and their generic attribution is used preliminarily. The best-known species is the Spanish *Hex. crusafonti* (Aguirre, 1963), but other samples from Spain and Italy have been described as separate species: *Hex. pantanellii* (Joleaud, 1920), *Hex. siculus* (Hooijer, 1946), and *Hex. primaevus* (Crusafont, Adrover and Golpe, 1963). All four may actually belong to a single species, that is *Hex. crusafonti*, which lived at the end of the Miocene and the beginning of the Pliocene (zones MN13–MN14) in the South of Europe (Spain, Italy, and South France—Coryndon 1978; Faure and Méon 1984; van der Made 1999). Another important migration wave occurred at the beginning of the Pleistocene, which involved tetraprotodont forms, morphologically very close to the extant *Hip. amphibius*. The European *Hippopotamus* became widespread in the continent, particularly in the south and west regions, and reached as north as England during the Last Interglacial. Taxonomically, they are usually attributed to two species: the very large *Hip. antiquus* Desmarest, 1822 (Early–Middle Pleistocene) and the generally smaller *Hip. amphibius* (Middle–Late Pleistocene), usually similar in size to the living individuals. The morphological and metrical variation exhibited throughout the genus' presence in Europe has led to the recognition of two additional species, *Hip. incognitus* Faure, 1984, and *Hip. tiberinus* Mazza, 1991, both of which are commonly considered as junior synonyms of one or both of the former species (Petronio 1986, 1995; Mazza 1995). On the other hand, several authors in the past, and more recently (e.g., Kahlke 1997, 2001), consider the observed variation of subspecific rank (i.e., normal for a mammal species of that body size). Here I follow the two-species taxonomy, principally because *Hip. antiquus* and *Hip. amphibius* are reported to be well distinct to each other on cranial characters (Caloi et al. 1980). However, given the fact that cranial specimens are quite rare, the results of Caloi et al. (1980) are difficult to corroborate, and the species-level determination is usually based on biometrical characters alone.

3 Historical Overview and Distribution

The oldest reference to a hippopotamid fossil in Greece was made by Forsyth Major (1887), who mentioned the presence of *Hippopotamus* on the SE Aegean island of Kos, in the vicinity of the village Antimáchia (Fig. 1). The specimens studied by the author can still be found in the collections of the Cantonal Museum of Geology in Lausanne, Switzerland (Roussiakis, pers. comm. 2018). Several years later, Airaghi (1928) and Desio (1931) referred to the same finds when mentioning the presence of *Hip. major* (= *Hip. antiquus*) at Antimáchia, Kos. The rest of the fossil samples collected in the area have a Villafranchian character and might be of Early Pleistocene age, though the collection is probably mixed.

Another early reference to hippo fossil remains (*Hippopotamus* sp.) was made by Deprat (1904), who discovered them at Panaghía sta Éria, Euboea Island, Central

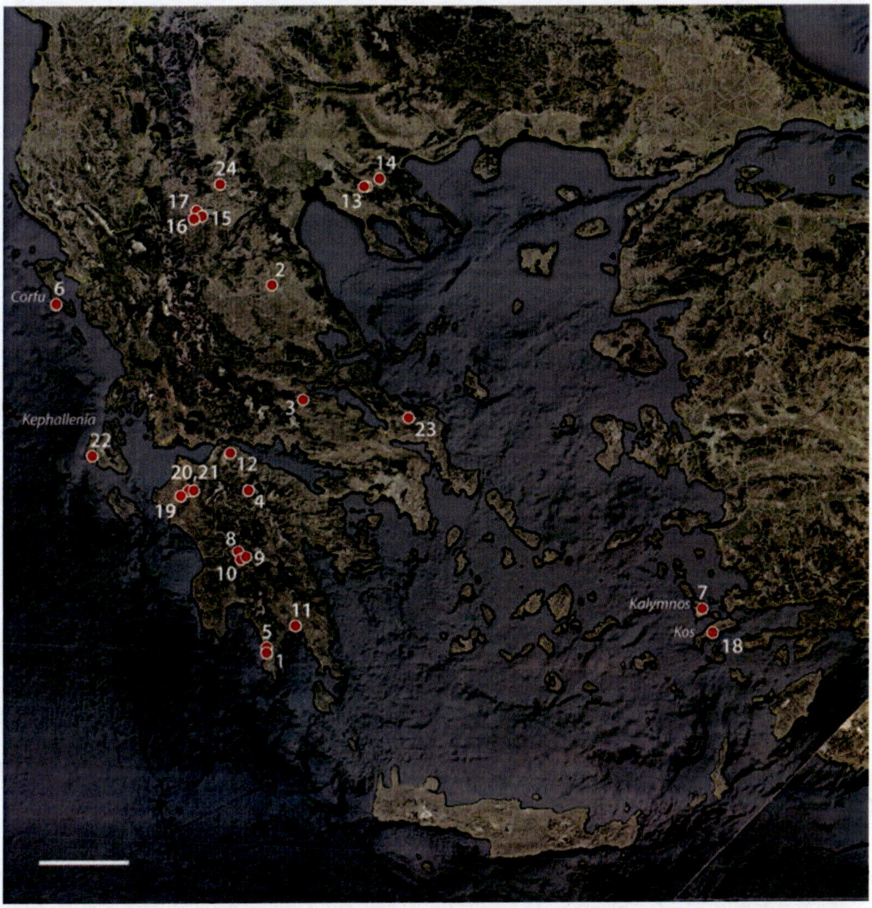

Fig. 1 Map of Greece showing the geographic distribution of localities with continental hippopotamids (including insecure ones). **1**, Dyrós Cave; **2**, Peniós Valley; **3**, Tíchos; **4**, Limnón Cave; **5**, Apídima Caves A, B; **6**, Paliávlako; **7**, Kálymnos; **8**, Marathoúsa 1, 2; **9**, Kyparíssia 1, 3, 4, T; **10**, Chorémi 3; **11**, Myrtiá; **12**, Káto Salmeníko; **13**, Kalamotó 1, 2; **14**, Ravin Voulgarákis; **15**, Kapetánios, **16**, Q-Profil; **17**, Libákos; **18**, Antimáchia; **19**, Hághios Demétrios; **20**, Aetorráchi; **21**, Elis; **22**, Mantzavináta, **23**, Panaghía sta Eria. Sites of unknown geographical position within the Megalopolis Basin are located in the area of 8–10. See Appendix for more information. Image exported from Google Earth Pro © 2019, map data from US Department of State Geographer, SIO, NOAA, US Navy, NGA, GEBCO, image from Landsat/Copernicus. Scale bar equals 80 km. North faces upward

Greece. The associated fossil assemblage given by the author (*Hipparion gracile*, *Sus erymanthius*, *Tragoceras amaltheus*, *Paleoreas pallasii*) is of Late Miocene age, which is in accordance with other dating proxies of the fossiliferous sedimentary sequence (Katsikatsos et al. 1981). In a much later publication on the Miocene faunas of Euboea, Mitzopoulos (1947) left *Hippopotamus* out of the locality's faunal list, which might mean that he considered Deprat's determination doubtful (he did not make any relevant comment, though). Nevertheless, an occurrence of a hippopotamid at this locality cannot be excluded, given the scarce presence of *Hexaprotodon*-like Late Miocene species in other Southern European countries. Neogene hippopotamids are otherwise unknown in Greece, so the presence of a hippo on Euboea, if true, would be quite exceptional. However, since the whereabouts of Deprat's material are unknown, this occurrence should be considered uncertain.

Psarianos (1954) described briefly a partly preserved upper dentition with P4–M3 (Fig. 2b) from the area of Mantzavináta, on Kephallenía Island. The author did not refer it to a species. This specimen was re-examined recently and is further reviewed herein (see below). A year later, Thenius (1955) reported on dental specimens (maxilla fragment with M2 and M3, three distal premolars, dental fragments) from marine deposits in the area of Elis, NW Peloponnesus, which he referred to *Hip.* aff. *amphibius major* (= *Hip.* aff. *antiquus*). These finds are important for their early geological age (documented using invertebrate biochronology of the marine layers), which is equivalent to what is now the earliest Pleistocene. More recent research in the same area produced scanty osteological and dental remains from Hághios Demétrios (Symeonidis and Theodorou 1986), as well as a partially preserved skull excavated at Aetorráchi (Fig. 2a; Reimann and Strauch 2008). All three localities in the area of Elis have yielded large-sized specimens consistent with an attribution to *Hip. antiquus* (see, e.g., Faure 1985), are closely located, and have probably similar age.

An interesting locality, because of the association of mammalian fossils with paleolithic artifacts, is Peniós Valley, Thessaly. The material, collected since the early 1960s from the Peniós River lower terrace deposits, comprises a rather poor hippopotamus sample which is not determinable to the species level (Boessneck in Milójčić et al. 1965; Athanassiou 2011). A radiometric dating of the fossiliferous strata to about 40 ka (Demitrack 1986; Athanassiou 2011) implies a plausible taxonomic attribution to the geochronologically younger species *Hip. amphibius*. One of the richest fossil hippopotamus-yielding areas in Greece is the Megalopolis Basin, central Peloponnesus, which is known to produce remains of this animal since the beginning of the twentieth century (Bürchner 1903). Historically old material, excavated by Professor Theodore Skouphos (University of Athens) in 1902, was described in detail by Melentis (1964, 1966a), who referred it to *Hip. antiquus*. This sample consists of numerous osteological and dental specimens, including a complete juvenile mandible (Fig. 2d; Melentis 1966a, pl. 78). Extensive surface collections carried out during the years 1960, 1962, and 1963 by a German team working on the mining potential of the basin's lignite deposits, yielded several specimens (mostly fragmentary), which were referred by Sickenberg (1976) to the same taxon. Recent fieldwork in the Megalopolis area corroborated with new material (Fig. 3) the prominent position of this species in the Middle Pleistocene ecosystem of the paleo-lake that covered this region (Athanassiou et al. 2018; Konidaris et al. 2018, 2019).

In a comprehensive study of the Early Pleistocene faunas from the Kastoriá–Grevená Basin (West Macedonia), Steensma (1988) described a dental and postcra-

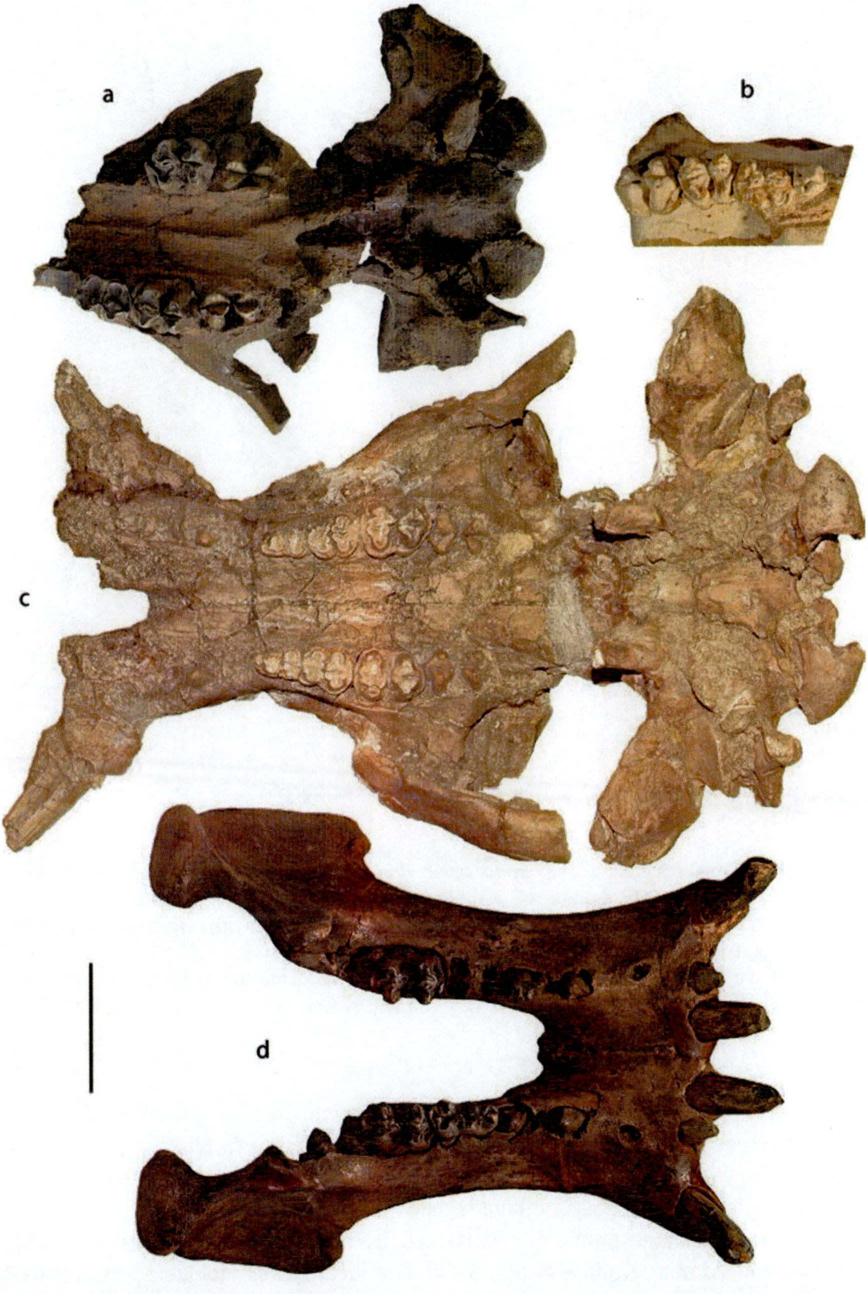

Fig. 2 Cranial specimens of *Hippopotamus* from Greek localities. (**a**) Juvenile skull from Aetorráchi, NW Peloponnesus (fossil collection of the University of Patras) (Reimann and Strauch 2008), (**b**) right maxilla fragment with P4–M3 from Mantzavináta, Kephallenía Island (Psarianos 1954), (**c**) juvenile skull RVL-88 from Ravin Voulgarákis, Central Macedonia (Museum of Geology–Paleontology, Aristotle University of Thessaloniki) (Kostopoulos 1996) (photo credit: D. Kostopoulos), (**d**) juvenile mandible from Megalopolis, Central Peloponnesus (Museum of Paleontology and Geology, National and Kapodistrian University of Athens) (Melentis 1966a). All specimens in occlusal view. Scale bar equals 10 cm

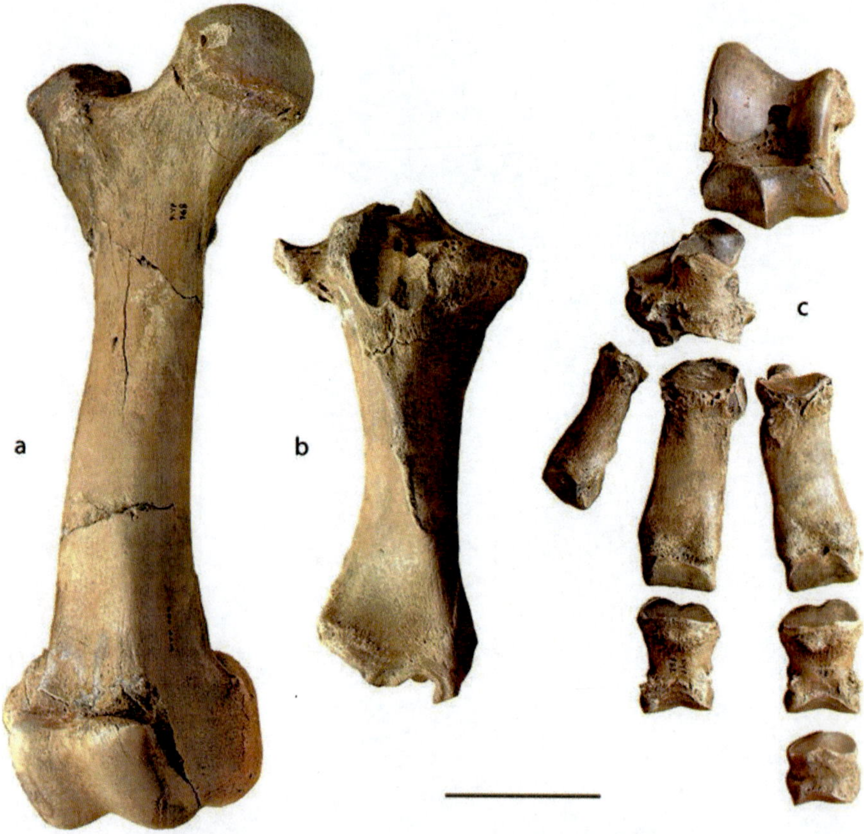

Fig. 3 Postcranial specimens of *Hippopotamus antiquus* from Kyparíssia, Megalopolis Basin, Central Peloponnesus (Ephorate of Palaeoanthropology–Speleology, Ministry of Culture) (Athanassiou et al. 2018). (**a**) Right femur KYP4-768, cranial view, (**b**) right tibia KYP4-314, cranial view, (**c**) right autopodium elements, presumably of the same individual (astragalus KYP1-794, fourth tarsal KYP1-786, third metatarsal KYP1-790, fourth metatarsal KYP1-781, fifth metatarsal KYP1-774, proximal phalanges KYP1-782 and KYP1-783, middle phalanx KYP1-787), dorsal view. Scale bar equals 10 cm

nial sample from the locality Libákos, which documents the presence of a large, though rather slender, hippopotamus. The author refers it to *Hip. amphibius antiquus* (= *Hip. antiquus*), and indeed, the studied specimens fit dimensionally to the range of this taxon. The assemblage is dated to the late Villafranchian.

Another important sample was excavated at Ravin Voulgarákis, Mygdonía Basin, Central Macedonia (Koufos et al. 1989). The locality has yielded several skeletal elements of hippopotamus, including a complete skull with associated mandible (Fig. 2c; Kostopoulos 1996). This nice specimen belongs to a juvenile individual, as its dentition still retains deciduous teeth. The Voulgarákis hippopotamus was a large-sized animal, despite its young ontogenetic age, metrically close to the maximal dimensions of *Hip. antiquus*, and it is referred to this species (*Hip. major* according to Koufos et al. 1989, *Hip. amphibius antiquus* according to Kostopoulos 1996). Biochronologically, the fauna is dated to the latest Early Pleistocene (MNQ19).

Giannopoulos (2000) and Symeonidis and Giannopoulos (2001) described another rich sample of more than a hundred complete dental and postcranial elements from Dyrós Cave, South Peloponnesus, which has been excavated mostly underwater. The presence of fossils in this cave (associated with smaller numbers of carnivore and cervid remains) had been already reported after the first speleological explorations of the cave (Petrochilos 1958). The authors described briefly the fossils, which they refer to *Hip. amphibius*, and published a radiometric date of 32 ka. The relatively small dimensions of the specimens recovered from the cave deposits and their rather recent age are consistent with their taxonomic attribution to *Hip. amphibius*. The cave is quite probably even richer in *Hippopotamus* material, but the high consolidation of the containing deposits, as well as the presence of a large body of water that floods most of the cave galleries, make the excavation demanding in resources and equipment, and excessively time consuming as well.

Hippopotamus is a particularly frequent faunal element in the Kalamotó sites, Chalkidikí Peninsula, central Macedonia. Tsoukala and Chatzopoulou (2005) referred the sample to *Hip. amphibius antiquus* (= *Hip. antiquus*), as the bones are large, though rather less robust than the largest individuals from Central and Western Europe. The authors dated the fauna (that also contains remains of carnivores, an elephant, a horse, a rhinoceros, cervids, bovins, and a vole) to the early Middle Pleistocene (zone MNQ20). On the basis of more recent stratigraphic work in the area, the age of Kalamotó is currently revised to the late Early Pleistocene (MNQ19) (Kostopoulos and Konidaris, pers. com. 2020).

Found in a Paleolithic archaeological context at the site of Paliávlako, Korissíon Lake, Corfu Island, a right mandibular fragment is briefly described and pictured by Darlas et al. (2006). The specimen lacks the ventral part of the dentary and preserves the fourth premolar and the three molars. No further information is available, since it was not described in detail, nor measured by the authors. According to Darlas (pers. com. 2019), the find is currently kept in the Archaeological Museum of Corfu.

A number of other publications, such as Melentis (1966b, 1969), Stratigopoulos (1986), Steensma (1988; localities Kapetánios and Q-Profil), Tsoukala (1999), Kranis (2003), Athanassiou and Bouzas (2010), and Masseti (2012, p. 20), deal with small dental and/or osteological samples of local importance, often insufficient for an attribution to a species. It should be noted that the dental specimen from Haliákmon Valley described by Melentis (1966b) comes from an unknown locality along the lower part of the 300-km-long valley. Melentis (1969) reports the presence of mostly juvenile postcranial bones from Limnón Cave, North Peloponnesus, which he referred to a Middle Pleistocene *Hip. antiquus*. However, the described material and the associated fauna are inadequate to support these taxonomic and biochronologic results. Stratigopoulos (1986) provided draft sketch-drawings of his studied specimens, which are not convincing about their hippopotamid identity. Unfortunately, there are no published photographs, and the whereabouts of the material are unknown. Nevertheless, the presence of hippopotamus in the wider area of the Ptolemaís Basin is documented from a semimandible kept in the Palaeontological and Historical Museum of Ptolemaís, which comes from Proásteion or Perdíkkas Formation (both of Early–Middle Pleistocene age) and is referrable to *Hip. antiquus* (Kostopoulos pers. comm. 2019). Tsoukala (1999) referred two small-sized specimens from Apídima Caves (an upper canine fragment from Cave A and a phalanx from Cave B) to female individuals, possibly young ones, of *Hip. antiquus*, but this poor sample is rather not referrable to a taxon further than the genus level. A comprehensive list of hippopotamid-bearing localities in Greece is given in Appendix.

4 Systematic Paleontology

This section addresses only the Quaternary non-endemic hippopotamids (genus *Hippopotamus*), excluding any Miocene taxa, whose presence in Greece is questionable (see previous section on the locality Panaghía sta Éria).

Hippopotamidae Gray, 1821

Type Genus *Hippopotamus* Linnaeus, 1758.

⊙*Hippopotamus* Linnaeus, 1758

Type Species *Hippopotamus amphibius* Linnaeus, 1758.

Included Taxa The genus is known with two species in Greece, *Hip. antiquus* and *Hip. amphibius* (extant in Africa), which are usually found in Pleistocene fluvial or lacustrine deposits.

Remarks *Hippopotamus* is a semiaquatic megaherbivore with very robust body and short and stout limbs. The skull is large, strongly constricted pre- and postorbitally, has short braincase, strong and protruding canine processes, and is characterized by highly elevated orbits and wide contact between the lacrymal and the nasal bones. The dentition is tetraprotodont (there exist four upper and four lower incisors). The distal groove of the upper canine is shallow, while the lower canine has prominent convergent enamel ridges. The genus is known in Greece from the beginning of the Pleistocene to the latest Pleistocene.

The Kephallenía Hippopotamus

A maxillary fragment with P4–M3 (Fig. 2b) was presented by Psarianos (1954) as a find from Mantzavináta, a village on the western part of Kephallenía Island. The specimen remains partially embedded in its original sandstone matrix. Despite its occurrence on an island, the Mantzavináta hippo is not a small-sized insular endemic. Its M1–M3 length of 144 mm is comparable to a small to medium *Hip. antiquus* or a large extant *Hip. amphibius* (compare, e.g., Faure 1985, p. 32). Note that the dimensions given by Psarianos (1954) contain errors; newly measured dimensions are given in Table 1. The lack of ecologically balanced Pleistocene faunas on the island may indicate that the sea strait between Paleo-Kephallenía and the mainland has posed an insurmountable barrier for most large mammals, except for those with advanced swimming abilities, such as hippos. The rather narrow sea channel may have allowed for a two-way migration route to the mainland, which precluded isolation and subsequent insular dwarfing. This supposed lack of genetic isolation leads to the inclusion of the Mantzavináta hippo with the continental populations, though its identification with *Hip. antiquus* or *Hip. amphibius* is not possible.

Table 1 Dental dimensions (mesiodistal × linguobuccal diameters, in mm) of the *Hippopotamus* sp. maxilla from Mantzavináta (Kephallenía Island, Western Greece; Psarianos 1954)

P4	M1	M2	M3	M1–M3
28 × 27	47 × 37	50 × 42	56 × 48	144

Hippopotamus antiquus Desmarest, 1822

Nomenclatural and Taxonomical History The large-sized fossil European hippopotamus was initially described in the beginning of the nineteenth century by Cuvier (1804), but was not given any name, except for the designation as "grand hippopotame fossile." The same author latinized this descriptive characterization as a formal species name 20 years later, coining *Hip. major* Cuvier, 1824. However, in the meantime, Desmarest (1822) published a short description of this fossil hippo under the species name *Hip. antiquus*, which has priority according to the International Code of Zoological Nomenclature. An additional species described in Europe, *Hip. tiberinus* Mazza, 1991, has been subsequently considered as a junior synonym of *Hip. antiquus* (Petronio 1986, 1995; Mazza 1995). The same is also true for *Hip. georgicus* Vekua, 1959, from Akhalkalaki, Georgia, which represents the easternmost recorded occurrence of *Hip. antiquus* (Faure 1985; Kahlke 1990). Depending on individual authors' views on intraspecific variation, *Hip. antiquus* may be regarded as a subspecies of *Hip. amphibius* as well.

Type Material The species was defined mainly on the sample from Upper Valdarno, Tuscany, Italy, without any type designation. Blandamura and Azzaroli (1977) designated the skull IGF-1043 from Figline (Valdarno)—mounted on a composite skeleton and exhibited in the Museum of Natural History of the University of Florence, Italy—as the lectotype of the species.

Type Locality Upper Valdarno, Tuscany, Italy, Early Pleistocene.

Distribution In Greece, this species has a widespread distribution, as it has been found in many localities from Central Macedonia in the North to Southern Peloponnesus and Kos Island in the South. More specifically, the *Hip. antiquus*-bearing localities are the following (North to South): Ravin Voulgarákis, Kalamotó 1 and 2, Libákos, Kapetánios, Q-Profil, Káto Salmeníko, Aetorráchi, Elis, Hághios Demétrios, Kyparíssia, Marathoúsa 1 and 2, Chorémi (and other sites of unknown exact location within the Megalopolis Basin), Myrtiá, and Antimáchia. Biostratigraphically, the species appears in the Greek fossil record at the beginning of the Pleistocene (area of Elis, North Peloponnesus; Thenius 1955; Reimann and Strauch 2008, possibly also Symeonidis and Theodorou 1986). Other Lower Pleistocene localities include Ravin Voulgarákis, Kalamotó, Libákos, Kapetánios, Q-Profil, and Antimáchia (Fig. 4). In the Middle Pleistocene, the species is well represented in several sites within the Megalopolis Basin (Central Peloponnesus; Melentis 1966a; Sickenberg 1976; van Vugt et al. 2000; Athanassiou et al. 2018; Konidaris et al. 2018, 2019). Marathousa 1 has been dated to 420–560 ka (Blackwell et al. 2018; Tourloukis et al. 2018), and other *Hippopotamus*-bearing sites situated

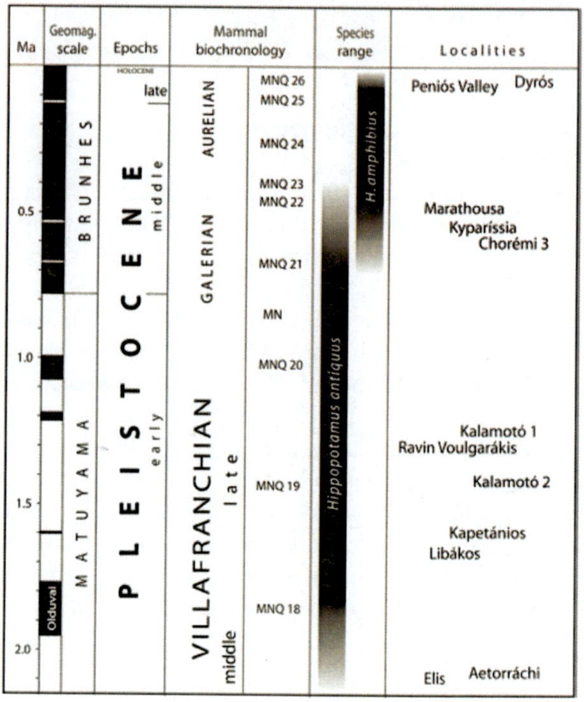

Fig. 4 Stratigraphic chart of well-dated *Hippopotamus*-bearing localities of Greece. Chronostratigraphic and magnetostratigraphic subdivisions according to Pillans and Gibbard (2012). Mammal biozonation according to Guérin (1990). The species ranges are based on biostratigraphic data from European localities (Faure 1985; Mazza 1995; Petronio et al. 2011; Pandolfi and Petronio 2015) and the present review

within the lacustrine layers of the basin are not expected to differ significantly in age (Fig. 4). Consequently, *Hip. antiquus* might have been present in Greece more recently than 500 ka, that is during MIS13 or even MIS12. This is generally in accordance with the species' upper biostratigraphic limit placed in lower Middle Pleistocene (Faure 1985), but postdates its last dated occurrence in other southern European regions, such as Italy (MIS15; Pandolfi and Petronio 2015).

Taxonomic Remarks This is the largest-bodied species of the genus *Hippopotamus*. Morphologically, it is very similar to the type species of the genus, from which it is distinguished dimensionally. Caloi et al. (1980) observed, however, several cranial characters that seem to effectively distinguish between the two species, which are mainly the following: *Hip. antiquus* has more elongate and slender skull, less laterally expanded zygomatic arches (with regard to the skull length), shorter sagittal crest, higher occipital, proportionally longer maxilla that is not concave ventrally, more caudally positioned choanae with regard to the third molar, and presents a diastema between the second and third premolars. Moreover, the mandible of *Hip. antiquus* has a rather concave, not convex ventral border, as is the case in *Hip. amphibius*. It also tends to be proportionally longer with less robust corpus, subparallel, not divergent, left and right corpora, and has a rather straight, not rostrally curving, angular process. Nevertheless, most fossil samples in Greece comprise scanty or isolated dental and postcranial specimens that are inadequate for the recognition of these diagnostic characters. Where cranial material is available (e.g., in Megalópolis, Ravin Voulgarákis, and Aetorráchi), the characteristic morphology of the species is observed (Melentis 1966a; Kostopoulos 1996; Reimann and Strauch 2008).

⊙*Hippopotamus amphibius* Linnaeus, 1758, the common hippopotamus

Nomenclatural and Taxonomical History This species name was coined by Linnaeus for the Nile hippopotamus. It is the genotype of *Hippopotamus* and the most characteristic species of the family Hippopotamidae. It has not been subjected to any nomenclatural changes. Nevertheless, it has been used by some authors to include all the Pleistocene *Hippopotamus* populations of Europe (see Sect. 2). One of the species founded much later on European material, *Hip. incognitus* Faure, 1984, has been subsequently considered a junior synonym of *Hip. amphibius* (Petronio 1986, 1995; Mazza 1995).

Distribution It is not well known in Greece, because the available material is scanty in all localities, except for Dyrós Cave (South Peloponnesus). As such, the attributions to this species are not certain. For the same reason, its biochronological distribution is also unknown. A radiometric date of about 32 ka for the Dyrós sample (Giannopoulos 2000; Symeonidis and Giannopoulos 2001) and another of about 40 ka for the Peniós Valley (Thessaly) sample (Demitrack 1986; Athanassiou 2011), which is referrable to the same species as well, point to a late survival of *Hip. amphibius* in Greece, in accordance to what is known for other Mediterranean region of Europe (Faure 1985; Pandolfi and Petronio 2015).

Taxonomic Remarks The European representatives of the species are very similar in size and morphology to the extant African populations. *Hippopotamus amphibius* is distinguished by older species *Hip. antiquus* by its smaller body size, as well as by a suite of morphological characters of the skull and mandible (Caloi et al. 1980; see the relevant section on *Hip. antiquus*).

5 Conclusions

The fossil record of Hippopotamidae in Greece spans biostratigraphically almost the entire Pleistocene. Moreover, a report of a hippopotamid in association with a late Miocene assemblage indicates the possible presence of a Neogene taxon (*Hexaprotodon*?). More than 20 localities have been recorded (see Fig. 1 and Appendix), but most of them have yielded scanty remains. Nevertheless, *Hippopotamus* constitutes one of the main faunal elements in Megalopolis Basin, Ravin Voulgarákis, Kalamotó, and Dyrós Cave. Most localities, stratigraphically placed in the Lower or Middle Pleistocene, have yielded material attributed or attributable to the species *Hip. antiquus*, whereas *Hip. amphibius* is certainly known only from Upper Pleistocene ones. The occurrence of *Hippopotamus* in these sites is a strong paleoecological indicator for the presence of a large freshwater body and a lacustrine or a riverside ecosystem, where the hippo was a key species.

Acknowledgments The author wishes to thank S. Roussiakis (National and Kapodistrian University of Athens) and D. Kostopoulos (Aristotle University of Thessaloniki) for useful discussions, as well as G. Iliopoulos (University of Patras) for providing access to the Aetorráchi skull. The comments of an anonymous reviewer greatly improved the original manuscript.

Appendix

List of fossil localities with occurrences of Hippopotamidae in Greece (including insecure ones), arranged in approximate geochronological order. The age of each locality is taken from the corresponding original publication cited below (in certain cases revised on the basis of more recent data), while the taxonomic identification follows the present revision. Locality numbers refer to the collection numbers of the PaleoBiology Database (PBDB)

Localities[PBDB No]	Age (ELMA; MNQ; chronometric in ka)	Taxon
Dyrós Cave	Late Pleistocene (≥31 ka)	*Hippopotamus amphibius*[1]
Peniós Valley[204397]	Late Pleistocene	*Hippopotamus* cf. *amphibius*[2]
Tíchos	Late Pleistocene	*Hippopotamus* sp.[3]
Limnón Cave	Middle Pleistocene	*Hippopotamus* sp.[4]

(continued)

Localities[PBDB No]	Age (ELMA; MNQ; chronometric in ka)	Taxon
Apídima Cave A[32059]	Middle–Late Pleistocene	*Hippopotamus* sp.[5]
Apídima Cave B[32062]	Middle–Late Pleistocene	*Hippopotamus* sp.[5]
Paliávlako (Corfu)	Middle–Late Pleistocene	*Hippopotamus* sp.[6]
Kálymnos Island	Middle–Late Pleistocene	*Hippopotamus* sp.[7]
Megalópolis Basin (undefined sites)[182721]	Middle Pleistocene	*Hippopotamus antiquus*[8]
Marathoúsa 1[187637]	Middle Pleistocene (~500 ka)	*Hippopotamus antiquus*[9]
Marathoúsa 2	Middle Pleistocene (~500 ka)	*Hippopotamus antiquus*[10]
Kyparíssia 1[194472]	Middle Pleistocene	*Hippopotamus antiquus*[11]
Kyparíssia 3[194474]	Middle Pleistocene	*Hippopotamus antiquus*[11]
Kyparíssia 4[194475]	Middle Pleistocene	*Hippopotamus antiquus*[11]
Kyparíssia T[194476]	Middle Pleistocene	*Hippopotamus antiquus*[11]
Chorémi 3[200081]	Middle Pleistocene (late Biharian)	*Hippopotamus* sp.[12]
Myrtiá[200093]	Early–Middle Pleistocene	*Hippopotamus antiquus*[13]
Ptolemaís Basin	Early–Middle Pleistocene?	*Hippopotamus antiquus*[14]
Káto Salmeníko[200094]	Early–Middle Pleistocene	*Hippopotamus antiquus*[15]
Kalamotó 1[200083]	late Early Pleistocene (MNQ19)	*Hippopotamus antiquus*[16]
Kalamotó 2[200084]	late Early Pleistocene (MNQ19)	*Hippopotamus antiquus*[16]
Ravin Voulgarákis[34783]	late Early Pleistocene (MNQ19)	*Hippopotamus antiquus*[17]
Kapetánios[34781]	late Villafranchian	*Hippopotamus antiquus*[18]
Q-Profil[34812]	late Villafranchian	*Hippopotamus antiquus*[18]
Libákos[34764]	late Villafranchian	*Hippopotamus antiquus*[18]
Antimáchia[207130]	Early Pleistocene	*Hippopotamus antiquus*[19]
Hághios Demétrios[200095]	Early–Middle Pleistocene	*Hippopotamus antiquus*[16]
Aetorráchi[200096]	middle Villafranchian	*Hippopotamus antiquus*[20]
Elis[200096]	earliest Pleistocene	*Hippopotamus antiquus*[21]
Haliákmon Valley	Pleistocene	*Hippopotamus* sp.[22]
Mantzavináta	Pleistocene	*Hippopotamus* sp.[23]
Panaghía sta Éria	late Miocene	*Hexaprotodon*? sp.[24]

ELMA European Land Mammal Age, *MNQ* Mammal Neogene–Quaternary Zone
[1]Giannopoulos (2000), [2]Boessneck in Milójčić et al. (1965), [3]Kranis (2003), [4]Melentis (1969), [5]Tsoukala (1999), [6]Darlas et al. (2006), [7]Masseti (2012), [8]Melentis (1966a), [9]Konidaris et al. (2018), [10]Konidaris et al. (2019), [11]Athanassiou et al. (2018), [12]van Vugt et al. (2000), [13]Athanassiou and Bouzas (2010), [14]Kostopoulos pers. com. (2019), [15]Symeonidis and Theodorou (1986), [16]Tsoukala and Chatzopoulou (2005), Kostopoulos and Konidaris, pers. com. (2020), [17]Kostopoulos (1996), [18]Steensma (1988), [19]Forsyth Major (1887), [20]Reimann and Strauch (2008), [21]Thenius (1955), [22]Melentis (1966b), [23]Psarianos (1954), [24]Deprat (1904)

References

Airaghi C (1928) Mammiferi pliocenici dell'isola di Coo (Dodecaneso). Atti Soc It Sci Nat Mus Civ St Nat Milano 67:125–135

Athanassiou A (2011) The Late Pleistocene fauna of Peneiós Valley (Lárissa, Thessaly, Greece): new collected material. In: van der Geer A, Athanassiou A (eds) 9th European Association of Vertebrate Palaeontologists Meeting, Heraklion, 2010. Abstracts, p 14

Athanassiou A, Bouzas D (2010) New hippopotamid finds in Eurotas Valley (Laconia, Greece). In: Christofides G, Kantiranis N, Kostopoulos DS, Chatzipetros AA (eds) Proceedings of the XIX Congress of the Carpathian-Balkan Geological Association, Thessaloniki, 2010. Sci Ann School Geol, Fac Sci, Aristotle Univ Thessaloniki, pp 57–62

Athanassiou A, Michailidis D, Vlachos E, Tourloukis V, Thompson N, Harvati K (2018) Pleistocene vertebrates from the Kyparíssia lignite mine, Megalopolis Basin, S. Greece: Testudines, Aves, Suiformes. Quat Int 497:178–197

Blackwell BAB et al (2018) ESR dating ungulate teeth and molluscs from the Paleolithic site Marathousa 1, Megalopolis Basin, Greece. Quaternary 1(22):1–25

Blandamura F, Azzaroli A (1977) L'"Ippopotamo Maggiore" di Filippo Nesti. Atti Accad Naz Lincei, Mem 14:169–188

Boisserie J-R (2005) The phylogeny and taxonomy of Hippopotamidae (Mammalia: Artiodactyla): a review based on morphology and cladistic analysis. Zool J Linnean Soc 143:1–26

Boisserie J-R (2007) Family Hippopotamidae. In: Prothero DR, Foss SE (eds) The evolution of artiodactyls. John Hopkins University Press, Baltimore, pp 106–119

Boisserie J-R, Lihoreau F, Brunet M (2005a) The position of Hippopotamidae within Cetartiodactyla. Proc Natl Acad Sci U S A 102:1537–1541

Boisserie J-R, Lihoreau F, Brunet M (2005b) Origins of Hippopotamidae (Mammalia, Cetartiodactyla): towards resolution. Zool Scr 34:119–143

Boisserie J-R, Lihoreau F, Orliac M, Fisher RE, Weston EM, Ducrocq S (2010) Morphology and phylogenetic relationships of the earliest known hippopotamids (Cetartiodactyla, Hippopotamidae, Kenyapotaminae). Zool J Linnean Soc 158:325–366

Boisserie J-R, Fisher RE, Lihoreau F, Weston EM (2011) Evolving between land and water: key questions on the emergence and history of the Hippopotamidae (Hippopotamoidea, Cetancodonta, Cetartiodactyla). Biol Rev 86:601–625

Bürchner L (1903) Wichtige Funde fossiler Knochen in Arkadien. Ber Naturwiss Ver Regensburg 9:119–123

Caloi L, Palombo MR, Petronio C (1980) Resti cranici di *Hippopotamus antiquus* (= *H. major*) e *Hippopotamus amphibius* conservati nel Museo di Paleontologia dell'Università di Roma. Geol Rom 19:91–119

Colbert EH (1935) Siwalik mammals in the American Museum of Natural History. Trans Am Philos Soc 26:1–401

Coryndon SC (1978) Hippopotamidae. In: Maglio VJ, Cooke HBS (eds) Evolution of African mammals. Harvard University Press, Cambridge, pp 483–495

Cuvier G (1804) Sur les ossements fossiles d'hippopotame. Ann Mus Hist Nat 4:99–122

Darlas A, Karkanas P, Palli O, Papadea A (2006) [Palaeolithic research on south-West Corfu]. Athens Ann Archaeol 39:11–30 [in Greek]

Demitrack A (1986) The late Quaternary geologic history of the Larissa plain, Thessaly, Greece: tectonic, climatic, and human impact on the landscape. PhD thesis, Stanford University

Deprat J (1904) Étude géologique et pétrographique de l'île d'Eubée. Thèse, Faculté des Sciences de l'Université de Paris, Besançon, Paris

Desio A (1931) Le isole Italiane dell'Egeo. Mem Descr Carta Geol Italia 24:1–534

Desmarest AG (1822) Mammalogie ou description des espèces de mammifères, seconde partie, contenant les ordres des rongeurs, des édentés, des pachydermes, des ruminans et des cétacés. Agasse, Paris

Faure M (1985) Les hippopotames quaternaires non-insulaires d'Europe occidentale. Nouv Arch Mus Hist Nat Lyon 23:13–79

Faure M, Méon H (1984) L'*Hippopotamus crusafonti* de la Mosson (près Montpellier). Première reconnaissance d'un hippopotame néogène en France. CR Séances Acad Sci 298:93–98

Forsyth Major CJ (1887) Faune mammalogiche delle isole di Kos e di Samos. Atti Soc Tosc Sc Nat, Proc Verb 5:272–275

Gatesy J (1997) More DNA support for a Cetacea/Hippopotamidae clade: the blood-clotting protein gene γ-fibrinogen. Mol Biol Evol 14:537–543

Gatesy J, Hayashi C, Cronin MA, Arctander P (1996) Evidence from milk casein genes that cetaceans are close relatives of hippopotamid artiodactyls. Mol Biol Evol 13:954–963

Giannopoulos VI (2000) [Contribution to the study of recent and old environments of the most important Greek caves]. PhD thesis, National and Kapodistrian University of Athens [in Greek]

Guérin C (1990) Biozones or Mammal Units? Methods and limits in biochronology. In: Lindsay EH, Fahlbusch V, Mein P (eds) European Neogene mammal chronology. NATO ASI Series (A: Life Sciences). Plenum, New York, pp 119–130

Kahlke R-D (1990) Zum Stand der Erforschung fossiler Hippopotamiden (Mammalia, Artiodactyla): eine Übersicht. Quartärpaläont 8:107–118

Kahlke R-D (1997) Die *Hippopotamus*-Reste aus dem Unterpleistozän von Untermaßfeld. In: Kahlke R-D (ed) Das Pleistozän von Untermaßfeld bei Meiningen (Thüringen), Teil 1, vol 40 (1). Monographien des Römisch-Germanischen Zentralmuseums Mainz. Habelt, Mainz, pp 277–374

Kahlke R-D (2001) Schädelreste von *Hippopotamus* aus dem Unterpleistozän von Untermaßfeld. In: Kahlke R-D (ed) Das Pleistozän von Untermaßfeld bei Meiningen (Thüringen), Teil 2, vol 40 (2). Monographien des Römisch-Germanischen Zentralmuseums Mainz. Habelt, Mainz, pp 483–500

Katsikatsos G, de Bruijn H, van der Meulen AJ (1981) The Neogene of the island of Euboea (Evia), a review. Geol Mijnb 60:509–526

Konidaris GE, Athanassiou A, Tourloukis V, Thompson N, Giusti D, Panagopoulou E, Harvati K (2018) The skeleton of a straight-tusked elephant (*Palaeoloxodon antiquus*) and other large mammals from the Middle Pleistocene butchering locality Marathousa 1 (Megalopolis Basin, Greece): preliminary results. Quat Int 497:65–84

Konidaris GE, Tourloukis V, Athanassiou A, Giusti D, Thompson N, Panagopoulou E, Karkanas P, Harvati K (2019) Marathousa 2: a new Middle Pleistocene locality in Megalopolis Basin (Greece) with evidence of human modifications on faunal remains. In: 9th Annual Meeting of the European Society for the Study of Human Evolution, Liège, Proc Eur Soc Study Human Evol, vol 8, p 82

Kostopoulos DS (1996) [The Plio-Pleistocene Artiodactyls of Macedonia (Greece): systematics, palaeoecology, biochronology, biostratigraphy]. PhD thesis, Aristotle University of Thessaloniki [in Greek]

Koufos G, Syrides G, Koliadimou K (1989) A new Pleistocene mammal locality from Macedonia (Greece). Contribution to the study of Villafranchian (Villanyian) in Central Macedonia. Bull Geol Soc Gr 23:113–124

Kranis HD (2003) [Neotectonic activity of fault zones in central-eastern mainland Greece (Lokris)]. PhD thesis (1999), National and Kapodistrian University of Athens, Gaia 10 [in Greek]

Lihoreau F, Boisserie J-R, Manthi FK, Ducrocq S (2015) Hippos stem from the longest sequence of terrestrial cetartiodactyl evolution in Africa. Nat Commun 6:6264

Lyras GA, Athanassiou A, van der Geer AAE (this volume) The fossil record of insular endemic mammals from Greece. In: Vlachos E (ed) The fossil vertebrates of Greece Vol. 2 – Laurasiatherians, artiodactyles, perissodactyles, carnivorans, and island endemics. Springer – Nature Publishing Group, Cham

Masseti M (2012) Atlas of terrestrial mammals of the Ionian and Aegean islands. de Gruyter, Berlin

Matthew WD (1929) Critical observations upon Siwalik mammals (exclusive of Proboscidea). Bull Am Mus Nat Hist 56:437–560
Mazza P (1995) New evidence on the Pleistocene hippopotamuses of Western Europe. Geol Rom 31:61–241
McKenna MC, Bell SK (1997) Classification of mammals above the species level. Columbia University Press, New York
Mein P (1999) European Miocene mammal biochronology. In: Rössner GE, Heissig K (eds) The Miocene land mammals of Europe. Dr. Friedrich Pfeil, München, pp 25–38
Melentis JK (1964) [Die fossilen Rhinocerotiden, Hippopotamiden und andere Säugetiere aus dem Becken von Megalopolis im Peloponnes (Griechenland)]. Prakt Akad Ath 39:388–400 [in Greek]
Melentis JK (1966a) Über *Hippopotamus antiquus* Desmarest aus dem Mittelpleistozän des Beckens von Megalopolis in Peloponnes (Griechenland). Ann Géol Pays Hellén 16:403–435
Melentis JK (1966b) Die pleistozäne Säugetierfauna des Beckens von Haliakmon (Griechenland). Ann Géol Pays Hellén 17:247–265
Melentis JK (1969) [Die quartären Vertebraten der Höhle der Seen von Kleitoria (im Gebiet der Aroania-Gebirge)]. Prakt Akad Ath 43:350–363 [in Greek]
Milójčić V, Boessneck J, Jung D, Schneider H (1965) Paläolithikum um Larissa in Thessalien. Beiträge zur Ur- und Frühgeschichtlichen Archäologie des Mittelmeerkulturräumes, vol 1. Rudolf Habelt, Bonn
Mitzopoulos MK (1947) Die Verbreitung der Pikermistufe auf der Insel Euböa. Ann Géol Pays Hellén 1:209–216
Nikaido M, Rooney AP, Okada N (1999) Phylogenetic relationships among cetartiodactyls based on insertions of short and long interpersed elements: hippopotamuses are the closest extant relatives of whales. Proc Natl Acad Sci U S A 96:10261–10266
Orliac M, Boisserie J-R, MacLatchy L, Lihoreau F (2010) Early Miocene hippopotamids (Cetartiodactyla) constrain the phylogenetic and spatiotemporal settings of hippopotamid origin. Proc Natl Acad Sci U S A 107:11871–11876
Pandolfi L, Petronio C (2015) A brief review of the occurrences of Pleistocene *Hippopotamus* (Mammalia, Hippopotamidae) in Italy. Geol Croat 68:313–319
Pearson HS (1927) On the skulls of early Tertiary Suidae, together with an account of the otic region in some other primitive Artiodactyla. Phil Trans R Soc London 215:389–460
Petrochilos J (1958) [Palaeontological findings in the cave "Glyphada"]. Bull Soc Spél Gr 4:119 [in Greek]
Petronio C (1986) Nuovi resti di ippopotamo del Pleistocene medio-inferiore dei dintorni di Roma e problemi di tassonomia e filogenesi del gruppo. Geol Rom 25:63–72
Petronio C (1995) Note on the taxonomy of Pleistocene hippopotamuses. Ibex 3:53–55
Petronio C, Bellucci L, Martinetto E, Pandolfi L, Salari L (2011) Biochronology and palaeoenvironmental changes from the Middle Pliocene to the Late Pleistocene in Central Italy. Geodiversitas 33:485–517
Pickford M (1983) On the origins of Hippopotamidae together with descriptions of two new species, a new genus and a new subfamily from the Miocene of Kenya. Geobios 16:193–217
Pickford M (1989) Update on hippo origins. CR Acad Sci Paris 309:163–168
Pickford M (2008) The myth of the hippo-like anthracothere: the eternal problem of homology and convergence. Rev Esp Paleont 23:31–90
Pickford M (2011) *Morotochoerus* from Uganda (17.5 Ma) and *Kenyapotamus* from Kenya (13–11 Ma): implications for hippopotamid origins. Estud Geol 67:523–540
Pillans B, Gibbard P (2012) The Quaternary Period. In: Gradstein FM, Ogg JG, Schmitz MD, Ogg GM (eds) The geologic time scale 2012. Elsevier, Oxford, pp 979–1010
Psarianos P (1954) Über das Vorkommen von *Hippopotamus* auf Kephallinia (Griechenland). Prakt Akad Ath 28:408–412
Reimann CK, Strauch F (2008) Ein *Hippopotamus*-Schädel aus dem Pliozän von Elis (Peloponnes, Griechenland). N Jb Geol Pal Abh 249:203–222

Sickenberg O (1976) Eine Säugetierfauna des tieferen Bihariums aus dem Becken von Megalopolis (Peloponnes, Griechenland). Ann Géol Pays Hellén 27:25–73

Steensma KJ (1988) Plio-/Pleistozäne Großsäugetiere (Mammalia) aus dem Becken von Kastoria/ Grevena, südlich von Neapolis – NW Griechenland. Inaugural-Dissertation, Technische Universität Clausthal

Stratigopoulos S (1986) [On the fossil remains of *Ursus etruscus* Cuv. and *Hippopotamus amphibius major* Cuv. in Ptolemais basin]. Bull Geol Soc Gr 18:183–197 [in Greek]

Symeonidis NK, Giannopoulos VI (2001) [Pleistocene faunas of Glyfada cave, Diros, Laconia]. Bull Geol Soc Gr 34:515–522 [in Greek]

Symeonidis NK, Theodorou GE (1986) [New localities with fossil *Hippopotamus* in northwestern Peloponnese]. Ann Géol Pays Hellén 33:51–67 [in Greek]

Thenius E (1955) *Hippopotamus* aus dem Astien von Elis (Peloponnes). Ann Géol Pays Hellén 6:206–212

Tourloukis V, Muttoni G, Karkanas P, Monesi E, Scardia G, Panagopoulou E, Harvati K (2018) Magnetostratigraphic and chronostratigraphic constraints on the Marathousa 1 Lower Palaeolithic site and the Middle Pleistocene deposits of the Megalopolis basin, Greece. Quat Int 497:154–169

Tsoukala E (1999) Quaternary large mammals from the Apidima caves (Lakonia, S. Peloponnese, Greece). Beitr Paläontol 24:207–229

Tsoukala E, Chatzopoulou K (2005) A new Early Pleistocene (latest Villafranchian) site with mammals in Kalamotó (Mygdonia Basin, Macedonia, Greece)—preliminary report. Mitt Komm Quartärforsch Österr Akad Wiss 14:213–233

van der Geer A, Lyras G, de Vos J, Dermitzakis M (2010) Evolution of island mammals: adaptation and extinction of placental mammals on islands. Wiley-Blackwell, Oxford

van der Made J (1999) Superfamily Hippopotamoidea. In: Rössner GE, Heissig K (eds) The Miocene land mammals of Europe. Dr. Friedrich Pfeil, München, pp 203–208

van Vugt N, de Bruijn H, van Kolfschoten T, Langereis CG (2000) Magneto- and cyclostratigraphy and mammal-fauna's of the Pleistocene lacustrine Megalopolis Basin, Peloponnesos, Greece. Geol Ultraject 189:69–92

Weston E, Boisserie J-R (2010) Hippopotamidae. In: Werdelin L, Sanders WJ (eds) Cenozoic mammals of Africa. University of California Press, Berkerley, pp 861–879

Zurano JP, Magalhães FM, Asato AE, Silva G, Bidau CJ, Mesquita DO, Costa GC (2019) Cetartiodactyla: updating a time-calibrated molecular phylogeny. Mol Phylogenet Evol 133:256–262

The Fossil Record of Giraffes (Mammalia: Giraffidae) in Greece

George Iliopoulos and Socrates Roussiakis

1 Introduction

Giraffidae consist a group of medium- to large-sized ruminant cetartiodactyl mammals which as typical pecorans are characterized by the presence of cranial appendages and more specifically by the presence of ossicones in the males of some taxa and in both males and females in some other taxa. Ossicones are epiphyseal osseous protuberances found either paired on the dorsal part of the frontoparietal suture, or paired or unpaired on the frontonasal of giraffid skulls (Churcher 1990) but their structure and development is completely distinct from that of horns and antlers. As epiphyseal growths they are deriving from the ossification of cartilage during the gradual development of the animal (Lankester 1907; Janis and Scott 1987; Solounias 1988; Prothero and Schoch 2002). At early stages, the ossicone cartilage is not attached to the skull, but gradually as the ossicones develop and get ossified, they are fused to the skull (Solounias 1988; Prothero and Schoch 2002). In addition, unlike horns and antlers, ossicones are permanently covered by skin and fur (Harris et al. 2010). Another typical feature of the giraffids is their bilobed lower canine (Hamilton 1978) in which, compared to Climacoceridae, the second lobe is notably enlarged (Solounias 2007; Harris et al. 2010). The elongation of the neck, typical for long necked giraffes, seems to be a general trait that can be seen in most giraffids (Danowitz et al. 2015), as it is the elongated skull as well. Similarly, another feature seen in most giraffids is the elongation of their limb bones (Solounias 2007).

G. Iliopoulos (✉)
Department of Geology, University of Patras, Rio Patras, Greece
e-mail: iliopoulosg@upatras.gr

S. Roussiakis
Faculty of Geology and Geoenvironment, National and Kapodistrian University of Athens, Athens, Greece
e-mail: srousiak@geol.uoa.gr

E. Vlachos (ed.), *Fossil Vertebrates of Greece Vol. 2*,
https://doi.org/10.1007/978-3-030-68442-6_10

Giraffids appear for the first time in the early Miocene of Africa and more specifically at Gebel Zelten, Libya (Churcher 1978) with *Canthumeryx sirtensis* as the first representative of the family. During the middle Miocene they migrate to Eurasia, and since then they consist a common faunal element in African and Eurasian faunas until the Early Pleistocene and particularly during the late Miocene when they present their maximum diversity with several species. Despite their significant contribution in the Neogene faunas of Eurasia, the last Eurasian giraffes became extinct in the interval 1.3–1.1 Ma (Lebatard et al. 2014). Until recently, two giraffid species were supposed to survive in Africa, *Okapia johstoni* and *Giraffa camelopardalis*, that belong to the subfamilies Palaeotraginae and Giraffinae, respectively. Concerning the latter species, a recent study by Fennessy et al. (2016) which applied multi-locus population genetic analyses on the genetic material of the 9 *G. camelopardalis* subspecies, indicated that notably four distinct giraffe species should be considered instead of one, namely *G. camelopardalis, G. tippelskirchi, G. giraffa*, and *G. reticulata*. However, two new studies that were only published this year using multiple methods of multi-locus DNA sequence analysis on genetic material from the different subspecies (Petzold and Hassanin 2020) and mitochondrial sequences from old museum specimens (Petzold et al. 2020), suggested that there are 10 extant subspecies of *Giraffa*, which belong to three different species, namely *G. camelopardalis, G. tippelskirchi,* and *G. giraffa*.

The Greek giraffid fossil record is fairly rich, containing 14 species ranging from the middle Miocene to the Early Pleistocene, of which 10 have been named based on material from Greek type localities. To date, more than 50 localities, mainly in the eastern and northern part of mainland Greece have revealed giraffid findings, nevertheless a number of these localities is found on islands of the Eastern Aegean Sea such as Lesvos, Chios, Samos, and Rhodes and surprisingly there is an additional report for a giraffid astragalus found on the island of Antikythera (Verikiou-Papaspyridakou 1986). Most of the localities as well as most of the species (11) are found in late Miocene deposits. The stratigraphically oldest Greek finding is *Georgiomeryx georgalasi* from the middle Miocene of Thymiana, Chios, whereas 7 localities bear the Plio-Pleistocene (Villafranchian) taxon *Palaeotragus inexspectatus*. The Greek late Miocene record includes the species *Helladotherium duvernoyi, Samotherium boissieri, Samotherium major, Palaeotragus rouenii, Palaeotragus coelophrys, Palaeotragus quadricornis, Palaeotragus* aff. *berislavicus, Palaeogiraffa macedoniae, Palaeogiraffa pamiri, Palaeogiraffa major, Bohlinia attica*, and *Bohlinia nikitiae*.

2 Historical Overview

In 1854, M. Chaeretis sent to the Natural History Museum in Paris fossil bones from Pikermi which in the same year were studied and published by Duvernoy (1854); in that paper, two alleged giraffid species, one "giraffe" and one large species of giraffe are mentioned. This was the first report of giraffid fossil material from Greece.

Since then, several giraffid findings, mainly of Late Miocene age, have been discovered in the Greek Peninsula. The most important findings and the most eminent giraffid workers will be reported herein.

Two years later, Gaudry and Lartet (1856) studied the "Chaeretis" material, as well as new material from excavations they contacted that year at Pikermi and described two new species: *Camelopardalis attica* and *Camelopardalis duvernoyi*. In 1860, Gaudry, after studying additional material from the excavations at Pikermi, proposed a new generic name for the robust giraffid, *Helladotherium duvernoyi*. A year later he (Gaudry 1861) erected the species *Palaeotragus rouenii,* a small and gracile giraffid, which he at first considered to be a large bovid. On the same year, Wagner (1861), after studying material also from Pikermi, described the taxon *Giraffa vetusta*, which today is considered as an invalid taxon. Before the end of the nineteenth century, Forsyth Major conducted excavations at the locality Mytilinii on Samos Island, where he collected bones that belonged to a large giraffid. He originally described this material as *H. duvernoyi*, but later he established a new species, *Samotherium boissieri* (Forsyth Major 1888). Four decades later, Bohlin (1926) after studying material from Samos considered that the large-sized *Samotherium* material presented certain different characters from the typical *S. boissieri* and thus he described it as *Samotherium boissieri* var. *major*. Weithofer (1888), after studying material from Pikermi, erected the new species *Giraffa parva*, which today is also considered as an invalid taxon. Despite the fact that in the following decades major excavation projects took place in Pikermi and Samos, no significant data concerning giraffids have been produced, except for the new findings from Axios valley reported by Arambourg and Piveteau in 1929. The next important giraffid finding came in 1940 by Paraskevaidis who described a mandibular fragment with P2–3 from the middle Miocene locality Thymiana in Chios which belonged to a primitive giraffid and which he named *Georgiomeryx georgalasi*. Sickenberg (1967) reported from the Villafranchian deposits of Volax the giraffid taxon *Macedonitherium martini*. Athanassiou (2014) in his study on the giraffid material from Sesklo, considered that this material as well as all Greek and Eurasian Villafranchian giraffid material belongs to *Palaeotragus inexspectatus*. In the early 1970s, Denis Geraads (1974) working on his PhD project studied the giraffid skeletal material that had been collected from the late Miocene localities of Axios Valley. In the years to come he published a number of papers on this material (Geraads 1978, 1979, 1989) and in one of them (1989) he erected the species *Decenatherium macedoniae* from the Vallesian locality of Pentalophos 1, which later Bonis and Bouvrain (2003) revised. They erected the new genus *Palaeogiraffa*, to accommodate *P. macedoniae* and determined material from the locality of Xirochori 1 as a second different species of the same genus, *P. pamiri*, and raised a third new species as well, *P. major*, from the Vallesian locality of Ravin de la Pluie.

Solounias is another important worker that reported giraffid material from Samos (Solounias 1981), and also has done among others a lot and significant work in determining the feeding and dietary habits of giraffids found in Greek localities and defining their paleoecological adaptations (e.g., Solounias et al. 2000, 2010). Kostopoulos et al. (1996) studied the giraffids from the localities of Nikiti and erected a new species of *Bohlinia*, *B. nikitiae*. Iliopoulos (2003) studied for his thesis the giraffid material from the different sites of the late Miocene locality of Kerassia. Kostopoulos and Koufos (2006) described the giraffid taxa that have been determined in the late Miocene locality of Perivolaki. Kostopoulos (2009) provided a detailed account on the giraffid findings from the late Miocene localities of Samos Island. A new generation of Greek vertebrate paleontologists in the last few years has added new data to the Greek giraffid record. Lazaridis (2015) for his thesis reported giraffid taxa from Kryopigi and Kassandra, Xafis et al. (2019) from Thermopigi and recently Laskos (2020) for his MSc thesis revised old *Palaeotragus* material from Vallesian localities of Macedonia, reporting for the first time the presence of *Palaeotragus* aff. *berislavicus* from Nikiti 1.

3 Phylogenetic Relationships

Giraffidae as ruminants are cetartiodactyl pecorans having ossicones as cranial appendages. Ruminants with ossicones located above the orbits have been placed to the clade Giraffomorpha, which includes two superfamilies, Palaeomerycoidea and Giraffoidea (Solounias 2007; Sánchez et al. 2015; Rios et al. 2017). According to Janis and Scott (1987), Giraffoidea consist a monophyletic group that originated from Gelocidae before the early Miocene. The main synapomorphies of the Giraffoidea that support their monophyly are: the bilobed lower canine (Hamilton 1978) and the lacking of first premolars (Harris et al. 2010). Two families are included in the Giraffoidea, Climacoceratidae, and Giraffidae. Climacoceratidae present a small accessory lobe and a plesiomorphic ruminant-like P4 (Hamilton 1978). The synapomorphies that characterize the Giraffidae are, on the other hand, mainly the large size of the second lobe of the bilobed lower canine (Hamilton 1978) (Fig. 1), the long slender limbs, the large body size, and the narrow occipital (Solounias 2007), indicating that Giraffidae consist a monophyletic clade.

Solounias (2007) based on metacarpal morphology proposed that the family consists of eight subfamilies: Canthumerycinae, Bohlininae, Okapiinae, Giraffokerycinae, Sivatheriinae, Samotheriinae, Palaeotraginae, and Giraffinae. Nevertheless, as the phylogeny of this group is still not very clear, herein a more conservative approach will be followed, taking under consideration what is widely accepted and more specifically that four subfamilies should be included in the family, Canthumerycinae (Hamilton 1978), Sivatheriinae, Palaeotraginae, and Giraffinae (Merceron et al. 2018). Recent phylogenetic analysis (Rios et al. 2017) has shown that Sivatheriinae and Giraffinae are monophyletic groups, whereas Palaeotraginae is paraphyletic.

The Greek fossil record contains giraffid fossils from all four subfamilies, such as the Canthumerycinae *Georgiomeryx georgalasi* from the middle Miocene, the Sivatheriinae *Helladotherium duvernoyi* from the late Miocene, the Giraffinae *Bohlinia attica* and *Bohlinia nikitiae* from the late Miocene as well, and the fairly diverse Palaeotraginae represented by *Samotherium boissieri, Samotherium major,*

Fig. 1 The left bilobed canine of *Giraffa camelopardalis* (lingual view) presenting the large-sized second lobe, a typical feature of the Giraffidae. Scale bar equals 10 mm

Palaeotragus rouenii, Palaeotragus coelophrys, Palaeotragus quadricornis, Palaeotragus aff. *berislavicus, Palaeogiraffa macedoniae, Palaeogiraffa pamiri* and *Palaeogiraffa major* from the late Miocene, and *Palaeotragus inexspectatus* from the Early Pleistocene.

4 Distribution

Taking in mind literature, as well as the most recent reports on the fossil giraffid findings of the Greek Peninsula, 56 localities bearing giraffid remains have been recorded until now (see Fig. 2 and Appendix). These localities are mainly found in the eastern and northern part of the peninsula, as well as on islands of the Eastern and South Aegean Sea such as Lesvos, Chios, Samos, Rhodes, Crete, and Antikythera (see Appendix). The two oldest localities are Thymiana B from the late Orleanian (MN5) of Chios Island (40), the type locality of *Georgiomeryx georgalasi* (Paraskevaidis 1940), and Melampes from the Astarachian (MN6 or MN7–8) of South Rethymnon on Crete (39), reported herein for the first time.

The latter consists mainly of a few long bone parts belonging possibly to a Palaeotraginae and which were collected in 1975 by Kuss (now stored in the collections of the Natural History Museum of Crete). Giraffids from the Vallesian and mainly from MN10 have been found in localities from Axios Valley (localities 35, 36, 38) and Chalkidiki (34, 37), represented by *Palaeogiraffa* spp., robust *Palaeotragus* species (*P. coelophrys, P.* aff. *berislavicus*), *Bohlinia* spp. (*B. attica, B. nikitiae*), and *Helladotherium duvernoyi*. Giraffids during the Turolian were more widespread expanding in the eastern and northern part of the peninsula, and have been found in several localities at Samos Island (14), as well as in Rhodes Island (10) (Appendix). More specifically, early Turolian (MN11) giraffids have been found in localities at Axios Valley (32), Chalkidiki (33), and Samos (14). The

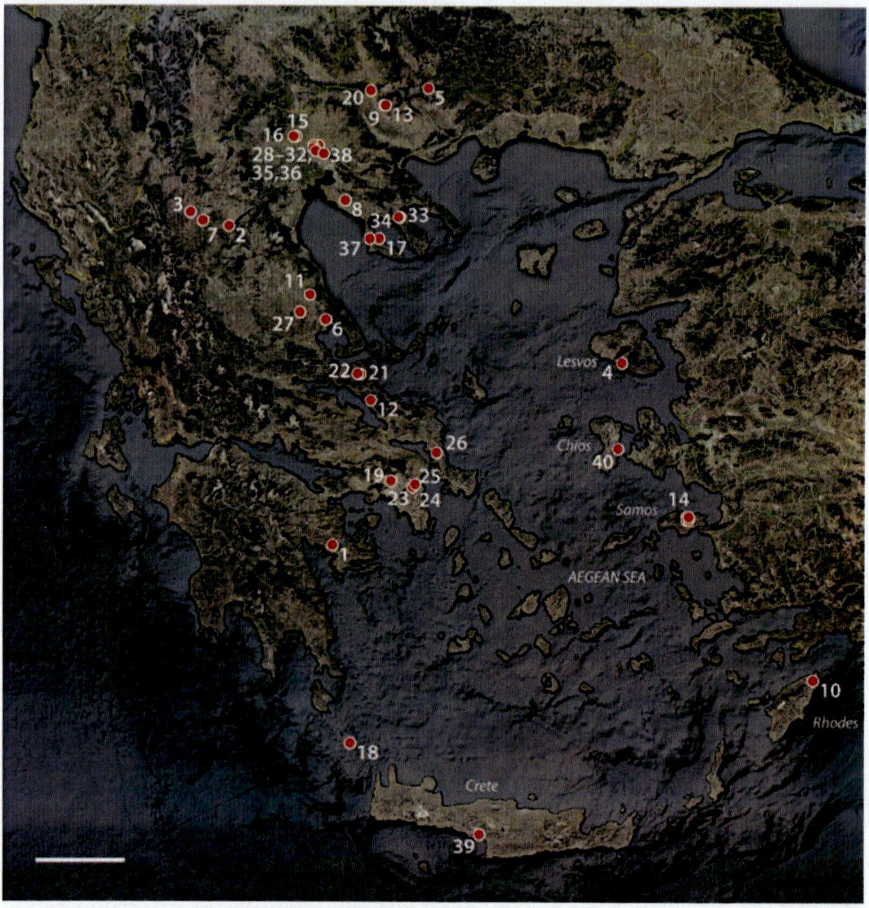

Fig. 2 Map of the occurrences of fossil giraffes from Greece. See Appendix for further information. **Pleistocene**: **1**, Karnezeika; **2**, Haliakmon; **3**, Libakos; **4**, Vatera-F; **5**, Volax; **6**, Sesklon; **7**, Dafnero. **Pliocene/Miocene**: **8**, Nea Silata; **9**, Maramena. **Miocene**: **10**, Rhodes Island; **11**, Alifakas; **12**, Drazi; **13**, Ano Metochi; **14**, various localities in Mytilinii basin on Samos Island; **15**, Dytiko 2; **16**, Dytiko 1; **17**, Kryopigi; **18**, Kamarella; **19**, Pyrgos Vassilissis; **20**, Thermopigi; **21**, Kerassia 1; **22**, Kerassia 3 and 4; **23**, Pikermi Valley 1; **24**, Pikermi; **25**, Chomateri; **26**, Halmyropotamos; **27**, Perivolaki; **28**, Ravin X; **29**, Prochoma; **30**, Vathylakkos 3; **31**, Vathylakkos 2; **32**, Ravin des Zouaves; **33**, Nikiti 2; **34**, Nikiti 1; **35**, Xirochori 1; **36**, Ravin de la Pluie; **37**, Fourka; **38**, Pentalophos 1; **39**, Melambes; **40**, Thymiana B. Data from the Paleobiology Database (PBDB). Image exported from Google Earth Pro © 2019, map data from US Dept. of State Geographer, SIO, NOAA, U.S. Navy, NGA, GEBCO, image from Landsat/Copernicus. Scale bar equals 80 Km, north faces upward

Early Turolian localities of Samos are the only Greek sites where *Samotherium boissieri* has been found. Furthermore, *H. duvernoyi, Palaeotragus rouenii, P. coelophrys,* and *B. attica* are also common early Turolian giraffids. The majority of the Greek late Miocene localities come from the late Turolian (MN12–13) and these localities are situated in Axios Valley (15, 16, 28, 29, 30, 31), Chalkidiki (8, 17), Strymon Valley (9, 13, 20) Euboea Island (12, 21, 22, 26), Attiki (19, 23, 24, 25), Samos (14), and Rhodes Island (10). The late Turolian giraffid fauna is characterized by the presence of *H. duvernoyi, Samotherium major, P. rouenii, P. coelophrys, Palaeotragus quadricornis*, and *B. attica*. *P. quadricornis* has been only reported from Samos Island. Finally, early Pleistocene giraffids have been discovered in the Haliakmon Valley (2, 3, 7), Drama (5), Magnesia (6), Lesvos Island (4), and Argolida (Karnezeika) (1), represented by the Villafranchian taxon *Palaeotragus inexspectatus*. The latter locality is reported for the first time herein (Sianis pers. comm.).

5 Systematic Paleontology

Giraffidae Gray, 1821

Type Genus *Giraffa* Brisson, 1762.

Canthumerycinae Hamilton, 1978

★***Georgiomeryx* Paraskevaidis,** 1940

Type Species *Georgiomeryx georgalasi* Paraskevaidis, 1940.

Distribution Middle Miocene (MN5) of Chios Island, Greece (Bonis et al. 1997).

Comments Primitive giraffid consisting probably the oldest European giraffid taxon.

★***Georgiomeryx georgalasi* Paraskevaidis,** 1940

Type Material Mandibular part with p2–3, Athens Museum of Paleontology and Geology, National and Kapodistrian University of Athens (figs. 4–5, pl. 13 in Paraskevaidis 1940).

Type Locality Thymiana B, Chios Island, Greece, late Orleanian, MN5.

Distribution Known only from the type locality.

Taxonomic Remarks *Georgiomeryx georgalasi* is a medium size primitive canthumerycine giraffid and, as a basal form, presents primitive features, such as the flat and wide roof of the skull, and a pair of flattened ossicones projecting laterally just above the orbits (Fig. 3). It presents a typical brachyodont giraffid dentition. Upper fourth premolars and molars have a clear and strong lingual cingulum. Upper P3 and P4 have clearly different shape and morphology (heteromorphic). *G. georg-*

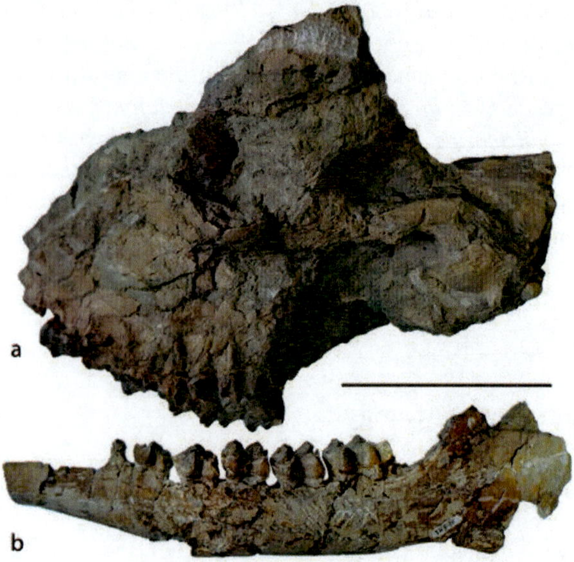

Fig. 3 Left lateral view of the cranium (THB 30) (**a**) and left mandible (THB 16) (**b**) of *Georgiomeryx georgalasi* from Thymiana B, Chios Island. Scale bar equals 10 cm (photos provided by DS Kostopoulos)

alasi as a canthumericine differs from Climacoceratidae in the shorter and more molarized lower P2 and P3. Nevertheless, its lower premolars are less molarized than *Canthumeryx*. Compared with *Injanatherium*, it is missing the small triangular anterior pair of ossicones found in front of the orbit and also the posterior pair projects laterally just above the orbits and not posteriorly behind the orbits, and its dentition is more brachyodont with stronger cingulums in the upper teeth (Bonis et al. 1997). *Giraffokeryx* also possessed two pairs of ossicones; however, the anterior one was placed anteriorly to the orbit above the frontals and the posterior one behind the orbits at the frontoparietal region (Bonis et al. 1997).

Comments When Paraskevaidis (1940) raised the new species for the first time, he considered that it belonged to Cervidae.

Palaeotraginae Pilgrim, 1911

★*Palaeotragus* Gaudry, 1861

Type Species *Palaeotragus rouenii* Gaudry, 1861.

Included Taxa 17 species are included in *Palaeotragus*: *Palaeotragus rouenii*; *Palaeotragus microdon* (Koken, 1885); *Palaeotragus pavlowae* Pavlow, 1913; *Palaeotragus coelophrys* (Rodler and Weithofer, 1890); *Palaeotragus quadricornis*

Bohlin, 1926; *Palaeotragus expectans* Borissiak, 1914; *Palaeotragus borissiaki* Alexeiev, 1930; *Palaeotragus hoffstetteri* Ozansoy, 1965; *Palaeotragus berislavicus* Korotkevich, 1957; *Palaeotragus moldavicus* Godina, 1979; *Palaeotragus asiaticus* Godina, 1975; *Palaeotragus tungurensis* Colbert, 1936; *Palaeotragus inexspectatus* (Samson and Radulesco, 1966); *Palaeotragus progressus* Tang and Ji, 1983; *Palaeotragus germaini* Arambourg, 1959; *Palaeotragus lavocati* Heintz, 1976; *Palaeotragus robinsoni* Crusafont-Pairó, 1979.

Distribution Middle Miocene of North Africa, Mongolia and China; late Miocene of North and East Africa, and Eurasia; late Pliocene and Early Pleistocene of Eurasia (Laskos 2020; Athanassiou 2014).

Comments This genus includes small- to middle-sized palaeotragines. Based on their relative size, the late Miocene species of *Palaeotragus* are separated into two groups: small and large (Geraads 1974, 1986; Kostopoulos et al. 1996; Iliopoulos 2003; Kostopoulos and Saraç 2005; Kostopoulos 2009). *P. rouenii*, *P. microdon,* and *P. pavlowae* are members of the small-sized group ("*P. rouenii*" group), which is characterized by slender and quite elongated limb bones. *P. coelophrys*, *P. expectans*, *P. borissiaki*, *P. hoffstetteri*, *P. moldavicus*, *P. asiaticus,* and *P. berislavicus* on the other hand are included in the large-sized group ("*P. coelophrys*" group), having shorter but more robust limb bones.

★*Palaeotragus rouenii* Gaudry, 1861

Type Material MNHN.F.PIK1670 (Fig. 4), almost complete skull with dentition, Muséum national d'Histoire naturelle, Paris (Pl. 1, Tabs. 1–5 in Gaudry 1861).

Type Locality Pikermi, Attiki, Greece, late Miocene, Turolian, MN12.

Distribution Late Miocene (Turolian, MN11–13) of Greece, Republic of North Macedonia, Bulgaria, Turkey, Moldova, Ukraine, and Afghanistan. In Greece, it has been found in Turolian (MN11–13) localities of Axios Valley (Ravin des Zouaves 5, Dytiko 2), Kryopigi (Chaldidiki), Thermopigi (Strymon valley), Perivolaki (Magnesia), Kerassia (Euboea Island), Pikermi and Chomateri (Attica), and Samos Island (Q1, Q2, QA, MGLS Stefana hill, MGLS Andrianos ravine, Mytilinii 4 MLN, Mytilinii-1A MTLA, Mytilinii-1B MTLB) (Fig. 2).

Taxonomic Remarks Small-sized palaeotragines with ossicones that are located above the orbits and are absent in females (Fig. 4), and slender and quite elongated postcranial elements (Fig. 5). Fourth lower premolar is molarized. It differs from *Palaeotragus microdon* in the shape of the ossicones, which are curved in *P. rouenii* but straight in *P. microdon*, and the presence of weak ossicones in the females in *P. microdon* (Bohlin 1926; Geraads 1974). The more robust representatives of the genus, the "*P. coelophrys*" group, have shorter but more robust elements and their teeth are significantly larger. In *B. attica,* the teeth and the toothrows are also slightly larger. The external tubercles and the stylids of the upper premolars are not so fully developed as in *B. attica* and in addition, their parastyle is not bifurcated.

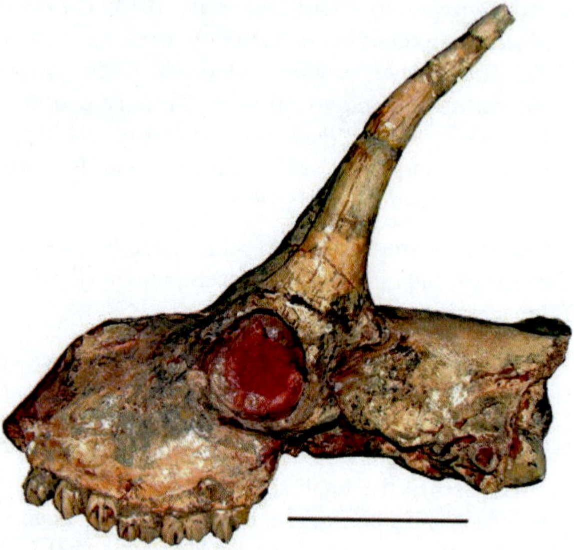

Fig. 4 Left lateral view of *Palaeotragus rouenii* cranium (MNHN.F.PIK1670) from Pikermi. Scale bar equals 10 cm

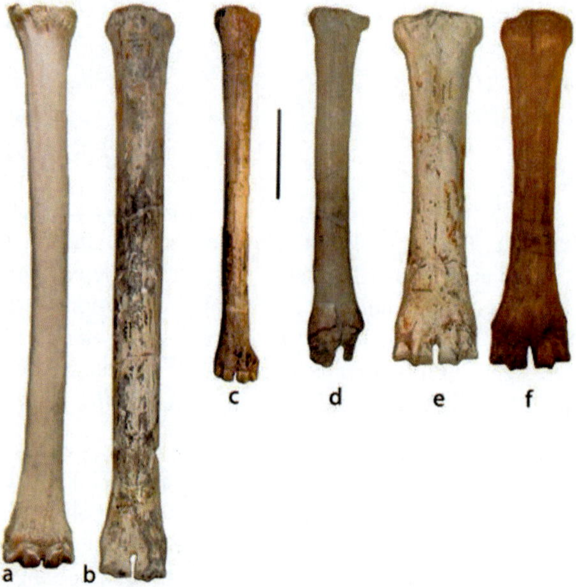

Fig. 5 Dorsal view of metacarpal bones of different Giraffidae taxa from Greek localities. (**a**) Extant *Giraffa reticulata* (NHMW 26429), (**b**) *Bohlinia attica* from Pikermi (AMPG(V) PA1923/91), (**c**) *Palaeotragus rouenii* from Pikermi (MNHN.F.PIK1692), (**d**) *Palaeotragus inexpectatus* from Volax (AMPG(V) 981), (**e**) *Helladotherium duvernoyi* from Pikermi (AMPG(V) PA1457/91), (**f**) *Samotherium major* from Samos (Andrianos) (MGL S535). Scale bar equals 10 cm

Comments *P. rouenii* is a very widespread palaeotragine in western Eurasia during the late Miocene. Based on masticatory morphology and tooth microwear, *P. rouenii* has been interpreted as a seasonal mixed feeder (Solounias et al. 1999, 2000, 2010; Merceron et al. 2018).

Palaeotragus coelophrys (Rodler and Weithofer, 1890)

Nomenclatural and Taxonomical History *Alcicephalus coelophrys* in Rodler and Weithofer 1890 (new species); *Achtiaria coelophrys* in de Mecquenem 1924 (new combination); *Palaeotragus coelophrys* in Bohlin 1926 (new combination).

Type Material Right cranium part with orbit and dentition, Naturhistorisches Museum Wien (Pl. 1, fig. 2 in Rodler and Weithofer 1890).

Type Locality Maragheh, Iran, late Miocene, Turolian, MN11.

Distribution Late Miocene (Vallesian and Turolian, MN9–13) of Turkey, Ukraine, Iraq, Iran, Georgia, and China. In Greece, it has been found in the Vallesian (MN9–10) of Axios Valley (Pentalophos 1, Ravin de la Pluie) and in the Turolian (MN11–12) of Samos Island (Qx, Q1, MGLS Andrianos ravine) (Fig. 2).

Taxonomic Remarks *P. coelophrys* is a medium-sized palaeotragine. Ossicones are absent in female skulls. The crowns of the teeth are relatively low compared to the type species (Fig. 6). Its long bones are slightly elongated and relatively robust. The morphology and the dimensions of the skull and the dentitions are quite similar with *P. expectans, P. borissiaki,* and *P. hoffstetteri* (Bohlin 1926; Alexeiev 1930; Geraads 1986). The long bones are less elongated and more robust than *P. rouenii* and *P. microdon*, and the teeth are significantly larger. On the other hand, the dimensions of its skeletal elements are clearly smaller than *S. boissieri*, which consists a much larger giraffid.

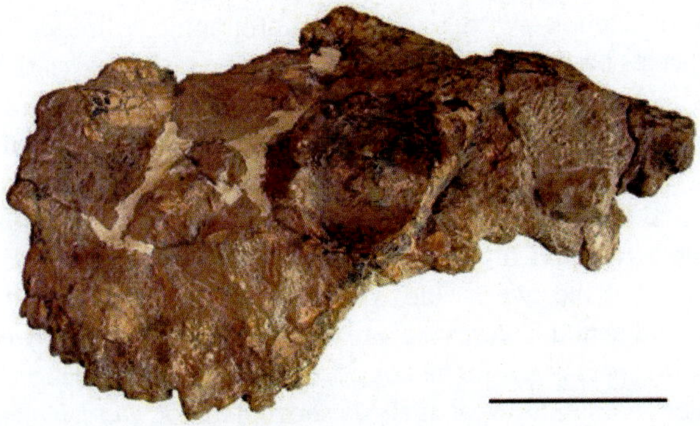

Fig. 6 Left lateral view of a female *Palaeotragus coelophrys* cranium (R.Pl 91) from Ravin de la Pluie. Scale bar equals 10 cm (photo provided by DS Kostopoulos)

Comments Geraads (1974, 1986) stated that the differences between *P. coelophrys, P. expectans, P. borissiaki*, and *P. hoffstetteri* are not significant and considered them all as synonyms of *P. coelophrys*. Tooth microwear analysis consistently showed that *P. coelophrys* was a grazer (Solounias and Moelleken 1993; Solounias et al. 1999, 2000) although more recent analyses have indicated browsing and mixed feeding diets (Danowitz et al. 2016; Merceron et al. 2018).

★***Palaeotragus quadricornis*** **Bohlin,** 1926

Type Material No catalog number, part of cranium with dentition and ossicone and related right mandible with dentition, Bayerische Staatssammlung für Paläontologie und historische Geologie-München (BSPM) (figs. 53–57 in Bohlin 1926)—permanently lost during World War II (Kostopoulos 2009).

Type Locality Samos, Greece, late Miocene, Turolian, MN12.

Distribution Besides its type locality (MN11–12), it has been found in the Turolian of Iran (MN11–13).

Taxonomic Remarks A medium-sized palaeotragine, with ossicones present in female skulls being small in size, conical, and straight. Male skulls have a subquadrangular orbit located behind the third molar, and their ossicones are long and slightly curved caudally found just above the orbit. Toothrows of *P. quadricornis* are longer than *P. coelophrys*, while ossicones are absent from the female skulls of *P. coelophrys*. The orbit of *P. coelophrys* is located in a more anterior position than *P. quadricornis*, found above M2–3 (Kostopoulos 2009). The slight but notable bilobation of P3 and P4 in *P. quadricornis* and *P. hoffstetteri* is not observed in *P. coelophrys and P. expectans*. Also, the first two species have a P2 with a paracone that differs from *P. coelophrys* and *P. expectans* in having a flat posterior flange and a reduced parastyle than a concave one.

Comments The available material is scarce and fragmentary coming from different localities of Samos—with uncertain stratigraphic affinities. Bohlin (1926) erected this species based on his observation that the taxon possessed an extra anterior pair of ossicones on the anterior margin of the orbits, characterizing as ossicone an osseous protuberance of the frontal sinuses. Several authors in the past have synonymized *P. quadricornis* with *P. coelophrys* (Bosscha-Erdbrink 1977; Geraads 1986; Gentry and Heizmann 1996). Nevertheless, Kostopoulos (2009), after studying the available material of *P. quadricornis* and *P. coelophrys*, considered that despite the fact that the two species present several common features, there are certain cranial and dental features that are different. Hence, herein we agree with his view that the two species should be considered as distinct until more suitable material is available. Tooth microwear analysis showed that *P. quadricornis* was a grazer (Solounias et al. 1999, 2000).

Palaeotragus aff. *berislavicus* Korotkevitch, 1957

Nomenclatural and Taxonomical History *Palaeotragus* cf. *rouenii* in Kostopoulos 1996 (initial identification).

Distribution Late Miocene (Vallesian, MN10) of Greece and Ukraine. In Greece it has been found in the Vallesian (MN10) of Nikiti 1 in Chalkidiki (Fig. 2).

Taxonomic Remarks *P. berislavicus* is a medium-sized palaeotragine, slightly smaller than *P. coelophrys* and *P. expectans* with a large dentition though. The diastema between the canine and p2 is short, less than the length of p2–m3. The teeth are brachyodont. Long bones are elongated and rather slender. It has a size in between *P. rouenii and P. coelophrys*, having slender metapodials with a length in between *P. rouenii and P. coelophrys* and a large skull (Laskos 2020). The skeletal elements of *P. berislavicus* are smaller than *P. expectans* (Korotkevich 1957). The length of the dentition is close to *P. coelophrys*, *P. expectans,* and *P. borissiaki.*

Comments The only locality that *P. berislavicus* had been described until recently was the Vallesian site of Berislav in Ukraine (Korotkevich 1957). Recently, Laskos (2020) studied *Palaeotragus* material from Nikiti 1 that Kostopoulos et al. (1996) had previously described as *P.* cf. *rouenii*, and assigned it to *P.* aff. *berislavicus,* based on morphological as well as metrical differences. Concerning the latter, he identified that the length of the metapodials is in between *P. rouenii* and *P. coelophrys* with values that fitted *P. berislavicus*. So he considered that the material belonged to a taxon of intermediate size. Thus, he proposed the presence of a third distinct late Miocene *Palaeotragus* group with dimensions of the limb bones in between *P. rouenii and P. coelophrys* and included *P. asiaticus* and *P. berislavicus*. Iliopoulos (2003) reported *Palaeotragus* sp. from Kerassia 4 and Xafis et al. (2019) *Palaeotragus* sp. from Thermopigi. Both agreed that the studied material belonged to a large-sized *Palaeotragus* that presented differences though from the typical *P. coelophrys*. Nevertheless, this material from both localities could possibly belong to the intermediate-sized group and could be affiliated to *P. berislavicus* as well.

Palaeotragus inexspectatus (Samson and Radulesco, 1966)

Nomenclatural and Taxonomical History *Mitilanotherium inexspectatum* Samson and Radulesco 1966 (new species); *Palaeotragus inexspectatus* Athanassiou 2014 (new combination). Considered as the senior synonym of *Macedonitherium martini* Sickenberg, 1967, *Sogdianotherium kuruksaense* Sharapov, 1974, and *Palaeotragus (Yuorlovia) priasovicus* Godina and Bajgusheva, 1985.

Type Material No 5227, ISERB, distal part of left lower M3, Institute of Speleology "Emil Racoviţă," Bucharest (Pl. 1–2 in Samson and Radulesco 1966).

Type Locality Fîntîna lui Mitilan, Oltenia, Romania, Early Pleistocene, Villafranchian, MN17.

Distribution Late Pliocene–Early Pleistocene (Villafranchian, MN16–MNQ19) of Eurasia. In Greece, it has been found in the middle Villafranchian (MN17–MNQ19) of Volax (Drama), Haliakmon Valley (Dafnero, Libakos, Haliakmon) Sesklo (Magnesia), Vatera (Lesbos Island), and Karnezeika (Argolida) (Fig. 2).

Taxonomic Remarks *P. inexspectatus* is a medium-sized and moderately brachyodont palaeotragine. It possesses simple, long and parallel ossicones, rostrally inclined and located above the orbit (Fig. 7). The orbit is found behind the M3. The parietal crests are developed. The premolar rows are relatively reduced. Its limb bones are slightly elongated and relatively robust (Fig. 5). Concerning its size this is comparable to that of the large "*P. coelophrys*" group. Unlike the typical representatives of the palaeotragines, the ossicones of which are inclined backward, in *P. inexspectatus* they are inclined rostrally (Athanassiou 2014).

Comments Athanassiou (2014) suggested that all the post-Miocene palaeotragines found in Western Eurasia belong to the genus *Palaeotragus* and more specifically should be attributed to *Palaeotragus inexspectatus*. This taxon represents the last member of the family in western Eurasia.

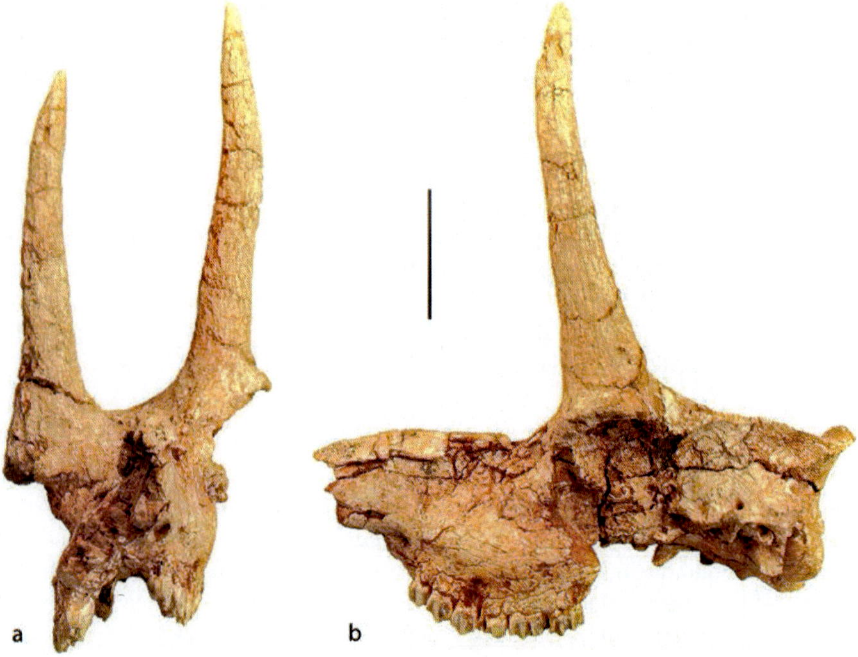

Fig. 7 Cranium of *Palaeotragus inexspectatus* (AMPG SE3–31) from Sésklo. (**a**) Rostral view, (**b**) left lateral view. Scale bar equals 10 cm (photos provided by A Athanassiou)

★*Palaeogiraffa* Bonis and Bouvrain, 2003

Type Species *Decennatherium macedoniae* Geraads, 1989.

Included Taxa *Palaeogiraffa macedoniae* (Geraads, 1989) Bonis and Bouvrain 2003; *Palaeogiraffa pamiri* (Ozansoy, 1965); *Palaeogiraffa major* Bonis and Bouvrain, 2003.

Distribution Late Miocene (Vallesian) of Greece and Turkey (Kostopoulos and Sen 2016).

Comments *Palaeogiraffa* is a genus with a very restricted distribution. Until now, the three species that have been included in the genus have been reported only from three localities in the Axios valley (Pentalophos 1, Xirochori, and Ravin de la Pluie) and two localities in Turkey (Yassiören and Sinap). Rios et al. (2016) in their phylogenetic analysis linked *Palaeogiraffa* with the samothere clade and thus herein we include this taxon in the Palaeotraginae subfamily.

★*Palaeogiraffa macedoniae* (Geraads, 1989)

Nomenclatural and Taxonomical History *Decennatherium macedoniae* Geraads, 1989 (new species); *Palaeogiraffa macedoniae* in Bonis and Bouvrain 2003 (new combination).

Type Material PNT 111, mandible, Museum of Geology, Paleontology, Palaeoanthropology, Univ. of Thessaloniki, Greece (Pl. II, fig. 1, 3 in Geraads 1989).

Type Locality Pentalophos 1, Axios Valley, Greece, late Miocene, Vallesian, MN9–10.

Distribution Besides its type locality, it is found in Turkey (Vallesian, MN9–10).

Taxonomic Remarks It is small in size, and relatively hypsodont *Palaeogiraffa*. The P3 and P4 have a sub-rectangular shape with enlarged lingual sides (Fig. 8). The styles in both molars and premolars are strong and prominent and particularly the metastyle of P3 that projects occlusally. The p4 is molarized with a continuous lingual wall. The posterior part of p3 forms a clear lobe. The canines are slightly bilobed. The limb bones are moderately elongated. Compared to the other two *Palaeogiraffa* species it is the smallest one in size. The short diastema differentiates it from other giraffids. *Samotherium* differentiates from *Palaeogiraffa* in having brachyodont teeth, a relatively long diastema, the shape of the incisors and the canines, the clearly different ulna and radius, the presence of acromion in the scapula (Geraads 1989). The morphology of p3 and particularly the non-isolated and strong mesolingual conid clearly differentiates *P. macedoniae* from *Decenatherium* and *Birgerbohlinia* (Geraads 1989; Rios et al. 2016).

Comments Geraads (1989) identified the differences between *Decenatherium* and the material from Pentalophos 1 and despite the fact that he affiliated the new spe-

Fig. 8 *Palaeogiraffa macedoniae*: (**a**) Occlusal view of left maxilla with P2–M3 (PNT 145), (**b**) Occlusal view of right mandible with p2–m3 (PNT 111, holotype), (**c**) Lingual view of right mandible with p2–m3 (PNT 111, holotype). Scale bars equal 10 cm

cies he erected to *Decenatherium,* he suggested that the material certainly deserves the erection of a new generic name. *P. macedoniae* was a browser that included also herbaceous monocotyledons in its diet as tooth microwear analysis has shown (Merceron et al. 2018).

Palaeogiraffa pamiri (Ozansoy, 1965)

Nomenclatural and Taxonomical History *Samotherium pamiri* Ozansoy, 1965 (new species); *Palaeogiraffa pamiri* in Bonis and Bouvrain (2003) (new combination).

Type Material MNHN.F.TRQ1031, Maxilla, Muséum national d'Histoire naturelle, Paris (France) (Plate VII, fig. 1 in Ozansoy 1965).

Type Locality Yassiören, Sinap, Turkey, late Miocene, Vallesian.

Distribution Late Miocene (Vallesian, MN10) of Greece and Turkey. In Greece, it has been found in the Vallesian (MN10) of Xirochori (Fig. 2).

Taxonomic Remarks A large-sized giraffid which is considered as relatively medium in size species of *Palaeogiraffa*, having very strong styles. It is larger than *P. macedoniae* and smaller than *P. major*. The lingual cingulum in the upper molars is weak or absent. The lingual wall of DP2 is not bilobed, whereas in DP3 the two lobes are not symmetrical with the anterior lobe being elongated (Fig. 9). A rounded fossa found on the posterior flange of the hypocone of the upper M2 is a feature shared by both *P. pamiri* and *P. major* and which is not present in *P. macedoniae*. The presence of a continuous eocristid in p4 differentiates *P. pamiri* from *Decenatherium* (Kostopoulos and Sen 2016). *Bohlinia attica* is smaller in size than *P. pamiri* having weaker styles and pillars in the upper molars (Bonis and Bouvrain 2003).

Fig. 9 *Palaeogiraffa pamiri*: (**a**) Occlusal view of left maxilla with DP2–M2 (XIR 131), (**b**) Occlusal view of left mandible with dp2–m1 (XIR 131), (**c**) Lingual view of left mandible with dp2–m1 (XIR 131). Scale bar equals 10 cm

Comments The *P. pamiri* material from Xirochori is rather scarce and consists of juvenile cranial material having the milk molars still attached. Tooth microwear analysis has shown that *P. pamiri* could be identified as a browser (Merceron et al. 2018).

★*Palaeogiraffa major* Bonis and Bouvrain, 2003

Nomenclatural and Taxonomical History *Decennatherium* cf. *pachecoi* Geraads 1979; *Palaeogiraffa major* in Bonis and Bouvrain 2003 (new species).

Type Material RPl 733, right Maxilla with DP2–M1, Museum of Geology, Paleontology & Palaeoanthropology of Thessaloniki University (Greece) (Plate 4 in Bonis and Bouvrain 2003).

Type Locality Ravin de la Pluie, Axios Valley, Greece, late Miocene, Vallesian, MN10.

Distribution Late Miocene (Vallesian, MN10) of Greece (Fig. 2).

Taxonomic Remarks A large-sized giraffid, being the largest of the three species of *Palaeogiraffa*, being even larger than *P. pamiri*. It possesses strong styles both in molars and milk molars. The lingual wall of DP2 is bilobed, whereas in DP3 the two lobes are not symmetrical with the anterior lobe fairly enlarged (Fig. 10). Except for its larger size, *P. major* differs from *P. pamiri* in having a bilobed DP2, a DP3 with a larger anterior lobe, low cingulums, molars that are higher, and the first lobe of dp4 is larger and more symmetrical (Bonis and Bouvrain 2003). The morphology of p3 and particularly the non-isolated and strong mesolingual conid clearly differenti-

Fig. 10 *Palaeogiraffa major*: (**a**) Occlusal view of right maxilla with DP2-M1 (RPL 733), (**b**) Buccal view of left mandible with dp3–m3 (RPL 734, holotype). Scale bar equals 10 cm

ates *P. major* from *Decenatherium* and *Birgerbohlinia* (Geraads 1989; Rios et al. 2016). *P. major* presents a larger second lobe in the deciduous canines than *Decenatherium pachecoi* (Bonis and Bouvrain 2003). *Samotherium* species possess more derived deciduous molars than the *Palaeogiraffa* species, having a much shorter DP3 and a more molarized dp3 (Bonis and Bouvrain 2003). The members of the large "*P. coelophrys*" group differentiate from *Palaeogiraffa* in having weaker styles and stylids in their upper and lower molars and deciduous molars (Bonis and Bouvrain 2003).

Comments The material is scarce and represents mainly subadult individuals. The absence of limb bones makes comparisons with other giraffid taxa rather difficult. Geraads (1979) described this material from Ravin de la Pluie as *Decennatherium* cf. *pachecoi*. Tooth microwear analysis has shown that *P. major* was a browser that included also herbaceous monocotyledons in its diet (Merceron et al. 2018).

★*Samotherium* Forsyth Major, 1888

Type Species *Samotherium boissieri* Forsyth Major, 1888.

Included Taxa *Samotherium boissieri* Forsyth Major, 1888; *Samotherium major* Bohlin, 1926; *Samotherium neumayri* (Rodler and Weithofer, 1890); *Samotherium sinense* (Schlosser, 1903); *Samotherium africanum* Churcher, 1970; *Samotherium eminens* (Alexejev, 1916) Bohlin 1926; *Samotherium mongoliense* Godina, 1954; *Samotherium maeoticum* Korotkevitch, 1978; *Samotherium borissiaki* Godina, 2002; *Samotherium irtyshense* Godina, 1962; *Samotherium korotkevichae* Godina, 2002.

Distribution Late Miocene (Vallesian and Turolian) of Algeria, Libya, Kenya, Greece, Turkey, Moldova, Ukraine, Georgia, Iran, Iraq, Kazakhstan, Kyrgyzstan, Tajikistan, Thailand, and China (NOW database).

Comments The genus includes large-sized palaeotragines.

★*Samotherium boissieri* Forsyth Major, 1888

Type Material M4215, cranium with mandible, Natural History Museum of London (NHML) (fig. 1, p. 318 in Forsyth-Major 1891).

Type Locality MGLS Stefana Hill Samos, Greece, late Miocene, Turolian, MN11.

Distribution Late Miocene (Turolian, MN11–12) of Greece, Turkey, Moldova, and Iraq. In Greece, it has been found in the Turolian (MN11–12) of Samos Island (Qx, Q2, MGLS Stefana hill, Vrysoula, Mytilinii-4 MLN) (Fig. 2).

Taxonomic Remarks *S. boissieri* is a large-sized palaeotragine, nevertheless it can be considered as a medium-sized *Samotherium*. Ossicones are absent from females and which are long and curved caudally in males, located above the orbit (Fig. 11). The orbit is placed rather high above the middle of M3. It has a relatively brachyodont dentition, nevertheless the molars have a relatively higher crown with narrow but well-marked styles. The premolars compared to the molars are smaller in size. The p3 presents a low degree of molarization. The limb bones are weakly elongated. *S. boissieri* is smaller in size than *S. major* and *S. sinense*. According to Geraads (1974), the differences of *S. boissieri* with *S. neumayri* are not quite significant, nevertheless the latter species is considered for the moment a separate taxon. *B. attica* has relatively longer premolar rows and lower tooth crowns and far longer limb bones (Geraads 1974). *Palaeotragus* has shorter premolar rows and except for size differences and the more brachyodont teeth, usually in the upper premolars the paracone is separated from the metacone (Geraads 1974).

Comments Based on tooth microwear, *S. boissieri* has been identified as a seasonal mixed feeder that included herbaceous monocotyledons in its diet (Solounias et al. 2000; Merceron et al. 2018).

Fig. 11 Right lateral view of *Samotherium boissieri* cranium (M4215, holotype) from MGLAS Stefana Hill Samos. Scale bar equals 10 cm (photo provided by DS Kostopoulos)

★*Samotherium major* Bohlin, 1926

Type Material S90 and S87, Incomplete cranium and left mandible, Senckenbergisches Naturhistorisches Museum-Frankfurt (text fig. 140, Pl. IX, figs. 8–11 in Bohlin 1926).

Type Locality Andrianos ravine Samos, Greece, late Miocene, Turolian, MN12.

Distribution Late Miocene (Turolian, MN12) of Greece and Turkey. In Greece, it has been found in the Turolian (MN12) of Samos (Mytilinii-1B, Mytilinii-1 BC, Mytilinii-3, Mytilinii-6, Q1, Q4, MGLS Andrianos ravine), Axios valley (Vathylakkos 3), Thermopigi (Strymon valley), and Kerassia 1, 3 and 4 (Euboea Island) (Fig. 2).

Taxonomic Remarks *S. major* is a *Samotherium* of great size with long and robust long bones (Fig. 5) but less robust than *H. duvernoyi* and shorter than *S. sinense*. Ossicones are absent from females and are long and inclined caudally in males at different angles, located above the posterior part of the orbit. The orbit is located behind the M3. The p3 is always molarized resembling p4 (Iliopoulos 2003) (Fig. 12). *S. major* was an intermediate-necked giraffid (Danowitz et al. 2015). The always molarized p3 of *S. major* differentiates it clearly from *S. boissieri* and *H. duvernoyi* (Geraads 1994). The fairly larger third lobe of m3 is a character that indicates *H. duvernoyi*. The postcranial elements of *S. sinense* are generally longer than *S. major* (Geraads 1994), while the upper and lower toothrows are evidently longer

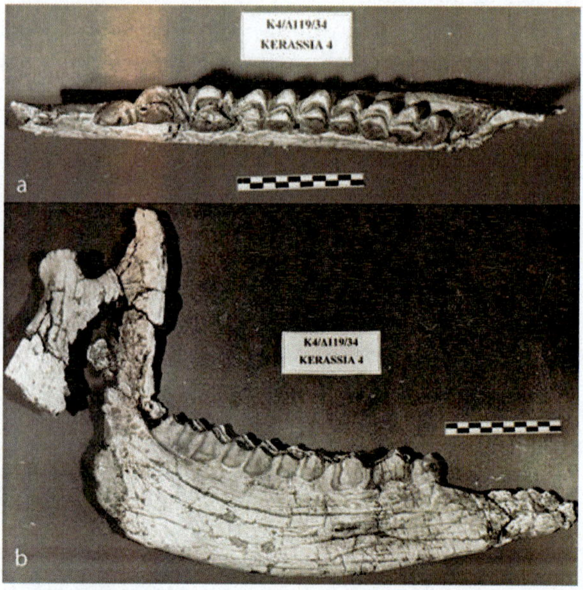

Fig. 12 Occlusal view (**a**) and buccal view (**b**) of a right mandible of *Samotherium major* (K4/Δ119/34) from Kerassia K4. Scale bar equals 10 cm

in *S. major*, as well as the diastema. *S. major* is clearly larger in size and more robust than *S. boissieri* and *S. neumayri*.

Comments *S. major* is a species with a very restricted distribution in Western Eurasia as it has been reported until now only from localities in Greece and Turkey. Both dental mesowear and tooth microwear analysis clearly indicated that *S. major* was a mixed feeder (Solounias et al. 2010; Danowitz et al. 2016).

Sivatheriinae Zittel, 1893

★*Helladotherium* Gaudry, 1860

Type Species *Camelopardalis duvernoyi* Gaudry and Lartet, 1856.

Included Taxa *Helladotherium duvernoyi*; *Helladotherium gaudryi* de Mecquenem, 1924.

Distribution Late Miocene (Vallesian and Turolian) of Greece, Turkey, Bulgaria, Republic of North Macedonia, Hungary, Moldova, Ukraine, Georgia, Iran (The NOW Community 2020), Italy, and France.

★*Helladotherium duvernoyi* (Gaudry and Lartet 1856)

Nomenclatural and Taxonomical History *Camelopardalis duvernoyi* Gaudry and Lartet, 1856 (new species); *Helladotherium duvernoyi* in Gaudry 1860 (new combination).

Type Material MNHN.F.PIK1500 (lectotype), cranium with dentition, Muséum national d'Histoire naturelle, Paris (Pl. XL, fig. 1, Pl. XLII fig. 1 in Gaudry 1862–67).

Type Locality Pikermi, Attiki, Greece, late Miocene, Turolian, MN12.

Distribution Late Miocene (Vallesian, MN10, and Turolian, MN11–13) of Greece, Turkey, Bulgaria, Republic of N. Macedonia, Hungary, Moldova, Ukraine, Georgia, Iran, and France. In Greece, it has been found in the Vallesian (MN10) of Nikiti 1 (Chalkidiki) and in the Turolian (MN11–13) of Samos, Axios valley (Ravin des Zouaves 5, Prochoma, Ravin X), Nikiti 2, Kryopigi (Chalkidiki), Ano Metochi, Thermopigi (Strymon valley), Perivolaki, Alifakas (Magnesia), Pikermi (Attiki), Halmyropotamos, Kerassia (Euboea Island) and Rhodes (Fig. 2).

Taxonomic Remarks *H. duvernoyi* is a large-sized sivatherine having long and fairly robust long bones (Fig. 5). It is a typically brachyodont taxon with relatively long premolar rows (Fig. 13). Styles and pillars of the teeth are fairly strong. The skull is fairly long and the presence of ossicones is still unknown as the only complete skull found until now in Pikermi, belonging presumably to a female individual, has no ossicones. The p3 is always not molarized. It is a very robust animal and

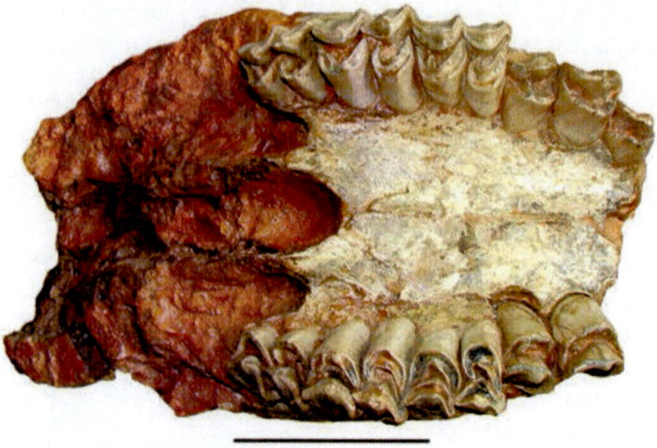

Fig. 13 Occlusal view of a *Helladotherium duvernoyi* cranium (MNHN.F.PIK1501) from Pikermi with P2–M3 on the right maxilla and P3–M3 on the left. Scale bar equals 10 cm

compared to *S. major*, an almost similar in size taxon, its postcranial elements clearly indicate a far more robust giraffid.

Comments *H. duvernoyi* is one of the most common giraffid representatives in the late Miocene faunas of western Eurasia. Postcranial elements show clear sexual dimorphism (Roussiakis and Iliopoulos 2004). Based on tooth microwear analysis *H. duvernoyi* has been identified as a browser (Solounias et al. 2000; Merceron et al. 2018).

Giraffinae Zittel, 1893

★*Bohlinia* Matthew, 1929

Type Species *Camelopardalis attica* Gaudry and Lartet, 1856.

Included Taxa *Bohlinia attica*; *Bohlinia nikitiae* Kostopoulos et al., 1996.

Distribution Late Miocene (Vallesian and Turolian) of Greece, Turkey, Bulgaria, Republic of North Macedonia, Iraq and Iran (The NOW Community 2020).

★*Bohlinia attica* (Gaudry and Lartet, 1856)

Nomenclatural and Taxonomical History *Camelopardalis attica* Gaudry and Lartet, 1856 (new species); *Orasius atticus* in Bohlin 1929 (new combination); *Bohlinia attica* in Geraads 1974 (new combination).

Type Material MNHN.F.PIK2757 (lectotype), almost complete left hind limb, Muséum national d'Histoire naturelle, Paris (France) (Plate XL, fig. 1 left, 5, 6, 8 in Gaudry 1862–67). Lectotype was designated by Geraads in 1979 (p. 378); however,

Gentry (2003) commented that the selected specimen was not part of the 1856 type series as it was collected years later, after 1860, therefore the above mentioned lectotype should be reconsidered as a neotype.

Type Locality Pikermi, Attiki, Greece, late Miocene, Turolian, MN12.

Distribution Late Miocene (Vallesian, MN10, and Turolian, MN11–13) of Greece, Turkey, Bulgaria, Republic of North Macedonia, Iraq, and Iran. In Greece, it has been found in the Vallesian (MN10) of Nikiti 1 (Chalkidiki) and Ravin de la Pluie (Axios valley) and in the Turolian (MN11–13) of Axios valley (Ravin des Zouaves 5, Vathylakkos 2, Vathylakkos 3, Dytiko 1, Dytiko 2), Pikermi, Pikermi Valley, Pyrgos Vassilissis (Attiki), and Kerassia 1 (Euboea Island) (Fig. 2).

Taxonomic Remarks *B. attica* is a giraffine of large size with a long neck and very long but slender long bones (Fig. 5). The orbit is placed above the middle of M3. At least the skulls of the males bear a pair of robust ossicones with a wide base, found above the posterior half of the orbits and extending up to the parietals (Fig. 14a). They are slightly curved and inclined caudally. Pneumatization of the frontal bone is observed which extends up to the parietal bones. The profile of the nasal bones is convex. It is a typically brachyodont taxon with comparatively small in size teeth having a low crown. The parastyle of the upper premolars is bifurcated, and the upper molars are wider than long having a weak eperon hypoconal. The upper deciduous molars are long with primitive features. In *Honnanotherium*, the ossicones are straight and rather vertical to the roof of the skull (Kostopoulos et al. 1996).

Comments Postcranial elements of *B. attica* present clearly sexual dimorphism (Roussiakis and Iliopoulos 2004). *B. attica* was characterized as a browser based on its masticatory morphology and tooth microwear analysis (Solounias et al. 1999, 2000; Merceron et al. 2018). A maxilla (CM458) from Samos that has been recently determined as *B. attica* by Parizad et al. (2019) consists probably an erroneous determination. The specimen depicted in their fig. 5 (page 28) most likely belongs to *Palaeotragus*.

★*Bohlinia nikitiae* Kostopoulos, Koliadimou and Koufos, 1996

Type Material NKT-147, cranium part with maxilla and both toothrows, Museum of Geology, Paleontology, Palaeoanthropology, Univ. of Thessaloniki, Greece (fig. 13 in Kostopoulos et al. 1996).

Type Locality Nikiti 1, Chalkidiki, late Miocene, Vallesian, MN10.

Distribution Late Miocene (Vallesian, MN10) of Greece.

Taxonomic Remarks *B. nikitiae* has a skull relatively shorter than *B. attica*. The orbit is found high far from the alveolar level. The ossicones are robust located behind the orbit toward the parietals (Fig. 14b). They are inclined and curved caudally having an elliptical to rounded transverse section, and present strong and deep

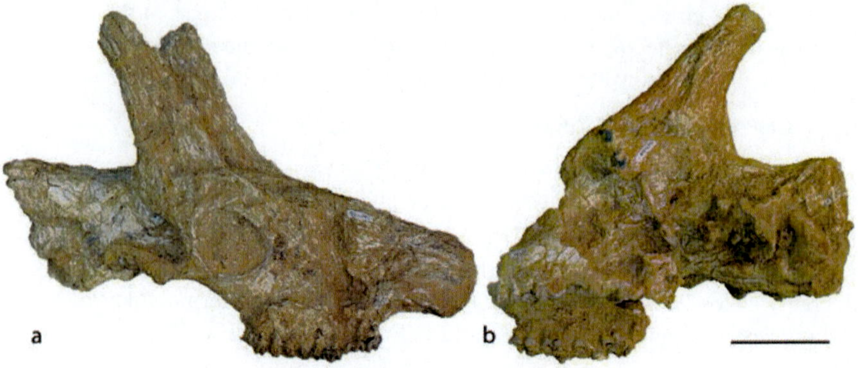

Fig. 14 (**a**) Right lateral view of a *Bohlinia attica* cranium (NKT-148) from Nikiti 1, (**b**) Left lateral view of a *Bohlinia nikitiae* cranium (NKT-147, holotype) from Nikiti 1. Scale bar equals 10 cm

grooves along the external surface. The anterior and posterior carines of the ossicones are weak. Compared to the skull the toothrow is considered long, having rather short premolars relative to the molars. According to Kostopoulos et al. (1996), *B. nikitiae* differs from *B. attica* in the position of the orbit which is placed higher above the alveolar level, the shape of the ossicones' transverse section which is elliptical at the base for *B. nikitiae* and rhomboid for *B. attica*, which becomes in the middle rounded and squarish respectively, the flat posterior end and fine and shallow grooves on the ossicones of *B. attica*, and the relatively short toothrow accompanied by a moderate to long premolar row of *B. attica*. Kostopoulos et al. (1996) observed similarities in the dentition between *B. nikitiae* and the female skull of *P.* cf. *coelophrys* from Ravin de la Pluie.

Comments *B. nikitiae* to date is known only from Nikiti 1. The material referred to this taxon consists a skull and a proximal part of a metacarpal bone. The skull presents a certain degree of damage, being distorted, the front part of the maxilla is broken, the left toothrow is incomplete and a number of teeth of the right one are broken. We believe that more material is needed, including postcranial material, to verify the attribution of this material to *B. nikitiae*, nevertheless we believe that for the time being, *B. nikitiae* should be considered a valid species.

Giraffidae indet.

Except for the determined material that has been assigned to all the taxa mentioned above, there are still a number of giraffid findings from six localities (Melampes, Fourka, Nikiti 2, Kamarella, Drazi, and Nea Silata) that has not been possible to be determined to the species or even to the genus level. Nevertheless, since this is the first comprehensive review of the Greek Giraffidae, it is considered important to discuss these findings further.

The giraffid material that was collected by Kuss in 1975 from the middle Miocene (Astaracian MN6 or MN7–8) locality of Melampes from south Crete Island consists of 5 hindlimb bone parts and a few bone fragments, including a rather well preserved diaphysis and lower epiphysis of a left metatarsal. The long bone parts are elongated and slender indicative of a small-sized giraffid that shows affinities with Palaeotraginae, and taking in mind the age of the locality this giraffid can be considered as a very primitive representative of this group. The material is reported for the first time herein and due to its fragmentary nature it is determined as Palaeotraginae indet.

Lazaridis (2015) in his PhD thesis described giraffid remains from the late Miocene (Vallesian) locality Fourka in Chalkidiki. The studied material consists of two large ossicones, a lower second molar and three long bone parts. The dimensions of the postcranial material are comparable to *S. boissieri* and smaller than those of *H. duvernoyi*. Nevertheless, the morphology of the ossicones indicates similarity with the ossicones of Sivatherinae. As no ossicones of the common west Eurasian *H. duvernoyi* have been found until now, the similarities with the uncommon sivatherine for western Eurasia *Bramatherium*, and the small number of specimens of the studied material, Lazaridis (2015) suggested that Sivatherinae indet. should be considered a suitable assignment for the time being.

Among the giraffid material from the late Miocene (Turolian MN11) locality Nikiti 2 in Chalkidiki, two parts of first phalanges and two juvenile second phalanges have been referred by Kostopoulos (2016) as Palaeotraginae indet., as they are larger than *P. rouenii*, with dimensions similar to *B. attica*, but with a morphology resembling the phalanges of palaeotragines.

Verikiou-Papaspyridakou (1986) in her PhD thesis reported the presence of a left incomplete astragalus belonging to a large giraffid, which she found in a karstic cavity located in Paleocene limestones at the locality Kamarella in Antikythera Island. Verikiou-Papaspyridakou (1986) suggests that this astragalus has been reworked from older deposits. The specimen had been originally determined by E. Thenius as *Camelopardalis* sp. or its synonym *Giraffa* sp. Herein, as the specimen is only known to us by a photograph of the anterior view of the astragalus as it is presented in the PhD thesis of Verikiou-Papaspyridakou (1986), we consider that the specimen should be referred as Giraffidae indet.

In 1947, Mitzopoulos presented in his study on the localities of Euboea Island with Pikermian fauna, fossil mammal bones from the late Miocene (Turolian MN11–13) locality of Drazi that had been collected by A.S.Woodward in 1901. Among the material there were some giraffid remains which Woodward suggested that they should belong to a form related to *Samotherium*. However, due to the fragmented character of the material we consider that this material should be referred as Giraffidae indet. as well.

In 2006, Koufos described a distal part of a metacarpal bone with only the internal condyle still attached on from the latest Miocene (Miocene/Pliocene boundary, late Turolian–early Ruscinian, MN13–14) locality of Nea Silata in Chalkidiki. Koufos (2006) compared the specimen with *Helladotherium* and *Samotherium* metacarpals and he suggested that metrically it could belong either to a *Helladotherium* or a *Samotherium*. Therefore, we consider more appropriate for the specimen to be referred as a large Giraffidae indet.

Acknowledgments We would like to thank Dimitris Kostopoulos and Athanassios Athanassiou for providing the photographs for some of the species mentioned in the text. We would like to thank A. Xafis for valuable comments that helped improving this manuscript.

Appendix

Table with a detail list of the occurrences of giraffid fossils in Greece. Type localities are marked with bold. Locality numbers refer to the collection numbers of the PaleoBiology Database (PBDB)

Localities[PBDB No]	Age (MN)	Taxon
Karnezeika[202122]	Early Pleistocene, Villafranchian (MN17–MNQ18)	*Palaeotragus inexspectatus*[a]
Haliakmon	Early Pleistocene, Villafranchian (MN17)	*Palaeotragus inexspectatus*[1,2]
Libakos[34764]	Early Pleistocene, Villafranchian (MN17–MNQ19)	*Palaeotragus inexspectatus*[1,2]
Vatera F[183341]	Early Pleistocene, Villafranchian (MN17)	*Palaeotragus inexspectatus*[3,2]
Volax[34593]	Early Pleistocene, Villafranchian (MN17)	*Palaeotragus inexspectatus*[4,2]
Sesklon[34614]	Early Pleistocene, Villafranchian (MN17)	*Palaeotragus inexspectatus*[2]
Dafnero[34594]	Early Pleistocene, Villafranchian (MN17)	*Palaeotragus inexspectatus*[5]
Nea Silata[191612]	Late Miocene, Late Turolian–Early Ruscinian (MN13/14)	Large Giraffidae indet.[6]
Maramena[32189]	Late Miocene, Late Turolian–Early Ruscinian (MN13/14)	*Samotherium* sp.[7]
Rhodes	Late Miocene, Turolian (MN9–13)	*Helladotherium duvernoyi*[8,9]
Alifakas[207134]	Late Miocene, Turolian (MN11–13)	*Helladotherium duvernoyi*[10]
Drazi, Euboea	Late Miocene, Turolian (MN11–13)	Giraffidae indet.[11]
Ano Metochi-2[31924], 3[31928]	Late Miocene, Turolian (MN13)	*Helladotherium* cf. *duvernoyi*[12]
Samos Q5[95690]	Late Miocene, Turolian (MN13)	*Palaeotragus rouenii*[13], *Samotherium major*[13], *Helladotherium duvernoyi*[13]
Dytiko 2[32375]	Late Miocene, Turolian (MN13)	*Palaeotragus rouenii*[14,15] *Bohlinia attica*[14,15]
Dytiko 1[32374]	Late Miocene, Turolian (MN13)	*Bohlinia attica*[15]
Kryopigi[157582]	Late Miocene, Turolian (MN12–13)	*Palaeotragus rouenii*[16], *Helladotherium duvernoyi*[16]
Kamarella, Antikythera	Late Miocene, Turolian	Giraffidae indet.[17]

(continued)

Localities[PBDB No]	Age (MN)	Taxon
Pyrgos Vassilissis[195555]	Late Miocene, Turolian (MN12)	cf. *Palaeotragus* sp.[18], *Bohlinia attica*[18]
Thermopigi[73553]	Late Miocene, Turolian (MN12)	*Palaeotragus rouenii*[19], *Palaeotragus* sp.[19], *Samotherium major*[19], *Helladotherium duvernoyi*[19]
Kerassia 1 Upper[195432]	Late Miocene, Turolian (MN12)	*Palaeotragus rouenii*[20], *Samotherium major*[20], *Helladotherium duvernoyi*[20], *Bohlinia attica*[20]
Kerassia 3[195434], 4 Lower[195435]	Late Miocene, Turolian (MN12)	*Palaeotragus rouenii*[20], *Palaeotragus* sp.[20], *Samotherium major*[20], *Helladotherium duvernoyi*[20]
Pikermi Valley (PV1) [202630]	Late Miocene, Turolian (MN12)	*Bohlinia* cf. *attica*[21]
Pikermi[182754]	Late Miocene, Turolian (MN12)	***Palaeotragus rouenii*[23,24], *Helladotherium duvernoyi*[22], *Bohlinia attica*[22]**
Chomateri[195562]	Late Miocene, Turolian (MN12)	*Palaeotragus rouenii*[25]
Halmyropotamos[202213]	Late Miocene, Turolian (MN12)	*Helladotherium duvernoyi*[26]
Perivolaki[194879]	Late Miocene, Turolian (MN12)	*Palaeotragus rouenii*[27], *Helladotherium duvernoyi*[27]
Samos[207137]	Late Miocene, Turolian (MN12)	***Palaeotragus quadricornis*[28,13]**
MGLS Andrianos ravine, Samos	Late Miocene, Turolian (MN12)	*Palaeotragus coelophrys*[13], *Palaeotragus rouenii*[13], ***Samotherium major*[28]**, *Helladotherium duvernoyi*[13]
Samos Q4[95689]	Late Miocene, Turolian (MN12)	*Samotherium major*[13]
Samos QA[206462]	Late Miocene, Turolian (MN12)	*Palaeotragus rouenii*[13]
Samos Q1[95691]	Late Miocene, Turolian (MN12)	*Palaeotragus coelophrys*[13], *Palaeotragus rouenii*[13], *Samotherium major*[13], *Helladotherium duvernoyi*[13]
Mytilinii 1C[202217]	Late Miocene, Turolian (MN12)	*Samotherium major*[13]
Mytilinii 1B[202216]	Late Miocene, Turolian (MN12)	*Palaeotragus rouenii*[13] *Palaeotragus* sp.[13], *Samotherium major*[13]
Mytilinii 1A[202215]	Late Miocene, Turolian (MN12)	*Palaeotragus rouenii*[13], *Samotherium major*[13], *Helladotherium duvernoyi*[13]
Mytilinii 6	Late Miocene, Turolian (MN12)	*Samotherium major*[13]
Mytilinii 3[202218]	Late Miocene, Turolian (MN12)	*Samotherium major*[13]
Mytilinii 4[202219]	Late Miocene, Turolian (MN12)	*Palaeotragus* sp.[13], *Palaeotragus rouenii*[13], *Samotherium boissieri*[13]
Ravin X[182745]	Late Miocene, Turolian (MN12)	? *Helladotherium duvernoyi*[29]
Prochoma[202222]	Late Miocene, Turolian (MN12)	*Helladotherium duvernoyi*[30]

(continued)

Localities[PBDB No]	Age (MN)	Taxon
Vathylakkos 3[182750]	Late Miocene, Turolian (MN12)	*Samotherium major*[14,15], *Bohlinia attica*[14,15]
Vathylakkos 2[202703]	Late Miocene, Turolian (MN12)	*Bohlinia attica*[15]
Vrysoula, Samos	Late Miocene, Turolian (MN11)	*Samotherium boissieri*[13]
MGLS Stefana hill, Samos	Late Miocene, Turolian (MN11)	*Palaeotragus rouenii*[13], ***Samotherium boissieri***[31,13]
Samos Q2[206460]	Late Miocene, Turolian (MN11)	*Palaeotragus rouenii*[13], *Samotherium boissieri*[13]
Samos Q6[206462]	Late Miocene, Turolian (MN11)	*Samotherium* sp.[32]
Samos Qx[211913]	Late Miocene, Turolian (MN11)	*Samotherium boissieri*[13,32], *Palaeotragus coelophrys*[13,32], *Palaeotragus* sp.[13,32]
Ravin des Zouaves 5[195489]	Late Miocene, Turolian (MN11)	*Palaeotragus rouenii*[33], *Helladotherium duvernoyi*[33], *Bohlinia attica*[33]
Nikiti 2[73869]	Late Miocene, Turolian (MN11)	*Helladotherium duvernoyi*[34], *Palaeotragus rouenii*[34], Palaeotraginae indet.[34]
Nikiti 1[202729]	Late Miocene, Vallesian (MN10)	*Palaeotragus* aff. *berislavicus*[36], *Helladotherium duvernoyi*[35,36], *Bohlinia attica*[35,36], ***Bohlinia nikitiae***[35,36]
Xirochori 1[195490]	Late Miocene, Vallesian (MN10)	*Palaeotragus* sp.[36,37], *Palaeogiraffa pamiri*[36,37]
Ravin de la Pluie[191070]	Late Miocene, Vallesian (MN10)	*Palaeotragus coelophrys*[14,15,37], ***Palaeogiraffa major***[14,15,37], *Bohlinia attica*[14,15,37]
Fourka[202330]	Late Miocene, Vallesian	Sivatherinae indet.[16]
Pentalophos 1[202119]	Late Miocene, Vallesian (MN9–10)	*Palaeotragus coelophrys*[36,38], ***Palaeogiraffa macedoniae***[36,38]
Melampes[208183]	Middle Miocene, Astaracian (MN6 or MN7–8)	Palaeotraginae indet[b]
Thymiana B [202728]	Middle Miocene, Late Orleanian (MN5)	***Georgiomeryx georgalasi***[39,40]

MN Mammal Neogene

[1]Steensma (1988), [2]Athanassiou (2014), [3]de Vos et al. (2002), [4]Sickenberg (1967), [5]Kostopoulos (1996), [6]Koufos 2006, [7]Köhler et al. (1995), [8]Boni (1943), [9]de Bruijn (1976), [10]Melentis and Schneider (1966), [11]Mitzopoulos (1947), [12]de Bruijn (1989), [13]Kostopoulos (2009), [14]Geraads (1978), [15]Geraads (1979), [16]Lazaridis (2015), [17]Verikiou-Papaspyridakou (Verikiou-Papaspyridakou 1986), [18]Böhme et al. (2017), [19]Xafis et al. (2019), [20]Iliopoulos (2003), [21]Theodorou et al. (2010), [22]Gaudry and Lartet (1856), [23]Gaudry (1860), [24]Gaudry (1861), [25]de Bruijn et al. (1999), [26]Melentis (1970), [27]Kostopoulos and Koufos (2006), [28]Bohlin (1926), [29]Arambourg and Piveteau (1929), [30]Bonis et al. (1988), [31]Forsyth Major (1888), [32]Solounias (1981), [33]Koufos (2006), [34]Kostopoulos (2016), [35]Kostopoulos et al. (1996), [36]Laskos (2020), [37]Bonis and Bouvrain (2003), [38]Geraads (1989), [39]Paraskevaidis (1940), [40]Bonis et al. (1997)

[a]Sianis pers. com

[b]This study

References

Alexeiev A (1930) Die obersarmatische Fauna von Eldar. I. *Achtiaria borissiaki* nov. sp. Trav Mus Geol Acad Sci URSS 8:167–204

Alexejev A (1916) Animaux fossiles du village Novo-Elizavetovka. Tipografiya "Technik". pp 453

Arambourg C (1959) Vertébrés continentaux du Miocène supérieur de l'Afrique du Nord. Mém Serv Carte Géol Algérie, (n.s.). Paléontology 4:1–161

Arambourg C, Piveteau J (1929) Les Vertébrés du Pontien de Salonique. Ann Paléontol 18:59–138

Athanassiou A (2014) New giraffid (Artiodactyla) material from the lower Pleistocene locality of Sésklo (SE Thessaly, Greece): evidence for an extension of the genus *Palaeotragus* into the Pleistocene. Zitteliana B 32:71–89

Bohlin B (1926) Die Familie Giraffidae. Palaeontologica Sinica, Series C 4:178

Böhme M, Spassov N, Ebner M, Geraads D, Hristova L, Kirscher U, Kötter S, Linnemann U, Prieto J, Roussiakis S, Theodorou G, Uhlig G, Winklhofer M (2017) Messinian age and savannah environment of the possible hominin *Graecopithecus* from Europe. PLoS One 12(5):e0177347

Boni A (1943) Fauna di *Hipparion* a Rodi. Palaeontographia It 41:23–36

Bonis LD, Bouvrain G (2003) Nouveaux Giraffidae du Miocène supérieur de Macédoine (Grèce). In: Petculescu A, Stiuca E (eds) Advances in vertebrate paleontology "hen to Panta". Romanian Academy, Bucharest, pp 5–16

Bonis LD, Bouvrain G, Koufos GD (1988) Late Miocene mammal localities of the lower Axios valley (Macedonia, Greece) and their stratigraphical significance. Modern Geol 13:141–147

Bonis LD, Koufos GD, Sen S (1997) A giraffid from the middle Miocene of the island of Chios, Greece. Palaeontology 40:121–134

Borissiak A (1914) Mammifères fossiles de Sébastopol. I: Mem Com Géol 87:105–154

Bosscha-Erdbrink DP (1977) On the distribution in time and space of three giraffid genera with Turolian representatives at Maragheh in N.W. Iran. Proc K Neder Akad Weten B 80(5):337–355

Brisson MJ (1762) Regnum animale, in classes IX distributum sive synopsis methodical. Theodorum Haak, Leiden, 294 pp

Churcher CS (1970) Two new upper Miocene giraffids from fort Ternan, Kenya, East Africa: *Palaeotragus primaevus* n sp and *Samotherium africanum* n sp. Fossil Vertebrates of Africa 2:1–106

Churcher CS (1978) Giraffidae. In: Maglio VJ, Cooke HBS (eds) Evolution of African Mammals. Princeton, Princeton University Press, pp 509–535

Churcher CS (1990) Cranial appendages of Giraffoidea. In: Bubenik GA, Bubenik AB (eds) Horns, pronghorns, and antlers: evolution, morphology, physiology, and social significance. Springer, New York, pp 180–196

Colbert EH (1936) *Palaeotragus* from the Tung Gur formation of Mongolia. Am Mus Novit 874:1–17

Crusafont-Pairó M (1979) Les Girafidés des gisements du Bled Douarah (W. de Gafsa, Tunisie). Not Service Geol 44:7–75

Danowitz M, Domalski R, Solounias N (2015) The cervical anatomy of *Samotherium*, an intermediate-necked giraffid. R Soc Open Sci 2(11):150521

Danowitz M, Sukuan H, Mihlbachler M, Hastings V, Solounias N (2016) A combined-mesowear analysis of late Miocene giraffids from north Chinese and Greek localities of the Pikermian biome. Palaeogeogr Palaeoclimatol Palaeoecol 449:194–204

de Bruijn H (1976) Vallesian and Turolian rodents from Biotia, Attica and Rhodes (Greece) I and II. Proc K Neder Akad Weten, Ser B 79:361–384

de Bruijn H (1989) Smaller mammals from the upper Miocene and lower Pliocene of the Strimon basin, Greece. Rodentia and Lagomorpha. Boll Soc Paleontol It 28:189–195

de Bruijn H, Saraç G, van den Hoek Ostende LW, Roussiakis S (1999) The status of the genus name *Parapodemus* Schaub, 1938; new data bearing on an old controversy. DEINSEA 7:95–112

de Mecquenem R (1924) Contribution à l'étude des fossiles de Maragha. Ann Paléontol 13:135–160 de Vos J, van der Made J, Athanassiou A, Lyras G, Sondaar P, Dermitzakis M (2002) Preliminary note on the late Pliocene fauna from Vatera (Lesvos, Greece). Ann Géol Pays Helléniques 39:37–69

Duvernoy M (1854) Sur des ossements de mammifères fossiles découverts à Pikermi, village près d'Athènes, au pied du mont Pentélique. C R Acad Sci, Paris 38:251–257

Fennessy J, Bidon T, Reuss F, Kumar V, Elkan P, Nilsson MA, Vamberger M, Fritz U, Janke A (2016) Multi-locus analyses reveal four giraffe species instead of one. Curr Biol 26(18):2543–2549

Forsyth Major CI (1888) Sur un gisement d'ossements fossiles dans l'ile de Samos, contemporain de l'âge de Pikermi. C R Acad Sci, Paris 107:1178–1182

Forsyth-Major CJ (1891) On the fossil remains of species of the family Giraffidae. Proc Zool Soc London 59:315–326

Gaudry A (1860) Résultats des fouilles exécutées en Grèce sous les auspices de l'Académie. C R Acad Sci, Paris 51:802–804

Gaudry A (1861) Note sur la Girafe et l' Helladotherium trouvés à Pikermi (Grèce). Bull Soc Géol Fr, Ser. 2 18:587–598

Gaudry A (1862–67) Animaux Fossiles et Géologie de l'Attique. Recherches faites en 1855–1856, 1860. Savy F, Paris

Gaudry A, Lartet E (1856) Sur les résultats de recherches paléontologiques entreprises dans l'Attique sous les auspices de l' Académie. C R Acad Sci, Paris 43:271–274

Gentry AW (2003) Ruminantia (Artiodactyla). In: Fortelius M, Kappelman J, Sen S, Bernor RL (eds) Geology and paleontology of the Miocene Sinap formation, Turkey. Columbia University Press, New York, pp 332–379

Gentry A, Heizmann EPJ (1996) Miocene ruminants of the central and eastern Tethys and Paratethys. In: Bernor RL, Fahlbusch V, Mittmann HW (eds) The evolution of Western Eurasian Neogene mammal faunas. Columbia University Press, New York, pp 378–391

Geraads D (1974) Les giraffidés du Miocène Supérieur de la région de Thessalonique (Grece) [PhD thesis]; Université de Paris VI, Paris

Geraads D (1978) Les Palaeotraginae (Giraffidae, Mammalia) du Miocène supérieur de la région de Thessalonique (Grèce). Géol Méditerranéenne 5:269–276

Geraads D (1979) Le Giraffinae (Artiodactyla, Mammalia) du Miocène supérieur de la région de Thessalonique (Grèce). Bull Mus National d'Hist Nat Paris 4e ser C 1(4):377–389

Geraads D (1986) Remarques sur la systématique et la phylogénie des Giraffidae (Artiodactyla, Mammalia). Geobios 19(4):465–477

Geraads D (1989) Un nouveau giraffidé du Miocène supérieur de Macédoine (Grèce). Bull Mus National d'Hist Nat Paris 4e Ser 11:189–199

Geraads D (1994) Les gisements de mammifères du Miocène supérieur de Kemiklitepe, Turquie. VIII: Giraffidae. Bull Mus National d'Hist Nat Paris 4e Ser C 16(1):159–173

Godina AY (1954) A new extinct giraffe from Mongolia. Tr Paleontol Inst (Moscow) 47:172–180

Godina AY (1962) A new *Samotherium* species from Kazakhstan. Paleontol Zh 1:131–139

Godina AY (1975) *Palaeotragus* from Neogene deposits of Mongolia and middle Asia. In: Academia Nauk USSR (ed) The fossilized fauna and flora of Mongolia. Transactions of unified Soviet-Mongolian Palaeontology, vol 2, pp 67–76

Godina AY (1979) History of fossil giraffes of the genus *Palaeotragus*. Trudy Paleontol Inst Akad Nauk USSR 177

Godina AY (2002) On the taxonomy and evolution of *Samotherium* (Giraffidae, Artiodactyla). Paleontol J 36:395–402

Godina AY, Bajgusheva VS (1985) A new *Palaeotragus* species from the late Pliocene of the peri-Azov region. Paleontol Zh 3:84–89

Gray JE (1821) On the natural arrangement of vertebrate animals. London Medical Repository 15:296–310

Hamilton WR (1978) Fossil giraffes from the Miocene of Africa and a revision of the phylogeny of the Giraffoidea. Phil Trans R Soc London B 283:165–229

Harris J, Solounias N, Geraads D (2010) Giraffoidea. In: Werdelin L, Sanders WJ (eds) Cenozoic mammals of Africa. University of California Press, Berkley, CA, pp 797–811

Heintz E (1976) Les Giraffidae (Artiodactyla, Mammalia) du Miocène de Beni Mellal, Marocco. Geologie Méditerranéenne 3:91–104

Iliopoulos G (2003, Published 2015) the Giraffidae (Mammalia, Artiodactyla) and the study of the histology and chemistry of fossil mammal bone from the late Miocene of Kerassia (Euboea Island, Greece). University of Leicester, UK. ProQuest LLC, UMI195587, Ann Arbor

Janis CM, Scott KM (1987) The interrelationships of higher ruminant families: with special emphasis on the members of the Cervoidea. Am Mus Novit 2893:1–85

Köhler M, Moyà-Solà S, Morales J (1995) The vertebrate locality Maramena (Macedonia, Greece) at the Turolian-Ruscinian boundary (Neogene), 15. Bovidae and Giraffidae (Artiodactyla, Mammalia). Münch Geowiss Abh A Geol Paläontol 28:167–180

Koken E (1885) Über fossile Saugetiere aus China. Paläeontologische Abhand Bd III 2:85

Korotkevich EL (1957) Giraffes in Hipparion fauna of Berislav. Trudy. Zool Inst Acad Nauk USSR 14:129–140

Korotkevitch EL (1978) A new *Samotherium* species from the Meotian of the northern part of the Black Sea region. Vestnik Zoologii 4:9–18

Kostopoulos DS (1996) The Plio-Pleistocene Artiodactyls of Macedonia (N. Greece). Systematics, Palaeoecology, Biochronology, Biostratigraphy. Unpublished PhD Thesis, Aristotle University of Thessaloniki, Greece

Kostopoulos DS (2009) The late Miocene mammal faunas of the Mytilinii Basin, Samos Island, Greece: new collection: 13. Giraffidae. In: Koufos GD, Nagel D (eds) the late Miocene mammal faunas of Samos. Beit Paläontol 31:299–343

Kostopoulos DS (2016) Artiodactyla. In: Koufos GD, Kostopoulos DS (eds) Palaeontology of the upper Miocene vertebrate localities of Nikiti (Chalkidiki peninsula, Macedonia, Greece). Geobios 49(1–2):119–134

Kostopoulos DS, Koufos GD (2006) The late Miocene vertebrate fauna from Perivolaki (Thessaly, Greece). 8. Giraffidae. Palaeontographica, Abt A 276:135–149

Kostopoulos DS, Saraç G (2005) The late Miocene mammal locality of Akkaşdaği (Central Anatolia, Turkey). Giraffidae Geodiversitas 27(4):735–745

Kostopoulos DS, Sen S (2016) Suidae, Tragulidae, Giraffidae, and Bovidae, in Sen S. (ed.), late Miocene mam-mal locality of Küçükçekmece, European Turkey. Geodiversitas 38(2):273–298

Kostopoulos DS, Koliadimou K, Koufos GD (1996) The Giraffids (Mammalia, Artiodactyla) from the late Miocene mammalian localities of Nikiti (Macedonia, Greece). Palaeontographica Abt A 239:61–88

Koufos GD (2006) The Neogene mammal localities of Greece: faunas, chronology and biostratigraphy. Hellenic J Geosci 41:183–214

Lankester ER (1907) The origin of the lateral horns of the giraffe in foetal life on the area of the parietal bones. Proc Zool Soc London 1907:100–115

Laskos K (2020) The Vallesian large-sized *Palaeotragus* (Giraffidae, Mammalia) from northern Greece. Master thesis, interinstitutional program of postgraduate studies in Palaeontology-Geobiology. School of Geology, Aristotle University of Thessaloniki

Lazaridis G (2015) Study of the Late Miocene Vertebrate Locality of Kryopigi and Other Localities of Kassandra Peninsula, Chalkidiki (Greece). Systematics, Taphonomy, Paleoecology, Biochronology. PhD Thesis, Aristotle University of Thessaloniki. Scientific Annals of the School of Geology 174:350

Lebatard AE, Alçiçek MC, Rochette P, Khatib S, Vialet A, Boulbes N, Bourlès DL, Demory F, Guipert G, Mayda S, Titov VV, Vidal L, de Lumley H (2014) Dating the *Homo erectus* bearing travertine from Kocabaş (Denizli, Turkey) at least 1.1 ma. Earth Planet Sci Lett 390:8–18

Matthew WD (1929) Critical observations upon Siwalik mammals. Bull Am Mus Nat Hist 56:437–560

Melentis JK (1970) Die Pikermifauna von Halmyropotamos (Euböa, Griechenland)—1. Teil: Odontologie und Kraniologie. Ann Géol Pays Helléniques 24:283–411

Melentis JK, Schneider H (1966) Eine neue Pikermifauna in der Niihe der Ortschaft Alifaka in Thessalien. Ann Géol Pays Helléniques 17:267–288
Merceron G, Colyn M, Geraads D (2018) Browsing and non-browsing extant and extinct giraffids: evidence from dental microwear textural analysis. Palaeogeogr Palaeoclimat Palaeoecol 505:128–139
Mitzopoulos MK (1947) Die Verbreitung der Pikermistufe auf der Insel Euboea. Ann Géol. Pays Helléniques 12:209–216
Ozansoy F (1965) Étude des gisements continentaux et des mammifères du Cénozoïque de Turquie. Mem Soc Géol. Fr N Ser 44:1–92
Paraskevaidis I (1940) Eine obermiocäne Fauna von Chios. N Jahrb Mineral Geol Paläontol 83:363–442
Parizad E, Ataabadi MM, Mashkour M, Solounias N (2019) First giraffid skulls (*Bohlinia attica*) from the late Miocene Maragheh fauna, Northwest Iran. Geobios 53:23–34
Pavlow M (1913) Mammifères tertiaires de la Nouvelle Russie. 1re partie. N Mem Soc imp Nat Moscou 17(3):1–72
Petzold A, Hassanin A (2020) A comparative approach for species delimitation based on multiple methods of multi-locus DNA sequence analysis: a case study of the genus Giraffa (Mammalia, Cetartiodactyla). PLoS One 15(2):e0217956
Petzold A, Magnant AS, Edderai D, Chardonnet B, Rigoulet J, Saint-Jalme M, Hassanin A (2020) First insights into past biodiversity of giraffes based on mitochondrial sequences from museum specimens. Eur J Taxon 703:1–33
Pilgrim GE (1911) The fossil Giraffidae of India. Memoirs of the Geological Survey of India 4:1–29
Prothero DR, Schoch RM (2002) Horns, tusks, and flippers: the evolution of hoofed mammals. The Johns Hopkins University Press, Baltimore, MD
Rios M, Sanchez IM, Morales J (2016) Comparative anatomy, phylogeny, and systematics of the Miocene giraffid Decennatheriumpachecoi Crusafont, 1952 (Mammalia, Ruminantia, Pecora): state of the art. J Vertebr Paleontol 36(5):e1187624
Rios M, Sanchez IM, Morales J (2017) A new giraffid (Mammalia, Ruminantia, Pecora) from the late Miocene of Spain, and the evolution of the sivathere-samothere lineage. PLoS One 12(11):e0185139
Rodler A, Weithofer KA (1890) Die Wiederkäuer der Fauna von Maragha. Denks Math Naturwissen Cl K Akad Wissen 57:753–771
Roussiakis S, Iliopoulos G (2004) Preliminary observations on the metrical variation of *Helladotherium duvernoyi* and *Bohlinia attica*. In: 5th International Symposium on Eastern Mediterranean Geology, Thessaloniki, pp 343–346
Samson P, Radulesco C (1966) Sur la présence des Girafidés dans le Villafranchien supérieur de Roumanie. N Jahrb Geol Paläontol Monats 10:588–594
Sánchez IM, Cantalapiedra JL, Ríos M, Quiralte V, Morales J (2015) Systematics and evolution of the Miocene three-horned palaeomerycid ruminants (Mammalia, Cetartiodactyla). PLoS One 10(12):e0143034
Schlosser M (1903) Die fossilen Säugetiere Chinas nebst einer odontographie der rezenten Antilopens. Abh Bayer Akad Wiss Math-Phys Kl 22:220 pp
Sharapov S (1974) *Sogdianotherium*—a new genus of the family Giraffidae from the upper Pliocene of Tadzhikistan. Paleontol J 8(4):517–521
Sickenberg O (1967) Die unterpleistozäne Fauna von Wolaks (Griech.-Mazedonien): I. Eine neue Giraffe (*Macedonitherium martinii* nov. gen. nov. spec.) aus dem Unteren Pleistozän von Griechenland. Ann Géol Pays Helléniques 18:314–330
Solounias N (1981) The Turolian fauna from the island of Samos, Greece, with special emphasis on the Hyaenids and the Bovids. Contributions to Vertebrate Evolution 6:1–232
Solounias N (1988) Prevalence of ossicones in Giraffidae (Artiodactyla, Mammalia). J Mammal 69:845–848

Solounias N (2007) Family Giraffidae. In: Prothero DR, Foss SE (eds) The evolution of artiodactyls. Johns Hopkins University Press, Baltimore, MD, pp 257–277

Solounias N, Moelleken SMC (1993) Dietary adaptation of some extinct ruminants determined by premaxillary shape. J Mammal 74:1059–1074

Solounias N, Plavcan JM, Quade J, Witmer L (1999) The paleoecology of the Pikermian biome and the savanna myth. In: Agusti J, Rook L, Andrews P (eds) The evolution of Neogene terrestrial ecosystems in Europe. Hominoid evolution and climatic change in Europe, vol 1. Cambridge, Cambridge University Press, pp 436–453

Solounias N, McGraw WS, Hayek LAC, Werdelin L (2000) The paleodiet of the Giraffidae. In: Vrba ES, Schaller GB (eds) Antelopes, deer, and relatives. Yale University Press, New Haven, CT, pp 84–95

Solounias N, Rivals F, Semprebon GM (2010) Dietary interpretation and paleoecology of herbivores from Pikermi and Samos (late Miocene of Greece). Paleobiol 36:113–136

Steensma KJ (1988) Plio-/Pleistozäne Großsäugetiere (Mammalia) aus dem Becken von Kastoria/Grevena, südlich von Neapolis—NW Griechenland. Inaugural-Dissertation, Technische Universität Clausthal

Tang Y, Ji H (1983) A Pliocene–Pleistocene transitional fauna from Yuxian, northern Hebei. Vertebrata PalAsiatica 21(3):245–254

The NOW Community (2020) New and Old Worlds Database of Fossil Mammals (NOW). Licensed under CC BY 4.0. Release [09/2017]. http://www.helsinki.fi/science/now/. Accessed 16 Jun 2020

Theodorou GE, Roussiakis S, Athanassiou A, Filippidi A (2010) Mammalian remains from a new site near the classical locality of Pikermi (Attica, Greece). In: Proceedings of the XIX CBGA congress, Thessaloniki, Greece. Scientific Annals, School of Geology, Aristotle University of Thessaloniki, vol 99, pp 109–119

Verikiou-Papaspyridakou E (1986) Geomorphological study of the area of Maleas Cape-Kythera-Antikythera-Gramvousa. Unpublished PhD Thesis, National and Kapodistrian Univ Athens, Greece

Wagner R (1861) Nachträge zur Kenntniss der fossilen Hufthier-Ueberreste von Pikermi. Sitzunberichte Bayer Akad Wiss 2:78–82

Weithofer A (1888) Beiträge zur Kenntniss der Fauna von Pikermi bei Athen. Beit Paläontol Oesterreich-Ungarns 6(3):225–292

Xafis A, Tsoukala E, Solounias N, Mandic O, Harzhauser M, Grimsson F, Nagel D (2019) Fossil Giraffidae (Mammalia, Artiodactyla) from the late Miocene of Thermopigi (Macedonia, Greece). Palaeontol Electronica 22(3.68):1–38

Zittel KA (1893) Handbuch der Paläontologie. Vertebrata (Mammalia), vol 4. R. Oldenbourg, Munich and Leipzig

The Fossil Record of Tragulids (Mammalia: Artiodactyla: Tragulidae) in Greece

Socrates Roussiakis

1 Introduction

The family Tragulidae Milne-Edward, 1864 is a family of the suborder Ruminantia Scopoli, 1777 that includes the smallest living artiodactyls, also known as chevrotains or mouse deer. The modern representatives of the family are secretive, mostly solitary, nocturnal to crepuscular artiodactyls, with a shoulder height of about 20–40 cm and body mass from 1.0 kg to 15 kg. Their neck is short, their eyes large and their body is characterized by an arched back that rises toward the rears. Their limbs are short and slender and possess four-toed feet with hooves. They live in forestry humid environments of Africa, India, Sri Lanka, Southeast Asia, and southern China. They feed on fruits fallen on the ground, shoots, flowers, leaves, and mushrooms, whereas some species are also known to include in their diet arthropods, fishes, and other small-sized animals. As ruminants, the tragulids chew the cud and they possess a complex four-chambered stomach. However, the third chamber, the omasum, is poorly developed to absent.

Concerning their skeletal anatomy, the skull lacks cranial appendages, the orbital fossa is very large and has only one lachrymal foramen, there is a completely formed postorbital bar, there is no postglenoid process, the external exposure of the mastoid bone is very small, and the auditory bulla is cancellous (except in *Moschiola* where it is hollow). The tragulids lack upper incisors, as well as P1 and p1 (p1 can be present however in some fossil species), and the lower canine is incisiform and closely rooted to the third lower incisor. The upper canines in male individuals are large, housed in long alveoli and used in intraspecific fights. The P4 is triangular in occlusal view. The upper molars have labially projecting styles and cones. The lower

S. Roussiakis (✉)
Faculty of Geology and Geoenvironment, National and Kapodistrian University of Athens, Athens, Greece
e-mail: srousiak@geol.uoa.gr

E. Vlachos (ed.), *Fossil Vertebrates of Greece Vol. 2*,
https://doi.org/10.1007/978-3-030-68442-6_11

molars are characterized by the presence of a Σ (or M) structure formed of the cristids of the distal half of the first lobe.

On the axis, the odontoid process is peg-like and not crescentic. On the appendicular skeleton, the radius and the ulna are unfused, there is a capitatotrapezoid bone in the carpus formed by the fusion of the capitate and the trapezoid, trapezium sometimes absent. The fibula is reduced to a malleolar bone which sometimes is fused to the distal part of the tibia. The astragalus is narrow, elongated, and not parallel sided. Instead, the mediolateral axes of the proximal and distal pulleys converge medially. The fibular facet of the calcaneus is rather flat. There is a cubonavicular bone formed by the fusion of the cuboid and navicular bones. The ecto- and mesocuneiform bones are also fused and sometimes also fused with the cubonavicular. The first metapodial has been lost. The median metatarsals (III and IV) are fused, whereas the median metacarpals are either unfused or partly fused. The distal keel on the metapodials is incompletely developed and does not extend dorsally (Rössner 2007; Geraads 2010; Meijaard 2011; Guzmán-Sandoval 2018).

2 Phylogenetic Relationships and Taxonomy

The Tragulidae are considered the most primitive family and sister group of all other Ruminantia that consist the infraorder Pecora Linnaeus, 1758 (Hernández Fernández and Vrba 2005). The Tragulidae, together with some other primitive ruminants (e.g., archaeomerycids, hypertragulids, lophiomerycids, leptomerycids, bachitheriids), are variably grouped in the infraorder Tragulina Flower, 1883. However, there is no consensus for the context of this infraorder (or parvorder according to some authors) (Geraads et al. 1987; Hernández Fernández and Vrba 2005; Rössner 2007; Mennecart et al. 2011), whereas other authors consider Tragulina paraphyletic. As ruminants, the tragulids have the ability to ruminate, and share with other members of the Ruminantia the fused cuboid and navicular bones in the tarsus, the absence of upper incisors and the incisiform lower canine.

Some frequent tragulid characters are also shared with the Pecora, as the absence of P1, the fusions of the capitate with the trapezoid and the ecto- with the mesocuneiform, and the reduction of the fibula to a malleolar bone. However, tragulids are considered primitive compared to the pecorans because the omasum is poorly developed, the skull lacks appendages (lacking also on members of the pecoran family Moschidae), there is no postglenoid process, the mastoid exposure is small to absent, the odontoid process of the axis is peg-like, the radius and ulna remain unfused, the astragalus is narrow and not parallel sided, the lateral metapodials are fully developed and free, the median metapodials are unfused or partly fused, and the metapodial distal keels do not extant dorsally. Derived characters could be considered the cancellous auditory bulla, several characters of their dentition, the rather flat fibular facet of the calcaneus, the frequent fusion of the cubonavicular with the ecto-mesocuneiform and that of the malleolar bone with the distal part of the tibia.

The extant Tragulidae can be considered as one of the least diversified families among the living ruminants, with three existing genera currently recognized: *Tragulus* Brisson, 1762 with 6 species living in Southeast Asia (Myanmar, China, Thailand, Vietnam, Laos, Cambodia, Malaysia, Borneo, Sumatra, Indonesia, Philippines), *Moschiola* Gray, 1852, with one species living in India and 2 (potentially 3) living in Sri Lanka, and *Hyemoschus* Gray, 1845 with one species living in West and Central Africa (Guinea, Gabon, Sierra Leone, Ghana, Nigeria, Angola, Cameroon, Congo, Uganda) (Rössner 2007; Meijaard 2011; Clauss and Rössner 2014).

Despite the small diversity of living tragulids, in the past the family had greater biodiversity and, besides Asia and Africa, it was also present in Europe. The most ancient tragulid genus is considered to be *Archaeotragulus* Métais et al., 2001 from the late Eocene of Thailand (Métais et al. 2001; Métais and Vislobokova 2007; Rössner 2007; Sánchez et al. 2011, 2015, 2018), though is not accepted as tragulid by Geraads (2010). Other fossil tragulid genera include *Iberomeryx* Gabounia, 1964 from the early Oligocene of Europe and Georgia and the late Oligocene of Turkey; *Afrotragulus* Sánchez et al., 2010 from the latest early Miocene of Kenya; *Yunnanotherium* Han, 1986 from the late Miocene of China; *Siamotragulus* Thomas et al., 1990 from the middle Miocene of Thailand, early Miocene of Pakistan and latest early Miocene of Africa; *Dorcabune* Pilgrim, 1910 from the Miocene of South Asia, Miocene and Pleistocene of China, and Miocene of eastern Mediterranean; and *Dorcatherium* Kaup, 1833 from the late early to middle Miocene of Europe, early to middle Miocene of Africa, middle to late Miocene of South Asia, middle Miocene of China, and late Miocene of Turkey (list based on Rössner 2007 and supplemented with data from Sánchez et al. 2010, 2018, Pickford 2001, Kostopoulos and Sen 2016, Guzmán-Sandoval 2018, Koufos 2020).

Krabitherium Métais et al., 2007, another primitive ruminant genus from the late Eocene of Thailand, was tentatively referred to the Tragulidae, but further confirmation is needed for a definite conclusion (Métais et al. 2007; Sánchez et al. 2010). Some authors also include the genus *Nalameryx* Métais et al., 2009 from the Oligocene of India and Pakistan, (originally attributed to Lophiomerycidae) to Tragulidae (Mennecart et al. 2011). If this is correct, *Nalameryx* and the poorly known *Iberomeryx* could be the only Oligocene tragulids, filling the stratigraphic gap between the Eocene and Miocene occurrences of the family.

3 Historical Overview and Distribution

The oldest reference of fossil tragulids in Greece was made by Arambourg and Piveteau (1929) who described tragulid remains from the late Miocene of "Ravin de Vatilük or Ravin G" (currently Vathýlakkos 3 (VAT), Central Macedonia). Subsequently, the paleontological research in Greece has unearthed fossil tragulids in Chios Island (Paraskevaidis 1940; Lehmann and Tobien 1995; de Bonis et al. 1998), in Mélambes in Crete (Bonneau and Ginsburg 1974) and Kastéllios Hill also in Crete (van der Made 1996), in Antónios in Chalkidiki (Koufos and Syrides 1997; Koufos 2020), in Kerassiá in Euboea Island (van der Made and Moyà-Solà 1989), and in Dytikó in Axios Valley (Bouvrain and de Bonis 2007).

In most cases, the remains are few and fragmentary. Most specimens are referred to *Dorcatherium* and few to *Dorcabune*. However, and as it will be obvious from the following sections, the alpha-taxonomy of *Dorcatherium* is considered highly problematic, and currently *Dorcatherium* is considered probably paraphyletic (Qiu and Gu 1991; Rössner 2007; Sánchez et al. 2010, 2018; Rössner and Heissig 2013; Alba et al. 2014).

A list of localities in Greece that have yielded tragulid remains is given in the Appendix, and Fig. 1 shows their geographical distribution.

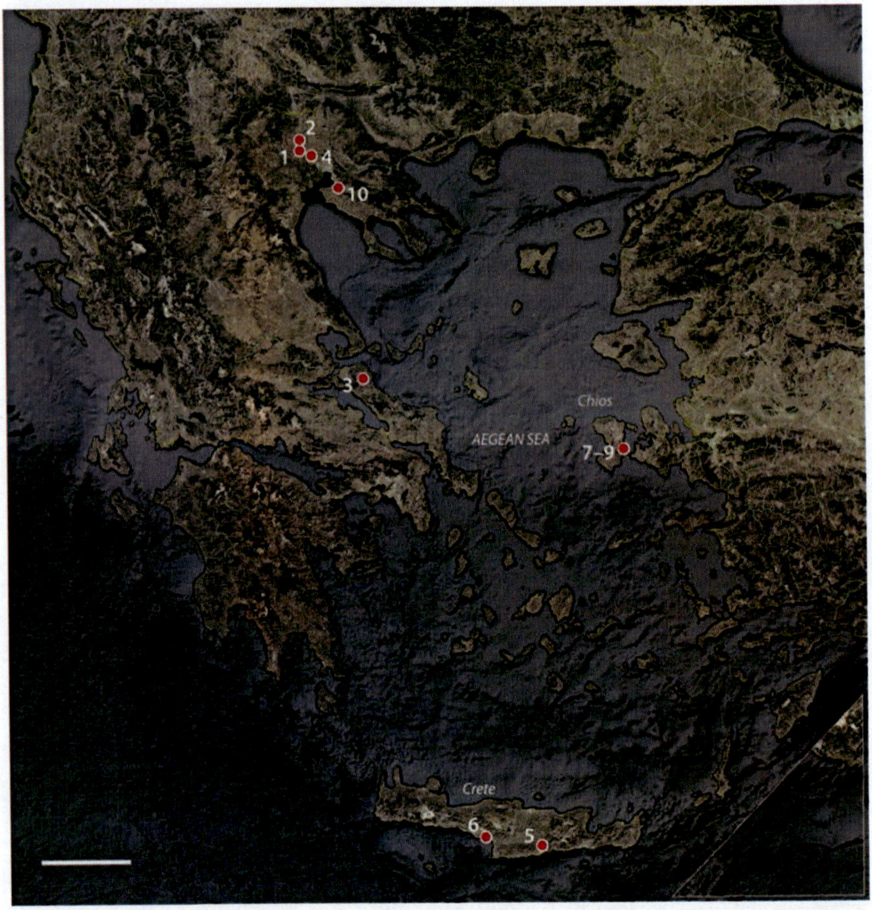

Fig. 1 Map of the occurrences of fossil tragulids from Greece. See Appendix for further information and text for details. **Late Miocene: 1**, Dytikó 3, **2**, Dytikó 2, **3**, Kerassiá, **4**, Vathylakkos 3. **Middle Miocene: 5**, Kastéllios Hill, **6**, Mélambes, **7–9**, Thymianá localities, **10**, Antonios. Data from the Appendix. Image exported from Google Earth Pro © 2020, map data from US Dept. of State Geographer, SIO, NOAA, U.S. Navy, NGA, GEBCO, image from Landsat/Copernicus. Scale bar equals 100 km, north faces upward

4 Systematic Paleontology

Tragulidae Milne-Edwards, 1864

Dorcatherium Kaup, 1833

Type Species *Dorcatherium naui* Kaup and Scholl, 1834.

Included Species *Dt. pigotti* Whitworth, 1958 from the early to middle Miocene of Africa; *Dt. chappuisi* Arambourg, 1933 from the early to middle Miocene of Africa; *Dt. crassum* (Lartet, 1851) from the late early to middle Miocene of Europe; *Dt. iririensis* Pickford, 2002 from the early Miocene of Africa; *Dt. vindebonense* von Meyer, 1846 from the late early to middle and probably early late Miocene of Europe; *Dt. guntianum* von Meyer, 1846 from the late early to middle Miocene of Europe and middle Miocene of Pakistan; *Dt. minus* Lydekker, 1876 from the middle to late Miocene of Pakistan; *Dt. peneckei* (Hofmann, 1893) from the early middle Miocene of Europe; *Dt. namaquensis* Sánchez et al., 2018 from the early middle Miocene of Namibia; *Dt. orientale* Qiu and Gu, 1991 from the middle Miocene of China; *Dt. minimus* West, 1980 from the middle Miocene of South Asia; *Dt. majus* Lydekker, 1876 from the middle and late Miocene of Pakistan; *Dt. nagrii* Prasad, 1970 from the middle and late Miocene of Pakistan and India; *Dt. dehmi* Guzmán-Sandoval and Rössner, 2019 from the middle and late Miocene of Pakistan; *Dt. naui* Kaup and Scholl, 1834 from the late middle Miocene to late Miocene of Europe and the middle Miocene of Pakistan; *Dt. jourdani* (Depéret, 1887) from the late Miocene of Europe; *Dt. puyhauberti* Arambourg and Piveteau, 1829 from the late Miocene of Europe; *Dt. maliki* Kostopoulos and Sen, 2016 from the late Miocene of Turkey (Pickford 2001; Rössner 2007; Sánchez et al. 2010, 2018; Aiglstorfer et al. 2014; Kostopoulos and Sen 2016; Rössner 2017; Guzmán-Sandoval and Rössner 2019).

Remarks Currently, the genus *Dorcatherium* is considered paraphyletic, an artificial assemblage of various unrelated bunoselenodont and selenodont tragulid forms that needs taxonomical revision (Rössner 2007; Sánchez et al. 2010, 2018). In that sense, the specific context of *Dorcatherium* is highly uncertain and the list provided above is provisional. According to Qiu and Gu (1991) some European *Dorcatherium* species (e.g., *Dorcatherium crassus* [sic], *Dorcatherium vindobonensis* [sic], and *Dorcatherium penekei* [sic]) should be included to *Dorcabune*. Until recently, *Dorcatherium* was considered as the only extinct African tragulid genus (e.g., Rössner 2007; Geraads 2010). However, this view changed recently with the assignment of the Early Miocene African species *Dorcatherium moruorotensis* Pickford, 2001 and *Dorcatherium parvum* Whitworth, 1958 to the genus *Afrotragulus* Sánchez et al., 2010, and the discovery of a *Siamotragulus* species, genus until recently known only from Asia—*Siamotragulus songhorensis* Sánchez et al. 2015 to the early Miocene of Africa (Sánchez et al. 2010, 2015, 2018). Additionally, the generic assignment of some species to *Dorcatherium* has been challenged. *Dorcatherium bulgaricum* Bakalov and Nikolov, 1962 was originally described from Bulgaria and considered of probably Pliocene age (Rössner 2007). However, Morales et al. (2012) noted that this form has similarities with Oligocene genera, whereas Aiglstorfer et al. (2014) consider its attribution to *Dorcatherium* quite uncertain and additionally they noted that it could be of Paleogene age. Furthermore, Sánchez et al. (2018) performed a preliminary phylogenetic analysis that shows that *Dt. crassum*, *Dt. namaquensis*, *Dt. pigotti*, and *Dt. iririensis* could belong in different genera/genus than *Dorcatherium*. However, as the authors note, such a hypothesis requires further support to be confirmed.

Dorcatherium guntianum von Meyer, 1846

Type Material Upper and lower permanent and deciduous teeth and postcranial bones (catalog number 1881 IX) curated in the Staatliche Naturwissenschaftliche Sammlungen Bayerns – Bayerische Staatssammlung für Paläontologie und Geologie, Munich, Germany (Guzmán-Sandoval and Rössner 2019).

Type Locality Early Miocene (MN4) of Reisensburg (Günzburg, Bavaria, Germany).

Distribution In Greece, this species is documented from the uppermost part of the early Miocene (Orleanian, MN4/5 boundary) of Antónios (ANT) in Chalkidiki (Central Macedonia) (Koufos 2020).

Remarks The species name *Dorcatherium guntianum* was made available by von Meyer (1846, p. 472) who considered it smaller than *Dt. naui* from Eppelsheim. *Dorcatherium guntianum* is a small-sized selenodont Miocene tragulid, intermediate in size between the larger *Dt. minus* and the smaller *Dt. nagri*. Its body mass ranges from 9.6 and 14.5 kg (mean 12.4 kg) and is reconstructed as a grass-dominated mixed feeder. Its postcanine teeth are rather brachyodont compared to *Dt. majus*, *Dt. naui*, *Dt. minus*, and *Dt. dehmi*, but rather hypsodont compared to *Dt. nagri* and the bunoselenodont European *Dt. crassum*, *Dt. vindebonense*, and *Dt. peneckei* (Rössner and Heissig 2013; Guzmán-Sandoval 2018; Guzmán-Sandoval and Rössner 2019).

The presence of *Dt. guntianum* in Antónios is based on maxillary and mandibular fragments as well as isolated upper and lower teeth described recently by Koufos (2020, figs. 2a–2f, figs. 3a–3o; Fig. 2a).

Dorcatherium cf. *crassum* (Lartet, 1851)

Neotype Left hemimandible preserving the p2–p3 (MNHN Sa9950), designated by Morales et al. (2012, figs. 1–3).

Type Locality Middle Miocene (MN6) of Sansan (Gers, France).

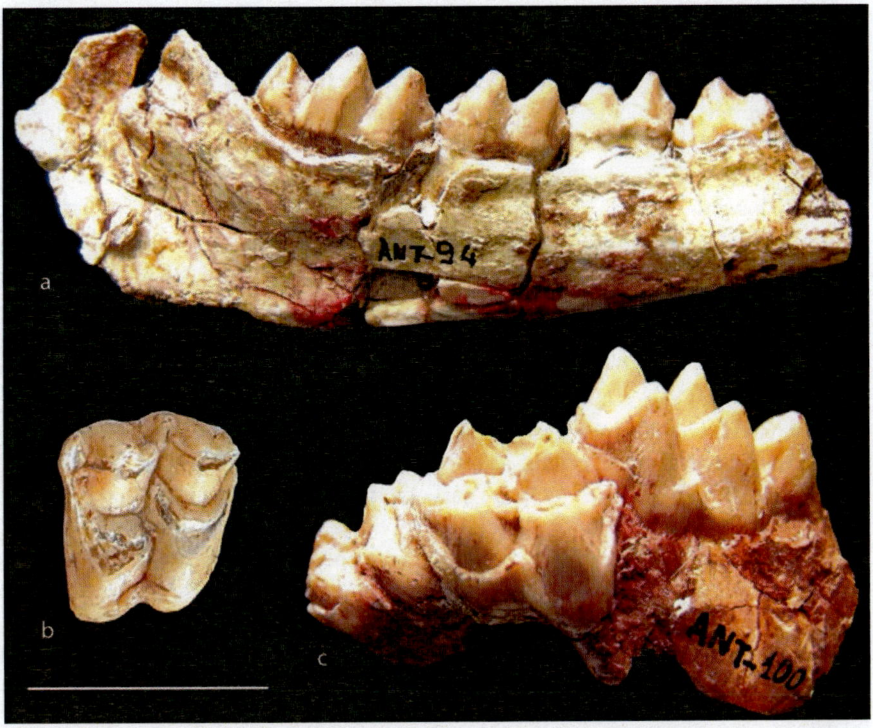

Fig. 2 Selected specimens of tragulids from the early/middle Miocene of Antonios. (**a**) LGPUT ANT 94, left mandible with p4–m3 of *Dorcatherium guntianum* in lingual view; (**b**) LGPUT ANT 98, left ?M3 of *Dorcatherium* cf. *crassum* in occlusal view; (**c**) LGPUT ANT 100, left mandible with m1–m3 of *Dorcabune* cf. *anthracotheroides* in labial view. Photographs courtesy of G.D. Koufos. Scale bar equals 2 cm

Distribution In Greece, *Dt.* cf. *crassum* is referred from the uppermost part of the early Miocene (Orleanian, MN4/5 boundary) of Antónios (ANT) in Chalkidiki (Central Macedonia).

Remarks As has been noted in the past (Rössner and Heissig 2013; Alba et al. 2014), *Dt. crassum* has been considered as a synonym of *Dt. naui* by some authors. At present *Dt. crassum* is considered a valid species, different in both dental and postcranial characters from *Dt. crassum* (Sánchez et al. 2012). Furthermore, as previously noted, some authors consider the attribution of this species to *Dorcatherium* questionable (Qiu and Gu 1991; Sánchez et al. 2018).

Dorcatherium crassum is a medium-sized bunoselenodont *Dorcatherium* with a body mass that ranges from 18.6 to 34.0 kg (mean 24.9 kg). It is considered as a browser with the ability, however, to consume harder food items as roots and seeds (Guzmán-Sandoval 2018).

The presence of *Dt.* cf. *crassum* in Antónios is based on poor material, two fragmentary preserved upper molars (Koufos 2020, figs. 2n–2r; Fig. 2b).

★*Dorcatherium puyhauberti* Arambourg and Piveteau, 1929

Type Material The type material of the species is housed in MNHN; however, there are certain problems. The type material includes a right maxillary fragment (MNHN F.SLQ784) preserving the much worn M1–M3, two upper molars (M2 and M3) (MNHN F.SLQ785) fixed on a plaster base, a right m3 (MNHN F.SLQ786), two right lower molars (m1 and m2) (MNHN F.SLQ787), two left lower molars (m2 and m3) fixed on a plaster base (MNHN F.SLQ788), and a right m1 (MNHN F. SLQ789). Arambourg and Piveteau (1929, p. 90) also mentioned a mandibular fragment preserving the p4–m3, though I could not locate any specimen fitting such a description in my various visits in MNHN. From the type series housed in MNHN, the specimen F.SLQ785 that preserves the left M2 and M3 could be considered the specimen figured by Arambourg and Piveteau (1929, pl. 5(or 7), fig. 1). However, as already mentioned by Aiglstorfer et al. (2014), the molars are fitted on a plaster base and in inverse order (M2 behind M3), whereas in the original photo provided by Arambourg and Piveteau the teeth are part of a maxillary fragment and in the correct order. Additionally, Aiglstorfer et al. (2014) noticed that the enamel color of the two molars is different and suggested that the teeth could originate from different individuals. It is also important to mention that the material plausibly was not collected by Arambourg himself, but from Dr. Puyhaubert (and possibly others following him) (Arambourg and Piveteau 1929, p. 60).

Type Locality "Ravin de Vatilük or Ravin G" (Arambourg and Piveteau 1929, p. 70). The locality corresponds to Vathýlakkos 3 (VAT) dated to the late Miocene (middle Turolian, MN12) with an estimated age of about 7.3 Ma (Koufos 2012, 2013, 2020).

Remarks *Dorcatherium puyhauberti* is a poorly known tragulid species and its validity is considered questionable by some authors. According to Morales et al. (2012), *Dt. puyhauberti* could be included in *Dt. naui*, whereas Aiglstorfer et al. (2014) consider that *Dt. puyhauberti* could be synonym of *Dt. guntianum*. It has an estimated body mass of about 19.3 kg (Guzmán-Sandoval 2018). It is characterized by the presence of an elongated postmetacristid on the lower molars that tends to reach the *Tragulus*-fold (Geraads et al. 2005; Kostopoulos and Sen 2016). Additionally, its upper molars are characterized from a lingual cingulum well developed on the mesial lobe but only slightly developed on the distal one.

Malik and Nafiz (1933) referred some specimens from Küçükçekmece (Turkey) to *Dt. puyhauberti*, but later on Kostopoulos and Sen (2016) referred these to *Dt. maliki*. Nevertheless, Guzmán-Sandoval (2018) and Guzmán-Sandoval and Rössner (2019) considered the validity of the latter species questionable assuming a potential synonym with *Dt. naui*. Geraads et al. (2011) tentatively referred some specimens from the late Miocene (middle Turolian) locality of Strumyani (Bulgaria) to *Dt.* cf. *puyhauberti*, though Aiglstorfer et al. (2014) referred to that these remains could belong to *Dt. naui*.

Dorcabune Pilgrim, 1910

Type Species *Dorcabune anthracotheroides* Pilgrim, 1910.

Included Species *Db. welcommi* Ginsburg, Morales and Soria, 2001 from the early Miocene of Pakistan, *Db. anthracotheroides* Pilgrim, 1910 from the late Miocene of Pakistan, *Db. nagrii* Pilgrim, 1915 from the late Miocene of Pakistan, *Db. sindiense* Pilgrim, 1915 from the late Miocene of Pakistan, *Db. progressus* (Yan, 1978) from the late Miocene of China, *Db. liuchengense* Han, 1974 from the Early Pleistocene of China (Rössner 2007; Guzmán-Sandoval 2018; Guzmán-Sandoval and Rössner 2019).

Remarks As previously mentioned (see *Dorcatherium* section), some European species commonly included to *Dorcatherium* could belong to *Dorcabune* or other genera.

Dorcabune cf. *anthracotheroides* Pilgrim, 1910

Type Material Geological Survey of India B 580, a left maxilla with M1 to M3 (Pilgrim 1915: pl. 21, figs. 1 and 1a) (holotype) and various isolated upper and lower teeth as well as a fragmentary mandible figured by Pilgrim 1915 (pl. 21, figs. 2, 7, 7a, 8; fig; pl. 22, figs. 4, 4a, 5) (paratypes) (Guzmán-Sandoval 2018; Guzmán-Sandoval and Rössner 2019).

Type Locality Chinji, Chakwal, Punjab Province of Pakistan, Lower Siwalik Subgroup, Siwalik Group (Guzmán-Sandoval 2018; Guzmán-Sandoval and Rössner 2019).

Distribution In Greece, *Db.* cf. *anthracotheroides* is referred from the uppermost part of the early Miocene (Orleanian, MN4/5 boundary) of Antónios (ANT) in Chalkidiki (Central Macedonia) (Koufos 2020).

Remarks *Dorcabune anthracotheroides* is a tragulid with bunoselenodont dentition, a *Dorcatherium* platform on the lower molars, prominent and isolated parastyle and mesostyle and rugose enamel on the upper molars. It is the largest tragulid with a body mass from 72.6 to 139.4 kg (mean 111.5 kg) (Guzmán-Sandoval 2018; Guzmán-Sandoval and Rössner 2019).

Dorcabune cf. *anthracotheroides* has recently referred in Antónios and is based on two mandibular fragments (Koufos 2020, figs. 6a–6g; Fig. 2c). The possible presence of the species in Kastéllios Hill (Crete) (van der Made 1996) is discussed in the following section.

Miscellaneous Tragulid Occurrences

Additional tragulid specimens have been recorded from various Greek localities. However, in most cases these remains are fragmentary and/or poorly informative for a precise taxonomic identification and are here referred to as *Dorcatherium* sp.

The early Miocene locality of Antónios in Chalkidiki, apart from *Dt. guntianum*, *Dt.* cf. *crassum* and *Db.* cf. *anthracotheroides*, has also yielded a fragmentary left upper canine that is referred to *Dorcatherium* sp. (Koufos 2020, figs. 2g–h). According to Koufos (2020), this canine corresponds to a small- or medium-sized tragulid better referrable as *Dorcatherium* sp.

Middle Miocene tragulids are known from Chios Island (North Aegean). The first remains from Chios were described by Paraskevaidis (1940). The fossils come from an area near to a ceramic factory close to the Thymianá village south of the Chios town, and the locality is commonly met to the bibliography as "Ziegelei Michalos," "Keramiaschichten südlich der Stadt," or "Kerameia." In his study, Paraskevaidis (1940) included fossils collected by himself in 1937, as well as fossils collected earlier on by local people. Paraskevaidis (1940, table 13, figs. 1–3) described and referred a tragulid mandibular fragment preserving the d4 and m1 to *D.* aff. *naui*. Later on, Paraskevaidis (1977) referred a tragulid astragalus from "Keramiaschichten südlich der Stadt" to *Dorcatherium* sp., but he did not provide any description, photo, or measurements. In the late 1960s, further campaigns investigating the middle Miocene mammal localities of Chios took part. The first period was during March of 1967 through collaboration of the National and Kapodistrian University of Athens and the University of Mainz, whereas later field campaigns were carried on by the University of Mainz in October of 1967, April and May of 1968, and September of 1968 (Tobien 1977). Nine fossiliferous localities were spotted north of Thymianá, namely Thy-1 to Thy-9 (Rothausen 1977). Lehmann and Tobien (1995) referred to a tragulid premolar fragment from Thy-1 as ? *Dorcatherium* sp. (aff. *D. crassum*), but they did not provide any picture of the specimen. The authors proposed an early Astaracian (MN6) age for the fauna.

In 1990s, excavations in Thymianá were carried out by a Hellenic-French team. Three fossiliferous levels were located: a lower one namely Thymianá A (THA), an intermediate one (Thymianá B, THB), and an upper one (Thymianá C, THC). From these levels, only Thymianá B yielded large mammal remains (de Bonis et al. 1998; Koufos 2013, 2016). The tragulid remains included a skull fragment as well as a left and a right hemimandible. The specimens are poorly preserved with very worn teeth and details about their morphology are not observable. All specimens are referred by the authors to *Dorcatherium* sp. Biostratigraphical and magnetostratigraphical data indicate that the Thymianá fauna is slightly younger than the Antónios one and can be referred to middle Miocene (late Orleanian) with an age of about 15.5 Ma and can be correlated to MN5 (Kondopoulou et al. 1993; de Bonis et al. 1998; Koufos 2013, 2016). Keeping in mind that the old fossiliferous sites have not been correlated with the new ones, but based on the newest data, all Thymianá sites are correlated to MN5.

Additional middle Miocene tragulid remains are known from the island of Crete, from the locality Mélambes (Rethymnon prefecture), firstly referred by Bonneau and Ginsburg (1974) and then by Kuss (1976). Bonneau and Ginsburg (1974) referred the presence of both dental and postcranial elements, representing according to their interpretation the remains of an almost complete tragulid skeleton of a not fully grown individual. The authors described an M1 and an M2 as well as a cubonavicular. They attributed these remains to *Dt. puyhauberti* and considered the locality of Vallesian age. Van der Made (1996) referred the Mélambes tragulid remains to *Dt. naui* claiming that an older age, MN6 or early MN7/8 age, could be more possible, whereas later on he suggested an early Astaracian (MN6) age for the Mélambes fauna (van der Made 1999). Later on, Koufos (2006) assigned the Mélambes tragulid to *Dt. naui*, whereas Rössner and Heissig (2013: online resource 2 and based on affiliations from the NOW database) listed the Mélambes tragulid as *D. crassum*.

Another presence of tragulids in Crete is in the Kastéllios Hill (Heraklion prefecture) where fossils of micromammals and large mammals were collected from four different stratigraphic levels, from the "eastern slope" of Kastéllios Hill as well as from the surface (de Bruijn et al. 1971; de Bruijn and Zachariasse 1979; van der Made 1996). The Kastéllios Hill localities are considered of late Vallesian (MN10) age (van der Made 1999; Koufos 2003, 2006). However, paleomagnetic data indicate for the Kastéllios section an age between 10.99 and 10.30 Ma (Sen et al. 1986) suggesting an early Vallesian (MN9) age. The presence of a tragulid is originally reported from the "eastern slope" of Kastéllios Hill by de Bruijn et al. (1971) who referred two isolated teeth as *Dorcatherium* sp., but no measurements, figures, or descriptions were provided. This was questioned by van der Made (1996) who studied the relative collections. Whereas he failed to relocate any tragulid teeth and considered that some bovid teeth could have incorrectly considered as tragulid ones, he found a large-sized tragulid astragalus stored with the K2 material. Van der Made (1996) attributed this tragulid to cf. *Dorcabune anthracotheroides*. Geraads et al. (2005) considered such an interpretation unlikely in paleogeographical terms considering *Dorcabune* a south-eastern Asian form. Though, as previously noted, the presence of *Dorcabune* in Europe cannot be excluded since Qiu and Gu (1991) and Sánchez et al. (2018) have suggested that some European tragulid species, traditionally included to *Dorcatherium*, could belong to *Dorcabune*. Moreover, the presence of *Dorcabune* in Europe is also supported by Koufos (2020) who described *Dorcabune* remains from the early Miocene of Antónios as referred previously. Based on its dimensions, the astragalus from Kastéllios also fits to the larger dimensions provided by Kostopoulos and Sen (2016, fig. 4) for *Dorcatherium maliki* from the Vallesian locality of Küçükçekmece (Turkey).

In Central Greece, tragulids are known only from the late Miocene locality of Kerassiá in Northern Euboea. The first excavation in the area was carried out by Hans de Bruijn, Albert van der Meulen (University of Utrecht), and Constantin Doukas (National and Kapodistrian University of Athens) during 1982. Since 1992, excavations were followed by the National and Kapodistrian University of Athens under the supervision of George Theodorou. During these later excavations seven fossiliferous sites (namely K1–K7) were discovered, grouped in two main fossiliferous levels an upper and a lower one. The only tragulid remain from Kerassiá comes from the 1982 excavation, from a fossiliferous level (namely KE) that most possibly corresponds to the upper fossiliferous level, and is firstly referred by van der Made and Moyà-Solà (1989) to *Dorcatherium* sp. It is a skull fragment and preliminarily we can mention that its molars are slightly larger than those of *Dt.*

puyhauberti and close to the overlapping size zone of *Dt. crassum* and *Dt. naui*, though they have in common with *Dt. puyhauberti* the slightly developed cingulum around the distal lobe. Based on their faunal content both Kerassiá levels are considered of middle Turolian (MN12) age, though an early Turolian age (MN11) cannot be rejected to the moment (van der Made and Moyà-Solà 1989; Iliopoulos 2003; Roussiakis and Theodorou 2003; Theodorou et al. 2003; Athanassiou et al. 2014; Kampouridis et al. 2019).

The late Miocene localities of Dytikó 2 (DIT) and Dytikó 3 (DKO) in Central Macedonia (Axios Valley) have also yielded tragulid fossils. Both localities are of late Turolian (MN13) age between 7.0 and 6.0 Ma (Koufos and Vasileiadou 2015). Bouvrain and de Bonis (Bouvrain and de Bonis 2007, figs. 15–16) described an upper maxilla preserving the M1–M3 from Dytikó 2 (Fig. 3) as well as a left mandible with p3–m3 from Dytikó 3 and referred both specimens to *Dorcatherium* sp. Koufos (2006) initially referred the Dytikó 2 and Dytikó 3 tragulid specimens to *Dt. puyhauberti*, though subsequently (Koufos 2020) considered the material poor for a specific determination. Rössner and Heissig (2013: online resource 2, listing affiliations given on NOW database) also assigned the Dytiko 2 tragulid to *Dt. puyhauberti*. However, the upper teeth of the Dytikó 2 tragulid are characterized by strong lingual cingulum, well developed around the mesial and distal lobes of the upper molars (Bouvrain and de Bonis 2007: fig. 15), whereas in *Dt. puyhauberti* the lingual cingulum is well developed around the mesial lobes of the molars but it is weakly developed around the distal lobes. Unfortunately, the Dytikó 3 mandibular specimen has strongly worn out teeth and their morphological characters are barely observable.

Finally, tragulid remains have also been referred from the famous late Miocene localities of Samos (North Aegean) and Pikermi (Attica). Solounias (1981, table 4) referred the presence of *Dt. naui* in Samos (unknown exact locality) and Koufos (2006, appendix 1) referred *Dorcatherium* sp. in his faunal list for Pikermi. As far as I know, however, no *Dorcatherium* specimen has been described either from Samos or from Pikermi, and the existence of tragulids in these localities is herein considered uncertain.

Fig. 3 *Dorcatherium* sp. Right maxilla with M1–3 (LGPUT DIT 83) from the late Miocene of Dytiko 2, in occlusal view, originally described by Bouvrain and de Bonis (2007). Scale bar equals 1 cm

5 Conclusions

In Greece, the localities that have yielded fossil Tragulidae are few and in all cases the remains are scanty and/or fragmentary. These localities span from the uppermost part of the early Miocene (Orleanian, MN4/5) to the late Miocene (late Turolian, MN13) and geographically they are distributed from Northern Greece (e.g., Dytikó) to Crete (e.g., Mélambes) in the south as well as to the east Aegean (Chios). The few and poorly preserved findings do not enable always a specific determination and in some cases the findings are here referred as *Dorcatherium* sp., pending also for a revision of the genus.

Acknowledgments I would like to deeply thank G. D. Koufos for providing photographs of tragulid material from Antonios and G. Rössner for valuable comments that improved this manuscript.

Appendix

List of localities with occurrences of fossil Tragulidae in Greece arranged in geochronological order. The age of each locality is based on the most recent data available as explained on the main text. The taxonomy follows the present revision and the superscript numbers correspond to the first original publication. Type localities are marked with bold. Locality numbers refer to the collection numbers of the PaleoBiology Database (PBDB)

Localities[PBDB]	Age (ELMA; MN)	Taxon
Dytikó 3[32376]	Late Miocene (late Turolian, MN13)	*Dorcatherium* sp.[1]
Dytikó 2[32373]	Late Miocene (late Turolian, MN13)	*Dorcatherium* sp.[1]
Kerassiá[195431]	Late Miocene (middle Turolian, MN12)	*Dorcatherium* sp.[2]
Vathýlakkos 3[182750]	**Late Miocene (middle Turolian, MN12)**	***Dorcatherium puyhauberti*[3]**
Kastéllios Hill[205364]	Late Miocene (early Vallesian, MN9)	cf. *Dorcabune anthracotheroides*[4]
Mélambes[208183]	Middle Miocene (Astaracian) MN6 or early MN7/8	*Dorcatherium* sp.[5]
Thymianá (Michalos)[182,752]	Middle Miocene (late Orleanian, MN5)	*Dorcatherium* sp.[6]
Thymianá-1	Middle Miocene (late Orleanian, MN5)	*Dorcatherium* sp.[7]
Thymianá B[202728]	Middle Miocene (late Orleanian, MN5)	*Dorcatherium* sp.[8]
Antónios[73861]	Early Miocene (Orleanian, MN4/MN5)	*Dorcatherium guntianum*[9] *Dorcatherium* cf. *crassum*[9] *Dorcabune* cf. *anthracotheroides*[9] *Dorcatherium* sp.[9]

[1]Bouvrain and de Bonis (2007), [2]van der Made and Moyà-Solà (1989), [3]Arambourg and Piveteau (1929), [4]van der Made (1996), [5]Bonneau and Ginsburg (1974), [6]Paraskevaidis (1940), [7]Lehmann and Tobien (1995), [8]de Bonis et al. (1998), [9]Koufos (2020)

References

Aiglstorfer M, Rössner GE, Böhme M (2014) *Dorcatherium naui* and pecoran ruminants from the late middle Miocene Gratkorn locality (Austria). Palaeobio Palaeoenv 94:83–123

Alba DM, DeMiguel D, Morales J, Sánchez IM, Moyà-Solà S (2014) New remains of *Dorcatherium crassum* (Artiodactyla: Tragulidae) from the early Miocene (MN4) of Els Casots (Subirats, Vallès-Penedès Basin). C R Palevol 13:73–86

Arambourg C (1933) Mammifères Miocènes du Turkana (Afrique orientale). Ann Paléontol 22:123–146

Arambourg C, Piveteau J (1929) Les Vertébrés du Pontien de Salonique. Ann Paléontol 18:59–138

Athanassiou A, Roussiakis SJ, Giaourtsakis IX, Theodorou GE, Iliopoulos G (2014) A new hornless rhinoceros of the genus *Acerorhinus* (Perissodactyla: Rhinocerotidae) from the upper Miocene of Kerassiá (Euboea, Greece), with a revision of related forms. Palaeontographica 303:23–59

Bakalov P, Nikolov I (1962) Les fossiles de Bulgarie. X. Mammifères Tertiaires. Acad Sci Bulgarie 10:1–162

Bonneau M, Ginsburg L (1974) Découverte de *Dorcatherium puyhauberti* Arambourg et Piveteau (Mammalia) dans les faciès continentaux de base de la molasse tertiaire de Crète (Grèce): Conséquences stratigraphiques et tectoniques. C R somm S G F 1974(1):11–12

Bouvrain G, de Bonis L (2007) Ruminants (Mammalia, Artiodactyla: Tragulidae, Cervidae, Bovidae) des gisements du Miocène supérieur (Turolien) de Dytiko (Grèce). Ann Paléontol 93:121–147

Clauss M, Rössner GE (2014) Old world ruminant morphophysiology, life history, and fossil record: exploring key innovations of a diversification sequence. Ann Zool Fennici 51:80–94

de Bonis L, Koufos GD, Sen S (1998) Ruminants (Bovidae and Tragulidae) from the middle Miocene (MN 5) of the island of Chios, Aegean Sea (Greece). N Jb Geol Paläontol Abh 210:399–420

de Bruijn H, Zachariasse WJ (1979) The correlation of marine and continental biozones of Kastellios Hill reconsidered. Ann Géol Pays Helléniques 1:219–226

de Bruijn H, Sondaar PY, Zachariasse WJ (1971) Mammalia and foraminifera from the Neogene of Kastellios hill, (Crete), a correlation of continental and marine biozones. Proc Koninkl Nederl Akad Wetenschappen B 74(5):1–22

Geraads D (2010) Tragulidae. In: Werdelin L, Sanders WJ (eds) Cenozoic mammals of Africa, University of California Press, Berkeley, CA pp. 723–729

Geraads D, Bouvrain G, Sudre J (1987) Relations phyletiques de *Bachitherium* Filhol, ruminant de l' Oligocene d' Europe occidentale. Palaeovertebrata 17(2):43–73

Geraads D, Kaya T, Mayda S (2005) Late Miocene large mammals from Yulafli, Thrace region, Turkey, and their biogeographic implications. Acta Palaeontol Pol 50:523–544

Geraads D, Spassov N, Hristova L, Markov GN, Tzankov T (2011) Upper Miocene mammals from Strumyani, South-Western Bulgaria. Geodiversitas 33:451–484

Guzmán-Sandoval JA (2018) Palaeobiology of Tragulids (Mammalia: Artiodactyla: Ruminantia). Dissertation zur Erlangung des Doktorgrades an der Fakultät für Geowissenschaften der Ludwig-Maximilians-Universität München, 1–236

Guzmán-Sandoval JA, Rössner GE (2019) Miocene chevrotains (Mammalia, Artiodactyla, Tragulidae) from Pakistan. Hist Biol:1. https://doi.org/10.1080/08912963.2019.1661405

Han D (1986) Fossils of Tragulidae from Lufeng, Yunnan. Acta Anthropologica Sinica 5:68–78

Hernández Fernández M, Vrba ES (2005) A complete estimate of the phylogenetic relationships in Ruminantia: a dated species-level supertree of the extant ruminants. Biol Rev 80:269–302

Iliopoulos G (2003) The Giraffidae (Mammalia, Artiodactyla) and the study of the histology and chemistry of fossil mammal bone from the Late Miocene of Kerassia (Euboea Island, Greece). PhD thesis, University of Leicester

Kampouridis P, Roussiakis S, Kargopoulos N, Giaourtsakis I, Dimakopoulos G, Iliopoulos G, Svorligkou G, Theodorou G (2019) Faunal diversity at the Turolian locality of Kerassia (northern Euboea, Greece). In: Proceedings of the 15th International Congress of the Geological Society of Greece, Athens, 22–24 May 2019. Bull Geol Soc Gr, Sp. Publ. No 7, pp. 52–53

Kondopoulou D, de Bonis L, Koufos GD, Sen S (1993) Palaeomagnetic and biostratigraphic data from the middle Miocene vertebrate locality of Thymiana (Chios Island, Greece). In: Proceedings of the 2nd Congress of the Geophysical Society of Greece 1:626–635

Kostopoulos DS, Sen S (2016) Suidae, Tragulidae, Giraffidae, and Bovidae. Geodiversitas 38:273–298

Koufos GD (2003) Late Miocene mammal events and biostratigraphy in the eastern Mediterranean. Deinsea 10:343–371

Koufos GD (2006) The Neogene mammal localities of Greece: faunas, chronology and biostratigraphy. Hellenic J Geosci 41:183–214

Koufos G (2012) New material of Chalicotheriidae (Perissodactyla, Mammalia) from the late Miocene of Axios Valley, Macedonia (Greece) with the description of a new species. Ann Paléontol 98:203–224

Koufos G (2013) Neogene mammal biostratigraphy and chronology of Greece. In: Wang X, Flynn LJ, Fortelius M (eds) Fossil mammals of Asia: neogene biostratigraphy and chronology, Columbia University Press, New York pp. 595–625

Koufos GD (2016) Neogene and quaternary continental biostratigraphy of Greece based on mammals. Bull Geol Soc Gr 50:55–64

Koufos GD (2020) New tragulid remains from the early/middle Miocene and a revision of their occurrence in Greece. Hist Biol:1. https://doi.org/10.1080/08912963.2020.1795650

Koufos GD, Syrides GE (1997) A new early/middle Miocene mammal locality from Macedonia, Greece. C R Acad Sci 325:511–516

Koufos GD, Vasileiadou K (2015) Miocene/Pliocene mammal faunas of southern Balkans: implications for biostratigraphy and palaeoecology. Palaeobio Palaeoenv 95:285–303

Kuss SE (1976) Ein erster Fund von *Pliohyrax* aus dem Vallesian von Kreta/Griechenland. N Jb Geol Pal Mh 3:157–162

Lehmann U, Tobien H (1995) Artiodactyle Fossilien (Mammalia) aus dem Miozän von Thymiana, Chios. Ann Géol Pays Helléniques 36:403–414

Malik A, Nafiz H (1933) Vertébrés fossiles de Küçükçekmece. Publ Inst Géol Univ Istanbul 8:1–119

Meijaard E (2011) Family Tragulidae (chevrotains). In: Wilson DE, Mittermeier RA (eds) Handbook of the mammals of the world. 2. Hoofed mammals, Lynx Edicions, Barcelona, pp. 320–334

Mennecart B, Damien Becker D, Berger J-P (2011) *Iberomeryx minor* (Mammalia, Artiodactyla) from the early Oligocene of Soulce (Canton Jura, NW Switzerland): systematics and palaeodiet. Swiss J Geosci 104(Suppl 1):S115–S132

Métais G, Vislobokova IA (2007) Basal ruminants. In: Prothero DR, Foss SC (eds) The evolution of artiodactyls, pp. 189–212

Métais G, Chaimanee Y, Jaeger J-J, Ducrocq S (2001) New remains of primitive ruminants from Thailand: evidence of the early evolution of the Ruminantia in Asia. Zool Scripta 30:231–248

Métais G, Chaimanee Y, Jaeger J-J, Ducrocq S (2007) Eocene bunoselenodont Artiodactyla from southern Thailand and the early evolution of Ruminantia in South Asia. Naturwiss 94:493–498

Métais G, Welcomme J-L, Ducrocq S (2009) New lophiomerycid ruminants from the Oligocene of the Bugti Hills (Balochistan, Pakistan). J Vertebr Paleontol 29:231–241

Morales J, Sánchez IM, Quiralte V (2012) Les Tragulidae (Artiodactyla de Sansan). In: Peigné S, Sen S (eds) Mammifères de Sansan, vol 203. Mém Mus nat d'Hist nature, Paris, pp. 225–247

Paraskevaidis I (1940) Eine obermiocäne Fauna von Chios. N Jahr Mineral Geol Paläontol 83:363–442

Paraskevaidis I (1977) Säugetierreste aus Griechenland. In: Proc. VI Coll. Geol. Aegean Region, III, pp 1143–1154

Pickford M (2001) Africa's smallest ruminant: a new tragulid from the Miocene of Kenya and the biostratigraphy of east African Tragulidae. Geobios 34:437–447

Qiu Z, Gu Y (1991) The Aragonian vertebrate fauna of Xiacaowan, Jiangsu - 8. *Dorcatherium* (Tragulidae, Artiodactyla). Vertebrata Palasiatica 29:21–37

Rössner GE (2007) Family Tragulidae. In: Prothero DR, Foss SE (eds) The evolution of artiodactyls. Johns Hopkins University Press, Baltimore, pp 213–220

Rössner GE (2017) New palaeontological and geological age constraints for the arrival and dispersal of Miocene tragulids (Tragulidae, Artiodactyla) in Europe. Zitteliana 89:359–367

Rössner GE, Heissig K (2013) New records of *Dorcatherium guntianum* (Tragulidae), stratigraphical framework, and diphyletic origin of Miocene European tragulids. Swiss J Geosci 106:335–347

Rothausen K (1977) Die mittelmiozänen Wirbeltierfundstellen südlich Thymiana (Insel Chios, Ägäis, Griechenland). 2 Teil: Geologie: Die Fundstellen und ihre Abfolge. Ann Géol Pays Helléniques 28:495–515

Roussiakis SJ, Theodorou GE (2003) Carnivora from the late Miocene of Kerassia (northern Euboea, Greece). Deinsea 10:469–497

Sánchez IM, Quiralte V, Morales J, Pickford M (2010) A new genus of tragulid ruminant from the early Miocene of Kenya. Acta Palaeontol Pol 55:177–187

Sánchez IM, Quiralte V, Morales J (2011) Solving and old dispute: anatomical differences between the European Miocene chevrotains *Dorcatherium crassum* Lartet, 1839 and *Dorcatherium naui* Kaup & Scholl, 1834 (Mammalia, Ruminantia, Tragulidae). Paleontol Evol 5:343–347

Sánchez IM, Quiralte V, Ríos M, Morales J, Pickford M (2015) First African record of the Miocene Asian mouse-deer *Siamotragulus* (Mammalia, Ruminantia, Tragulidae): implications for the phylogeny and evolutionary history of the advanced selenodont tragulids. J Syst Palaeontol 13:543–556

Sánchez IM, Morales J, Cantalapiedra JL, Quiralte V, Pickford M (2018) Preliminary phylogenetic analysis of the Tragulidae (Mammalia, Cetartiodactyla, Ruminantia) from Arrisdrift: implications for the African Miocene tragulids. Comm Geol Surv Namibia 19:110–122

Sen S, Valet J-P, Ioakim C (1986) Magnetostratigraphy and biostratigraphy of the Neogene deposits of Kastellios hill (Central Crete, Greece). Palaeogeogr Palaeoclimatol Palaeoecol 53:321–334

Solounias N (1981) The Turolian fauna from the island of Samos, Greece, with special emphasis on the hyaenids and bovids. Contrib Vertebr Evol 6:1–232

Theodorou G, Athanassiou A, Roussiakis S, Iliopoulos G (2003) Preliminary remarks on the late Miocene herbivores of Kerassiá (northern Euboea, Greece). Deinsea 10:519–530

Thomas H, Ginsburg L, Hintong C, Suteethorn V (1990) A new tragulid, *Siamotragulus sanyathanai* n. gen. n. Sp. (Artiodactyla, Mammalia) from the Miocene of Thailand (Amphoe pong, Phayao Province). C R Acad Sci Paris II 310:989–995

Tobien H (1977) Die mittelmiozänen Wirbeltierfundstellen südlich Thymiana (Insel Chios, Ägäis, Griechenland). 1 Teil: Einleitung. Ann Géol Pays Helléniques 28:489–494

van der Made J (1996) Pre-Pleistocene land mammals from Crete. In: Reese DS (ed) Pleistocene and Holocene Fauna of Crete and its first settlers. Prehistory Press, Madison, WI, pp 69–79

van der Made J (1999) Intercontinental relationship Europe—Africa and the Indian subcontinent. In: Rössner GE, Heissig K (eds) The Miocene land mammals of Europe. Dr. Friedrich Pfeil, München, pp 457–472

van der Made J, Moyà-Solà S (1989) European Suinae (Artiodactyla) from the late Miocene onwards. Bol Soc Pal It 28:329–339

von Meyer H (1846) Mittheilungen an professor Bronn. Neues Jahrb Mineral Geol Geognoise Petrefaktenkund Jg 1846:462–476

Whitworth T (1958) Fossil mammals of Africa. No 15: Miocene ruminants of East Africa. Bull British Mus (Nat Hist) Geol 15:1–50

The Fossil Record of Equids (Mammalia: Perissodactyla: Equidae) in Greece

George D. Koufos, Theodora D. Vlachou, and Anastasia G. Gkeme

1 Introduction

Equidae is a large family of perissodactyls (odd-toed ungulate mammals), including all horses, donkeys, zebras, and related extinct taxa. Although they were widely distributed in the past, their current distribution is significantly reduced and wild equids are encountered in Eastern and Southern Africa, as well as in some regions of Asia. The equids inhabit various open habitats, from lush grasslands and savannah to sandy and stony deserts. The hypsodont (high-crowned) teeth with complex enamel plication help them to consume hard food. The elongated and thin legs with a single toe are adapted to running in the open habitats. The skull of the equids is long, with elongated muzzle, long and narrow nasal bones and orbits situated far behind the teeth. The cheek teeth are squarish (upper) and rectangular (lower) with various enamel folds and much cement. All equids are herbivorous grazers to browsers, alert runners and prefer to escape from danger rather than fight (Ballenger and Myers 2001; MacFadden 2005).

The modern equids are all included in the single genus *Equus* with the species: *E. ferus* (wild horse and Przewalskii's horse), *E. zebra* (mountain zebra), *E. quagga* (plain zebra), *E. grevyi* (Gravy's zebra), *E. africanus* (African wild donkey), *E. kiang* (kiang), and *E. hemionus* (onager, hemione, Asiatic wild donkey).

G. D. Koufos (✉) · T. D. Vlachou · A. G. Gkeme
Department of Geology, Laboratory of Geology and Palaeontology, Aristotle University of Thessaloniki, Thessaloniki, Greece
e-mail: koufos@geo.auth.gr

E. Vlachos (ed.), *Fossil Vertebrates of Greece Vol. 2*,
https://doi.org/10.1007/978-3-030-68442-6_12

2 Historical Overview

The first equid-like animal (Equoidea) appeared in Greece is *Palaeotherium*, recognized near Ferres (Alexandroupolis, Western Thrace); some postcranial remains have been described as *Palaeotherium* cf. *magnum*, dated to the Middle-Late Eocene (Metais and Sen 2017). During the end of Miocene *Anchitherium* appeared in Greece. The migration of the three-toed horses from the New to the Old World marks the beginning of the Vallesian (early late Miocene) epoch and is extensively used in the biostratigraphy and biochronology of the Eurasian faunas. The first appearance of the hipparion horses in the Old World is known as "*Hipparion* Datum" and is thought to have occurred around 11.2 Ma (Vlachou 2013 and references). A great number of Late Miocene mammal localities with fossil equids are known from Greece. Approximately 40 of these are correlated with the Turolian (MN11–MN13) and the remaining with the Vallesian (MN9–MN10) (Appendix 1). *Hipparion* remains are very common in the late Miocene faunas and usually constitute more than 50% of the collected material. The large number of specimens makes the Equidae family particularly important for the extraction of biostratigraphical, paleoecological, and paleobiogeographical conclusions. Pikermi is possibly the locality where equids are originally recognized in Greece. Wagner (1848) described a hipparion skull associated with the mandible as *Equus primigenius* but later was transferred to *Hippotherium primigenium* var. *mediterraneum* by Roth and Wagner (1854). Few years later, Gaudry (1862–67) refers the Pikermi hipparionine horse's material as *H. gracile*. Then, the hipparions were recognized in Samos, Axios Valley and in any Late Miocene and Pliocene Greek localities (Vlachou 2013 and references therein), while recently their presence was recorded in the Early Pleistocene (Athanassiou 2018).

The equids are also common in the Plio-Pleistocene localities of Greece with the genus *Equus* (Appendix 2). The earliest evidence of the genus was recognized in the locality Damatria, Rhodes Island (van der Meulen and van Kolfschoten 1986), which is probably correlated with the beginning of the Early Pleistocene (Koufos and Kostopoulos 2016). The genus is still living in the country as a domestic animal. It is worth mentioning that a small-sized breed of *Equus* is living in Skyros Island, having a mean height at acromion of about 110 cm.

3 Phylogenetic Relationships

The equids are the most frequent group in the fossil mammal faunas with many remains. The rich collections, the presence of several intermediate forms, and the extended studies on equids provided many data to demonstrate the principles of evolution. The earliest known equid is *Hyracotherium* (= *Eohippus*) known from the early Eocene (54.0 Ma). *Hyracotherium* was a fox-sized animal with squarish, bunodont teeth, four toes in the front legs and three in the hind ones with remains of the first and fifth toe; it was a browser feeding on soft plants and adapted to running. *Hyracotherium* evolved gradually increasing the crown height of the teeth and the molarization of the premolars, while the teeth became gradually lophodont. This change is clear in the middle Eocene with the genus *Orohippus* and integrated at the end of the Eocene with *Epihippus*. The first tridactyl equids with the lateral metapodials, being still in contact with the ground, appeared in the Oligocene with the genus *Mesohippus*. This leg type is also present in the primitive anchitheres, migrated to Eurasia at the beginning of the Miocene. During that time a strong increasing in the height of the third metapodial, tips up the third toe and the lateral metapodials lost their contact with the ground. This is clear in the genus *Merychippus*, from which *Hipparion* originates. The hipparions migrated to Asia through Beringian Bridge during the beginning of the late Miocene, at ~11.2 Ma and rapidly dispersed in Europe and Africa with many species until the Early Pleistocene. *Merychippus* gave another branch, which gradually gave the genus *Equus* in America, which migrated to Eurasia and Africa at the beginning of the Pleistocene (~2.6 Ma), and still exists in these regions. *Equus* preserves the central metapodial with a single toe, while the lateral metapodials decreased remarkably and remain as small spine-like bones, reaching to about the middle of the third metapodial. In America, *Equus* disappeared at ~8.000 years B.P., but it was re-introduced there a few hundred years ago by the European immigrants.

Three main trends have been recognized in the fossil evolution:

- An overall size increase. In fact, the size of *Eohippus/Hyracotherium* is significantly smaller than *Equus*. Although this character is used as a general trend in Equidae evolution, it is not typical. There are evolutionary lines in which the size decreases with the time as adaptation to a new paleoecological conditions.
- Premolar's moralization and crown height (hypsodonty) increase.
- Increase running adaptation. Elongation of the limb bones, reduction of the metapodial's number and size and tip-toed legs; these trends led gradually to more elastic legs and thus better adaptation for running.

4 Distribution

As mentioned above, the equids are widely distributed, covering all world except Australia where they were carried by the Europeans before some hundreds of years. The main Greek mammal localities with equids are demonstrated in Apendices 1 and 2. The earliest presence of the equids in the Greek fossil record is known from the early Miocene locality Aliveri (Evia Island), where *Anchitherium* was recognized (van Hoek Ostende et al. 2015). Hipparionines data are abundant in all Greek late Miocene localities but rare in the Pliocene ones, because the Pliocene localities are rare in Greece. In the Quaternary fossil mammal localities, the genus *Equus* is the most common and abundant taxon (see Appendix 1, 2).

5 Systematic Palaeontology

Hipparions
Equidae Gray, 1821
***Anchitherium* von Meyer, 1844**

Type Species *Palaeotherium aurelianense* Cuvier, 1825.

***Anchitherium aurelianense* (Cuvier, 1825)**

Nomenclatural and Taxonomical History *Palaeotherium aurelianense* in Cuvier 1825 (new species); *Anchitherium aurelianense* in von Meyer 1844 (new genus and new combination).

Type Material MOR-201, right maxillary fragment with P2–P4 (neotype, designated by Sánchez et al. 1998). No type specimen has been originally designated, and the specimen figured in Abusch (1983) was chosen as the neotype for this species.

Type Locality Montabuzard, France; early Miocene, MN4 (Sánchez et al. 1998).

Distribution The species is known from several fossiliferous sites of Eurasia (Spain, France, Germany, Turkey, Kazakhstan, Mongolia, China). In Greece, the species is only known from Aliveri.

Remarks *Anchitherium* was recently recognized in Greece by a single p4 (van Hoek Ostende et al. 2015). The species is characterized by: small size; brachyodonty; canines larger than the incisors; presence of the prefossette only in the M3; absence of postfossette in the upper teeth; elongated facet for the second and fourth metatarsals; presence of a non-articular pit on the facet for the cuneiform in the third metatarsal (Sánchez et al. 1998).

***Hipparion* de Christol, 1832**

Type Species *Hipparion prostylum* de Christol, 1832.

Remarks The Old Word hipparionines with tridactyl limbs and isolated protocone have been originally included within the genus *Hipparion*. According to MacFadden (1980), the name *Hipparion* was given to some material from Mt. Luberon (France); later Gervais (1849) referred a syntypic series from Mt. Luberon, while Osborn (1918) proposed *H. prostylum* as the genotype of *Hipparion*. Woodburne and Bernor (1980) gave emphasis to the cranial characters, especially those of the preorbital fossa, and divided the hipparionines to morphotypes; these morphotypes, years later, were recognized by Bernor et al. (1996) as different evolutionary lineages of supraspecific rank. Nowadays, both nomenclatures are in use. In this paper, we are following the old one, while Appendix 3, and 4 depict the Greek hipparionine species in morphotypes and/or genera according to Vlachou (2013).

★*Hipparion brachypus* Hensel, 1862

Nomenclatural and Taxonomical History *Equus primigenius* in Wagner 1848 (initial identification; in part); *Hipparion gracile* var. *mediterraneum* in Roth and Wagner 1854 (new combination; in part); *Hipparion brachypus* in Hensel 1862 (new species).

Type Specimen The species was originally erected on some metapodials from Pikermi (Hensel 1862), whose location is unknown; there is a cast in the Bayerische Staatssamlung für Paläontologie und Geologie, München, figured by Abel (1926), which can be considered a suitable lectotype for the species (Heissig, pers. comm. to Koufos). As Hensel's specimens are not found, and taking in mind that the original distinction of the two Pikermi hipparions was based on the metapodials, Koufos (1987a) proposed the metacarpal NHML-PIK-M.11240, housed in the National History Museum of London, as a suitable neotype and the metatarsal NHML-PIK-M.11265 and some metapodials of the Gaudry collection (MNHNP-PIK-42, 46, 48, 52, 54, 59, 104), housed in the Museum national d'Histoire naturelle of Paris, as potential topotypes of the species.

Type Locality Pikermi (PIK), Attica, Greece; late Miocene, MN12.

Distribution The geographic distribution of *H. brachypus* is restricted in the Eastern Mediterranean region (MN12–13; 7.3–6.5 Ma). Besides Greece, it was also recorded in Bulgaria, FYROM, and Turkey, while recently Bernor et al. (2016) described as aff. *H. brachypus* material from the early Turolian of the Middle Maragheh levels (Iran). Especially in Greece, and except its type locality (Fig. 1), *H. brachypus* is known from the localities Mytilinii 1A, B, Quarry 1, and Quarry 4 of Samos Island. Similar hipparionine forms have been also recognized in Pikermi Valley-1 and Pyrgos Vassilisis (Attica, close to the type locality Pikermi), in Kryopigi (Chalkidiki), in Kerassia (Northern Evia Island) and maybe in Quarry 6 of Samos Island. The questionable recognition of the species in the latter localities is based on scanty material, mostly metapodials.

Remarks The main characters of the species are: large size (~285 kg); elongated and wide muzzle; deep narial opening, nasal notch retracted above either the mesostyle of the P2 or the mesial margin of the P3; deep, elliptical-subtriangular, well-delineated and pocketed posteriorly preorbital fossa situated far from the orbit; rich enamel plication in the upper cheek teeth; usually elliptical protocone; double and sometimes triple pli caballin; occasionally presence of the dP1 (30.7% in the Pikermi); crenulated or plicated flexid's borders in the lower cheek teeth; pli caballinid rarely present; relatively short and robust metapodials (Koufos 1987a; Bernor et al. 1996; Vlachou 2013).

The comparison of the Greek *H. brachypus* samples reveals two different size forms. The first one, in general terms, follows the dimensions of the type material and have been found in the localities of southern Balkans dated up to 7.2 Ma (Appendix 3). The other one has larger size, without other significant morphological differences and has been recorded in Samos and Turkey localities dated to 7.0–7.2 Ma (Appendix 4). This is a spatiotemporal difference that is often observed between the taxa in the two regions on either side of the Aegean Sea (Vlachou 2013, and cited references therein) and probably related with the different paleoecology of the two regions at the second half of MN12 (Koufos et al. 2009a, b, 2011).

★*Hipparion proboscideum* Studer, 1911

Nomenclatural and Taxonomical History *Hipparion proboscideum* in Studer 1911 (new species); *Hemihipparion proboscideum* in Wehrli 1941 (new genus and new combination).

Type Specimen NHMB-no 46, cranium (holotype), housed in the Naturhistorisches Museum Bern (Studer 1911: figs 1, 3). According to Sondaar (1971), the mandible figured with the holotype belongs to another individual.

Type Locality The holotype originates from Samos Island (Greece), but the exact locality is unknown. All known mammal localities of Samos have been dated to Turolian, between 8.0–6.7 Ma (Koufos et al. 2009a, b).

Distribution *Hipparion proboscideum* is also referred from FYROM and possibly from Romania and Turkey. In Greece, and especially in Samos Island, *H. proboscideum* has been recorded in Mytilinii 1A, 1B, 3, and Quarry X, 1, 4 localities, but none of the available skulls are alike the type one; in continental Greece it is certainly present in the localities Ravin des Zouaves 5 in Axios Valley (Fig. 2) and probably Perivolaki in Thessaly. The possible presence of this large-sized species is also referred from other mammal localities of Greece (Mytilinii 4, Prochoma, Vathylakkos) but the specimens are not enough for definite identification (see also Appendix 4).

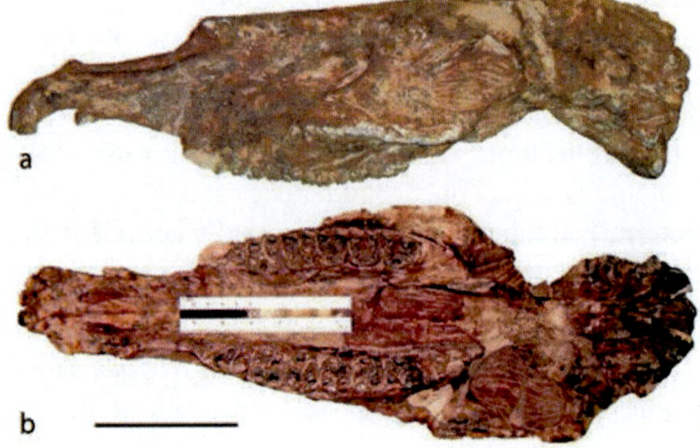

Fig. 1 The cranium of *Hipparion brachypus* (NHML-PIK-M.11191) from the middle Turolian (MN12) of Pikermi, Attica in **a** left lateral and **b** palatal views. Scale bar equals 10 cm

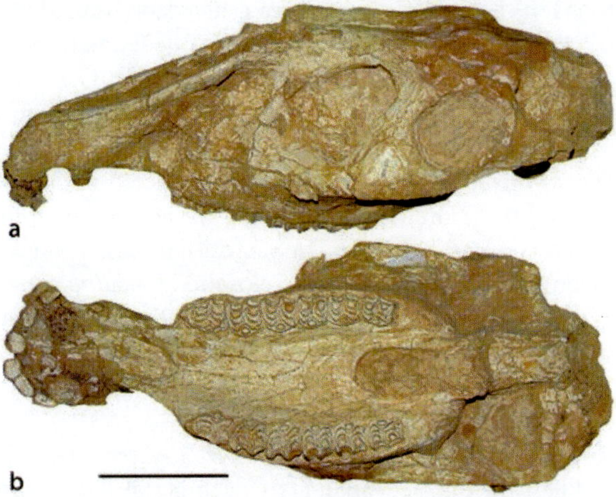

Fig. 2 The almost complete cranium of *Hipparion proboscideum* (LGPUT-RZO-172) from the early Turolian (MN11) of Ravin de Zouaves 5, Axios Valley in **a** left lateral and **b** palatal views. Scale bar equals 10 cm

Remarks The main characters of *H. proboscideum* are: large size; relatively elongated and narrow muzzle; deep narial opening, the nasal notch is variably retracted from the distal end of the P2 (Samos Quarry 1 and Quarry X) to the P4 (Samos holotype, Ravin des Zouaves 5); the large preorbital fossa which is subtriangular, well delineated, very deep and wide, antero-ventrally oriented, deeply pocketed posteriorly and situated very close to the orbit; presence of a large, subtriangular, deep, well-delineated and antero-ventrally oriented canine fossa; presence of the dP1; rich enamel plication in the upper cheek teeth; elliptical-oval protocone; double and rarely triple pli caballin; relatively short and robust metapodials (Vlachou 2013).

★***Hipparion mediterraneum*** **(Roth and Wagner, 1854)**

Nomenclatural and Taxonomical History *Equus primigenius* in Wagner 1848 (initial identification; in part); *Hipparion gracile* var. *mediterraneum* in Roth and Wagner 1854 (new species-group name as new variation; in part); *Hipparion mediterraneum* in Hensel 1862 (elevation to the species level).

Type Specimen The holotype of the species is unknown. Forstén (1968) proposed the skull and mandible described by Wagner (1848: fig. 1) and later by Roth and Wagner (1854) as the lectotype of the species; however, Vlachou (2013) considers that this skull belongs to *H. brachypus*. Based on the metapodial's sample from Pikermi, Hensel (1862) considered that there are two hipparions, the slender *H. mediterraneum* and the robust *H. brachypus*. Unfortunately, this material has been lost. Bernor (1985) proposed the skull MNHNP-PIK-259, housed in the Muséum national d'Histoire naturelle, Paris as the neotype of the species, while Koufos (1987a), based on the Hensel's (1862) distinction of the metapodials, proposed the associated metacarpals NHML-M.11240, housed in the Natural History Museum of London as the neotype. Nowadays, the skull and associated mandible

MNHNP-PIK-259 is considered as the neotype of *H. mediterraneum*; it is housed in the Muséum national d'Histoire naturelle, Paris (Fig. 3).

Type Locality Pikermi, Attica, Greece; late Miocene, MN12.

Distribution *H. mediterraneum* is restricted in the area of southern Balkans. It is referred from Greece and Bulgaria while, for the moment, there are no remains of the species eastwards, i.e., in Samos (Greece), Turkey, and Iran. In Greece, besides Pikermi, a hipparionine horse close to *H. mediterraneum* has been recently recognized in Ravin des Zouaves 5 (Axios Valley), while, based on few metapodials, the species is possibly present in Nikiti 2 (Chalkidiki Peninsula) and Perivolaki in Thessaly.

Remarks The main characters of *H. mediterraneum* are: moderate size; elongated and narrow muzzle; moderately deep narial opening; nasal notch retracted at least above the P2; moderately deep, well delineated, usually invading lacrymal bone and posteriorly pocketed preorbital fossa situated close to the orbit; occasionally presence of a small, subtriangular and variably in depth canine fossa with weakly-moderately developed posterior margin; moderate enamel plication in the upper cheek teeth with rounded and shallow plis; subtriangular-elliptical protocone; small, single and rarely double pli caballin; elongated and slender metapodials (Koufos 1987a; Bernor et al. 1996; Vlachou 2013; Koufos and Vlachou 2019).

★***Hipparion philippus* Koufos and Vlachou**, 2016

Nomenclatural and Taxonomical History *Hipparion dietrichi* in Koufos 1980 (initial identification); *Hipparion* cf. *dietrichi* in Vlachou 2013 (open specific identification); *Hipparion philippus* Koufos and Vlachou 2016 (new species).

Type Specimen LGPUT-NIK-894, skull and associated mandible (holotype) (Koufos and Vlachou 2016: figs. 18a, b, 19a, b) stored in the Laboratory of Geology and Palaeontology, University of Thessaloniki.

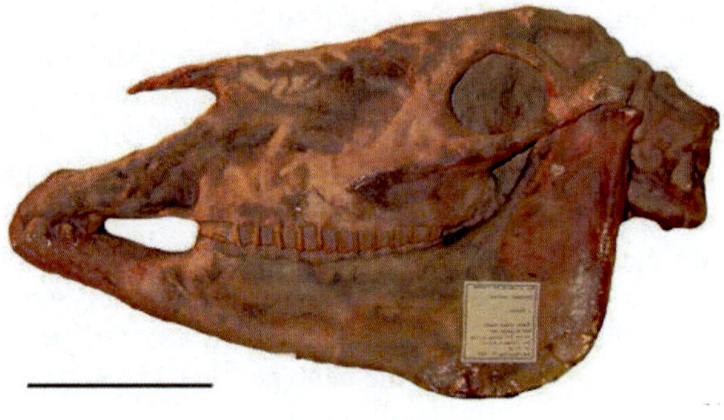

Fig. 3 The cranium and mandible of *Hipparion mediterraneum* (NHML-PIK-M.1883, neotype, cast housed in Paris) from the middle Turolian (MN12) of Pikermi, Attica in left lateral view. Scale bar equals 10 cm

Type Locality Nikiti 2, Chalkidiki Peninsula, Macedonia, Greece; late Miocene, MN11.

Distribution *Hipparion philippus* is certainly known from Greece and probably in the neighboring countries. In Greece, it is recognized from the localities Ravin des Zouaves 5, Prochoma 1, Vathylakkos 1, 2, 3 in Axios Valley, Nikiti-2 in Chalkidiki (Fig. 4), and Perivolaki in Thessaly.

Remarks The species shares common cranial morphology with *H. dietrichi* from Samos, but it has significantly smaller size and remarkably slenderer skeleton. For many years, the Axios Valley, Nikiti 2, and Perivolaki samples were described under the name *H. dietrichi*. Koufos and Vlachou (2016) decided the creation of the new species *H. philippus* based on the rich Nikiti 2 medium-sized hipparion sample. The different species name shed light on the different morphometry of the two hipparion forms, *H. dietrichi* and *H. philippus,* which was an unnoticed but important information related to the different paleoecological conditions on either side of the Aegean Sea (Appendices 1 and 2).

The main diagnostic features of the species are: moderate size; short and wide muzzle; short narial opening, nasal notch retracted above the middle of the diastema P2–C; weak, shallow, elliptical-subtriangular, antero-ventrally oriented and not posteriorly pocketed POF situated far from the orbit; wide and oval choanae with their anterior margin at the contact between M1 and M2; short and wide snout; high and thick mandibular corpus; short symphysis; moderate enamel plication in the upper cheek teeth; protocone elliptical-oval and isolated, except in the worn teeth; simple and large pli caballin and pli caballinid; low plicated flexid borders in the lower cheek teeth; elongated and slender metapodials (Koufos and Vlachou 2016).

★***Hipparion phlegrae* Lazaridis and Tsoukala,** 2014

Nomenclatural and Taxonomical History *Hipparion dietrichi* in Lazarid is 2010 (initial identification); *Hipparion phlegrae* in Lazaridis and Tsoukala 2014 (new species).

Type Specimen LGPUT-KRY 2800, skull (holotype), housed in the Laboratory of Geology and Palaeontology, Aristotle University of Thessaloniki, Greece (Fig. 5).

Type Locality Kryopigi, Chalkidiki Peninsula, Macedonia, Greece; middle-late Turolian, MN12–13.

Distribution It is only known from the type locality.

Remarks *H. phlegrae* is similar to *H. philippus*. The creation of the new species *H. phlegrae* relied on three main characters: the short and extremely broad muzzle, the reduced body mass, and the vestigial preorbital fossa. Other characters of this species are: the long distance between preorbital fossa and orbit; the shallow narial

Fig. 4 (**a**) The cranium and mandible of *Hipparion philippus* (LGPUT-NIK-909) from the early Turolian (MN11) of Nikiti 2, Chalkidiki in left lateral view, and **b** The cranium of *Hipparion philippus* (LGPUT-NIK-894, holotype) from the early Turolian (MN11) of Nikiti 2 in palatal view. Scale bar equals 10 cm

Fig. 5 The cranium of *Hipparion phlegrae* (LGPUT-KRY-2800, holotype, from the middle-late Turolian (MN12–13) of Kryopigi, Chalkidiki in **a** right lateral and **b** palatal views. Scale bar equals 10 cm. Photographs courtesy of G. Lazaridis

opening; the retraction of nasal notch above the P2; the simple enamel plication in the upper cheek teeth; the single pli caballin; and the small and slender metapodials (Lazaridis and Tsoukala 2014).

★*Hipparion dietrichi* (Wehrli, 1941)

Nomenclatural and Taxonomical History *Hipparion mediterraneum* in Forsyth Major 1892 (initial identification); *Hemihipparion dietrichi* in Wehrli 1941 (new genus, new combination); *Hipparion dietrichi* in Sondaar 1971 (new combination).

Type Specimen 1/7 GIM skull (holotype), housed in the Geological Institute of the University of Münster, Germany (Wehrli 1941: taf. 17, fig. 4; tafs 19, 20; taf. 22, figs. 2–4; taf. 23, figs. 5–7).

Type Locality The holotype originates from Samos Island, but the exact locality is unknown. All known mammal localities of Samos are of Turolian age, between 8.0 and 6.7 Ma (Koufos et al. 2009a, b).

Distribution Besides Greece, *H. dietrichi* is referred from Turkey and Iran; it is also mentioned from the Balkans with a smaller form, to *H. philippus*. In Greece is recognized in the localities Mytilinii 1A, B, Quarry 1, and Quarry 4 of Samos Island.

Remarks The main characters of *H. dietrichi* are: medium size; short and wide muzzle; weak, elliptical-subtriangular, shallow, open anteriorly but well defined posteriorly, and situated far from the orbit preorbital fossa; reduced enamel plication; elliptical to oval protocone; simple pli caballin and relatively elongated and slender metapodials (Vlachou 2013).

★*Hipparion macedonicum* Koufos, 1984

Nomenclatural and Taxonomical History *Hipparion* sp. in Koufos 1980 (initial identification); *Hipparion macedonicum* in Koufos 1984 (new species).

Type Specimen LGPUT-RPl-21, mandible with right i1–m3 and left i1–m2 (holotype), housed in the Laboratory of Geology and Palaeontology, University of Thessaloniki (Koufos 1984: figs 2a, 3a; Fig. 6).

Type Locality Ravin de la Pluie, Axios Valley, Macedonia, Greece, late Vallesian, MN10.

Distribution The species has been certainly identified in continental Greece. Some postcranial remains, however, close to *H. macedonicum*, are also referred from the localities Strumyani 1, Bulgaria (Geraads et al. 2011), Sumeg, Hungary (Bernor et al. 1999), and Montredon, France (Eisenmann 1988). In Greece, it has been fully described in the localities Ravin de la Pluie (type locality), Ravin des Zouaves 5, Prochoma, Vathylakkos, 1, 3 in Axios Valley, Nikiti 1, 2 in Chalkidiki Peninsula, and Perivolaki in Thessaly, Central Greece. It is also referred in the locality Pentalophos 1 and Ravin des Zouaves 1 in Axios Valley, but the recognition is based on few isolated metapodials (Appendix 3).

Remarks The species was originally based on mandibular remains from the late Vallesian locality Ravin de la Pluie (Koufos 1984). Later, the discovery of more cranial and postcranial material from RPl and younger levels added some more information for the species morphology. The main characters of the species are: small size; relatively elongated and narrow muzzle; shallow narial opening, the

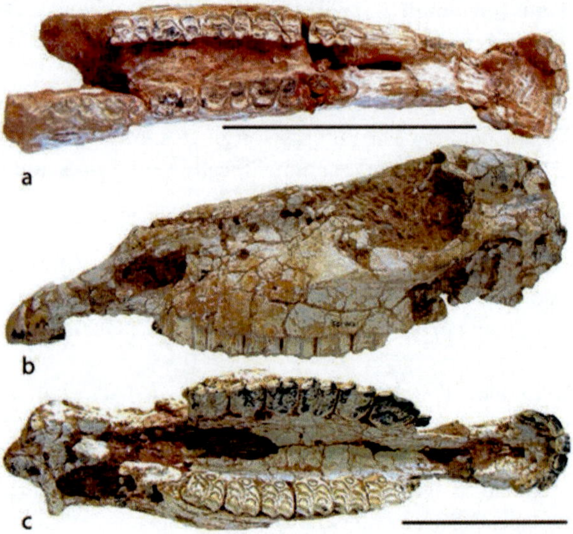

Fig. 6 (**a**) The mandible of *Hipparion macedonicum* (LGPUT-RPl-21, holotype) from the late Vallesian (MN10) of Ravin de la Pluie, Axios Valley in palatal view; (**b**, **c**) The cranium of *Hipparion macedonicum* (LGPUT-RPl-125) from the late Vallesian (MN10) of Ravin de la Pluie, Axios Valley in **b** left lateral, and **c** palatal views. Scale bar equals 10 cm

nasal notch is retracted above the middle of the C–P2 diastema; elliptical, shallow, antero-posteriorly oriented, not pocketed posteriorly with moderately developed margin and open anteriorly preorbital fossa, situated moderately far from the orbit; infraorbital foramen encroaches upon the antero-ventral border of the preorbital fossa; moderate enamel plication in the upper cheek teeth with narrow and deep plis; simple-double pli caballin; elliptical-oval and isolated protocone; often connection of the fossettes; very rare presence of a weak lingual hypoconal groove; often presence of a functional dP1; elongated and slender metapodials (Koufos 2016).

★*Hipparion matthewi* Abel, 1926

Nomenclatural and Taxonomical History *Hipparion minus* in Forsyth Major 1894 (initial identification); *Hipparion matthewi* in Abel 1926 (new species).

Type Specimen UGR OK/557, skull associated with the mandible (holotype) figured by Kormos (1911) and Abel (1926); it is housed in the Ungarische Geologische Reichsanstalt, Budapest (Fig. 7).

Type Locality The holotype originates from Samos Island, but the exact locality is unknown. All known mammal localities of Samos are of Turolian age, between 8.0 and 6.7 Ma (Koufos et al. 2009a, b).

Distribution Several small-sized hipparions from Samos and adjacent areas are referred as belonging to this species. *H. matthewi* has been mentioned in Spain, Bulgaria, Turkey, and Iran. However, except the Kemiklitepe and Sazak samples from Turkey, the rest material in most cases is identified as *H.* cf. *matthewi*

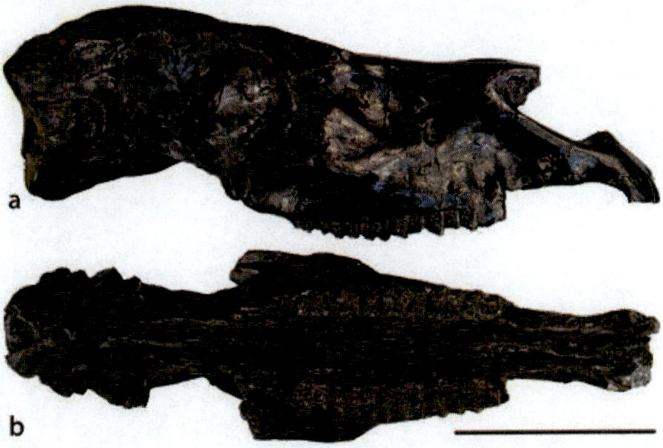

Fig. 7 The cranium of *Hipparion matthewi* (UGR OK/557, holotype, cast, housed in the Laboratory of Geology and Palaeontology, University of Thessaloniki) from an unknown locality of Samos Island in **a** right lateral and **b** palatal views. Scale bar equals 10 cm

(Vlachou 2013 and references therein). In our opinion, these determinations are under question.

Remarks The main characters of the species are: small size; short and narrow muzzle; shallow narial opening; nasal notch retracted above the parastyle of P2; single, oval, subtriangular, dorso-ventrally oriented and placed far from the orbit preorbital fossa; simple enamel plication in the upper cheek teeth; small and simple pli caballin; and elongated and slender metapodials (Bernor et al. 1996).

★***Hipparion sithonis*** **Koufos and Vlachou,** 2016

Nomenclatural and Taxonomical History *Hipparion macedonicum* in Koufos 1987 (initial identification; in part); *Hipparion sithonis* in Koufos and Vlachou 2016 (new species).

Type Specimen LGPUT-NIK-1707, skull (holotype), housed in the Laboratory of Geology and Palaeontology, University of Thessaloniki (Fig. 8).

Type Locality Nikiti 2, NIK, Chalkidiki, Macedonia, Greece; early Turolian, MN11.

Distribution It is only known from Greece; besides Nikiti 2, it was recently recognized in the locality Ravin des Zouaves 5 of Axios Valley. Even though there is not cranial material to verify the presence of *H. sithonis* in the neighboring localities, it is quite possible to be present in other early–middle Turolian localities not only in Greece but also in the Balkans.

Remarks *H. sithonis* and *H. macedonicum* share almost the same body mass and postcranial morphometry and thus they were referred under the name *H. macedonicum* (Koufos 1987b). The finding of some cranial remains of small hipparions in Nikiti 2 allow us to recognize two different cranial morphologies and to separate the small-sized material in two species (Koufos and Vlachou 2016; Koufos and Vlachou

Fig. 8 The cranium of *Hipparion sithonis* (LGPUT-NIK-1707, holotype) from the early Turolian (MN11) of Nikiti 2, Chalkidiki in **a** left lateral and **b** palatal views. Scale bar equals 10 cm

2019). The proposed diagnostic features of the *H. sithonis* are: small size; short and relatively wide muzzle; moderately deep narial opening; nasal notch retracting well behind the mesostyle of the P2; shallow, oval-subtriangular, not posteriorly pocketed and antero-ventrally oriented POF, placed close to the orbit; presence of a small and weak canine fossa; wide and deep palate; wide choanae with their anterior border at the middle of the M2; short upper tooth row; small and elliptical-oval protocone; very small and simple pli caballin; low-moderate enamel plication in the upper cheek teeth; no plicated or crenulated lower cheek teeth; rudimentary or absent pli caballinid; elongated and slender metapodials. *H. sithonis* differs from the small-sized *H. macedonicum*, in having slightly larger size, nasal notch retracting well behind the mesostyle of the P2, a small but well-defined canine fossa, and relatively larger and more robust metapodials (Appendix 3) (Koufos and Vlachou 2016).

★*Hipparion nikosi* Bernor and Tobien, 1989

Nomenclatural and Taxonomical History *Hipparion matthewi* in Tobien 1938 (initial identification); *Hipparion nikosi* in Bernor and Tobien 1989 (new species).

Type Specimen BSPM-1899 VII 31b, cranial fragment (holotype), housed in the Bayerische Staatssamlung für Paläontologie und Historische Geologie, München.

Type Locality The holotype originates from Samos Island, but the exact locality is unknown. All known mammal localities of Samos are of Turolian age, between 8.0 and 6.7 Ma (Koufos et al. 2009a, b).

Distribution It is mainly known from Samos Island, Greece; besides the unknown type locality, it was recognized in the localities Q5 in Samos Island (Fig. 9) and Kryopigi in Chalkidiki where probably coexists (in our opinion) with a form similar to *H. macedonicum* (Vlachou and Koufos 2009; Lazaridis 2015).

Fig. 9 The partial cranium of *Hipparion nikosi* (AMNH-Q5–22908) from the late Turolian (MN13) of Quarry 5, Samos Island in **a** right lateral and **b** palatal views. Scale bar equals 10 cm

Remarks *Hipparion nikosi* is characterized by small size, not different than that of *H. matthewi* and *H. macedonicum*; elongated and narrow muzzle; deep narial opening, nasal notch retracting above P3-P4; elliptical, shallow and antero-posteriorly oriented preorbital fossa; simple-moderate enamel plication in the upper cheek teeth; simple and small pli caballin; rounded-oval protocone; elongated and slender metapodials (Bernor and Tobien 1989; Vlachou 2013).

Hipparion longipes **Gromova**, 1952

Type Specimen PIN 2413/5030, third metatarsal (holotype), housed in the Palaeontological Institute of the Russian Academy of Sciences.

Type Locality Pavlodar, Irtych, Kazakhstan; Late Miocene–Early Pliocene.

Distribution Besides Kazakhstan, the species was recorded in Hungary and Turkey. In Greece, a few postcranial remains are known from the locality Megalon Emvolon, near Thessaloniki.

Remarks The creation of the species *H. longipes* is only based on postcranial material. Relying on the original diagnosis of Gromova (1952), *H. longipes* is characterized by: large size; very long and slender extremities; elongated metapodials; moderate enamel plication in the upper cheek teeth; short and wide protocone.

Hipparion crassum **Gervais,** 1859

Type Specimen According to Forstén (1968), the skull described and figured by Deperét (1890) is the lectotype of the species, and it is possibly housed in the museum of Lyon.

Type Locality Perpignan, Roussillon Basin, France; Pliocene, MN15 (Mein 1990).

Distribution Besides France, *H. crassum* is found in Spain, Hungary, and Bulgaria. In Greece, it was found in Ptolemais Basin (Western Macedonia), while it is also referred from the locality of Apolakkia in Rhodes Island.

Remarks The known material of *H. crassum* is poor. The morphology of the facial region is unknown; it is possibly a large hipparion (cheek tooth length = 170 mm) with rich enamel plication in the upper cheek teeth and very short and robust metapodials (Bernor et al. 1996). A maxillary fragment and some postcranial remains from Ptolemais Basin have been described as *H. crassum*; the described third metacarpal has similar morphology (very short and robust) and dimensions with this species (Koufos 1982).

Greek Hipparions with Questionable Determination

The Greek hipparionine sample is abundant, including material from several localities. In some cases, however, the available material is poor and/or fragmentary (some isolated teeth and/or bones), not allowing definite identifications. In these cases, the occurrence is identified as *Hipparion* sp. and they are not included in this paper. In the following lines, we will refer hipparion species that either they have not been fully recognized in the Greek fauna, but their presence (or a relative form) is quite possible (cf.) or their species name is under study (aff.) or their old species name is under study, so they are included in quote marks.

Hipparion cf. *sebastopolitanum*

Taxonomical History *Hipparion primigenium* in Koufos 1980 (initial identification); *Hipparion* cf. *depereti* in Koufos 2000a (new identification); *Hipparion* cf. *sebastopolitanum* in Vlachou 2013 (new identification).

Localities Pentalophos 1 (Fig. 10), Ravin de la Pluie in Axios Valley, Macedonia, Greece; Vallesian, MN9–10.

Remarks *Hipparion gracile sebastopolitanum* was erected as a subspecies on some material from Sebastopol (Crimea, Russia) by Borissiak (1914) and later upgraded to species level by Gromova (1952). The Greek material has the following characters: moderate to large size; relatively elongated and wide muzzle; shallow narial opening, nasal notch retracted above the middle of the diastema C–P2; oval, deep, posteriorly pocketed, moderately delineated, antero-ventrally oriented and situated far from the orbit preorbital fossa; invasion of the lacrymal to the preorbital fossa; moderate enamel plication in the upper cheek teeth; oval-subtriangular protocone; usually double pli caballin; presence of the dP1; strong protostylid in all lower

Fig. 10 (**a**) The cranium of *Hipparion* cf. *sebastopolitanum* (LGPUT-PNT-118) from the early Vallesian (MN9) of Pentalophos 1, Axios Valley in right lateral view; (**b**, **c**) The partial cranium of *Hipparion* cf. *sebastopolitanum* (LGPUT-PNT-136) from the early Vallesian (MN9) of Pentalophos 1, Axios Valley in **b** right lateral, and **c** palatal views. Scale bar equals 10 cm

cheek teeth; weak pli caballinid in the molars; short and robust metapodials (Koufos 2000a; Vlachou 2013).

H. cf. *sebastopolitanum* is considered as the most primitive hipparion in Greece. Together with *Cormohipparion sinapensis* and *H. kecigibi* from Sinap (Turkey) (Bernor et al. 2003), as well as *Hipparion* sp. from Yulafli (Turkey) (Geraads et al. 2005), they are probably remains of the first migrated hipparion clade that dispersed in Asia and Eastern Mediterranean region during early Vallesian.

Hipparion* aff. *giganteum

Taxonomical History *Hipparion primigenium* in Koufos 2000b (initial identification); *Hipparion* aff. *giganteum* in Vlachou 2013 (new identification).

Locality Nikiti 1, Chalkidiki, Macedonia, Greece; terminal Vallesian, MN10.

Remarks The species *H. giganteum* was erected by Gromova (1952) on some material from the late Miocene locality Grebeniki (Ukraine), correlated with the zone MN10 (Vangengeim and Tesakov 2013). The main characters of the Nikiti 1 hipparion (Fig. 11) are: large size, similar to that of *H. giganteum* from Grebeniki; moderately long and wide muzzle; shallow narial opening; oval, moderately deep and antero-posteriorly oriented preorbital fossa situated far from the orbit; moderate enamel plication in the upper cheek teeth; oval-subtriangular protocone; double pli caballin; short and robust metapodials with similar dimensions and proportions to those of *H. intrans* and *H.* aff. *intrans* of Central Paratethys (Hungary) (Koufos 2000b; Vlachou 2013).

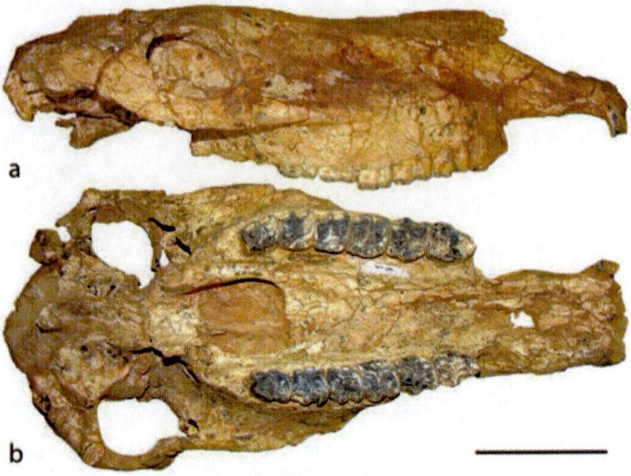

Fig. 11 The almost complete cranium of *Hipparion* cf. *giganteum* (LGPUT-NKT-183) from the terminal Vallesian (MN10) of Nikiti 1, Chalkidiki in **a** right lateral and **b** palatal views. Scale bar equals 10 cm

Hipparion aff. *platygenys*

Taxonomical History *Hipparion mediterraneum* in Koufos 1980 (initial identification); *Hipparion* aff. *platygenys* in Vlachou 2013 (new identification).

Localities Dytiko 1, 2, 3, Axios Valley, Macedonia, Greece; late Turolian, MN13.

Remarks Gromova (1952) erected *H. platygenys* on some material from the locality Taraklia (Moldavia); the locality is of late Miocene age and correlated to the biozone MN12 (Vangengeim and Tesakov 2013). The Dytiko remains of this hipparion (Fig. 12) are characterized by: relatively elongated and narrow muzzle; shallow narial opening, nasal notch situated above the mesial margin of the P2; oval and moderately deep preorbital fossa situated far from the orbit; moderate enamel plication in the upper cheek teeth; round-oval protocone; single pli caballin; relatively elongated and slender metapodials (Koufos 1988; Vlachou 2013). The large-sized hipparion from Dytiko localities, though it shares common cranial characters with *H. platygenys* sample from Hadjidimovo, its postcranial remains are almost identical to *H. mediterraneum*.

Hipparion aff. *moldavicum*

Taxonomical History *Hipparion matthewi* in Koufos 1988 (initial identification); *Hipparion* aff. *moldavicum* in Vlachou 2013 (new identification).

Localities Dytiko 1, 2, 3, Axios Valley, Macedonia, Greece; late Miocene, MN13.

Distribution It is only known from the three Dytiko localities.

Remarks Gromova (1952) erected the species *H. moldavicum*, based on material found in the locality Taraklia (Ukraine). According to Vlachou (2013), some postcranial remains from the Dytiko localities have similarities with *H. moldavicum*, but

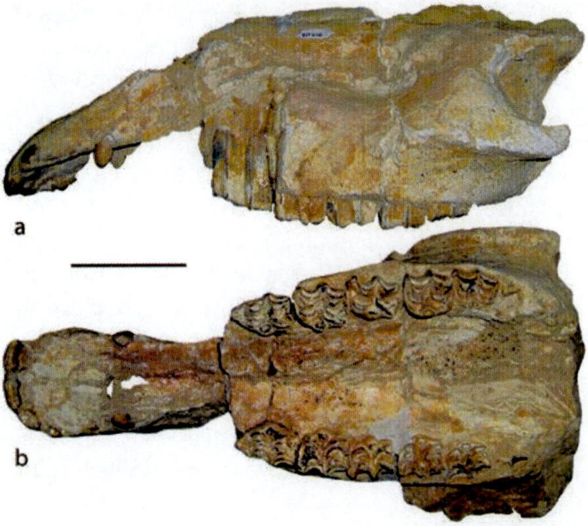

Fig. 12 The partial cranium of *Hipparion* aff. *platygenys* (LGPUT-DTK-6) from the late Turolian (MN13) of Dytiko 1, Axios Valley in **a** left lateral and **b** palatal views. Scale bar equals 10 cm

the absence of crania and the limited material cannot allow definite determination and thus they are referred to as *H.* aff. *moldavicum*.

Hipparion "prostylum" Gervais, 1849

Nomenclatural and Taxonomical History *Hipparion prostylum* in Gervais 1849 (new species).

Type Specimen Sondaar (1974) refers a right maxilla with P4–M2, figured by Gervais (1859: pl 19, fig. 2), as the type specimen of the species; this specimen is probably housed in the Muséum Requien, Avignon, France. The partial skull, NHML-M33603, stored in the Natural History Museum of London is referred as the lectotype of the species (Bernor 1985).

Type Locality Mont Lubéron (or Mont Lebéron or Cucuron), France; middle Turolian, MN12.

Distribution In Greece, it is certainly referred from the locality Quarry 6 in Samos Island. Vlachou (2013) refers the same morphometry also in the locality Mytilinii 4 in Samos and the localities of Kemiklitepe D and Çorakyerler, in Turkey.

Remarks The species is characterized by moderate size; short and wide muzzle; shallow narial opening (nasal notch retracts in front of the P2); moderate, oval-shaped, well-delineated and antero-posteriorly oriented preorbital fossa; moderate enamel plication in the upper cheek teeth; elliptical-oval protocone; simple or rarely double pli caballin; relatively elongated and slender metapodials (Vlachou 2013).

The Quarry 6 sample (Fig. 13) strongly reminds *H. philippus* in size and cranial morphology but the preorbital fossa is more oval, almost antero-posteriorly oriented and certainly deeper, like in *H. prostylum*. Furthermore, the metapodials seem to be more robust, closer to *H. prostylum* rather than to *H. philippus* and under this point of view the Quarry 6 sample was attributed to *H. prostylum*. It is important to say that Bernor et al. (2011) attribute at least the postcranials from Q6 to *H. brachypus*. Old references described *H. prostylum* from Mt. Lubéron as the ancestor of *H. dietrichi*. According to our opinion, based on the new available material, the two species probably evolved in different bioprovinces following different lineages. For this reason, the Samos samples must be referred under different species name.

Hipparion* aff. *forstenae

Taxonomical History *Hipparion* cf. *matthewi* in Forstén and Kaya 1995; cf. *Hipparion mediterraneum* in Forstén 1999; aff. *H. mediterraneum* in Vlachou and Koufos 2004; *Hipparion* aff. *forstenae* in Vlachou 2013.

Localities Mytilinii 1A, 1B, 3 and Quarry 1, Samos Island, Greece; middle Turolian, MN12.

Remarks The species *H. forstenae* was erected on some material from Shanxi, China by Zhegallo (1971). The main characters of the material from Samos and Turkey are: moderate size; elongated and narrow muzzle; moderately deep narial opening; nasal notch situated above the P2; elliptical-subtriangular, shallow, well-delineated and antero-ventrally situated preorbital fossa; presence of a weak but well-defined canine fossa; moderate enamel plication in the upper cheek teeth; elliptical-oval protocone; single pli caballin; elongated and slender metapodials (Vlachou 2013). *H.* aff. *forstenae* is represented in Quarry 1 and Mytilinii 1A (Fig. 14) of Samos by a significant number of postcranials and more than six almost complete skulls. The sample was referred by Vlachou (2013) to *H.* aff. *forstenae* because the skull morphology described above was known only in China. Deng

Fig. 13 The cranium of *Hipparion "frostylum"* (AMNH-Q6–22990) from the early Turolian (MN11) of Quarry 6, Samos Island in **a** right lateral and **b** palatal views. Scale bar equals 10 cm

Fig. 14 The almost complete cranium of *Hipparion* aff. *forstenae* (NHMA-MTLA-338) from the middle Turolian (MN12) of Mytilinii 1A, Samos Island in **a** left lateral and **b** palatal views. Scale bar equals 10 cm

(2009) refers that the faunal exchange or dispersal between the two regions was gradually increased from the early Vallesian to the middle Turolian, but the common species, apart from the suid *Microstonyx major,* were mainly carnivores. Nowadays, the recognition of *H. sithonis* in the early Turolian fauna of Nikiti 2 led us to presume that *H.* aff. *H. forstenae* is probably an evolved form of *H. sithonis* adapted to the new paleoecological conditions of Asia Minor, its resemblance with the Chinese *H. forstenae* is a homoplasy.

Plesiohipparion* cf. *shanxiense

Taxonomical History *Equus (Hippotigris) stenonis* in Symeonidis 1992 (initial identification); *Plesiohipparion* cf. *shanxiense* in Athanassiou 2018 (new identification).

Locality Sesklon, Volos, Thessaly, Greece; late Pliocene, MN16.

Remarks *Plesiohipparion shanxiense* was described from Shanxi, China (Bernor et al. 2015). The Sesklon material includes some mandibular and postcranial remains. The lower teeth are large with U-shaped linguaflexid, angular metaconid and metastylid, simple enamel plication and wide and short ectoflexid; the metapodials are relatively elongated and slender (Athanassiou 2018).

The Evolution of Greek Hipparions

Hipparionine horses arrived in Eurasia through Beringian Bridge from North America at about 11.2 Ma marking the beginning of the Late Miocene. During early Vallesian MN9, the *Hipparion* fauna indicates that there were no natural barriers preventing the dispersal of the North American *Hipparion* ancestor throughout

Eurasia. As soon as hipparions inhabited Eurasia, they were restricted in biogeographic provinces possibly having temporary communication each other, where evolved under the pressure of local paleoecological conditions, firstly in the skeleton morphometry and later (especially during Turolian) in the cranial morphology (Eisenmann 1995; Bernor et al. 1996; Vlachou 2013).

In the Greco-Iranian province, right after the arrival of the first medium-sized hipparions, two main evolutionary trends-branches appeared. The first one is characterized by an increase in the body size and skeleton robustness, while the other one by a relative lengthening of the limbs compared to the body and later by a decrease in the body mass and a gradual change of the cranial morphology (Fig. 15). In Greece, *H.* cf. *sebastopolitanum* is almost directly connected to the first hipparion immigrant and includes the most primitive characters of the Greek hipparionines, while *H. giganteum* and the younger Turolian form *H. brachypus* belong to the lineage with the large and robust build hipparions (Fig. 15). All members of this lineage have conservative evolution and share more or less similar cranial characters (*primigenium* morphotype) (Appendices 3 and 4).

Although primitive members of the second branch (with more slender limbs), such as the medium-sized *H. uzunagizli*, have been recognized in Anatolia (Sinap) during early Vallesian (MN9), (Bernor et al. 2003), in Greece the slender build *H. macedonicum* from the early Vallesian (MN9), (Koufos 2000a), is quite smaller than *H. uzunagizli* or the derived *H. moldavicum*, and it is characterized by even more derived cranial morphology. From this point of view, *H. macedonicum* probably represents the first member of an early divergence clade from the second branch (Fig. 15).

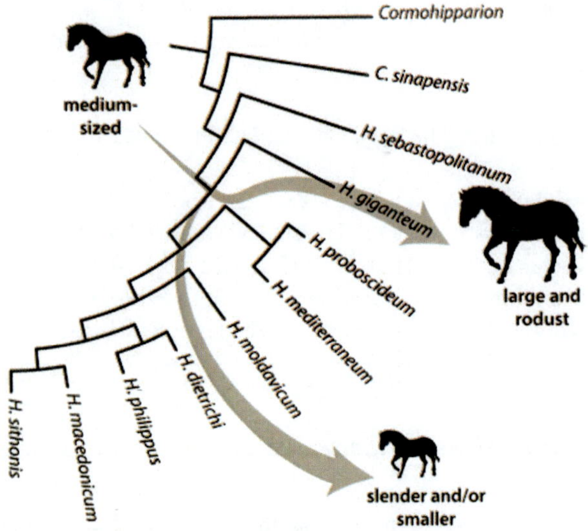

Fig. 15 Cladogram corresponding to the single tree obtained in the cladistic analysis of the Greek hipparione taxa by Koufos and Vlachou (2019), with a simplified indication of the changes in the body size of these hipparione horses in Greece

During Turolian, the increased seasonality, contrary to the more humid conditions of the Vallesian, led to more open landscapes, especially in the Eastern Mediterranean region, where even from the beginning of the Vallesian the conditions were open and dry (de Bonis et al. 1992; Koufos 2006c). Thus, slender forms had the advantage of having further differentiation/diversification (Fig. 15). It is worth mentioning that the number of species per locality increased from two or three in the Vallesian to five or more in the Turolian (Appencices 3 and 4).

H. sithonis and *H. philippus,* originally recognized from the early Turolian fauna (MN11) of Nikiti 2, are probably related to *H. macedonicum* as they share a lot of common characters. On the other hand, *H. proboscideum* and *H. mediterraneum,* although members of the second branch, retain some primitive characters in their morphology, suggesting probably a divergence of this sister clade from members of the second branch, most probably relatives of *H. moldavicum* (Fig. 15).

H. dietrichi and *H.* aff. *forstenae* occurred quite later, during the middle of MN12, in Samos and Asia Minor faunas, at the opposite side of the Aegean Sea. Both species morphologies are known from the MN11 hipparion faunas of Continental Greece as well. Spassov et al. (2018) consider *H. philippus* of Nikiti 2 to be a synonym of *H. dietrichi* downplaying the importance of their differences, while Koufos and Vlachou (2016) regard *H.* aff. *forstenae* as relative of *H. sithonis* based on their morphological affinities.

As a matter of fact, *H. dietrichi* arguably differs from *H. philippus* and these differences are not random but due to different populations of different times. So, ascribing their differences to paleoecological changes (apomorphic characters), *H. dietrichi* could be a descendant of *H. philippus* (Fig. 15). Similarly, *H.* aff. *forstenae* could be a descendant of *H. sithonis*. However, presently too little material is known from the MN11 hipparion faunas from Asia Minor to substantiate the above phylogenetic relationships. Especially for the *H. dietrichi* evolution, the convergence to *H. philippus* evolution will be a possibility, if the presence of *H. "prostylum"* was certified in the late MN11 fauna of Samos and Asia Minor. In such a case, *H. philippus* probably will be a paraphyletic form of *H. dietrichi* in respect to the first hipparion immigrant that appeared and evolved separately to *H. dietrichi*, while their morphological affinities are result of homoplasy. In any case, we believe that more extensive systematic and biogeographic analyses are necessary for the *dietrichi* morphotype.

H. matthewi and *H. nikosi* belong to the small hipparions clade that evolved parallel to *H. macedonicum* in Asia Minor. It is plausible that *H. nikosi* is a derived member of this clade probably firstly appeared in Asia Minor during MN11 and later migrated to the Balkans at ~7.2 Ma where evolved and gave *H.* aff. *H. nikosi* (Kryopigi) or *H.* cf. *macedonicum* of Dytiko reducing its size. *H. macedonicum* and *H. matthewi* share a common ancestor and they are probably parts of the same lineage that evolved in parallel on each side of the Aegean Sea.

At the end of Miocene, these species extincted and were succeeded by larger, higher-crowned horses: *H. longipes*, *H. crassum,* and *Plesiohipparion* cf. *shanxiense,* all members of "*Sivalhippus*" group derived from East and South Asia (Bernor et al. 1996).

The Dispersal of Hipparion *Throughout the Greco-Iranian Bioprovince*

The active tectonics of the Aegean region caused the isolation of the Southern Balkans from the Near East during MN10 (Kostopoulos 2009). This is reflected in the hipparion assemblages from both sides of Aegean Sea through the absence of Turkish hipparion species from the Greek localities. However, the presence of *H. macedonicum* in Southern Balkans during Vallesian and the recognition of the small-sized *H. minus* in Sebastopol (Ukraine), and *Hipparion* sp., with affinities to *H. macedonicum,* in SE France and Hungary, at the same time, make possible the contemporaneous communication between the Southern Balkans, the Black Sea area, the Central and West Europe during Vallesian (Fig. 16).

According to Kostopoulos (2009), the regions on both sides of the Aegean Sea remain isolated throughout the Turolian except for a short period of time, around 7.3–7.2 Ma, when changes in the Mediterranean Sea hydrography led to temporary connections of areas across the Aegean allowing mammalian species exchanges. Contrary to the well-recognized early Turolian (MN11) hipparion fauna of the Continental Greece, little are known about the early Turolian hipparion fauna of Asia Minor. The skulls of *H. proboscideum* recognized from the locality Quarry X of Samos Island and the fragmentary material of *H. matthewi* from the locality Kemiklitepe D of Turkey, attest that at least two hipparion morphologies were common on both sides of the Aegean Sea during early Turolian. On the other hand, the presence of *H. "prostylum"* in the Quarry 6 of Samos Island and the absence of *H. mediterraneum* from the Samos and Turkish localities reinforce a kind of isolation, but the available material is too scarce to fully support this possibility. However, the fossil records indicate that hipparion species of similar body mass (100–170 kg) and morphologies were living on both areas during the early Turolian, suggesting similar paleoecological conditions on both sides of the Aegean Sea at least during MN11, open landscapes with increased seasonality (Vlachou 2013 and references therein) (Fig. 17).

During the middle Turolian (MN12; ~7.2 Ma), the conditions changed and the hipparion species in each side of the Aegean Sea completely changed as well. In the Southern Balkans, *H. proboscideum*, *H. macedonicum, H. sithonis*, and *H. philippus* disappeared and their ecological niche was occupied by the relatively large-sized *H. mediterraneum* and the newly appeared in the Balkans *H. brachypus* (Fig. 18). This faunal turnover is coeval with a dramatic cooling in the Mediterranean region at the Tortonian-Messinian transition. Paleobotanic proxies demonstrate C4-grass dominated wooded grassland-to-woodland habitats of a savannah biome for the Pikermi Formation (Böhme et al. 2017). On the other side of the Aegean Sea, based on data from the localities Mytilinii 3 (~7.3 Ma) and Mytilinii 1 (~7.1 Ma) in Samos Island, an increase of the open and dry characters of the environment is quite possible (Koufos et al. 2009a, b) (Fig. 20b). In evidence of this change five hipparion species, of different body mass participate in the composition of the middle Turolian mammal fauna *Hipparion proboscideum* and *H. matthewi* seem to have a

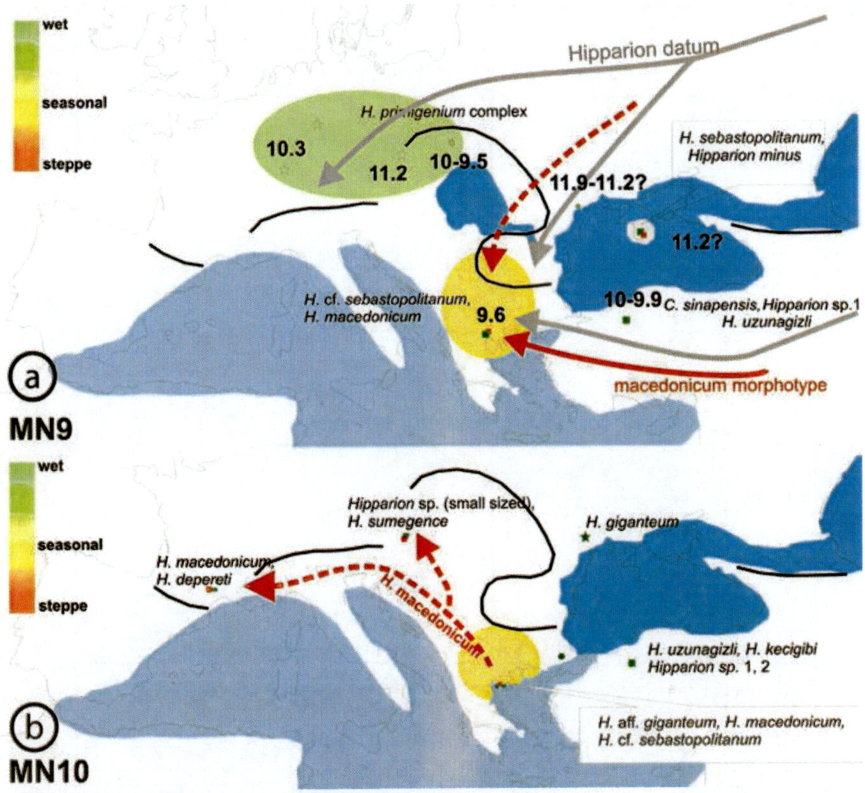

Fig. 16 Paleogeographic map of the Eastern Mediterranean region, indicating the *Hipparion* exchanges between the Balkans and the neighboring areas during the **a** early (MN9) and **b** late (MN10), Vallesian, (Vlachou 2013 and references therein)

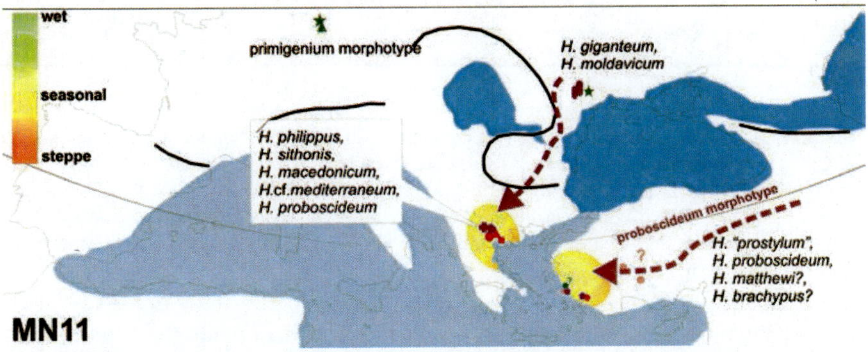

Fig. 17 Paleogeographic map of the Eastern Mediterranean region, indicating the *Hipparion* exchanges between the Balkans and the neighboring areas during the lower part of the middle Turolian, MN11 (Vlachou 2013 and references therein)

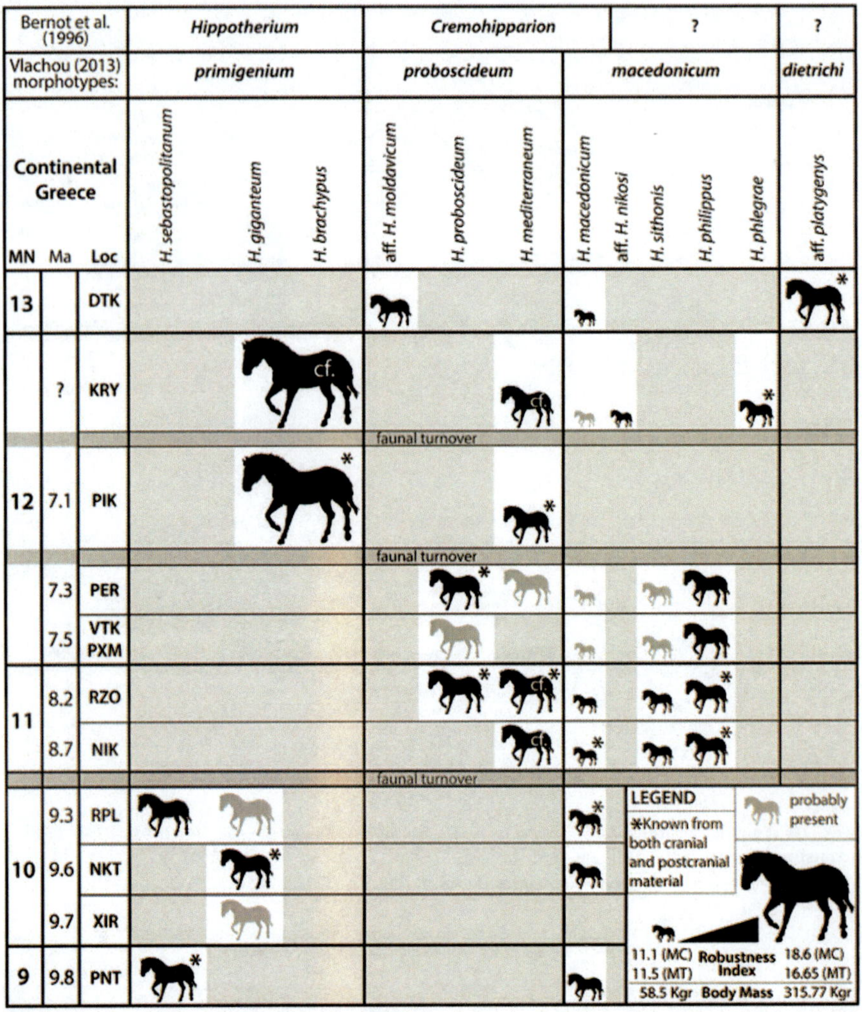

Fig. 18 Biostratigraphic distribution of the different *Hipparion* morphotypes and species found in Continental Greece. Asterisk indicates that the identification is based on both cranial and postcranial material; body-mass and robustness index are given in 95% CI. Silhouette from phylopic.org, made by M. Yrayzoz and M. Keesey. Locality abbreviations: **PNT** Pentalophos 1, **XIR** Xirochori 1, **NKT** Nikiti 1, **RPl** Ravin de la Pluie, **RZO** Ravin des Zouaves 5, **PXM** Prochoma 1, **VTK** Vathylakkos 2, **PER** Perivolaki, **PIK** Pikermi, **KRY** Kryopigi, **DTK** Dytiko 1. See Appendix 3 for further information

continuous presence in the area throughout the Turolian, while the appearance of *H. dietrichi* and *H.* aff. *forstenae,* which are probably descendants of *H. philippus* (or *H. "prostylum"*) and *H. sithonis* respectively, in addition to the increasing hipparion body mass, in some cases, up to 50%, indicate paleoecological changes (Figs. 19 and 20). *Hipparion brachypus* was firstly recognized in the Q4 fauna of Samos at ~7.2 Ma (Koufos et al. 2009a, b). Although larger, it shares common

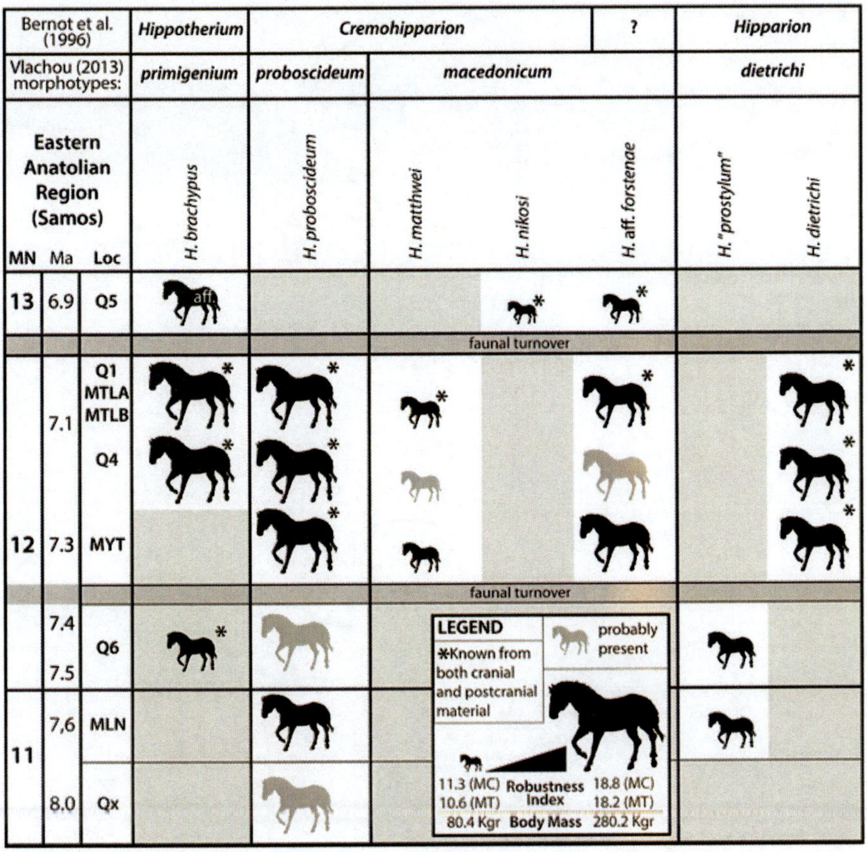

Fig. 19 Biostratigraphic distribution of the different *Hipparion* morphotypes and species found in Samos Island, Greece. Asterisk indicates that the identification is based on both cranial and post-cranial material; body-mass and robustness index are given in 95% CI. Silhouette from phylopic.org, made by M. Yrayzoz and M. Keesey. Locality abbreviations: **Qx** Quarry X, **MLN** Mytilinii 4, **Q6** Quarry 6, **MYT** Mytilinii 3, **Q4** Quarry 4, **Q1** Quarry 1, **MTL** Mytilinii 1, **Q5** Quarry 5. See Appendix 4 for detailed information

characters with the sample of Pikermi without real differences in their skeleton morphometry. The small-sized *H. nikosi* has not been recorded in the Samos faunas older than 6.9 Ma, but Bernor and Tobien (1989) refer that the type skull of the species is probably coming from the locality Q2 of Samos, recently dated to 7.4 Ma (Koufos et al. 2009a, b).

So, the different hipparion faunas and evolutionary trends, recorded to each side of the Aegean Sea after 7.2 Ma, agree with the assumption of Kostopoulos (2009), indicating isolation of the areas. On the other hand, the morphological affinities of *H.* aff. *forstenae* and *H. dietrichi* from Samos with *H. sithonis* and *H. philippus* from Balkans, respectively, in addition to the appearance of *H. brachypus* in Pikermi and Hadjidimovo faunas and the presence of *H. nikosi* in Titov Veles fauna as well,

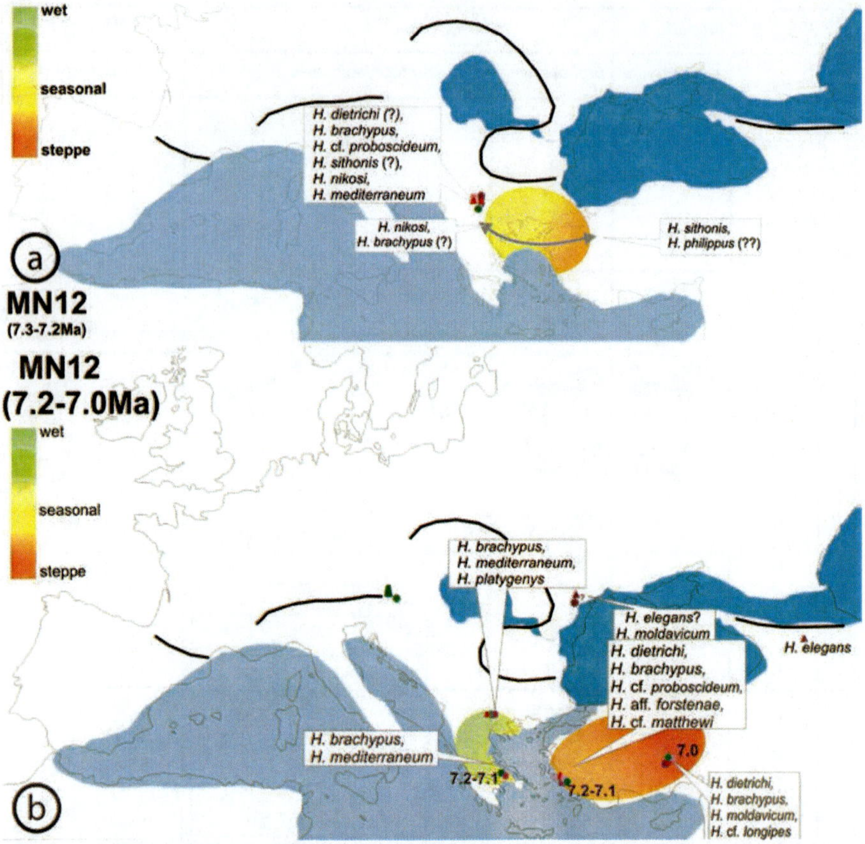

Fig. 20 Paleogeographic map of the Eastern Mediterranean region, indicating the *Hipparion* exchanges between the Balkans and the neighboring areas during the **a** middle and **b** uppermost part of the middle Turolian, MN12 (Vlachou 2013 and references therein)

indicate a new dispersal wave from east toward the Balkans and vice versa. This suggests a short time communication between both regions on each side of the Aegean. *Hipparion* data placed this communication prior to 7.2 Ma and probably right after 7.5 Ma (Fig. 20a).

At the beginning of the late Turolian (MN13), the paleoecological transition to more humid conditions (de Bonis et al. 1992; Koufos 2006c; Koufos and Vasileiadou 2015) caused significant changes in the composition of the hipparion fauna on both sides of the Aegean Sea and important reduction of the hipparion's weight at about 100 kg on average (Figs. 18 and 19). Southern Balkans and Anatolia have few mammal localities with informative fauna from this time interval. Hence, hipparionines are not fully determined and in consequence the inferred information is limited. The possible presence of species with affinities to *H. platygenys* and *H. moldavicum* in Dytiko fauna indicates that both species moved southwards during the early MN13,

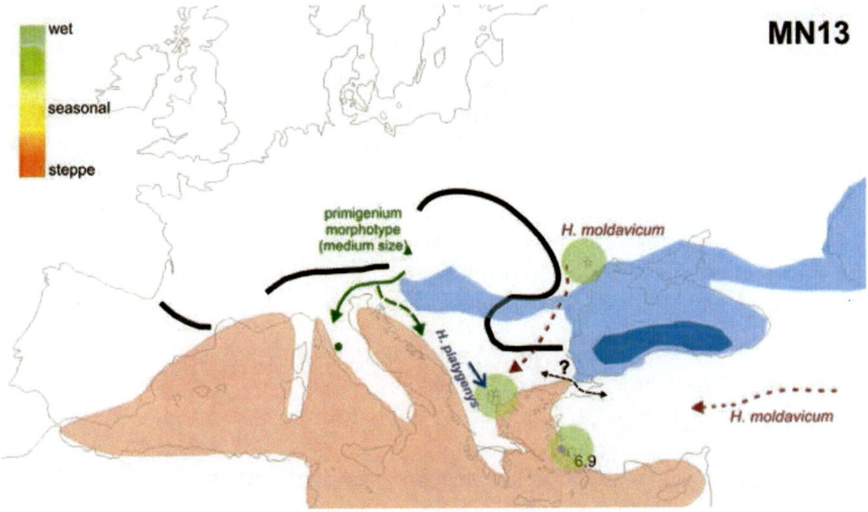

Fig. 21 Paleogeographic map of the Eastern Mediterranean region, indicating the *Hipparion* exchanges between the Balkans and the neighboring areas during the late Turolian, MN13 (Vlachou 2013 and references therein)

extending their territory and taking advantage from the reduction of the open landscapes (Fig. 21). But it is not secure based only on postcranial similarities to infer definite species relationships, since postcranials could belong either to unrelated hipparion forms of similar posture or even more to similar, but not directly related, hipparion forms. Old references, for example, refer the dwarf hipparion from Dytiko as a relative of *H. periafricanum* from Spain and part of the same evolutionary lineage. Orlandi-Oliveras et al. (2019) suggest a different growth strategy between Eastern and Western dwarf hipparions resulting in body mass reduction, which probably indicates different ancestors.

Horses

Equus Linnaeus, 1758

Type Species *Equus caballus* Linnaeus, 1758.

Equus (Allohippus) stenonis Cocchi, 1867

Nomenclatural and Taxonomical History There are several taxonomical interpretations regarding the position of the European stenonoid horses (in this case *Equus stenonis*) relative to the extant horses. The name *Allohippus* Kretzoi, 1938, originally has been used on the genus level, now it is used at different levels of the taxonomic rank. Gromova (1949a, b) and Azzaroli (1982) use *Allohippus* as a subgenus of *Equus*. Eisenmann and Baylac (2000) and Eisenmann (2002) separated the primitive horses (*Allohippus*) from the modern ones (true *Equus*) and consider *Allohippus* to represent a genus comprising the stenonoid species *Allohippus stenonis*. There is still no consensus on the taxonomic status of the stenonoid equids.

Horses from several Eurasian local faunas, dating from the middle to late Villafranchian, have been mostly referred to different subspecies of *E. stenonis*. The main differences among them are in size and proportions. Several subspecies have been proposed, containing *E. s. stenonis* Cocchi, 1867 from Upper Valdarno (Italy), *E. s. vireti* Prat, 1964 from Saint-Vallier (France), *E. s. guthi* Boeuf, 1986 from Chilhac (France), *E. s. pueblensis* Caloi, 1997 from La Puebla de Valverde (Spain), and *E. s. olivolanus* Caloi, 1997 from Olivola (Italy). The validity of these subspecies is debated (Forstén 1999). Other subspecies: *E. s. pamirensis* (Sharapov, 1986), *E. s. bactrianus* (Zhegallo 1988), even from North America *Equus stenonis anguinus* Azzaroli and Voorhies, 1993 from Idaho.

Type Specimen IGF 560 (holotype), skull and mandible, belonging to a male adult individual, housed in the Geological Institute of Florence, Italy (Azzaroli 1965: p. 5; tab. I, fig. 3; tab. II, fig. 1, 1a; tab. V, fig. 1, 1a).

Type Locality Upper Valdarno (Terranova), Italy, Early Pleistocene, MNQ18–20.

Distribution The species is widely distributed in Eurasia from Spain to China, known by several subspecies. In Greece, it is referred from the localities Vassiloudi-1, Ravin of Voulgarakis, Kalamoto-1, 2 and Riza-1 in Mygdonia Basin, Sésklon in Thessaly, Kos Island and probably from Pyrgos in Peloponnese, Vatera-E in Lesvos Island, and Aliakmon Q-Profil in Western Macedonia.

Remarks *Equus stenonis* is a large-sized and robust equid. A typical feature of this species is the deep indentation of the narial notch, which extends caudally above the third premolar. The great elongation of the narial notch is always apparent. *Equus stenonis* has a slender and elongated snout and the diastema is very long. The braincase is rather small in relation to the length of the face and is strongly deflected and the forehead is transversally undulated. *Equus stenonis* shares some common features of the forehead and braincase with *Equus livenzovensis*, but the preorbital pit is poorly developed and the posterior palatine foramina are shifted caudally, at the level of the M3 in the type skull. The protocone on the upper cheek teeth is usually short, similar to *Equus livenzovensis*, but it tends to be more elongated in some subspecies. The enamel of the inner fossettes is richly plicated. The lower cheek teeth are typically stenonoid (zebrine) with V-shaped linguaflexid and the ectoflexid is either deep or shallow; in the type skull is shallow (Azzaroli 1965, 1982). *Equus stenonis* followed *E. livenzovensis* in time and suggests an intermediate form between *E. livenzovensis* and *E. s. senezensis* (Azzaroli, 1989). *Equus stenonis* is believed to have been replaced by the gracile *Equus altidens* sometime during the latest Villafranchian (Fig. 22).

***Equus altidens* von Reichenau, 1915**

Nomenclatural and Taxonomical History *Equus altidens* in von Reichenau 1915 (new species).

Fig. 22 The partial cranium of *Equus stenonis* cf. *vireti* (LGPUT-DFN-112) from the middle Villafranchian of Dafnero 1 in **a** right lateral and **b** palatal views. Scale bar equals 10 cm

Type Specimen The lectotype of this species is, according to Alberdi et al. (1998), the right P_2 figured by von Reichenau (1915: pl. 6, fig. 17), and designated by Schwarz (1928: 437), stored in the Universität Halle a.d. Saale Collection (Germany).

Type Locality Süssenborn, Germany, Middle Pleistocene, MNQ22.

Distribution The species is recognized from Italy, Spain, Germany, and France, while it is possibly known from Romania, Georgia, Great Britain, Turkey, Tadjikistan, Hungary, and Russia. In Greece, it is known from the localities Gerakarou-1 in Mygdonia Basin (Fig. 23), as well as from Libakos, Polylakkos, and Aliakmon Basin in Western Macedonia.

Remarks This species was originally described by von Reichenau (1915), who described a few isolated teeth from Süssenborn, Germany. Later, Musil (1969) included several teeth and slender metapodials from the same locality to this species, part of which are certainly belonging to *E. altidens*. Alberdi and Ruiz-Bustos (1989) revised *E. granatensis* and referred it as a subspecies of *E. altidens,* named as *E. a. granatensis*.

Equus altidens is a medium-sized stenonoid horse with dental morphology like *E. stenonis* (usually short protocone, V-shaped double knot). The shape of the protocone is variable and the linguaflexid is either shallow or deep depending on the level of the attrition. The limb bones of *Equus altidens* are more elongated and very slender (especially the metapodials) than any other stenonoid horse. The most closely related species to *E. altidens* is *E. s. senezensis*, but it differs from the latter having slenderer metapodials and phalanges (Alberdi et al. 1998).

Fig. 23 The cranium of *Equus altidens* (LGPUT-GER-8) from the late Villafranchian of Gerakarou 1 in **a** dorsal and **b** palatal views. Scale bar equals 10 cm

★*Equus apolloniensis* Koufos, Kostopoulos, Sylvestrou, 1997

Type Specimen LGPUT-APL-148 (holotype) skull, stored in the Laboratory of Geology and Paleontology, University of Thessaloniki, described and figured by Koufos et al. (1997: pl. I, figs. 1, 2).

Type Locality Apollonia 1, Mygdonia Basin, Macedonia, Early Pleistocene, MNQ20.

Distribution Besides the type locality, the species is recognized in the Greek locality Platanochori-1 in Mygdonia Basin and possibly in Denizli (Turkey).

Remarks *Equus apolloniensis* is a large-sized horse with both caballoid and stenonoid features. Unlike *E. stenonis*, *E. apolloniensis* has a short narial notch that ends above the distal lobe of the P2. The buccinator fossa is well developed and the braincase is rather small like *E. stenonis*. The facial crest is shorter than *E. stenonis*. The choanae are more elongated than any other stenonoid horse and the palate is relatively shorter. The protocone is more elongated than in *E. stenonis* and the double knot is typically stenonoid (V-shaped). The metapodials are larger and slenderer than *E. stenonis*. *Equus apolloniensis* possibly represents a transitional form from the typical (archaic) stenonoid horses (*E. stenonis*) to the advanced *E. suessenbornensis* (Koufos et al. 1997) (Fig. 24).

Equus "caballus" Linnaeus, 1758

Distribution The modern widespread distribution of *E. caballus* is due to its domestication, and as such, its geographic distribution during the last glacial period (9,500–15,000 years ago) is considered as the typical one. During that time wild horse populations were living in Eurasia, North America, and North Africa (Bennet

Fig. 24 The cranium of *Equus apolloniensis* (LGPUT-GER-8) from the Epivillafranchian (latest Villafranchian) of Apollonia 1 in **a** right lateral and **b** palatal views. Scale bar equals 10 cm

and Hoffmann 1999). Several subspecies are known from the different geographic areas. In Greece, the species is referred from the localities of Aggitis in Eastern Macedonia, Penios Valley in Thessaly, as well as in several prehistoric places. The subspecies *E. c. piveteaui* and *E. c.* cf. *germanicus* are referred from the cave Agios Georgios (Kilkis area, Macedonia).

Remarks This name is used for several samples described from the Greek sites but the material needs more detailed studies for better determination and thus the name caballus is used within brackets. The main characters of *E. caballus* are: heavy body build, elongated skull, large and dorso-laterally oriented orbits, elongated molar row (>80 mm), wide incisor's cup, usually long and bipartite protocone, deep hypoconal groove, more richly plicated fossettes of the upper cheek teeth than in the other species of the genus, and caballoid lower cheek teeth (Bennet and Hoffmann 1999).

Equus hydruntinus **Regalia,** 1907

Nomenclatural and Taxonomical History Regalia (1904) described some equid's remains from Romanelli caves as *Equus (Asinus)* sp.; a few years later the same author established *Equus (Asinus) hydruntinus* as a new subspecies (Regalia 1907).

Type Material One upper and one lower molar, distal fragment of metacarpal and three phalanges (first second and third) described and figured by Regalia (1904, 1907).

Type Locality Romanelli Caves, Castello Italy, Paleolithic.

Distribution The species is recognized in Iberian Peninsula, France, Germany, Czech Republic, Bulgaria, Azerbaijan, Jordan, Syria, and Iran; it is possibly present in North Africa (Libya and Tunisia). In Greece, it is referred from the cave of Agios

Georgios in Central Macedonia, Penios Valley in Thessaly and Franchthi Cave (H1B faunal series) in Argolis Peninsula.

Remarks *Equus hydruntinus* is characterized by small-size, slender limb bones, cursorial limb proportions resembling those of the hemiones, short narial opening (relatively shorter than that of asses and hemiones), relatively wide muzzle, relatively small cheek teeth (the cheek teeth pattern mostly resembles modern plains zebra or the primitive *Equus stenonis*, short protocone on the upper cheek teeth, primitive enamel pattern on the lower cheek teeth resembling that of *E. stenonis*, and deep ectoflexid on the lower molars (Gromova 1949a, b; Davis 1980; Forstén 1986; Burke et al. 2003).

Greek Horses with Questionable Identification

Equus stenonis cf. *vireti*

Localities In Greece, a form similar to *Equus stenonis vireti* is described from the localities Dafnero-1 in Western Macedonia, and Volax in Eastern Macedonia.

Remarks *Equus stenonis vireti* is a large-sized horse, named by Prat (1964). The size of the skull is like *Equus stenonis*, but *E. s. vireti* has larger and more robust limbs. The Greek material is characterized by elongated skull, deep narial opening like *E. stenonis* from Saint-Vallier, typically stenonoid cheek teeth, and short and robust metapodials like *E. s. vireti* from Saint-Vallier (France) and *E. s. guthi* from La Puebla de Valverde (Spain) (Koufos and Kostopoulos 1993; Koufos and Vlachou 1997).

Equus cf. *stehlini*

Localities *Equus stehlini* Azzaroli, 1965 was described from Upper Valdarno (Terranova, Italy) as a new species (Azzaroli 1965). In Greece, a similar form is referred without description from the area of Pyrgos in Peloponnese, Early Pleistocene?, MNQ 18 (Van der Meulen and Van Kolfschoten 1986).

Remarks *Equus stehlini* is characterized by its smaller size relative to *E. stenonis* and even to *E. s. senezensis*. The skull is smaller in size than *E. stenonis*, it has slenderer muzzle, but shows the same specialization in the deep narial notch. On the upper teeth, the protocones are very short, and the ectoflexid is deep in the lower molars. The metapodials are short with a moderate slenderness index (Azzaroli 1965, 1982; Alberdi et al. 1998).

★"*Equus petraloniensis*" Tsoukala, 1989

Nomenclatural and Taxonomical History *Equus petraloniensis* in Tsoukala 1989 (new species), *Equus* cf. *altidens* (Forstén 1999), *Equus hydruntinus petraloniensis* (Eisenmann et al. 2008).

Type Specimen PEC-500 (holotype) second and third metacarpal, stored in the Laboratory of Geology and Palaeontology, University of Thessaloniki, described and figured by Tsoukala (1989: pl. XXVII, fig. a).

Type Locality Petralona Cave, Chalkidiki, Central Macedonia, Middle-Late Pleistocene.

Remarks *Equus petraloniensis* is possibly an intermediate form of *E. stenonis-hydruntinus* group (Tsoukala 1989). The species is characterized by its gracile and slender metapodials. The lower teeth are typical stenonoid (V-shaped ectoflexid with variable depth). The taxonomic position of *Equus petraloniensis* is still debated. According to Forstén (1999), the gracile equid from Petralona resembles the equids from Libakos, Krimni and Gerakarou, and they are all referred as *E.* cf. *altidens*. Eisenmann et al. (2008) include the gracile equid from Petralona, within the variation of *Equus hydruntinus* and refer it as "*E. hydruntinus petraloniensis*."

***Equus "asinus"* Linnaeus, 1758**

Distribution Several Greek samples were referred as *E. asinus*, but as we have not seen the material we are not sure about the determinations and thus the name asinus used into brackets. Domesticated asses are living worldwide, although in most of the developed countries they are in the zoos or in special parks. On the other hand, in the underdeveloped countries they are used by humans for different works. Feral populations of *E. asinus* exist in many places, especially in Western Australia and western United States. In Greece, it was recognized in Lerna, Crete Island. In the fossil record, the ancestral forms of *E. asinus* have been traced in the early Pleistocene of northern Africa and then dispersed in the peri-Mediterranean region and Red-Sea area by the end of the Pleistocene (Churcher 1982; Azzaroli 1995).

Remarks The species includes the asses which are known by several subspecies living in different regions. *E. asinus* is the smallest living species of the genus. The morphology of the skull of *E. asinus* is similar to *E. hemionus*. *E. asinus* is differentiated from *E. hemionus* by the lower enamel plication in the fossette's borders, and the shorter and more robust metapodials (Grubb 1993; Grinder et al. 2006).

***Equus ferus* Boddaert, 1785**

Nomenclatural and Taxonomical History *Equus ferus* in Boddaert 1785 (new species); *Equus gmellini* in Antonius 1912 (junior synonym); *Equus sylvestris* in Brinken 1826 (junior synonym); *Equus sylvaticus* in Vetulani 1928 (junior synonym); *E. tarpan* in Pidoplichko 1951 (junior synonym).

Holotype Unknown.

Type Locality Woronesk, Russia, Asia.

Distribution This species represents the primitive wild horse of Europe, ranging from Iberian Peninsula to Central Russia. In Greece, it was recognized in Penios Valley, Thessaly.

Remarks The wild horse was living in steppe and forest environments of Eurasia. Based on archeological evidences, the wild horses lived in Eurasia for a long time; ancient paintings found in caves of Southern France and Northern Spain and dated from 17–39 Ky include several horses. After the last glaciation, the increase of the humans and forests had as a result the decrease of the wild horse which slowly-slowly disappeared. During the last decades, biologists tried to create a breed resembling that of the wild horse, using selective breeding. The breeds they got have some primitive features but genetically were not a wild horse.

***Equus abeli* Antonius, 1913**

Type Specimen Antonius (1913) did not designate a holotype for this species.

Type Locality Heiligenstädter, Wien, Austria.

Distribution It is referred from different European countries. In Greece, the species is referred from Kos Island, (Aegean Sea) and Megalopolis (Peloponnese).

Remarks It is closely related to *E. mosbachensis,* while some researchers synonymized it with *E. caballus*. A mandibular fragment preserving the right tooth row i2–p3 and some bone fragments from Megalopolis (Peloponnese) have been described as *E. abeli* by Melentis (1966). The presence of this genus is questionable as the available material from Megalopolis is very poor, while the determinations for the Kos Island material (Forsyth-Mayor 1887) are outdated and need revision.

Appendix 1

List of the Greek localities containing hipparionine's fossils. Type localities are marked in bold. Locality numbers refer to the collection numbers of the PaleoBiology Database (PBDB).

Localities[PBDB]	Age (ELMA; MN)	Taxon
Alifakas[207134]	Late Miocene (MN11–13)	*Hipparion mediterraneum* [1]
Achladi[207133]	Late Miocene (MN11–13)	*Hipparion mediterraneum* [2]
Rhodes	Late Miocene (MN11–13)	*Hipparion dietrichi* [3]
Tanagra	Late Miocene (MN11–13)	*Hipparion mediterraneum* [4]
Triada[207135]	Late Miocene (MN11–13)	*Hipparion mediterraneum* [2]
Sesklon[205043]	Early Villafranchian (MN16)	*Plesiohipparion* cf. *shanxianse* [5]
Milia[185859–62, 195290, 195298, 195299]	Early Villafranchian (MN16)	*Hipparion* sp.[6]
Megalon Emvolon[36579]	Ruscinian (MN15)	*Hipparion longipes* [7]
Ptolemais[35086]	Ruscinian (MN14)	*Hipparion crassum* [8]
Nea Silata[191612]	Turolian/Ruscinian (MN13/14)	*Hipparion mediterraneum* [9]
Thermopigi[73553]	Turolian (MN11–13)	*Hipparion* cf. *matthewi* [10] *Hipparion* cf. *dietrichi* [10]
Samos, Loc. Unknown	**Early-Middle Turolian (MN13; 6.9–6.7)**	***Hipparion nikosi*** **[11]** ***Hipparion matthewi*** **[12]**
Samos, Quarry 5[95690]	**Late Turolian (MN13; 6.9–6.7)**	Hipparion nikosi [13], *Hipparion* aff. *forstenae* [13] *Hipparion* aff. *dietrichi* [13]
Dytiko 1, 2, 3[32374–6]	Late Turolian (MN13; 7.0–6.5)	*Hipparion* aff. *platygenys* [13] *Hipparion* aff. *moldavicum* [13] *Hipparion* cf. *macedonicum* [13]
Kryopigi[157582]	**Middle-Late Turolian (MN12–13; 7.3–6.4)**	***Hipparion phlegrae*** **[14]** *Hipparion* cf. *nikosi* [15] cf. *Hipparion brachypus* [15] *Hipparion* sp. [15]
Halmyropotamos[202213]	Middle Turolian (MN12)	*Hipparion mediterraneum* [16] *Hipparion dietrichi* [13] *Hipparion brachypus* [13] *Hipparion* cf. *proboscideum* [13] *Hipparion* cf. *matthewi* [13] *Hipparion* aff. *forstenae* [13]
Chomateres[195562]	Middle Turolian (MN12; ~7.1)	*Hipparion mediterraneum* [17]
Pikermi[182754]	**Middle Turolian (MN12; 7.1–7.25)**	***Hipparion brachypus*** **[18]** ***Hipparion mediterraneum*** **[19]**

(continued)

Localities[PBDB]	Age (ELMA; MN)	Taxon
Perivolaki[194879]	Middle Turolian (MN12; ~7.3–7.1)	*Hipparion macedonicum* [13,20] *Hipparion philippus* [13,20] *Hipparion* cf. *proboscideum* [13] ?*Hipparion* cf. *mediterraneum* [13,20]
Samos, Quarry 4[95689]	Middle Turolian (MN12; ~7.3–7.2)	*Hipparion dietrichi* [13] *Hipparion brachypus* [13] *Hipparion* cf. *proboscideum* [13] *Hipparion* cf. *matthewi* [13] *Hipparion* aff. *forstenae* [13]
Pyrgos Vassilissis[195555]	Middle Turolian (MN12; ~7.3)	*Hipparion brachypus* [21]
Vathylakkos 1[182750], 2[202702], 3[202702]	Middle Turolian (MN12; ~7.3)	*Hipparion philippus* [13,20] *Hipparion macedonicum* [13,20]
Mytilinii 3[202218]	Middle Turolian (MN12; ~7.3)	*Hipparion* sp. (large size) [13] *Hipparion* cf. *matthewi* [13] *Hipparion* aff. *forstenae* [13] ?*Hipparion* aff. *prostylum* [13]
Prochoma[202222]	Early Turolian (MN11; ~7.4)	*Hipparion* sp. [13,20] *Hipparion phillipus* [13,20] *Hipparion macedonicum* [13,20]
Kerassia[195432]	Early-Middle Turolian (MN11–12)	*Hipparion brachypus* [22] ?*Hipparion mediterraneum* [22] ?*Hipparion dietrichi* [22]
Mytilinii 4[202219]	Early Turolian (MN11; 7.5–7.7)	*Hipparion* aff. *proboscideum* [13] *Hipparion* aff. *prostylum* [13]
Samos, Quarry X[211913]	Early Turolian (MN11; 8.0–7.6)	*Hipparion proboscideum* [13]
Samos, Loc. Unknown	**Early-Middle Turolian (8.0–6.7)**	*Hipparion dietrichi* [23] ***Hipparion proboscideum*** **[24]**
Ravin des Zouaves 5[195489]	Early Turolian (MN11; ~8.2)	*Hipparion proboscideum* [25] *Hipparion* cf. *mediterraneum* [25] *Hipparion philippus* [25] *Hipparion sithonis* [25] *Hipparion macedonicum* [25]
Nikiti 2[73869]	**Early Turolian (MN11; 8.7–8.2)**	*Hipparion* sp. (large sized) [20] ***Hipparion philippus*** **[20]** ***Hipparion sithonis*** **[20]** *Hipparion macedonicum* [20]
Nikiti 1[202729]	Terminal Vallesian (MN10; 9.3–8.7)	*Hipparion* cf. *giganteum* [13] *Hipparion macedonicum* [13]
Ravin des Zouaves 1[182746]	Late Vallesian (MN10; ~9.3)	*Hipparion macedonicum* [13]
Ravin de la Pluie[191070]	**Late Vallesian (MN10; ~9.3)**	*Hipparion* cf. *sebastopolitanum* [13] ***Hipparion macedonicum*** **[26]**
Xirochori 1[195490]	Late Vallesian (MN10; ~9.6)	*Hipparion* sp. (large-sized) [27]
Diavata[210627]	?Early Vallesian (MN10)	*Hipparion* indet. [28]

(continued)

Localities[PBDB]	Age (ELMA; MN)	Taxon
Pentalophos 1[202119]	Early Vallesian (MN9)	*Hipparion macedonicum* [13] *Hipparion* cf. *sebastopolitanum* [13]
Kastellios K2[205361], K5[205364]	Vallesian (MN9–10)	*Hipparion* cf. *primigenium* [29]
Aliveri[68032]	Middle Orleanian (MN4)	*Anchitherium aurelianense*30

MN Mammal Neogene Zone
[1]Melentis and Schneider (1966), [2]Mitzopoulos (1947), [3]Boni (1943), [4]Mitzopoulos (1961), [5]Athanassiou (2018), [6]Vlachos et al. (2018), [7]Steffens et al. (1979), [8]Koufos (1982), [9]Koufos (2000a), [10]Geraads et al. (2007), [11]Bernor and Tobien (1989), [12]Abel (1926), [13]Vlachou (2013), [14]Lazaridis and Tsoukala (2014), [15]Lazaridis (2015), [16]Melentis (1966), [17]Marinos and Symeonidis (1974), [18]Hensel (1862), [19]Roth and Wagner (1854), [20]Koufos and Vlachou (2016), [21]Böhme et al. (2017), [22]Kampouridis et al. (2019), [23]Wehrli (1941), [24]Studer (1911), [25]Koufos and Vlachou (2019), [26]Koufos (1984), [27]Koufos (2006b), [28]Koufos (1984), [29]de Bruijn et al. (1971), [30]van Hoek Ostende et al. (2015)

Appendix 2

List of the Greek localities containing *Equus* fossils. Type localities are marked in bold. Locality numbers refer to the collection numbers of the PaleoBiology Database (PBDB).

Localities[PBDB]	Age (MN)	Taxon
Penios Valley[204397]	Late Pleistocene	*Equus ferus* [1] *Equus hydruntinus* [2] *Equus caballus* [2]
Aggitis Cave[204391]	Late Pleistocene	*Equus caballus* [3]
Franchthi Cave[182709]	Late Pleistocene	*Equus hydruntinus* [4]
Lerna	Late Pleistocene	*Equus asinus* [5]
Megalopolis[34809]	Middle-Late Pleistocene	*Equus abeli* [6]
Petralona Cave[183123]	**Middle-Late Pleistocene**	***Equus petraloniensis*** [7] *Equus caballus piveteaui* [7]
Kyparissia[194472]	Middle Pleistocene	*Equus* sp. [8]
Marathousa[34765]	Middle Pleistocene	*Equus* sp. [9]
Kos	?Villafranchian	*Equus abeli* [10] *Equus stenonis* [10]
Q-Profil[34812]	Villafranchian	*Equus stenonis* [11]
Volos[34808]	Villafranchian	*Equus stenonis* [12]
Riza 1[219198]	Villafranchian	*Equus stenonis* [13]
Platanochori[204658]	Latest Villafranchian (MNQ20)	*Equus apolloniensis* [14]
Kalamoto 1[200083]	Latest Villafranchian (MNQ20)	*Equus stenonis* [15]
Ravin Voulgarakis[34783]	Latest Villafranchian (MNQ20)	*Equus stenonis* [13]

(continued)

Localities[PBDB]	Age (MN)	Taxon
Apollonia[34784]	**Latest Villafranchian (MNQ20)**	***Equus apolloniensis*** **[16]**
Alikes[34782]	Late Villafranchian (MNQ19)	*Equus stenonis* [2]
Kapetanios[34781]	?Late Villafranchian (?MNQ19)	*Equus* sp. [17]
Polylakkos[34763]	Late Villafranchian (MNQ18–19)	*Equus altidens* [17] *Equus* cf. *altidens* [17]
Libakos[34764]	Late Villafranchian (MNQ18–19)	*Equus altidens* [17]
Krimni1[34762]	Late Villafranchian (MNQ18–19)	*Equus stenonis* [13]
Tsiotra Vrissi[197943]	Late Villafranchian (MNQ18–19)	*Equus* sp. (large-sized) [14] *Equus* sp. (small-sized) [14]
Pyrgos[34655]	?Late Villafranchian (?MNQ18)	*Equus* cf. *stehlini* [12] *Equus* cf. *stenonis* [12]
Vassiloudi[34651]	Late Villafranchian (MNQ18)	*Equus stenonis* [18]
Gerakarou 1[34617]	Late Villafranchian (MNQ18)	*Equus altidens* [13, 19]
Vatera E[183342]	Middle/Late Villafranchian (MNQ17/18)	*Equus* sp. [20] *Equus* cf. *stenonis* [20]
Volax[34593]	Middle/Late Villafranchian (MNQ17/18)	*Equus stenonis* cf. *vireti* [21]
Sesklon[34614]	Middle/Late Villafranchian (MNQ17/18)	*Equus stenonis* [22]
Dafnero[34594]	Middle/Late Villafranchian (MNQ17/18)	*Equus stenonis* cf. *vireti* [23]
Tourkovounia 3–5[34592]	?Early Middle Villafranchian (MN17)	*Equus* indet. [24]
Epanomi[204660]	Early Villafranchian (MN16)	*Equus* sp. [25]
Damatria[34591]	?Early Villafranchian (MN 16)	*Equus* sp. [26]

MN/MNQ Mammal Neogene–Quaternary Zone
[1]Athanassiou (2011), [2]Athanassiou (2002), [3]Koufos (1981), [4]Stiner and Munro (2011), [5]Reese (2008), [6]Melentis (1966), [7]Tsoukala (1989), [8]Athanassiou (2018), [9]Konidaris et al. (2018), [10]Forsyth-Mayor (1887), [11]Steensma (1988), [12]Van der Meulen and Van Kolfschoten (1986), [13]Koufos (1992), [14]Konidaris et al. (2015), 15Tsoukala and Chatzopoulou (2005), [16]Koufos et al. (1997), [17]Gkeme (2016), 18Koufos et al. (1995), [19]Gkeme et al. (2017), [20]de Vos et al. (2002), [21]Koufos and Vlachou (1997), [22]Athanassiou (2018), [23]Koufos and Kostopoulos (1993), [24]Symeonidis and de Vos (1976), [25]Athanassiou and Kostopoulos (2010), [26]Benda et al. (1977)

Appendix 3

Continental Greece		Bernor et al. (1996)	***Hippotherium***			***Cremohipparion***				?				?
		Vlachou (2013)	*primigenium* **morphotype**			*proboscideum* **morphotype**			*macedonicum* **morphotype**					*dietrichi* **morphotype (?)**
		Basic morphology	Medium- to large-sized hipparions with short to deep narial opening, deep and large POF placed far from the orbit, robust postcranials			Medium-sized hipparion with short narial opening, deep and large POF placed moderately far from the orbit, relatively robust metapodials	Large-sized hipparions with deep narial opening, deep POF placed very close to the orbit and to the facial crest, canine fossa usually well distinct, relatively robust postcranials		Medium- to small-sized hipparions with short to deep narial opening, medium to shallow POF placed the facial crest but either close to or far from the orbit, elongared postcranials					Large-sized hipparions with short narial opening, rounded shallow POF placed far from the orbit and the facial crest, robust postcranials
MN Biozones	**Ma**	**Localities**	*H. sebastopolitanum*	*H. giganteum*	*H. brachypus*	**aff.** *H. moldavicum*	*H. proboscideum*	*H. mediterraneum*	*H. macedonicum*	**aff.** *H. nikosi*	*H. sithonis*	*H. philippus*	*H. phlegrae*	**aff.** *platygenys*
MN13		**DTK**				12.2–13.2(MT) ($n = 2$) 112.8–124.6 kgr			(aff.) 11.5–11.7(MT) ($n = 2$) 11.1–13(MC) 58.5–65.2 kgr					* 13.3–14.8(MT) 14.6–15.8(MC) 148.9–164.7 kgr

(continued)

MN12	?	**KRY**			(cf.) 15.53(MT) 315.77 kgr			(cf.) 13.9(MT) (*n* = 1) 145.14 kgr (*n* = 2)	Possibly present	* 11.4 (MT) 12.6(MC) 59.7 kgr			* 11.7(MT) 13.2(MC) 94.3 kgr	
	Faunal turnover													
	7.1	**PIK**			* 15.7– 16.65(MT) 17.7– 18.6(MC) 227.8– 259.3 kgr			* 12.1–13.5(MT) 13.9–14.9(MC) 131.8– 152.2 kgr						
	Faunal turnover													
	7.2													
	7.3													
	7.3	**PER**					* 13.7– 13.8(MT) (*n* = 3) 17.2– 17.3(MC) (*n* = 2) 143.6– 177.6 kgr	Possibly present	The statisticals might have a bias because the sample possibly includes also specimens of *H. sithonis*		Possibly present	* 11.6– 12.3(MT) 13.5– 14.1(MC) 126.9– 140.9 kgr		
	7.4 7.5	**VTK, PXM**					Possibly present (fragmentary postcranials)							

MN11	8 8.2	**RZO**					* (?)15.5(MC) (*n* = 1) (?)14.61– 14.65(MT) (*n* = 2) (?)152.6– 240.8 kgr (*n* = 3)	* (cf.) 14.4–14.6(MT) (*n* = 2) 14.4–16.6(MC) 155.8– 199.9 kgr	* 11.1– 11.5(MT) 12.2– 12.8(MC) 66.6–73.9 kgr		114– 11.9(MT) 12.5– 13.1(MC) 82.8– 88.3 kgr	* 11.9– 12.7(MT) 13.1– 14.1(MC) 116.3– 128.6 kgr		
	8.7	**NIK**						(cf.) 14.5–14.6(MC) (*n* = 2) 144.1– 145.3 kgr (*n* = 2)						
	Faunal turnover													
MN10	~9.3	**RPI**	(?) 15.4 _MT 149.98 kgr	Possibly present (fragmentary material)					* 11.1(MT) (*n* = 1) 89.8–97.6 kgr (*n* = 2)					
	9.6	**NKT**	?	* 14.9–15.6 _MT 18.2–18.4_ MC (*n* = 2) 141.77– 158.6 kgr										
	9.7	**XIR**		Possibly present (fragmentary material)					?					
MN9	9.8	**PNT**	* 14.02–15.1 (MT) 156.04–176.12 kgr						96.1 kgr (*n* = 1)					

Appendix 4

	Bernor et al. (1996)		***Hippotherium***	***Cremohipparion***				**?**	***Hipparion***	
	Vlachou (2013)		***primigenium* morphotype**	***proboscideum* morphotype**		***macedonicum* morphotype**			***dietrichi* morphotype**	
Eastern Anatolian Region (Samos)	**Basic morphology**		Medium- to large-sized hipparions with deep narial opening, deep and large POF placed far from the orbit, robust postcranials	Medium- to large-sized hipparions with short to deep narial opening, deep POF placed close to the orbit and to the facial crest, robust (to elongate?) postcranials		Small-sized hipparions with short to deep narial opening, medium to shallow POF placed close to or far from the orbit and far from the facial crest, elongared postcranials			Medium- to large-sized hipparions with short narial opening, medium in depth POF placed far from the orbit and to the facial crest, relatively robust postcranials	
MN Biozones	**Ma**	**Localities**	***H. brachypus***	***H. proboscideum***		***H. matthwei***	***H. nikosi***	**aff. *forstenae***	***H. "prostylum"***	***H. dietrichi***
MN13			?				?	?		
	6.9	**Q5**	aff				10.6–11.7(MT)	12.07–14.8(MT) (*n* = 2)		
			15.9–16.5(MC) (*n* = 2)				11.3–14.5(MC)	14.9(MC) (*n* = 1)		
			200.1–205.4 kgr (*n* = 2)				80.4–99.8 kgr *	104.7–133.1 kgr *		

	Faunal turnover								
MN12	**7.1**	**Q1, MTLA, MTLB**	* 15.3–18.2(MT)		11.1–11.9(MT) (n = 2) 11.3–14.4(MC) 108.3–111.2 kgr *		* (?) 12.9–14.4(MT) 14.5–15.6(MC) 182.6–201.9 kgr		*
			16.7–18.8(MC) (n = 3)	* 14.1–15.6(MT)					12.9–13.4(MT)
		Q4	237.7–280.2 kgr *	18.2–20.3(MC) (n = 3)	Fragmentary material		Fragmentary material		14.4–15.3(MC)
	7.3	**MYT**	?	224.1–266.5 kgr	12.5(MT) (n = 1) 103.2–108.4 kgr (n = 2)		15.3MC (n = 1) 200.9 kgr (n = 1)		189.6–191.2 kgr
	Faunal turnover								
	7.4	**Q6**	15.5(MC) (n = 1) 138.2 kgr (n = 1) (Bernor et al. 2013) *	(Only a mandible)				* 15.5(MC) (n = 1)	
	7.5								
MN11	**7.6**	**MLN**		15.51(MT) (n = 1) 191.7–201 kgr (n = 2)				129.4–157 kgr	
	8	**Qx**		No postcranials *					

References

Abel O (1926) Die Geschichte der Equiden auf dem boden Nordamerikas. Verh Zool-Bot Gesell Wien 74–75:159–164

Abusch-Siewert S (1983) Gebissmorphologische Untersuchungen an eurasiatischen Anchitherien (Equidae, Mammalia) unter besonderer Berücksichtigung der Fundstelle Sandelzhausen. Cour Forschungsinst Senck 62:1–401

Alberdi MT, Ruiz-Bustos A (1989) Taxonomia y Biostratigrafia de Equidae (Mammalia, Perissodactyla) en la Cuenca de Guadix-Baza (Granada). Trabajos sobre el Neógeno-Cuaternario 11:239–270

Alberdi MT, Ortiz-Jaureguizar E, Prado JL (1998) A quantitative review of European stenonoid horses. J Paleo 72(2):371–387

Antonius O (1913) *Equus abeli* nov. spec. Beitr z Paläontol Geol Österreich-Ungarns und des Orients 26:241–294

Athanassiou A (2018) A Villafranchian *Hipparion*-bearing mammal fauna from Sésklo (E. Thessaly, Greece): implications for the question of *Hipparion–Equus* sympatry in Europe. Quaternary 1. https://doi.org/10.3390/quat1020012

Azzaroli A (1965) The two Villafranchian horses of the upper Valdarno. Palaeontogr Ital 59:1–12

Azzaroli A (1982) On Villafranchian palaearctic equids and their allies. Industrie Grafiche V, Lischi & Figli

Azzaroli A (1995) A synopsis of the Quaternary species of *Equus* in North America. Boll Soc Paleontol It 34:205–221

Azzaroli A (1989) The genus Equus in Europe. In: Lindsay EH, Fahlbusch V, Mein P (eds), European Neogene Mammal Chronology. NATO ASI series, Life sciences, 180:339-356.

Athanassiou A (2011) The Late Pleistocene fauna of Peneiós Valley (Lárissa, Thessaly, Greece): new collected material. In: van der Geer A, Athanassiou A (eds) 9th European Association of Vertebrate Palaeontologists Meeting, Heraklion, 2010. Abstracts, p 14

Athanassiou A (2002) Neogene and Quaternary faunas of Thessaly. Ann Géol Pays Helléniques 39:279–293

Athanassiou A, Kostopoulos DS (2010) On mammal remains in the area of Epanomi (Macedonia, Greece). In: Christofides G, Kantiranis N, Kostopoulos DS, Chatzipetros AA (eds) Proceedings of the XIX Congress of the Carpathian-Balkan Geological Association, Thessaloniki, 2010. Sci Ann School Geol, Fac Sci, Aristotle Univ Thessaloniki, pp 63–68

Athanassiou A (1998) [Contribution to the study of the fossil mammals of Thessaly]. PhD thesis (1996), National and Kapodistrian University of Athens. Gaia 5:1–393 (in Greek)

Ballenger L, Myers P (2001) Equidae. Animal diversity web. https://animaldiversity.org/accounts/Equidae/. Accessed 2 Feb 2020

Bennet D, Hoffmann R (1999) *Equus caballus* Linneaus, 1758. Mamm Species 628:1–54

Bernor RL (1985) Systematic and evolutionary relationships of the hipparionine horses from Maragheh, Iran (late Miocene, Turolian age). Palaeovertebrata 15:173–269

Bernor RL, Tobien H (1989) Two small species of *Cremohipparion* (Equidae, Mammalia) from Samos, Greece. Mitt Bayer Staatssammlung Paläontol Hist Geol 29:207–226

Bernor RL, Koufos GD, Woodburne M, Fortelius M (1996) The evolutionary history and biochronology of European and southeastern Asian late Miocene and Pliocene hipparionine horses. In: Bernor RL, Fahlbusch V, Mittman HW (eds) The evolution of Western Eurasian Neogene mammal faunas. Columbia University Press, New York, pp 7–46

Bernor LR, Kaiser MT, Kordos L, Scott R (1999) Stratigraphic context, systematic position and paleoecology of *Hippotherium sumegense* Kretzoi 1984 from MN10 (late Vallesian of the Pannonian Basin). Mitt Bayer Staatssammlung Paläontol Hist Geol 39:115–149

Bernor RL, Scott RS, Fortelius M, Kappelman J, Sen S (2003) Systematics and evolution of the late Miocene Hipparions from Sinap, Turkey. In: Fortelius M, Kappelman J, Sen S, Bernor RL (eds) The geology and paleontology of the miocene sinap formation. Columbia University Press, Turkey, New York

Bernor RL, Kaiser TM, Nelson SV, Rook L (2011) Systematics and Paleobiology of *Hippotherium malpassii* n. sp. (Equidae, Mammalia) from the latest Miocene of Baccinello V3 (Tuscany, Italy). Boll Soc Paleontol It 50:175–208

Bernor RL, Sun B, Chen Y (2015) *Plesiohipparion shanxiense* n. sp. from the early Pleistocene (Nihowanian) of E. Shanxi, China. Boll Soc Paleontol It 54:197–210

Benda L, Meulenkamp J, van der Weerd A (1977) Biostratigraphic correlations in the eastern Mediterranean Neogene 3. Correlation between mammal, sporomorph and marine microfossil assemblages from the upper Cenozoic of Rhodes, Greece. Newsl Stratigr 6:117–130

Bernor RL, Mirzaie AM, Meshida K, Wolf D (2016) The Maragheh hipparions, late Miocene of Azarbaijan, Iran. Palaeobiod Palaeoenvir 96:453–488

Böhme M, Spassov N, Ebner M, Geraads D, Hristova L, Kirscher U et al (2017) Messinian age and savannah environment of the possible hominin *Graecopithecus* from Europe. PLoS One 12(5):e0177347. https://doi.org/10.1371/journal.pone.0177347

Borissiak AA (1914) Sevastopolskaya fauna melkopitayaushikh. Trudy geol komiteta 87:1–154

Burke A, Eisenmann V, Ambler GK (2003) The systematic position of *Equus hydruntinus*, an extinct species of Pleistocene equid. Quatern Res 59:459–469

Boni A (1943) Geologia della regione fra il Sebino e l'Eridio: introduzione. Bibliografia tettonica. Parte prima: la porzione centrale. Atti Ist Geol Univ Pavia 1:1–41

Churcher CS (1982) Oldest ass recovered from Olduvai Gorge, Tanzania, and the origin of asses. J Paleo 56:1124–1132

Davis SJ (1980) Late Pleistocene and Holocene equid remains from Israel. Zool J Linn Soc 70:289–312

de Bonis L, Bouvrain G, Geraads D, Koufos GD (1992) Diversity and palaeoecology of Greek late Miocene mammalian faunas. Pal Pal Pal 91:99–121

de Bruijn H, Sondaar P, Zachariasse WJ (1971) Mammalia and Foraminifera from the Neogene of Kastellios hill (Crete): correlations of continental and marine biozones. Proc Kon Neder Akad Weten B7:1–22

de Bruijn H, Sondaar P, Zachariasse WJ (1971) Mammalia and Foraminifera from the Neogene of Kastellios hill (Crete): correlations of continental and marine biozones. Proc Kon Neder Akad Weten Ser B 74(5):1–22

de Vos J, van der Made J, Athanassiou A, Lyras G, Sondaar PY, Dermitzakis MD (2002) Preliminary note on the Late Pliocene fauna from Vatera (Lesvos, Greece). Ann Géol Pays Helléniques 39:37–70

Deng T (2009) Paleoecological comparison between late Miocene localities of China and Greece based on *Hipparion* faunas. Geodiversitas 28:499–516

Deperét C (1890) Les Animaux Pliocènes du Roussillon. Mèm Soc Géol Fr Paléontol 3:1–194

Eisenmann V (1988) Contributions à l ètude du gisement Miocènes supérieur de Montredon (Herault). Les grands mammifères. 5. Les Perissodactyles, Equidae. Palaeovertebrata 1988:65–96

Eisenmann V (1995) What metapodial morphometry has to say about some Miocene hipparions. In: Vrba ES, Denton GH, Pattride T, Burckle LH (eds) Paleoclimate and evolution, with emphasis on human origin. Yale University Press, New Haven and London, pp 149–164

Eisenmann V (2002) Discriminating *Equus* skulls: the Franck's index and the new palatal index. In: Mashkour M (ed) Equids in time and space: papers in honour of Véra Eisenmann. Oxbow Books, Oxford, pp 172–182

Eisenmann V, Baylac M (2000) Extant and fossil *Equus* (Mammalia, Perissodactyla) skulls: a morphometric definition of the subgenus *Equus*. Zool Scripta 29:89–100

Eisenmann V, Howe J, Pichardo M (2008) Old world hemiones and new world slender species (Mammalia, Equidae). Palaeovertebrata 36:159–233

Forstén AM (1968) Revision of the Palearctic *Hipparion*. Acta Zool Fennica 119:1–134

Forstén AM (1986) A review of the Süssenborn horse and the origin of *Equus hydruntinus* Regalia. Quartärpaläontol 6:43–52

Forstén AM (1999) A review of *Equus stenonis* Cocchi (Perissodactyla, Equidae) and related forms. Quat Sci Rev 18:1373–1408
Forstén AM, Kaya T (1995) The hipparions (Mammalia, Equidae) from Gülpinar (Çanakkale, Turkey). Paläontol Z 69:491–501
Forsyth-Mayor CJ (1887) Fauna mammalogische delle isole di Kos e di Samos. Att Soc Ital Sci Nat 5:272–275
Gkeme A, Koufos GD, Kostopoulos DS (2017) The Early Pleistocene stenonoid horse from Libakos (Western Macedonia, Greece): Biochronological and palaeoecological implications and dispesal events. 15th RCMNS Congress, Athens
Gaudry A (1862–67) Animaux fossiles et géologie de l'Attique. Savy, Paris
Geraads D, Kaya T, Mayda S (2005) Late Miocene large mammals from Yulaflı, Thrace region, Turkey, and their biogeographic implications. Acta Palaeontol Pol 50:523–544
Geraads D, Spassov N, Hristova L, Markov G, Tzankov T (2011) Upper Miocene mammals from Strumyani, South-Western Bulgaria. Geodiversitas 33(3):451–484
Gervais PM (1849) Zoologie et Paleontologie Françaises. Arthur Bertrand, Paris
Gervais PM (1859) Zoologie et Paleontologie Francaise. Nouvelles recherches sur les animaux vertebres. Arthur Bertrand, Paris
Grinder MI, Krausman PR, Hoffmann RS (2006) *Equus asinus*. Mamm Species 794:1–9
Gromova V (1949a) Histoire des chevaux (genre *Equus*) dans l'Ancien Monde. 1[re]partie: Revue et description des formes. Trudy Paleontol Inst Akad Nauk USSR 27:1–373
Gromova V (1949b) Histoire des chevaux (genre *Equus*) dans l'Ancien Monde. 2[e]partie: Evolution et classification du genre. Trudy Paleontol Inst Akad Nauk USSR 27:1–162
Gromova V (1952) Le genre *Hipparion*. Bureau de Recherches géologiques et Minières CEDP 12:1–288
Grubb P (1993) Order Perissodactyla. In: Wilson DE, Reider DM (eds) Mammal species of the world, 2nd edn. Smithsonian Institution Press, Washington, DC, pp 369–372
Geraads D, Tsoukala E, Spassov N (2007) A skull of Ancylotherium (Chalicotheriidae, Mammalia) from the late Miocene of Thermopigi (Serres, N. Greece) and the relationships of the genus. J Vertebr Paleontol 27:461–466
Hensel RF (1862) Über die Reste einiger Säugetierarten von Pikermi in der Münchener Sammlung. Monats K Prussischen Akad Wisse Berlin 14:560–569
Hensel RF (1862) Über die Reste einiger Säugetierarten von Pikermi in der Münchener Sammlung. Monats Akad Wiss 27:560–569
Kormos T (1911) Der Pliozäne Knochenfund bei Polgardi. Foldtani Kozlony Budapest 41:171–189
Kostopoulos DS (2009) The Pikermian event: temporal and spatial resolution of the Turolian large mammal fauna in SE Europe. Pal Pal Pal 274:82–95
Kostopoulos DS, Koufos GD (2016) The evolution of living world: Chordata. Vertebrate Palaeontology. Association of Greek Academic Libraries, Athens, pp 1–285
Koufos GD (1980) Palaeontological and stratigraphical study of the Neogene continental deposits of the Axios Valley basin. PhD thesis, Aristotle University of Thessaloniki
Koufos GD (1982) *Hipparion crassum* Gervais, 1859 from the lignites of Ptolemais (Macedonia-Greece). Proc K Nederlandse Akad Wet B 85:229–239
Koufos GD (1984) A new hipparion (Mammalia, Perissodactyla) from the Vallesian (late Miocene) of Greece. Paläontol Z 58:307–317
Koufos GD (1987a) Study of the Pikermi hipparions. Part I: Generalities and taxonomy. Bull Mus Nat Hist Nat Paris 4 ser 9C 2:197–252 ; Part II: Comparisons and odontograms. Bull Mus Nat Hist Nat Paris 4 Ser 9C 3:327–363
Koufos GD (1987b) Study of the Turolian hipparions of the lower Axios Valley (Macedonia, Greece). 1. Locality "Ravin des Zouaves-5" (RZO). Geobios 20:293–312
Koufos GD (1988) Study of the Turolian hipparions of the lower Axios Valley (Macedonia, Greece). 4. Localities of Dytiko. Palaeovertebrata 18:187–239
Koufos GD (2000a) New material of Vallesian hipparions (Mammalia, Perissodactyla) from the lower Axios Valley, Macedonia, Greece. Senckenb Lethaea 80:231–255
Koufos GD (2000b) The hipparions of the late Miocene locality "Nikiti 1", Chalkidiki, Macedonia, Greece. Rev Palaeobiol 19:47–77

Koufos GD (2004) Vertebrate palaeontology. Zitis Publications, Thessaloniki

Koufos GD (2006a) The large mammals from the Miocene/Pliocene locality of Silata, Macedonia, Greece with implications about the latest Miocene palaeoecology. Beitr f Paläont 30:293–313

Koufos GD (2006b) The Neogene mammal localities of Greece: faunas, chronology and biostratigraphy. Hellenic J Geosci 41:183–214

Koufos GD (2006c) Palaeoecology and chronology of the Vallesian (late Miocene) in the eastern Mediterranean region. Pal Pal Pal 234:127–145

Koufos GD (2016) *Hipparion macedonicum* revisited: new data on evolution of hipparionine horses from the late Miocene of Greece. Acta Palaeontol Pol 61:519–536

Koufos GD, Kostopoulos DS (1993) A stenonoid horse (Equidae, Mammalia) from the Villafranchian of Western Macedonia (Greece). Bull Geol Soc Greece 28:131–143

Koufos GD, Kostopoulos DS (1994) The late Miocene mammal localities of Kemiklitepe (Turkey). 3. Equidae. Bull Mus Nat Hist Natur Paris 4e Ser Sect C 16:41–80

Koufos GD, Kostopoulos DS (2016) The Plio-Pleistocene large mammal record of Greece: implications for early human dispersals into Europe. In: Harvati K, Roksanic M (eds) Palaeoanthropology of the Balkans and Anatolia: human evolution and its context. Springer, Dordrecht, pp 269–280

Koufos GD, Vasileiadou K (2015) Miocene/Pliocene mammal faunas of southern Balkans: implications for biostratigraphy and palaeoecology. Palaeobiodiv Palaeoenviron 95:285–303

Koufos GD, Vlachou TD (1997) *Equus stenonis* from the middle Villafranchian locality of Volax (Macedonia, Greece). Geodiversitas 19:641–657

Koufos GD, Vlachou TD (2016) Equidae. In: Koufos GD, Kostopoulos DS (eds) Palaeontology of the upper Miocene vertebrate localities of Nikiti (Chalkidiki peninsula, Macedonia, Greece). Geobios 49:85–118

Koufos GD, Vlachou TD (2019) New material and revision of hipparions from the upper Miocene locality Ravin des Zouaves-5, RZO of Axios Valley (Macedonia, Greece). Geobios 57:33

Koufos GD, Kostopoulos DS, Sylvestrou I (1997) *Equus apolloniensis* n. sp. (Mammalia, Equidae) from the latest Villafranchian locality of Apollonia, Macedonia, Greece. Paleontol Evol 30:4976

Koufos GD, Kostopoulos DS, Merceron G (2009a) The late Miocene mammal faunas of the Mytilinii Basin, Samos Island, Greece: new collection. 17. Palaeoecology – Palaeobiogeography. Beit z Paläontol 31:409–430

Koufos GD, Kostopoulos DS, Vlachou TD (2009b) The late Miocene mammal faunas of the Mytilinii Basin, Samos Island, Greece: new collection. 16. Chronology. Beit z Paläontol 31:397–408

Koufos GD, Kostopoulos DS, Vlachou TD, Konidaris GE (2011) A synopsis of the late Miocene mammal fauna of Samos Island, Aegean Sea, Greece. Geobios 44:237–251

Kretzoi M (1938) Die raubtiere von Gombaszog nebst einer Übersicht der Gesamtfauna. Annali Musei Nationalis Hungarici, Pars Mineralogii, Geologii, Paleontologii 31:88–157

Kampouridis P, Roussiakis S, Kargopoulos N, Giaourtsakis I, Dimakopoulos G, Iliopoulos G, Svorligkou G, Theodorou G (2019) Faunal diversity at the Turolian locality of Kerassia (Northern Euboea, Greece). Bull Geol Soc Greece Sp Pub 7 122:52–53

Koufos GD (1985) *Hipparion* sp. (Equidae, Perissodactyla) from Diavata (Thessaloniki, northern Greece). Bull Brit Mus Nat Hist (Geol) 38:335–345

Koufos GD (1981) A new Late Pleistocene (Würmian) mammal locality from the basin of Drama (Northern Greece). Sci Annals, Fac Phys & Mathem, Univ Thessaloniki 21:129–148

Konidaris GE, Athanassiou A, Tourloukis V, Thompson N, Giusti D, Panagopoulou E, Harvati K (2018) The skeleton of a straight-tusked elephant (Palaeoloxodon antiquus) and other large mammals from the Middle Pleistocene butchering locality Marathousa 1 (Megalopolis Basin, Greece): preliminary results. Quat Int 497:65–84

Koufos GD (1992) Early Pleistocene equids from Mygdonia basin (Macedonia, Greece). Palaeontographia It 79:167–199

Konidaris GE, Tourloukis V, Kostopoulos DS, Thompson N, Giusti D, Michailidis D, Koufos GD, Harvati K (2015) Two new vertebrate localities from the Early Pleistocene of Mygdonia Basin (Macedonia, Greece): Preliminary results. C R Palevol 14:353–362

Koufos GD, Syrides GE, Kostopoulos DS, Koliadimou KK (1995) Preliminary results about thestratigraphy and the palaeoenvironment of Mygdonia Basin, Macedonia, Greece. Geobios 28:243–249

Lazaridis G (2015) Study of the Late Miocene vertebrate locality of Kryopigi and other localities of Kassandra Peninsula, Chalkidiki (Greece). Systematics, Taphonomy, Paleoecology, Biochronology. PhD Thesis, Aristotle University of Thessaloniki. Sci Ann School Geol 174:1–351

Lazaridis G, Tsoukala E (2014) *Hipparion phlegrae*, sp. nov. (Mammalia, Perissodactyla): a new species from the Turolian locality of Kryopigi (Kassandra, Chalkidiki, Greece). J Vertebr Paleontol 34:164–178

MacFadden BJ (1980) The Miocene horse *Hipparion* from North America and from the type locality in southern France. Palaeontology 23:617–635

MacFadden BJ (2005) Fossil horses-evidence for evolution. Science 307:1728–1730

Mein P (1990) Updating of MN zones. In: Lindsay E, Fahlbusch V, Mein P (eds) European Neogene mammal chronology. Plenum Press, New York, pp 73–90

Melentis JK (1966) Studien über Vertebraten Griechenlands. 8.Über *Equus abeli* aus dem Mittepleisozän des beckens von Megalopolis imPeloponnes (Griechenland). Ann Geol Pays Helleniques 17:158–168

Metais G, Sen S (2017) First occurrence of Palaeotheriidae (Perissodactyla) from the late–middle Eocene of eastern Thrace (Greece). C R Palevol 16:382–396

Musil R (1969) Die Equiden reste aus dem Pleistozän von Süssenborn bei Weimar. Paläont Abh A III:317–364

Melentis JK, Schneider H (1966) Studien über fossile Vertebraten Griechenlands. 17. Eine neue Pikermifauna in der Nähe der Ortschaft Alifaka in Thessalien (Griechenland). Ann Géol Pays Helléniques 17:267–288

Mitzopoulos MK (1947) Die Verbreitung der Pikermistufe auf der insel Euboea. Ann Géol Pays Helléniques 1:209–216

Mitzopoulos MK (1961) Die Hipparionfauna von Tanagra bei Theben. Ann Géol Pays Helléniques 12:301–304

Melentis JK (1967) Die Pikermi-Fauna von Halmyropotamos (Euböa, Griechenland). Pract Akad Athens 42:147–152

Marinos G, Symeonidis N (1974) Neue Funde aus Pikermi (Attika, Griechenland) und eine allgemeine geologische Übersicht dieses paläontologischen Raumes. Ann Géol Pays Helléniques 26:1–27

Orlandi-Oliveras G, Nacarino-Meneses G, Koufos GD, Kohler M (2019) Bone histology provides insights into the life history mechanisms underlying dwarfing in hipparionines. Sci Rep 8(17203):1–15. https://doi.org/10.1038/s41598-018-35347-x

Osborn HF (1918) Equidae of the Oligocene, Miocene and Pliocene of North America. Iconographic type revision. Am Mus Novit 2:1–326

Prat F (1964) Contribution à la classification des Équidés Villafranchiens. Procès-Verbaux de la Societé Linnéenne de Bourdeaux 101:14–32

Regalia E (1904) Sulla fauna della grotta di Romanelli (Castro, Lecce). Archivo per l' Anthropologie e la Etnologia 34(1):33–79

Regalia E (1907) Sull'*Equus* (*Asinus*) *hydruntinus* Regalia della grotta di Romanelli (Castro, Lecce). Archivio per l'Antropologia e l'Etnologia 37:375–390

Roth J, Wagner A (1854) Die fossilen Knochenüberreste von Pikermi in Griechenland. Abh Bayerische Akad Wiss München Math Phys Kl 7:371–464

Reese DS (2008) Faunal remains from Late Helladic Lerna (Argolid, Greece). Med Archaeol Archaeometry 8(1):5–25

Sánchez IM, Salesa MJ, Morales J (1998) Revisión sistemática del género *Anchitherium* Meyer 1834 (Equidae; Perissodactyla) en España. Estudios Geol 54:39–63

Schwarz E (1928) Über diluviale Pferde der *Equus caballus*-Gruppe. Jahrb Preuss Geol Landesanst 48:429–476

Sondaar PY (1971) The Samos *Hipparion*. Proc K Nederlandse Akad Wet B 74:417–441

Sondaar PY (1974) The *Hipparion* of the Rhone valley. Geobios 7(4):289–306

Spassov N, Geraads D, Hristova L, Markov GN, Garevska B, Garevski R (2018) The late Miocene mammal faunas of the Republic of Macedonia (FYROM). Palaeontographica Abt A 311:1–85

Steffens P, de Bruijn H, Meulenkamp J, Benda L (1979) Field guide to the Neogene of Northern Greece (Thessaloniki area and Strimon basin). Publ Dept Geol Palaeont Univ Athens ser A 35:1–14

Studer T (1911) Eine neue Equidenform aus dem Obermiocän von Samos. Verhand Deutschen Zool Gesell 1911:192–200

Stiner MC, Munro ND (2011) On the evolution of diet and landscape during the Upper Paleolithic through Mesolithic at Franchthi Cave (Peloponnese, Greece). J Hum Evol 60(5):618–636

Steensma KJ (1988) Plio-/Pleistozäne Großsäugetiere (Mammalia) aus dem Becken von Kastoria / Grevena, südlich von Neapolis — NW Griechenland. Inaugural-Dissertation, Technische Universität Clausthal

Symeonidis N, de Vos J (1976) Grosssäuger-funde aus den Plistozänen Spaltenfullungen von Turkovunia in Athen. Ann Geol Pays Helléniques 28:135–144

Tsoukala E, Chatzopoulou K (2005) A new Early Pleistocene (latest Villafranchian) site with mammals in Kalamotó (Mygdonia Basin, Macedonia, Greece) – preliminary report. Mitt Komm Quartärforsch Österr Akad Wiss 14:213–233

Tsoukala ES (1989) Contribution to the study of the Pleistocene fauna of large mammals (Carnivora, Perissodactyla, Artiodactyla) from Petralona cave Chalkidiki (N. Greece). PhD thesis, Aristotle University of Thessaloniki.

Van der Meulen A, Van Kolfschoten T (1986) Review of the LateTurolian to early Biharian mammal faunas from Greece and Turkey. Mem Soc Geol It 31:201–211

van Hoek Ostende LW, Mayda S, Oliver A, Madern A, Hernández-Ballarín V, Peláez-Campomanes P (2015) Aliveri revisited a biogeographical appraisal of the early Miocene mammals from the eastern Mediterranean. Palaeobiodiver Palaeoenviron 95:271–284

Vangengeim E, Tesakov A (2013) Late Miocene mammal localities of Eastern Europe and Western Asia. In: Wang X, Flynn LJ, Fortelius M (eds) Fossil mammals of Asia-Neogene biostratigraphy and chronology. Columbia University Press, New York, pp 521–545

van den Hoek Ostende LW, Mayda S, Oliver A, Madern A, Hernández–Ballarín V, Peláez–Campomanes P (2015) Aliveri revisited, a biogeographical appraisal of the early Miocene mammals from the eastern Mediterranean. Palaeobiodiv Palaeoenviron 95(3):271–284

von Reichenau W (1915) Beiträge zur Näheren Kenntnis Fossiler Pferde aus Deutschem Pleistozän, insbesondere über die Entwicklung und die Abkaustadien des Gebisses vom Hochterrassenpferd (*Equus mosbachensis* v. R.). Abh Geol Landsanstalt Darmstadt 7:1–155

Vlachou TD (2013) Palaeontological, biostratigraphical and palaeoecological study of the Greek hipparions. Ph.D. Thesis, Aristotle Univ Thessaloniki. Sci Ann School Geol 154:1–580

Vlachou TD, Koufos GD (2009) Equidae. In: Koufos GD, Nagel D (eds.), The Late Miocene Mammal Faunas of the Mytilinii Basin, Samos Island, Greece: New Collection. Beitr Paläont 31:207–281.

Vlachos E, Tsoukala E, Crégut-Bonnoure E, Guérin C, Mol D (2018) The Paradise Lost of Milia (Grevena, Greece; Late Pliocene, Early Villafranchian, MN15/MN16a): Faunal Composition and Diversity. Quaternary 1(2):13

Wagner A (1848) UrweltlicheSäugetier-UeberresteausGriechenland. Abh Bayer Akad Wiss 5(2):155–170

Wehrli H (1941) Beitrag zur Kenntnis der “Hipparionen” von Samos. Paläontol Z 22:321–386

Woodburne MO, Bernor RL (1980) On superspecific groups of some Old World hipparionine horses. J Paleo 54:1319–1348

Zhegallo VI (1971) Gipparionoy iz neogenovykh otolozhenii Zapadnoi Mongolii I Tuvy. Trudy Sovmestnoi Sovetsko-Mongolskoi Nauchno-Issledovatelskoi Gelogicheski Ekspeditesii 3:98–119

The Fossil Record of Tapirs (Mammalia: Perissodactyla: Tapiridae) in Greece

Evangelia Tsoukala

1 Introduction

Tapirs (Mammalia: Perissodactyla: Tapiridae) are odd-toed ungulates. From the short and robust limbs, the forefoot also preserves the fifth digit, sometimes used, the dental formula is 3-1-4-3 / 3-1-3/4-3 with low-crowned, lophodont cheek teeth and molarized premolars, and possess a short and flexible proboscis (Rustioni and Mazza 2001). Due to their particular lifestyle in forests, significant water sources are required and given that fossil tapirids do not differ significantly from living tapirs (Kurtén 1968) (living fossils), they are considered to be useful indicators of the paleoenvironment. This group is known at least since the Early Eocene of Kyrgyzstan (Asia) and has always been associated with tropical-subtropical forestry environments. Besides the extant genus—which is known by five extant and more than 15 extinct species (Cozzuol et al. 2013 and references therein)—Tapiridae includes several other extinct genera and species whose phylogenetic relationships are not well known. After their appearance in the Ypresian (Eocene), tapirids quickly achieved their maximum diversity in the Lutetian with at least 14 species; since then, the diversity of this group has been gradually declining (PaleoBiology Database 2019). Molecular analyses corroborate the monophyly of crown tapirs and recover them as the sister-group of the rhinoceroses, forming together the perissodactyl clade called Ceratomorpha Wood, 1937 (Steiner and Ryder 2011). The inclusion of extinct tapir species in phylogenetic analyses has been slow, but Cozzuol

E. Tsoukala (✉)
Laboratory of Geology and Paleontology, Aristotle University of Thessaloniki, Thessaloniki, Greece
e-mail: lilits@geo.auth.gr

E. Vlachos (ed.), *Fossil Vertebrates of Greece Vol. 2*,
https://doi.org/10.1007/978-3-030-68442-6_13

et al. (2013) presented a phylogeny containing all extant tapirs and several extinct species. Based on this analysis, it seems that crown Tapiridae is divided into two main clades, one defined by the extant Baird's and Malayan tapir and the other by the extant Brazilian, Mountain, and Kabomani tapirs (Cozzuol et al. 2013). Extinct tapirs are placed within those clades, as well as along the stem of crown Tapiridae (Cozzuol et al. 2013).

In Europe, the Plio-Pleistocene fossil record is represented by two species of tapirs: *Tapirus arvernensis* Croizet and Jobert, 1828 and *T. jeanpiveteaui* Boeuf, 1991 (Boeuf 1991; Guérin and Eisenmann 1994). The former is much more common than the latter and closer to the extant *T. terrestris* (Linnaeus, 1758) from South America. *Tapirus arvernensis* includes two sub-species: *T. a. minor* Michaux et al., 1976 and *T. a. arvernensis* Croizet and Jobert, 1828.

2 Historical Overview

Tapir fossils are rare in the Greek fossil record (Table 1 and Fig. 1). Paraskevaidis (1977) referred, but without description and measurements, three upper cheek teeth of *Tapirus* sp. found during research in the late Miocene lignite of Servia (Kozani, W Macedonia), and without a known host location of the specimens; as such, this occurrence is cited herein with caution. The only two confirmed occurrences of a tapir are known from the research and the systematic excavations in Milia (Grevena) area in the last 25 years. The Milia association is also unique in Greece and very rare in Europe, defining the MN16a biozone of Late Pliocene. The Milia vertebrate paleo-fauna, besides *Tapirus arvernensis*, consists of mastodons, gomphotheres, rhinoceroses, suids, carnivorans, bovids, cervids, hipparions, micromammals, and turtles (Guérin and Tsoukala 2013; Tsoukala 2000; Tsoukala and Mol 2016; Crégut-Bonnoure and Tsoukala 2017; Vlachos and Tsoukala 2016; Vlachos et al. 2018). The tapir remains from Milia are the southernmost occurrences of the family in the Pliocene of Europe.

Table 1 List of the Greek localities containing tapir fossils. Locality numbers refer to the collection numbers of the PaleoBiology Database (PBDB)

Localities[PBDB No]	Age (MN)	Taxon
Milia-2[195290]	Late Pliocene (MN16a)	*Tapirus a. arvernensis*[1]
Milia-5[185859]	Late Pliocene (MN16a)	*Tapirus a. arvernensis*[1]
Servia[202257]	Late Miocene	*Tapirus* sp.[2]

MN Mammal Neogene

[1]Guérin and Tsoukala (2013), [2]Paraskevaidis (1977)

Fig. 1 Map of Greece showing the geographic distribution of the localities with tapir fossils. **1**, Milia-2; **2**, Milia-5; **3**, Servia. See Table 1 for more information. Image exported from Google Earth Pro © 2019, map data from US Dept. of State Geographer, SIO, NOAA, U.S. Navy, NGA, GEBCO, image from Landsat/ Copernicus. Scale bar equals 10 km, North faces upward

3 Systematic Paleontology

Perissodactyla Owen, 1848

Tapiridae Burnett, 1830

Tapirus Brisson, 1762

Type Species *Hippopotamus terrestris* Linnaeus, 1758.

Included Taxa Extant tapirs comprise five species, distributed in the Central and South America and Southeast Asia (Cozzuol et al. 2013 and references therein). Their fossil record is more extensive and widely distributed, including at least 15 species distributed in Europe, Americas, and Asia.

Remarks The mentioned material from the late Miocene of Servia (W Macedonia) by Paraskevaidis (1977: pl. II.4), comprising three upper cheek teeth, has never been described in detail. As we do not know if and where this material is preserved/ stored, it is not possible to confirm this identification. As such, this material is tentatively included herein with its original identification as *Tapirus* sp.

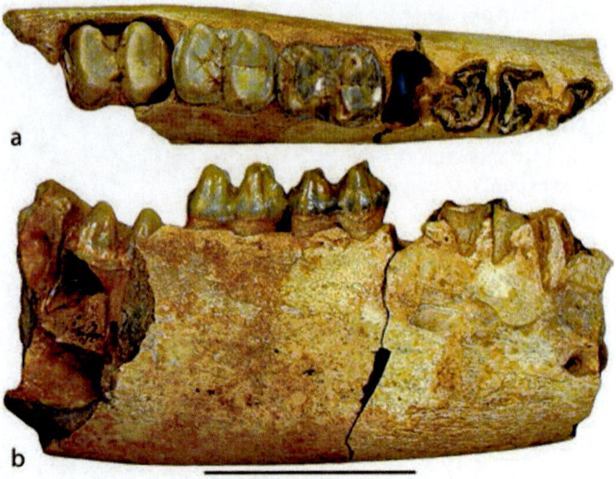

Fig. 2 A selected fossil of tapir from Greece. The right hemi-mandible (LGPUT MIL 649) of *Tapirus a. arvernensis* from the Late Pliocene (MN16a) of Milia, in **a** occlusal, and **b** buccal view. Scale bar equals 5 cm

Tapirus arvernensis arvernensis **Croizet and Jobert, 1828**

Type Material Mandible with p3–m3, left; mandible with d1–d4, m1; I2 right; upper molar; atlas (Croizet and Jobert 1828:pl. 2, figs. 1, 3, 5 and pl. 12, figs. 4–6).

Type Locality Perrier-les-Étouaires, France, late Pliocene, late Ruscinian, MN16a.

Distribution This species is known from several localities in the Neogene of Europe. In Greece, this species is known only from Milia (N Greece).

Remarks The most complete material of this species is known from the Late Pliocene of Camp del Ninots (Spain; Gómez de Soler et al. 2012), with one complete skeleton of *T. arvernensis* in articulation; this occurrence is roughly coeval with the occurrences of this species in Greece. The referred material from Greece consists of a complete, well-preserved right calcaneum of an adult and a partial right hemi-mandible, with the alveoli of d2 and d3, with d4 and m1; the m2 is unerupted, indicating that this specimen belonged to a juvenile individual (Fig. 2). The Milia fossils are clearly larger than those of *T. a. minor*, and are safely attributed to *T. a. arvernensis* (Guérin and Tsoukala 2013). Based on the known fossils, the estimated size of *T. arvernensis* is similar to the extant Mountain Tapir from South America, with a length between 1.8 and 2.0 m, a shoulder height of 75 and 80 cm, and weight over 200 kg (Guérin and Tsoukala 2013, and references therein).

4 Concluding Remarks

Tapirs are among the least common elements of the Neogene faunas of Greece. Given that earliest report of tapirs in Greece (Paraskevaidis 1977) cannot be evaluated, the two occurrences in Milia (Guérin and Tsoukala 2013) confirm the presence of tapirs in Greece with the species *Tapirus a. arvernensis*. The association of the *Tapirus a.*

arvernensis with the other species from Milia in the Early Villafranchian (~3,5 Ma, MN16a, Late Pliocene), bearing anatomical characteristics similar to those of extant species, suggests paleoecological implications on similar ecology: a warm and humid environment, probably a dense wooded area, and access to sufficient water resources. These are the southernmost occurrences of this species in the Pliocene of Europe.

References

Boeuf O (1991) *Tapirus jeanpiveteaui* nov. sp., nouvelle espèce de Tapiridae (Perissodactyla) du Pliocène de Charente (France). Palaeontographica A 217:177–194

Brisson MJ (1762) Le regnum animale in classes IX distributum, sive synopsis methodica sistens generalem cetaceorum, particularem divisionem in ordines, sectiones, genera & species. Editio altera auctior, Lugduni Batavorum, Theodorum Haak Paris 296 pp

Cozzuol MA, Clozato CL, Holanda EC, Rodrigues FH, Nienow S, de Thoisy B, Redondo RA, Santos FR (2013) A new species of tapir from the Amazon. J Mammology 94:1331–1345

Crégut-Bonnoure E, Tsoukala E (2017) The late Pliocene Bovidae and Cervidae (Mammalia) of Milia (Grevena, Macedonia, Greece). Quat Int 445:215–249

Croizet JB, Jobert A (1828) Recherches sur les ossements fossiles du département du Puy-de-Dôme. Adolphe Delahays, Paris

Gómez de Soler B, Campeny Vall-Llosera G, van der Made J, Oms O, Agustí J, Sala R, Blain HA, Burjachs F, Claude J, García Catalán S, Riba D (2012) A new key locality for the Pliocene vertebrate record of Europe: the camp dels Ninots maar (NE Spain). Geol Acta 10. https://doi.org/10.1344/105.000001702

Guérin C, Eisenmann V (1994) Les tapirs (Mammalia, Perissodactyla) du Miocène supérieur d'Europe occidentale. Geobios 27:113–127

Guérin C, Tsoukala E (2013) The Tapiridae, Rhinocerotidae and Suidae (Mammalia) of the early Villafranchian site of Milia (Grevena, Macedonia, Greece). Geodiversitas 35:447–489

Kurtén B (1968) Pleistocene mammals of Eyrope. Aldine Publishing Company, Chicago

Linnaeus C (1758) Systema naturae per regna tria naturae, secundum classes, ordines, genera, species, cum characteribus, differentiis, synonymis, locis. Laurentii Salvii, Stockholm

Michaux J, Sigé B, Sudre J (1976) Présence de *Tapirus arvernensis* (Mammalia, Perissodactyla) dans le Pliocène supérieur de Bassan (Hérault). C R Soc Geol Fr 1:9–11

PaleoBiology Database (2019) Sampled-in-bin diversity in the species level for Tapiridae. https://paleobiodb.org. Accesed 18 Jun 2019

Paraskevaidis E (1977) Säugetierreste aus Griechenland. In: Proceedings of the VIth Colloquium on the Geology of the Aegean Region, Athens, pp 1143–1159

Rustioni M, Mazza P (2001) Taphonomic analysis of *Tapirus arvernensis* remains from the lower Valdarno (Tuscany, Central Italy). Geobios 34:469–474

Steiner CC, Ryder OA (2011) Molecular phylogeny and evolution of the Perissodactyla. Zool J Linn Soc 163:1289–1303

Tsoukala E (2000) Remains of a Pliocene *Mammut borsoni* (Hays, 1834) (Proboscidea, Mammalia), from Milia (Grevena, W. Macedonia, Greece). Ann Paleontol 86:165–191

Tsoukala E, Mol D (2016) The Proboscidea of the early Villafranchian site of Milia (Grevena, Macedonia, Greece). Quat Int 406:4–24

Vlachos E, Tsoukala E (2016) The diverse fossil chelonians from Milia (late Pliocene, Grevena, Greece) with a new species of *Testudo* Linnaeus, 1758 (Testudines: Testudinidae). Pap Palaeontol 2:71–86

Vlachos E, Tsoukala E, Crégut-Bonnoure E, Guérin C, Mol D (2018) The paradise lost of Milia (Grevena, Greece; late Pliocene, early Villafranchian, MN15/MN16a): faunal composition and diversity. Quaternary 1. https://doi.org/10.3390/quat1020013

The Fossil Record of Rhinocerotids (Mammalia: Perissodactyla: Rhinocerotidae) in Greece

Ioannis X. Giaourtsakis

1 Introduction

Rhinocerotoids (superfamily Rhinocerotoidea) comprise the largest and most ecologically diverse group of perissodactyls. Since their initial diversification during the Eocene, large populations of numerous and evolutionary diverse lineages comprising hundreds of species were once widespread in Europe, Asia, Africa, and North America (Prothero et al. 1989; Heissig 1989; Lucas and Sobus 1989; Wall 1989). Most notably, during the Oligocene and Miocene, many fossiliferous localities have often yielded more than three different sympatric genera, each one with unique anatomical and dietary adaptations. This exceptional richness of rhinocerotoid taxa is explained by diverse herbivorous adaptations and successful feeding strategies within a wide range of ecological niches (Prothero et al. 1989; Prothero and Schoch 2002). The systematic assessment of fossil rhinoceroses provides therefore a significant biogeographical and ecological component for the reconstruction of terrestrial ecosystems during the past and their response to the environmental changes.

This work and the nomenclatural acts it contains has been registered in the Zoobank under the LSID urn:lsid:zoobank.org:pub:C87AE7D0-4410-4D19-8DF9-ECDCD0A2B9A4

Electronic Supplementary Material The online version of this chapter (https://doi.org/10.1007/978-3-030-68442-6_14) contains supplementary material, which is available to authorized users.

I. X. Giaourtsakis (✉)
Department of Earth and Environmental Sciences, Section of Paleontology and Geobiology, Ludwig-Maximilians-Universität München, Munich, Germany
e-mail: giaourtsakis@snsb.de

E. Vlachos (ed.), *Fossil Vertebrates of Greece Vol. 2*,
https://doi.org/10.1007/978-3-030-68442-6_14

2 Historical Overview

Fossil remains of animals were known to the ancient Greeks, as testified by surviving legends, myths, and artifacts (Mayor 2000). The systematic study of fossil vertebrates in Greece began during the middle of the nineteenth century. The classical Late Miocene locality of Pikermi was discovered incidentally by the renowned Scottish historian and philhellene George Finley in 1836, during a tour he had undertaken in the Mesogea region of Attica with the hope of revealing the remains of the temples of the Brauronian Artemis and the Oropian Amphiaraos. A few weeks later, Finlay presented a talk on his discovery at the newly established Physiographic Society of Athens, and donated the fossil findings to the collections of the society (Karadimas 2013). Sir William Wilde, a prominent Irish oto-ophthalmologist and antiquarian explorer, father of renowned poet and playwright Oscar Wilde, visited the locality of Pikermi and briefly examined the collection at the Physiographic Society; among the specimens, he did recognize the skull of a rhinoceros (Wilde 1840: 448). In a short note announcing the acquisition of some fossils from Pikermi, Goldfuss (1841: 358) mentioned also the occurrence of rhinoceros at the site. A few years later, Wagner (1848) described and illustrated for the first time in detail rhinocerotid specimens from Pikermi (Fig. 1), introducing a new species as *Rhinoceros pachygnathus*, along with a second one, *Rhinoceros schleiermacheri*, previously known from some Western European localities. In subsequent studies, Roth and Wagner (1854) and Wagner (1857) reported a few additional specimens from the locality. In his monumental monograph on the Pikermi fauna, Gaudry (1862–67) documented ample cranial and postcranial material, confirming the occurrence of two distinct horned rhinocerotid species. In addition, Gaudry described as *Acerotherium* sp. an adult mandible with a well-developed second lower incisor referable to a hornless taxon.

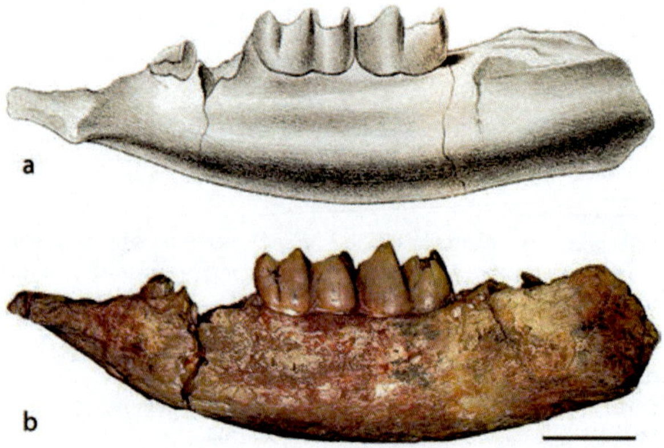

Fig. 1 (**a**) The first depicted fossil rhinocerotid specimen from Greece as figured by Wagner (1848: pl. 10, fig. 3), a juvenile mandible from the Late Miocene locality of Pikermi (type specimen of *Rhinoceros pachygnathus* Wagner, 1848). (**b**) The original specimen (BSPG: AS.II.612) in left lateral view, presently attributed to *Dihoplus pikermiensis* (Toula, 1906). Scale bar equals 4 cm

Inspired by the legends on the mythical beasts of Neades by Euphorion of Chalcis (*Commentaries*, C3rd B.C.) accounted by Aelian (*De Natura Animalium*, C2nd A.D.), as well as by the mythical battle between the God of Wine Dionysus and the female warrior tribe of Amazons narrated by Plutarch (*Moralia*, C1st A.D.), the English physician and naturalist C. I. Forsyth Major carried out the first systematic excavations on the island of Samos during the years 1885–1889 (Forsyth-Major 1888). In his descriptive catalog, Forsyth-Major (1894) attributed most of the rhinocerotid specimens to *Rhinoceros pachygnathus*, and a few of them potentially to *Rhinoceros schleiermacheri*. The first detailed studies on the Samos rhinoceroses were undertaken by Weber (1904, 1905), who described and illustrated a remarkably rich collection hosted at that time at the Paleontological Museum of Munich. In his first contribution, Weber (1904) documented the remains of the horned species *Rhinoceros pachygnathus* and *Rhinoceros schleiermacheri*, while in his second work Weber (1905) studied a remarkable sample of hornless rhinocerotid material and described two new species, *Aceratherium schlosseri* and *Aceratherium samium*. Unfortunately, the entire rhinocerid material from Samos at Munich was destroyed by a heavy bombing during the Second World War. Andrée (1921) described and named two additional hornless species based on material from Samos at the Paleontological Museum of Münster, *Aceratherium wegneri* and *Aceratherium angustifrons*.

Toula (1906) considered the specimens of *Rhinoceros schleiermacheri* from Pikermi and Samos to represent two different subspecies, which he named *R. schleiermacheri pikermiensis* and *R. schleiermacheri samius,* respectively. A great number of expeditions by scientists and fossil dealers spread the amply recovered material from Pikermi and Samos to numerous museums and institutional collections across Europe and the USA (Solounias 1981; Appendices 2–3). Arambourg and Piveteau (1929) documented the presence of *Rhinoceros pachygnathus* and *Aceratherium* sp. at the Neogene deposits of the Axios Valley, near Thessaloniki. Naturally, modern workers have gradually modified and refined the initial generic assignments and some of the species names applied by these first authors (e.g., Ringström 1924; Heissig 1975, 1996; Geraads 1988; Giaourtsakis 2003, 2009; Athanassiou et al. 2014; present study).

After the Second World War and up to the end of the previous century, little systematic work was undertaken on the fossil rhinoceroses of Greece. Nonetheless, during this period of time, a few occasional papers and many references in faunal lists and preliminary reports unveiled the presence of the family in numerous Neogene and Quaternary localities of Greece (Appendices 1–3; Supl.-Tables 1–3). The rhinoceroses of Pikermi and Samos continued to be frequently mentioned and compared in major reviews and general works on the family (e.g., Arambourg 1959; Hooijer 1968; Viret 1958; Guérin 1980; Heissig 1989, 1996), but in most cases neither new material was introduced nor was the old one properly revised. A notable exception was the significant contribution by Geraads (1988), who revised the cranial specimens of the horned rhinocerotids from Pikermi and Samos as *Ceratotherium neumayri* and *Dicerorhinus pikermiensis*, establishing reliable criteria for distinguishing their craniodental remains. Two years later, Geraads and Koufos (1990) studied the rich material discovered at the Late Miocene locality Pentalophos-1, near Thessaloniki, and created the new hornless species *Aceratherium kiliasi*.

During the past two decades, the systematic research on the Greek fossil rhinocerotids resumed with several dedicated studies. Giaourtsakis (2003) provided a detailed overview of the Neogene rhinocerotid record of Greece. Symeonidis et al. (2006) described the remains of *Stephonorhinus* cf. *etruscus* from the Pleistocene locality of Aivaliki, near Serres, and provided an updated overview of the Plio-Pleistocene rhinocerotid record of Greece. Giaourtsakis et al. (2006) studied the horned rhinocerotids from the Late Miocene locality of Kerassia on Euboea Island. From the same locality, Athanassiou et al. (2014) documented later the new hornless species *Acerorhinus neleus*. Giaourtsakis et al. (2009) studied the material from the new excavations on Samos Island and reassessed further the craniodental and postcranial differences between the Late Miocene horned species "*Diceros*" *neumayri* and *Dihoplus pikermiensis*. Koufos and Kostopoulos (2013) described remains of *Brachypotherium brachypus* from the Chryssavgi in Macedonia. Guérin and Tsoukala (2013) documented as *Dicerorhinus jeanvireti* a notable rhinocerotid sample from the Pliocene fossiliferous site of Milia in Grevena. Tsoukala and Guérin (2016) described specimens of *Dicerorhinus hemitoechus* from the Pleistocene of Petralona Cave, in Chalkidiki. Most recently, Tsoukala (2018) studied the rhinocerotid material from the Neogene localities of Platania, Thermopigi, and Angelochori in Northern Greece.

3 Phylogenetic Relationships

The superfamily Rhinocerotoidea Gray, 1821 is commonly considered to encompass three distinct families: the Hyracodontidae Cope, 1879, the Amynodontidae Scott and Osborn, 1883, and the Rhinocerotidae Gray, 1821. (Radinsky 1966; Prothero et al. 1986, 1989; Prothero and Schoch 1989). According to an alternative analysis by Wang et al. (2016), the Indricotheriidae Borissiak, 1923 (= Paraceratheriidae Osborn, 1923) and the Eggysodontidae Breuning, 1924 may be regarded as two separate families from the Hyracodontidae within the Rhinocerotoidea. Fossil rhinocerotoids are represented in Greece thus far only by the family Rhinocerotidae. Within the Rhinocerotoidea, the family Rhinocerotidae is defined by the unique specialization of a prolonged, chisel-shaped first upper incisor occluding with an enlarged and procumbent tusk-like second lower incisor; other key features include the reduction of the dorsally ascending process of the premaxilla, resulting to the loss of the premaxilla-nasal contact, and an upper third molar with extremely reduced or lost metastyle (Radinsky 1966; Prothero et al. 1986, 1989; Heissig 2012a). The primitive rhinocerotoid *Uintaceras radinskyi* Holbrook and Lucas, 1997, documented by a fairly complete skeleton from the Middle Eocene deposits of the Uinta Basin (Utah, USA), is considered to best represent the sister taxon of the Rhinocerotidae as currently defined (Holbrook and Lucas 1997; Prothero 2005).

The supra-generic classification and the phylogenetic relationships within the family Rhinocerotidae remain controversial, with several systematic arrangements proposed and debated during the past four decades (e.g., Heissig 1973, 1989, 2012a;

Groves 1983; Prothero et al. 1986; Guérin 1989; Prothero and Schoch 1989; Cerdeño 1995; McKenna and Bell 1997; Antoine 2002; Antoine et al. 2010; Becker et al. 2013; Pandolfi et al. (2019b). It has to be noted that the majority of these recent systematic arrangements recognize more or less the same contextual supra-generic rhinocerotid lineages (e.g., the rhinocerotines, the aceratheriines, the elasmotheriines, the teleoceratines, the dicerotines to name a few), which roughly correspond to the "eleven rhinocerotid subfamilies" of the most plain systematic scheme outlined by Guérin (1989). Although it is evident by most authors that a multistage systematic hierarchy among these broadly acknowledged supra-generic lineages is phylogenetically appropriate, it is the relative taxonomical position (i.e., as a subfamily or a tribe, etc.), as well as the precise taxonomical content of each lineage that remain the subject of persistent controversy. In particular, the increasing application of computer-based cladistic analyses, apart from offering valuable insights into decisive aspects of rhinocerotid interrelationships, has also rendered a plethora of frequently contradicting phylogenetic outcomes. The recurrent and perplexing convergences observed in key cranial, dental, and postcranial characters among rhinocerotids are common disconcerting issues acknowledged by all authors. For the supra-generic taxonomy of the Greek rhinocerotid taxa under consideration (see the box below), the systematic arrangement proposed by Heissig (2012a) will be followed, as it outlines with adequate clarity the fundamental phylogenetic interrelationships among most rhinocerotid lineages.

Systematic Taxonomy and Biostratigraphical Distribution of Rhinocerotidae in Greece

Family Rhinocerotidae Gray, 1821

Subfamily Aceratheriinae Dollo, 1885

Tribe Teleoceratini Hay, 1902

Brachypotherium brachypus (Lartet in Laurillard, 1848); Middle Miocene

Tribe Aceratheriini Dollo, 1885

Aceratherium simorrense (Lartet, 1851); Middle Miocene

Acerorhinus neleus Athanassiou et al., 2014; Late Miocene

Chilotherium schlosseri (Weber, 1905); Late Miocene

Chilotherium samium (Weber, 1905); Late Miocene

Subfamily Rhinocerotinae Gray, 1821

Tribe Rhinocerotini Gray, 1821

Subtribe Dicerotina Ringström, 1924

Miodiceros neumayri (Osborn, 1900); Late Miocene

Subtribe Dicerorhinina Ringström, 1924

Dihoplus pikermiensis (Toula, 1906); Late Miocene

Stephanorhinus jeanvireti (Guérin, 1972); Late Pliocene

Stephanorhinus etruscus (Falconer *in* Ansted, 1859); Late Pliocene – Middle Pleistocene

Stephanorhinus hemitoechus (Falconer *in* Gaudin, 1859); Middle – Late Pleistocene

Coelodonta antiquitatis (Blumenbach 1799); Late Pleistocene

4 Extant Record and Distribution

Today, five rhinocerotid species placed in four genera and one Subfamily survive in parts of the Afrotropical realm of Sub-Saharan Africa and the Indomalayan realm of Southern and Southeastern Asia. All extant species are considered either threatened or critically endangered, most of them clinging to existence due to the relentless poaching for their horn and the increasing habitat loss (Groves and Grubb 2011; Emslie et al. 2019).

The black or hook-lipped rhinoceros, *Diceros bicornis* (Linnaeus, 1758), is found in a very wide range of sub-Saharan habitats, from montane and lowland forest, through savanna woodland, bush and thicket, mixed grassland and woodland, scattered-tree grassland, to semi-desert and arid desert, pursuing a browsing diet (Hillman-Smith and Groves 1994). Its historical range was continuous, extending from Central, through Eastern, to Southern Africa, but today the species survives in a limited number of scattered pocket populations (Emslie et al. 2019; Rookmaaker and Antoine 2012). A number of at least seven subspecies of *D. bicornis* has been recognized during historical times (Groves 1967; Groves and Grubb 2011). Today, conservation efforts are focused on just three surviving ones with a total number of ca. 5495 individuals (Emslie et al. 2019).

The extant white or square-lipped rhinoceros, *Ceratotherium simum* (Burchell, 1817), is a dedicated grazer preferring grassland and savannah habitats; it is divided in two well-founded subspecies with a strikingly discontinuous range (Heller 1913; Groves 1972). The southern subspecies *C. s. simum* (Burchell, 1817) inhabits Southern Africa, whereas the virtually extinct northern *C. s. cottoni* (Lydekker, 1908) used to occupy parts of central Africa (Groves 1975; Rookmaaker and Antoine 2012). A recent vigorous study by Groves et al. (2010) proposed recognizing the two subspecies as separate species. Relentless hunting had reduced the population of *C. s. simum* to a handful of animals in the 1900s, but after tenuous conservation efforts the subspecies recovered to a remarkable number of ca. 21,316 individuals today (Emslie et al. 2019; Groves 1975). The northern subspecies *C. s. cottoni* was considered by 2006 extinct in the wild. Four remaining individuals from the Czech Dvůr Králové Zoo were moved to the Ol Pejeta Conservancy in Laikipia, Kenya for a last-chance protected breeding program (Ryder et al. 2020). Unfortunately, the death of the last known male named Sudan in 2018, has left only two surviving female individuals. Alternatives in artificial reproduction (in vitro fertilization and embryo transfer) are presently explored as the ultimate hope for revival (Hildebrandt et al. 2018; Ryder et al. 2020).

The monotypic Indian or greater one-horned rhinoceros, *Rhinoceros unicornis* Linnaeus, 1758, is today restricted in several reserves south of the outer Himalayan foothills, pursuing a mix-feeding diet in the flood-plains grasslands and the riverine forest tracts valleys of northern India, southern Nepal, and Bhutan (Laurie et al. 1983; Dinerstein 2015); the surviving population is estimated at ca. 3588 individuals (Emslie et al. 2019). During the Holocene, the distribution of the species was broader encompassing the entire outer Himalayan foothills from Pakistan to easternmost India, whereas during the Pleistocene it extended throughout southern Asia, including the Indochinese Province and south China (Antoine 2012; Filoux and Suteethorn 2018).

The Javan or lesser one-horned rhinoceros, *Rhinoceros sondaicus* Desmarest, 1822, is a dedicated browser and was widespread throughout the Indochinese Peninsula and the Sundaic subregion during historical times, comprising three different subspecies (Groves and Leslie 2011; Groves and Guérin 1980). Today though only the nominotypical subspecies *R. s. sondaicus* Desmarest, 1822 survives in a single population at the Ujung Kulon National Park on the island of Java, Indonesia, with a current total estimate of just 65 individuals (Emslie et al. 2019). Unfortunately, despite the strenuous conservation efforts, the last individual of the Vietnamese mainland subspecies *R. s. annamiticus* Heude, 1892 was poached at the Cat Tien National Park in 2010 (Brook et al. 2012, 2014), following thus the fate of the Sundarbans subspecies *R. s. inermis* Lesson, 1838 from eastern India, Bangladesh and Myanmar, which went extinct in the early 1900s (Rookmaaker 1997).

The smallest extant species, the Sumatran or Asiatic two-horned rhinoceros *Dicerorhinus sumatrensis* (Fischer, 1814), used to have a wide distribution in several parts of the Indomalayan region during the Holocene, primarily in the Indochinese Peninsula and the Sundaic subregion, and extending probably to south China (Groves and Kurt 1972; Rookmaaker 1980, 2006; Antoine 2012). *D. sumatrensis* is a selective browser, with a habitat preference for tropical rain forests and mountain moss forests. After a drastic demographic decline during historical times, the species is presently restricted only to four isolated sites in Indonesia (Emslie et al. 2019). On the island of Sumatra, three National Parks host an uncertain number of ca. 37–75 individuals of the nominotypical subspecies *D. s. sumatrensis* (Fischer, 1814). On the island of Borneo, a handful of just 3–10 individuals of the Bornean subspecies *D. s. harrissoni* Groves, 1965 may still survive in the Kutai Barat Forest at the Indonesian province of East Kalimantan, whereas the last few known individuals from the Tabin Wildlife Reserve in the Malaysian state of Sabah have recently deceased (Kretzschmar et al. 2016; Mukhlisi et al. 2016; Emslie et al. 2019). The northernmost Indochinese mainland subspecies, *D. s. lasiotis* (Sclater *in* Buckland, 1872) is sadly considered to be extinct several decades ago (Groves and Grubb 2011).

5 Fossil Record and Distribution in Greece

The presence of the rhinocerotoid families Amynodontidae and Hyracodontidae (including the Paraceratheriinae and Eggysodontinae) is known from the Eocene and Oligocene deposits of neighboring countries in the Balkan and Anatolian Peninsulas (e.g., Petronijevic and Thenius 1957; Nikolov and Heissig 1985; Antoine et al. 2008; Tissier et al. 2018). Nevertheless, the absence of both amynodontid and hyracodontid rhinocerotoids in Greece can be explained by the notable scarcity of pre-Miocene localities with fossil large mammals in the country (Koufos 2016a, b;

see also other chapters in this volume). Thus far, only the family Rhinocerotidae of the "true" rhinocerotid rhinoceroses is represented in Greece, comprising eight genera and 11 species. The occurrence of fossil rhinoceroses has been documented in over 80 Greek localities, ranging from the Middle Miocene to the Late Pleistocene (Appendices 1–3, Figs. 2 and 3, Tables S1–S3). The taxonomic context of the Greek

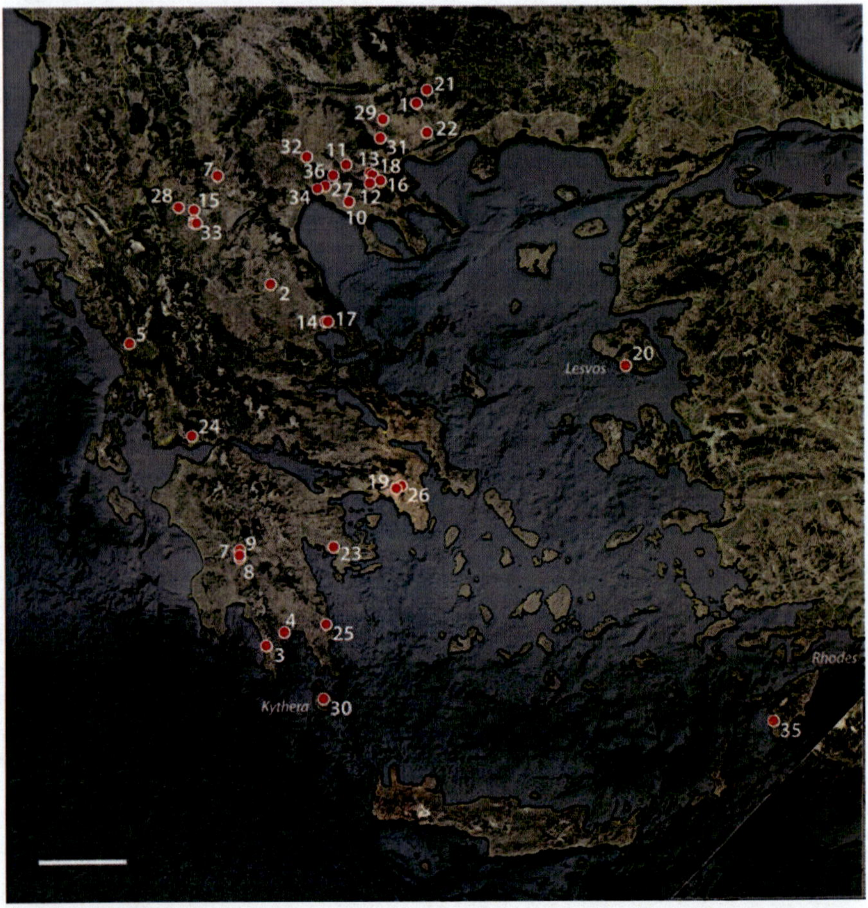

Fig. 2 Map of Greece showing the geographic distribution of the most important Plio–Pleistocene localities with rhinocerotid fossils. **1**, Aggitis Cave; **2**, Penios riverbank; **3**, Kalamakia; **4**, Lakonis I; **5**, Asprochaliko; **6**, Perdikkas; **7**, localities in Megalopolis; **8**, Kyparissia; **9**, Marathousa; **10**, Petralona Cave; **11**, Apollonia 1; **12**, Kalamoto 1 and 2; **13**, Platanochori; **14**, Alikes; **15**, Livakos; **16**, Krimni; **17**, localities in Sesklo; **18**, Tsiotra Vrisi; **19**, Tourkovounia 3–5; **20**, Vatera DS; **21**, Volax; **22**, Aivaliki; **23**, Karnezeika; **24**, Molikrio; **25**, Richea; **26**, Psychiko; **27**, Agia Triada; **28**, Neapolis; **29**, Serres basin; **30**, Kythera; **31**, Nigrita; **32**, Gephyra; **33**, localities in Milia; **34**, Angelochori; **35**, Apolakkia; **36**, Allatini. See Appendix 1 for more information. Image exported from Google Earth Pro © 2019, map data from US Dept. of State Geographer, SIO, NOAA, U.S. Navy, NGA, GEBCO, image from Landsat/Copernicus. Scale bar equals 80 km, North faces upward

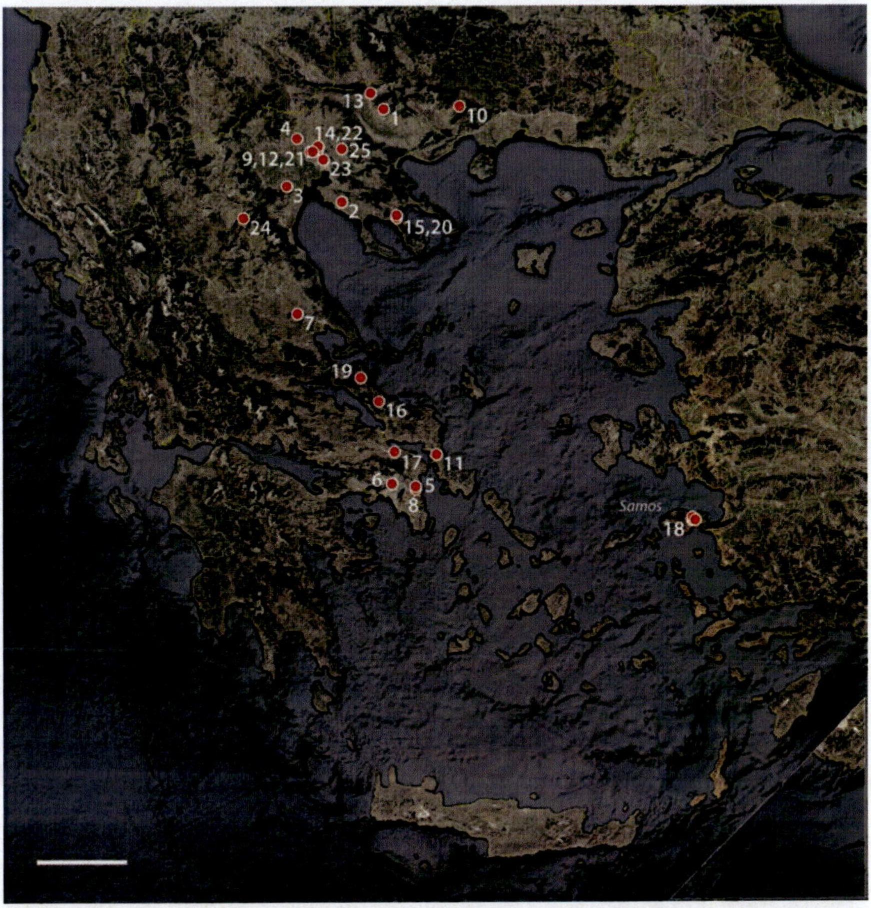

Fig. 3 Map of Greece showing the geographic distribution of the most important Miocene localities with rhinocerotid fossils. **1**, Maramena; **2**, Nea Silata; **3**, Asprogiannos 1; **4**, Dytiko 1; **5**, Chomateri; **6**, Pyrgos Vassilissis; **7**, Perivolaki; **8**, Pikermi localities; **9**, Vathylakkos localities; **10**, Platania; **11**, Halmyropotamos; **12**, Ravin X; **13**, Thermopigi; **14**, Ravin des Zouaves 5; **15**, Nikiti 2; **16**, Prokopi; **17**, Chalkoutsi; **18**, Samos localities; **19**, Kerassia localities; **20**, Nikiti 1; **21**, Ravin de la Pluie; **22**, Xirochori; **23**, Pentalophos 1; **24**, Servia; **25**, Chryssavgi. See Appendices 2 and 3 for more information. Image exported from Google Earth Pro © 2019, map data from US Dept. of State Geographer, SIO, NOAA, U.S. Navy, NGA, GEBCO, image from Landsat/Copernicus. Scale bar equals 80 km, North faces upward

fossil rhinocerotids allows their convenient presentation in a combined systematic and biostratigraphical order, offering the possibility to discuss some key biogeographical and paleoecological aspects associated with significant paleoenvironmental events.

Nomenclature and Abbreviations Anatomical terminology follows the sixth edition of Nomina Anatomica Veterinaria (NAV) as published by the International

Committee on Veterinary Gross Anatomical Nomenclature (2017); the works of König and Liebich (2004) and Schaller (2007) were also considered. Upper teeth are designated by upper-case letters and lower teeth by lower-case letters. References to provisions of the International Code of Zoological Nomenclature (ICZN) follow the current fourth edition, incorporating the Declarations 44 and 45 and the 2012 Amendment, as published by the International Commission on Zoological Nomenclature (1999, 2003, 2012, 2017). To facilitate nomenclatural opinions, whenever the authors' names and the year of publication are separated by a comma, they represent a citation of taxonomic authorship for the preceding taxon according to ICZN Art. 51. In changed combinations (species-group names combined with a generic name other than the original one), the use of parentheses around the authors' names and dates is applied according to ICZN Art. 51.3. The names of genera are abbreviated, only when it is contextually proper and clear.

Institutional Abbreviations **AMNH** American Museum of Natural History, New York; **AMPG** Athens Museum of Palaeontology and Geology, National and Kapodistrian University of Athens; **BMNH** British Museum of Natural History (= Natural History Museum), London; **BSPG** Bayerische Staatssammlung für Paläontologie und Geologie, Staatlichen Naturwissenschaftlichen Sammlungen Bayerns, München; **CM** Carnegie Museum of Natural History, Pittsburgh, Pennsylvania; **GPMM** Geologisch-Paläontologisches Museum (Geomuseum) der Westfälischen Wilhelms-Universität Münster; **GPIH** Geologisch-Paläontologisches Institut der Universität Hamburg (Centrum für Naturkunde, Hamburg); **HLMD** Hessisches Landesmuseum Darmstadt; **FSL:** Faculté des Sciences, Université Claude Bernard, Lyon; **IPUW** Institut für Paläontologie der Universität Wien; **IRSNB** Institut Royal des Sciences Naturelles de Belgique, Brussels; **KMMA** Koninklijk Museum voor Midden-Afrika, Tervuren (AfricaMuseum, Belgium); **LGPUT** Laboratory of Geology and Palaeontology, Aristotle University of Thessaloniki; **MAFI** Magyar Állami Földtani Intézet, Budapest; **MCGL** Musée Cantonal de Géologie Lausanne; **MGPP** Museo di Geologia e Paleontologia dell'Università di Padova; **MNHB** Museum der Naturkunde für Humboldt Universität zu Berlin; **MNHN** Muséum National d'Histoire Naturelle, Paris; **NHMA** Natural History Museum of Aegean, Mytilinii, Samos; **NHMB** Naturhistorisches Museum Basel; **NHMM:** Natural History Museum of Milia, Grevena; **NHMW** Naturhistorisches Museum Wien; **NMBE** Naturhistorisches Museum Bern; **NME:** National Museum of Ethiopia, Addis Ababa; **NMP** Národní Museum Prague; **NRM** Naturhistoriska Riksmuseet Stockholm; **PMMS** Paleontological Museum of Mytilinii Samos (now integrated with NHMA); **RMNH** Rijksmuseum van Natuurlijke Historie, Leiden (Naturalis Biodiversity Center); **SMF** Forschungsinstitut und Naturmuseum Senckenberg, Frankfurt am Main; **SMNK** Staatliches Museum für Naturkunde Karlsruhe; **SMNS** Staatliches Museum für Naturkunde Stuttgart; **STIPB** Steinmann-Institut für Geologie, Mineralogie und Paläontologie (Goldfuss Museum), Rheinische Friedrich-Wilhelms-Universität Bonn; **UCMP** University of California Museum of Paleontology (Vertebrate Collection); **USNM** United States National Museum, Washington, D.C. (National

Museum of Natural History, Smithsonian Institution); **YPM** Yale Peabody Museum of Natural History; **ZMA** Zoölogisch Museum, Universiteit van Amsterdam (now part of Naturalis Biodiversity Center, Leiden) **ZMUC** Zoologisk Museum Universitet København (Statens Naturhistoriske Museum), Copenhagen.

6 Systematic Paleontology

Family Rhinocerotidae Gray, 1821

Type Genus *Rhinoceros* Linnaeus, 1758.

Subfamily Aceratheriinae Dollo, 1885

Type Genus *Aceratherium* Kaup, 1832.

Tribe Teleoceratini Hay, 1902

Type Genus *Teleoceras* Hatcher, 1894.

***Brachypotherium* Roger,** 1904

Type Species *Rhinoceros brachypus* Lartet *in* Laurillard, 1848. Type species fixed in the original publication (Roger 1904: 13) by original designation (ICZN, Art. 68.2). Subsequent citations or fixations of *Rhinoceros goldfussi* Kaup, 1834 as the type species of the genus *Brachypotherium* (e.g., Prothero 1998; Geraads 2010; Geraads & Miller 2013) are considered either incorrect or invalid respectively (ICZN, Art. 70.2), even in the case that the two species are to be regarded as subjective synonyms (ICZN, Art. 67.1.2).

Remarks *Brachypotherium*, as currently understood, is a long-lasting Miocene teleoceratine genus with a notably vast spatial distribution throughout Eurasia and Africa; a dedicated systematic revision is thus apparently required. The genus currently comprises all large-sized teleoceratine species of this region, which are characterized by a brachycephalic cranium, a markedly enlarged I1 occluding with a powerful tusk-like i2, as well as a notably robust and shortened appendicular skeleton resembling hippopotamid proportions. Apart from the aforementioned species *B. brachypus* and *B. goldfussi* from Europe (Heissig 1996, 1999), it does also comprise *B. perimense* (Falconer and Cautley, 1847) from Southern Asia (Heissig 1972) and perhaps *B. pugnator* (Matsumoto, 1921) from Japan (Fukuchi and Kawai 2011). In addition, the species *B. snowi* (Fourtau, 1918), *B. heinzelini* Hooijer, 1963, *B., minor* Geraads and Miller, 2013, and *B. lewisi* Hooijer and Patterson, 1972 have been described from Africa (Geraads 2010). The latter species is the only representative of the genus that seems to have persisted into the Early Pliocene, at the locality of Lothagam in Kenya (Harris and Leakey 2003). In North America, primitive teleoceratine material from the late Hemingfordian has been frequently attributed to the species *Brachypotherium americanum* Yatkola and Turner, 1979 (Prothero 1998). However, this species has been re-assigned to the more derived genus *Teleoceras* Hatcher, 1894, which presently comprises all North American teleoceratine species (Prothero 2005).

Brachypotherium brachypus **(Lartet *in* Laurillard, 1848)**

Nomenclatural Note The species-group name *brachypus* was first established for the species *Rhinoceros brachypus* by Lartet *in* Laurillard (1848: 101). The following quotation made by Lartet (1851: 29): "Rhinocéros brachypus. *Lart.* C. R. Ac. Se. 1837." (sic) seems to have been the source for referring to the date of authorship as 1837 by some authors (e.g., Cerdeño 1992; Heissig 1996, 2012b; Prothero 1998). However, this is apparently incorrect and the name *brachypus* was not introduced before 1848 as noted by Mestre (1934: 316). I have also verified the original references, and neither Lartet (1837a, b, c), nor (Lartet in) de Blainville (1837a, b), or Lartet (1836) mentioned the name *brachypus*. Sometimes, the species-group name *brachypus* has been cited with a later date of authorship as "Lartet, 1851" (e.g., Cerdeño 1993; Koufos and Kostopoulos 2013), which is also inaccurate. Based on the available evidence, and under the provisions of ICZN Art. 50.1.1, Lartet is considered as the author of the name in Laurillard (1848: 101).

Type Material To the best of our knowledge, neither a holotype, nor a lectotype or a neotype has been designated (Heissig 1976, 2012b; Prothero 1998). Following ICZN Arts. 72.1.1 and 73.2, all specimens based on which Lartet established *Rhinoceros brachypus* in the original publication of Laurillard (1848) are to be considered as the syntypes of the species.

Type Locality Simorre, Gers, Occitanie, France; Middle Miocene (MN7/8).

Distribution The species is a common element of Early and Middle Miocene faunas in Europe and neighboring Asian countries (Heissig 1976, 1996; Cerdeño 1993).

Remarks Psarianos (1958) initially described as "*Diceros pachygnathus*" two lower rhinocerotid premolars from a clay-pit near the village of Chryssavgi (Mygdonia basin, N. Greece). Dimopoulos (1972) studied the same teeth, including also some upper molars and a few rib fragments, and assigned them to "*Dicerorhinus orientalis.*" Both authors implied a Late Miocene age for the fossils. In a faunal review, de Bonis and Koufos (1999) referred to this material as Rhinocerotidae indet., suggesting though a Middle Miocene (MN7–8) age based on a micromammalian association collected later from an adjacent spot at Chryssavgi (Koliadimou 1996). In a review on the Neogene rhinoceroses of Greece, Giaourtsakis (2003) provisionally retained the determination Rhinocerotidae indet. For the material from Chryssavgi. During 2004, I had the opportunity to examine the material from Chryssavgi at the collections of LGPUT, updating its assignment to *Brachypotherium* sp.; this determination was also followed by Koufos (2006a).

Subsequently, Koufos and Kostopoulos (2013) described in detail the specimens from Chryssavgi as *Brachypotherium brachypus* and discussed thoroughly their systematic affinities, updating also the stratigraphical and biochronological context of the locality within the Mygdonia Basin.

The dental remains from Chryssavgi are characterized by their large size and thick enamel. The right M3 features a subtriangular occlusal outline with a continuous ectometaloph, a well-developed paracone fold, a deep and widely open medisinus valley (despite the advanced stage of dental wear) with a cingular bulge on its entrance, a relatively small crochet and antecrochet, as well as a constricted protocone with flattened lingual wall. The partially preserved left M2 features a concave occlusal surface and a notably sharp metacone cusp. These dental features are in accordance with an attrition-dominated occlusal wear and suggestive of a browsing-dominated diet (Fortelius 1982; Fortelius and Solounias 2000). The most significant feature of the lower molars from Chryssavgi is the shallow ectoflexid on the buccal aspect of the teeth, which is typical for the genus (Fig. 4). The detailed metrical comparisons provided by Koufos & Kostopoulos (2013: fig. 5) place the specimens from Chryssavgi within the known range of *B. brachypus*.

Tribe Aceratheriini Dollo, 1885

Type Genus *Aceratherium* Kaup, 1832

***Aceratherium* Kaup,** 1832

Type Species *Aceratherium incisivum* Kaup, 1832 (nec *Rhinoceros incisivus* Cuvier, 1822). Type species hereby fixed as *Aceratherium incisivum* Kaup, 1832 under Article 70.3 of the ICZN, misidentified as *Rhinoceros incisivus* Cuvier, 1822 in the original designation by Kaup (1832a: 904). Kaup's explicit and discrete employment of the binomen "*Aceratherium incisivum* Kaup" under the binomen "*Rhinoceros incisivus* Cuvier" is considered to have been deliberately cited as a junior synonym (Kaup, 1832a: Taf. XVII, Fig. 1; compare also the citation of Kaup 1832b: 419). As such, the binomen *Aceratherium incisivum* Kaup, 1832 was commonly treated as an available name before 1961 in accordance with the provisions

Fig. 4 Teeth of *Brachypotherium brachypus* (Lartet *in* Laurillard, 1848) from the Middle Miocene of Chryssavgi, Greece, in occlusal views, LGPUT collections: **a** right M3 (CHR-500); **b** left M2 (CHR-501); **c** left mandibular fragment with m2–m3 (CHR-502). Scale bar equals 3 cm

of Art. 11.6.1, and it was predominantly adopted as the name of this taxon in prevailing usage (e.g., Kaup 1854: 1; Brandt 1878: 27; Lydekker 1881: 4, 145; Osborn 1900: 248; Ringström 1924: 74 etc; *vide* Giaourtsakis and Heissig 2004). It was thereby made available under Art. 11.6.1 and dates from its first publication as a synonym by Kaup (1832a). In accordance with Art. 50.7, Kaup 1832a is considered as the author of the name.

Following the provisions of Art. 72.4.3, the type series of *Aceratherium incisivum* Kaup, 1832 consists of the specimens cited with that name in the published synonymy by Kaup (1832a). The partially preserved cranium of an adult individual from the Late Miocene (Vallesian) locality of Eppelsheim housed in the collections of the Hessisches Landesmuseum Darmstadt in Germany (HLMD: DIN 1927) is part of the type series described by Kaup (1832a); it is hereby formally designated as the lectotype of *Aceratherium incisivum* Kaup, 1832, in accordance with the provisions of the Art. 74. The lectotype cranium has been figured by Kaup (1834: tab. X, fig. 2a) and by Giaourtsakis & Heissig (2004: fig. 1.3). The lithographic drawing of the cranium depicted with a mandible and an enlarged first upper incisor by Kaup (1832a: Taf. XVIII, Fig. 1) and Kaup (1834: tab. X, fig. 2) is only partially based on the lectotype cranium; it is in fact a composite and thus imprecise reconstruction based on several specimens from different individuals and species (Giaourtsakis and Heissig 2004). The same is true for a fairly enlarged plaster reconstruction of a complete cranium (Kaup and Scholl 1834), which has been widely distributed as a cast of *Aceratherium incisivum* in several Natural History Museums, including London, Paris, and New York, and may have been frequently but uncritically used for systematic comparisons (*vide* discussion in Giaourtsakis and Heissig 2004). The species *A. incisivum* is best documented in the Vallesian (MN9) locality of Höwenegg (Hegau, Germany), where two nearly complete skeletons have been described in detail (Hünermann 1989).

Nomenclatural Notes *Acerotherium* (Kaup 1834: 49), *Acerathium* (Schubert 1835: 431), *Acrotherium* (Lartet 1851: 28), and *Acerothirium* (Gill 1872: 87) are incorrect subsequent spellings and under ICZN Art. 33.3 they are all to be considered as not available names.

Remarks *Aceratherium* was the third nominal genus of a fossil rhinocerotid to be ever described, following *Elasmotherium* Fischer, 1808 and *Coelodonta* Bronn, 1831. Such was the impact of this first discovery of a hornless rhinoceros in the literature, that for almost one century *Aceratherium* has been commonly used as the collective genus for nearly all hornless rhinocerotid and several rhinocerotoid species that were being discovered in the Old and the New World (e.g. Lydekker 1874: 3-5; Roger 1896: 186–189). Subsequent systematic revisions have gradually assigned most of these species to separate genera (cf. summaries in Heissig 1989, 1999, 2012b; Prothero 2005). In the writer's opinion, the taxonomic content of the genus *Aceratherium* must be restricted to the primitive smaller sized aceratheriini (i.e., non-chilothere) from the Middle and Late Miocene of Europe and adjacent

Asian countries, comprising the species *A. incisivum* Kaup, 1832, *A. simorrense* (Lartet, 1851), and *A. alfambrense* Cerdeño and Alcalá, 1989.

Ginsburg and Guérin (1979) created the subgenus *Aceratherium* (*Alicornops*) with type species *Rhinoceros simorrensis* Lartet, 1851 from the Simorre (Gers, France). Subsequently, *Alicornops* was elevated to genus level, and the taxon was sometimes reported as *Alicornops simorrensis* (e.g., Guérin 1989; Cerdeño and Sanchez 2000). However, this concept was primarily based on the misconception that *Aceratherium incisivum* (the nominotypical species of the genus *Aceratherium*) was a "large acerathere," and that the "smaller aceratheres" required a different taxonomic allocation. *Aceratherium incisivum* was inaccurately regarded by Guérin (1980) as the potential descendant of the significantly larger and long-limbed Middle Miocene species "*Aceratherium*" *tetradactylum* (Lartet, 1836). The latter taxon is currently placed in the separate genus *Hoploaceratherium* Ginsburg and Heissig 1989. *Aceratherium incisivum* is in fact associated with the smaller and more robust postcranial elements, which are scarce in the type locality of Eppelsheim (Giaourtsakis and Heissig, 2004), but they are well documented with two nearly complete skeletons in the locality of Höwenegg (Hünermann, 1989). *Aceratherium incisivum* must be thus regarded as the successor of *Aceratherium simorrense* and not of *Hoploaceratherium tetradactylum.*

***Aceratherium simorrense* (Lartet, 1851)**

Type Material Lectotype, a partially and fragmentarily preserved cranium (MNHN F.Si.48) and the associated mandible (MNHN F.Si.50) housed in the collections of the Muséum National d'Histoire Naturelle in Paris, France; subsequently fixed and figured by Ginsburg & Guérin (1979: 115, fig. 1). The dentition of the specimen was also figured by Osborn (1900: figs. 13b, 14b), but he did not designate a lectotype for this species. The species-group name *simorrensis* was first established for *Rhinoceros simorrensis* by Lartet (1851: 29), and not by Lartet in Laurillard (1848) as stated by Heissig (2012b: 344).

Type Locality Simorre - Malartic, Gers, Occitanie, France; Middle Miocene (Astaracian, MN7/8).

Distribution The species is a frequent element of primarily Middle Miocene and early Late Miocene (MN5–9) faunas across Europe and the neighboring eastern parts of Asia (Heissig 1976, 1996, 2012b; Cerdeño and Sanchez 2000).

Remarks Symeonidis (1974) described as *Aceratherium* sp. some interesting dental remains originating from a small lignite pit near the city of Atalanti. The material comprises the rostral tip of a thin, hornless nasal fragment, as well as portions of two functional and well-formed second lower incisors; a larger one attributed to a male individual (Fig. 5c), and a smaller one attributed to a female individual (Fig. 5b). The maxillary dentition is represented by several remains. The third and fourth premolars feature a lingual cingulum, a simple crochet, and a faintly expressed

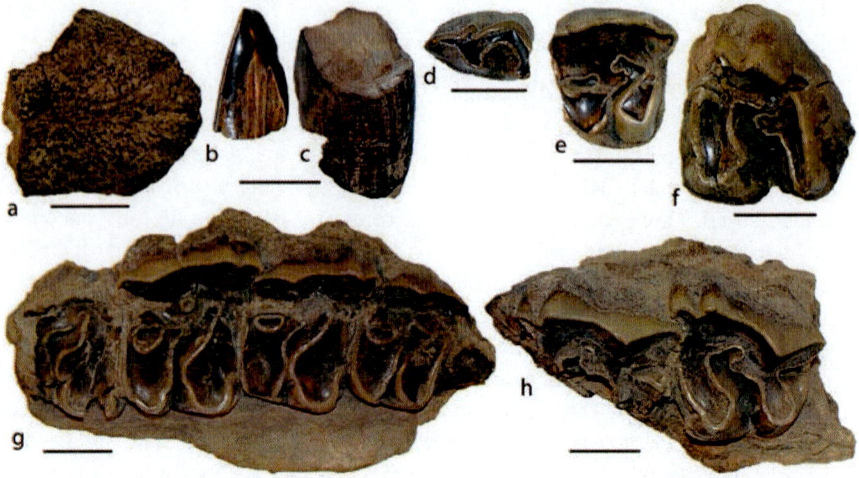

Fig. 5 *Aceratherium simorrense* (Lartet, 1851) from the Middle Miocene of Atalanti, Greece, AMPG collections: **a** rostral fragment of nasal tip; **b** fragment of left female i2 (ATL-15); **c** fragment of right male i2 (ATL-16); **d** right p2 (ATL-6); **e** right P2 (ATL-5); **f** left M3 (ATL-12); **g** maxillary fragment with P1–M1 (ATL-1); **h** maxillary fragment with left M1-M2 (ATL-2). Scale bar equals 2 cm

protocone constriction; a crista is not present, but a closed medifossette may be formed between the crochet and the worn ectoloph in older adult specimens during the late stages of occlusal wear. The molars are characterized by a moderately constricted protocone, a weak antecrochet, and a marked paracone fold on the buccal wall of the ectoloph; the metacone cusp is notably high and sharp (Fig. 5g, h). The morphology and the relative small size of the teeth readily exclude an ascription to *Brachypotherium*. Moreover, these specimens lack derived features such as the prominent antecrochet, the weaker paracone fold, the frequently blunt metacone, and the markedly constricted protocone observed in the molars of *Chilotherium*, and to a lesser extent in *Acerorhinus*. The affinities of this material are similar to the morphology of the "Western European" aceratheres, in particular to the smaller sized Middle Miocene species *A. simorrense*.

The dentition of *A. simorrense* is characterized by the relatively low-crowned molars featuring a concave surface of occlusal wear, notably sharp metacone apices, and a high relief in-between the buccal cutting cusps. These dental features are indicative of an exclusive browsing diet. The appendicular skeleton is shortened with relatively robust epiphyses, especially on the metapodials (Cerdeño and Sanchez 2000). The species is considered to have preferred forested areas or open woodlands in warm and moist habitats (Guérin 1980).

Biogeographical and Paleoecological Remarks Resembling *Brachypotherium*, the genus *Aceratherium* (sensu stricto) seems to have retained a continuous biogeographical distribution during the Middle Miocene across Europe and the neighboring western Asian countries. Both genera frequently appear sympatric, particularly in the well-sampled localities of Central and Western Europe (Heissig 1996, 1999). The transition toward the Late Miocene in Europe was characterized by a significant faunal turnover known as the Vallesian Crisis. It was caused by a major global climatic change that favored the gradual expansion of more open and arid habitats in several regions (Casanovas-Vilar et al. 2005; Koufos 2006b; Eronen et al. 2009). These environmental changes affected also the rhinocerotid distribution. In Central and Western Europe, where more temperate and closed habitats persisted, both *Brachypotherium* and *Aceratherium* continued uninterrupted into the Late Miocene (Heissig 1996, 1999). However, both taxa disappeared from the Eastern Mediterranean region and they were gradually replaced by the more specialized aceratheriine genera *Acerorhinus* and *Chilotherium*, which have migrated from Asia during the Late Miocene (Heissig 1996; Giaourtsakis 2003).

***Acerorhinus* Kretzoi,** 1942

Type Species *Aceratherium zernowi* Borissiak, 1914 by monotypy (Kretzoi 1942a: 311).

Remarks A detailed overview on the systematic affinities and the context of the genus *Acerorhinus* is provided by Athanassiou et al. (2014). Specimens of *Acerorhinus* have been frequently misidentified as *Aceratherium* or *Chilotherium*. Based on some of the most salient differences, *Acerorhinus* differs from *Aceratherium* by the more widened frontal region with relatively enlarged supraorbital processes and more constricted nasal base, the frequent presence of a strong facial crest, the notably concave ventral side of the mandibular symphysis, and the absolutely longer and relatively more slender metapodials. *Chilotherium* differs from *Acerorhinus* by the more flattened and frequently depressed frontal region, the sub-trapezoid occipital outline, the reduced size of the premolar series, the markedly widened mandibular symphysis, and the significantly more shortened and robust appendicular skeleton.

The oldest specimens with evident *Acerorhinus* affinities have been documented from the Middle Miocene (MN8) localities of the Tung-gur Formation in Inner Mongolia, China (Cerdeño 1996; Athanassiou et al. 2014). The majority of *Acerorhinus* species feature a Late Miocene (MN9–13) temporal distribution and include: *A. neleus* Athanassiou et al., 2014 from the Balkan Peninsula and perhaps Italy and Turkey; *A. zernowi* (Borissiak, 1914) and its potential junior synonyms *A. simplex* (Krokos, 1914), *A. asiaticus* (Borissiak, 1915), *A. ggadzhievi* (Dzhafarov, 1958), *A. dgadzhievi* (Dzhafarov, 1958), and *A. beleavi* (Korotkevich, 1974) from the Northern Peri-Pontic region. Further, *A. eldaricus* (Tsiskarishvili, 1987) from the locality of Eldari-2 in Azerbaijan, and its potential junior synonym *A. rodleri* (Pandolfi, 2016) from the nearby locality of Maragheh in the East Azerbaijan province of Iran. In addition, with more or less (?= questionable) certainty: *A.? hipparionus* (Koken, 1885), *A. palaeosinensis* (Ringström, 1924), *A.? minus* (Ringström, 1924), *A. tsaidamensis* (Bohlin, 1937), *A. huadeensis* (Qiu, 1979), *A.? xiaoheensis*

Gao and Ma, 1997, *A. yuanmouensis* Zong, 1998, *A. hezhengensis* Qiu et al., 1988, *A. fuguensis* Deng 2000, *A.? lufengensis* Deng and Qi, 2009, and *A. porpani* (Deng et al. 2013) from Central and Eastern Asia (China and neighboring countries). The Asian taxa are in need of a dedicated revision, and several synonymies are to be expected (Lu 2013; Athanassiou et al. 2014).

★*Acerorhinus neleus* Athanassiou, Roussiakis, Giaourtsakis, Theodorou, Iliopoulos, 2014

Type Material Holotype (AMPG K4/119.37), a well-preserved female adult cranium with associated mandible at the collections of the Athens Museum of Palaeontology and Geology of the National and Kapodistrian University of Athens, Greece; presently on display at the local Fossil Mammal Museum of Kerassia, Euboea Island. Figured by Athanassiou et al. (2014: pls. 1, 2), and herein (Fig. 6).

Type Locality and Age Kerassia site K4, near the village of Kerassia, Euboea Island, Greece; Late Miocene, Turolian.

Distribution Several Late Miocene localities of primarily Turolian age (MN11–13) in the Balkan Peninsula (Greece, North Macedonia, Bulgaria). The occurrence of the species in Italy and Turkey is possible, but requires additional comparative studies.

Remarks *Acerorhinus neleus* features a relatively dolichocephalic cranium (Fig. 6a) with relatively short hornless nasals, a gently concave dorsal cranial profile, a subvertical occipital plane with a bell-shaped occipital outline, as well as closely converging parietal crests in females that may form a sagittal crest in males. The mandible is characterized by a narrow and ventrally concave symphysis with a relatively short diastema (Fig. 6b). The well-developed second lower incisors exhibit a marked sexual dimorphism. The dentition is relatively low-crowned featuring a concave occlusal surface and sharp metacone cusps on the ectoloph profile (Fig. 6c). The upper premolars feature a continuous lingual cingulum, a faint protocone constriction, no crista, a well-developed crochet, and a weak paracone fold. The upper molars feature a moderate protocone constriction, a weak to moderate antecrochet, a well-developed crochet, no crista, and a marked paracone fold. The lower premolars are characterized by a continuous crenulated labial cingulum. The metapodials of *A. neleus* feature significantly more slender proportions with respect to the frequently sympatric species of *Dihoplus*, *Miodiceros,* or *Chilotherium*. They are absolutely shorter than the metapodials of the horned species, but significantly longer than the metapodials of *Chilotherium* (Fig. 7).

Athanassiou et al. (2014) attributed to *A. neleus* the aceratheriine rhinocerotid specimens from the Greek Turolian localities of Kerassia, Pikermi, and Chomateri. A partially preserved juvenile cranium with associated mandible from Pikermi has been also referred to this species (Giaourtsakis et al. 2018). A well-preserved adult cranium with associated mandible from the Turolian locality of Karaslari in North

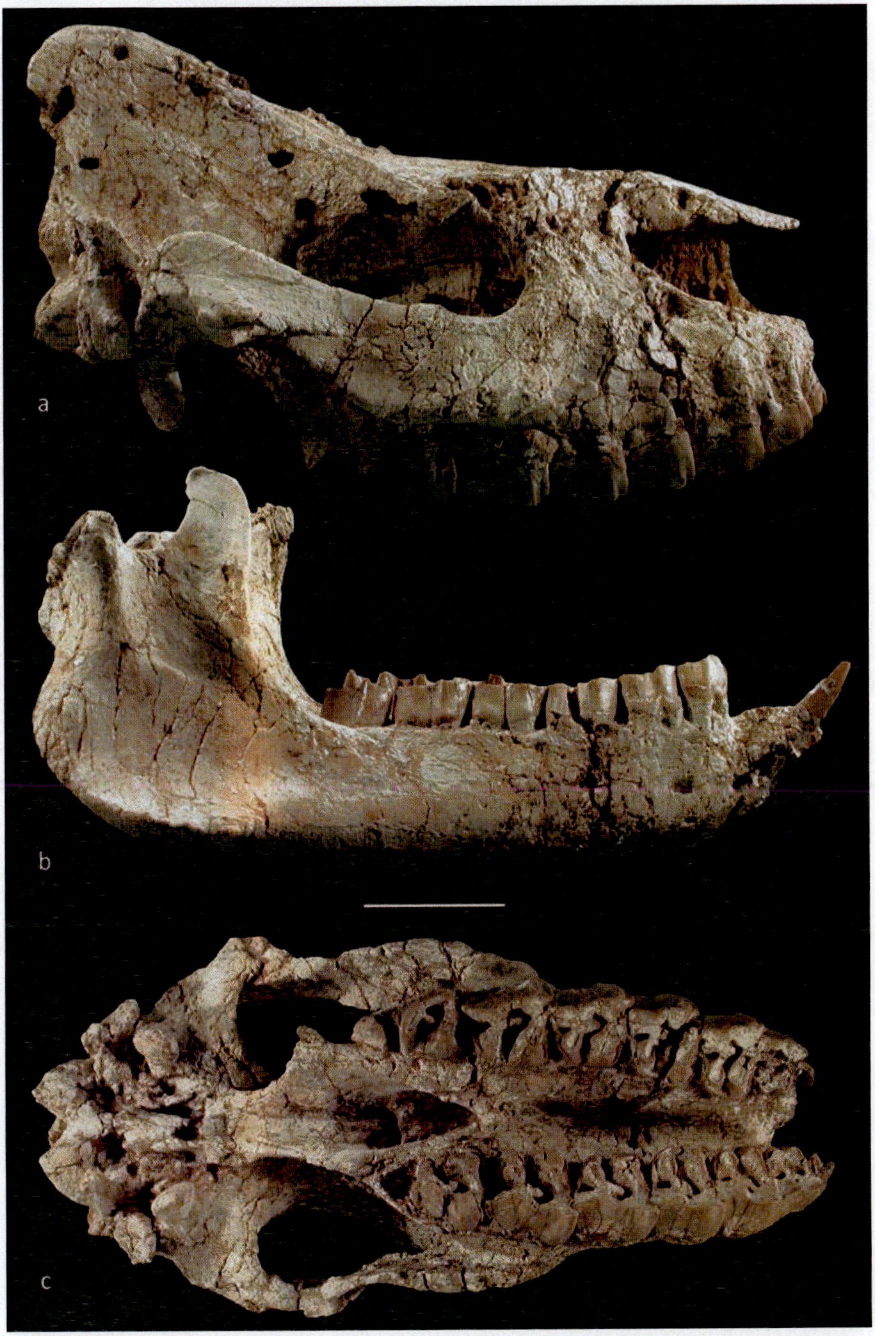

Fig. 6 *Acerorhinus neleus* Athanassiou et al., 2014 from the Late Miocene of Kerassia, Greece. Holotype adult cranium and associated mandible (AMPG: K4/119.37): **a** cranium in right lateral view; **b** mandible in right lateral view; **c** cranium in ventral view, with upper dentition in occlusal view. Scale bar equals 10 cm

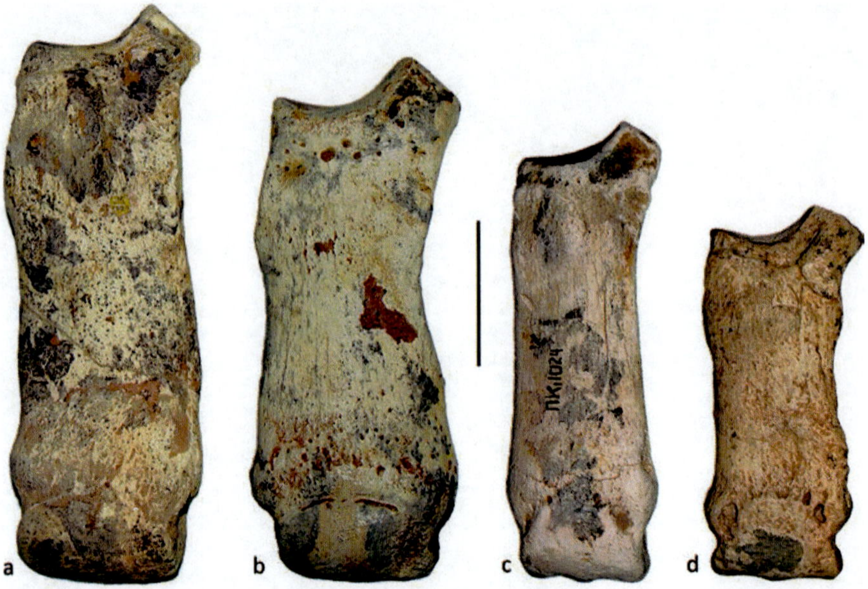

Fig. 7 Comparison between the third metacarpals (Mc-3) of Late Miocene rhinoceroses from the Eastern Mediterranean. **a** *Dihoplus pikermiensis* from Pikermi, Greece (BMNH: M.11297), mirrored image; **b** *Miodiceros neumayri* from Pikermi (MNHN: PIK-1010d), mirrored image; **c** *Acerorhinus neleus* from Pikermi (AMPG: ΠK1–1024); **d** *Chilotherium* sp. from Garkin, Turkey (BSPG: 1968.VI.389). Scale bar equals 5 cm

Macedonia (Spassov et al. 2018: pl. 11, fig. 5) is morphologically very similar with the holotype skull of Kerassia and it is also included in *A. neleus*. The same is true for the material referred to *Acerorhinus* sp. (Geraads and Spassov 2009: pl. 2) from the Turolian locality of Kalimantsi in Bulgaria. The well-preserved frontal, parietal and occipital regions, the short diastema of the mandibular symphysis, and the morphology of the upper dentition are very similar with the Kerassia type specimen. The more densely converging parietal crests forming virtually a sagittal crest, and the stronger second lower incisors observed in the Kalimantsi and Karaslari specimens may be attributed to the sexual dimorphism of male individuals.

Material described as *Acerorhinus zernowi* from several Turolian, and possibly late Vallesian localities in Turkey (Heissig 1975, 1999; Kaya and Heissig 2001; Fortelius et al. 2003) may also belong *A. neleus*, but a dedicated revision is necessary. A mandible from the Vallesian locality of Pentalophos (LGPUT: PNT-142), originally included in the hypodigm of *Chilotherium kiliasi* by Geraads & Koufos (1990: pl. 3, figs. 2, 3), has been confidently re-assigned to *Acerorhinus* sp. (Heissig 1999; Fortelius et al. 2003; Athanassiou et al. 2014: fig. 6). The fragmentarilly preserved cranium PNT-32, may also belong to *Acerorhinus*. An incompletely preserved mandible discovered in Late Miocene reworked sediments at Monte delle Piche (Rome, Italy) shows the typical morphological traits of *Acerorhinus* (Pandolfi et al. 2013); it currently marks the western-most occurrence of the genus, and perhaps of *A. neleus*.

The conservative dental morphology of *A. neleus* features a low-crowned and functionally brachyodont dentition (sensu Fortelius 1982), a simple and uncomplicated enamel pattern with reduced thickness in the secondary folds, a thin cement coating, well-developed cinguli on the premolars, and concave occlusal surfaces on the upper teeth with sharp metacone apices. All these features are indicative of a predominately browsing diet based on non-fibrous and nonabrasive vegetation and suggest a preference for the denser niches of the mosaic open woodland regimes within the Pikermian Biome (Athanassiou et al. 2014).

Chilotherium Ringström, 1924
(= ★_Eochilotherium_ Geraads and Spassov, 2009)

Type Species *Chilotherium anderssoni* Ringström, 1924 by original designation (Ringström 1924: 26).

Amended Diagnosis *Chilotherium* is a highly specialized genus of the tribe Aceratheriini characterized by a moderately dolichocephalic cranium with hornless nasals and frontals, a marked and dorsally unfused internasal suture, a deep nasal incision, and a flattened and frequently depressed frontal region with well-developed postorbital processes. The dorsal cranial profile is gently concave or flattened; the parietal crests are moderately to widely separated. The nuchal crests are wide featuring a nearly straight or slightly concave caudal border, and the occipital plain is subvertical with a roughly sub-trapezoid occipital outline. The mandible is characterized by a notably widened mandibular symphysis with a concave ventral side. The permanent upper incisors are lost; the second permanent lower incisors are strong and tusk-like, and they are characterized by a notable sexual dimorphism; the first lower incisors are diminutive and peg-like. The relative length of the premolars with respect to the molars is reduced, primarily due to the reduction of the size of the second upper and lower premolars. The first upper deciduous molar persists almost always during the adulthood; the first lower deciduous premolar is single-rooted and does not persist in adulthood. The molars feature a strong antecrochet and a marked protocone constriction. The appendicular skeleton is notably shortened and relatively robust.

Remarks The oldest species with evident *Chilotherium* affinities is *Ch. primigenium* Deng, 2006, documented from the early Late Miocene locality of Zhongmajia, in the Linxia Basin of the Gansu Province, China (Deng 2006). The biostratigraphical range of the genus includes the Late Miocene, but it also extends to the Early Pliocene (MN9–14). The paleobiogeographical dispersal of *Chilotherium* is fairly similar to *Acerorhinus*, including the Eastern Palearctic subregion of Asia, the Northern Peri-Pontic and Caucasus regions, as well as the Anatolian and Balkan Peninsulas.

Despite the readily identifiable morphological features and the well-defined systematic context, the specific content of *Chilotherium* remains unsettled. An anecdotal euphemism among rhinocerotid paleontologists refers to the chilotheres as the "hippariones of rhinos," being a reference to the large numbers of species that this genus has been associated with. Indeed, more than forty species and local subspecific variants may have been described as, or attributed to *Chilotherium* since the establishment of the genus. Several of these taxa have been rightfully transferred to other genera, including *Acerorhinus* Kretzoi, 1942, *Shansirhinus* Kretzoi, 1942, *Subchilotherium* Heissig, 1975, and *Aprotodon* Forster-Cooper, 1915 (e.g., Heissig 1975, 1999; Deng 2005, 2006; Athanassiou et al. 2014). For the purpose of this chapter we shall focus on the occurrences of the genus in Greece and adjacent regions, pending though future studies to clarify several of the existing systematic uncertainties. Regarding the purported occurrence of the genus *Chilotherium* in Western Europe, it must be noted that the hypodigm of *Dicerorhinus orientalis hungaricus* Gaál, 1938 from the Late Miocene Hungarian locality of Hatvan (Gaál 1943), which was subsequently assigned to as Chilotherium anderssoni hungaricum by Gaál (1953, 1954), does not belong to the genus. Further, the hypodigms of two species described as *Chilotherium quintanelensis* Zbyszewsky, 1952 and *Chilotherium ibericum* Antunes, 1972 from Portugal have been reassigned to the elasmotheriine rhinocerotid *Hispanotherium matritense* Lartet *in* de Prado, 1864 (Villalta and Crusafont 1955, Heissig 1975, Deng 2006). It is also to be noted that in China, apart from the nominotypical species *Ch. anderssoni* Ringström, 1924, the species *Ch. wimani* Ringström, 1924 is the most adequately documented; as currently recognized, it is represented by abundant material and is suitable as a basis for systematic comparisons (Deng 2001, 2002b; Chen et al. 2010). For the numerous taxa from Central and Eastern Asia that have been associated with *Chilotherium*, a dedicated revision is necessary.

★*Chilotherium schlosseri* (Weber, 1905)
(= ★*Aceratherium wegneri* Andrée, 1921 = ★*Aceratherium angustifrons* Andrée, 1921)

Type Material Lectotype by present designation under the provisions of ICZN Art. 74, a relatively well-preserved adult cranium with associated mandible described and figured as *Aceratherium schlosseri* by Weber (1905: Taf. VIII, figs 1-4; Taf. IX, fig. 1 *pars*). It was collected by Kommerzienrat Th. Stützel or C. Hentschel on the island of Samos (Greece) and was formerly stored at the collections of the Bayerische Staatssammlung für Paläontologie und Geologie München, in Germany. The specimen is believed to have been destroyed during the Second World War; the fact that the specimen no longer exists or cannot be traced does not of itself invalidate the designation (ICZN Art. 74.4), and the available descriptions and figures may be used as a reference.

Note The species name *Aceratherium schlosseri* Weber *in* Schlosser, 1904 (as published by Schlosser 1904: 113) is a *nomen nudum*, because it fails to conform to the requirements of ICZN Art. 12. The species name is valid as published by Weber (1905).

Type Locality and Age Samos Island, Greece; Late Miocene fluviolacustrine volcanoclastic deposits of the Mytilinii formation.

Distribution Several Late Miocene localities in the Balkan and Anatolian Peninsulas; extending perhaps to the Northern Peri-Pontic region, if a subjective synonymy with *Chilotherium kowalevskii* (Pavlow, 1913) is pursued.

Remarks The hornless rhinocerotid material from Samos at the BSPG collections has been adequately documented by Weber (1905), before its destruction during the Second World War. Despite their aesthetic appeal, classic lithographs must be always critically examined, since a certain level of artistic enhancement is to be expected; in particular incomplete or deformed regions of the specimen may have been retouched, or complex cranial regions may have been simplified. Weber (1905) states that the nasal incision of the lectotype cranium terminates caudally over the middle of the M1, but in the depiction it appears somewhat more retracted and terminates over its distal border; in addition, the orbital and auditory regions are certainly somewhat simplified and over-shaded. Apart from such known limitations, the description, measurement tables, and figures provided by Weber (1905) for the lectotype of *Ch. schlosseri* are more than sufficient to evaluate its key cranial and dental morphological features and identify them straightforwardly with other published and unpublished cranial specimens from Samos.

Some of the salient diagnostic feature of the lectotype cranium include: the frontal bones are markedly flattened and depressed; this notable depression of the dorsal cranial surface extents caudally in-between the parietal crests, which are widely separated; the nuchal crests are wide, and their caudal border is straight; the occipital plane is nearly vertical, and the occipital outline is sub-trapezoid. The depiction of the orbital region appears smoothened and over-shaded; presumably this region may have been slightly damaged and thus artistically somewhat enhanced in the drawing. Nevertheless, the dorsal border of the orbit is evidently placed near the cranial roof, and a shaded hint for the presence of a postorbital process is indicated. According to Weber's description, the rostral margin of the orbit opens above the mesial border of the M3. The zygomatic bone is strong featuring a small postorbital process; the zygomatic process of the temporal bone is thinner and relatively uplifted; the relationship between the postglenoid and posttympanic process cannot be securely evaluated because the auditory region is simplified and shaded. The nasal bones are thin, featuring a deep and unfused internasal suture; the rostral nasal tip is unfortunately broken. According to Weber's description, the nasal notch terminates above the middle of the M1. The mandibular angle of the mandible is rounded and widened, featuring rugocities on the ventral border; the caudal border of the mandibular symphysis is situated below the distal border of p3; the rostral portion of the symphysis is broken. The upper dentition depicted by Weber (1905: pl. 9, fig. 1) is much worn and indicates an old adult individual (Fig. 8a). It is to be noted that the M1 featured within the depicted dentition belongs to another younger individual, because this teeth was much worn in the original specimen; the P4 and M2 of the younger individual are depicted in separate figures (Weber 1905:pl. 9, figs 2, 3). For this reason, the M1 is excluded from our Fig. 8a. Despite the old age, the DP1 is present; the size of the P2 is reduced. Depending on the level of occlusal wear, the

Fig. 8 *Chilotherium schlosseri* (Weber, 1905) from the Late Miocene of Samos, Greece. **a** Upper dentition of lectotype cranium with left P1–P4 and M2–M3 in occlusal view, as figured by Weber (1905: pl. 9, fig. 1; the M1 of a different individual is excluded); formerly at the BSPG collections, destroyed during WWII. **b** Left upper dentition with P1-M3 of the cranium NHMW: 1911/0005/0128 (see also Fig. 9). Scale bar equals 4 cm

inner secondary folds of each tooth (crochet, antecrochet, cristae) and the cinguli are variably expressed, forming multiple small fossettes on the occlusal surface; some of the inner folds even bear a few but notable enamel plications. The molars are characterized by a very strong antecrochet and a marked protocone constriction. The M3 features a roughly subtriangular outline; the ectometaloph is essentially continuous, but it is characterized by a markedly angular disto-buccal corner ("stark pontenzirtes Schmelztalon an der Hinterseite," sensu Weber 1905).

The adult cranium with associated mandible described as *Aceratherium wegneri* and depicted in photographs by Andrée (1921: pl. 1) exhibits exactly the same set of significant cranial features as *Ch. schlosseri*, e.g., the flattened and notably depressed frontal and parietal regions, the widely separated parietal crests, and the high and caudally uplifted zygomatic arc; the nasals are complete and hornless, featuring a marked internasal suture. The morphology of the associated mandible with the markedly widen and ventrally concave mandibular symphysis is typical of *Chilotherium*, while the strong i2 indicate a male individual. This cranium and associated mandible constitute the holotype by monotypy of *Aceratherium wegneri* Andrée,1921 (a juvenile partial mandible doubtfully attributed to the taxon by Andrée 1921: 201-202 is excluded from the type series, according to the provisions of ICZN Art. 72.4.1). Unfortunately, only the adult mandible (Fig. 10b), can be presently traced in the collections of the Geologisch-Paläontologisches Museum der Universität Münster (*vide* Meiburg and Siegfried 1970). It must be noted that all mandibles found associated with a cranium in Weber's material did not preserve the rostral portion of the symphysis intact. An isolated mandible, featuring a complete and characteristically widened mandibular symphysis (Weber 1905: pl. 10, fig. 3), was not assigned to a species and was referred to as *Aceratherium* sp. by Weber (1905, 1907). The main features reported by Andrée 1921 to differentiate *A. wegneri* from *Ch. schlosseri* were basically the slightly less retracted caudal border of the nasal incision (distal part of P4 vs. middle of M1) and the less retracted opening of the orbit (distal part of M2 vs. mesial border of M3), which are clearly well within the limits of the expected intra-specific variability as documented in several rhinocerotid species (e.g., Guérin 1980, for the extant species). The reported difference in the occipital outline (sub-rectangular vs. sub-trapezoid) is obviously exaggerated and represents the result of a mild dorso-ventral compression that characterizes Andrée's specimen (Andrée 1921: text fig. 2) in conjunction with an apparent artistic enhancement in the symmetry of the occipital outline in Weber's depiction (Weber 1905: pl. 8, fig. 3). Therefore, *A. wegneri* is considered a subjective junior synonym of *Ch. schlosseri* as previously suggested (Heissig 1975; Giaourtsakis 2003, 2009; Deng 2006).

Fortunately, a virtually identical cranium of a similar ontogenetical stage as the lectotype of *Ch. schlosseri* is preserved in the collections of the Natural History Museum of Vienna and confirms all major morphological observations discussed above (NHMW 1911/0005/0128; Fig. 9). It belongs to a somewhat older adult individual than the lectotype, as it can be deduced by its more worn dentition (Fig. 8b). Apart from the notable cranial similarity, the dental similarity between the specimens is also evident (Fig. 8), as exemplified by the form and strength of the antecrochet and the shape of the markedly constricted protocone on the less worn P4, M2, and M3; the secondary folds and the cinguli are worn and variably confluent, forming small fossettes on the occlusal surface and occasionally also some small enamel plications. Apparently, some of the traits appear differently expressed, simply because of the different stage of their relative occlusal wear. For example, the absence of the postfossette on the P3 of the Vienna specimen is due to the obliteration of this feature as a result of the more advanced stage of occlusal wear; on the contrary, the formation of a small fossette in-between the protocone and the hypocone on the same tooth was caused by the wear of a small cingular pillar that was present in front of the entrance of the medisinus and which is now confluent with the aforementioned cusps.

This cingular pillar is for instance evident on the less worn P3 of an adult cranium, originally described and figured as *Aceratherium angustifrons* by Andrée (1921: pl. 3, fig 3). This partially preserved cranium was collected in 1909 by Prof. Dr. Wegner in the Late Miocene fluviolacustrine volcanoclastic deposits of the Samos Island, and it is hereby designated as the lectotype of *Aceratherium angustifrons* Andrée, 1921 under the provisions of ICZN Art. 74. The specimen was housed at the collections of the Geologisch-Paläontologisches Museum der Universität Münster in Germany before the Second World War, but it is presently considered to be lost (Meiburg and Siegfried 1970). As previously suggested by Giaourtsakis (2009), *A. angustifrons* is also to be considered as a subjective junior synonym of

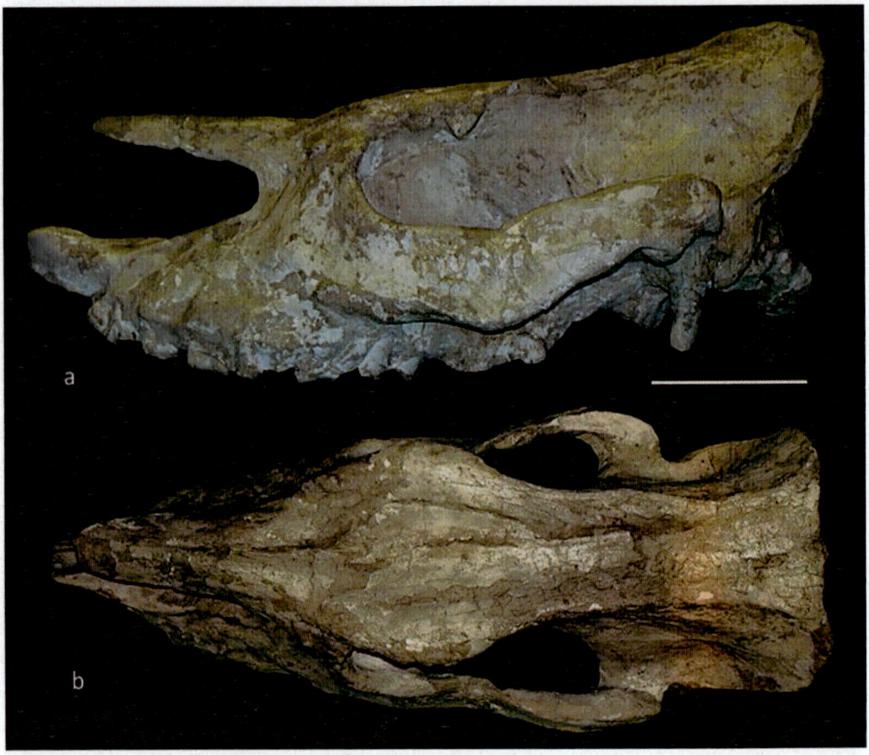

Fig. 9 *Chilotherium schlosseri* (Weber, 1905) from the Late Miocene of Samos, Greece. Adult cranium (NHMW: 1911/0005/0128) in **a** left lateral view and **b** dorsal view. Scale bar equals 10 cm

Ch. schlosseri. The main cranial morphological features of this cranium are very similar to the types of *Ch. schlosseri* and *A. wegneri*, including the markedly flattened and depressed frontal region, the strong and caudally uplifted zygomatic arc, and the hornless nasals with a marked internasal suture. Further, according to Andrée (1921), it was discovered in the same deposits as the cranium of *A. wegneri*. The dentition of this ontogenetically younger specimen (M3 not fully erupted) bears the same morphological features, expressed though at an earlier stage of occlusal wear (Andrée 1921: pl. 3, fig. 3).

A partially preserved and partly restored cranium of *Ch. schlosseri* at the collections of the Hessische Landesmuseum Darmstadt (HLMD: SS-192) features a remarkably well-preserved dentition, which is at the same ontogenetical stage as the last specimen (Fig. 10a). An exceptionally well-preserved cranium in the collections of the Istituto Geologico dell'Università di Padova in Italy features a slightly more worn dentition. This specimen was briefly described and adequately figured as *Ch. wegneri* by Leonardi (1947: pl. 1, figs. 1-3; pl. 2, fig. 12), and features all the typical craniodental characters of *Ch. schlosseri* described above. A well-preserved

Fig. 10 *Chilotherium schlosseri* (Weber, 1905) from the Late Miocene of Samos, Greece. (**a**) Upper dentition with right P1–M3 of a young adult individual in occlusal view (HLMD: SS-192). (**b**) Adult mandible (GPMM: Samos-567) in dorsal view (holotype of *Aceratherium wegneri* Andrée, 1921), image courtesy of M. Bertling (GPMM); note the characteristically widened mandibular symphysis. Scale bar equals 10 cm

cranium of *Ch. schlosseri* (GPIH Nr. 3015) with a seemingly associated mandible (GPIH Nr. 3015a) was briefly documented and figured by Lehman (Lehmann 1984: pl. 3, figs. 2–3); they are stored at the paleontological collections of the Centrum für Naturkunde der Universität Hamburg in Germany, and they also do share the same key features. A moderately well-preserved cranium with associated mandible at the Naturmuseum Senckenberg in Frankfurt (SMF M.3601) was figured as *Ch. schlosseri* by Drevermann (1930: fig. 6); it features the most worn dentition from all specimens discussed, and it is provisionally best referable to as *Chilotherium* sp.

A full assessment of the intra-specific and ontogenetic variability of the Sub-Parathethyan chilotheres is currently under study. Within this context, the spatio-temporal evolutionary patterns between successive or adjacent populations, as well as the potential influence of various environmental factors are essential parameters to be considered. Seasonality, for example, may affect the dental occlusal wear pattern, influencing the development and relative expression of some occlusal morphological features. During extended dry seasons, for instance, the increased consumption of forage with higher fracture toughness may act as a significant abrasive agent resulting to an increase of high-pressure occlusal strokes on the dental surface.

The importance of assessing the dental intra-specific variability has been also stressed by Pavlow (1913), who described as *Aceratherium kowalevskii* a large sample of *Chilotherium* specimens from the Maeotian locality of Grebeniki in Ukraine (correlated with MN10 after Vangengeim and Tesakov 2008a). Pavlow (1913: pl. VI) illustrated an ontogenetic series of seven specimens and discussed the development and variation of several dental and cranial morphological features. Krokos (1917) extended the studies on the variability of the Grebeniki material documenting in detail several additional specimens, and suggested a synonymy between *Ch. kowalevskii* and *Ch. schlosseri*. This synonymy may further include a cranial fragment described as *Teleoceras ponticus* Lubicz-Niezabitowski, 1912 from Odessa, Ukraine, which is very similar with both taxa (Lubicz-Niezabitowski 1913: pls. 24–25; Khomenko 1913; Kiernik 1914: pl. 45). Provisionally though, it may be preferable to maintain the two established species names for the material from Samos and Grebeniki as different, pending for a contemporary comparative study between the two populations.

Among the Late Miocene rhinocerotids of Greece, the genus *Chilotherium* features without doubt the most specialized combination of cranial, dental, and postcranial features. The dentition is relatively high-crowned, featuring a gently concave occlusal surface with mostly rounded paracone and metacone apices on the buccal wall of the ectoloph and a relatively low but not flat intermediate relief. The increased crown height is associated with increased resistance to abrasion, whereas the more flattened occlusal surface, with respect to *Acerorhinus*, enhances the crushing chewing face of the high-pressure occlusal strokes that are required for the comminution of thin and tough fibrous vegetation such as grasses (Fortelius 1982, 1985). In addition, the dentition features a thin to moderate cement coating (frequently though removed during preparation). The remarkably strong antecrochet and the increased complexity of the inner folds and fossettes within the medisinus are associated with improved abrasion efficiency, because they increase the effective length of the enamel ridges perpendicular to the occlusal motion, and they may also support the growing demands for additional cement investment in the crown.

Unlike other rhinocerotids that utilize low-growing vegetation, the cranium of chilotheres did not develop a backwards-inclined occipital plane with caudally extending nuchal crests. The preservation of a relatively subvertical head posture with a correspondingly subvertically oriented occipital plane was necessitated by the efficiency of the principal defensive mechanism of this hornless animal: the strong and sharp tusk-like second lower incisors. On the contrary, incisorless taxa equipped with strong horns, such as *Ceratotherium*, *Coelodonta,* and *Elasmotherium* adapted a sloping head posture with a markedly backwards-inclined occipital plane in order to facilitate both the consumption of lower vegetation and also optimize the effectiveness of their principal defensive mechanism, the horns. Since *Chilotherium* retained a relatively less inclined head posture, its mobilization closer to the ground was facilitated by the significant shortening of the appendicular skeleton in order to exploit lower-growing vegetation. Further, in order to maintain and handle the relative weight of the head and the trunk, the robustness of the postcranial elements increased accordingly. An additional benefit gained by this overall height reduction may further have been the more efficient utilization of available cover against potential predators in relatively open habitats. The markedly widened mandibular symphysis with a ventrally concave surface can be interpreted as an additional assistive adaptation for the feeding activities closer to the ground, which were further supported by the strong lower incisors with strikingly sharp cutting edges.

Based on these craniodental and postcranial adaptations, *Ch. schlosseri* can be interpreted as a flexible mixed-feeder, being capable of utilizing increased quantities of tougher and nutritionally inferior forage, including C3 grasses, and exploit the relatively more open and drier niches within the Pikermian Biome (sensu Solounias et al. 1999). The fact that the occurrence of *Chilotherium* gradually increases toward the eastern localities of the Sub-Paratethyan bioprovince (sensu Bernor 1983) is in accordance with the more rapid expansion of open and seasonally arid habitats in these areas (Giaourtsakis 2009; Koufos et al. 2009a; Eronen et al. 2009).

★*Chilotherium samium* (Weber, 1905)
(= ★*Aceratherium kiliasi* Geraads and Koufos, 1990)

Type Material Lectotype, a partially preserved cranium with associated mandible described and figured by Weber (1905: Taf. IX. fig. 5 *pars*; Taf. X, figs. 1, 2), formerly stored at the collections of the Bayerische Staatssammlung für Paläontologie und Geologie München, in Germany. Subsequently designated and fixed by Geraads and Koufos (1990: 163). The specimen is believed to have been destroyed during the Second World War; the fact that the specimen no longer exists or cannot be traced does not of itself invalidate the designation (ICZN Art. 74.4), and the available descriptions and figures may be used as a reference.

Note The species name *Aceratherium samium* Weber *in* Schlosser, 1904 (as published by Schlosser 1904: 113) is a *nomen nudum*, because it fails to conform to the requirements of ICZN Art. 12. The species name is valid as published by Weber (1905).

Type Locality and Age Samos Island, Greece; Late Miocene fluviolacustrine volcanoclastic deposits of the Mytilinii formation.

Distribution Several Late Miocene localities in Greece, Bulgaria, and Turkey.

Remarks Unlike the lectotype skull of *Ch. schlosseri*, the lectotype skull of *Ch. samium* is less complete; it was apparently more fragmentarily preserved, so that its lithographic depictions may have been somewhat more retouched (Weber 1905: pl. 10, fig. 1, 2). The frontal region is described as slightly depressed, but the shading of this feature in the drawing is not that as evident as in *Ch. schlosseri*. Some salient

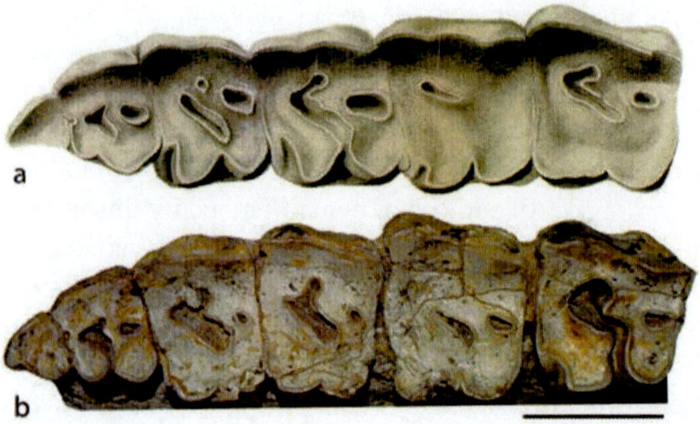

Fig. 11 *Chilotherium samium* (Weber, 1905) from the Late Miocene of Greece: (**a**) Upper dentition of lectotype cranium with left P1–M2 in occlusal view, figured by Weber (1905: pl. 9, fig. 5; the M3 of a different individual has been excluded); formerly at the BSPG collections, destroyed during WWII. (**b**) Upper dentition with left P1–M2 in occlusal view, of the cranium LGPUT: PNT-135 (holotype of *Aceratherium kiliasi* Geraads and Koufos, 1990). Scale bar equals 10 cm

morphological features supported by the description and the illustration alike, include the relatively more slender zygomatic arch, the termination of the nasal incision above the level of P4 (may be though more rostrally depicted), and the rostral border of the orbit placed above the level of the mesial part of the M2. The mandible features a gentle mandibular angle without depicted rugocities, and the incompletely preserved symphysis is reported to have less strong and caudally extending alveoli suggesting a female individual. The most notable differences discussed by Weber were reported in the dentition, despite the fact that the latter was quite worn representing a very old individual. The relative proportions and the outline of the teeth of *Ch. samium* are described and depicted as more square-shaped; the development of the secondary folds inside the medisinus appears to be less complex, the antecrochet is less prominent, and the protocone constriction is less marked (Fig. 11a). According to Schlosser (1904) and Weber (1905), *Ch. schlosseri* and *Ch. samium* may have been recovered from different deposits, based on the sediment matrix adherent to the fossils.

Subsequent reviews and studies postulated the presence of at least two *Chilotherium* species in the Eastern Mediterranean region based primarily on dental differences. Specimens with more pronounced and complex dental features (i.e., very strong antecrochet, marked protocone constriction, more complex development of secondary folds and fossettes within the medisinus) were readily assigned to *Ch. schlosseri* or *Ch. kowalevskii*. Conversely, specimens with less complex and pronounced dental features (i.e., less marked antecrochet and protocone constrictions, more simple structures within medisinus) were deemed as more primitive and were tentatively related with *Ch. samium*, *Ch. kiliasi*, or even the Chinese species

Ch. habereri (Schlosser, 1903) (e.g. Heissig 1975, 1996, 1999; Geraads and Koufos 1990; Fortelius et al. 2003). For instance, Heissig (1996) and Fortelius et al. (2003) referred to the latter group as "mesodont forms" and "plesiomorphic" or "primitive" *Chilotherium*; in contrast, *Ch. schlosseri* from Samos, *Ch. kowalevskii* from Grebeniki, and *Ch. persiae* Pohlig, 1886 from Maragheh were deemed as more "hypsodont" and "derived" *Chilotherium* species. Nonetheless, all authors acknowledge these assessments as preliminary pending further investigation.

The taxon *Aceratherium kiliasi* was initially established by Geraads and Koufos (1990) based on a mixed aceratheriine sample from the Vallesian locality of Pentalophos, which was collectively referred to as a new species. The sample comprises several cranial and mandibular specimens, mostly fragmentarily preserved, housed at the collections of the LGPUT in Thessaloniki. The holotype cranium PNT-135 (Geraads and Koufos 1990: pl. 2, figs. 1–2, pl. 3, fig. 4) is lacking most part of the right side and the occipital region, and it is heavily skewed and compressed relative to the sagittal plane. With respect to *Ch. samium*, PNT-135 features a similarly weak zygomatic arc, a V-shaped nasal incision, and an orbit with a rostral border opening above the middle of the M2 and an upper border that is same level as the frontals. The frontals are flattened, and the nasals are weak and rostrally incompletely preserved. Among the three adult mandibles from Pentalophos, only specimen PNT-142 preserved intact the symphyseal region, featuring a notably narrow symphysis with a rostrally concave ventral surface in-between the i2 alveoli (Geraads and Koufos 1990: pl. 3, fig. 2). The peculiar combination of a chilothere-like craniodental morphology with a narrow mandibular symphysis was regarded as primitive by Geraads and Koufos (1990). Systematic comparisons with the genus *Acerorhinus* were not assessed by the authors, and the Pentalophos material was collectively assigned to a new species of the genus *Aceratherium*. Subsequently though, the holotype cranium PNT-135 was properly recognized as *Chilotherium*, whereas the mandible PNT-142 as belonging to sympatric *Acerorhinus* (Heissig 1999; Fortelius et al. 2003; Giaourtsakis 2003; Geraads and Spassov 2009). The systematic affinities of the Pentalophos aceratheriine material have been further revised and updated by Athanassiou et al. (2014: 42–44; figs. 3–6).

The upper dentition of the chilothere cranium PNT-135 from Pentalophos is well worn, but it seems to represent a somewhat younger adult individual than the lectotype dentition of *Ch. samium*. Both dentitions share a similar dental morphology expressed at different stages of occlusal wear (Fig. 11), demonstrating the traits regarded as characteristic for the "plesiomorphic" chilotheres as discussed above. An incomplete but well-preserved dentition from Locality 49 of the Sinap Formation in Turkey shows essentially the same features expressed at a similar stage of occlusal wear (Fortelius et al. 2003: fig. 12.7). Apart from the presumed sexually dimorphic widening of the mandibular symphysis, due to the inclusion of the *Acerorhinus* mandible in the hypodigm another unusual feature cited in the initial diagnosis of *A. kiliasi* was the presence of "a long persisting dI/2, resulting in a hexaprotodon dentition even in old individuals" (Geraads and Koufos 1990: 166; Geraads and Spassov 2009: 103). The occasional presence of an additional smaller alveolus, which is more dorso-laterally placed with respect to the alveolus of an erupting permanent i2, corresponds to the remnant alveolus of a di2, as correctly discussed by Geraads and Koufos (1990). This is the condition normally documented in rhinocerotids replacing their deciduous incisors with the permanent ones, and these remnant alveoli are commonly integrated within the permanent ones when the di2 are lost and the i2 are fully exposed. The ontogenetic stage of the younger adult mandible PNT-12 may still represent this arrangement. On the other hand, the condition observed in the older adult specimen PNT-31 may represent an unusual pathological case of a persisting di2 or a supernumerary malformed I2 (Geraads and Koufos 1990: pl. 3, fig. 1). Similar pathological oddities of persisting deciduous and misplaced or supernumerary teeth have been occasionally reported for several rhinocerotid taxa in general (e.g., Guérin 1980; von Koenigswald et al. 2007; Giaourtsakis et al. 2009), and for *Chilotherium* in particular (Ringström 1924; Chen et al. 2011). Apparently they do not qualify as valid specific differences, unless they are validated by significantly larger samples, a point also outlined by Geraads and Koufos (1990: 159-160).

The occurrence of two different temporal chilothere populations within the same region has been more convincingly demonstrated in the Northern Peri-Pontic region. Apart from the well-documented *Chilotherium kowalevskii* from the Maeotian locality of Grebeniki (~MN10) (Pavlow 1913; Krokos 1917), another species based on ample material was described as *Chilotherium sarmaticum* Korotkevich, 1958 from the Late Sarmatian locality of Berislav in Ukraine (correlated with MN9 after Vangengeim and Tesakov 2008b). Korotkevich (1970) compared in detail the substantial available material from both localities. The most notable differences were documented in the relative size and some morphological specializations of the appendicular skeleton. *Ch. sarmaticum* was shown to be slightly but distinctively smaller than *Ch. kowalevskii*, with minimally overlapping morphometric values; it also featured a more flexible autopodium. Apart from the biostratigraphical separation, Korotkevich (1970) suggested also a potential niche differentiation, with *Ch. sarmaticum* preferring more closed habitats with softer ground than *Ch. kowalevskii.*

A similar temporal and perhaps spatial and niche differentiation could explain the diversification of two or even more chilothere species in the Eastern Mediterranean region, but the evaluation of additional cranial and postcranial material is certainly required to clarify the systematic affinities and the level of distinctiveness between them. One may even question their value of specific or subspecific importance, considering for instance the notable morphological and metrical variation observed among the extant subspecies of *D. bicornis* (Groves 1967; Rookmaaker and Groves 1978; Giaourtsakis et al. 2009). Therefore, the context of the "plesiomorphic" chilotheres of the Eastern Mediterranean region as discussed herein must be regarded as tentative. In case that a subjective synonymy between *Ch. schlosseri* and *Ch. samium* is ever pursued at specific level, then *Ch. schlosseri* must be regarded as the valid name (synonym) by present selection (precedence fixed under ICZN Art. 24.2), because its type material has been more adequately documented and figured by Weber (1905), and it can be thus undoubtedly identified with existing specimens from Samos. From a morphofunctional and paleoecological perspective, the principal craniodental and postcranial features of *Ch. samium* as currently understood are not significantly different from *Ch. schlosseri.* The presumed more low-crowned dentition with less pronounced secondary folds may only indicate an increase in the browsing component of an otherwise rather similar mix-feeding diet.

Subfamily Rhinocerotinae Gray, 1821

Type Genus *Rhinoceros* Linnaeus, 1758

Tribe Rhinocerotini Gray, 1821

Subtribe Dicerotina Ringström, 1924

Type Genus *Diceros* Gray, 1821.

***Miodiceros* gen. nov.**

LSID urn:lsid:zoobank.org:act:1C0CECD0-D342-4A1A-9BE9-3CB31B4250A5.

Etymology A combination of Mio- (prefix of Miocene) and -diceros, a compound word based on the Greek words "δύο" and "κέρας" (meaning: two horns). In reference to the presence of two distinct horn bosses on the cranium, serving as attachment sites for a frontal and a nasal horn, as well as to the biostratigraphical distribution of the genus during the upper part of the Miocene.

Type Species *Atelodus neumayri* Osborn, 1900 by present designation.

Referred Species Only the type species.

Distribution Numerous localities in the Balkan and Anatolia Peninsulas, the northern part of Iran, and the Southern Caucasus region during the Late Miocene (Giaourtsakis 2009; Geraads 1988; Heissig 1996). Its presence in the northern Peri-Pontic region cannot be excluded, but requires further investigation.

Descriptive Diagnosis A fairly robust dicerotine rhinocerotid of large size. Cranium dolichocephalic (Fig. 12); large nasal and frontal horns present; nasal bones rostrally rounded with abrupt and broad termination; nasal septum not ossified; premaxillary bones reduced and edentulous in adult individuals; remnant rudimentary upper deciduous incisors infrequently present, rarely preserved in fossil specimens. Contact between nasal and lachrymal bones lost or reduced to a point; rostral orbital border mostly placed over M2; dorsal surface of zygomatic bone sloping laterally downwards; supraorbital processes very strong, postorbital processes on frontals virtually absent. Cranial roof markedly convex in-between the supraorbital processes; dorsal cranial profile gently to moderately concave; parietal crests always separated, never forming a sagittal crest. Strong nuchal crests always forming a marked V-shaped occipital notch, and extending caudally over the level of the occipital condyles; occipital plane backward inclined; occipital outline in caudal view nearly square or sub-rectangular; postglenoid and posttympanic processes in close proximity, but without contact. Mandible with moderately shortened

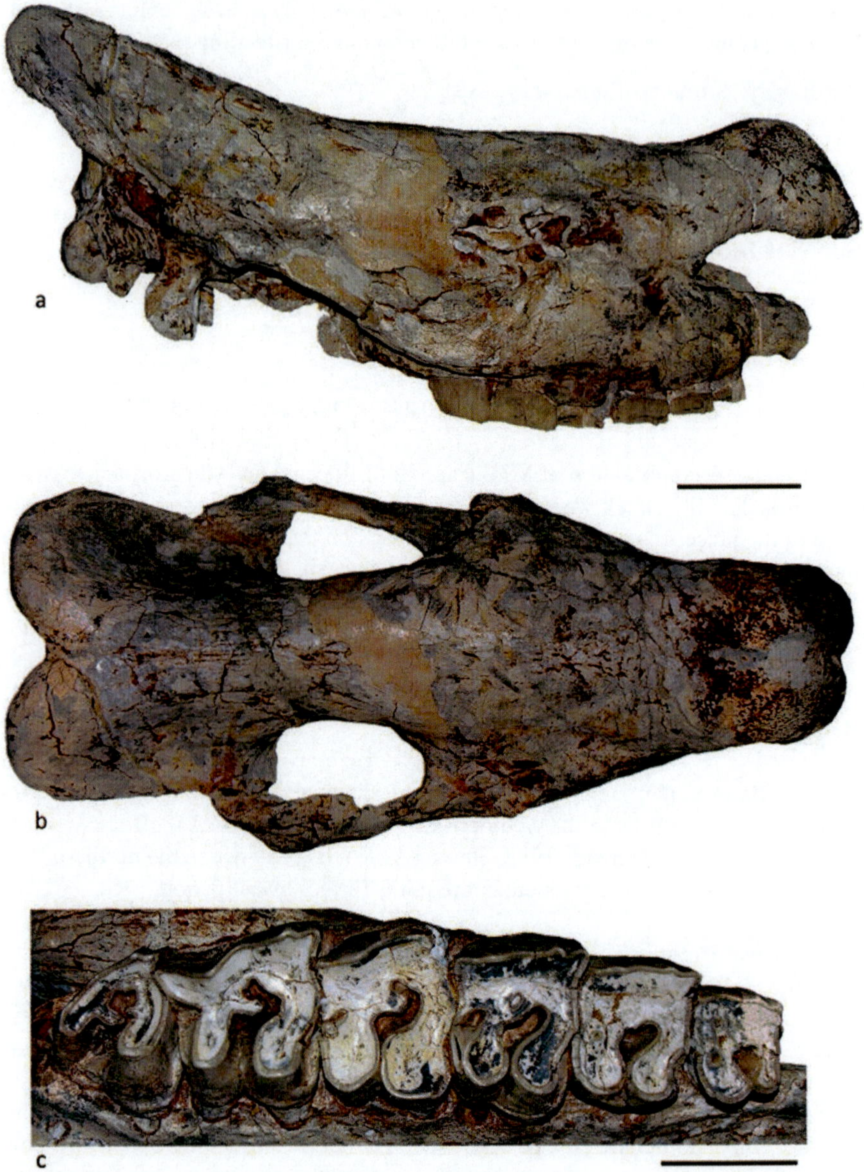

Fig. 12 *Miodiceros neumayri* (Osborn, 1900) from the Late Miocene of Pikermi, Greece. Adult cranium (AMPG: PA 4721/91) in (**a**) right lateral view and (**b**) dorsal view. Scale bar equals 10 cm; (**c**) *idem*, right upper dentition with P2–M3 in occlusal view. Scale bar equals 5 cm

and narrow symphysis, edentulous in adult individuals; rudimentary deciduous lower incisors may have infrequently occurred as indicated by diminutive remnant alveoli, though rarely preserved in fossil specimens.

Upper permanent dentition moderately high-crowned with concave occlusal surface and weak cement coating. Upper permanent premolars molariform, featuring always a continuous lingual cingulum. Upper permanent molars with a marked paracone fold, a weakly expressed mesostyle, and without protocone constriction and antecrochet; crochet mostly weakly developed; crista and gegencrochet occasionally present; a closed medifossette is rarely formed and only at advanced stages of occlusal wear. M3 with continuous ectometaloph and subtriangular outline. D1 with long protoloph and short metaloph; D2 with lingual cingular pillar present in the entrance of the medisinus; paracone fold strong and often double; D3 and D4 with crista always present and medifossette mostly closed; d1 double-rooted; d2 with frequently closed talonid valley forming a fossettid; d3 with double paralophid frequently forming a closed fossetid; d3 without or with only faintly expressed mesial groove on the buccal wall of the trigonid.

Atlas featuring an alar incision and no alar foramen. Appendicular skeleton with relatively heavy mediportal proportions. Humerus with the cranial part of the lesser tubercle stronger and more prominent than the cranial part of the greater tubercle; intermediate tubercle always present and well developed. Radius with a depressed attachment surface for the brachial biceps on the cranial aspect of the proximal epiphysis, and a strong proximal tubercle on the lateral aspect. Ulna featuring an articular facet for the semilunar. Scaphoid with a reduced palmar tuber and three articular surfaces for the semilunar. Semilunar with dorso-proximal and palmar articular facets for the scaphoid always separated. Pyramidal with an asymmetrical distal articular surface for the semilunar. Magnum with a nearly straight palmar process. Unciform with articular facets for the pyramidal and the fifth metacarpal always in contact.

Second metacarpal with an articular facet for the trapezium infrequently present and small; palmar articular facet for the third metacarpal absent or faintly expressed as a minute linear continuance of the large dorsal one. Third metacarpal robust; articular facet for the second metacarpal laterally shifted; single articular surface for the unciform large; shaft with a markedly concave lateral side. Fourth metacarpal with dorsal and ventral articular facets for the third metacarpal mostly in contact but not fused; medial rugosity on the shaft for the attachment of the interosseous ligament strong and pronounced. Fifth metacarpal reduced to a rudiment, resulting to a tridactyl manus.

Fovea of femoral head always present; third trochanter of femur well developed, featuring a cranially curved lateral border with an upturned proximolateral rim. Tibia with a relatively wide and shallow *sulcus* on the tibial tuberosity; *sulcus malleolaris* on the medial malleolus always present and well-marked. Fibula with a robust lateral malleolus, featuring a marked groove for the *m. peronaeus longus*. Astragalus with an ectal calcaneal facet featuring a low distal extension; a sustentacular calcaneal facet featuring a nearly high-oval outline, and mostly without or seldom with a minimal contact to the distal calcaneal facet; a bowed-shaped articular facet for the lateral malleolus of the fibula, gradually narrowing toward its proximal tip, and frequently not contacting the ectal calcaneal facet. Calcaneus with a relatively robust but not massive calcaneal tuber; an articular facet for the tibia always present and large; and three talar articular facets without any contact among them. Entocuneiform with a well-defined and notched distal process (co-ossified remnant of the halux). Mesocuneiforme with a subtriangular outline and without a markedly projecting ventral tubercle. Navicular flattened with confluent proximoventral and distoventral articular facets for the cuboid. Cuboid with a dorsal face always higher than wide and with an asymmetrical proximal border; ventral tuber of the cuboid very large, bearing a pair of shallow ligamentous depressions on the proximoventral aspect.

Second metatarsal with separated dorsal and ventral articular facets for the ectocuneiform. Articular head of the third metatarsal with a reduced and weakly expressed ventral sagittal keel, not surpassing the level of the medial rim. Fourth metatarsal with dorsal and ventral articular facets for the third metatarsal almost always separated, and also featuring a well-developed bulbous protuberance on the ventrolateral corner of the proximal end.

Differential Diagnosis *Miodiceros* differs from the genus *Diceros* Gray, 1821 by its absolutely longer and markedly more dolichocephalic cranium (Fig. S1). In *Diceros*, the anterior border of the orbit is less retracted and mostly placed over the mesial and distal limits of the M1; the dorsal cranial profile is more concave; the nuchal crests are less pronounced and do not markedly extend backwards beyond the occipital condyles; a V-shaped caudal notch on the nuchal crests is always absent, and the occipital plane is subvertical or even slightly rostro-dorsally inclined. The mandibular symphysis of *Diceros* is more abbreviated with respect to *Miodiceros*. The appendicular skeleton of *Diceros* is characterized by significantly more slender proportions with respect to *Miodiceros* (Fig. 13).The femur of *Diceros* is less massive and the fovea of the femoral head is absent or faintly expressed; on the unciform there is no contact between the articular facets for the pyramidal and the fifth metacarpal; on the cuboid of *Diceros*, the proximal and distal articular facets for the navicular are always separated.

Miodiceros differs from the genus *Ceratotherium* Gray, 1868 (including *Serengeticeros* Dietrich, 1942, which is available for *C. efficax* Dietrich, 1942) by its less dolichocephalic cranium, and the significantly less specialized and lower crowned dentition featuring a concave occlusal surface with a saw-toothed ectoloph wear profile. The dorsal cranial profile of *Ceratotherium* is more flattened and the anterior border of the orbit is mostly retracted behind the middle of M2. In addition, the nuchal crest extends further beyond the occipital condyles and the occipital plane inclines more markedly backward (Fig. S1). The dentition of *Ceratotherium* is markedly different (Fig. S2); it is more hypsodont featuring a notably flattened occlusal surface with a nearly straight ectoloph wear profile; the enamel roughness and thickness in *Ceratotherium* is increased and it is evenly distributed in the inner valleys of the teeth; the cement investment is significantly thicker in Plio-Pleistocene

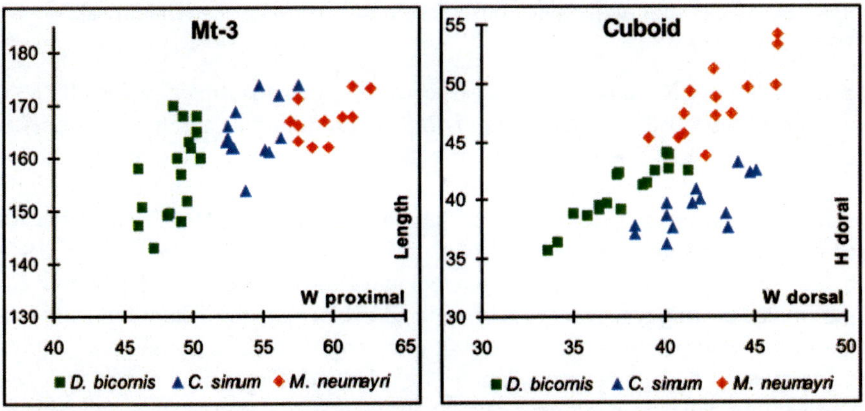

Fig. 13 On the left, bivariate plot of Mt-3 maximum length versus proximal maximum width comparing *Diceros bicornis* (*n* = 17), *Ceratotherium simum* (*n* = 13), and *Miodiceros neumayri* (*n* = 11). On the right, bivariate plot of Cuboid dorsal height versus dorsal width, comparing *Diceros bicornis* (*n* = 19), *Ceratotherium simum* (*n* = 14), and *Miodiceros neumayri* (*n* = 14)

Ceratotherium, and abundant in Pleistocene and extant *C. simum*. In the upper molars of *Ceratotherium*, the protoloph and metaloph are more oblique, bending markedly distolingually; a crista is mostly present forming a closed medifossette with the crochet; the paracone fold on the ectoloph is reduced or faintly expressed. The mandibular symphysis of *Ceratotherium* is absolutely wider and relatively shorter with respect to *Miodiceros*. In *Ceratotherium*, there is no contact between the articular facets for the pyramidal and the fifth metacarpal on the unciform, and the cuboid features a dorsal face with a symmetrical proximal border and subequal sides. Primitive Plio-Pleistocene species of *Ceratotherium*, *C. efficax* (Dietrich, 1942) and *C. mauritanicum* (Pomel, 1895), are characterized by absolutely longer and significantly more slender metapodials than *Miodiceros* (Giaourtsakis et al. 2009: tab. 14.4). In derived Pleistocene and extant *C. simum*, the metapodials are shortened with respect to *C. efficax* and *C. mauritanicum*, but they follow a divergent pattern and degree of relative robustness with respect to the Late Miocene *Miodiceros* (Fig. 13).

The position of the genus *Paradiceros* Hooijer, 1968 within the subtribe Dicerotina has been questioned by Giaourtsakis et al., (2009) and Hernesniemi et al. (2011); a similar reluctance was also expressed by Geraads (2010). The genus comprises only its type species, *Paradiceros mukirii* Hooijer, 1968, which was described from the Fort Ternan in Kenya (ca. 13–14 Ma). In any case, *Miodiceros* differs from the genus *Paradiceros* by the significantly larger and markedly more dolichocephalic cranium, the larger frontal and nasal horns bosses, the more abrupt and broad termination of the nasal bones, the lower border of the orbit that is sloping laterally downwards, and the markedly convex cranial roof in-between the supraorbital processes. The size of the metapodials assigned to *Paradiceros* by Hooijer (1968: tab. 9) is significantly smaller than *Miodiceros*; it is even smaller with respect to the majority of the minimum values observed in *Diceros bicornis* (Giaourtsakis et al. 2009: tab. 14.4).

Nomenclatural Remarks The type species of the genus *Atelodus* Pomel, 1853, originally described as a subgenus of *Rhinoceros* Linnaeus, 1758, is *Rhinoceros elatus*, Croizet and Jobert, 1828 by subsequent designation (Sclater 1900: 297; *vide* Palmer 1904: 127), a species currently assigned to the dicerorhinine rhinocerotid genus *Stephanorhinus* Kretzoi, 1942 (Ballatore and Breda 2016). The type species of the genus *Pliodiceros* Kretzoi, 1942 is *Rhinoceros pachygnathus* Wagner, 1848 by monotypy (Kretzoi 1942a: 314; Kretzoi and Kretzoi 2000), a species whose eligible name bearing type (Fig. 1) is considered to belong to the dicerorhinine rhinocerotid *Dihoplus pikermiensis* (Toula 1906) (Heissig 1975; Giaourtsakis 2003, 2009). Consequently, both genus-group names *Atelodus* and *Pliodiceros* are *ipso facto* not pertinent for a dicerotine rhinocerotid taxon.

Remarks *Diceros gansuensis* Deng and Qiu 2007 from the Late Miocene of Gansu, China differs from *Miodiceros* by exactly the same cranial features that differentiate the new genus from *Diceros*; its postcranial elements are not known. The close resemblance of most craniodental features of the Chinese Late Miocene *Diceros gansuensis* with the Pleistocene and extant African *Diceros* is biogeographically and phylogenetically remarkable, and merits further investigation (Deng and Qiu 2007; Deng 2010: fig. 2).

Dicerorhinus primaevus Arambourg, 1959 is an early Late Miocene species described from the Vallesian site of Bou Hanifia (Oued el Hammam) in Algeria. The Bou Hanifia Tuff, found below the mammal horizon, has provided a radiometric date of 12.18 ± 1.03 Ma (Ameur et al. 1976). The species was subsequently recognized as a Dicerotini and it was allocated to *Diceros* by Geraads (1986) and Giaourtsakis et al. (2009), or to *Ceratotherium* by Geraads (2005, 2010). The holotype specimen (MNHN 1951–9-222) is a partially preserved and somewhat crushed juvenile cranium (Arambourg 1959: tab. 6, figs. 1–3). The preserved portion of the cranium bears the typical Dicerotini features, but the absence of the occipital region prohibits the evaluation of significant characters (Geraads 1986; Giaourtsakis et al. 2009). The unworn M1 and M2 of a juvenile maxilla (MNHN 1951–9-219) are morphologically identical to extant *Diceros* but somewhat less high-crowned (Arambourg 1959: tab. 7, fig. 3; Giaourtsakis et al. 2009: 455–456). This primitive dental morphology definitely excludes the ascription to *Ceratotherium* proposed by Geraads (2005, 2010). An ascription to *Miodiceros* is also excluded due to the smaller size and the significantly more slender proportions of the postcranial elements, which are very similar to *Diceros* (Giaourtsakis et al. 2009: tab. 14.4); thus *Diceros primaevus* is retained within the latter genus.

Diceros australis Guérin, 2000 from the Arrisdrift in the Orange River Valley of Namibia (ca. 17,5–17 Ma) is based on several isolated postcranial and dental specimens. It is excluded from *Miodiceros*, due to its significantly larger postcranial elements with relatively more slender proportions (including the type specimen, a third metacarpal), as well as by the purported presence of a small but functional second lower incisor (an isolated finding, assigned to the hypodigm). The generic affinities of this species and even its allocation within the subtribe Dicerotina are currently considered uncertain until additional material is made available (Giaourtsakis et al. 2009; Geraads 2010).

The species *Diceros douariensis* Guérin, 1966 was described from the Late Miocene locality of Douaria, Tunisia. The holotype comprises a partially preserved adult cranium with an almost complete dentition and the associated mandible (FSL-16749 and FSL-16750). A partially preserved juvenile cranium with incomplete dentition from the same locality (FSL-16752) was originally included in the hypodigm of the species (Guérin 1966: fig. 2, 6), but it belongs to a different rhinocerotid, most likely a teleoceratine (Giaourtsakis et al. 2009; Geraads 2010). The dentition of the type adult cranium (Guérin 1966: figs. 5, 8) is morphologically more similar to *Diceros* and *Miodiceros*, and prohibits an attribution to *Ceratotherium* (Giaourtsakis et al. 2009: 453–455). However, the first upper molar features a notable lingual protocone groove and a quite oblique protoloph, two derived characters that are very frequently expressed in some primitive Pliocene representatives of *Ceratotherium* (Giaourtsakis et al. 2009). Unfortunately, the incomplete preservation of the neurocranium (Guérin 1966: figs. 5, 8) obstructs the evaluation of important features related to the development of the posterior cranial region. A partially preserved cranium (NME: KUS-VP-1/20) from the Ethiopian locality of Kuseralee dated close to the Miocene-Pliocene boundary was assigned to *Diceros douariensis* by Giaourtsakis et al. (2009), but it is better referable as *Diceros* aff. *douariensis*.

A very interesting Dicerotini specimen has been recently reported by Pandolfi et al. c from the late Miocene site of Cava Gentile in Vibo Valentia (Italy) under the name "*Ceratotherium*" *advenientis*, sp. nov. The material consists of a partially preserved cranium retaining primarily the frontal, parietal, occipital, and basioccipital regions. The specimen has been described as having a concave dorsal profile, a concave nuchal crest, and a vertical occipital face that is very wide and laterally expanded. According to the biogeographical interpretation provided by Pandolfi et al. (2019b), the locality of Cava Gentile was part of the Calabrian-Peloritan arc, which is considered to have been a northern extension of the African continental shelf during the late Miocene. As such, the affinities with the North African "*Diceros*" *douariensis* need to be further assessed. However, the preserved portions of the Douaria and the Cava Gentile crania are largely incompatible to allow definite conclusions. Unfortunately, any discussion regarding the generic allocation of the material from Cava Gentile is presently not possible due to a prohibiting technicality. The electronic version of Pandolfi et al. (2019b), as it is currently posted in the Latest Articles section of the journal *Historical Biology*, is to be regarded as unpublished for the purposes of zoological nomenclature, because it does not comply with the requirements for electronic publications of the ICZN Art. 8.5.3. Consequently, the species-group name *advenientis* as published in the binomen "*Ceratotherium*" *advenientis* by Pandolfi et al. (2019b) is considered unavailable according to ICZN Art. 11.1 (*vide* Krell and Pape 2015). The name will be made available, when a printed version of the journal is published with the article, in accordance with the provisions of the code. The authors have been promptly notified in order to take appropriate actions.

***Miodiceros neumayri* (Osborn, 1900)**
(= *Diceros gabuniae* Tsiskarishvili, 1987)

Type Material Holotype by original designation (Osborn 1900: 263), a partially preserved adult cranium with associated mandible housed in the collections of the Naturhistorisches Museum Wien in Austria (NHMW 2014/0424/0001). The second left upper molar of the type cranium has been figured by Osborn (1900: fig. 16); the specimen has been depicted by Thenius (1955: Abb. 1), and by Pandolfi et al. (2015: figs 3b, 4c, 5b); a figure of the right upper dentition is provided in Fig. S2a.

Type Locality and Age Kopran Mescha, near Maragheh, East Azerbaijan province, Iran; Lower horizon of the Maragheh Formation, Late Miocene, Vallesian (MN10).

Distribution Same as for the genus. The reported occurrence of the species in the Vienna Basin (Thenius 1956) was revised as a *Brachypotherium* by Giaourtsakis et al. (2006). Some specimens from Spain reported as *Diceros pachygnathus* by Guérin (1980) were correctly assigned to *Dihoplus schleiermacheri* by Cerdeño (1992). Specimens referred to as *Rhinoceros pachygnathus* from Mont Léberon, France (Gaudry 1873) and Baltavar, Hungary (Pethő 1884), also belong to *Dihoplus schleiermacheri* (Giaourtsakis et al. 2009). A much worn upper second premolar from Sahabi, Libya, reported as *Diceros neumayri* by Heissig (1982) and Bernor et al. (1987: fig. 15), is best referred to as Rhinocerotidae indet., since it is much worn and does not bear any diagnostic features (Giaourtsakis et al. 2009). The scanty material reported as *Diceros* cf. *pachygnathus* from Tugen Hills, Kenya by Guérin (2011) belongs to a different dicerotine rhinocerotid.

Revised Diagnosis As for the genus.

Remarks The most important morphological features of the species are detailed in the descriptive and differential diagnoses of the genus. *Miodiceros neumayri* is the dominant rhinocerotid in the well-sampled Turolian localities on the Island of Samos (Weber 1904; Giaourtsakis 2009), and it is also well represented in the classical Turolian locality of Pikermi (Gaudry 1862–67; Geraads 1988; Giaourtsakis 2009). The species has been documented in several other Late Miocene Greek localities (Appendix 2), with notable occurrences at the sites of Pentalophos-1 in Axios Valley (Geraads and Koufos 1990), Kerassia on Euboea Island (Giaourtsakis et al. 2006), and Platania in Drama (Tsoukala 2018).

Thus far, the Vallesian locality of Pentalophos-1 marks the earliest occurrence of *M. neumayri* in Greece and the Balkans (Geraads and Koufos 1990). In other Balkan countries, the species is known from Turolian localities (Geraads and Spassov 2009; Spassov et al. 2018). In Turkey, the species has been reported from some Vallesian, but primarily from several Turolian localities (Heissig 1975, 1996; Kaya 1994; Fortelius et al. 2003; Antoine and Saraç 2005; Geraads 2017). At the late Valesian and Turolian localities of Maragheh, Iran, *M. neumayri* is the second most frequent rhinocerotid after *Chilotherium persiae*. The occurrence of *M. neumayri* at the locality of Eldari in Azerbaijan, where it has been initially described as *Diceros gabuniae* by Tsiskarishvili 1987, is currently dated at late Vallesian to early Turolian, at ca. MN10–MN11 (Bukhsianidze and Kakhaber 2018).

With respect to the two extant diceotine genera, *Miodiceros* seems to form a separate, monophyletic, extra-African evolutionary lineage (Giaourtsakis et al. 2009; Hooijer and Patterson 1972; Guérin 1980; Heissig 1999; Deng & Qiu 2007), rather than being the common ancestor of both *Diceros* and *Ceratotherium* (Geraads 2005, 2010). The dispersal of this lineage during the course of the Late Miocene in the Eastern Mediterranean and adjacent regions of the Sub-Paratethyan bioprovince was followed by the gradual establishment of a unique combination of primitive and derived craniodental and postcranial features, as detailed in the descriptive and differential diagnoses. Based on the available African fossil record, the split between the two extant Dicerotini genera *Diceros* and *Ceratotherium* seems to have taken place during the Miocene (Giaourtsakis et al. 2009). The highly derived anatomical specializations of *Ceratotherium* toward an exclusive grazing diet were gradually established during the course of the Plio-Pleistocene, following the rapid expansion of open grasslands in Africa (Giaourtsakis et al. 2009; Hernesniemi et al. 2011).

The available record of *M. neumayri* shows a series of several gradual adaptations throughout the Vallesian and Turolian, to cope with the increasingly dry and seasonal environmental conditions that characterized the expansion of the Pikermian chronofauna (Giaourtsakis 2009; Heissig 1975). The dentition of *M. neumayri* is morphologically and functionally very similar to the dentition of the extant browsing *D. bicornis*, and is missing nearly all derived features of the highly specialized dentition of the extant grazing *C. simum* (Fig. S2). In particular, it closely resembles the dentition of the larger and more arid-adapted extant subspecies of *D. bicornis*. However, the derived cranial morphology and the robust postcranial proportions of *M. neumayri* are evidently more specialized with respect to any of the extant subspecies of *D. bicornis* (Giaourtsakis 2009; Giaourtsakis et al. 2009). The markedly dolichocephalic and downwards inclined skull is interpreted as a key adaptation to the consumption of lower growing vegetation. This may be further correlated with the ability to dwell in more open and seasonally arid habitats and utilize more abrasive forage such as lower bushy vegetation or soft C3 herbaceous grasses, when other resources became occasionally or seasonally unavailable (Giaourtsakis 2009). The increased robustness of the appendicular skeleton is also suggestive of a locomotion in more open habitats, including the mosaic open woodland or perhaps open bushland regimes with increased seasonality that have been proposed for the Pikermian Biome (Solounias et al. 1999; Eronen et al. 2009; Koufos et al. 2009a). Based on the combination of its cranial, dental, and postcranial adaptations, *Miodiceros neumayri* may be interpreted as a competent browsing-dominated mixed-feeder that maintained a certain level of dietary flexibility to adjust its nutritional preferences according to the seasonal environmental conditions and the forage availability.

Subtribe Dicerorhinina Ringström, 1924

Type Genus *Dicerorhinus* Gloger, 1841.

***Dihoplus* Brandt, 1878**

Type Species *Rhinoceros schleiermacheri* Kaup, 1832. Dihoplus schleiermacheri Kaup, 1832" reffers to Kaup (1832a). Type species fixed in the original publication (Brandt 1878: 49) by monotypy (ICZN Art. 68.3). Brandt (1878) also included a second species with question marks within the genus, quoted as: "?Spec. 2. *Dihoplus sansaniensis* Lartet Nob.?" (sic). His main concern was, whether it constituted a separate species. Because it was doubtfully included, it is deemed not to have been originally included for the purpose of the type species fixation (ICZN Art 67.2.5).

Remarks *Dihoplus* is a large-sized dicerorhinini, significantly larger than the extant nominotypical representative of the subtribe, *Dicerorhinus sumatrensis*. *Dihoplus* is characterized by the presence of well-developed nasal and frontal horn bosses, the strong and elongated nasals without an ossified nasal septum, and the presence of gradually reduced but still functional and sexually dimorphic second lower incisors in the mandibular symphysis. Upper first incisors may be present, but small or rudimentary. The appendicular skeleton is long with generally mediportal proportions. The dentition is relatively low-crowned with concave occlusal surface; the upper premolars are paramolariform (early fusion between protocone and hypocone) and they are characterized by the presence of a well-developed paracone fold along with a notable metacone fold on the buccal wall of the ectoloph; the molars feature a weakly constricted protocone, a faint antecrochet and a weak to moderate crochet, as well as sharp metacone cusps on the ectoloph.

The biostratigraphical distribution of the genus extends from the Late Miocene to the second half of the Pliocene (MN9–MN16). Late Miocene species include *D. schleiermacheri* (Kaup, 1832) from Central and Western Europe, *D. pikermiensis* (Toula, 1906) from the Balkan and Eastern Mediterranean region, *D. longirostris* (Krokos, 1915) from the Northern Peri-Pontic region, perhaps *D. kurmetiensis* (Bayshashov, 1983) from Eastern Kazakhstan, and *D. ringstroemi* (Arambourg, 1959) from China. In Europe, *Dihoplus* is the only Late Miocene rhinocerotid genus that survived into the Pliocene, with the well-documented large-sized species *Dihoplus megarhinus* (de Christol, 1834). Other Pliocene representatives of the genus may include the newly described species *D. bethlehemsis* Pandolfi et al., 2020 and perhaps *D. vekuai* (Tsiskarishvili, 1987) from Georgia. Morphological and systematic studies suggest a sister taxon relationship of *Dihoplus* with the group of *Stephanorhinus* and *Coelodonta* (Guérin 1980; Groves 1983; Giaourtsakis 2009; Deng et al. 2011).

★*Dihoplus pikermiensis* (Toula, 1906)
(= ★*Rhinoceros pachygnathus* Wagner, 1848 = ★*Rhinoceros schleiermacheri samius* Toula, 1906 = *Ceratorhinus schleiermacheri orientalis* Schlosser, 1921)

Type Material Lectotype by present designation under the provisions of ICZN Art. 74, a moderately well-preserved subadult cranium with left P2-P3, DP4, M1-M2 and right DP4, M1-M2 (MNHN.F.PIK-972), housed in the collections of the Muséum National d'Histoire Naturelle Paris, France. Toula (1906: 34) explicitly states that the name *Rhinoceros schleiermacheri pikermiensis* is established for the crania from Pikermi recognized and discussed as *Rhinoceros schleiermacheri* by Gaudry. Specimen MNHN.F.PIK-972 is the best preserved, and it was the cranium comprehensively discussed and figured as *Rhinoceros schleiermacheri* by Gaudry (1862–67: pl. XXXII, figs. 1, 2). Hence it constitutes without doubt part of the type series (syntype) in accordance with ICZN Art. 72.4.1, and is thus eligible to be herby designated as the lectotype.

Type Locality and Age Pikermi, Attica, Greece. Late Miocene, Turolian (MN12) sediments of reddish silts with clastic channels of conglomerates and sandstones belonging to the lower Red Conglomeratic Member of the Pikermi Formation (Böhme et al. 2017).

Distribution Several Late Miocene localities, of primarily Turolian age (MN11–13), in the Balkan and Anatolian Peninsulas, and perhaps in the Northern Peri-Pontic region (Heissig 1999; Giaourtsakis et al. 2006; Geraads and Spassov 2009; Giaourtsakis 2009).

Restricted Synonymy The species *Rhinoceros pachygnathus*, and the subspecies *Rhinoceros schleiermacheri samius* and *Ceratorhinus schleiermacheri orientalis*, are considered subjective synonyms of *Dihoplus pikermiensis*. Their nomenclatural status has been thoroughly discussed and prevailing usage is maintained (Kretzoi 1942b; Heissig 1975, 1999; Geraads 1988; Giaourtsakis et al. 2006; Giaourtsakis 2009).

Remarks *D. pikermiensis* is the dominant rhinocerotid at the classical Turolian locality of Pikermi (Geraads 1988; Giaourtsakis et al. 2006), whereas it is rather scarcely represented in the well-sampled Turolian localities of Samos (Weber 1904; Giaourtsakis 2009). Other notable occurrences in Greece include the localities of Halmyropotamos and Kerassia on Euboea Island (Giaourtsakis et al. 2006), Nikiti-1 in Chalkidiki (Koufos et al. 2016) and Thermopigi in Serres (Tsoukala 2018). In neighboring Balkan countries, the species emerges as the dominant horned rhinocerotid (Geraads and Spassov 2009; Spassov et al. 2018). In Anatolian localities the species is infrequent (Heissig 1996; Antoine and Saraç 2005), whereas in the Iranian locality of Maragheh it is completely absent (Giaourtsakis 2009).

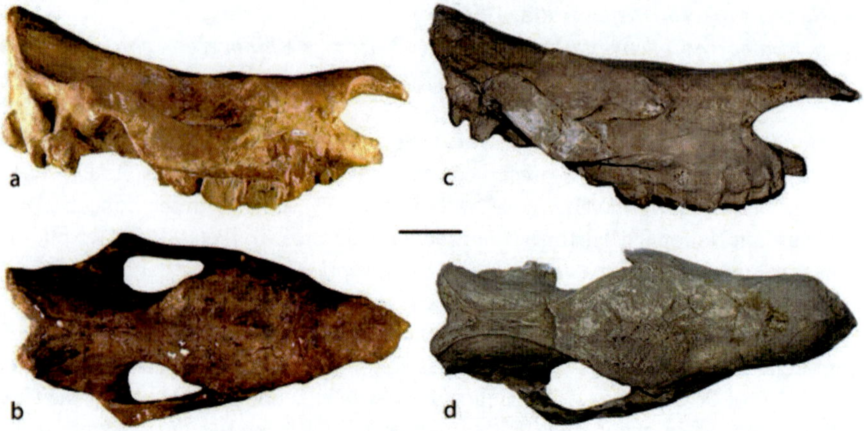

Fig. 14 *Dihoplus pikermiensis* (Toula, 1906) from the Late Miocene of Greece. (**a**) Lectotype, subadult cranium from Pikermi (MNHN: F.PIK972) in right lateral view and (**b**) in dorsal view (credit: MNHN, Elodie Lerat 2018, used under CC-BY 4.0, background color modified, http://coldb.mnhn.fr/catalognumber/mnhn/f/pik972). (**c**) Adult cranium from Samos (MAFI: Ok-546) in right lateral view and (**d**) in dorsal view. Scale bar equals 10 cm

The most significant craniodental and postcranial differences between *Dihoplus pikermiensis* and the frequently sympatric *Miodiceros neumayri* have been discussed in detail (Geraads 1988; Giaourtsakis et al. 2006; Giaourtsakis 2009). *D. pikermiensis* is a large-sized dicerorhinine featuring a dolichocephalic cranium with a relatively vertically oriented occipital plane (Fig. 14), and a comparatively more slender appendicular skeleton (Fig. 7). The dentition is relatively lower crowned with a concave occlusal surface, featuring unspecialized paramolariform premolars with a notably strong pair of paracone and metacone folds on the ectoloph; the molars have a weak protocone constriction, a well-marked paracone fold, and a sharp metacone apex. These dental features are in accordance with an attrition-dominated occlusal wear profile. As a dedicated browser, *D. pikermiensis* must have preferred the denser niches offered by the woodland habitats of the Pikermian Biome, especially on the western parts of the Sub-Paratethyan region (Giaourtsakis 2009; Solounias et al. 2010; Koufos et al. 2009a).

Biogeographical and Paleoecological Remarks The Vallesian locality of Pentalophos-1 is the only Greek site that has yielded both *Acerorhinus* and *Chilotherium*, along with the horned rhinocerotid *M. neumayri*. Thus far, Greek Turolian localities have yielded only one of the two hornless genera. In well-sampled Turolian localities with ample material, where both horned rhinocerotids *D. pikermiensis* and *M. neumayri* appear sympatric, one of them is more abundantly represented, signifying a clear interspecific dominance. For instance, at the locality Pikermi, *D. pikermiensis* is the dominant horned species, *M. neumayri* is less frequent, and hornless *Acerorhinus* is present. In contrast at Samos, *M. neumayri* emerges as the dominant horned species, *Dihoplus* is infrequent, and *Chilotherium* is the sole hornless taxon present. In the cases of sympatry, a resource partitioning with limited dietary competition among the species has been suggested due to variable dietary preferences (Giaourtsakis et al. 2006; Giaourtsakis 2009).

Attrition dominated	Taxa	PIK	KER	SAM	ANA	MAR	Closed habitats
	Acerorhinus	x	x	-	x	x	
	Dihoplus	xxx	x	x	x	-	
	Miodiceros	xx	xxx	xxx	xxx	xx	
	Chilotherium	-	-	xx	xx	xxx	
	Iranotherium	-	-	-	-	x	
Abrasion dominated		WEST →				EAST	Open habitats

Fig. 15 Distribution and relative abundance of rhinocerotid genera during the Late Miocene in the Sub-Paratethyan bioprovince. Pikermi (**PIK**), Kerassia (**KER**), Samos (**SAM**), Anatolia (**ANA**), Maragheh (**MAR**). Relative Frequency: Dominant (**xxx**), Frequent (**xx**), Infrequent (**x**), Absent (**–**) Abrasion and Attrition dental wear gradient based on occlusal functional properties (Fortelius 1982, 1985) and relative mesowear profiles (Fortelius and Solounias 2000)

The marked differences observed in the relative distribution and abundance of rhinocerotid taxa among the Turolian localities of the Sub-Paratethyan bioprovince (Fig. 15) seem to have been primarily influenced by environmentally controlled provincial differences (Giaourtsakis 2009). The lower crowned and relatively slenderer *Dihoplus* and *Acerorhinus* seem to have favored more closed and temperate niches, whereas the craniodentally more specialized and postcranially more robust *Miodiceros* and *Chilotherium* appear to have preferred more open and dry habitats. At the locality of Maragheh (Iran), the occurrence of the massive elasmotheriine *Iranotherium morgani* (de Mecquenem, 1908) must be included, as the only true grazing rhinocerotid of the region with a highly specialized and flattened hypsodont dentition. The gradual changes observed are markedly associated with the more rapid expansion of open and arid habitats in the central and eastern parts of the Sub-Paratethyan bioprovince (Eronen et al. 2009; Koufos et al. 2009a).

The end of the Miocene was marked by a significant faunal turnover caused by major climatic changes, also known as the Messinian Salinity Crisis. Throughout the terminal part of the Messinian, the Mediterranean Sea went into a cycle of partly or nearly complete desiccation caused by the closure of the precursor of the Strait of Gibraltar (Gautier et al. 1994). The environmental changes were detrimental to the large land mammals. According to Eronen et al. (2009), a possible explanation for the ultimate extinction of the Pikermian chronofauna in its core area could have been the double impact of increased seasonality and regional differentiation, followed by significantly increased humidity and forestation during the Pliocene. At the end of the Miocene, most Turolian rhinocerotid species became extinct in Greece, unable to survive the paleoenvironmental changes at the Miocene-Pliocene boundary. The aceratheriine rhinocerotids disappeared completely from Europe and Western Asia, whereas the dicerotine rhinocerotids were restricted to Africa, where they survive until today. In Europe, only the dicerorhinine genus *Dihoplus* managed to survive into the Pliocene. During the second half of the Pliocene, it was replaced by its successor *Stephanorhinus*.

***Stephanorhinus* Kretzoi, 1942**
(= *Brandtorhinus* Guérin 1980 = *Procerorhinus* Kretzoi, 1942)

Type Species *Rhinoceros etruscus* Falconer *in* Ansted, 1859 by original designation (Kretzoi 1942b: 348).

Remarks The fossil genus *Stephanorhinus* comprises large-sized dolichocephalic two-horned rhinoceroses, which are characterized by the unique combination of a nasal and frontal horn, a partially ossified nasal septum in adults connecting rostrally the nasals with the elongated premaxillary bones, and the significant size reduction or complete loss of functional incisors. *Stephanorhinus* succeeded *Dihoplus* during the course of the Pliocene, and quickly expanded across the Palearctic until its diminishment during the Late Pleistocene.

After almost two centuries of nomenclatural controversies over dozens of specific names and subspecific variants, a general consensus for the discrimination of five species has been reached in Europe, whose biogeographical range may further extend in Western and North-Central Asia: *S. jeanvireti* (Guérin, 1972), *S. etruscus* (Falconer *in* Ansted, 1859), *S. hundsheimensis* (Toula, 1901), *S. hemitoechus* (Falconer *in* Gaudin, 1859), and *S. kirchbergensis* (Jäger, 1839) (Guérin 1980; Fortelius et al. 1993; Lacombat 2005, 2007). In Asia, craniodental material with evident *Stephanorhinus* affinities has been locally attributed to the species *S. binagadensis* (Dzhafarov, 1958) in Azerbaijan, *S. yunchuchenensis* (Chow, 1963), *S. lantianensis* (Hu and Qi, 1978), and *S. choukoutiensis* (Wang, 1931) in China, as well as *S. niponicus* (Shikama et al. 1967) in Japan. The close systematic affinities between the latter two species and *S. kirchbergensis* have been often discussed (Tong 2012; Handa and Pandolfi 2016). *Stephanorhinus* has also migrated into North Africa, whenever a Gibraltar land bridge permitted it; it has been locally described as *Dicerorhinus africanus* Arambourg, 1970 from Pliocene deposits, and *Rhinoceros subinermis* Pomel, 1895 from Pleistocene ones (Guérin 1980; Geraads 2010).

Morphological studies suggest an evident evolutionary lineage between the three large-sized dicerorhinine genera *Dihoplus, Stephanorhinus,* and *Coelodonta* as currently understood (e.g., Guérin 1980; Groves 1983; Deng et al. 2011). Recent molecular studies unambiguously place *Stephanorhinus* as a sister taxon of *Coelodonta*, and also confirm the strong phylogenetic affinities of this group with the extant *Dicerorhinus sumatrensis* among the five surviving rhinocerotid species (Kirillova et al. 2017; Welker et al. 2017; Cappellini et al. 2019).

***Stephanorhinus jeanvireti* (Guérin, 1972)**

Type Material Holotype by original designation, (Guérin 1972: 135), a moderately well-preserved and partially restored adult cranium with associated mandible (NHMB: Vt-627), originally described and figured as *Dicerorhinus jeanvireti* by

Guérin (1972: Pl. 1); housed in the collections of the Naturhistorisches Museum of Basel in Switzerland.

Type Locality and Age Vialette (Haute-Loire, France), Late Pliocene (MN16).

Distribution Several European Late Pliocene localities, primarily during the MN16 (Guérin 1972, 1980; Lacombat and Mörs 2008; Guérin and Tsoukala 2013); last occurrence reported from the early Pleistocene (MN17/18) in Romania (Pandolfi et al. 2019a).

Nomenclatural Note Recently, Ballatore and Breda (2016) designated in a dedicated study a lectotype for the rarely used species *Rhinoceros elatus* Croizet and Jobert, 1828 and proposed a subjective synonymy with *Dicerorhinus jeanvireti* Guérin 1972. Prior to this designation, the former syntypes of *R. elatus* were considered to represent a mixture of two different species, and the taxon was effectively treated as *nomen dubium* and purportedly *oblitum* (*vide* Guérin 1972: 58, 1980: 444; Guérin and Tsoukala 2013: 454). Since this is a case of subjective and not objective synonymy, and the recently designated lectotype of *R. elatus* is restricted to postcranial elements, we shall maintain the prevailing usage of *Stephanorhinus jeanvireti* (see also Pandolfi et al. 2017, 2019a).

Remarks The Pliocene rhinocerotid record of Greece is very scanty (Appendix 1). A notable exception are the early Villafranchian (MN16) sites of Milia at the municipality of Grevena, which have yielded almost 200 rhinocerotid specimens. This significant material has been recently documented in detail by Guérin and Tsoukala (2013), and was attributed to *Stephanorhinus jeanvireti*. The well-preserved cranium MIL-162 from Milia (Fig. 16) is in fact the only complete rhinocerotid cranium recovered thus far from the Pliocene and Pleistocene deposits of Greece.

S. jeanvireti is a large tandem-horned rhinoceros. The facial area is massive, featuring very long and relatively broad nasal bones. In lateral view, the nasal bones form a pronounced, domelike, nasal horn boss, which is covered by extensive and rugose vascular impressions. The partially ossified nasal septum occupies the cranial half of the nasal cavity and it is ventrally fused with the tips of the long premaxillary bones, providing additional support for the nasal horn during lifetime. The Milia cranium features a minute alveolus on the ventral tip of the premaxillary bone that accommodates a vestige peg-like first upper incisor (Guérin and Tsoukala 2013: fig. 3e). On the dorsal surface of the frontal bones, the insertion of the second horn boss is also broad and well marked. The upper premolars are paramolariform, showing a fusion between the protocone and hypocone after early to middle stages of occlusal wear that closes the lingual side of the medisinus valley (Fig. 16c). The internal folds of the premolars always present a crochet and very frequently a crista, which may occasionally fuse to form a closed medifossette during the middle-late

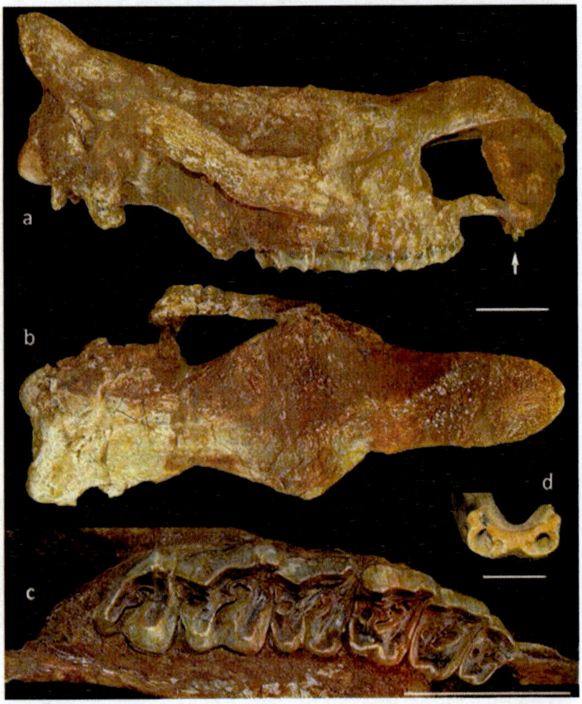

Fig. 16 *Stephanorhinus jeanvireti* (Guérin, 1972) from the Late Pliocene of Milia, Greece (Guérin and Tsoukala 2013). (**a**) Adult cranium (NHMM: MIL-162) in right lateral view, arrow indicates vestige I1; (**b**) *idem*, in dorsal view; (**c**) *idem*, left upper dentition with P2–M3 in occlusal view. Scale bars **a-c** equal 10 cm. (**d**) Adult mandible (NHMM: MIL-244) in rostral view; Scale bar equals 5 cm. Images courtesy of E. Tsoukala

stage of occlusal wear; the protocone is not constricted and a weakly express lingual cingulum is mostly present. The upper molars always feature a crochet, but a crista is rarely observed; the protocone is slightly constricted, and there is no lingual cingulum developed. A notable feature of the adult, presumably male, mandible MIL-244 is the presence of 2 diminutive alveoli for minuscle i1, along with two well-formed alveoli for reduced second lower incisors (Fig. 16d; also Guérin and Tsoukala 2013: figs. 5c, d). The size of these alveoli seems smaller with respect to preceding *Dihoplus* species, but their presence suggests that the acquisition of the partially ossified nasal septum in early *Stephanorhinus* species was achieved before the complete functional loss of the permanent second lower incisors. The appendicular skeleton of *S. jeanvireti* is somewhat smaller, but distinctly more slender than the Ruscinian *Dihoplus megarhinus*. On the contrary, it is clearly larger and somewhat more robust than the typical Villafranchian species *S. etruscus* (Guérin 1980; Guérin and Tsoukala 2013; Ballatore and Breda 2016, 2019).

Apart from the Milia sites and the nearby locality of Saint George Priporos at Grevena, some postcranial specimens referable to *S. jeanvireti* have been also documented from the early Villafranchian (MN16) site of Angelochori at the municipality of Thermaikos (Tsoukala 2018). According to Guérin (1980), *S. jeanvireti* probably favored a browsing-dominated diet in a woodland habitat with grassy areas that consist of graminae and ferns in a humid environment.

Stephanorhinus etruscus **(Falconer *in* Ansted, 1859)**

Type Material Lectotype, a nearly complete cranium (IGF 756) from Val'd'Arno (Tuscany, Italy) housed in the collections of the Museo di Storia Naturale, sezione di Geologia e Paleontologia, Florence, Italy. Subsequently designated and fixed by Toula (1906: 36) in accordance with the provisions of ICZN Art. 74. According to ICZN Art. 50.1.1, Falconer is considered as the author of the name in Ansted (1859: 602). Sufficient evidence is provided in Ansted (1859) that the Val'd'Arno cranium was explicitly recognized by Falconer as *Rhinoceros etruscus* when the nominal species was established, and thus it constituted part of the type series (syntype) in accordance with ICZN Art. 72.4.1.1. The specimen has been figured by Falconer (1868: pl. 27, figs. 1–3), and by Mazza (1988: pl. 1, fig. 1a–1d).

Type Locality and Age Upper Valdarno Basin (Val'd'Arno) in Tuscany, Italy; Early Pleistocene fluviolacustrine deposits of the Upper Valdarno Basin.

Distribution Numerous European localities ranging from Late Pliocene until the end of the Early Pleistocene (ca. MN16a – MNQ20), estimated at ca. 3.5–0.9 Ma (Guérin 1980; Fortelius et al. 1993; Lacombat 2005; Pandolfi et al. 2017). During this period of time, the species is also present in adjacent regions of Western Asia. A comprehensive overview is provided by Pandolfi et al. (2017).

Remarks *Stephanorhinus etruscus* is the typical Villafranchian rhinocerotid of Europe. In Greece, the species has been reported with more or less certainty from several Villafranchian sites (Appendix 1), but the available material is scarce and restricted to a few dental, mandibular, or postcranial remains. The most adequately documented occurrence comes from the locality of Aivaliki (Symeonidis et al. 2006), which has yielded a partially preserved adult mandible (Fig. 17), as well as a juvenile one. The most striking feature of the Aivaliki mandible is the small size of

Fig. 17 *Stephanorhinus etruscus* (Falconer, 1859) from the Early Pleistocene locality of Aivaliki near Alistrati, Greece. Adult mandible in dorsal view, with p2–m3 sin (AMPG: AVL-1) and m1–m3 dex (AMPG: AVL-2). Scale bar equals 5 cm

the dentition with relative low-crowned teeth, especially the molars. The size of the lower molars from Aivaliki is close to the minimum values observed for *S. etruscus* (Symeonidis et al. 2006: Tab. 1). The second lower premolar is well developed, with a deep ectoflexid and a constricted prominent paralophid, suggesting a conservative morphology. The presence of a labial cingulum on the lower premolars has been frequently documented in *S. etruscus* (Fortelius et al. 1993; Lacombat 2005). According to Mazza et al. (1993) and Fortelius et al. (1993), the small-sized populations of the etruscus-group may be regarded as transitional forms showing various intermediate stages between *S. etruscus* and *S. hundsheimensis*, or they may be considered as locally adapted populations associated with the environmental stress caused by the altering climatic conditions during the Early Pleistocene.

S. etruscus was a long-limbed animal with a rather low-crowned dentition and a head posture suggestive of a primarily browsing diet on vegetation of intermediate height. The limb proportions are more cursorial with respect to other *Stephanorhinus* species, which together with the relatively shallow joints suggests a locomotion pattern in open woodland habitats (Fortelius et al. 1993; Loose 1975).

***Stephanorhinus hemitoechus* (Falconer *in* Gaudin, 1859)**

Type Material Lectotype, a partially preserved cranium (BMNH: M.27836) from Clacton on Sea (Essex, Great Britain) housed in the collections of the Natural History Museum of London in the United Kingdom; subsequently designated and fixed by Azzaroli (1962: 22) in accordance with the provisions of ICZN Art. 74. According to ICZN Art. 50.1.1, Falconer is considered as the author of the name in Gaudin (1859: 131). Sufficient evidence is provided in the posthumous publication of Falconer (1868), presented by Murchison (1868), that the Clacton skull constituted part of the type series (syntype) and was explicitly recognized as *Rhinoceros hemitoechus* by Falconer when the nominal species was established in 1859 (in accordance with ICZN Art. 72.4.1.1; *contra* van der Made 2010: 478). The lectotype specimen has been figured by Owen (1846: figs. 131 and 138–140), Falconer (1868: pl. 15, figs. 1–3), and Azzaroli (1962: pl. 16, fig. 4).

Type Locality and Age Clacton on Sea, Essex, Great Britain; Clacton Channel deposits, Middle Pleistocene (MNQ23).

Distribution Numerous European localities ranging from Middle to Late Pleistocene (ca. MNQ23 – MNQ26), estimated at ca. 0.500–0.040 Ma (Guérin 1980, 1982; Fortelius et al. 1993; Lacombat 2005; van der Made 2010; Pandolfi and Tagliacozzo 2015). During this period of time, the species appears to have also migrated in Western Asia, where it has been reported as *Rhinoceros binagadensis* Dzhafarov, 1955, and in North-western Africa, where it has been referred to as *Rhinoceros subinermis* Pomel, 1895 (Guérin 1980; Geraads 2010).

Remarks In Greece, the species is best known from the Middle Pleistocene Petralona Cave in the peninsula of Chalkidiki, a celebrated locality for its renowned

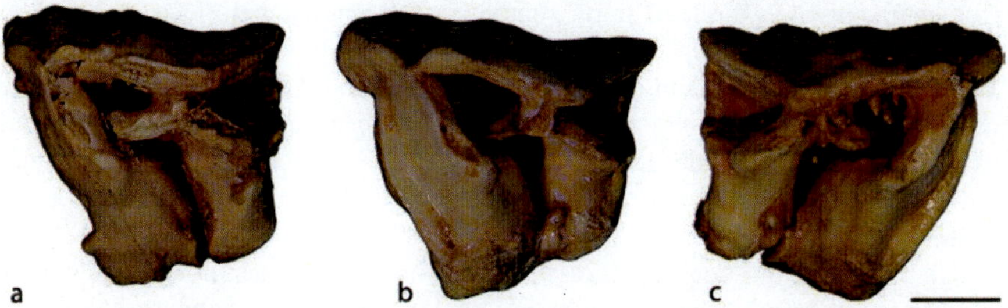

Fig. 18 *Stephanorhinus hemitoechus* (Falconer *in* Gaudin, 1859) from the Middle Pleistocene of Petralona Cave, Greece. (**a**) Left M1 (LGPUT: PEC 963), (**b**) left M2 (LGPUT: PEC 964), (**c**) right M2 (LGPUT: PEC 965); all in occlusal views. Images courtesy of E. Tsoukala. Scale bar equals 2 cm

archaic hominid skull. The fossil material from Petralona Cave is divided between the old collection, which is housed in the Paleontological Museum of the Geology School at the Aristotle University of Thessaloniki (LGPUT), and the new collection, which is stored in an exhibitional building next to the cave (Tsoukala and Guérin 2016). The rhinocerotid material of the old collection has been documented by Tsoukala (1989, 1991a) and Tsoukala & Guérin (2016) whereas the material from the new collection has been previously studied by Fortelius and Poulianos (1979, 1980) and Poulianos (1981). A partially and fragmentarily preserved cranium maintains only the skull roof from the posterior part of the second horn base to the crest, and the posterior parts of the maxillae with fragments of the zygomatic arches, including the left P4–M3 and the right M1–M3 (Fortelius and Poulianos 1979: pl. 1–4). The remaining material from both the old and the new collection comprises numerous dental and postcranial specimens, which have been described and figured by Tsoukala and Guérin (2016) and Fortelius and Poulianos (1979), respectively (e.g. Fig. 18). Their morphological and morphometrical features are in accordance with the known range of variation established for the species (Guérin 1980; Fortelius et al. 1993; Lacombat 2005).

Apart from the Petralona Cave, a single tooth and a partially preserved mandible from the Late Pleistocene deposits at the banks of the Penios River, near the city of Larissa (Thessaly), have been potentially attributed to *S. hemitoechus*. (Boessneck 1965; Athanassiou 2001, 2011). Melentis (1965) described a few dental and postcranial rhinocerotid specimens from presumably Middle Pleistocene deposits in the Megalopolis basin (Peloponnese) as *Dicerorhinus kirchbergensis* and *Dicerorhinus hemitoechus*. After a brief re-examination in the collections of the Athens Museum of Palaeontology and Geology (AMPG), they are preliminary referred to as *Stephanorhinus* sp.

The species *S. hemitoechus* is characterized by a low-slung, downward directed cranium, with an occipital plane that inclines notably backward (compare Loose 1975: tab. 7, pls. 5, 6; van der Made and Grube 2010: fig. 5). This feature has been interpreted as a key adaptation to the consumption of lower growing vegetation in rather open habitats. The face is long and massive, with relatively narrow nasals featuring a partially ossified nasal septum in the anterior half. The dentition of *S. hemitoechus* is comparatively more high-crowned with respect to other *Stephanorhinus* species, and it is further characterized by a relatively reduced premolar segment and an enlarged M3. Postcranial elements are characterized by relatively shallow limb joints and short and broad proportions, suggesting a more graviportal locomotion adapted to more open habitats than the preceding species *S. jeanvireti*, *S. etruscus*, and *S. hundsheimensis* (Guérin 1980; Fortelius et al. 1993; Lacombat 2005; van der Made 2010). Based on its craniodental and postcranial adaptations, *S. hemitoechus* probably preferred temperate open habitats rich in low-growing vegetation (Fortelius et al. 1993). A recent mesowear analysis of dental wear patterns has interpreted *S. hemitoechus* as a flexible mixed-feeder that was able to adjust its dietary composition within a range of suitable habitats (van Asperen and Kahlke 2015).

***Coelodonta* Bronn, 1831**

Type Species *Coelodonta boiei* Bronn, 1831 by monotypy (Bronn 1831: 61), now regarded as a subjective junior synonym of *Rhinoceros antiquitatis* Blumenbach, 1799. The species name "*bojei,*" as cited in some subsequent publications (e.g., Brandt 1877; Roger 1896), is an incorrect subsequent spelling of "*boiei*" (ICZN Art. 33.3).

Remarks Over the past two centuries, numerous specific and subspecific names have been proposed and debated for Eurasian material related to the woolly rhinoceros. Currently, only four of them are generally recognized at species level (Deng et al., 2011). The earliest known member of the woolly rhinoceros lineage, *Coelodonta thibetana* Deng et al. 2011 dates to the end of the Early Pliocene (ca. 3.7 Ma) of the Tibetan plateau in China. This suggests that the genus may have gained its adaptations to cold and open habitats as a result of living at high altitudes in lower latitudes, before migrating to lower altitudes in higher latitudes (Deng et al. 2011). During the Late Pliocene, at ca. 2.5 Ma, a more derived species, *Coelodonta nihowanensis* Kahlke, 1969 migrated north of the Himalayan–Tibetan uplift at Longdan in Linxia Basin (Gansu, China) and expanded in several adjacent northern Chinese localities dating to 2.4–1.0 Ma (Deng 2002a; Kahlke and Lacombat 2008).

During the Early-Middle Pleistocene, a more evolved form, *C. tologoijensis* Belyaeva in Vangengeym et al., 1966, spread further north- and westward in Asia and presumably may have also reached Europe (Kahlke and Lacombat 2008; but for a different opinion compare Guérin 2010). All through the subsequent Middle to Late Pleistocene cold stages, the typical species *Coelodonta antiquitatis* (Blumenbach, 1799) dominated Northern Asia and expanded rapidly across Europe (Guérin 1980, 2010; Kahlke and Lacombat 2008; Stuart and Lister 2012). Unlike mammoth, the woolly rhinoceros did not cross the Bering land bridge to North America.

Several molecular studies place *Coelodonta* as a sister taxon of *Stephanorhinus*; further, all molecular studies unambiguously concur that, among the extant rhinocerotids, these two taxa are closely related to the Sumatran rhino *Dicerorhinus sumatrensis* (Orlando et al. 2003; Kirillova et al. 2017; Willerslev et al. 2009; Yuan et al. 2013; Welker et al. 2017; Cappellini et al. 2019).

***Coelodonta antiquitatis* (Blumenbach, 1799)**

Type Material Lectotype hereby designated under the provisions of ICZN Art. 74, the anterior half of an adult cranium preserving the frontals, nasals, premaxillary bones, and fragments of the maxilla and the right zycomatic bone, housed in the historical collections of the Museum of the Geowissenschaftliches Zentrum of the Georg-August-Universität Göttingen, Germany. This material was donated to the University by Baron Georg Thomas von Asch before the end of eighteenth century (Cuvier 1806; Gehler et al. 2007, 2008). The acquisition of rhinocerotid material from Siberia was briefly mentioned for the first time by Blumenbach (1791: 698). The species name was established as *Rhinoceros antiquitatis* by Blumenbach (1799: 697) in accordance with the provisions of the ICZN Arts. 11 and 12, and in particular Art. 12.2.1 for the definition of the species by an indication to a bibliographic reference of a previously published description (Kalandadze et al. 2009: 101). The specimen was figured for the first time by Cuvier (1806: pl. 3, figs. 4, 5), as well as in the first and subsequent editions of his grand oeuvre *Recherches sur les ossemens fossiles de quadrupeds* (Cuvier 1812: pl. 3, figs. 4, 5; Cuvier 1822: pl. VII, figs. 4, 5, etc.). Among the existing syntypes of the species at the University of Göttingen (Gehler et al. 2007, 2008), the designated lectotype is the only cranial specimen that preserves the undisputable key morphological features of the species: the presence of well-defined frontal and nasal horn bosses combined with a completely ossified nasal septum; in addition, it has been adequately figured and documented by Cuvier (1806, 1812, 1822).

Type Locality and Age In the vicinity of the River Kurtamysh (Река Куртамыш), located within the current Kurtamyshsky District (Куртамы́шский райо́н) of the Kurgan Oblast (Курга́нская о́бласть) in the Russian Federation, but the exact provenance was not precisely documented. Late Pleistocene fluvial deposits; an absolute age based on AMS ^{14}C dating of the Siberian material at the University of Göttingen was reported as 40,610 ± 340 years before present (Gehler et al. 2008). According to Cuvier (1806: 38) the specimen was found "près du fleuve Kartamisch dans le gouvernement d'Ufa en Sibérie" (sic). Presently the source of the River Kurtamysh (Река Куртамыш) is placed at ca. 500 km west of the city of Ufa (Republic of Bashkortostan), within the Kurtamyshsky District (Kurgan Oblast), and flows south east through the town of Kurtamysh into the River Tobol (Река Тобол), for a distance of ca.70 km in straight line (or ca. 125 km in curves).

Distribution Numerous Middle to Late Pleistocene European and Asian localities, mostly with a latitude of over 40°N (Guérin 1980, 2010; Kahlke and Lacombat 2008; Stuart and Lister 2012). The last representatives of *C. antiquitatis* have survived until the beginning of the Holocene, at ca. 0.01 Ma (Orlova et al. 2008; Stuart and Lister 2012).

Remarks The presence of the woolly rhinoceros in Greece remains currently scarce and tentative. Melentis (1965) described thoroughly a well-preserved cranium of *Coelodonta antiquitatis*, which he believed to have been excavated by Th. Skouphos in 1902 from Late Pleistocene deposits at the basin of Megalopolis in Peloponnese. Thereafter, this reference has been frequently reproduced as the southernmost occurrence of the woolly rhinoceros in Europe (e.g., Alvarez-Lao and Garcia 2011; Pandolfi and Tagliacozzo 2013). However, Lyras (2007) re-examined the provenance of the cranium at the AMPG collections and provided adequate archival evidence that the specimen did not belong to the Megalopolis material excavated by Skoufos in 1902, but it was rather part of a few fossil remains acquired from Ukraine in 1904 (Telichka region, near Kiev). Because these fossils share a similar brownish-black color, they must have been unintentionally mixed during the post-war transportation of the Museum's collections to a new building. Iliopoulos et al. (2010) performed a geochemical rare earth elements (REE) analysis on related specimens from Megalopolis and the small Ukrainian collection and confirmed this case.

Tsoukala (1991b) described as *Coelodonta antiquitatis praecursor* Guérin, 1980 several postcranial specimens belonging to a partial left foreleg of a single individual from a Pleistocene locality close to the village of Gephyra in the valley of Axios River, 25 km west of Thessaloniki. The morphometric and morphological evaluation of the specimens suggests some mosaic features, which obscure a straightforward attribution to *Stephanorhinus* or early *Coelodonta*. In particular, the reported absolute length values of the Gephyra metapodials are equal to, or surpass the maximum values of a very large sample of *C. antiquitatis* documented in Europe by Guérin (1980, 2010). The relative proportions and some morphological features of the specimens from Gephyra indicate that they may correspond to a longer limbed and slightly slender species of *Stephanorhinus*, an alternative option that was also discussed by Tsoukala (1991b: 78).

Koufos (1981) reported in a preliminary study the presence of *C. antiquitatis* in a Late Pleistocene fossiliferous site at the cave-mouth through which the Angitis River flows, at the NW edge of the Drama basin in Northern Greece. Subsequent excavations in the cave revealed more material, and Trandalidou (2013: 60–62; figs. 11–13) documented additional dental and postcranial specimens attributable to *C. antiquitatis*. The fauna of the Angitis Cave has yielded typical taxa of the Late Pleistocene megafauna, such as *Mammuthus primigenius* and *Megaloceros giganteus*. Absolute dating obtained on rhinoceros tooth enamel examined with the ESR method and complemented with systematic natural radioactivity, produced measurements ranging between ca 0.034 and 0.027 Ma. In neighboring countries, the presence of *C. antiquitatis* has been documented with complete skulls in Serbia (Markovic 1998: pls 2, 3) and Kroatia (Malez and Brajkovic 1991: pl. 1, fig. 2). Although scarce in low latitudes (~40°N), the species has been also reported in Italy (Billia 2010; Pandolfi and Tagliacozzo 2013) and northern Iberia (Alvarez-Lao and Garcia 2011).

C. antiquitatis is characterized by several highly specialized features, including a low-slung head, a relatively heavy mediportal stance, and a high-crowned plagiolophodont dentition with mostly flattened occlusal surface and thickened enamel suggestive of an abrasion-dominated diet. These adaptations enabled the species to explore huge belts of the open tundra–steppe and boreal habitats during the Pleistocene glacial cycles, pursuing a predominantly grazing diet of low-growing herbaceous vegetation. The external physiology of the animal is well documented, since complete or partially mummified carcases have been discovered in the permafrost of Siberia in Russia (e.g. Pallas 1773; Boeskorov et al. 2011), as well as in an ozokerite mine near the village of Starunia in Ukraine (Nowak et al. 1930; Kotarba 2009). Fossil horns of *C. antiquitatis* have been also thoroughly documented (Pallas 1769; Fortelius 1983; Shidlovskiy et al. 2011). The woolly rhinoceros was hunted by Paleolithic humans along with other members of the Pleistocene megafauna, and it has been frequently depicted in numerous cave-paintings (Nougier and Robert 1957; Braun and Zessin 2009). Apart from the human interference, the significant stress caused by the environmental changes toward the end of the Last Glacial led to the gradual population decline and the final extinction of the woolly rhinoceros in Eurasia (Orlova et al. 2008; Stuart and Lister 2012).

Acknowledgments For facilitating access to specimens under their care, as well as constructive discussions, I am greatly indebted to G. Theodorou, M. Dermitzakis, S. Roussiakis, A. Athanassiou (AMPG); J. Hooker, A. Currant, R. Sabin (BMNH); K. Heissig, G. Rössner (BSPG); M. Bertling (GPMM); O. Sandrock (HLMD); D. Nagel, G. Rabeder (IPUW); G. Lenglet (IRSNB); E. Gilissen (KMMA); G. Koufos, E. Tsoukala, D. Kostopoulos (LGPUT); L. Kordos, K. Palotás (MAFI); P. Tassy, C. Sagne, C. Lefevre, J. Cuisin, J.-G. Michard (MNHN); L. van de Hoek Ostende, J. de Vos, N. den Ouden, H.J. van Grouw, P. Kamminga (Naturalis); O. Hampe, N. Klein (NHMB); U. Göhlich, G. Höck, F. Zachos, F. Spitzenberger, B. Herzig, A. Bibl (NHMW); P. Benda (NMP); Berhane Asfaw (MNE); N. Solounias (NYIT); T. White, R. Bernor (RHOI); F. Schrenk, G. Plodowski, O. Kullmer, J.L. Franzen (SMF); E. Frey, H.W. Mittmann, W. Munk, D. Schreiber (SMNK); E. Heizmann (SMNS); W. von Koenigswald, M. Sander (STIB); H. Baagøe (ZMUC). Bibliographical research has been greatly facilitated by the Rhino Resource Center and the Biodiversity Heritage Library. Financial support for comparative studies was provided by the Deutscher Akademischer Austausch Dienst (DAAD), the European Commission's Research Infrastructure Action (EU-SYNTHESYS: GB-TAF-574, FR-TAF-1226, NL-TAF-2513, FR-TAF-2545, BE-TAF-5664, CZ-TAF-4227, DK-TAF-5242, AU-TAF 5710), the European Science Foundation (ESF-EEDEN/2003/EX05), and the Revealing Hominid Origins Initiative (RHOI, under NSF award BCS-0321893). The present contribution is dedicated to memory of our beloved colleague and friend Katerina Vasileiadou.

Appendix 1

List of the Greek localities containing Pliocene and Pleistocene rhinocerotid fossils. Locality numbers refer to the collection numbers of the PaleoBiology Database (PBDB). Biochronological data (age) are primarily based on Koufos (2013, 2016b) and Koufos and Kostopoulos (2016), as well as on evaluation of the biostratigraphical and faunal context of each locality. A detailed taxonomic history of the rhinocerotid record of each locality is provided in chronological order.

Localities	Age	Taxonomic history	Key references
Aggitis Cave (204391)	Late Pleistocene Aurelian MNQ 26 34.0–27.9 Ka	*Coelodonta antiquitatis*[1,2]	[2]Trantalidou (2013) [1]Koufos (1981)
Agia Triada (204745)	Pleistocene (imprecise)	Rhinocerotidae indet.[2] = *Dicerorhinus* sp.[1]	[2]Symeonidis et al. (2006) [1]Marinos (1965)
Aivaliki (195881)	Early Pleistocene Villafranchian	*Stephanorhinus etruscus*[2] = *Stephanorhinus* cf. *etruscus*[1]	[2]Present study [1]Symeonidis et al. (2006)
Aliakmon	Pleistocene (imprecise)	*Stephanorhinus* sp.[3] = Rhinocerotidae indet.[2] = *Dicerorhinus etruscus*[1]	[3]Tsoukala (2018) [2]Symeonidis et al. (2006) [1]Psarianos (1958)
Alikes (34782)	Early Pleistocene L. Villafranchian MNQ 19	Rhinocerotidae indet.[1,2,3]	[3]Symeonidis et al. (2006) [2]Athanassiou (1998) [1]Athanassiou (1996)
Allatini (182683)	Early Pliocene Ruscinian	Rhinocerotidae indet.[2] = *Rhinoceros* sp.[1]	[2]Symeonidis et al. (2006) [1]Marinos (1965)
Angelochori (211694)	Late Pliocene E. Villafranchian MN16	*Stephanorhinus jeanvireti* [1]	[1]Tsoukala (2018)
Apolakkia, Rhodes (34607)	Early Pliocene Ruscinian MN15	Rhinocerotidae indet.[1,2,3]	[3]Symeonidis et al. (2006) [2]Koufos (2006a) [1]Van de Weerd et al. (1982)
Apollonia (34784)	Early Pleistocene Epivillafranchian MNQ 19 1.3–1.0 Ma	Rhinocerotidae indet.[1,2,3,4]	[4]Symeonidis et al. (2006) [3]Koufos (2001) [2]Koufos and Kostopoulos (1997) [1]Koufos et al. (1992)
Asprochaliko (208564)	Late Pleistocene Aurelian MNQ 26	Rhinocerotidae indet.[2] = *Dicerorhinus kirchbergensis*[1]	[2]Present study [1]Bailey et al. (1983)
Gephyra	?Pleistocene	Rhinocerotidae indet.[3] = *Coelodonta antiquitatis praecursor*[1,2]	[3]Present study [2]Tsoukala (1992) [1]Tsoukala (1991a)
Kalamakia (184245)	Late Pleistocene Aurelian MNQ 26 100–39 Ka	*Stephanorhinus* sp.[4] = *Rhinoceros* sp.[1,2,3]	[4]Harvati et al. (2013) [3]Gardeisen et al. (2001) [2]Darlas and Lumley (2000) [1]Gardeisen and Trantalidou (1998)

(continued)

Localities	Age	Taxonomic history	Key references
Kalamoto 1 (200083)	Early Pleistocene Epivillafranchian MNQ 19	*Stephanorhinus etruscus*[2] = *Dicerorhinus etruscus*[1]	[2]Tsoukala (2018) [1]Tsoukala and Chatzopoulou (2005)
Kalamoto 2 (200084)	Early Pleistocene Epivillafranchian MNQ 19	*Stephanorhinus etruscus*[2] = *Dicerorhinus etruscus*[1]	[2]Tsoukala (2018) [1]Tsoukala and Chatzopoulou (2005)
Karnezeika (202122)	Early Pleistocene Villafranchian	*Stephanorhinus* sp.[1]	[1]Kokotini et al. (2019)
Krimni (34762)	Early Pleistocene L. Villafranchian MNQ 18	*Stephanorhinus etruscus*[2,3,4] = *Stephanorhinus* cf. *etruscus*[5] = *Dicerorhinus etruscus*[1]	[5]Symeonidis et al. (2006) [4]Koufos (2001) [3]Koufos and Kostopoulos (1997) [2]Tsoukala (1992) [1]Sakellariou-Mane et al. (1979)
Kyparissia (194477)	Mid. Pleistocene Galerian MNQ 20-21	*Stephanorhinus* sp.[2] = Rhinocerotidae indet.[1]	[2]Athanassiou (2018b) [1]Athanassiou (2016)
Kythera Island	?Pleistocene (imprecise)	Rhinocerotidae indet.[1,2]	[2]Symeonidis et al. (2006) [1]Bartsiokas (1998)
Lakonis I (207778)	Late Pleistocene Aurelian MNQ 26 100–40 Ka	Rhinocerotidae indet.[1]	[1]Panagopoulou et al. (2004)
Livakos (34764)	Early Pleistocene L. Villafranchian MNQ 18 1.8–1.6 Ma	*Stephanorhinus etruscus*[2,4] = *Stephanorhinus* cf. *etruscus*[3] = *Dicerorhinus etruscus*[1] = *Dicerorhinus* sp.[1] = *Dicerorhinus*? sp.[1]	[4]Present study [3]Symeonidis et al. (2006) [2]Tsoukala (1992) [1]Steensma (1988)
Megalopolis, Marathousa (34809)	Mid. Pleistocene Galerian MNQ 20–21	Rhinocerotidae indet.[4] = *Stephanorhinus etruscus*[1,2,3]	[4]Symeonidis et al. (2006) [3]Koufos and Kostopoulos (1997) [2]Tsoukala (1992) [1]Sickenberg (1975)
Megalopolis (old col.)	Middle–Late Pleistocene (imprecise)	*Stephanorhinus* sp.[6,10] = *Stephanorhinus hemitoechus*[5,9] = *Stephanorhinus kirchbergensis*[5,9] = *Dicerorhinus hemitoechus*[3,4] = *Dicerorhinus merckii* [3,4] = *Rhinoceros* sp.[2] = Rhinocerotidae indet.[1] Non *Coelodonta antiquitatis*[3–10]	[10]Present study [9]Tsoukala (2018) [8]Iliopoulos et al. (2010) [7]Lyras (2007) [6]Symeonidis et al. (2006) [5]Tsoukala (1992) [4]Melentis (1965) [3]Melentis (1964) [2]Skouphos (1905) [1]Bürchner (1903)

(continued)

Localities	Age	Taxonomic history	Key references
Milia, Grevena (185859–62, 195290–99)	Late Pliocene E. Villafranchian MN16	*Stephanorhinus jeanvireti* [4,5,6] = *Dicerorhinus jeanvireti* [3] = *Dicerorhinus etruscus*[2] = Rhinocerotidae indet.[1]	[6]Present study [5]Tsoukala (2018) [4]Vlachos et al. (2018) [3]Guérin and Tsoukala (2013) [2]Tsoukala (2005) [1]Tsoukala (2000)
Molikrio	Early Pleistocene Villafranchian	*Stephanorhinus* cf. *etruscus*[2,3,4] = *Dicerorhinus cf. etruscus*[1]	[4]Symeonidis et al. (2006) [3]Koufos and Kostopoulos (1997) [2]Tsoukala (1992) [1]Symeonidis et al. (1986)
Neapolis	Pleistocene (imprecise)	Rhinocerotidae indet.[2] = *Rhinoceros* sp.[1]	[2]Symeonidis et al. (2006) [1]Brunn (1956)
Nigrita (202533)	Plio-Pleistocene Villafranchian (imprecise)	Rhinocerotidae indet.[1]	[1]Athanassiou (2017)
Perdikas	Middle–Late Pleistocene	Rhinocerotidae indet.[3] = *Coelodonta antiquitatis*[1,2]	[3]Present study [2]Koufos and Pavlides (1988) [1]Pavlides (1985)
Penios (old col.)	Late Pleistocene Aurelian MNQ 26	*Stephanorhinus* cf. *hemitoechus*[2,4] = *Dicerorhinus* cf. *hemitoechus*[1,3]	[4]Symeonidis et al. (2006) [3]Athanassiou (2002) [2]Tsoukala (1992) [1]Boessneck (1965)
Penios riverbank (204397)	Late Pleistocene Aurelian MNQ 26 45–35 Ka	*Stephanorhinus hemitoechus*[1]	[1]Athanassiou (2011)
Petralona Cave (183123)	Mid. Pleistocene Galerian110	*Stephanorhinus hemitoechus*[9,10,11,13,14] = *Dicerorhinus hemitoechus*[7,8,12] = *Dicerorhinus* cf. *hemitoechus*[5,6] = *Stephanorhinus* sp.[4] = *Didermoceros* sp.[3] = *Didermoceros kirchbergensis*[2] = *Rhinoceros mercki*[1]	[14]Present study [13]Tsoukala (2018) [12]Tsoukala and Guérin (2016) [11]Symeonidis et al. (2006) [10]Fortelius et al. (1993) [9]Tsoukala (1992) [8]Tsoukala (1991a) [7]Tsoukala (1989) [6]Poulianos (1981) [5]Fortelius and Poulianos (1979) [4]Kretzoi (1977) [3]Sickenberg (1971) [2]Sickenberg (1964) [1]Kanellis (1962)
Platanochori (204658)	Early Pleistocene Epivillafranchian MNQ 19	*Stephanorhinus* sp.[2] = *Stephanorhinus hundsheimensis*[1]	[2]Present study [1]Konidaris et al. (2015)
Psychiko	Early Pleistocene (imprecise)	*Stephanorhinus* cf. *etruscus*[2] = *Rhinoceros* cf. *etruscus*[1]	[2]Symeonidis et al. (2006) [1]Paraskevaidis (1953)
Richea (195310)	Early Pleistocene Villafranchian	*Stephanorhinus etruscus*[2] = *Stephanorhinus* cf. *etruscus*[1]	[2]Present study [1]Kampouridis et al. (2018)

(continued)

Localities	Age	Taxonomic history	Key references
Saint George Priporos	Late Pliocene E. Villafranchian MN16	*Stephanorhinus jeanvireti* [3] = *Dicerorhinus jeanvireti* [2] = *Dicerorhinus etruscus*[1]	[3]Present study [2]Guérin and Tsoukala (2013) [1]Tsoukala (2005)
Serres Basin	Pleistocene (imprecise)	Rhinocerotidae indet.[2] = *Dicerorhinus* cf. *etruscus*[1]	[2]Symeonidis et al. (2006) [1]Marinos (1965)
Sesklon (34614)	Early Pleistocene L. Villafranchian MN 17	*Stephanorhinus* sp.[2,3,4,6] = *Stephanorhinus* cf. *etruscus*[5] = *Dicerorhinus* cf. *etruscus*[1]	[6]Athanassiou (2018a) [5]Symeonidis et al. (2006) [4]Koufos (2006a) [3]Athanassiou (1998) [2]Athanassiou (1996) [1]Symeonidis (1992)
Sesklon (lower) (205043)	Late Pliocene E. Villafranchian MN16	*Stephanorhinus* sp.[2] = *Dicerorhinus* cf. *etruscus*[1]	[2]Athanassiou (2018a) [1]Symeonidis (1992)
Tourkovounia 3-5 (34592)	Early Pleistocene M. Villafranchian MN 17	*Stephanorhinus* cf. *etruscus*[2-5] = *Dicerorhinus cf. etruscus*[1]	[5]Present study [4]Symeonidis et al. (2006) [3]Koufos (2006a) [2]Tsoukala (1992) [1]Symeonidis and de Vos (1976)
Tsiotra Vrisi (197943)	Early Pleistocene L. Villafranchian	*Stephanorhinus* sp.[1]	[1]Konidaris et al. (2015)
Vatera DS (183344)	Early Pleistocene M. Villafranchian MN 17	*Stephanorhinus* cf. *etruscus*[1,2,3]	[3]Symeonidis et al. (2006) [2]Koufos (2006a) [1]de Vos et al. (2002)
Volax (34593)	Early Pleistocene M. Villafranchian MN 17	Rhinocerotidae indet.[3,4,5] = *Stephanorhinus* cf. *etruscus*[2] = *Didermoceros* sp.[1]	[5]Symeonidis et al. (2006) [4]Koufos (2006a) [3]Koufos and Kostopoulos (1997) [2]Sickenberg (1975) [1]Sickenberg (1968)

Appendix 2

List of the Greek localities containing Miocene rhinocerotid fossils. Locality numbers refer to the collection numbers of the PaleoBiology Database (PBDB). Biochronological data (age) are primarily based on Koufos (2013, 2016b), as well as on evaluation of the biostratigraphical and faunal context of each locality. A detailed taxonomic history of the rhinocerotid record of each locality is provided in chronological order.

Localities	Age	Taxonomic history	Key references
Asprogiannos 1 (205337)	Late Miocene Turolian MN 13 5.96–5.33 Ma	Rhinocerotidae indet.[1,2]	[2]Koufos and Vasileiadou (2015) [1]Sylvestrou (2002)

(continued)

Localities	Age	Taxonomic history	Key references
Atalanti	Middle Miocene Astaracian MN 7+8	*Aceratherium simorrense* [3] = Aceratheriini indet.[2] = *Aceratherium* sp.[1]	[3]Present study [2]Giaourtsakis (2003) [1]Symeonidis (1974)
Chalkoutsi (205344)	Late Miocene Turolian	Rhinocerotidae indet.[1,2]	[2]Giaourtsakis (2003) [1]Theodorou et al. (1995)
Chrysavgi (205366)	Middle Miocene Astaracian MN 7 + 8	*Brachypotherium brachypus* [6,7] = ?*Brachypotherium* sp.[5] = Rhinocerotidae indet.[3,4] = *Dicerorhinus orientalis*[2] = *Diceros pachygnathus*[1]	[7]Present study [6]Koufos and Kostopoulos (2013) [5]Koufos (2006a) [4]Giaourtsakis (2003) [3]Bonis and Koufos (1999) [2]Dimopoulos (1972) [1]Psarianos (1958)
Dytiko 1 (32374)	Late Miocene Turolian MN 13 6.81–5.96 Ma	*Miodiceros neumayri* [6] = *Ceratotherium neumayri* [3,4,5] = *Dicerorhinus orientalis*[1,2]	[6]Present study [5]Koufos (2006a) [4]Giaourtsakis (2003) [3]Bonis and Koufos (1999) [2]Bonis et al. (1988) [1]Koufos (1980)
Halmyropotamos (202213)	Late Miocene Turolian MN 12	*Dihoplus pikermiensis* [5–8] = *Dicerorhinus pikermiensis* [3,4] = *Dicerorhinus orientalis* [1,2]	[8]Present study [7]Giaourtsakis et al. (2006) [6]Koufos (2006a) [5]Giaourtsakis (2003) [4]Bonis and Koufos (1999) [3]Geraads (1988) [2]Melentis (1969) [1]Melentis (1968b)
Kerassia (Ke) (old collection) (195431)	Late Miocene Turolian MN 11–12	*Miodiceros neumayri* [3] = "*Diceros*" *neumayri* [1,2]	[3]Present study [2]Giaourtsakis et al. (2020) [1]Kampouridis et al. (2019)
Kerassia 1 (K1) (195432)	Late Miocene Turolian MN 11–12	*Miodiceros neumayri* [3] = "*Diceros*" *neumayri* [1,2]	[3]Present study [2]Giaourtsakis et al. (2020) [1]Kampouridis et al. (2019)
Kerassia 3 (K3) (195434)	Late Miocene Turolian MN 11–12	*Miodiceros neumayri* [12] = "*Diceros*" *neumayri* [8,9,10,11] = *Ceratotherium neumayri* [3,5,6,7] = Rhinocerotidae indet.[1,2,4]	[12]Present study [11]Giaourtsakis et al. (2020) [10]Kampouridis et al. (2019) [9]Athanassiou et al. (2014) [8]Giaourtsakis (2009) [7]Giaourtsakis et al. (2006) [6]Koufos (2006a) [5]Theodorou et al. (2003) [4]Giaourtsakis (2003) [3]Theodorou et al. (2001) [2]Theodorou et al. (1998) [1]Theodorou et al. (1995)

(continued)

Localities	Age	Taxonomic history	Key references
Kerassia 4 (K4) (195435)	Late Miocene Turolian MN 11–12	*Miodiceros neumayri* [12] = "*Diceros*" *neumayri* [8,9,10,11] = *Ceratotherium neumayri* [3,5,6,7] *Dihoplus pikermiensis* [7-12] = *Dihoplus* cf. *pikermiensis* [6] = "*Dicerorhinus*" cf. *pikermiensis* [5] = *Dicerorhinus pikermiensis* [3] = Rhinocerotidae indet.[1,2] *Acerorhinus neleus* [9,10,11,12] = Rhinocerotidae n.sp.[5,6] = Rhinocerotidae indet.[2,3]	[12]Present study [11]Giaourtsakis et al. (2020) [10]Kampouridis et al. (2019) [9]Athanassiou et al. (2014) [8]Giaourtsakis (2009) [7]Giaourtsakis et al. (2006) [6]Koufos (2006a) [5]Theodorou et al. (2003) [4]Giaourtsakis (2003) [3]Theodorou et al. (2001) [2]Theodorou et al. (1998) [1]Theodorou et al. (1995)
Kerassia 5 (K5) (195436)	Late Miocene Turolian MN 11–12	*Miodiceros neumayri* [3,4,5] = *Dihoplus* cf. *pikermiensis* [2] = "*Dicerorhinus*" cf. *pikermiensis*[1]	[5]Present study [4]Giaourtsakis et al. (2020) [3]Kampouridis et al. (2019) [2]Koufos (2006) [1]Theodorou et al. (2003)
Maramena (32189)	Mio-Pliocene boundary MN 13–14 5.40–5.23 Ma	Rhinocerotidae indet.[3-7] = *Diceros neumayri* [2] = *Diceros neymari* (sic)[1]	[7]Present study [6]Koufos (2006a) [5]Giaourtsakis (2003) [4]Bonis and Koufos (1999) [3]Schmidt-Kittler et al. (1995) [2]Meulen and Kolfschoten (1986) [1]Karistineos (1984)
Nea Silata (191612)	Mio-Pliocene boundary MN 13–14 5.40–5.23 Ma	Rhinocerotidae indet.[1,2,3,4]	[4]Koufos and Vasileiadou (2015) [3]Koufos (2006c) [2]Giaourtsakis (2003) [1]Syrides (1990)
Nikiti 1 (202729)	Late Miocene Vallesian MN 10 9.3–8.7 Ma	*Miodiceros neumayri* [6] = "*Diceros*" *neumayri* [5] = Rhinocerotidae indet.[1,2,3,4] *Dihoplus pikermiensis* [5,6]	[6]Present study [5]Koufos et al. (2016) [4]Koufos (2006a) [3]Giaourtsakis (2003) [2]Bonis and Koufos (1999) [1]Kostopoulos (1994)
Nikiti 2 (73869)	Late Miocene Turolian MN 11 8.7–8.3 Ma	*Miodiceros neumayri* [3] = "*Diceros*" *neumayri* [2] = *Rhinocerotidae* indet.[1]	[3]Present study [2]Koufos (2016a) [1]Kostopoulos et al. (1996)

(continued)

Localities	Age	Taxonomic history	Key references
Pentalophos 1 (202119)	Late Miocene Vallesian MN 9–10	*Miodiceros neumayri* [10] = "*Diceros*" *neumayri* [7,9] = *Ceratotherium neumayri* [1–6,8] *Acerorhinus* sp.[7,8,9,10] = Acerorhinus cf. *zernowi* [6] = *Acerorhinus zernowi* [4] =p. *Aceratherium kiliasi* [1,2,3] ***Chilotherium samium*** [10] = ***Chilotherium*** cf. *samium* [9] = ***Chilotherium (Eochilotherium) kiliasi*** [8] = ***Chilotherium*** sp.[7] = ***Chilotherium kiliasi*** [2,4,6] =p. *Aceratherium kiliasi* [1,3,5]	[10]Present study [9]Athanassiou et al. (2014) [8]Geraads and Spassov (2009) [7]Giaourtsakis (2009) [6]Koufos (2006a) [5]Giaourtsakis (2003) [4]Heissig (1999) [3]Bonis and Koufos (1999) [2]Heissig (1996) [1]Geraads and Koufos (1990)
Perivolaki (194879)	Late Miocene Turolian MN12 7.3–7.1 Ma	*Miodiceros neumayri* [3] = *Ceratotherium neumayri* [2] = *Ceratotherium* sp.[1]	[3]Present study [2]Koufos (2006d) [1]Koufos (2006a)

(continued)

Localities	Age	Taxonomic history	Key references
Pikermi (182754) Classic collections: MNHN, BSPG, AMPG, BMNH, NHMW, IPUW, SMNS, MNHB, STIBP, NRM, USNM, YPM, UCMP	Late Miocene Turolian MN 12 7.33–7.29 Ma	*Miodiceros neumayri* [33] = *"Diceros" neumayri* [29–32] = *Ceratotherium neumayri* [23–28] = *Diceros neumayri* [20] = *Diceros pachygnathus pachygnathus* [19] = *Pliodiceros pachygnathus* [17] = *Diceros pachygnathus* [15,16,21,22] = *Atelodus pachygnathus* [11] = *Rhinoceros (Atelodus) pachygnathus* [10] = *Atelodus (Colodus) pachygnathus* [9] =p. *Rhinoceros schleiermacheri* [7] =p. *Rhinoceros pachygnathus* [7] = *Rhinoceros pachygnathus* [6,8,12] Non *Chalicotherium (Colodus) pachygnathus* [5] non *Rhinoceros pachygnathus* [4] = *Rhinoceros schleiermacheri* [2] Non *Rhinoceros pachygnathus* [2]	[33]Present study [32]Giaourtsakis et al. (2018) [31]Athanassiou et al. (2014) [30]Giaourtsakis (2009) [29]Giaourtsakis et al. (2009) [28]Koufos (2006a) [27]Giaourtsakis et al. (2006) [26]Giaourtsakis (2003) [25]Heissig (1999) [24]Heissig (1996) [23]Geraads (1988) [22]Solounias (1981) [21]Guérin (1980) [20]Heissig (1975) [19]Thenius (1955) [18]Kretzoi (1942b) [17]Kretzoi (1942a) [16]Ringström (1924) (continued below)

(continued)

Localities	Age	Taxonomic history	Key references
Pikermi (continued)		*Dihoplus pikermiensis* [26–33] = *Stephanorhinus pikermiensis*[24,25] = *Dicerorhinus schleiermacheri* [22] = *Stephanorhinus pachygnathus* [20] = *Dicerorhinus pikermiensis*[17,18,23] = *Dicerorhinus orientalis* [16,21] = *Ceratorhinus* aff. *schleiermacheri*[15] = *Rhinoceros (Ceratorhinus) schleiermacheri orientalis* [14] = *Rhinoceros schleiermacheri pikermiensis* [13] = *Ceratorhinus schleiermacheri*[11] = *Rhinoceros (Dihoplus) schleiermacheri* [10] = *Dihoplus schleiermacheri* [9] =p. *Rhinoceros pachygnathus* [7] =p. *Rhinoceros schleiermacheri* [7] = *Rhinoceros schleiermacheri* [3,4,6,8,12] = *Rhinoceros pachygnathus* [2] = *Rhinoceros* sp.[1] *Acerorhinus neleus* [31,32,33] = *Acerorhinus* sp.[30] = *Aceratherium* sp.[28] = Aceratheriini indet.[26] = *Aceratherium* cf. *incisivum* [22] = *Aceratherium incisivum* [10] = *Acerotherium* sp.[7]	[15]Abel (1922) [14]Schlosser (1921) [13]Toula (1906) [12]Weber (1904) [11]Osborn (1900) [10]Zittel (1893) [9]Brandt (1878) [8]Weithofer (1888) [7]Gaudry (1862–1867) [6]Gaudry (1864) [5]Wagner (1861) [4]Wagner (1857) [3]Roth and Wagner (1854) [2]Wagner (1848) [1]Goldfuss (1841)

(continued)

Localities	Age	Taxonomic history	Key references
Pikermi Chomateri (195562)	Late Miocene Turolian MN 12 7.17–7.11 Ma	*Dihoplus pikermiensis* [7] *Acerorhinus neleus* [6,7] = Aceratheriini indet.[4] = *Diceros* aff. *pachygnathus* [2] = *Aceratherium* sp.[1,2,3,5]	[7]Present study [6]Athanassiou et al. (2014) [5]Koufos (2006a) [4]Giaourtsakis (2003) [3]Bonis and Koufos (1999) [2]Marinos and Symeonidis (1974) [1]Marinos and Symeonidis (1973)
Pikermi Valley-1 (202630)	Late Miocene Turolian MN 12 ~7.27 Ma	*Miodiceros neumayri* [3] = "*Diceros*" *neumayri* [2] *Dihoplus pikermiensis* [2,3] = Rhinocerotidae indet.[1] *Acerorhinus neleus* [3]	[3]Present study [2]Filis et al. (2019) [1]Roussiakis et al. (2014)
Pikermi Valley-3 (202631)	Late Miocene Turolian MN 12 7.33–7.29 Ma	*Miodiceros neumayri* [2] *Dihoplus pikermiensis* [1,2] *Acerorhinus neleus* [2]	[2]Present study [1]Roussiakis et al. (2014)
Platania (182682)	Late Miocene Turolian MN12	*Miodiceros neumayri* [3] = *Ceratotherium neumayri* [2] = Rhinocerotidae indet.[1]	[3]Present study [2]Tsoukala (2018) [1]Vlachos and Tsoukala (2014)
Prokopi (=Achmet Aga) Euboea (214079)	Late Miocene Turolian	Rhinocerotidae indet.[1,2,3]	[3]Koufos (2006a) [2]Giaourtsakis (2003) [1]Woodward (1901)
Pyrgos Vassilissis (=Tour la Reine) (195555)	Late Miocene Turolian MN 12 7.18–7.17 Ma	*Miodiceros neumayri* [7] = ?*Ceratotherium neumayri* [5,6] = Rhinocerotidae indet.[4] = *Stephanorhinus pachygnathus*[2] = *Dicerorhinus* ?*orientalis* [1,3]	[7]Present study [6]Böhme et al. (2017) [5]Koufos (2006a) [4]Giaourtsakis (2003) [3]Bonis and Koufos (1999) [2]Paraskevaidis (1977) [1]Freyberg (1951)
Ravin de la Pluie (191070)	Late Miocene Vallesian MN10 ~9.3 Ma	Rhinocerotidae indet.[1,2,3,4]	[4]Koufos (2006a) [3]Giaourtsakis (2003) [2]Bonis and Koufos (1999) [1]Bonis et al. (1988)
Ravin des Zouaves 5 (195489)	Late Miocene Turolian MN 11 ~8.2 Ma	*Miodiceros neumayri* [6] = *Ceratotherium neumayri* [3,4,5] = *Dicerorhinus orientalis* [1,2]	[6]Present study [5]Koufos (2006a) [4]Giaourtsakis (2003) [3]Bonis and Koufos (1999) [2]Bonis et al. (1988) [1]Koufos (1980)

(continued)

Localities	Age	Taxonomic history	Key references
Ravin X (182745)	Late Miocene Turolian MN 12	*Miodiceros neumayri* [5] = *Ceratotherium neumayri* [2,3,4] = *Rhinoceros pachygnathus*[1]	[5]Present study [4]Koufos (2006a) [3]Giaourtsakis (2003) [2]Bonis and Koufos (1999) [1]Arambourg and Piveteau (1929)
Samos (various localities and collections)	Late Miocene Turolian MN11–13	*Miodiceros neumayri* [1] *Dihoplus pikermiensis*[1] *Chilotherium schlosseri* [1] *Chilotherium samium*[1]	[1]Present study (see separate Appendix 3)
Servia (202257)	Late Miocene	Rhinocerotidae indet.[2,3] = *Rhinoceros* sp.[1]	[3]Koufos (2006a) [2]Giaourtsakis (2003) [1]Paraskevaidis (1977)
Tanagra	Late Miocene Turolian	Rhinocerotidae indet.[3,4,5] = Rhinocerotidarum gen. indet.[1,2]	[5]Koufos (2006a) [4]Giaourtsakis (2003) [3]Bonis and Koufos (1999) [2]Mitzopoulos (1961) [1]Mitzopoulos (1960)
Thermopigi (73553)	Late Miocene Turolian MN 12	*Dihoplus pikermiensis* [1,2]	[2]Tsoukala (2018) [1]Geraads et al. (2007)
Vathylakkos 2 (202703)	Late Miocene Turolian MN 12 ~7.3 Ma	Rhinocerotidae indet.[1,2,3]	[3]Koufos (2006a) [2]Giaourtsakis (2003) [1]Bonis and Koufos (1999)
Vathylakkos 3 (182750)	Late Miocene Turolian MN 12 ~7.3 Ma	*Miodiceros neumayri* [10] = *Stephanorhinus pikermiensis* [7] = *Ceratotherium neumayri* [5,6,8,9] = *Dicerorhinus orientalis* [3,4] = *Diceros pachygnathus pachygnathus* [2] = *Rhinoceros pachygnathus* [1] Aceratheriini indet.[8,10] = Rhinocerotidae indet.[6] = *Aceratherium* sp.[1]	[10]Present study [9]Koufos (2006a) [8]Giaourtsakis (2003) [7]Heissig (1999) [6]Bonis and Koufos (1999) [5]Geraads (1988) [4]Bonis et al. (1988) [3]Koufos (1980) [2]Thenius (1955) [1]Arambourg and Piveteau (1929)
Xirochori (195490)	Late Miocene Vallesian MN 10 9.64–9.58 Ma	Rhinocerotidae indet.[1,2,3]	[3]Koufos (2006a) [2]Giaourtsakis (2003) [1]Bonis and Koufos (1999)

(continued)

Appendix 3

List of the various collections of rhinocerotid fossils on Samos Island. Some records correspond to precise and biochronologically well-calibrated points of collections (localities). Other records refer to old fossil collections housed in various institutions, whose exact point of excavation may have not been documented. Locality numbers refer to the collection numbers of the PaleoBiology Database (PBDB). Biochronological data (age) are based on Koufos et al. (2009b) and Koufos (2013, 2016b). A detailed taxonomic history of the rhinocerotid record of each collection is provided in chronological order.

Localities	Age	Taxonomic history	Key references
Samos BSPG col. (207137) (horned taxa)	Late Miocene Turolian	*Miodiceros neumayri* [23] = *"Diceros" neumayri* [21,22] = *Ceratotherium neumayri* [14–20] = *Diceros neumayri* [11] = *Diceros pachygnathus pachygnathus*[10] = *Pliodiceros pachygnathus*[8] = *Diceros pachygnathus*[7,12,13] = *Rhinoceros pachygnathus*[3,5,6] =p. *Rhinoceros pachygnathus*[1] *Dihoplus pikermiensis*[19–23] = *Stephanorhinus pikermiensis*[17,18] = *"Dicerorhinus" schleiermacheri*[16] = *Dicerorhinus schleiermacheri* [13] = *Stephanorhinus pachygnathus*[11] = *Dicerorhinus pikermiensis*[8,9,14,15] = *Dicerorhinus orientalis*[7,12] = *Rhinoceros (Ceratorhinus) schleiermacheri orientalis*[5] = *Rhinoceros schleiermacheri samius*[4] = *Rhinoceros* aff *schleiermacheri*[3,6] =p. *Rhinoceros pachygnathus*[1] =p. *Rhinoceros schleiermacheri* [1]	[23]Present study [22]Giaourtsakis (2009) [21]Giaourtsakis et al. (2009) [20]Giaourtsakis et al. (2006) [19]Giaourtsakis (2003) [18]Heissig (1999) [17]Heissig (1996) [16]Bernor et al. (1996) [15]Geraads and Koufos (1990) [14]Geraads (1988) [13]Solounias (1981) [12]Guérin (1980) [11]Heissig (1975) [10]Thenius (1955) [9]Kretzoi (1942b) [8]Kretzoi (1942a) [7]Ringström (1924) [6]Andrée (1921) [5]Schlosser (1921) [4]Toula (1906) [3]Weber (1907) [2]Weber (1905) [1]Weber (1904)

(continued)

Localities	Age	Taxonomic history	Key references
Samos BSPG col. (207137) (hornless taxa)	Late Miocene Turolian	*Chilotherium schlosseri* [6,7,9,12–14,16,17] = *Chilotherium (Chilotherium) schlosseri* [8,11,15] = *Aceratherium schlosseri* [1–5] *Chilotherium samium*[6,7,9,12–14,16,17] = *Chilotherium (Eochilotherium) samium*[15] = *Chilotherium (Chilotherium) samium*[8,11] = *Aceratherium sp.?* [2,3] = *Aceratherium samium*[1,2,3,5,10]	[17]Present study [16]Giaourtsakis (2009) [15]Geraads and Spassov (2009) [14]Giaourtsakis (2003) [13]Heissig (1999) [12]Bernor et al. (1996) [11]Heissig (1996) [10]Geraads and Koufos (1990) [9]Solounias (1981) [8]Heissig (1975) [7]Borsuk-Bialynicka (1969) [6]Ringström (1924) [5]Andrée (1921) [4]Krokos (1917) [3]Kiernik (1914) [2]Weber (1907) [1]Weber (1905)
Samos GPMM col. (207193)	Late Miocene Turolian	*Miodiceros neumayri* [12] = *"Diceros" neumayri* [11] = *Ceratotherium neumayri* [6,9,10] = *Diceros neumayri* [6] = *Diceros pachygnathus*[4,7] = *Rhinoceros pachygnathus*[1] *Dihoplus pikermiensis*[10,11,12] = *Stephanorhinus pikermiensis*[9] = *Dicerorhinus pikermiensis*[8] = *Stephanorhinus pachygnathus*[6] = *Dicerorhinus orientalis*[4] = *Rhinoceros schleiermacheri* [1] *Chilotherium schlosseri* [4,7,9,10,11,12] = *Chilotherium kowalevskii* [9,10] = *Chilotherium(Chilotherium) schlosseri* [6] = *Chilotherium (Chilotherium) kowalevskii* [6] = *Chilotherium angustifrons*[4] = *Chilotherium wegneri* [4] = *Aceratherium habereri*[2,3] = *Aceratherium schlosseri* [1] = *Aceratherium angustifrons*[1] = *Aceratherium wegneri* [1,5]	[12]Present study [11]Giaourtsakis (2009) [10]Giaourtsakis (2003) [9]Heissig (1999) [8]Geraads (1988) [7]Solounias (1981) [6]Heissig (1975) [5]Leonardi (1947) [4]Ringström (1924) [3]Killgus (1922b) [2]Killgus (1922a) [1]Andrée (1921)

(continued)

Localities	Age	Taxonomic history	Key references
Samos Mytilinii-1A NHMA col. (202215)	Late Miocene Turolian MN 12 ~7.1 Ma	*Miodiceros neumayri* [4] = "*Diceros*" *neumayri* [3] *Dihoplus pikermiensis*[2,3,4] = *Dicerorhinus pikermiensis*[1]	[4]Present study [3]Giaourtsakis (2009) [2]Koufos (2006a) [1]Koufos et al. (1997)
Samos Mytilinii-1B NHMA col. (202216)	Late Miocene Turolian MN 12 ~7.1 Ma	*Miodiceros neumayri* [4] = "*Diceros*" *neumayri* [3] = *Dihoplus* cf. *pikermiensis*[2] = *Dicerorhinus* cf. *pikermiensis*[1]	[4]Present study [3]Giaourtsakis (2009) [2]Koufos (2006a) [1]Koufos et al. (1997)
Samos Mytilinii-1C NHMA col. (202217)	Late Miocene Turolian MN 12 ~7.1 Ma	?Rhinocerotidae indet.[3] = Rhinocerotidae indet[2] = *Dicerorhinus* sp.[1]	[3]Present study [2]Koufos (2006a) [1]Koufos et al. (1997)
Samos Mytilinii-1D NHMA col. (202220)	Late Miocene Turolian MN 12 ~7.1 Ma	*Miodiceros neumayri* [2] = "*Diceros*" *neumayri* [1]	[2]Present study [1]Giaourtsakis (2009)
Samos Mytilinii-3 NHMA col. (202218)	Late Miocene Turolian MN 12 ~7.3 Ma	*Miodiceros neumayri* [3] = "*Diceros*" *neumayri* [2] *Dihoplus pikermiensis*[2,3] = Rhinocerotidae indet.[1]	[3]Present study [2]Giaourtsakis (2009) [1]Koufos (2006a)
Samos Mytilinii-4 NHMA col. (202219)	Late Miocene Turolian MN 12 ~7.5 Ma	*Miodiceros neumayri* [3] = "*Diceros*" *neumayri* [2] = Rhinocerotidae indet.[1]	[3]Present study [2]Giaourtsakis (2009) [1]Koufos (2006a)
Samos (PMMS) NHMA col.	Late Miocene Turolian MN 12 ~7.1 Ma	*Miodiceros neumayri* [3] = "*Diceros*" *neumayri* [2] = *Dicerorhinus orientalis*[1]	[3]Present study [2]Giaourtsakis (2009) [1]Melentis (1968a)
Samos Q1 AMNH col. (95691)	Late Miocene Turolian MN 12 ~7.1 Ma	*Miodiceros neumayri* [2] = *Diceros pachygnathus*[1] ?*Dihoplus pikermiensis*[2] = *Dicerorhinus schleiermacheri*[1] *Chilotherium* sp.[2] = *Chilotherium samium*[1]	[2]Present study [1]Solounias (1981)
Samos Q2 AMNH col. (206460)	Late Miocene Turolian MN 12 ~7.5 Ma	*Chilotherium* sp.[2] = *Chilotherium samium*[1]	[2]Present study [1]Solounias (1981)
Samos Q4 AMNH col. (95689)	Late Miocene Turolian MN 11 8.0–7.6 Ma	*Miodiceros neumayri* [2] = *Diceros pachygnathus*[1] *Chilotherium* sp.[2] = *Chilotherium samium*[1]	[2]Present study [1]Solounias (1981)
Samos Q5 AMNH col. (95690)	Late Miocene Turolian MN 13 6.9–6.7 Ma	*Miodiceros neumayri* [2] = *Diceros pachygnathus*[1] ?*Dihoplus pikermiensis*[2] = *Dicerorhinus schleiermacheri* [1]	[3]Present study [2]Koufos (2006a) [1]Solounias (1981)

(continued)

Localities	Age	Taxonomic history	Key references
Samos S4 CM col. (205353)	Late Miocene Turolian MN 12 ~7.1 Ma	*Chilotherium* sp.[3] = *Chilotherium samium*[1,2]	[3]Present study [2]Solounias (1981) [1]Black et al. (1980)
Samos CM col.	Late Miocene Turolian	*Miodiceros neumayri* [1] *Chilotherium* sp.[1]	[1]Present study
Samos UCMP col.	Late Miocene Turolian	Rhinocerotidae indet.[1]	[1]Present study
Samos SMF col. (202725)	Late Miocene Turolian	*Miodiceros neumayri* [6] = *"Diceros" neumayri* [5] = *Ceratotherium neumayri* [3] = *Atelodus pachygnathus*[1] *Dihoplus pikermiensis*[6] *Chilotherium* sp.[5,6] = *Chilotherium (Eochilotherium) samium*[4] = *Chilotherium schlosseri* [1,2]	[6]Present study [5]Giaourtsakis (2009) [4]Geraads and Spassov (2009) [3]Giaourtsakis et al. (2006) [2]Heissig (1999) [1]Drevermann (1930)
Samos HLMD col. (202724)	Late Miocene Turolian	*Miodiceros neumayri* [4] = *"Diceros" neumayri* [3] = *Ceratotherium neumayri* [2] *Dihoplus pikermiensis*[4] *Chilotherium schlosseri* [1,4]	[4]Present study [3]Giaourtsakis (2009) [2]Giaourtsakis et al. (2006) [1]Heissig (1999)
Samos SMNS col.	Late Miocene Turolian	*Miodiceros neumayri* [3] = *"Diceros" neumayri* [2] = *Ceratotherium neumayri* [1] *Chilotherium* sp.[2,3]	[3]Present study [2]Giaourtsakis (2009) [1]Geraads (1988)
Samos GPIH col. (43007)	Late Miocene Turolian	*Miodiceros neumayri* [6] = *"Diceros" neumayri* [5] = *Ceratotherium neumayri* [3,4] = *Diceros neumayri* [2] *Chilotherium schlosseri* [2–6] = *Aceratherium* sp.[1]	[6]Present study [5]Giaourtsakis (2009) [4]Giaourtsakis (2003) [3]Heissig (1999) [2]Lehmann (1984) [1]Gürich (1912)
Samos NHMW col. (182751)	Late Miocene Turolian	*Miodiceros neumayri* [3] = *"Diceros" neumayri* [2] = *Ceratotherium neumayri* [1] *Dihoplus pikermiensis*[1,2,3] *Chilotherium schlosseri* [3] = *Chilotherium* sp.[2]	[3]Present study [2]Giaourtsakis (2009) [1]Giaourtsakis et al. (2006)
Samos IPUW col.	Late Miocene Turolian	*Miodiceros neumayri* [1]	[1]Present study

(continued)

Localities	Age	Taxonomic history	Key references
Samos MCGL col. (202120)	Late Miocene Turolian	*Miodiceros neumayri* [7] = *"Diceros" neumayri* [6] = *Ceratotherium neumayri* [5] = *Diceros pachygnathus*[4] = *Rhinoceros pachygnathus*[1,2,3] *Dihoplus pikermiensis*[6,7] = *Dicerorhinus schleiermacheri* [4] = *Rhinoceros schleiermacheri* [3] =?*Rhinoceros schleiermacheri* [2]	[7]Present study [6]Giaourtsakis (2009) [5]Heissig (1999) [4]Solounias (1981) [3]Weber (1904) [2]Forsyth-Major (1894) [1]Forsyth-Major (1888)
Samos NHMB col.	Late Miocene Turolian	*Miodiceros neumayri* [3] = *Ceratotherium neumayri* [2] = *Diceros pachygnathus*[1] *Chilotherium* sp.[3]	[3]Present study [2]Geraads (1988) [1]Viret (1958)
Samos NMBE col.	Late Miocene Turolian	*Miodiceros neumayri* [2] = *Rhinoceros (Atelodus) pachygnathus*[1] *Dihoplus pikermiensis*[2] = *Rhinoceros (Ceratorhinus) schleiermacheri* [1]	[2]Present study [1]Studer (1911)
Samos AMPG col. (208570)	Late Miocene Turolian	*Miodiceros neumayri* [3] = *"Diceros" neumayri* [1,2] *Dihoplus pikermiensis*[1,2,3] *Chilotherium* sp.[1,2,3]	[3]Present study [2]Svorligkou et al. (2019) [1]Giaourtsakis (2009)
Samos MNHN col.	Late Miocene Turolian	*Miodiceros neumayri*[2] = *"Diceros" neumayri*[1]	[2]Present study [1]Giaourtsakis (2009)
Samos BMNH col.	Late Miocene Turolian	*Miodiceros neumayri* [2] = *"Diceros" neumayri* [1] *Chilotherium* sp.[1,2]	[2]Present study [1]Giaourtsakis (2009)
Samos MAFI col.	Late Miocene Turolian	*Miodiceros neumayri* [2] =p. *Rhinoceros* sp.[1] *Dihoplus pikermiensis*[2] =p. *Rhinoceros* sp.[1]	[2]Present study [1]Kormos (1913)
Samos MGPP col.	Late Miocene Turolian	*Chilotherium schlosseri* [3,4] = *Chilotherium wegneri* [1,2]	[4]Present study [3]Giaourtsakis (2009) [2]Piccoli et al. (1975) [1]Leonardi (1947)

References

Abel O (1922) Lebensbilder aus der Tierwelt der Vorzeit. Verlag von Gustav Fischer, Jena

Alvarez-Lao DJ, Garcia N (2011) Southern dispersal and palaeoecological implications of woolly rhinoceros (*Coelodonta antiquitatis*): review of the Iberian occurrences. Quat Sci Rev 30:2002–2017

Ameur RC, Jaeger J-J, Michaux J (1976) Radiometric age of early *Hipparion* fauna in Northwest Africa. Nature 261:38–39

Andrée J (1921) Rhinocerotiden aus dem Unterpliozän von Samos. Paläontol Z 20:189–212

Ansted DT (1859) On the geology of Malaga and the southern part of Andalusia. Quart J Geol Soc London 15:585–604

Antoine P-O (2002) Phylogénie et évolution des Elasmotheriina (Mammalia, Rhinocerotidae). Mém Mus nat d'Hist natur 188:1–359

Antoine P-O (2012) Pleistocene and Holocene rhinocerotids (Mammalia, Perissodactyla) from the Indochinese peninsula. C R Palevol 11:159–168

Antoine P-O, Saraç G (2005) Rhinocerotidae (Mammalia, Perissodactyla) from the late Miocene of Akkaşdaği, Turkey. In: Sen S (ed) geology, mammals and environments at Akkaşdaği, late Miocene of Central Anatolia. Geodiversitas 27(4):601–632

Antoine P-O, Karadenizli L, Saraç G, Sen S (2008) A giant rhinocerotoid (Mammalia, Perissodactyla) from the late Oligocene of north-Central Anatolia (Turkey). Zool J Linnean Soc 152:581–592

Antoine P-O, Downing KF, Crochet JY, Duranthon F, Flynn LJ, Marivaux L, Metais G, Rajpar AR, Roohi G (2010) A revision of *Aceratherium blanfordi* Lydekker, 1884 (Mammalia: Rhinocerotidae) from the early Miocene of Pakistan: postcranials as a key. Zool J Linnean Soc 160:139–194

Antunes MT (1972) Un nouveau Rhinocerotidé, *Chilotherium ibericus* n. sp. Bol Mus Lab Mineral Geol Fac Ciênc 13(1): 25–33

Arambourg C (1959) Vertébrés continentaux du Miocène supérieur de l'Afrique du Nord. Publications du service de la carte géologique de l'Algérie (nouvelle série). Paléontol, Mem 4:1–159

Arambourg C, Piveteau J (1929) Les vertébrés du Pontien de Salonique. Ann Paléontol 18:1–82

Arambourg C (1970) Les Vertébrés du Pléistocène de l'Afrique du Nord. Arch Mus Hist Nat Paris (7e sér) 10:1–126

Athanassiou A (1996) Contribution to the study of the fossil mammals of Thessaly [in Greek]. Dissertation, National and Kapodistrian University of Athens

Athanassiou A (1998) Contribution to the study of the fossil mammals of Thessaly [in Greek]. Gaia 5:1–393

Athanassiou A (2001) New findings of fossil large mammal remains in the Penios valley (area of Larissa, Thessaly, Greece). Bull Geol Soc Greece 34:533–539

Athanassiou A (2002) Neogene and Quaternary faunas of Thessaly. Ann Géol Pays Helléniques 39(A): 279–293

Athanassiou A (2011) The late Pleistocene fauna of Peneiós Valley (Lárissa, Thessaly, Greece): new collected material. In: Program and abstracts of the 9th annual meeting of the European Association of Vertebrate Palaeontologists, Heraklion (Crete), p 14

Athanassiou A (2016) Pleistocene vertebrates from the Kyparíssia lignite mine, Megalópolis, S. Greece. In: Closing symposium, ERC starting Grant project "PaGE", Paleoanthropology at the Gates of Europe, Tübingen, Abstracts, pp 14–15

Athanassiou A (2017) A cranial specimen of *Anancus arvernensis* from Nigríta, northern Greece. In: VII international conference of mammoths and their relatives, Taichung, Abstracts, p 13

Athanassiou A (2018a) A Villafranchian hipparion-bearing mammal fauna from Sésklo (E. Thessaly, Greece): implications for the question of hipparion–Equus sympatry in Europe. Quaternary 1(2):1–25

Athanassiou A (2018b) Pleistocene vertebrates from the Kyparíssia lignite mine, Megalopolis Basin, S. Greece: Rodentia, Carnivora, Proboscidea, Perissodactyla, Ruminantia. Quat Int 497:198–221

Athanassiou A, Roussiakis SJ, Giaourtsakis IX, Theodorou GE, Iliopoulos G (2014) A new hornless rhinoceros of the genus *Acerorhinus* (Perissodactyla: Rhinocerotidae) from the upper Miocene of Kerassiá (Euboea, Greece), with a revision of related forms. Palaeontographica A 303(1–3):23–59

Azzaroli A (1962) Validita della specie *Rhinoceros hemitoechus* Falconer. Palaeontographia It 57:21–34

Bailey GN, Carter P, Gamble C, Higgs HP (1983) Asprochaliko and Kastritsa: further investigations of Paleolithic settlement and economy in Epirus (North-West Greece). Proc Prehist Soc 49:15–42

Ballatore M, Breda M (2016) *Stephanorhinus elatus* (Rhinocerotidae, Mammalia): proposal for the conservation of the earlier specific name and designation of a lectotype. Geodiversitas 38(4):579–594

Ballatore M, Breda M (2019) Revision of the rhinoceros remains (Rhinocerotidae, Mammalia) from the late Pliocene of Etouaires (Auvergne, France) and the morphological distinction between the postcranial bones of *Stephanorhinus elatus* and *S. etruscus*. C R Palevol 18(2):191–208

Bartsiokas A (1998) The paleontology of Kythera island [in Greek]. Publ Soc Kytherian Stud 9:1–96

Bayshashov BU (1983) Novyy nosorog iz neogena Kazakhstana. (A new rhinoceros from the Neogene of Kazakhstan) [in Russian]. Paleontol Zh 1983(1):131–135

Becker D, Antoine PO, Maridet O (2013) A new genus of Rhinocerotidae (Mammalia, Perissodactyla) from the Oligocene of Europe. J Syst Palaeontol 11(8):947–972

Bernor RL (1983) Geochronology and zoogeographic relationships of Miocene Hominoidea. In: Ciochon RL, Corruccini RS (eds) New interpretations of ape and human ancestry. Plenum Press, New York, pp 21–64

Bernor RL, Heissig K, Tobien H (1987) Early Pliocene Perissodactyla from Sahabi, Libya. In: Boaz N, El Arnauti A, Gaziry A, Heinzelin De J, Boaz DD (eds) Neogene paleontology and geology of Sahabi. Alan Liss, New York, pp 233–254

Bernor RL, Solounias N, Swisher CC, Van Couvering JA (1996) The corellation of three classical "pikermian" mammal faunas - Maragheh, Samos and Pikermi -with the European MN unit system. In: Bernor RL, Fahlbush V, Mittman H-W (eds) The evolution of Western Eurasian Neogene mammal faunas. Columbia University Press, New York, pp 137–154

Billia EME (2010) A brief review of the "woolly rhino" *Coelodonta antiquitatis* (Blumenbach, 1799) (Mammalia, Rhinocerotidae) in Italy (southern Europe). In: Podobina VM (ed) Proceedings of the IV IPA international symposium "evolution of life on the earth", Tomsk, 10–12 November, 2010, vol 4, pp 529–532

Black CC, Krishtalka L, Solounias N (1980) Mammalian fossils of Samos and Pikermi, part 1. The Turolian rodents and insectivores of Samos. Ann Carnegie Mus 49:359–378

Blumenbach JF (1791) Handbuch der Naturgeschichte. Vierte sehr verbesserte Ausgabe, 4th edn. Johann Christian Dieterich, Göttingen

Blumenbach JF (1799) Handbuch der Naturgeschichte. Sechste Ausgabe, 6th edn. Johann Christian Dieterich, Göttingen

Boeskorov GG, Lazarev PA, Sher AV, Davydov SP, Bakulina NT, Shchelchkova MV, Binladen J, Willerslev E, Buigues B, Tikhonov AN (2011) Woolly rhino discovery in the lower Kolyma River. Quat Sci Rev 30:2262–2272

Boessneck J (1965) Die Jungpleistozänen Tierknochenfunde aus dem Peneiostal bei Larissa in Thessalien. In: Milojčić V (ed) Paläolithikum um Larissa in Thessalien. Beiträge zur Ur–und Frühgeschichtlichen Archäologie des Mittelmeer–Kulturraumes, vol 1. R. Habelt Verlag, Bonn, pp 42–60

Böhme M, Spassov N, Ebner M, Geraads D, Hristova L, Kirscher U, Kötter S, Linnemann U, Prieto J, Roussiakis S, Theodorou G, Uhlig G, Winklhofer M (2017) Messinian age and savannah environment of the possible hominin *Graecopithecus* from Europe. PLoS One 12(5):e0177347. https://doi.org/10.1371/journal.pone.0177347

Bohlin B (1937) Eine tertiäre Säugetier-Fauna aus Tsaidam. Palæont Sin 14(1):1–111

Borissiak A (1914) Mammifères fossiles de Sébastopol, I. Mém Com Géol, nouv sér 87:1–154

Borissiak A (1915) Mammifères fossiles de Sébastopol, II. Mém Com Géol, nouv sér 137(2):1–47

Borissiak A (1923) O rod *Indricotherium* n.g. (sem. Rhinocerotidae). Zap Ross Akad Nauk (Ser 8) Fiz Mat Otd 35(6):1–128

Borsuk-Bialynicka M (1969) Lower Pliocene rhinocerotids from Altan Teli, Western Mongolia. Palaeontol Polonica 21:73–92

Brandt JF (1877) Versuch einer Monographie der Tichorhinen Nashörner nebst Bemerkungen über *Rhinoceros leptorhinus* Cuvier. Mém Acad Imp Sci St Petersbourg 7e Serie 24(4):1–135

Brandt J (1878) Tentamen synopseos rhinocerotidum viventium et fossilium. Mém Acad imp Sci St Petersbourg 7e Serie 26(5):1–66

Braun I, Zessin W (2009) Paläolithische Nashorndarstellungen und der Versuch ihrer zoologisch-ethologischen interpretation. Ursus Mitt Zoov Zoos Schwerin 15(1):3–19

Breuning S (1924) Beiträge zur Stammesgeschichte der Rhinocerotidae. Verh Zool Bot Ges Wien 73:5–46

Bronn HG (1831) Über die fossilen Zahne eines neuen Geschlechtes aus der Dickhauter-Ordung *Coelodonta* Hohlenzahn. Jahrb Min Geognos Geol Petrefakt 2:51–61

Brook SM, van Coevorden de Groot P, Scott C, Boag P, Long B, Ley RE, Reischer GH, Williams AC, Mahood SP, Hien TM, Polet G, Cox N, Bach Thanh H (2012) Integrated and novel survey methods for rhinoceros populations confirm the extinction of *Rhinoceros sondaicus annamiticus* from Vietnam. Biol Conserv 155:59–67

Brook SM, Dudley N, Mahood SP, Polet G, Williams AC, Duckworth JW, Van Ngoc T, Long B (2014) Lessons learned from the loss of a flagship: the extinction of the Javan rhinoceros *Rhinoceros sondaicus annamiticus* from Vietnam. Biol Conserv 174:21–29

Brunn JH (1956) Contribution a l'étude géologique du Pinde septentrional et d'une partie de la Macédoine occidentale. Ann Géol Pays Helléniques 7:1–358

Bukhsianidze M, Kakhaber K (2018) Synopsis of the terrestrial vertebrate faunas from the middle Kura Basin (eastern Georgia and Western Azerbaijan, South Caucasus). Acta Palaeontol Pol 63(3):441–461

Buckland F (1872) A new rhinoceros at the Zoological Gardens. Land and Water 1872:89

Burchell WJ (1817) Note sur une nouvelle espèce de rhinocéros. Bull Sci Soc Philomat Paris 1817:96–97

Bürchner L (1903) Wichtige Funde fossiler Knochen in Arkadien. Ber Naturwiss Ver Regensburg 9:119–123

Cappellini E, Welker F, Pandolfi L, Ramos-Madrigal J, Samodova D, Rüther PL, Fotakis AK, Lyon D, Moreno-Mayar JV, Bukhsianidze M et al (2019) Early Pleistocene enamel proteome from Dmanisi resolves *Stephanorhinus* phylogeny. Nature 574:103–107

Casanovas-Vilar I, Moyà-Solà S, Agustí J, Köhler M (2005) The geography of a faunal turnover: tracking the Vallesian Crisis. In: Elewa AMT (ed) Migration of organisms. Climate, geography, ecology. Springer, Heidelberg, pp 247–300

Cerdeño E (1992) Spanish neogene rhinoceroses. Paléo 35(2):297–308

Cerdeño E (1993) Étude sur *Diaceratherium aurelianense* et *Brachypotherium brachypus* du Miocène Moyen de France. Bull Mus Nat Hist Natur Paris 15(1–4):25–77

Cerdeño E (1995) Cladistic analysis of the family Rhinocerotidae. Am Mus Novit 3143:1–25

Cerdeño E (1996) Rhinocerotidae from the middle Miocene of the Tung-gur formation, Inner Mongolia (China). Am Mus Novit 3184:1–43

Cerdeño E, Alcalá B (1989) *Aceratherium alfambrense* sp. n., nueva especie de rinoceróntido del Vallesiense superior de Teruel (España). Rev Esp Paleontol 4:39–51

Cerdeño E, Sanchez B (2000) Intraspecific variation and evolutionary trends of *Alicornops simorrense* (Rhinocerotidae) in Spain. Zool Scripta 29(4):275–305

Chen S, Deng T, Hou S, Shi Q, Pang L (2010) Sexual dimorphism in perissodactyl rhinocerotid *Chilotherium wimani* from the late Miocene of the Linxia Basin (Gansu, China). Acta Palaeontol Pol 55(4):587–597

Chen S, Deng T, He W, Chen S-Q (2011) A dental pathological deformity of *Chilotherium wimani* from the Linxia Basin of Gansu, China. Vertebrata PalAsiatica 49(2):223–228

Chow BS (1963) A new species of *Dicerorhinus* from Yushe, Shansi, China [in Chinese with English summary]. Verteb PalAsiatica 7(4):325–329

Cope ED (1879) On the extinct american rhinoceroses. Am Nat 13(12):771a–771j

Croizet JB, Jobert A (1828) Recherches sur les ossements fossiles du département du Puy de Dôme. Delahays, Paris

Cuvier G (1806) Sur les rhinocéros fossiles. Ann Mus Hist Natur 7:19–52
Cuvier G (1812) Recherches sur les ossemens fossiles de quadrupèdes. Deterville, Paris
Cuvier G (1822) Recherches sur les ossemens fossiles, nouvelle édition (deuxième éd.), tome 2. G. Doufor et E. d'Ocagne (libraires), Paris
Darlas A, de Lumley H (2000) Fouilles franco-helléniques de la grotte de Kalamakia (Aréopolis, Péloponnèse). Bull Corresp Hellénique 124(2):659–663
de Blainville HMD (1837a) Rapport sur la découverte de plusieurs ossements fossiles de quadrumanes dans le dépôt tertiaire de Sansan, près d'Auch, par Lartet. C R hebd séances Acad Sci Paris 4:981–998
de Blainville HMD (1837b) Sur un nouvel envoi de fossiles provenant du dépôt de Sansan. C R hebd séances Acad Sci Paris 5:417–427
de Bonis L, Koufos GD (1999) The Miocene large mammal succession in Greece. In: Agusti J, Rook L, Andrews P (eds) Hominoid evolution and climatic change in Europe. The evolution of the Neogene terrestrial ecosystems in Europe, vol 1. Cambridge University Press, London, pp 205–237
de Bonis L, Bouvrain G, Koufos GD (1988) Late Miocene mammal localities of the lower Axios valley (Macedonia, Greece) and their stratigraphical significance. Mod Geol 13:141–147
de Christol J (1834) Recherches sur les caractères des grandes espèces de rhinocéros fossiles. Jean Martel Ainé (imprimeur de la faculté de Médecine), Montpellier
de Mecquenem R (1908) Contribution à l'étude du gisement des vertébrés de Maragha et de ses environs. In: de Morgan (ed) Annales d'Histoire Naturelle. Ministère de l'instruction publique et des beaux-arts (Délégation en Perse). Ernest Leroux (éditeur), Paris, Tome 1(1):27–79
de Prado C (1864) Descripción física y geológica de la provincia de Madrid. Junta General de Estadística, Imprenta Nacional, Madrid, pp 1–219
de Vos J, van der Made J, Athanassiou A, Lyras G, Sondaar PY, Dermitzakis M (2001) Preliminary note on the late Pliocene fauna from Vatera (Lesvos, Greece). Ann Géol Pays Helléniques 39:37–69
Deng T (2000) A new species of *Acerorhinus* (Perissodactyla, Rhinocerotidae) from the Late Miocene of Fugu, Shaanxi, China. Verteb PalAsiatica 38(3):203–217
Deng T (2001) New materials of *Chilotherium wimani* (Perrisodactyla, Rhinocerotidae) from the late Miocene of Fugu, Shaanxi. Vertebr PalAsiatica 39:129–138
Deng T (2002a) The earliest known wooly rhino discovered in the Linxia basin, Gansu Province, China. Geol Bull China 21:604–608
Deng T (2002b) Limb bones of *Chilotherium wimani* (Perissodactyla, Rhinocerotidae) from the late Miocene of the Linxia Basin in Gansu, China. Vertebr Pal Asiatica 40:305–316
Deng T (2005) New cranial material of *Shansirhinus* (Rhinocerotidae, Perissodactyla) from the lower pliocene of the Linxia basin in Gansu, China. Geobios 38:301–313
Deng T (2006) A primitive species of *Chilotherium* (Perissodactyla, Rhinocerotidae) from the late Miocene of the Linxia Basin (Gansu, China). Cainozoic Res 5(1/2):93–102
Deng T (2010) Linxia basin: an ancient paradise for late Cenozoic rhinoceroses in North China. Bull Chin Acad Sci 24(2):103–106
Deng T, Qiu Z-X (2007) First discovery of *Diceros* (Perissodactyla, Rhinocerotidae) in China. Vertebr PalAsiatica 45:287–306
Deng T, Qi G-Q (2009) Rhinocerotids (Mammalia, Perissodactyla) from *Lufengpithecus* site, Lufeng, Yunnan. Verteb PalAsiatica 47(2):135–152
Deng T, Wang X, Fortelius M, Li Q, Wang Y, Tseng ZJ, Takeuchi GT, Saylor JE, Saila LK, Xie G (2011) Out of Tibet: Pliocene woolly rhino suggests high-plateau origin of ice age Megaherbivores. Science 333:1285–1288
Deng T, Hanta R, Jintasakul P (2013) A new species of *Aceratherium* (Rhinocerotidae, Perissodactyla) from the Late Miocene of Nakhon Ratchasima, northeastern Thailand. J Vertebr Paleontol 33(4):977–985
Desmarest AG (1822) Mammalogie, ou description des espèces de Mammifères, seconde partie. Veuve Agasse (imprimeur), Paris, pp 277–555

Dietrich WO (1942) Zur Entwicklungsmechanik des Gebisses der afrikanischen Nashörner. Zentralbl Min Geol Paläontol Abt A 1942:297–300

Dimopoulos G (1972) *Dicerorhinus orientalis* aus dem Obermiozän (Pont.) des Beckens von Lagadas (Mazedonien/Griechenland). Folia Biochaem Biol Graeca 9:47–60

Dinerstein E (2015) Greater one-horned rhinoceros, *Rhinoceros unicornis*. In: Johnsingh AJT, Manjrekar N (eds) Mammals of South Asia, vol 2. Universities Press (India), Hyderbad, pp 95–111

Dollo L (1885) Rhinocéros vivants et fossiles. Rev Quest Sci 17:293–299

Drevermann F (1930) Aus der Zeit des dreizehigen Pferdes. Nat Mus 60:2–13

Dzhafarov RD (1955) Novy predstavitel' chetvertichnykh nosorogov (*Rhinoceros binagadensis* sp. nova) binagadinskoy fauny. (A new representative of the Quaternary rhinoceros (*Rhinoceros binagadensis* sp. nova) in the Binagady fauna) [in Russian]. In: Binagadinskoe Mestonakhozhdenie Chetvertichnoy Fauny i Flory IV. Trudy Estestvenno-Istoricheskogo Muzeya G. Zardaby, Izdatel'stvo Akademii Nauk Azerbaydzhanskoy SSR, Baku, pp 65–88

Dzhafarov RD (1958) Two new species of *Aceratherium* from upper Sarmatian deposits in the Caucasus [in Russian, Azerbaijani]. Izv Akad Nauk Azerbaidzhnsk SSR, Ser Geol Geogr Sci 1958(2):25–34

Emslie RH, Milliken T, Talukdar B, Burgess G, Adcock K, Balfour D, Knight MH (2019) African and Asian rhinoceroses - status, conservation and trade. CITES 17th meeting (Colombo, June 2019), CoP 18 doc.83.1 annex 3, p 1–38

Eronen JT, Mirzaie Ataabadi M, Micheels A, Karme A, Bernor RL, Fortelius M (2009) Distribution history and climatic controls of the late Miocene Pikermian chronofauna. Proc Nat Acad Sci 106(29):11867–11871

Falconer H (1868) On the European Pliocene and post-Pliocene species of the genus *Rhinoceros*. In: Murchison C (ed) Paleontological memoires and notes of the late Hugh Falconer, vol 2. Hardwicke R, London, pp 309–403

Falconer H, Cautley PT (1847) Fauna antiqua sivalensis, being the fossil zoology of the Sewalik Hills, in the North of India. Illustrations, part VIII: Suidae and Rhinocerotidae. Smith, Elder and Co (publishers), London, pls 69–80

Filis P, Roussiakis S, Giaourtsakis I, Kampouridis P, Kargopoulos N, Skandalos P, Sklavounou S, Koumoutsea I, Svorligkou G, Tsoukalas N, Kirdis S, Theodorou G (2019) A fieldwork update on the new palaeontological excavations at the classical Turolian locality of Pikermi (Attica, Greece). In: Kranis H (ed) proceedings of the 15th International Congress of the Geological Society of Greece, Harokopio University of Athens, 22-24 May 2019. Bull Geol Soc Greece Sp Pub 7:63–64

Filoux A, Suteethorn V (2018) A late Pleistocene skeleton of *Rhinoceros unicornis* (Mammalia, Rhinocerotidae) from western part of Thailand (Kanchanaburi Province). Geobios 51(1):31–49

Fischer G (1808) Notice d'un animal fossile de Sibérie inconnu aux naturalistes. In: Programme d'invitation à la Séance publique de la Société Impériale des Naturalistes, Imprimerie de l'Université Impériale, Moscou, pp 1–28

Fischer G (1814) Zoognosia tabulis synopticis illustrata. Nicolai Sergeidis Vsevolozsky (typis), Mosquae,vol 3, pp 1–734

Forster-Cooper C (1915) New genera and species of mammals from the Miocene deposits of Baluchistan. Preliminary notice. Ann Mag Nat Hist (Ser 8) 16:404–410

Forsyth-Major CI (1888) Sur un gisement d'ossements fossiles dans l'île de Samos contemporains de l'âge de Pikermi. C R Hebd Séanc Acad Sci Paris 107:1178–1181

Forsyth-Major CI (1894) Le gisement ossifère de Mytilini et catalogue d'ossemens fossiles recueillis à Mitylini, île de Samos, et déposés au Collège Galliard, à Lausanne. G Bridel et Cie (éditeurs), Lausanne

Fortelius M (1982) Ecological aspects of dental functional morphology in the Plio-Pleistocene rhinoceroses of Europe. In: Kurtén B (ed) Teeth: form, function and evolution. Columbia University Press, New York, pp 163–181

Fortelius M (1983) The morphology and paleobiological significance of the horns of *Coelodonta antiquitatis* (Mammalia: Rhinocerotidae). J Vertebr Paleontol 3(2):125–135

Fortelius M (1985) Ungulate cheek teeth: developmental, functional, and evolutionary interrelations. Acta Zool Fennica 180:1–76

Fortelius M, Poulianos NA (1979) *Dicerorhinus* cf. *hemitoechus* (Mammalia, Perissodactyla) from the middle Pleistocene cave at Petralona Chalkidiki, N. Greece. Preliminary report. Anthropos 6:15–43

Fortelius M, Poulianos NA (1980) Errata of: *Dicerorhinus* cf. *hemitoechus* (Mammalia Perissodactyla) from the middle Pleistocene cave at Petralona. Anthropos 7:324–325

Fortelius M, Solounias N (2000) Functional characterization of ungulate molars using the abrasion-attrition wear gradient: a new method for reconstructing paleodiets. Am Mus Nov 3301:1–36

Fortelius M, Mazza P, Sala B (1993) *Stephanorhinus* (Mammalia, Rhinocerotidae) of the western European Pleistocene, with a revision of *Stephanorhinus etruscus* (Falconer, 1868). Palaeontogr It 80:63–155

Fortelius M, Heissig K, Sarac G, Sen S (2003) Rhinocerotidae (Perissodactyla). In: Fortelius M, Kappelman J, Sen S, Bernor RL (eds) Geology and paleontology of the Miocene Sinap formation, Turkey. Columbia University Press, New York, pp 282–307

Fourtau R (1918) Contribution à l'étude des vertébrés miocènes de l'Egypte. Survey Department, Ministry of Finance Governement Press, Cairo

Freyberg BV (1951) Die Pikermi-Fauna von Tour la Reine (Attika). Ann Géol Pays Helléniques 3:7–10

Fukuchi A, Kawai K (2011) Revision of fossil rhinoceroses from the Miocene Mizunami group, Japan. Paleontol Res 15(4):247–257

Gao F, Ma B (1997) Perissodactyla [in Chinese with English summ]. In: He ZQ (ed), Yuanmou Hominoid Fauna. Yunnan Sci. Tech. Press, Yunan, pp 94–114

Gaál I (1938) A keleti orrszarvú – Dicerorhinus orientalis Schloss. – új alakjának csontmaradványai Magyarországon. Természettudományi Közlöny Pótfüzete 1938(10–12):130–138

Gaál I (1943) Alsó-pliocén emlősmaradványok Hatvanból (Unterpliozäne Säugetierreste aus Hatvan in Ungarn). Geol Hungarica (Ser Palaeontol) 20:1–120

Gaál I (1953) Újra megvizsgált néhány Hatvani és gödöllői pliocén emlősmaradványról és a pliocén tagozódásáról. Föld Közlöny 83:263–272

Gaál I (1954) Über einige neuerlich Untersuchte pliozäne Säugetierreste aus Hatvan und Gödölö. Föld Közlöny 84:79–81

Gardeisen A, Trantalidou C (1998) La faune du Pléistocène supérieur de la grotte de Kalamakia (Péloponnèse, Grèce): premiers résultats. In: proceedings of XIII congress International Union of Prehistoric and Protohistoric Sciences, 8-14 September 1996, Forlì (Italia), a.B.a.C.O. Edizioni 1(3):537–542

Gardeisen A, Trantalidou K, Darlas A (2001) Faunal remains from the Kalamakia Cave (Peloponnese, Greece). In: Cruz AR et al (eds) Human population origins in the circum-Mediterranean area: adaptations of the hunter-gatherer groups to environmental modifications. Ark Dent 5:111–120

Gaudin C-T (1859) Modifcations apportés par Mr. Falconer a la faune du Val d'Arno. Bull Soc Vaudoise Sci Natur 6(44):130–131

Gaudry A (1862–67) Animaux fossiles et géologie de l'Attique. F. Savy Editions, Paris

Gaudry A (1864) Des liens qui semblent unir plusieurs rhinocéros fossiles aux rhinocéros vivants. Bull Soc Géol France 21(2):233–236

Gaudry A (1873) Animaux fossiles du Mont Léberon (Vaucluse): étude sur les vertébrés. F. Savy Éditions, Paris

Gautier F, Clauzon G, Suc J-P, Cravatte J, Violanti D (1994) Age and duration of the Messinian salinity crisis. C R Acad Sci 318:1103–1109

Gehler A, Reich M, Mol D, van der Plicht H (2007) The type material of *Coelodonta antiquitatis* (Blumenbach) (Mammalia: Perissodactyla: Rhinocerotidae). In: Boeskorov GG (ed) Abstracts of the 4th International Mammoth Conference, Yakutsk, 18–22 June 2007, pp 178–179

Gehler A, Reich M, Mol D, van der Plicht H (2008) The type material of *Coelodonta antiquitatis* (Mammalia: Perissodactyla: Rhinocerotidae). In: Robbert Gradstein SR, Klatt S, Felix Normann F, Weigelt P, Willmann R, Wilson R (eds) Program and abstracts, systematics 2008 conference, 7–11 April 2008. Universitäts-verlag Göttingen, Göttingen, p 206

Geraads D (1986) Sur les relations phylétiques de *Dicerorhinus primaevus* Arambourg, 1959, rhinocéros du Vallésien d'Algérie. C R Acad Sci 302:835–837
Geraads D (1988) Révision des Rhinocerotidae (Mammalia) du Turolien de Pikermi. Comparaison avec les formes voisines. Ann Paléontol 74:13–41
Geraads D (2005) Pliocene Rhinocerotidae (Mammalia) from Hadar and Dikika (lower awash, Ethiopia), and a revision of the origin of modern African rhinos. J Vertebr Paleontol 25(2):451–461
Geraads D (2010) Rhinocerotidae. In: Werdelin L, Sanders WJ (eds) Cenozoic mammals of Africa. University of California Press, Berkeley, pp 669–683
Geraads D (2017) Late Miocene large mammals from Mahmutgazi, Denizli province, Western Turkey. N Jb Geol Paläont Abh 284(3):241–257
Geraads D, Koufos GD (1990) Upper Miocene Rhinocerotidae (Mammalia) from Pentalophos-1, Macedonia, Greece. Palaeontographica 210:151–168
Geraads D, Miller E (2013) *Brachypotherium minor* n. sp., and other Rhinocerotidae from the Early Miocene of Buluk, Northern Kenya. Geodiversitas 35(2):359–375
Geraads D, Spassov N (2009) Rhinocerotidae (Mammalia) from the late Miocene of Bulgaria. Palaeontographica A 287:99–122
Geraads D, Tsoukala E, Spassov N (2007) A skull of Ancylotherium (Chalicotheriidae, Mammalia) from the late Miocene of Thermopigi (Serres, N. Greece) and the relationships of the genus. J Vertebr Palaeont 27(2):461–466
Giaourtsakis IX (2003) Late Neogene Rhinocerotidae of Greece: distribution, diversity and stratigraphical range. In: Reumer JWF, Wessels W (eds) distribution and migration of tertiary mammals in Eurasia. Deinsea 10:235–253
Giaourtsakis IX (2009) The late Miocene mammal faunas of the Mytilinii Basin, Samos Island, Greece: new collection, 9. Rhinocerotidae. Beit Paläontol 31:157–187
Giaourtsakis IX, Heissig K (2004) On the nomenclatural status of *Aceratherium incisivum* (Rhinocerotidae, Mammalia). In: Chatzipetros AA, Pavlides SB (eds) Proceedings of the 5th International Symposium on Eastern Mediterranean Geology, vol 1, pp 314–317
Giaourtsakis IX, Theodorou G, Roussiakis S, Athanassiou A, Iliopoulos G (2006) Late Miocene horned rhinoceroses (Rhinocerotinae, Mammalia) from Kerassia (Euboea, Greece). N Jahrb Geol Paläontol Abh 239:367–398
Giaourtsakis IX, Pahlevan C, Haile-Selassie Y (2009) Chapter 14: Rhinocerotidae. In: Haile-Selassie Y, Wolde Gabriel G (eds) *Ardipithecus kadabba*: late Miocene evidence from middle awash, Ethiopia. University of California Press, Berkeley, pp 429–468
Giaourtsakis IX, Svorligkou G, Roussiakis S (2018) A juvenile skull of the hornless rhinocerotid *Acerorhinus neleus* (Rhinocerotidae, Mammalia) from the late Miocene locality of Pikermi (Attica, Greece). In: Abstracts of the 62nd annual meeting of the Palaeontological association, University of Bristol, Bristol, 14–17 December 2018
Giaourtsakis IX, Kampouridis P, Roussiakis S, Svorligkou G, Kargopoulos N, Alifieri E, Theodorou G (2020) Diversity assessment of the family Rhinocerotidae (Mammalia, Perissodactyla) at the Late Miocene locality of Kerassia (Euboea Island, Greece). In: Vlachos E et al (eds) Book of Abstracts, 2nd Palaeontological Virtual Congress, May 1–15th, 2020, Palaeontology in the virtual era, p 170
Gill T (1872) Arrangement of the families of mammals: with analytical tables. Smithsonian Misc Coll 230, vol. 11(1):1–98
Ginsburg L, Guérin C (1979) Sur l'origine et l'extension stratigraphique du petit Rhinocérotidé Miocène *Aceratherium (Alicornops) simorrense* (Lartet, 1851), nov. subgen. C R Somm Soc géol France 1979:114–116
Ginsburg L, Heissig K (1989) *Hoploaceratherium*, a new generic name for "*Aceratherium*" *tetradactylum*. In: Prothero DR, Schoch RM (eds) The evolution of Perissodactyls. Oxford Monographs on Geology and Geophysics, vol. 15, Clarendon Press: Oxford University Press, New York, pp 418–421
Gloger CWL (1841) Gemeinnütziges Hand- und Hilfsbuch der Naturgeschichte (erster Band). Verlag von Aug. Schulz and Co, Breslau, pp 1–496

Goldfuss GA (1841) Correspondence. N Jahr Mineral, Geol Paläontol 1841:357–358
Gray JE (1821) On the natural arrangement of vertebrose animals. London Medical Repository 15:297–310
Gray JE (1868) Observations on the preserved specimens and skeletons of the Rhinocerotidae in the collection of the British museum and Royal College of surgeons, including the description of three new species. Proc Zool Soc London 1867:1003–1032
Groves CP (1965) Description of a new subspecies of rhinoceros from Borneo, *Didermocerus sumatrensis harrissoni*. Säugetierkundliche Mitt 13(3):128–131
Groves CP (1967) Geographic variation in the black rhinoceros, *Diceros bicornis* (L, 1758). Zeit Saugetierkunde 32(5):267–276
Groves CP (1972) *Ceratotherium simum*. Mamm Species 8:1–6
Groves CP (1975) Taxonomic notes on the white rhinoceros *Ceratotherium simum* (Burchell, 1817). Säugetierkund Mitt 23(3):200–212
Groves CP (1983) Phylogeny of the living species of rhinoceros. Zeit Zool System Evolutionsforschung 21:293–313
Groves CP, Grubb P (2011) Ungulate taxonomy. John Hopkins University Press, Baltimore
Groves CP, Guérin C (1980) Le *Rhinoceros sondaicus annamiticus* (Mammalia, Perissodactyla) d'Indochine: distinction taxinomique et anatomique; relations phylétiques. Geobios 13(2):199–208
Groves CP, Kurt F (1972) *Dicerorhinus sumatrensis*. Mamm Species 21:1–6
Groves CP, Leslie DM (2011) *Rhinoceros sondaicus* (Perissodactyla: Rhinocerotidae). Mamm Species 43:190–208
Groves CP, Fernando P, Robovsky J (2010) The sixth rhino: a taxonomic re-assessment of the critically endangered northern white rhinoceros. PLoS One 5(4):e9703. https://doi.org/10.1371/journal.pone.0009703
Guérin C (1966) *Diceros douariensis* nov. sp., un Rhinocéros du MioPliocène de Tunisie du Nord. Doc Lab Géol Fac Sci Lyon 16:1–50
Guérin C (1972) Une nouvelle espèce de rhinocéros à Vialette et dans d'autres gisements du Villafranchien inférieur européen: *Dicerorhinus jeanvireti* n. sp. Doc Lab Géol Fac Sci Lyon 49:53–150
Guérin C (1980) Les Rhinocéros (Mammalia, Perissodactyla) du Miocène terminal au Pléistocène supérieur en Europe occidentale. Comparaison avec les espèces actuelles. Doc Lab Géol Fac Sci Lyon 79:1–1185
Guérin C (1989) La famille des Rhinocerotidae (Mammalia, Perissodactyla): systématique, histoire, évolution, paléoécologie. Cranium 6(2):3–14
Guérin C (2000) The Neogene rhinoceroses of Namibia. Palaeontol Africana 36:119–138
Guérin C (2010) *Coelodonta antiquitatis praecursor* (Rhinocerotidae) du Pléistocène moyen final de l'aven de Romain-la-Roche (Doubs, France). Revue Paléobiol 29:697–746
Guérin C (2011) Les Rhinocerotidae (Mammalia, Perissodactyla) miocènes et pliocènes des Tugen Hills (Kénya). Estud Geol 67:333–362
Guérin C, Tsoukala E (2013) The Tapiridae, Rhinocerotidae and Suidae (Mammalia) of the early Villafranchian site of Milia (Grevena, Macedonia, Greece). Geodiversitas 35:447–489
Gürich G (1912) Fossile Säugetierreste aus Samos. Verh naturwiss Ver Hamburg (3e F.) 19:LXXIX
Handa N, Pandolfi L (2016) Reassessment of the middle Pleistocene Japanese rhinoceroses (Mammalia, Rhinocerotidae) and paleobiogeographic implications. Paleontol Res 20:247–260
Harris JM, Leakey MG (2003) Lothagam Rhinocerotidae. In: Leakey MG, Harris JM (eds) Lothagam: the dawn of humanity in eastern Africa. Columbia University Press, New York, pp 371–386
Harvati K, Darlas A, Bailey SE, Rein TR, El Zaatari S, Fiorenza L, Kullmer O, Psathi E (2013) New Neanderthal remains from Mani peninsula, southern Greece: the Kalamakia middle Paleolithic cave site. J Human Evol 64(6):486–499
Hatcher JB (1894) A median horned rhinoceros from the Loup Fork Beds of Nebraska. Am Geol 13(3):149–150
Hay OP (1902) Bibliography and catalogue of the fossil vertebrata of North America. Bull US Geol Surv 179:1–868

Heissig K (1972) Paläontologische und geologische Untersuchungen im Tertiär von Pakistan – 5. Rhinocerotidae (Mammalia) aus den unteren und mittleren Siwalik-Schichten. Abh Bayerischen Akad Wiss. Math-naturwiss Kl München 152:1–112

Heissig K (1973) Die Unterfamilien und Tribus der rezenten und fossilen Rhinocerotidae (Mammalia). Säugetierkundliche Mitt 21:25–30

Heissig K (1975) Rhinocerotidae aus dem Jungtertiär Anatoliens. Geol Jahrb B 15:145–151

Heissig K (1976) Rhinocerotidae (Mammalia) aus der Anchitherium-Fauna Anatoliens. Geol Jahrb B 19:3–121

Heissig K (1982) Note on Sahabi Rhinocerotidae. In: Boaz NT, De Heinzelin J, Gaziry AW, El-Arnauti A (eds) Results from the international Sahabi research project, Garyounis scientific bulletin, special issue 4. University of Garyounis, Benghazi, p 85

Heissig K (1989) The Rhinocerotidae. In: Prothero DR, Schoch RM (eds) The evolution of Perissodactyls. Oxford Monographs on Geology and Geophysics, vol 15. Clarendon Press, Oxford University Press, New York, pp 399–417

Heissig K (1996) The stratigraphical range of fossil rhinoceroses in the late Neogene of Europe and eastern Mediterranean. In: Bernor RL, Fahlbush V, Mittman H-W (eds) The evolution of Western Eurasian Neogene mammal faunas. Columbia University Press, New York, pp 339–347

Heissig, K. (1999) Family Rhinocerotidae. In: Rössner G, Heissig, K. (eds): The Miocene land mammals of Europe. Dr. Friedrich Pfeil Verlag, Munich, pp 175–188

Heissig K (2012a) The American Genus *Penetrigonias* Tanner & Martin, 1976 (Mammalia: Rhinocerotidae) as a stem group elasmothere and ancestor of *Menoceras* Troxell, 1921. Zitteliana A 52:79–95

Heissig K (2012b) Les Rhinocerotidae (Perissodactyla) de Sansan. In: Peigné, S., & Sen, S., (eds.) Mammifères de Sansan. Mém Mus nat Hist natur 203:317–485

Heller E (1913) The white rhinoceros. Smithsonian Miscellaneous Collection 61(1):1–56

Hernesniemi E, Giaourtsakis IX, Evans AR, Fortelius M (2011) Chapter 11 Rhinocerotidae. In: Harrison T (ed) Paleontology and geology of Laetoli, Tanzania: human evolution in context. Fossil hominins and the associated fauna, vol 2. Springer, Dordrecht, pp 275–294

Heude PM (1892) Études Odontologiques 1. Herbivores trizygodontes et dizygodontes. Mém Hist Nat Empire Chinois 2(2):65–84

Hildebrandt TB, Hermes R, Colleoni S, Diecke S, Holtze S, Renfree MB, Stejskal J, Hayashi K, Drukker M, Loi P, Goerlitz F, Lazzari G, Galli C (2018) Embryos and embryonic stem cells from the white rhinoceros. Nat Commun 9:2589. https://doi.org/10.1038/s41467-018-04959-2

Hillman-Smith K, Groves CP (1994) *Diceros bicornis*. Mamm Species 455:1–8

Holbrook LT, Lucas SG (1997) A new genus of rhinocerotoid from the Eocene of Utah and the status of north American "*Forstercooperia*". J Vertebr Paleontol 17(2):384–396

Hooijer DA (1963) Miocene Mammalia of Congo. Ann K Mus Midden Afr, Geol Wet 46:1–150

Hooijer DA (1968) A rhinoceros from the late Miocene of fort Ternan, Kenya. Zool Mededelingen 43:77–92

Hooijer DA, Patterson B (1972) Rhinoceroses from the Pliocene of Northwestern Kenya. Bull Mus Comp Zool 144:1–26

Hu CK, Qi T (1978) Gongwangling Pleistocene mammalian fauna of Lantian, Shaanxi [in Chinese with English summary]. Palaeont Sinica (ser C) 21:36–39

Hünermann KA (1989) Die Nashornskelette (*Aceratherium incisivum* Kaup 1832) aus dem Jungtertiär vom Höwenegg im Hegau (Südwestdeutschland). Andrias 6:1–116

Iliopoulos G, Lyras G, Roussiakis S, Athanassiou A (2010) Woolly rhinos and woolly mammoths in southern Greece? Using REE elements to trace the provenance of old museum collections. In: 8th Annual meeting of the European Association of Vertebrate Palaeontologists, 7-12 June 2010, Aix-en-Provence, France, p 46

International Commission on Zoological Nomenclature (1999) International code of zoological nomenclature, 4th edn. International Trust for Zoological Nomenclature, London

International Commission on Zoological Nomenclature (2003) Declaration 44 - Amendment of Article 74.7.3. Bull Zool Nomenclature 60(4):263

International Commission on Zoological Nomenclature (2012) Amendment of articles 8, 9, 10, 21 and 78 of the international code of zoological nomenclature to expand and refine methods of publication. Bull Zool Nomenclature 9(3):161–169

International Commission on Zoological Nomenclature (2017) Declaration 45 – addition of recommendations to article 73 and of the term "specimen, preserved" to the glossary. Bull Zool Nomenclature 73(2–4):96–97

International Committee on Veterinary Gross Anatomical Nomenclature (2017) Nomina anatomica veterinaria, 6th edn. World Association of Veterinary Anatomists, Editorial Committee, Hanover, Ghent, Columbia, MO, Rio de Janeiro

Jäger GF (1839) Über die fossilen Säugethiere welche in Württemberg in verschiedenen Formationen aufgefunden worden sind, nebst geognostischen Bemerkungen über diese Formationen. Verlag Carl Erhard, Stuttgart, pp 1–214

Kahlke HD (1969) Die Rhinocerotiden-Reste aus den Kiesen von Süssenborn bei Weimar. Paläontolog Abh A 3(3/4):667–709

Kahlke RD, Lacombat F (2008) The earliest immigration of woolly rhinoceros (*Coelodonta tologoijensis*, Rhinocerotidae, Mammalia) into Europe and its adaptive evolution in Palaearctic cold stage mammal faunas. Quat Sci Rev 27(21–22):1951–1961

Kalandadze NN, Shapovalov AV, Tesakova EM (2009) On nomenclatural problems concerning the woolly rhinoceros *Coelodonta antiquitatis* (Blumenbach, 1799) [in Russian]. In: Shishkin MA, Tverdokhlebov VP (eds) Researches on paleontology and biostratigraphy of ancient continental deposits. Nauchnaya Kniga, Saratov, pp 98–111

Kampouridis P, Kargopoulos N, Svorligkou G, Giaourtsakis I, Roussiakis S, Theodorou G (2018) Rhinocerotidae (Mammalia, Perissodactyla) remains from the new Pleistocene locality of Richea (Laconia, Greece). In: Abstract Book of the 5th International Paleontological Congress, 9-13 July 2018, Paris, France, pp 1049

Kampouridis P, Roussiakis S, Kargopoulos N, Giaourtsakis I, Dimakopoulos G, Iliopoulos G, Svorligkou G, Theodorou G (2019) Faunal diversity at the Turolian locality of Kerassia (northern Euboea, Greece). In: Kranis H (ed) Proceedings of the 15th International Congress of the Geological Society of Greece, Harokopio University of Athens, 22-24 May 2019. Bull Geol Soc Greece Sp Pub 7:49–50

Kanellis A (1962) Die Höhle von Petralona. To Vouno (Das Gebirge). Zeitsch Deutsche-Griech Bergsteiger Verb Athen 227:91–102

Karadimas N (2013) George Finlay as a prehistoric archaeologist and the first paleontological research in Greece (Pikermi, 1836) [in Greek]. In: Stefanis A (ed) Proceedings of the 14th Scientific Meeting of Southeastern Attica. Society of Southeastern Attica Studies, Kalyvia Thorikou, pp 127–138

Karistineos NK (1984) Paleogeographic evolution of the Serres Basin. Lithostratigraphy, biostratigraphy and tectonics [in Greek]. Aristotle University of Thessaloniki. Fac Phys Math Sci Ann 23(sup. 10):1–230

Kaup J (1832a) Über *Rhinoceros incisivus* Cuv., und eine neue Art, *Rhinoceros schleiermacheri*. Isis von Oken Jg 1832(8):898–904

Kaup J (1832b) Fossile Säugethiere um Maynz. N Jahrb Geol Paläontol, Jg 1832:419–420

Kaup J (1834) Description d'ossements fossiles de mammifères inconnus jusqu'à présent, qui se trouvent au Muséum grand-ducal de Darmstadt; troisième cahier. JG Heyer, Darmstadt

Kaup J (1854) Beiträge zur näheren Kenntniss der urweltlichen Säugethiere, Erstes Heft. V. Leske Verlag, Darmstadt

Kaup J, Scholl J (1834) Verzeichnis der Gypsabgüsse von ausgezeichnesten urweltlichen. Thierresten des Grossherzoglichen Museums zu Darmstadt, Darmstadt, pp 1–28

Kaya T (1994) *Ceratotherium neumayri* (Rhinocerotidae, Mammalia) in the upper Miocene of Western Anatolia. Turkish J Earth Sci 3:13–22

Kaya T, Heissig K (2001) Late Miocene rhinocerotids (Mammalia) from Yulaı (Çorlu-Trace/Turkey). Geobios 34(4):457–467

Khomenko I (1913) Zamietka ob atseraterii iz okrestnoste Odessy (Notice on aceratheres from the surroundings of Odessa) [in Russian]. Ezheg Geol Mineral Rossii 15(7):202–203

Kiernik E (1914) Über einen Aceratheriumschädel aus der Umgebung von Odessa. Bull Int Acad Sci Cracovie, Classe Sci Math Nat (ser B, Sci Nat) 1913(9):808–864

Killgus H (1922a) Unterpliozäne Säuger aus China. Paläontol Z 5(3):251–257

Killgus H (1922b) Die Unterpliocänen chinesischen Säugetierreste der Tafelschen Sammlung zu Tübingen. Dissertation, Eberhard Karls Universität Tübingen

Kirillova IV, Chernova OF, van der Made J, Kukarskih VV, Shapiro B, van der Plicht J, Shidlovskiy FK, Heintzman PD, van Kolfschoten T, Zanina OG (2017) Discovery of the skull of *Stephanorhinus kirchbergensis* (Jäger, 1839) above the Arctic circle. Quat Res 88(3):537–550

Kokotini M, Kargopoulos N, Iliopoulos G, Roussiakis S, Skandalos P, Michailidis D, Svorligkou G, Kampouridis P, Theodorou G (2019) Karnezeika (Argolis, Peloponnese): preliminary data concerning a new Villafranchian locality of southern Greece. In: Kranis H (ed) proceedings of the 15th International Congress of the Geological Society of Greece, Harokopio University of Athens, 22-24 May 2019. Bull Geol Soc Greece Sp Pub 7:39–40

Koliadimou KK (1996) Paleontological and biostratigraphical study of the Neogene-quaternary micromammals from Mygdonia basin [in Greek]. Dissertation, School of Geology, Aristotle University, Thessaloniki, Greece

Konidaris GE, Tourloukis V, Kostopoulos DS, Thompson N, Giustia D, Michailidis D, Koufos GD, Harvati K (2015) Two new vertebrate localities from the early Pleistocene of Mygdonia Basin (Macedonia, Greece): preliminary results. Comptes Rendus Palevol 14:353–362

König HE, Liebich H-G (2004) Veterinary anatomy of domestic mammals: textbook and colour atlas, 4th edition. Schattauer Verlag, Stuttgart

Kormos T (1913) Bericht über meine ausländische Studienreise im Jahre 1911. Jahresber Ungar Geol Anst Jg 1911:272–291

Korotkevich EL (1958) A new *Chilotherium* species from the Sarmatian deposits of Ukraine [in Ukrainian]. Dopov Nac Akad Nauk Ukr RSR 12:1372–1376

Korotkevich EL (1970) The mammals of the Berislav late Sarmatian hipparion-fauna [in Ukrainian]. In: Pidoplichko I (ed) The natural environment and the fauna of the past, vol 5. Naukova Dumka, Kiev, pp 24–121

Korotkevich EL (1974) Novyy vid atserateriya iz rannego sarmata Ukrainy (A new species of acerathere from the Early Sarmatian of the Ukraine). [in Russian with English summary]. Vestn Zool 8:38–44

Kostopoulos DS (1994) *Microstonyx major* (Suidae, Artiodactyla) from the late Miocene locality of "Nikiti-1", Macedonia, Greece; some remarks about the species. Bull Geolog Soc Greece 30(1):341–355

Kostopoulos DS, Koliadimou KK, Koufos GD (1996) The giraffids (Mammalia, Artiodactyla) from the late Miocene mammalian localities of Nikiti (Macedonia, Greece). Palaeontographica Abt A 239(1–3):61–88

Kotarba MJ (2009) Interdisciplinary studies at Starunia palaeontological site and vicinity (Carpathian region, Ukraine) in the years 2006-2009: previous discoveries and research, purposes, results and perspectives. Ann Soc Geol Poloniae 79(3):219–241

Koufos GD (1980) Paleontological and stratigraphical study of the Neogene continental deposits of Axios Valley (Macedonia, Greece) [in Greek]. Dissertation, Aristotle Univ Thessaloniki. Fac Phys Math Sci Ann 19(sup. 11):1–322

Koufos GD (1981) A new late Pleistocene (Würmian) mammal locality from the basin of Drama (N. Greece). Aristotle University of Thessaloniki. Fac Phys Math Sci Ann 21:129–148

Koufos (2001) The Villafranchian mammalian faunas and biochronology of Greece. Boll Soc Paleontol Ital 40:217–223

Koufos GD (2006a) The Neogene mammal localities of Greece: faunas, chronology and biostratigraphy. Hellenic J Geosci 41:183–214

Koufos GD (2006b) Palaeoecology and chronology of the Vallesian (late Miocene) in the eastern Mediterranean region. Palaeogeogr Palaeoclimatol Palaeoecol 234:127–145

Koufos GD (2006c) The large mammals from the Miocene/Pliocene locality of Silata, Macedonia, Greece with implications about the latest Miocene palaeoecology. Beitr Paläontol 30:293–313

Koufos GD (2006d) The late Miocene vertebrate locality of Perivolaki, Thessaly, Greece. 5. Proboscidea, Rhinocerotidae. Palaeontogr Abt A 276:75–80

Koufos GD (2013) Neogene mammal biostratigraphy and chronology of Greece. In: Wang X, Flynn LJ, Fortelius M (eds) Fossil mammals of Asia. Neogene biostratigraphy and chronology. Columbia University Press, New York, pp 595–621

Koufos GD (2016a) Rhinocerotidae. In: Koufos GD, Kostopoulos DS (eds) Palaeontology of the upper Miocene vertebrate localities of Nikiti (Chalkidiki peninsula, Macedonia, Greece). Geobios 49:69–73

Koufos G (2016b) Neogene and quaternary continental biostratigraphy of Greece based on mammals. Bull Geol Soc Greece 50(1):55–64

Koufos GD, Kostopoulos DS (1997) Biochronology and succession of the Plio-Pleistocene macromammalian localities of Greece. In: Aguilar J-P, Legendre S, Michaux J (eds) Actes du Congrès BiochroM'97. Mémoires et travaux de l'Institut de Montpellier de l'Ecole Pratique des Hautes Etudes, vol 21, pp 619–634

Koufos GD, Kostopoulos DS (2013) First report of *Brachypotherium* Roger, 1904 (Rhinocerotidae, Mammalia) in the middle Miocene of Greece. Geodiversitas 35(3):629–641

Koufos GD, Kostopoulos DS (2016) The Plio-Pleistocene large mammal record of Greece: implications for early human dispersals into Europe. In: Harvati K, Roksandic M (eds) Paleoanthropology of the Balkans and Anatolia. Vertebrate paleobiology and paleoanthropology. Springer, Dordrecht, pp 269–280

Koufos G, Pavlides S (1988) Correlation between the continental deposits of the lower Axiosvalley and Ptolemais basin. Bull Geol Soc Greece 20:9–19

Koufos GD, Vasileiadou K (2015) Miocene/Pliocene mammal faunas of southern Balkans: implications for biostratigraphy and palaeoecology. Palaeobio Palaeoenv 95:285–303

Koufos GD, Syrides GE, Kostopoulos DS, Koliadimou K (1992) Apollonia, a new vertebrate site in the Pleistocene of the Mygdonia basin (Macedonia, Greece); the first fossil freshwater mollusks in the area. C R Seanc Acad Sci Paris (II) 315:1041–1046

Koufos GD, Syrides GE, Kostopoulos DS, Koliadimou KK, Sylvestrou IA, Seitanidis GC, Vlachos TD (1997) New excavations in the Neogene mammalian localities of Mytilinii, Samos Island, Greece. Geodiversitas 19(4):877–885

Koufos GD, Kostopoulos DS, Merceron G (2009a) The late Miocene mammal faunas of Mytilinii Basin, Samos Island, Greece: new collection. 17. Palaeoecology. Beit Paläontol 31:409–430

Koufos GD, Kostopoulos DS, Vlachou TD (2009b) The late Miocene mammal faunas of Mytilinii Basin, Samos Island, Greece: new collection. 16. Biochronology. Beit Paläontol 31:397–408

Koufos GD, Kostopoulos DS, Vlachou TD (2016) Revision of the Nikiti 1 (NKT) fauna with description of new material. In: Koufos GD, Kostopoulos DS (eds) Palaeontology of the upper Miocene vertebrate localities of Nikiti (Chalkidiki peninsula, Macedonia, Greece). Geobios 49:11–22

Krell F-T, Pape T (2015) Electronic publications need registration in ZooBank to be available. Bull Zool Nomenclature 72(3):245–251

Kretzoi M (1942a) Bemerkungen zum System der nachmiozänen Nashorn Gattungen. Földtani Közlöny 72:309–318

Kretzoi M (1942b) Präokkupierte und durch ältere zu ersetzende Säugetiernamen. Földtani Közlöny 72:345–349

Kretzoi (1977) The Fauna of Small Vertebrates of the Middle Pleistocene at Petralona. Anthropos 4:131–143

Kretzoi M, Kretzoi M (2000) Index generum et subgenerum Mammalium. In: Riegraf W (ed) Fossilium Catalogus I, Animalia pars 137. Sect. 1–2. Backhuys Publishers, Leide

Kretzschmar P, Kramer-Schadt S, Ambu L, Bender J, Bohm T, Ernsing M, Goeritz F, Hermes R, Payne J, Schaffer N, Thayaparan ST, Zainuddin ZZ, Hildebrandt TB, Hofer H (2016) The catastrophic decline of the Sumatran rhino (*Dicerorhinus sumatrensis harrissoni*) in Sabah: historic exploitation, reduced female reproductive performance and population viability. Global Ecol Conserv 6:257–275

Krokos VI (1914) Aceratherium simplex nov. sp. aus mäotischen Schichten in der Nähe des Dorfes Tudorowo im Bessarabischen Gouvernement, Ackermanischen Bezirks. [in Russian with German summary]. Separatum (Отдельный оттискъ) Zap Novorossijsk Obsc Estestvoisp 41, Kommercheskaya Typohrafiya BY Sapozhnykova, Odessa, pp 1–15

Krokos VI (1915) Rhinoceros longirostris nov. sp. from Pliocen south of Bessarabia [in Russian with English summary]. Separatum (Отдельный оттискъ) Zap Novorossijsk Obsc Estestvoisp 41, Kommercheskaya Typohrafiya BY Sapozhnykova, Odessa, pp 1–19

Krokos VI (1917) *Aceratherium Schlosseri* Web. du village de Grebeniki du gouvernement de Kherson. Zap Obsc Sel Choz Juz Rossij 87(2):1–96

Lacombat F (2005) Les Rhinocéros fossiles des sites préhistoriques de l'Europe méditerranéenne et du Massif Central, Paléontologie et implications biochronologiques. Br Archaeol Res Int Ser 1419:1–175

Lacombat F (2007) Phylogeny of the genus *Stephanorhinus* in the Plio-Pleistocene of Europe. Hallesches Jb Geowiss 23:63–64

Lacombat F, Mörs T (2008) The northernmost occurrence of the rare late Pleistocene rhinoceros *Stephanorhinus jeanvireti* (Mammalia, Perissodactyla). N Jahrb Geol Paläontol 249(2):157–165

Lartet E (1836) Nomenclature des mammifères et des coquilles qu'il a trouvés dans un terrain d'eau douce près de Simorre et de Sansan (Gers). Bull Soc Géol Fr 7:217–220

Lartet E (1837a) Note sur les ossements fossiles des terrains tertiaires de Simorre, de Sansan, etc., dans le département du Gers, et sur la découverte recente d'une mâchoire de singe fossile. C R hebd séances Acad sci Paris 4:85–93

Lartet E (1837b) Nouvelles observations sur une mâchoire inférieure fossile, crue d'un singe voisin du gibbon, et sur quelques dents et ossements attribués à d'autres quadrumanes. C R hebd séances Acad sci Paris 4:583–584

Lartet E (1837c) Sur les débris, fossiles trouvés à Sansan, et sur les animaux antédiluviens en général. C R hebd séances Acad sci Paris 5:158–159

Lartet E (1851) Notice sur la colline de Sansan, suivie d'une récapitulation des diverses espèces d'animaux vertébrés fossiles trouvés soit à Sansan, soit dans d'autres gisements du terrain Miocène dans le bassin sous-pyrénéen. JA Portes, Auch

Laurie WA, Lang EM, Groves CP (1983) *Rhinoceros unicornis*. Mamm Species 211:1–6

Laurillard F (1848) Rhinocéros fósiles. In: Orbigny CDD (ed) Dictionnaire universel d'Histoire naturelle, vol 11. Renard, Martinet & Cie édit, Paris, pp 99–102

Lehmann U (1984) Notiz über Säugetierreste von der Insel Samos in der Sammlung des Geologisch-Palaontologischen Instituts und Museums Hamburg. Mitt Geol-Paläontol Inst Univ Hamburg 57:147–156

Leonardi P (1947) Resti fossili inediti di rinoceronti conservati nelle collezioni dell' Istituto Geologico dell' Università di Padova. Mem Inst Geol Univ Padova 15:3–30

Lesson RP (1838) Compléments de Buffon, Races humaines et mammifères, 2ème ed., P. Pourrat Freres (editeurs), Paris, vol 1, pp 1–660

Linnaeus C (1758) Systema Naturae. Editio decima, reformata, tomus I. Impensis Direct. Laurentii Salvii, Holmiae

Loose H (1975) Pleistocene Rhinocerotidae of W. Europe with reference to the recent two-horned species of Africa and S. E. Asia. Scripta Geol 33:1–59

Lu X (2013) A juvenile skull of *Acerorhinus yuanmouensis* (Mammalia: Rhinocerotidae) from the late Miocene hominoid fauna of the Yuanmou Basin (Yunnan, China). Geobios 46:539–548

Lubicz-Niezabitowski E (1912) *Teleoceras ponticus* n. sp. Vorläufge Notiz. Nowy Targ, pp 1–2

Lubicz-Niezabitowski E (1913) Über das Schädelfragment eines Rhinocerotiden (*Teleoceras ponticus* Niez.) von Odessa. Bull Int Acad Sci Cracovie, Classe Sci Math Nat (ser B, Sci Nat) 1913(5):223–235

Lucas SG, Sobus JC (1989) The systematics of Indricotheres. In: Prothero DR, Schoch RM (eds) The evolution of Perissodactyls. Oxford monographs on geology and geophysics, vol 15. Clarendon Press, Oxford University Press, New York, pp 358–378

Lydekker R (1881) Siwalik Rhinocerotidae (Indian Tertiary and Post-Tertiary Vertebrata). Mem Geol Surv India, Palaeont Indica Ser 10, vol 2(1):1–62

Lydekker R (1886) Catalogue of the fossil Mammalia in the British museum (natural history), vol 3. Taylor and Francis, London

Lydekker R (1908) The white rhinoceros. The Field (London) 111(2878):319

Lyras G (2007) Museum labels and fieldnotes. Cranium 28(2):45–47

Malez M, Brajkovic D (1991) Late Pleistocene mammals from the alluvium of the Drava River (Croatia) [in Croatian]. Inst Geol Istrazivanja Zagreb, Geol Vjesnik (Paleontol Stratigr) 44:31–37

Marinos G (1965) Beiträge zur Kenntnis der Verbreitung des Pleistozäns in Mazedonien. Aristotle Univ Thessaloniki. Fac Phys Math Sci Ann 9:95–111

Marinos G, Symeonidis NK (1973) Neue Fossilfundstellen und neue Ergebnisse aus Pikermi (Attika) [in Greek]. Bull Geol Soc Greece 9(2):160–176

Marinos G, Symeonidis N (1974) Neue Funde aus Pikermi (Attika, Griechenland) und eine allgemeine geologische Übersicht dieses paläontologischen Raumes [in Greek]. Ann Géol Pays Helléniques 26:1–27

Markovic A (1998) Woolly rhinoceros *Coelodonta antiquitatis* (Blumenbach, 1803) from Pleistocene of Serbia. Bull Natur Hist Mus Belgrade A 47–50:217–237

Matsumoto H (1921) Descriptions of some new fossil mammals from Kani District, Prov. of Mino, with revisions of some Asiatic fossil Rhinocerotids. Sci Rep Tohoku Imp Univ, 2nd Ser, Geol 5(3):75–91

Mayor A (2000) The first fossil hunters: paleontology in Greek and Roman times. Princeton University Press, Princeton

Mazza P (1988) The Tuscan early Pleistocene rhinoceros *Dicerorhinus etruscus*. Palaeontogr It 75:1–87

Mazza P, Sala B, Fortelius M (1993) A small latest Villafranchian (late Early Pleistocene) rhinoceros from Pietrafitta (Perugia, Umbria, Central Italy), with notes on the Pirro and Westerhoven rhinoceroses. Palaeontographia Italica 80: 25–50

McKenna MC, Bell SK (1997) Classification of mammals above the species level. Columbia University Press, New York

Meiburg P, Siegfried P (1970) Katalog der Typen und Belegstücke zur Paläozoologie im Geologisch-Paläontologischen Institut der Westfälischen Wilhelms-Universität Münster. II. Teil: Vertebrata. Münstersche Forsch Geol Paläontol 15:1–84

Melentis JK (1964) Die fossilen Rhinocerotiden, Hippopotamiden und andere Säugetiere aus dem Becken von Megalopolis im Peloponnes [in Greek]. Proc Athens Akad 39:388–400

Melentis JK (1965) Die pleistozänen Nashörner des Beckens von Megalopolis in Peloponnes (Griechenland). Ann Géol Pays Helléniques 16:363–402

Melentis JK (1968a) Palaeontological excavations at Samos Island, a preliminary report [in Greek]. Proc Athens Academy 43:344–349

Melentis JK (1968b) Die Pikermifauna von Halmyropotamos (Euboea - Griechenland) 1. Teil: Odontologie und Craniologie. Ann Géol Pays Helléniques 19:83–341

Melentis JK (1969) Die Pikermifauna von Halmyropotamos (Euboea - Griechenland) 2. Teil: Osteologie. Ann Géol Pays Helléniques 21:217–306

Mestre L (1934) *Rhinoceros brachypus* de Simorre. Bull Soc Hist Natur Toulouse 66:309–318

Mitzopoulos MK (1960) Die Hipparionfauna von Tanagra bei Theben. Proc Athens Acad 34:295–309

Mitzopoulos MK (1961) Die Hipparionfauna von Tanagra bei Theben. Ann Géol Pays Helléniques 12:301–314

Mukhlisi M, Ningsih TS, Sari U, Kurniawan Y, Setiawan R, Muslim A (2016) Habitat utilization of the Sumatran rhinos (*Dicerorhinus sumatrensis harrissoni*) in Kutai Barat forest, East Kalimantan, Indonesia. Biodiversitas 19(5):1842–1850

Murchison C (ed) (1868) Paleontological memoires and notes of the late Hugh Falconer, vol 2. R. Hardwicke, London

Nikolov I, Heissig K (1985) Fossile Säugetiere aus dem Obereozän und Unteroligozän Bulgariens und ihre Bedeutung für die Palaogeographie. Mitt Bayerische Staats Paläontol Hist Geol 25:61–79

Nougier L-R, Robert R (1957) Le Rhinocéros dans l'art franco-cantabrique occidental (in: Le Rhinocéros dans l'Art Quaternaire). Bull Soc préhist Ariège, Tarascon sur Ariège 7:16–54
Nowak J, Panow E, Tokarski J, Szafer W, Stach J (1930) The second woolly rhinoceros (*Coelodonta antiquitatis* Blum.) from Starunia. Poland. Bull Int Acad polonaise Sci Lett, Cl Sci Math Natur B 1930:1–47
Orlando L, Leonard JA, Thenot A, Laudet V, Guérin C, Hännia C (2003) Ancient DNA analysis reveals woolly rhino evolutionary relationships. Mol Phylogenet Evol 28:485–499
Orlova LA, Vasilev SK, Kuzmin YV, Kosintsev PA (2008) New data on the time and place of extinction of the woolly rhinoceros *Coelodonta antiquitatis* Blumenbach, 1799. Doklady Biol Sci 423:403–405
Osborn HF (1900) Phylogeny of the rhinoceroses of Europe. Bull Am Mus Natur Hist 12:229–267
Osborn HF (1923) Baluchiterium grangeri, a giant hornless rhinoceros from Mongolia. Am Mus Nov 78:1–15
Owen R (1846) A history of British fossil mammals and birds. John van Hoorst, London
Pallas PS (1769) De ossibus Sibiriae fossilibus craniis praesertim rhinocerotum atque buffalorum, observationes. N Comm Acad Sci Imp Petropolitanae 13:436–477
Pallas PS (1773) De reliquis animalium exoticorum per Asiam borealem repertis. N Comm Acad Sci Imp Petropolitanae 17:576–606
Palmer TS (1904) Index generum mammalium: a list of the genera and families of mammals. US Department of Agriculture, Division of Biological Survey: north American Fauna, no. 23. Government Printing Office, Washington
Panagopoulou E, Karkanas P, Tsartsidou G, Kotjabopoulou E, Harvati K, Ntinou M (2004) Late Pleistocene archaeological and fossil human evidence from Lakonis cave, southern Greece. J Field Archaeol 29(3/4):323–349
Pandolfi L, Tagliacozzo A (2013) Earliest occurrence of woolly rhino (*Coelodonta antiquitatis*) in Italy (late Pleistocene, Grotta Romanelli site). Riv It Paleontol Stratigr, Milano 119(1):125–129
Pandolfi L, Tagliacozzo A (2015) *Stephanorhinus hemitoechus* (Mammalia, Rhinocerotidae) from the late Pleistocene of Valle Radice (Sora, Central Italy) and re-evaluation of the morphometric variability of the species in Europe. Geobios 48(2):169–191
Pandolfi L, Grossi F, Frezza V (2013) A Miocene Aceratheriine Rhinocerotid (Mammalia, Perissodactyla) from early Pleistocene marine deposits at Monte delle Piche (Rome, Central Italy). Riv It Paleontol Stratigr 119(3):401–405
Pandolfi L, Gasparik M, Piras P (2015) Earliest occurrence of "*Dihoplus*" *megarhinus* (Mammalia, Rhinocerotidae) in Europe (late Miocene, Pannonian Basin, Hungary): Palaeobiogeographical and biochronological implications. Ann Paléontol 101:325–339
Pandolfi L (2016) *Persiatherium rodleri*, gen. et sp. nov. (Mammalia, Rhinocerotidae) from the upper Miocene of Maragheh (northwestern Iran). J Vertebr Paleontol 36(1), https://doi.org/10.1080/02724634.2015.1040118
Pandolfi L, Cerdeno E, Codrea V, Kotsakis T (2017) Biogeography and chronology of the Eurasian extinct rhinoceros *Stephanorhinus etruscus* (Mammalia, Rhinocerotidae). C R Palevol 16:762–773
Pandolfi L, Codrea VA, Popescu A (2019a) *Stephanorhinus jeanvireti* (Mammalia, Rhinocerotidae) from the early Pleistocene of Coltesti (southwestern Romania). C R Palevol 18(8):1041–1056
Pandolfi L, Marra AC, Carone G, Maiorino L, Rook L (2019b) A new rhinocerotid (Mammalia, Rhinocerotidae) from the latest Miocene of southern Italy. Hist Biol:1. https://doi.org/10.1080/08912963.2019.1602615
Paraskevaidis E (1953) Canis und Equus aus Attika. N Jahrb Geol Paläontol Abh 96(3):453–496
Paraskevaidis E (1977) Säugetierreste aus Griechenland. In: Kallergis G (ed) Proceedings of the VIth Colloquium on the Geology of the Aegean Region, vol 3, pp 1143–1154
Pavlides SB (1985) Neotectonic evolution of Florina-Vegoritida-Ptolemais Basin (E. Macedonia) [in Greek]. Dissertation, Aristotle University of Thessaloniki. Fac Phys Math Sci Ann 23(sup 43):1–265
Pavlow MW (1913) Mammifères tertiaires de la nouvelle Russie, 1. Partie: Artiodactyla, Perissodactyla (*Aceratherium kowalevskii* n.s.). N Mém Soc imp Natur Moscou 17:1–68

Pethő G (1884) Über die Fossilien Säugethier-Überreste von Baltavár. Földtani Intézet Évi Jelentése 1884:63–73
Petronijevic Z, Thenius E (1957) Über den ersten Nachweis von Indricotherien (Baluchitherien, Rhinocerotidae, Mammalia) im Tertiär von Europa. Anz Math Naturwiss Kl, Österr Akad Wiss 9:153–155
Piccoli G, Franco F, Bertolotti O, Bimbatti M, Buja P, Cesga L, Gradenico M (1975) I resti di mammiferi del Neogene di Samos e Pikermi (Grecia) conservationel museo geologico e paleonto logico dell'università di Padova. Mem Inst Geol Univ Padova 31:1–39
Pohlig H (1886) On the Pliocene of Maragha, Persia, and its resemblance to that of Pikermi in Greece; on fossil elephant remains of Caucasia and Persia; and on the results of a monograph of the fossil elephants of Germany and Italy. Quart J Geol Soc London 42:177–182
Pomel NA (1853) Catalogue methodique et descriptif des vertebres fossiles decouvertes dans le bassin hydrographique superieur de la Loire, et surtout dans la vallee de son affluent principal. Ann Sci Litt Ind Auvergne 26:81–176
Pomel A (1895) Les rhinocéros quaternaires. Carte Géologique de l'Algérie Paléontologie Monographies), Imprimerie P Fontana & Co, Alger, pp 1–49
Poulianos NA (1981) New rhino findings from the Crenian period of Petralona cave. Anthropos 8:84–90
Prothero DR (1998) Rhinocerotidae. In: Janis CM, Scott KM, Jacobs LL (eds) Evolution of the tertiary mammals of North America. Terrestrial carnivores, ungulates, and Ungulatelike mammals, vol 1. Cambridge University Press, Cambridge, pp 595–605
Prothero DR (2005) The evolution of north American rhinoceroses. Cambridge University Press, Cambridge
Prothero DR, Schoch RM (1989) Classification of the Perissodactyla. In: Prothero DR, Schoch RM (eds) The evolution of Perissodactyls. Oxford Monographs on Geology and Geophysics, vol 15. Clarendon Press, Oxford University Press, New York, pp 530–537
Prothero DR, Schoch RM (2002) Horns, tusks & flippers: the evolution of hoofed mammals. John Hopkins University Press, Baltimore & London
Prothero DR, Manning E, Hanson BCM (1986) The phylogeny of the Rhinocerotoidea (Mammalia, Perissodactyla). Zool J Linnean Soc 87:341–366
Prothero DR, Guérin C, Manning E (1989) The history of the rhinocerotoidea. In: Prothero DR, Schoch RM (eds) The evolution of Perissodactyls. Oxford monographs on geology and geophysics, vol 15. Clarendon Press, Oxford University Press, New York, pp 321–340
Psarianos P (1958) Neue Rhinocerotidenfunde aus dem Tertiär und Quartär von Mazedonien (Griechenland). Proc Athens Acad 33:303–312
Qiu ZD (1979) Some mammalian fossils from the Pliocene of Inner Mongolia and Gansu (Kansu) [in Chinese with English summary]. Vertebr PalAsiatica 17:222–235
Qiu Z, Xie J, Yan D (1988) A new chilothere skull from Hezheng Gansu, China – with special reference to the Chinese "*Diceratherium*". Sci Sin 31(4):493–502
Radinsky LB (1966) The families of the Rhinocerotoidea (Mammalia, Perissodactyla). J Mammal 47(4):631–639
Ringström T (1924) Nashörner der Hipparion-Fauna Nord-Chinas. Palaeontol Sinica C 1(4):1–156
Roger O (1896) Verzeichniss der bisher bekannten fossilen Säugethiere. Berichte Naturwiss Vereins für Schwaben und Neuburg 32:1–272
Roger O (1904) Wirbeltierreste aus dem Obermiocän der bayerisch-schwäbischen Hochebene. V Teil Berichte Naturwiss Vereins für Schwaben und Neuburg 36:1–21
Rookmaaker LC (1980) The distribution of the rhinoceros in eastern India, Bangladesh, China and the indo-Chinese region. Zool Anz 205(3/4):253–268
Rookmaaker LC (1997) Records of the Sundarbans rhinoceros (*Rhinoceros sondaicus inermis*) in India and Bangladesh. Pachyderm 24:37–45
Rookmaaker LC (2006) Distribution and extinction of the rhinoceros in China: review of recent Chinese publications. Pachyderm 40:102–106

Rookmaaker LC, Antoine PO (2012) New maps representing the historical and recent distribution of the African species of rhinoceros: *Diceros bicornis*, *Ceratotherium simum* and *Ceratotherium cottoni*. Pachyderm 52:91–96

Rookmaaker LC, Groves CP (1978) The extinct cape Rhinoceros, *Diceros bicornis bicornis* (Linnaeus, 1758). Säugetierkundliche Mitt 26(2):117–126

Roth J, Wagner A (1854) Die fossilen Knochenüberreste von Pikermi in Griechenland. Abh Bayer Akad Wiss (Math-Phys Kl) 7(2):371–464

Ryder OA, Friese C, Greely HT, Sandler R, Saragusty J, Durrant BS, Redford KH (2020) Exploring the limits of saving a subspecies: The ethics and social dynamics of restoring northern white rhinos (*Ceratotherium simum cottoni*). Conserv Sci Pract 2(8):e241. https://doi.org/10.1111/csp2.241

Sakellariou-Mane H, Psilovikos A, Koufos GD (1979) Contribution to the study of Villafranchian in N. Chalkidiki. Aristotle Univ Thessaloniki. Fac Phys Math Sci Ann 19:279–296

Schaller O (2007) Illustrated veterinary anatomical nomenclature, 2nd edn. Enke Verlag, Stuttgart

Schlosser M (1903) Die fossilen Säugethiere Chinas nebst einer Odontographie der recenten Antilopen. Abh Bayer Akad Wissen (Math-Phys Kl) 22(1):1–222

Schlosser M (1904) Die fossilen Cavicornia von Samos. Beitr Paläontol Geol Österreich-Ungarns 17(1-2):21–118

Schlosser M (1921) Die Hipparionenfauna von Veles in Mazedonien. Abh Bayerischen Akad Wiss 29(4):1–55

Schmidt-Kittler N, de Bruijn H, Doukas C (1995) The vertebrate locality Maramena (Macedonia, Greece) at the Turolian/Ruscinian boundary (Neogene). 1. General introduction. Munch Geowiss Abh 28:9–18

Schubert GH (1835) Die Geschichte der Natur: als zweite, gänzlich umgearbeitete Auflage der allgemeinen Naturgeschichte. JJ Palm and Ernst Ente, Erlangen, vol 1:1–593

Sclater WL (1900) The mammals of South Africa: primates, carnivora and ungulata, vol 1. RH Porter, London

Scott WB, Osborn HF (1883) On the skull of the Eocene rhinoceros, *Orthocynodon*, and the relation of this genus to other members of the group. Contr EM Mus Geol Archaeol Princeton College Bull 3(1):1–22

Shidlovskiy FK, Kirillova IV, Wood J (2011) Horns of the woolly rhinoceros *Coelodonta antiquitatis* (Blumenbach, 1799) in the ice age museum collection (Moscow, Russia). Quat Int 255:125–129

Shikama T, Hasegawa Y, Okafuji G (1967) On a rhinocerid skull from Isa, Yamaguchi Prefecture, Japan. Bull National Sci Mus Tokyo 10:455–462

Sickenberg O (1964) Die Säugetierfauna der Höhle Petralona bei Thessaloniki (Vor. Mit.). Geol Geophys Res (Athens) 9:1–16

Sickenberg O (1968) Die Pleistozänen Knochenbrezien von Volax (Griech.-Mazedonien). Geol Jahrb 85:33–54

Sickenberg O (1971) Revision der Wirbeltier fauna der Höhle Petralona (Griech., Mazedonien). Ann Géol Pays Helléniques 23:230–264

Sickenberg O (1975) Eine Säugetierfauna des Tieferen Bihariums aus dem Becken von Megalopolis (Peloponnes, Griechenland). Ann Géol Pays Helléniques 27:25–73

Skouphos TG (1905) Über die palæontologischen Ausgrabungen in Griechenland in Beziehung auf das Vorhandensein des Menschen. In: Comptes Rendus du Congrès International d'Archéologie, 1re session. Athènes 1905:231–236

Solounias N (1981) Mammalian fossils of Samos and Pikermi. Part 2. Resurrection of a classic Turolian fauna. Ann Carnegie Mus 50:231–270

Solounias N, Plavcan M, Quade J, Witmer L (1999) The Pikermian biome and the savanna myth. In: Agusti J, Andrews P, Rook L (eds) Evolution of the Neogene terrestrial ecosystems in Europe. Cambridge University Press, Cambridge, pp 427–444

Solounias N, Rivals F, Semprebon GM (2010) Dietary interpretation and paleoecology of herbivores from Pikermi and Samos (late Miocene of Greece). Paleobiol 36:113–136

Spassov N, Geraads D, Hristova L, Markov GN, Garevska B, Garevski R (2018) The late Miocene mammal faunas of the republic of Macedonia (FYROM). Palaeontographica Abt A 311(1–6):1–85

Steensma KJ (1988) Plio-Pleistozäne Grosssäugetiere (Mammalia) aus dem Becken von Kastoria/Grevena, südlich von Neapolis NW Griechenland. Dissertation, Mathematisch. Naturwissenschaftlichen Fakultät der Technischen Universität, Clausthal, pp 1–315

Stuart AJ, Lister AM (2012) Extinction chronology of the woolly rhinoceros *Coelodonta antiquitatis* in the context of late quaternary megafaunal extinctions in northern Eurasia. Quat Sci Rev 51:1–17

Studer T (1911) Eine neue Equidenform aus dem Obermiocän von Samos. Verh deutsch Zool Ges 20–21:192–200

Svorligkou G, Giaourtsakis IX, Roussiakis S (2019) New material of the hornless rhinocerotid *Chilotherium* (Mammalia, Perissodactyla) from the Turolian Fauna of Samos Island, Greece. In: Kranis H (ed) Proceedings of the 15th International Congress of the Geological Society of Greece, Harokopio University of Athens, 22-24 May 2019. Bull Geol Soc Greece Sp Pub 7:61–62

Sylvestrou I (2002) Stratigraphic, palaeontological and palaeogeographic study of the Neogene-quaternary deposits of Katerini Basin, northern Greece [in Greek]. Dissertation, Aristotle University of Thessaloniki, School of Geology. Sci Ann 67:1–366

Symeonidis NK (1974) Ein bemerkenswerter Wirbeltierfund aus dem Lignit von Atalanti (Fthiotis, Griechenland). Ann Géol Pays Helléniques 26:306–313

Symeonidis NK (1992) Lower Pleistocene (Villafranchian) fossil mammals from the Sesklo basin (Volos) [in Greek]. Ann Géol Pays Helléniques 35:1–41

Symeonidis NK, de Vos J (1976) Großsäugerfunde aus den Pleistozänen Spaltenfüllungen von Turkovunia in Athen. Ann Géol Pays Helléniques 28:135–144

Symeonidis NK, Theodorou GE, Schütt H, Velitzelos E (1986) Paleontological and stratigraphical observations in the area of Achaea and Aetolia-Acarnania (W. Greece). Ann Géol Pays Helléniques 33(1):329–365

Symeonidis NK, Giaourtsakis IX, Seemann R, Giannopoulos VI (2006) Aivaliki, a new locality with fossil rhinoceroses near Alistrati (Serres, Greece). Beit Paläontol 30:437–451

Syrides GE (1990) Lithostratigraphic, biostratigraphic and palaeogeographic study of the Neogene-quaternary sedimentary deposits of Chalkidiki peninsula, Macedonia, Greece. Aristotle University of Thessaloniki, School of Geology. Sci Ann 11:1–243

Thenius E (1955) Zur Kenntnis der unterpliozänen *Diceros* Arten. Ann Naturhist Mus Wien 60:202–209

Thenius E (1956) Über das Vorkommen von *Diceros pachygnathus* (Wagner) im Pannon (Unter-Pliozaen) des Wiener Beckens. N Jahrb Geol Palaeontol Monats 1:35–39

Theodorou GE, Roussiakis SJ, Athanassiou A (1995) Contribution to the study of the terrestrial Neogene of Greece. Artiodactyla and Rhinocerotidae from the Kerassia and Chalkoutsi localities. In: 10th Congress of the Regional Committee on Mediterranean Neogene stratigraphy, Bucharest, 4-9 September 1995. Roman J Stratigr 76(sup. 7):129–130

Theodorou G, Athanassiou A, Roussiakis S, Iliopoulos G (1998) Preliminary results on the recent excavations of the Kerassia locality (Euboea). In: Abstracts of interim-colloquium/RCMNS "Mediterranean Neogene Cyclostratigraphy in marine-continental palaeoenvironments", Patras-Greece, 27-29 May 1998, p 66

Theodorou G, Athanassiou A, Roussiakis S, Iliopoulos G (2001) Remarks on the late Miocene vertebrates of Kerassia (northern Euboea, Greece). In: Reumer WF, van Dam JE, Doukas CS, van der Meulen AJ, Meulenkamp JE, Wessels W (eds) Distribution and migration of tertiary mammals in Eurasia. A three-day international conference in honour of Hans de Bruijn, University of Utrecht, 17-19 May 2001, abstracts, pp 44–45

Theodorou G, Athanassiou A, Roussiakis S, Iliopoulos G (2003) Preliminary remarks on the late Miocene herbivores of Kerassiá (northern Euboea, Greece). In: Reumer JWF, Wessels W (eds) Distribution and migration of tertiary mammals in Eurasia. Deinsea 10:519–530

Tissier J, Becker D, Codrea V, Costeur L, Farcas C, Solomon A, Venczel M, Maridet O (2018) New data on Amynodontidae (Mammalia, Perissodactyla) from Eastern Europe - phylogenetic and palaeobiogeographic implications around the Eocene-Oligocene transition. PLoS One 13(4):e0193774. https://doi.org/10.1371/journal.pone.0193774

Tong H (2012) Evolution of the non-*Coelodonta* dicerorhine lineage in China. C R Palevol 11(8):555–562

Toula F (1901) Das Nashorn von Hundsheim bei Deutsch-Altenburg in Niederösterreich. Verh k-k Geol Reichsanst 1901(15):309–311

Toula F (1906) Das Gebiss und Reste der Nasenbeine von *Rhinoceros* (*Ceratorhinus* Osborn) *hundscheimensis*. Abh Kaiser-König Geol Reichsanstalt 20(2):1–38

Trantalidou K (2013) On the banks of the river Angitis. Environmental data for the basin of Drama in prehistoric times [in Greek]. In: Proceeding of the 5th conference on Drama and its environment, history, and civilization, Drama, pp 41–156

Tsiskarishvili GV (1987) Pozdnetvetichnye nosorogi (Rhinocerotidae) Kavkaza (Late Tertiary rhinoceroses (Rhinocerotidae) of the Caucasus) [in Russian with English summary]. Akademya Nauk Gruzinskoy SSR, Gosudarstvennyy Muzey Gruzii, Izdatel'stvo Metsnierba, Tbilisi, pp 1–141

Tsoukala E (1989) Contribution to the study of the Pleistocene fauna of large mammals (Carnivora, Perissodactyla, Artiodactyla) from Petralona cave, Chalkidiki (N. Greece) [in Greek]. Dissertation. Aristotle University of Thessaloniki, School of Geology. Sci Ann 8:1–360

Tsoukala E (1991a) Contribution to the study of the Pleistocene fauna of large mammals (Carnivora, Perissodactyla, Artiodactyla) from Petralona cave, Chalkidiki (N. Greece). Preliminary report. C R Hebd Séances Acad Sci Paris 312(2):331–336

Tsoukala E (1991b) A *Coelodonta antiquitatis praecursor* (Mammalia, Rhinocerotidae, zone 24) from the lower Axios valley deposits (Gephyra, Macedonia, N. Greece). Bull Geol Soc Greece 25(2):473–485

Tsoukala E (1992) Quaternary faunas of Greece. Cour Forsch-Inst Senckenberg 153:79–92

Tsoukala E (2000) Remains of a Pliocene *Mammut borsoni* (Hays, 1834) (Proboscidea, Mammalia), from Milia (Grevena, W. Macedonia, Greece). Ann Paléont 86(3):65–191

Tsoukala E (2005) The pursuit of elephants in Grevena (W. Macedonia, Greece). In: Hellenic Speleological Society (publisher), Proceedings of the 14th international Congress of Speleology, 21-28 August 2005, Kalamos, Greece, vol 1: 114–117

Tsoukala E (2018) Rhinocerotidae from the late Miocene and late Pliocene of Macedonia, Greece: a revision of the Neogene-quaternary Rhinocerotidae of Greece. Rev Paléobiol 37(2):609–630

Tsoukala E, Chatzopoulou K (2005) A new early Pleistocene (latest Villafranchian) site with mammals in Kalamotó (Mygdonia Basin, Macedonia, Greece)—preliminary report. Mitt Komm Quartärforsch Österreich Akad Wiss 14:213–233

Tsoukala E, Guérin C (2016) The Rhinocerotidae and Suidae of the middle Pleistocene from Petralona cave (Macedonia, Greece). Acta Zool Bulgarica 68(2):243–264

van Asperen EN, Kahlke R-D (2015) Dietary variation and overlap in central and northwest European *Stephanorhinus kirchbergensis* and *S. hemitoechus* (Rhinocerotidae, Mammalia) influenced by habitat diversity. C R Palevol 18(2):191–208

van de Weerd A, Reumer JWF, de Vos J (1982) Pliocene mammals from Apolakkia formation (Rhodes, Greece). Proc Kon Ned Akad Wetensch B 85:89–112

van der Made J (2010) The rhinos from the middle Pleistocene of Neumark-Nord (Saxony-Anhalt). In: Mania D (ed) Ein interglaziales Ökosystem des mittelpaläolithischen Menschen Veröffentlichungen des Landesamtes für Denkmalpflege Archäol Sachsen-Anhalt, vol 62, pp 433–527

van der Made J, Grube R (2010) The rhinoceroses from Neumark-Nord and their nutrition. In: Meller H (ed) Elefantenreich–Eine Fossilwelt in Europa. Landesamt für Denkmalpflege und Archäologie Sachsen Anhalt. Landesmuseum für Vorgeschichte, Halle (Saale), pp 383–398

van der Meulen A, van Kolfschoten T (1986) Review of the late Turolian to early Biharian mammal faunas from Greece and Turkey. Mem Soc Geol Ital 31:201–211

Vangengeym EA, Belyaeva EI, Garutt VE, Dmitrieva EL. Zazhigin VS (1966) Mlekopitayushchie Eopleystozena Zapadnogo Zabaykal'ya (Eopleistocene Mammals of Western Transbaikalia) [In Russian]. Trudy Geolog Inst Akad Nauk SSSR 152:1–162

Vangengeim EA, Tesakov AS (2008a) Maeotian mammalian localities of eastern Paratethys: Magnetochronology and position in European continental scales. Stratigr Geol Correlation 16(4):437–450

Vangengeim EA, Tesakov AS (2008b) Late Sarmatian mammal localities of the eastern Paratethys: stratigraphic position, Magnetochronology, and correlation with the European continental scale. Stratigr Geol Correlation 16(1):92–103

Villalta JF, Crusafont M (1955) *Chilotherium quintanelensis* Zbysz., sinonimo de *Hispanotheriurn matritensis* (Prado). Notas Com Inst Geol Minero Esp 37:23–31

Viret J (1958) Perissodactyla. In: Piveteau J (ed) Traité de Paléontologie, vol 4. Masson et Cie, Paris, pp 368–475

Vlachos E, Tsoukala E (2014) *Testudo* cf. *graeca* from the new late Miocene locality of Platania (Drama basin, N. Greece) and a reappraisal of previously published specimens. Bull Geol Soc Greece 48:27–40

Vlachos E, Tsoukala E, Cregut-Bonnoure E, Guérin C, Mol D (2018) The paradise lost of Milia (Grevena, Greece; late Pliocene, early Villafranchian, MN15/MN16a): faunal composition and diversity. Quat MDPI 1(13):1–18

von Koenigswald W, Holly Smith B, Keller T (2007) Supernumerary teeth in a subadult rhino mandible (*Stephanorhinus hundsheimensis*) from the middle Pleistocene of Mosbach in Wiesbaden (Germany). Palaeontol Z 81(4):416–428

Wagner A (1848) Urweltliche Säugetiere-Überreste aus Griechenland. Abh Bayer Akad Wiss (Math-Phys Kl) 5(2):335–378

Wagner A (1857) Neue Beiträge zur Kenntniss der fossilen Säugthier-Überreste von Pikermi. Abh Bayer Akad Wiss (Math-Phys Kl) 8(1):109–158

Wagner A (1861) Nachträge zur Kenntnis der fossilen Hufthier- Überreste von Pikermi. Sitz Ber Bayer Akad Wiss, Jg 1861(2):78–82

Wall WP (1989) The phylogenetic history and adaptive radiation of the Amynodontidae (Perissodactyla, Rhinocerotoidea). In: Prothero DR, Schoch RM (eds) The evolution of Perissodactyls. Oxford Monographs on Geology and Geophysics, vol 15. Clarendon Press, Oxford University Press, New York, pp 341–354

Wang KM (1931) Die Fossilen Rhinocerotiden von Choukoutien. Contrib Nat Res Inst Geol Acad Sin (Nanking) 1:69–84

Wang H, Bai B, Meng J, Wang Y (2016) Earliest known unequivocal rhinocerotoid sheds new light on the origin of Giant Rhinos and phylogeny of early rhinocerotoids. Sci Rep 6:3960. https://doi.org/10.1038/srep39607

Weber M (1904) Über tertiäre Rhinocerotiden von der Insel Samos I. Bull Soc Imp Naturalistes Moscou 17(4):477–501

Weber M (1905) Über tertiäre Rhinocerotiden von der Insel Samos II. Bull Soc Imp Naturalistes Moscou 18(4):345–363

Weber M (1907) Nachtrag zu den Rhinocerotiden von Samos. Centralbl Min Geol Pal Jg 1907:29–30

Weithofer A (1888) Beiträge zur Kenntniss der Fauna von Pikermi bei Athen. Beitr Paläontol Österreich-Ungarns Orients 6:225–292

Welker F, Smith GM, Hutson JM, Kindler L, Garcia-Moreno A, Villaluenga A, Turner E, Gaudzinski-Windheuser S (2017) Middle Pleistocene protein sequences from the rhinoceros genus *Stephanorhinus* and the phylogeny of extant and extinct middle/late Pleistocene Rhinocerotidae. Peer J 5:e3033. https://doi.org/10.7717/peerj.3033

Wilde WR (1840) Narrative of a voyage to Madeira, Teneriffe, and along the shores of the Mediterranean. W. Curry, Jun, Dublin, pp 1–495

Willerslev E, Gilbert M, Binladen J, Ho S, Campos P, Ratan A, Tomsho L, da Fonseca R, Sher A, Kuznetsova T, Nowak-Kemp M, Roth T, Miller W, Schuster S (2009) Analysis of complete mitochondrial genomes from extinct and extant rhinoceroses reveals lack of phylogenetic resolution. BMC Evol Biol 9:1–30

Woodward AS (1901) On the bone-beds of Pikermi, Attica, and on similar deposits in northern Euboea. Geol Mag 8(4):481–486

Yatkola D, Tanner LG (1979) *Brachypotherium* from the tertiary of North America. Occas Pap Mus Nat Hist Univ Kansas 77:1–11

Yuan J, Sheng G, Hou X, Shuang X, Yi J, Yang H, Lai X (2013) Ancient DNA sequences from *Coelodonta antiquitatis* in China reveal its divergence and phylogeny. China Earth Sci 57(3):388–396

Zbyszewsky G (1952) Les mammifères miocènes de Quintanelas (Sabugo). Com Serv Géol Portugal 33:1–22

Zittel KA (1893) Handbuch der Palaeontoiogie. I. Abtheilung Palaeozoologie. IV. Band. Vertebrata (Mammalia). Verlag R. Oldenbourg, Munich & Leipzig

Zong GF (1998) A new evidence of dividing in the Neogene stratigraphy of Yuanmou Basin [in Chinese with English summary]. Mem Beijing Nat Hist Mus 56:159–178

The Fossil Record of Chalicotheres (Mammalia: Perissodactyla: Chalicotheriidae) in Greece

Evangelia Tsoukala

1 Introduction

Chalicotheres (Mammalia: Perissodactyla) are peculiar odd-toed animals that first appeared in the Eocene of Asia; the family Chalicotheriidae is first known from the Oligocene and diversified into two subfamilies, the Chalicotheriinae and Schizotheriinae, both of which survived into the Pleistocene. The identified species within Greece include: *Ancylotherium pentelicum* Gaudry and Lartet, 1856; *Ancylotherium hellenicum* Koufos, 2012; *Chalicotherium goldfussi* Kaup, 1833; and *Anisodon macedonicus* (de Bonis et al., 1995). The typical Turolian schizotheriine *Ancylotherium pentelicum* is a rather rare but characteristic taxon found in Upper Miocene localities such as the type locality of Pikermi and in adjacent areas of the Eastern Mediterranean. On the other hand, the typical Turolian chalicotheriine is *Anisodon macedonicus*, found in Dytiko (MN13), Vathylakkos, and Prochoma (MN12). The Schizotheriinae is commonly associated with more open woodlands, having higher-crowned and more elongated molars, compared to the Chalicotheriinae. It must be noted that *Ancylotherium pentelicum,* with the highest crowned molars of all chalicotheres, based on the microwear results, showed that it might have occupied a somewhat closed habitat and eaten primarily leaves with some twigs and bark (Semprebon et al. 2011). With these different adaptations the members of each subfamily apparently occupied different ecological niches, which in turn would allow their co-existence in some of the localities (Roussiakis and Theodorou 2001; Koufos 2012; but see Geraads et al. 2001). In addition, chalicotheriines having elongated forelimbs with a protractable humerus, hooked clawed digits and remarkably short

E. Tsoukala (✉)
Laboratory of Geology and Paleontology, Aristotle University of Thessaloniki, Thessaloniki, Greece
e-mail: lilits@geo.auth.gr

E. Vlachos (ed.), *Fossil Vertebrates of Greece Vol. 2*,
https://doi.org/10.1007/978-3-030-68442-6_15

hindlimbs, were capable of bearing their body weight in an erect posture (Giaourtsakis and Koufos 2009). More specifically, the posture of *Anisodon grande* was discussed by Zapfe (1979) and of *Ancylotherium pentelicum* by Schaub (1943).

2 Historical Overview

Apart from being peculiar animals, chalicotheres also have a peculiar history, as summarized by Anquetin et al. (2007 and references therein): The first chalicothere fossils from Eppelsheim (Germany) were found at the beginning of the nineteenth century, but as they were disarticulated they were attributed to different animals—ungual phalanges were thought to belong to a deinothere or a giant pangolin and the teeth to a large anoplothere (artiodactyl). Filhol (1891) concluded that the teeth and feet belonged to a single paradoxical animal. More disarticulated chalicothere fossils were discovered in Sansan (France), suggesting that they could belong to an edentate animal. Later, the discovery of a complete and articulated skeleton of the Sansan chalicothere proved that they belonged to the same peculiar animal and thus the anatomy of chalicotheres became better known (Anquetin et al. 2007; and references therein). Traces of this peculiar early history of chalicotheres can be found also in the history of their Greek representatives, as summarized by Roussiakis and Theodorou (2001 and references therein): the first fossils from Pikermi were considered to belong to an edentate species, whereas an upper maxilla of a chalicothere was described by Wagner (1857) as a rhinoceros fossil (Roussiakis and Theodorou 2001; and references therein). Chalicothere fossils were reported sporadically from Greece during the twentieth century, mostly from Pikermi and nearby coeval localities, suggesting the presence of at least two different species: *Ancylotherium pentelicum* and *Chalicotherium goldfussi* (Gaudry 1863; Arambourg and Piveteau 1929; Thenius 1953; Melentis 1969a, b; Symeonidis 1973; Symeonidis and Zapfe 1973). During recent decades several new fossils and specialized studies, focused on chalicotheres, have shed light on the anatomy, taxonomy, diversity, distribution, and evolution of Greek chalicotheres. In particular, important cranial fossils from Thermopigi (Geraads et al. 2007) and additional postcranial elements from Pikermi (Roussiakis and Theodorou 2001) confirmed the presence of *Anc. pentelicum* in Greece, whereas additional crania from Dytiko 3 (de Bonis et al. 1995) and Pentalophos 1 (Koufos 2012) allowed the establishment of two additional new species from Greece: *Anisodon macedonicus* and *Ancylotherium hellenicum*, respectivelly. At the same time, additional works established the presence of chalicotheres on Samos Island as well (Solounias 1981; Giaourtsakis and Koufos 2009; Coombs 2013).

3 Phylogenetic Relationships

Chalicotheres are peculiar odd-toed animals that were originally considered to be more closely related to horses (= Hippomorpha) than to rhinoceroses and tapirs (= Ceratomorpha) or along the stem of Perissodactyla as sister to the clade formed by both horses and rhinoceroses/tapirs. More recently, they have been considered to be close relatives of the Ceratomorpha based on shared characters of the dentition (Hooker 1984 and references therein). In Hooker's (1984) classification, which is largely followed today for the higher relationships of Chalicotherioidea, chalicotheres are placed within Ancylopoda (Chalicotherioidea + Lophiodontidae) as a sister group of the Ceratomorpha s.s.

The Greek chalicothere species have contributed to the reconstruction of the species-level phylogeny of the family (Anquetin et al. 2007; Fahlke et al. 2013). *Ancylotherium* belongs to the subfamily Schizotheriinae, whereas the other Greek species belong to the Chalicotheriinae. The Chalicotheriinae clade including *Anisodon* and its close relatives has been recently designated as the Tribe Anisodontini (Coombs and Göhlich (2020). In discussing various characters of schizotheriines and *Ancylotherium pentelicum* in particular, Geraads et al. (2007) supported the conclusion that *Anc. pentelicum* is more closely related to schizotheriines from Eurasia and Africa than to the North American *Tylocephalonyx*; these relationships would make the shared domed morphology of the crania of *Ancylotherium* and *Tylocephalonyx* a character that evolved in parallel in the two taxa. Derived Chalicotheriinae comprise two main clades, one defined by *Chalicotherium goldfussi* and the other by *Anisodon grande*; the Greek *Anisodon macedonicus* holds a derived position within the latter clade (Anquetin et al. 2007; Fahlke et al. 2013).

4 Distribution

Chalicothere fossils are found in at least 14 localities in Greece, concentrated in three main areas: in Attica and nearby regions in central Greece (e.g., Pikermi and Chomateri in Attica; Halmyropotamos and Kerassia in Euboea), in the Axios valley in Northern Greece (e.g., Vathylakkos 3, Prochoma 1, Dytiko 3, and Pentalophos 1), and on Samos Island. The occurrence in Thermopigi in northeastern Greece is the northernmost occurrence of the family in the country. All chalicothere occurrences come from the late Miocene. The Axios valley contains both the stratigraphically youngest (from the late Turolian, MN13, of Dytiko 3) and the oldest occurrence (from the early Vallesian, MN9, of Pentalophos 1) of the family in Greece. All other occurrences are dated to the middle Turolian (MN12). Based on the currently available evidence, the subfamily Schizotheriinae (represented by two species of *Ancylotherium*) was present during the entire temporal range of the family in Greece, whereas the subfamily Chalicotheriinae (represented by *Anisodon* and *Chalicotherium*) had a shorter, Turolian, temporal range. See Fig. 1 and Appendix for further details.

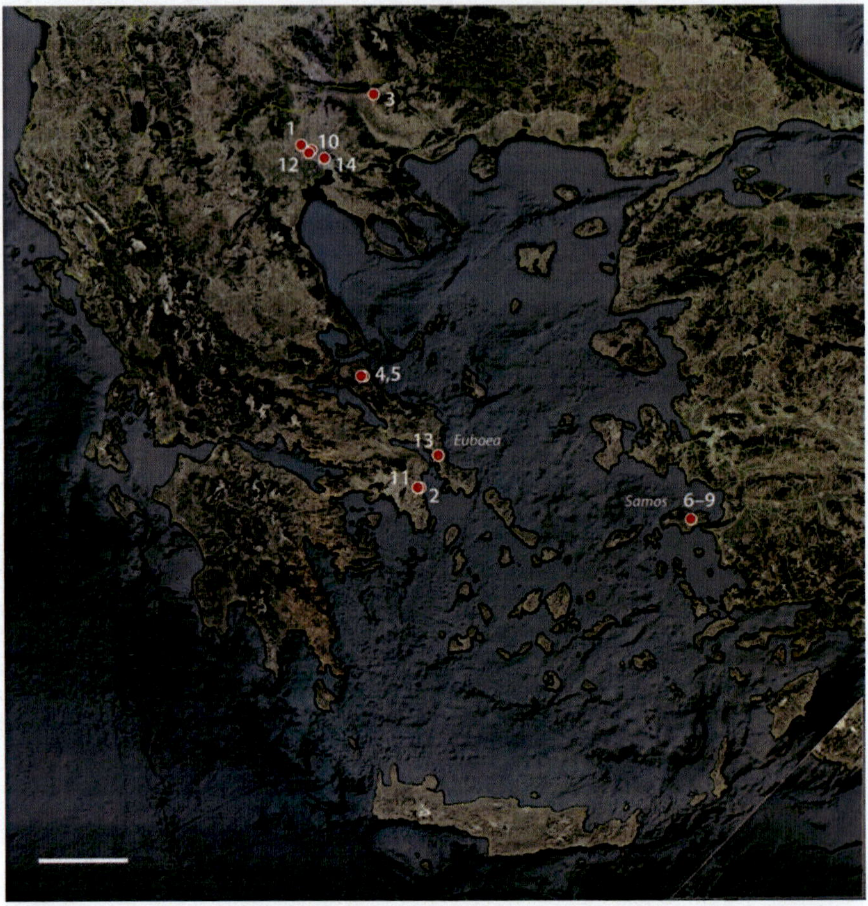

Fig. 1 Map of Greece showing the geographic distribution of the most important localities with chalicotheres in Greece. **1**, Dytiko 3; **2**, Chomateri; **3**, Thermopigi; **4**, Kerassia 1; **5**, Kerassia 4; **6**, Mytilinii 1A; **7**, Mytilinii 1B; **8**, Mytilinii 3; **9**, Samos Q1/QA; **10**, Vathylakkos 3; **11**, Pikermi; **12**, Prochoma 1; **13**, Halmyropotamos; **14**, Pentalophos 1. See Appendix for more information. Image exported from Google Earth Pro © 2019, map data from US Dept. of State Geographer, SIO, NOAA, U.S. Navy, NGA, GEBCO, image from Landsat/Copernicus. Scale bar equals 80 km, North faces upward

5 Systematic Paleontology

Perissodactyla Owen, 1848

Chalicotherioidea Gill, 1872

Chalicotheriidae Gill, 1872

Schizotheriinae Holland and Peterson, 1914

***Ancylotherium* Gaudry, 1863**

Type Species *Macrotherium pentelicum* Gaudry and Lartet, 1856.

Remarks The new genus *Ancylotherium* was erected by Gaudry in his volume on the fossil fauna and geology of Pikermi, which was issued in parts. The part that contains the name *Ancylotherium* was issued in 1863. The older genus *Metaschizotherium* von Koenigswald 1932 has been synonymized with *Ancylotherium.* There has been a discussion regarding the genera *Ancylotherium* and *Metaschizotherium* is whether they represent two distinct genera or the latter is a subgenus of the former. In this case, there are two subgenera: *Ancylotherium* (*Metaschizotherium*) and *Ancylotherium* (*Ancylotherium*). Coombs (1975) also refers to the size variation that may be due to sexual dimorphism, thus *Metaschizotherium fraasi* and *M. bavaricum* (von Koenigswald 1932) are synonymized with *Anc. (M.) fraasi.* Additional material recently described by Coombs (2009), Fahlke and Coombs (2009), argues that *M. bavaricum* and *M. fraasi* are regarded as separate species. Both of these papers remove the La Grive schizotheriine from Metaschizotherium (see p. 123 in Coombs 2009; p. 71 and p. 88 in Fahlke and Coombs 2009). These papers suggested that the La Grive schizotheriine is probably related to *Ancylotherium* and might be an early occurrence of that genus in southern Europe. Previously, Mein and Ginsburg (2002) had created a new species for the La Grive chalicothere, which they called *Phyllotillon grivensis.*

Some chalicothere specimens have been reported from the late Miocene of Kerassia, including a distally broken third metacarpal from Kerassia 1 and a complete second metatarsal from Kerassia 4 (Theodorou et al. 2003). All specimens are poorly preserved, and their morphological characters cannot be clearly observed, but they can be identified to the genus level; therefore they are referred to *Ancylotherium* sp. (Theodorou et al. 2003).

★*Ancylotherium pentelicum* (Gaudry and Lartet, 1856)

Nomenclatural and Taxonomical History *Macrotherium pentelicum* in Gaudry and Lartet 1856 (new species); *Ancylotherium pentelici* in Gaudry 1863 (new combination, subsequent spelling); *Nestoritherium pentelici* in Schlosser 1921 (new combination); *Colodus pentelicus* in Dietrich 1928 (new combination and mandatory change). See Roussiakis and Theodorou (2001) for further information.

Type Material MNHN.F.PIK 3549a (radius), 3549b (ulna), 3553 (right humerus) (syntypes), figured in Gaudry (1863: pl. 19). Although Gaudry and Lartet (1856) did not specify any type material, the MNHN database lists the above-mentioned specimens as the type material of this species. However, there are nearly 100 specimens of *Anc. pentelicum* in Gaudry's collection in MNHN. In the original publication, Gaudry and Lartet (1856) clearly described elements of the anterior limb (thus corresponding to the mentioned type material of the MNHN database) as well as elements of the carpus, altogether corresponding to the figured specimens of Gaudry (1863). Therefore, probably all the specimens in the Gaudry's collection in MNHN should be considered as the type series or the syntypes of *Anc. pentelicum*.

Type Locality Pikermi, Greece, late Miocene, MN12.

Distribution Besides its type locality, *Anc. pentelicum* has been found also at other Greek localities, including Halmyropotamos on Euboea Island, in Thermopigi (Northern Greece), and on Samos Island. Outside Greece it has been reported in Bulgaria and Turkey (Geraads et al. 2001, 2006; Kaya et al. 2012; Saraç et al. 2002; Geraads 2017; Giaourtsakis and Koufos 2009).

Remarks Besides the original material of Gaudry and Lartet (1856) and Gaudry (1863), several more fossils of *Ancylotherium pentelicum* have been unearthed from Pikermi. Thenius (1953) studied a left maxilla with P2–M3 housed in AMPG. From the same collection, Roussiakis and Theodorou (2001) gave an overview of the site and analyzed the rich postcranial material including rare or barely known skeletal parts (appendicular elements including a radius and ulna in life position, carpal and tarsal bones, metacarpals and metatarsals, and phalanges), many of which belonged to juvenile individuals. The excavations in the nearby site of Halmyropotamos (Euboea), in 1916, brought to light thousands of fossilized bones of late Miocene animals, similar to those of Pikermi. Since they are separated from each other through the narrow bay of Evoikos, at distance of only 45 km, it seems that there was free communication between Attiki and Euboea, during the upper Miocene (Melentis 1969a, b). The occurrence of *A. pentelicum* in Halmyropotamos was identified by a proximal phalanx of the third digit, associated with a proximal and intermediate phalanx, other phalanges, a left third metacarpal and a third cervical vertebra (Melentis 1969a, b). More specimens of *A. pentelicum* are reported from the historical paleontological localities of Samos. The fossil occurrences in this area have been well-known since 1888, when Major (1888, 1894) collected thousands of fossils, stored in the Geological Museum of Lausanne. In addition, other thousands of fossils have been collected from Samos Island, stored in paleontological collections of natural history museums all over the world (see Melentis 1969c; Koufos 2009 and references therein for further information). Among the chalicothere fossils of Samos, a complete left forelimb was discovered and documented by Schaub (1943) and therefore we know the morphology and the functional anatomy of this animal. A detailed overview of the Samos collection, including material from the relatively recent excavations of *Ancylotherium pentelicum* of middle Turolian (MN12, late Miocene) age, was presented in Giaourtsakis and Koufos (2009). The material was collected from the localities Mytilinii-1A (MTLA and B), ~7.1 Ma (MTLA: right third metatarsal, ungual phalanx; MTLB: left radius and ulna, right radius, and ulna, left femur) and from Mytilinii-3 (MYT), ~7.3 Ma (MYT: proximal phalanx and right third metatarsal). More recently, Coombs (2013) described a juvenile mandible with deciduous teeth of *Anc. pentelicum*, collected by Barnum Brown from the late Miocene of Samos.

The necessity of skull remains of *Anc. pentelicum* was mentioned as early as Gaudry (1863). The most complete and well-preserved skull of an ancylothere, ever reported, was excavated at Thermopigi (near Sidirokastro, Serres) (Fig. 2) among additional material of an upper molar (Fig. 3a–c), mandible (Fig. 3d, e), and postcranial bones. The frontal bone inflation is the most remarkable feature of the skull, which, though weaker, closely resembles the dome of the North American genus *Tylocephalonyx* Coombs 1979.

The analysis of some of the characters found in the Miocene Schizotheriinae and their relationship with other Eurasian and African members of *Ancylotherium,* as well as the evolution of the domes are discussed by Geraads et al. (2007). The locality has yielded a rich collection of more than 20 mammals, dating to the middle Turolian (MN12, late Miocene).

Additional dental material of *Anc. pentelicum* has been brought to light from the systematic excavations at Thermopigi since 1998, improving the description of the anatomy of the most important chalicothere from Greece; some of this new material is presented herein. The new M3 (LGPUT SIT 1072a; Fig. 3a–c) is moderately worn and is the least quadrate and least symmetrical upper molar because of the decrease of the metaloph and the hypocone: SIT 770 (length = 56.74 mm, breadth = 50.90 mm, breadth at the trigon = 52.50 mm; breadth at the talon = 41.32 mm).

It is high-crowned (the preserved heights of the strong parastyle and mesostyle are 40 and 41.3 mm, respectively). The ectoloph is intensely concave, whereas the metastyle has a height of 33.3 mm, and is weaker than the other two styles, though still well-developed. The metacone is less developed than the paracone. The well-developed protoloph joins the protocone anteriorly and meets a deep anterior valley, but weak anterior cingulum. The central valley is deep and open as the lingual cingulum stops between the protocone and hypocone. On the other hand, the disto-lingual cingulum closes the post-fossette. The right isolated P2 (LGPUT SIT 1515) is worn but the protocone and hypocone are clearly distinct and pointed (length = 27.3 mm, breadth = 21.3 mm). The new right mandible (LGPUT SIT 1482; Fig. 3f–h) is well-preserved with all cheek teeth (p2–m3). The length of the

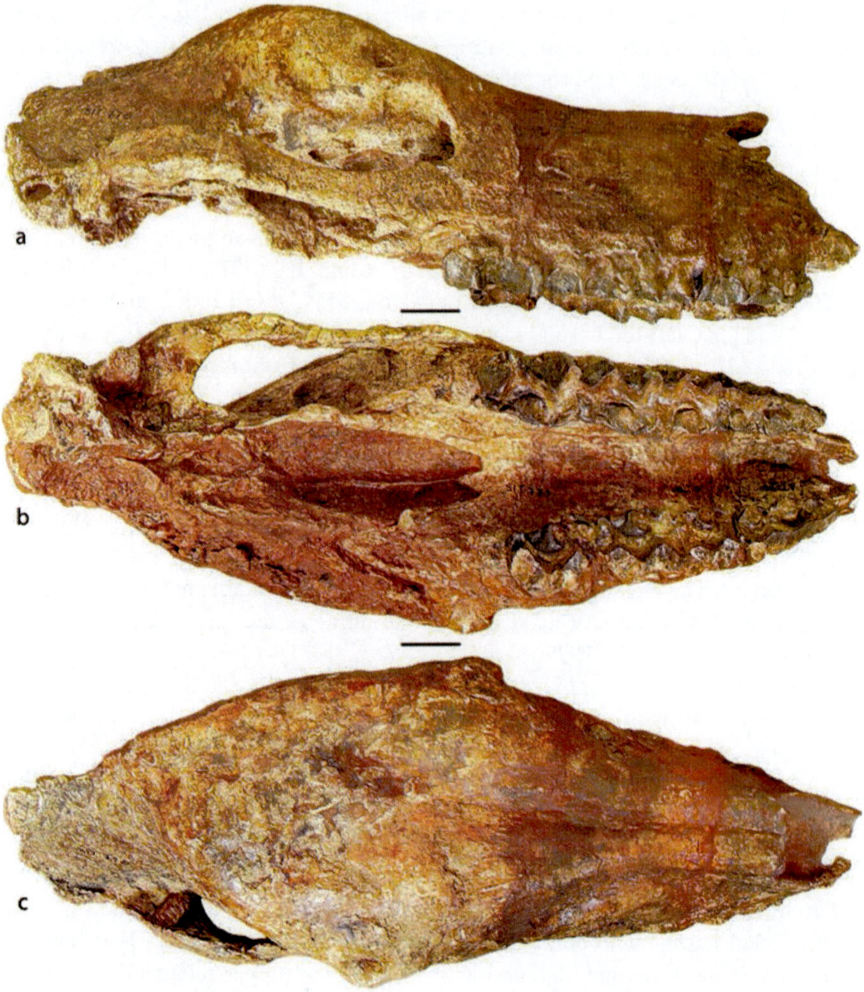

Fig. 2 The almost complete cranium (LGPUT SIT 770) of *Ancylotherium pentelicum* from the late Miocene of Thermopigi. **a** right lateral view, **b** occlusal view showing the left and right cheek teeth P2–M3, **c** cranial view. Scale bar equals 5 cm

premolars (p2–p4) is 82.42 mm, whereas that of the molars is much longer (m1–m3 = 182.00 mm); thus, the combined length of the cheek teeth is 262.60 mm. The height of the mandible at the diastema, between the p2 and the mental foramen, is 38.00 mm, which is clearly much shorter than the height below the p2 (48.50 mm) and the m3 (63.00 mm). Symphysis is rather short (40.49 × 20.03 mm) and ends about 10 mm in front of p2. The dimensions of each cheek tooth are given in Table 1. The premolars (especially p3 and p4) and molars are elongated. The least worn teeth are p2 and m3, whereas the intermediate teeth are moderately to very worn. The p2

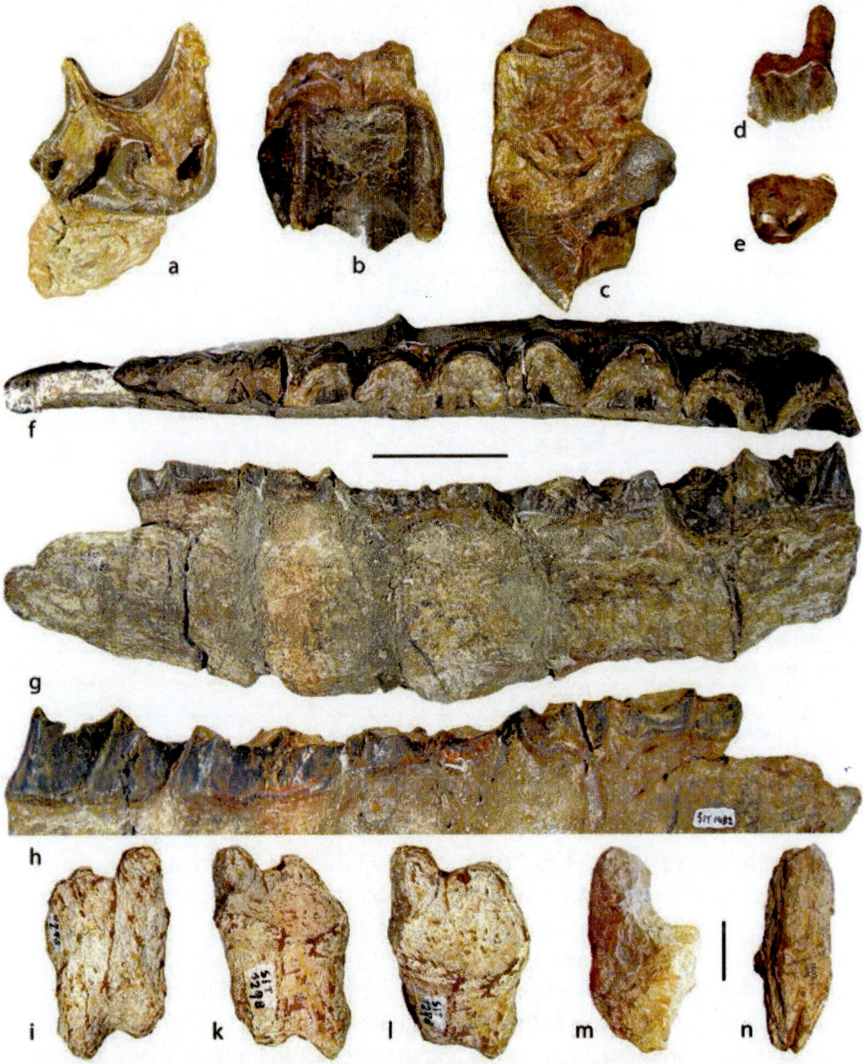

Fig. 3 Selected specimens of *Ancylotherium pentelicum* from the late Miocene of Thermopigi. (**a–c**) right M3 (LGPUT SIT 1072a) in **a** occlusal, **b** buccal, and **c** mesial views. (**d–e**) right P2 (LGPUT SIT 1515) in **d** occlusal and **e** buccal views. (**f–h**) right mandible with p2–m3 (LGPUT SIT 1482) in **f** occlusal, **g** lingual and **h** buccal views. (**i–l**) proximal phalanx (LGPUT SIT 1298) in **i** volar, **k** dorsal and **l** proximo-dorsal views. (**m–n**) distal phalanx (LGPUT SIT 2071) in **m** lateral and **n** dorsal views. Scale bars equal 5 cm

Table 1 Measurements of the various cheek teeth of the right mandible (LGPUT SIT 1482) and proximal (LGPUT SIT 1298) and distal (LGPUT SIT 2071) phalanx of *Ancylotherium pentelicum* from the late Miocene of Thermopigi (near Sidirokastro, Serres, Greece). Measurements in mm

Teeth	Length	Breadth	Breadth at trigonid	Breadth at talonid
p2	18.35	12.90	12.80	12.54
p3	28.83	17.00	17.39	17.66
p4	38.67	21.50	18.90	21.00
m1	51.24	24.00	23.82	23.94
m2	63.30	28.00	27.30	27.80
m3	65.97	29.97	29.97	27.83

	Length: 78.14	
Phalanx 1	Transversal diameter	Anteroposterior diameter
Proximal end	46.11	41.46
Proximal articulation	39.50	32.72
Minimum diaphysis	39.50	30.94
Distal end	40.50	31.95
Distal articulation	37.00	–

	Length: >160.00	
Phalanx 3[a]	Transversal diameter	Anteroposterior diameter
Proximal end	50.00	94.00
Maximum	62.00	–
Volar length	105.00	–
Articulation	42.20	–

[a]Measurements after Roussiakis and Theodorou (2001)

reveals a height of 16.20 mm. Only the premolars show a buccal cingulum and a distinctive lingual cingulum. On m3, the trigonid and talonid basins are open and the paraconid, metaconid/metastylid, and entoconid are well-developed. The m3 is rather hypsodont with a preserved height of 33.60 mm at the entoconid. Additional postcranial elements of *Anc. pentelicum* have been found at Thermopigi, including pelvic and scapular fragments, two complete well-preserved phalanges: a proximal phalanx with asymmetrical proximal and distal ends and a large distal phalanx of digit II of the manus (Fig. 3g–i). Morphology and dimensions correspond to those of *Anc. pentelicum* from Pikermi (see Roussiakis and Theodorou 2001).

★*Ancylotherium hellenicum* Koufos, 2012

Type Material Skull PNT-160 (holotype). A mandible (PNT-33), which is probably associated with the holotype skull, is the paratype (Koufos 2012)

Type Locality Pentalophos 1 (PNT), Axios Valley, Macedonia, late early Vallesian, late Miocene, late MN9.

Remarks The Axios valley has yielded thousands of fossilized mammalian remains from various localities since 1915–1916 and the initial excavations of C. Arambourg and afterwards, in 1972, by J. Melentis and L. de Bonis, continued by G. Koufos and his team. At least two different chalicothere species have been reported from the valley. One of them is *Ancylotherium hellenicum* and is characterized by small size and an elongated skull without the inflated frontal seen in the *Anc. pentelicum* skull from Thermopigi. A number of characters from the skull and mandible support its distinction, but the observed morphology is within the range of the genus *Ancylotherium* (Koufos 2012).

Chalicotheriinae Gill, 1872

Type Genus *Chalicotherium* Kaup, **1833**.

Remarks The subfamily Chalicotheriinae contains two derived clades defined by *Anisodon* and *Chalicotherium* (Anquetin et al. 2007), both of which are represented in the chalicothere fossil record of Greece. The main difference between the two clades is the brevirostry and derived cranial anatomy of *Anisodon* (Anquetin et al. 2007) and other Anisodontini (Coombs and Göhlich 2020).

Anisodon Lartet, 1851

Type Species *Anoplotherium grande* de Blainville, 1849.

Remarks Anquetin et al. (2007) revised the complicated nomenclature of chalicotheriines and made several proposals that have been largely followed ever since. First of all, the generic name *Macrotherium* Pictet, 1844, which has been largely used in the relevant literature up until recently and for most of the Greek taxa as well, is considered as a synonym of *Chalicotherium* (Anquetin et al. 2007). The species from Sansan is combined with *Anisodon*, and a lectotype is designated for *Anoplotherium grande* as the previous designation was judged to be invalid (Anquetin et al. 2007). As a result of these changes, the nomenclature of Greek chalicotheriines has also been modified.

★*Anisodon macedonicus* (de Bonis et al., 1995)

Nomenclatural and Taxonomical History *Macrotherium macedonicum* in de Bonis et al. 1995 (new species); *Anisodon macedonicus* in Anquetin et al. 2007 (new combination and mandatory spelling change).

Type Material LGPUT DKO-234 (holotype), skull with associated mandible (de Bonis et al. 1995; Fig. 4). A cast of the type material is also housed in MNHN under the collection number MNHN SLQ 1054a–c.

Type Locality Dytiko 3 (DKO), Axios Valley, late Turolian, MN13.

Fig. 4 The almost complete cranium of *Anisodon macedonicus* (LGPUT DKO-234) from the late Miocene of Dytiko 3 (Axios valley) in occlusal view. Scale bar equals 5 cm. Photograph courtesy of GD Koufos

Distribution Besides its type locality, *An. macedonicus* is also reported from the Late Miocene of Prochoma-1 and Vathylakkos-3 in the Axios Valley (Koufos 2012).

Remarks Initially, the skull and the associated mandible from Dytiko 3 were described as a new species of *Macrotherium* by de Bonis et al. (1995), an opinion tentatively followed by Geraads et al. (2001), who even argued that the species might be included in their new genus *Kalimantsia* Geraads et al. (2001). A species-level phylogeny of Chalicotheriinae by Anquetin et al. (2007) showed that the species from Dytiko is clearly distinct from *Kalimantsia* and is placed within the clade defined by *Anisodon grande*. Besides its type material, *An. macedonicus* from Dytiko 3 is also known from mandibles, and an atlas and epistrophaeus (de Bonis et al. 1995). Along with the description of *Anc. hellenicum*, Koufos (2012) assigned to *An. macedonicus* more material, including a right M3 from Prochoma 1 (middle Turolian MN12). Also, the chalicothere material from Vathylakkos 3—found by C. Arambourg in 1915–1916 and mentioned in Arambourg and Piveteau (1929) as "Ravin du Vatilük"—is also attributed to *Anisodon macedonicus* and is dated to the middle Turolian with a magnetostratigraphically estimated age of ~7.3 Ma (Koufos 2012). In Anquetin et al. (2007), the Vathylakkos chalicothere was identified as *Anisodon* sp. Antoine and Sen (2016) compared this specimen with *Kalimantsia* sp. and attributed it to the MN10 biozone. Based on the type material (de Bonis et al. 1995) and the referred material (Koufos 2012), *An. macedonicus* is clearly different from *An. grande* from the middle Miocene of Sansan and other material attributed to that species. The characters that diagnose the holotype (an almost complete cranium, atlas, and axis) of *An. macedonicus,* among others, are: medium sized animal with more shortened cranium than that of the type species, advanced upper molars, reduced paraloph not reaching by the protocone, P2 single-rooted and transversally stretched, p4 with distinct entoconid, long symphysis, the metastylid little detached from the metaconid in the lower molars, cervical vertebrae relatively short (from de Bonis et al. 1995).

Chalicotherium Kaup, 1833

Chalicotherium goldfussi Kaup, 1833

Type Material HLM Din. 3167 (lectotype), a right M3 described by Kaup (1833) and designated as a lectotype by Schaefer and Zapfe (1971); see Anquetin et al. (2007) for further information.

Type Locality Eppelsheim (Germany), late Miocene, Vallesian (MN9).

Distribution Besides its type locality, it is known from several Miocene localities from Germany, Austria, France, Hungary, and Turkey. In Greece, it is reported from the late Miocene of Pikermi and Chomateri, Attica.

Remarks Several of the chalicothere occurrences from Greece were initially identified as occurrences of *Chalicotherium goldfussi*, but are currently identified as *An. macedonicus* (see above). Arambourg and Piveteau (1929) identified the material from Vathylakkos 3 as *Ch.* cf. *goldfussi*, but the identification has changed several times until its present attribution to *An. macedonicus* by Koufos (2012). Similarly, the material from Prochoma 1 was listed by Kondopoulou et al. (1992) as Chalicotheriidae indet. and by Koufos (2006) as *Chalicotherium goldfussi*, but the final study of the material showed that it belongs to the Dytiko species, *An. macedonicus*. Other cited occurrences of *Chalicotherium goldfussi* come from the late Miocene of Pikermi and Chomateri (Attika)—both localities are very close to each other (Symeonidis and Zapfe 1973)—based on the study of a metacarpal and astragalus from Pikermi by Butler (1965), and a maxilla with right and left P3–M3 from Chomateri by Symeonidis (1973), respectively. In Veles (North Macedonia), a mandible assigned to *Ch. goldfussi* by Garevski and Zapfe (1983) was recently assigned to *Anisodon* sp. (Anquetin et al. 2007 and references therein). Roussiakis and Theodorou (1995, 2001) described numerous chalicothere specimens from Pikermi attributed to *Anc. pentelicum*, raising questions regarding the co-existence of both species in the late Miocene of the broader area of Pikermi. They argued that as both are representatives of different habitats and environments of about the same age (MN12), and the collected material from Pikermi could come from mixed stratigraphic levels, this co-existence cannot be excluded (Roussiakis and Theodorou 2001). Koufos (2006) cited the presence of *Ch. goldfussi* at Chomateri (= Chomateres therein), but the only chalicothere in Pikermi is *Anc. pentelicum*. Although the occurrences of *Ch. goldfussi* from Chomateri and Pikermi are included herein as originally identified, it is necessary to revise this material in order to confirm the presence of this species in Greece. Given that so much of the skull, jaw, and mandibular material that previously were referred to *C. goldfussi* are now assigned to *Anisodon macedonicus*, the lack of any postcranial comparative study makes the presence of *Ch. goldfussi* in Greece as tenuous for now (Coombs, pers. commun.)

6 Concluding Remarks

The presence of both chalicotheriid subfamilies in the late Miocene of the Balkan Peninsula (including Greece) and Turkey appears to be the western end of a broad distribution continuing into eastern Asia, especially China, where both the Schizotheriinae and Chalicotheriinae survived into the Early Pleistocene. The presence of *Anisodon* sp. at Dorn Dürkheim 1, MN11 (Fahlke et al. 2013) represents an unusually late occurrence of the Chalicotheriidae in central Europe.

Despite being relatively rare in the vertebrate fossil record of Greece, chalicotheres are known from some exceptional fossil remains that have allowed the description of a family having a notable diversity in the late Miocene of Greece. So far, both chalicotheriid subfamilies—namely Schizotheriinae and Chalicotheriinae—are identified in Greece, each represented by at least two different species. Schizotheriinae are represented by two species of *Ancylotherium*, *Anc. pentelicum*, and *Anc. hellenicum*, reaching the maximum of the geographical distribution (especially *Anc. pentelicum*, from the broader Pikermi area to Northern Greece and Samos Island) and temporal range (covering most of the Vallesian–Turolian). On the other hand, Chalicotheriinae are represented by two different species as representatives of the derived chalicotheriine clades, *An. macedonicus* and *Ch. goldfussi*. Chalicotheriines had both a more restricted geographical distribution (*Anisodon* in Northern Greece and *Chalicotherium* in Central Greece) and temporal range (found only in the Turolian). Chalicotheres have peculiar morphology, with no hooves, characteristic bifid claws, large size, and strange body proportions, which undoubtedly makes them some of the most extraordinary vertebrate fossils from Greece. The current available knowledge of the anatomy, taxonomy, diversity, and phylogenetic relationships of the group is a solid framework for future studies.

Acknowledgments I would like to thank Prof. Emer. GD Koufos for kindly providing the photograph of *Anisodon macedonicus* from Dytiko 3. Sincerest thanks to Emer. Professor M. Coombs for the useful comments that greatly helped to improve this manuscript, as well as to the anonymous reviewer.

Appendix

List of the Greek localities containing chalicothere fossils. Type localities are marked with bold. Locality numbers refer to the collection numbers of the PaleoBiology Database (PBDB)

Localities[PBDB No]	Age (MN; GPTS in Ma)	Taxon
Dytiko 3[32376]	**Late Turolian (MN13)**	***Anisodon macedonicus*[1]**
Chomateri[195562]	Middle Turolian (MN12)	*Chalicotherium goldfussi*[2,3]
Thermopigi[73553]	Middle Turolian (MN12)	*Ancylotherium pentelicum*[5]
Kerassia 1[195432]	Middle Turolian (MN12)	*Ancylotherium* sp.[5]
Kerassia 4[195435]	Middle Turolian (MN12)	*Ancylotherium* sp.[5]
Mytilinii 1A[202215]	Middle Turolian (MN12; ~7.1)	*Ancylotherium pentelicum*[6]
Mytilinii 1B[202216]	Middle Turolian (MN12; ~7.1)	*Ancylotherium pentelicum*[6]
Mytilinii 3[202218]	Middle Turolian (MN12; ~7.3)	*Ancylotherium pentelicum*[6]
Samos Q1/QA[95691]	Middle Turolian (MN12; 7.3–7.1)	*Ancylotherium pentelicum*[7,16]
Vathylakkos 3[182750]	Middle Turolian (MN12; ~7.3)	*Anisodon macedonicus*[8–9]
Pikermi[182754]	**Middle Turolian (MN12; 7.3–7.1)**	***Ancylotherium pentelicum*[10–12]** *Chalicotherium goldfussi*[13]
Prochoma 1[202222]	Middle Turolian (MN12; ~7.4)	*Anisodon macedonicus*[9]
Halmyropotamos[202213]	?Middle Turolian (MN12?)	*Ancylotherium pentelicum*[14–15]
Pentalophos 1[202119]	**Early Vallesian (MN9)**	***Ancylotherium hellenicum*[9]**

GPTS Geomagnetic Polarity Time Scale
[1]de Bonis et al. (1995), [2]Symeonidis and Zapfe (1973), [3]Symeonidis (1973), [4]Geraads et al. (2007), [5]Theodorou et al. (2003), [6]Giaourtsakis and Koufos (2009), [7]Coombs (2013), [8]Arambourg and Piveteau (1929), [9]Koufos (2012), [10]Gaudry (1863), [11]Thenius (1953), [12]Roussiakis and Theodorou (2001), [13]Butler (1965), [14]Melentis (1969a), [15]Melentis (1969b), [16]Solounias (1981)

References

Anquetin J, Antoine PO, Tassy P (2007) Middle Miocene Chalicotheriinae (Mammalia, Perissodactyla) from France, with a discussion of chalicotheriine phylogeny. Zool J Linn Soc 151:577–608

Antoine PO, Sen S (2016) Rhinocerotidae and Chalicotheriidae (Perissodactyla, Tapiromorpha), in Sen S. (ed.), late Miocene mammal locality of Küçükçekmece, European Turkey. Geodiversitas 38:245–259

Arambourg C, Piveteau J (1929) Les Vertébrés du Pontien de Salonique. Ann Paleontol 18:1–82

Butler PM (1965) Fossil mammals of Africa. **18.** East African Miocene and Pleistocene chalicotheres. Bull Br Mus (Nat Hist) Geol 10:165–237

Coombs MC (1975) Sexual dimorphism in Chalicotheres (Mammalia, Perissodactyla). Syst Zool 24:55–62

Coombs MC (1979) *Tylocephalonyx*, a new genus of north American dome-skulled Chalicotheres (Mammalia, Perissodactyla). Bull Am Mus Nat Hist 164:1–64

Coombs MC (2009) The chalicothere *Metaschizotherium bavaricum* (Perissodactyla, Chalicotheriidae, Schizotheriinae) from the Miocene (MN5) Lagerstätte of Sandelzhausen (Germany): description, comparison, and paleoecological significance. Palaeontol Z 83:85–129

Coombs MC (2013) A juvenile mandible with deciduous teeth of *Ancylotherium pentelicum* (Perissodactyla, Chalicotheriidae, Schizotheriinae), collected by Barnum Brown from the late Miocene of Samos (Greece). J Vertebr Paleontol 33:233–238

Coombs MC, Göhlich UB (2020) *Anisodon* (Perissodactyla, Chalicotheriinae) from the middle Miocene locality Gračanica (Bugojno Basin, Gornji Vakuf, Bosnia and Herzegovina). In Göhlich UB, Mandic O (eds.) the drowning swamp of Gračanica (Bosnia-Herzegovina)—a diversity hotspot from the middle Miocene in the Bugojno Basin. Palaeobiodiv Palaeoenviron 100:363–372

de Bonis L, Bouvrain G, Koufos GD, Tassy P (1995) Un crâne de chalicothère (Mammalia, Perissodactyla) du Miocène supérieur de Macédoine (Grèce): remarques sur la phylogénie des Chalicotheriinae. Palaeovertebrata 24:135–176

Dietrich WO (1928) Lassen sich *Chalicotherium* und verwandte Gattungen schon rekonstruieren? Centralbl Min Geol Paleont:366–376

Fahlke JM, Coombs MC (2009) Dentition and first postcranial description of *Metaschizotherium fraasi* Koenigswald, 1932 (Perissodactyla: Chalicotheriidae) and its occurrence on a karstic plateau—new insights into schizotheriine morphology, relationships and ecology. Palaeontogr Abt A 290:65–129

Fahlke JM, Coombs MC, Semprebon GM (2013) *Anisodon* sp. (Mammalia, Perissodactyla, Chalicotheriidae) from the Turolian of Dorn-Dürkheim 1 (Rheinhessen, Germany): morphology, phylogeny, and palaeoecology of the latest chalicothere in Central Europe. Palaeobiodiv Palaeoenviron 93:151–170

Filhol H (1891) Études sur les mammifères fossiles de Sansan. Libr Acad Méd Paris:1–319

Garevski R, Zapfe H (1983) Weitere Chalicotheriiden-Funde aus der Pikermi Fauna von Titov Veles (Mazedonien, Jugoslawien). Acta Mus Macedonici Sci Nat 17:1–20

Gaudry A (1863) Édentés. In: *Ancylotherium pentelici*. Animaux fossiles et géologie de l'Attique. Savy, Paris, pp 129–136

Gaudry A, Lartet EAIH (1856) Résultats des recherches paléontologiques entreprises dans l'Attique sous les auspices de l'Académie. C R Acad Sci Paris 43:271–274

Geraads D (2017) Late Miocene large mammals from Mahmutgazi, Denizli province, Western Turkey. N Jb Geol Paläont Abh 284(3):241–257

Geraads D, Spassov N, Kovachev D (2001) New Chalicotheriidae (Mammalia) from Bulgaria. J Vertebr Paleontol 21:596–606

Geraads D, Spassov N, Kovachev D (2006) The Bulgarian Chalicotheriidae (Mammalia): an update. Rev Paléobiol 25(2):429–437

Geraads D, Tsoukala E, Spassov N (2007) A skull of *Ancylotherium* (Chalicotheriidae, Mammalia) from the late Miocene of Thermopigi (Serres, N. Greece) and the relationships of the genus. J Vertebr Paleontol 27:461–466

Giaourtsakis IX, Koufos GD (2009) The late Miocene mammal faunas of the Mytilinii Basin, Samos Island, Greece: new collection. 10. Chalicotheriidae Beitr Palaeontol Wien 31:189–205

Hooker JJ (1984) A primitive ceratomorph (Perissodactyla, Mammalia) from the early tertiary of Europe. Zool J Linn Soc 82:229–244

Kaup JJ (1833) Description d'ossements fossiles de Mammifères inconnus jusqu'à-présent, qui se trouvent au Muséum grand-ducal de Darmstadt. JG Heyer, Darmstadt

Kaya T, Mayda S, Kostopoulos DS, Alcicek MC, Merceron G, Tan A, Karakutuk S, Gieslere AK, Scott RS (2012) Şerefköy-2, a new late Miocene mammal locality from the Yatağan formation, Muğla. SW Turk C R Palevol 11:5–12

Kondopoulou D, Sen S, Koufos GD, de Bonis L (1992) Magneto-and biostratigraphy of the late Miocene mammalian locality of Prochoma (Macedonia, Greece). Paleontol Evol 24–25:135–139

Koufos GD (2006) The Neogene mammal localities of Greece: faunas, chronology and biostratigraphy. Hellenic J Geosci 41:183–214

Koufos GD (2009) The late Miocene mammal faunas of the Mytilinii Basin, Samos Island, Greece: new collection. 1. History of the Samos fossil mammals. Beit Palaeontol 31:1–12

Koufos GD (2012) New material of Chalicotheriidae (Perissodactyla, Mammalia) from the late Miocene of Axios Valley, Macedonia (Greece) with the description of a new species. Ann Paleontol 98:203–224

Major CJF (1888) Sur un gisement d'ossements fossiles dans l'île de Samos, contemporain de l'âge de Pikermi. C R Séances Acad Sci Paris 107:1178–1182

Major CJF (1894) Le gisement ossifère de Mytilinii et catalogue d'ossements fossiles recueillis à Mitylinii, île de Samos, et déposés au Collège Galliard, à Lausanne. Georges Bridel & Cie, Lausanne

Mein P, Ginsburg L (2002) Sur l'âge relatif des différents dépôts karstiques miocènes de La Grive-Saint-Alban (Isère). Cahiers Scientifiques Mus d'Hist Nat Lyon 2:7–47

Melentis JK (1969a) Die Pikermi-Fauna von Halmyropotamos (Euböa, Griechenland), 2. Teil: Osteologie. Ann Geol Pays Helléniques 28:217–306

Melentis JK (1969b) *Ancylotherium pentelicum* Gaudry & Lartet aus dem Pont von Halmyropotamos (Euböa, Griechenland). Arch études d'Eubée 15:1–18

Melentis JK (1969c) Paläontologische Ausgrabungen auf der Insel Samos (eine vorläufige Mitteilung). Praktika Akad Wiss Athen 43:344–349

Roussiakis S, Theodorou G (1995) Fossil mammal collections of the Paleontological and Geological Museum of the University of Athens: *Ancylotherium pentelicum* Gaudry & Lartet (Mammalia, Perissodactyla) from Pikermi (Attica, Greece). Romanian J Stratigr, abstracts pp 99–100

Roussiakis S, Theodorou G (2001) *Ancylotherium pentelicum* (Gaudry & Lartet, 1856) (Perissodactyla, Mammalia) from the classic locality of Pikermi (Attica, Greece), stored in the Palaeontological and geological museum of Athens. Geobios 34:563–584

Saraç G, Kaya T, Geraads D (2002) *Ancylotherium pentelicum* (Perissodactyla, Mammalia) from the upper Miocene of central and western Turkey. Geobios 35:241–251

Schaefer H, Zapfe H (1971) *Chalicotherium grande* Blainv. Und *Chalicotherium goldfussi* Kaup. Verh Naturforsch Ges Basel 81:157–199

Schaub S (1943) Die Vorderextremität von *Ancylotherium pentelicum* Gaudry & Lartet. Schweiz Palaeontol Abh Basel 64:1–36

Schlosser M (1921) Die Hipparionenfauna von Veles in Mazedonien. Abh Bayer Akad Wiss Math Phys Kl 29:1–55

Semprebon GM, Sise PJ, Coombs MC (2011) Potential bark and fruit browsing as revealed by stereomicrowear analysis of the peculiar clawed herbivores known as chalicotheres (Perissodactyla, Chalicotherioidea). J Mamm Evol 18:33–55

Solounias N (1981) Mammalian fossils of Samos and Pikermi. Part 2. Resurrection of a classic Turolian fauna. Ann Carnegie Mus 50:231–270

Symeonidis NK (1973) *Chalicotherium goldfussi* Kaup (Perissodactyla, Mammalia) aus dem Altpliozän von Pikermi (Griechenland). Ann Geol Pays Helléniques 25:301–307

Symeonidis NK, Zapfe H (1973) Ein bedeutender Fund von *Chalicotherium goldfussi* Kaup (Perissodactyla, Mammalia) aus dem Altpliozän von Pikermi (Griechenland). Anz Österr Akad Wiss Math Natwiss Kl Wien:132–134

Thenius E (1953) Studien über fossile Vertebraten Griechenlands. III. Das Maxillargebiss von *Ancylotherium pentelicum* Gaudry und Lartet. Ann Geol Pays Helléniques 5:97–106

Theodorou G, Athanassiou A, Roussiakis S, Iliopoulos G (2003) Preliminary remarks on the late Miocene herbivores of Kerassia (N. Euboea, Greece). Deinsea 10:519–530

von Koenigswald GHR (1932) *Metaschizotherium fraasi* n.g. n.sp. ein Chalicotheriide aus dem Obermiozän von Steinheim a Alb. Palaeontographica Suppl 8:3–24

Wagner A (1857) Neue Beiträge zur Kenntnis der fossilen Säugthier-Ueberreste von Pikermi. Abh Bayer Akad Wiss 8:109–158

Zapfe H (1979) *Chalicotherium grande* (Blainv.) aus der miozänen Spaltenfüllung von Neudorf an der March (Devinská Nová Ves), Tschechoslowakei. Neue Denkschr Naturh Mus Wien 2:1–282

The Fossil Record of Felids (Mammalia: Carnivora: Felidae) in Greece

George D. Koufos

1 Introduction

Family Felidae includes several carnivoran species, which generally referred as "cats." The first felids appear in the fossil record at the end of the Oligocene at ~25 Ma with the genera *Proailurus* and *Pseudaelurus*, which are considered as the ancestral forms of the felids. The family is characterized by high, broad and short skull, with a short muzzle. The endotympanic bones form the auditory bullae and the auditory meatus is very short. The dental formula varies in the different forms as the number of the premolars reduces; except P1 and p1 some forms lack the second or third premolars as well. The canines are elongated and pointed with grooves in their surface; in some felids (Machairodontinae) the upper canines are huge with cutting edges. The upper carnassial is similar to that of hyaenids and the M1 is relatively small. In most cases, the lower carnassial lacks metaconid and the talonid is rudimentary or absent; the m2 is only present in primitive species, or as an atavism in extant ones. The extremities are elongated, equal-sized and digitigrades with retractile claws. The felids are morphologically adapted to hunting and they can kill animals as large as themselves or sometimes larger than them. They are dispersed worldwide, except Australia, New Zealand, Antarctica, and some islands; however, the domestic cats are living everywhere following the humans. Regarding the number of the extant felid taxa, although there are different opinions, most of the scientists recognize 18 genera and 36 species (Orlov 1968; Etnyre et al. 2011).

G. D. Koufos (✉)
Department of Geology, Laboratory of Geology and Paleontology,
Aristotle University of Thessaloniki, Thessaloniki, Greece
e-mail: koufos@geo.auth.gr

E. Vlachos (ed.), *Fossil Vertebrates of Greece Vol. 2*,
https://doi.org/10.1007/978-3-030-68442-6_16

2 Historical Overview

The felids are well known in the Greek fossil record including several taxa. They are early recognized in the first collected material from the Greek localities of Pikermi and Samos; several new genera and species have been described from this material (e.g., Gaudry 1862–1867; Forsyth Major 1894). Later, felid remains have been described from various Greek areas (Axios Valley, Chalkidiki Peninsula, Evia Island, Thessaly), which enriched remarkably the Greek fossil record (e.g., Arambourg and Piveteau 1929; Melentis 1967; Koufos 2000, 2006, 2009; Roussiakis 2001, 2002). The earliest evidence of the felids is recognized in the early/middle Miocene locality of Antonios with the presence of *Pseudaelurus* (Koufos 2008). During the late Miocene, several felid taxa have been described from Greece; in the Pliocene the felids are scarce (but this is due to the very small known number of localities), but their number increases in the Quaternary with several species. The known fossil Felidae of Greece include ten valid genera and 13 species, two of which are new; there is also another taxon referred to as "*Metailurus parvulus*" the systematic position of which is debated. The species *Felis silvestris* (European Wildcat) and possibly *Lynx lynx* (Balkan Lynx) are still living in the high mountains of Greece.

3 Phylogenetic Relationships

Several articles are referred to the phylogenetic relationships of Felidae; in this chapter, a short brief of their evolution and relationships is given. The felids are recognized in the Oligocene with their earliest representative *Proailurus lemanensis* dated at ~30.0 Ma. The primitive felids are characterized by short skull, not completely ossified auditory bullae, large-sized and sectorial canines, reduction of the cheek teeth and retractile claws. The true cats with the modern type of auditory bullae are included in the family Felidae (Carroll 1988). The family appeared at the base of the Miocene (~20.0 Ma) with the genus *Pseudaelurus*, the members of which are considered ancestral forms of the extant and living conical-toothed cats (subfamily Felinae) and the extinct saber-toothed cats (subfamily Machairodontinae) (Turner and Antón 1997). Modern molecular studies suggest that the extant cats diverged from the fossil species in eight lineages of related species: *Panthera* lineage at ~10.8 Ma, Bay Cat (*Catopuma* and *Pardofelis*) lineage at 9.4 Ma, *Caracal* lineage at ~8.5 Ma, Ocelot or *Leopardus* lineage at ~8.0 Ma, *Lynx* lineage at ~7.2 Ma, *Puma* lineage at ~6.7 Ma, Leopard Cat lineage at ~6.2 Ma, and *Felis* lineage at ~3.4 Ma (O'Brien and Johnson 2007).

4 Distribution

The fossil felids have a worldwide distribution covering the Old and New World. In Greece, fossil felids have been found in the entire country from Miocene to Holocene both in open area localities and in caves (see Fig. 1 and Appendix).

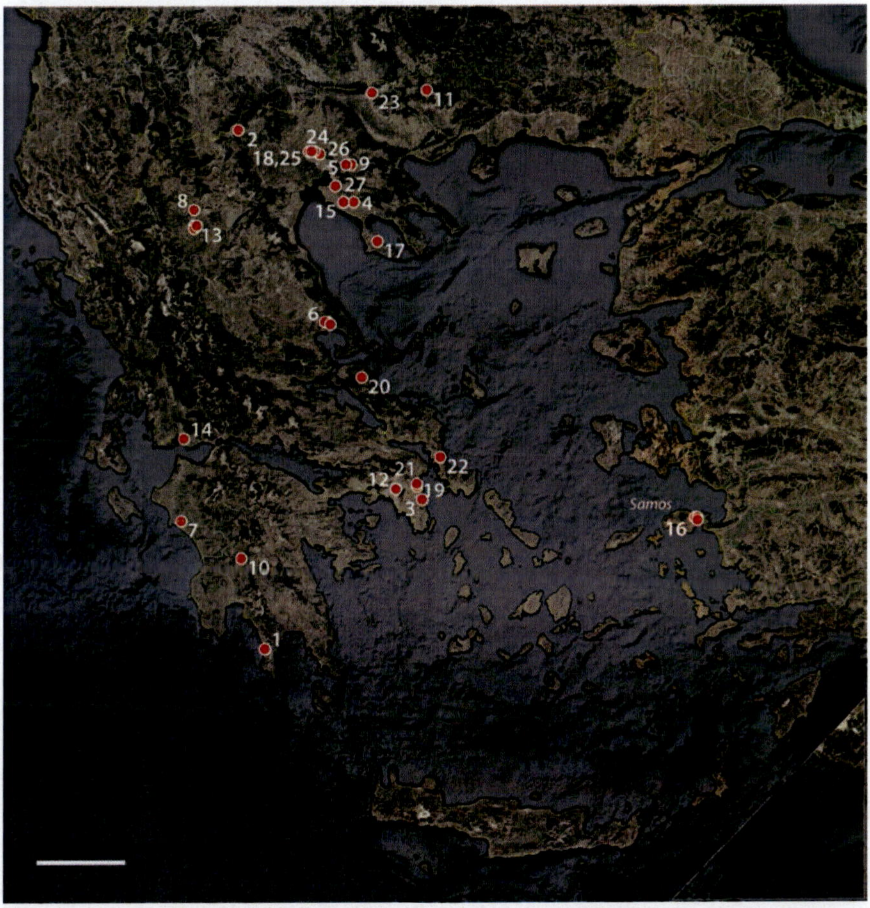

Fig. 1 Map of Greece showing the geographic distribution of the localities with viverrid fossils. **1**, Apidima Caves; **2**, Loutra Almopias Cave; **3**, Vraona Cave; **4**, Petralona Cave; **5**, Apollonia 1; **6**, localities near Volos (Alikes, Volos, Sesklon); **7**, Pyrgos; **8**, Libakos; **9**, Gerakarou 1; **10**, Marathousa 1; **11**, Volax; **12**, Tourkobounia 3–5; **13**, Milia localities; **14**, Makinia; **15**, Nea Silata; **16**, various localities in Mytilinii basin, in Samos Island; **17**, Kryopigi; **18**, Vathylakkos 3; **19**, Pikermi; **20**, Kerassia localities; **21**, Chomateres; **22**, Halmyropotamos; **23**, Thermopigi; **24**, Ravin des Zouaves 5; **25**, Ravin X; **26**, Ravin de la Pluie; **27**, Antonios. See Appendix for more information. Image exported from Google Earth Pro © 2019, map data from US Dept. of State Geographer, SIO, NOAA, U.S. Navy, NGA, GEBCO, image from Landsat/Copernicus. Scale bar equals 8 km, North faces upward

5 Systematic Paleontology

Felidae Fischer von Walheim, 1817

Felinae Fischer von Walheim, 1817

***Pseudaelurus* Gervais, 1850**

Type Species *Felis quadridentata* Blainville, 1843.

***Pseudaelurus romieviensis* Roman and Viret, 1934**

Nomenclatural and Taxonomical History *Pseudaelurus quadridentatus* mutation *romieviensis* in Roman and Viret 1934 (named as a new mutation); *Pseudaelurus romieviensis* in Heizmann 1973 (referral to the species level). More information about the taxonomy of the early-middle Miocene Felinae of Europe is given in Salesa et al. (2011).

Type Material FSL-1779 (holotype), left mandibular fragment with p3–m1, housed at the University of Lyon, France.

Type Locality La Romieu, France, middle Miocene, MN4.

Distribution In Europe, this species is reported from the middle Miocene of Spain, France, and Germany; in Greece, it is only known from Antonios.

Remarks The Greek material (Fig. 2a) includes a mandibular fragment with the p4 and the anterior part of the m1. The shallow mandibular corpus, the mental foramen below the c–p3 diastema, the relatively elongated and narrow p4 with high main cuspid, the strong anterior and large posterior accessory cuspid situated on a strong distal cingular projection, the strong buccal and lingual cingulum, the relatively large and high paraconid in the m1, and its size fits well with *P. romieviensis* (Koufos 2008).

***Metailurus* Zdansky, 1924**

Type Species *Metailurus major* Zdansky, 1924.

***Metailurus major* Zdansky, 1924**

Type Material Cranium and mandible (holotype; number unknown) described and figured by Zdansky (1924; taf. 29, figs 1–5).

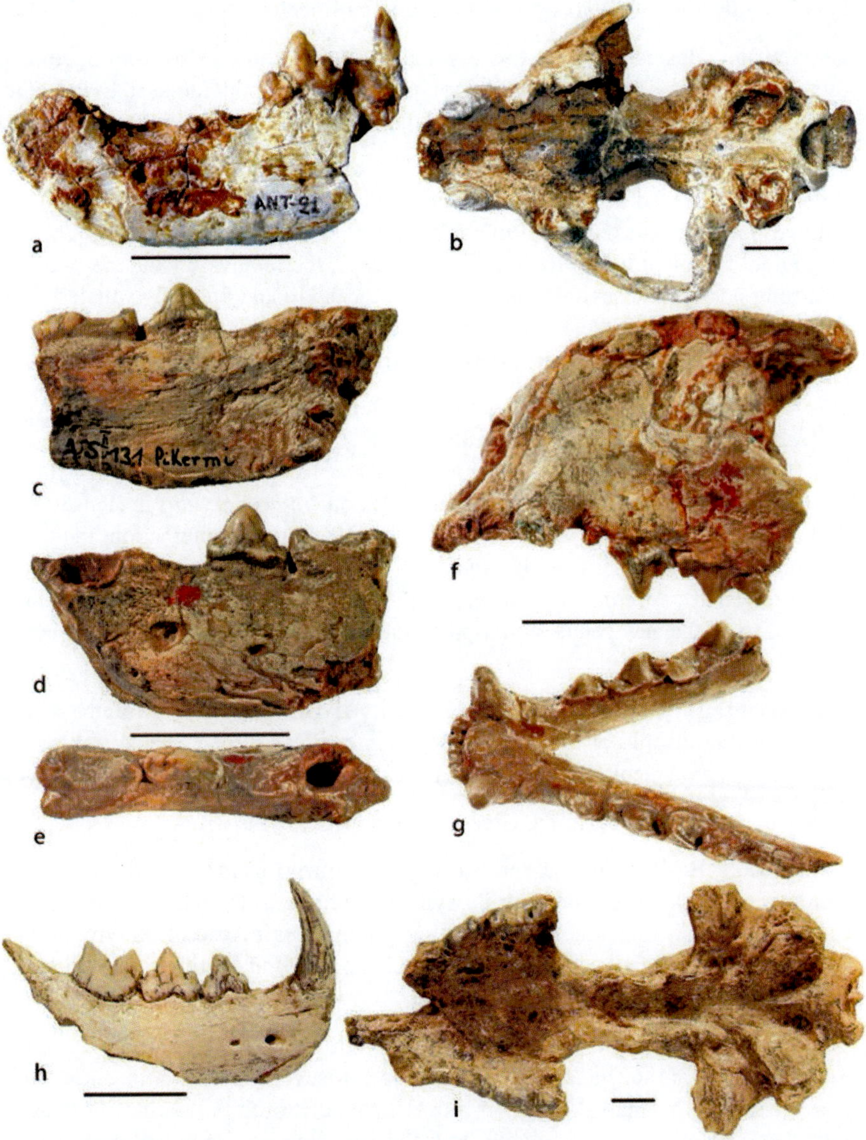

Fig. 2 Selected specimens of Felinae and Pantherinae (Felidae) from Greece. **a** mandible of *Pseudaelurus romieviensis* from Antonios in buccal view, **b** cranium of *Metailurus major* from Pikermi in palatal view, **c–e** left mandible (holotype) of "*Metailurus parvulus*" from Pikermi in **c** lingual, **d** buccal, and **e** occlusal views, **f–g** cranium and mandible (holotype) of *Pristifelis attica* from Pikermi in **f** lateral, and **g** occlusal views, **h** mandible of *Lynx issiodorensis* from Apollonia 1 in buccal view, **i** cranium of *Panthera gombaszogensis* from Gerakarou in palatal view. Scale bar equals 2 cm

Type Locality Locality 30, Shansi, China, late Miocene, MN13.

Distribution Besides the type locality, the species is known also from Spain and Italy. In Greece, it is known from Pikermi in Attica, Halmyropotamos in Evia Island, and Q5 and A in Mytilinii Basin, Samos Island.

Remarks The taxon is quite rare in Greece and the known material is scarce. Melentis (1967) described some cranial and mandibular remains from Halmyropotamos and recently Roussiakis (2001) described a skull (Fig. 2b) and one cranial fragment from Pikermi. The large size, the relatively wide mesially P3, the strong posterior accessory cusp in the P3, the small and lingually situated anterior accessory cusp in the P3, and the relatively large M1 of the Greek material allow its attribution to this species (Roussiakis 2001).

★"*Metailurus parvulus*" (Hensel, 1862)

Nomenclatural and Taxonomical History *Machairodus parvulus* in Hensel 1862 (new species); *Metailurus parvulus* in Thenius 1951 (new combination, including its Chinese junior synonyms *M. minor* of Zdansky 1924 and *Felis leiodon* of Weithofer 1888); "*Metailurus parvulus*" in Koufos 2018 (new combination, after the erection of the new genus *Yoshi* Spassov and Geraads 2014; see Remarks below for further information).

Type Material BSPM-PIK-AS II. 131 (holotype), left mandibular fragment with p3–p4 and the alveole of the canine, stored in the Bayerische Staadtsamlung fur Paläontologie und Geologie, München (Hensel 1862: p. 568; fig. 6; Fig. 2c–e).

Type Locality Pikermi, Attica, Greece, late Miocene, MN12.

Distribution Besides Greece, the taxon is recognized from Spain, France, Italy, Moldova, and China, considering its synonymized taxa as well. In Greece, it is known from Ravin de la Pluie in Axios Valley, Chomateres in Attica, Halmyropotamos and Kerassia 1 in Evia Island, Kryopigi in Chalkidiki, and Mytilinii-1A, Q5 and A in Samos Island.

Remarks The species was originally described from Pikermi as *Machairodus parvulus* by Hensel (1862) and later it was found in several Eurasian localities. It is characterized by a small skull, short muzzle, rounded orbit, voluminous braincase with flattened upper surface, ovoid bullae, wide choanae, short and wide palate, presence of posterior accessory cusp in the P3, very small and low protocone in the upper carnassial, protocone aligned with the mesial margin of the parastyle, small M1 situated vertically to the P4, shallow mandibular corpus, short symphysis almost vertical to the ramus, deep masseteric fossa with its anterior margin well behind the m1, absence of anterior accessory cuspid in the p3, small posterior accessory cuspid in the p3 situated on a distal cingular projection, presence of both accessory cuspids in the p4, and presence of a small talonid and tiny metaconid in the m1 (Roussiakis et al. 2006; Koufos 2009, 2012). It is worth mentioning that the recent discovery of a skull in the locality Karaslari (FYROM) allowed the erection of the new genus *Yoshi* with two species: *Y. garevskii*, including the Karaslari skull and *Y. minor*, including the Chinese material referred to as *M. minor*. As the distinction of these two species is only based on cranial characters and they have similar size and dental morphology, it is difficult to determine the fragmentary material referred as *M. parvulus*, which can belong to the one or the other species or to none of them (Spassov and Geraads 2014). Moreover, the synonymy between the two genera (*Metailurus* and *Yoshi*) is also possible until the finding of a complete skull from the type locality Pikermi, clarifying the taxonomic value of the various taxa. Therefore, the fragmentary material from Pikermi, although it could represent a valid distinct species, it also has uncertain generic affinities, and it could be referred to either as *Y.* cf. *garevskii* or *Y.* cf. *minor*, or *M.* cf. *parvulus* or *"M. parvulus"* (Koufos et al. 2018). Here, I used the last name; the brackets indicate the questionable taxonomic position of this species.

Pristifelis Salesa et al., 2012

Type Species *Felis attica* Wagner, 1857.

Remarks A felid cranium (MNHN-SLQ-935) of Arambourg's collection from Ravin du Vatiluk in Axios Valley (this locality is correlated to the new one Vathylakkos-3) has been described as *Felis* cf. *attica* by Arambourg and Piveteau (1929: p. 125; pl. 11, figs 1, 1a). Later, it is reported as *Felis attica* by Pilgrim (1931), *Felis* sp. by Koufos (2000, 2011) and recently as *Pristifelis attica* by Salesa et al. (2012). According to Arambourg and Piveteau (1929), the skull differs from the typical *F. attica* from Pikermi displaying larger size, ectoparastyle, and weaker protocone in the P4. Pilgrim (1931) added that MNHN-SLQ-935 has longer facial region than that of *F. attica*. A more recent comparison of MNHN-SLQ-935, except of the above-mentioned differences, indicates that the teeth are generally longer and narrower than the typical *F. attica* (Koufos 2000). Recently, Salesa et al. (2012) include this skull to the new genus *Pristifelis* and attributed it as *P. attica*. However, the limited material of these small felids cannot allow certain comparisons and definition of the variability in their characters and thus it is better to leave it separately without specific determination, waiting for more material. It is noteworthy that there is a maxillary fragment with p3–m1 from Maragheh (de Mequenem 1924–1925: pl. 9, fig. 5) displaying the same characters with the Axios Valley specimen (Arambourg and Piveteau 1929).

★*Pristifelis attica* (Wagner, 1857)

Nomenclatural and Taxonomical History *Felis attica* in Wagner 1857 (new species); *Pristifelis attica* in Salesa et al. 2012 (new genus and new combination).

Type Material BSPM-PIK-AS II.116 (holotype), cranial fragment associated with the mandible housed in the Bayerische Staadtsamlung für Paläontologie und Geologie, München (Wagner 1857: p. 122; tab. 3, fig. 4).

Type Locality Pikermi (PIK), Attica, Greece, late Miocene, MN12.

Distribution Besides Pikermi, the species is mentioned without description from Thermopigi in Serres Basin and from locality A and some unknown locality(–ies) of Samos Island. Besides Greece, it is mentioned from Spain, Germany, Hungary, Moldova, Georgia, Russia, Turkey, Iran, and Afghanistan.

Remarks Except the holotype (Fig. 2f–g), some more material is known from Pikermi, and has been described by Gaudry (1862–1867) and Roussiakis (2002). A cranium associated with the mandible from an unknown locality of Samos is referred to a new species *Felis neas* without description (Forsyth Major 1891, 1894), that is considered as a *nomen nudum* (de Beaumont 1961). Another cranium (NHMW-SAM-A.4751) from an unknown locality of Samos, housed at the Naturhistorisches Museum Wien, is also attributed to *P. attica* (Koufos 2000). Some postcranial remains from Samos referred to as *Felis* sp. by Forsyth Major (1894), attributed to "*Metailurus parvulus*" but later transferred to *Felis attica* (de Beaumont 1961, 1986). All these specimens are referred herein to *P. attica*. The species is characterized by elongated cranium, relatively long muzzle, inflated auditory bullae, oval-shaped orbits, oval nasal cavity, well-developed mastoid process, double-rooted P2, P3 lacking anterior but bearing a small posterior accessory cusp, upper carnassial with small protocone aligned with the mesial margin of the parastyle, absence of ectoparastyle, triangular M1 with clear paracone and metacone, wide and short symphysis, high mandibular corpus, elongated masseteric fossa with its anterior margin below the posterior half of the m1, presence of a small anterior and posterior accessory cuspid in the p3 and p4, and m1 with a small metaconid and absence of talonid (Pilgrim 1931; Salesa et al. 2012).

Felis Linneaus, 1758

Type Species *Felis catus* Linneaus, 1758.

⊙*Felis silvestris* Schreber, 1777—the Wildcat

Distribution *Felis silvestris* is an extant species living in Eurasia, Africa, and America, and is also known from several fossiliferous sites. In Greece, the species is recognized from Petralona Cave in Chalkidiki, Vraona Cave in Attica, and Apidima Cave-C in Peloponnese.

Remarks Fossils of this species are rare in Greece, found only in cave deposits. The earliest evidence of its presence comes possibly from Petralona Cave, where Sickenberg (1971) described some material as? *F. silvestris*. Later, its presence was confirmed by new material described as a new subspecies *F. s. hamadryas* by Kurtén and Poulianos (1977, 1981). Later on, the revision of the material from Petralona cave, housed at the Laboratory of Geology and Paleontology, Aristotle University of Thessaloniki, confirmed the presence of the species in the cave (Tsoukala 1989;

Baryshnikov and Tsoukala 2010). The Late Pleistocene forms of this taxon are generally similar to the modern ones, although there are some minor differences due to age and/or geographical reasons, allowing the distinction of several subspecies.

Lynx Kerr, 1792

Type Species *Felis lynx* Linnaeus, 1758.

Remarks Some felid material from the Early Pleistocene locality Volax in Drama Basin is referred to as? *Lynx issiodorensis* (Sickenberg 1968: pl. 3, fig. 4) and from Tourkobounia 3–5 as *L.* cf. *issiodorensis* by Symeonidis and de Vos (1976).

Lynx issiodorensis (Croizet and Jobert, 1828)

Nomenclatural and Taxonomical History *Felis issiodorensis* in Croizet and Jobert 1828 (new species); *Felis (Lynx) issiodorensis* in Viret 1954 (new combination); Morales et al. (2003) transferred this taxon to *Caracal* and they mentioned it as *Caracal issiodorensis.*

Type Material MNHB Prr 200 (holotype), cranium (Croizet and Jobert 1828).

Type Locality Étouaires, Perrier, France, Early Pleistocene, MN16.

Distribution In Greece, it is recognized from Apollonia 1 in Mygdonia Basin and Pyrgos in Peloponnese, as well as in the fissure fillings Tourkobounia 3–5 in Attica. The species is also known from several Eurasian countries, e.g., Spain, France, Italy, Romania, Turkey, Russia, Kazakhstan, and Mongolia.

Remarks The species is rare in Greece and the known material quite scarce, except the material from Apollonia 1 that is quite rich, including two hemimandibles and one upper canine (Fig. 2h). The main characters of *L. issiodorensis* are the relatively narrow skull, the robust zygomatic arches, the low sagittal and nuchal crests, the very large and rounded auditory bullae, the large and flat palate, the absence of the P2, the long and slender P3 relatively to that of *L. lynx* with posterior accessory cusp, the presence of a small ectoparastyle in the upper carnassial, the slender protocone with its mesial margin in front to that of the parastyle, the massive mandible, the high and posteriorly-oriented coronoid process, the small angular process, the very wide masseteric fossa extending far in front, the relatively long c–p3 diastema, the larger p3 than that of *L. lynx* displaying a small posterior accessory cuspid, the absence or vestigial metaconid in the m1, the vestigial talonid in the m1, the concave occlusal aspect of the lingual border of the m1 (Kurtén 1978; Koufos 1992; Cherin et al. 2013).

⊙*Lynx lynx* (Linneaus, 1758)—the Balkan Lynx

Distribution The modern lynx is well known from Europe and North America but it is rare in Asia. In Greece, fossil remains of this species are known from the Late Pleistocene deposits of Vraona Cave and Apidima Cave-B, C, D.

Remarks A single calcaneum is reported from Vraona Cave, while two upper canines, one p3 and some postcranial remains are known from Apidima Cave (Symeonidis et al. 1980; Tsoukala 1999).

Pantherinae Pocock, 1917

Panthera Oken, 1816

Type Species *Felis pardus* Linneaus, 1758.

Remarks The possible presence of *Panthera leo fossilis* is referred from Megalopolis, Marathousa (Sickenberg 1975).

Panthera gombaszogensis (Kretzoi, 1938)

Nomenclatural and Taxonomical History *Leo gombaszogensis* in Kretzoi 1938 (new species); *Panthera gombaszogensis* in Hemmer 1971 (new combination); *Panthera onca gombaszogensis* in Hemmer 2001 (new combination and referral to the subspecies level); *Panthera gombaszogensis* in Argant and Argant 2011 (referral to the species level).

Type Material MNM 291 (holotype), upper carnassial, described and figured by Kretzoi (1938: p. 10; taf. 1; figs 2, 3).

Type Locality Gombaszög, Hungary, Early Pleistocene.

Distribution This species is known from several Eurasian localities; Argant and Argant (2011) refer that, up to now, it is known from more than 50 Eurasian sites. In Greece, the taxon is known from Gerakarou 1 and Apollonia 1 in Mygdonia Basin, as well as from Alykes and Volos in Thessaly.

Remarks There is a debate concerning the taxonomy of the Villafranchian jaguars and different proposals have been formulated. Hemmer et al. (2010) classified all material in the modern species *Panthera onca* and separated three subspecies, *P. o. toscana, P. o. gombaszogensis,* and *P. o. georgica*. Argant and Argant (2011) proposed that *P. gombaszogensis* is a valid species, appeared in Africa at ~1.90 Ma and migrated to Europe; the Upper Valdarno and Olivola material represent the oldest European evidence of the species and reported to as *P. g. toscana*. This subspecies gave two branches: the one of *P. g. gombaszogensis* that survived in Europe until 0.3 Ma, and the other of *P. g. georgica* from which *P. onca* originates. The Gerakarou

1 skull (Fig. 2i) is closer to the Upper Valdarno and Olivola material and attributed as *P. g. toscana*; the material from the localities Apollonia 1 and Alykes is insufficient for subspecific determinations and thus it is attributed to *P. gombaszogensis*. There is only one reference without description for the presence of *P. gombaszogensis* in Volos (van der Meulen and van Kolfschoten 1986). The taxon is characterized by elongated and high skull, large infraorbital foramen, very high sagittal crest, very strong and canine like I3, strong and relatively small upper canine, elongated P3 without anterior and with strong posterior accessory cusps, elongated and narrow upper carnassial, relatively strong protocone with its mesial margin aligned to that of the parastyle, absence of ectoparastyle, small M1, relatively weak p3 with large anterior accessory cuspid and well-developed distal cingulum, p4 with large anterior and posterior accessory cuspids, and m1 without talonid (Kretzoi 1938; Koufos 1992, 2018; Argant and Argant 2011).

⊙*Panthera pardus* (Linneaus, 1758)—the Leopard

Distribution This is an extant species living in different countries of Asia and Africa; it is also known from several fossil localities as well. In Greece, it is known from Loutra Almopias Cave in Western Macedonia, Apidima Cave-B, C in Peloponnese, and Vraona Cave in Attica.

Remarks The taxon is found only in various caves of Greece and it is similar to the extant species.

⊙*Panthera leo* (Linneaus, 1758)—the Lion

Distribution This extant species is distributed in the Africa and Asia, with numerous occurrences in the fossil record as well. Fossil remains of the species are known from Marathousa-Megalopolis, in Peloponnese.

Remarks It is similar to the modern species. For the Greek fossils found in Megalopolis, there is not detailed description.

Panthera leo spelaea (Goldfuss, 1810)—the Cave Lion

Type Material MB.Ma.50948 (holotype), cranium of an adult male individual, housed at the Museum für Naturkunde, Berlin (Diedrich 2008: fig. 5).

Type Locality Zoolithen Cave No. D 106, Geilenreuth, Bavaria, Germany, Upper Pleistocene, Weichselian/Würmian (probably OIS 3; Diedrich 2008)

Distribution *Panthera leo spelaea* is well distributed in Eurasia, Canada and Alaska. The oldest known fossils of the taxon in Europe are known from Italy (0.7 Ma; Sala 1990) and from England (0.75–0.65 Ma, Lewis et al. 2010), and in

Asia from Siberia (0.62 Ma), while the youngest evidences dated at 13,000 years ago (Stuart and Lister 2011).

Remarks The material from Petralona Cave is the best known in Greece, including a skull and a mandibular fragment. The material is characterized by elongated and wide nasal bone, large infraorbital foramen, strong sagittal crest, strong occipital condyles, high and wide occipital foramen, well-developed posterior accessory cusp in the P3, strong parastyle and protocone in the P4, relatively very small M1, high mandibular corpus with almost vertical symphysis, strong lower canine, weak anterior and strong posterior accessory cuspid in the p3, large accessory cuspid in the p4, and posterior accessory cuspid situated on a distal cingular projection in both premolars (Tsoukala 1989).

Machairodontinae Gill, 1872

Amphimachairodus Kretzoi, 1929

Type Species *Machairodus palanderi* Zdansky, 1924.

Remarks The genus is mentioned without description from Thermopigi in Serres Basin, referred to as *Machairodus* sp. (Geraads et al. 2007).

★*Amphimachairodus giganteus* (Wagner, 1848)

Nomenclatural and Taxonomical History *Felis gigantea* in Wagner 1848 (new species); *Machairodus giganteus* in de Beaumont 1975 (new combination and mandatory change); *Amphimachairodus giganteus* in Salesa et al. 2012 (new combination).

Type Material Olecranon (holotype; number unknown) (Wagner 1848:375, fig. 6).

Type Locality Pikermi, Attica, Greece, late Miocene, MN12.

Distribution Besides the type locality, the species is recognized in the late Miocene of Spain, France, Hungary, Bulgaria, Turkey, and Tajikistan. In Greece, it is known from Ravin des Zouaves-5, Ravin X in Axios Valley, Kryopigi in Chalkidiki, Halmyropotamos (Fig. 3a, b), and Kerassia 4 in Evia Island, as well as from Mytilinii-1 and locality A in Samos Island.

Remarks According to de Beaumont (1975) *A. giganteus* differs from *Machairodus aphanistus* by its narrower upper canine, the more pronounced ectoparastyle in the P4, the reduced and more posteriorly situated protocone in the P4 and the longer and narrower upper teeth. In the lower teeth, the accessory cuspids of the p4 are higher and slenderer, the protoconid of the m1 is longer, and the talonid of the m1 is strongly reduced.

Paramachaerodus Pilgrim, 1913

Type Species *Machairodus orientalis* Kittl, 1887, by the subsequent designation of Pilgrim 1931. Originally, Pilgrim (1913) had not defined the type species of *Paramachaerodus*, which originally included the species *M. orientalis* Kittl, 1887 and *M. schlosseri* Weithofer, 1888. In a latter publication (Pilgrim 1931) fixed the former as the type species of *Paramachaerodus* (see also Salesa et al. 2010).

Remarks The genus is reported from the late Miocene locality Thermopigi in Serres Basin without description (Geraads et al. 2007).

Paramachaerodus orientalis (Kittl, 1887) (= ★*Machairodus schlosseri* Weithofer, 1888)

Nomenclatural and Taxonomical History *Machairodus orientalis* in Kittl 1887 (new species); *Paramachaerodus orientalis* in Pilgrim 1913 (new combination). More information for the taxonomy of this species are given by Salesa et al. (2010). Shortly after the erection of *Machairodus orientalis*, Weithofer (1888) described *Machairodus schlosseri* from the late Miocene of Pikermi. However, this species was subsequently considered as a junior synonym of *Machairodus orientalis*, an opinion that still holds up to now.

Type Material NHMW-2007z0172/0001 (holotype), frontal part of the cranium with the tooth rows, housed in the Naturhistorisches Museum Wien (Kittl 1887: p. 329; taf. 14, figs 1–5).

Type Locality Maragheh, Iran, late Miocene, MN11–12.

Distribution In Greece, it is described from Pikermi in Attica (Fig. 3c) and Nea Silata in Chalkidiki. Besides Iran and Greece, the species is also known from Spain, Germany, Hungary, Bulgaria, FYROM, Ukraine, and Turkey.

Remarks The species was erected based on a cranial fragment from Maragheh, Iran (Kittl 1887) as *Machairodus* and later transferred to *Paramachaerodus* by Pilgrim (1913). The main characters of the taxon, modified from the emended diagnosis of Salesa et al. (2010), are: serrated upper and lower canines; markedly larger I3 than I1 and I2; small posterior accessory cusp and absence of anterior on the P3; upper carnassial with well-developed parastyle and small ectoparastyle; reduced protocone; p3 and p4 wider mesially than distally; p3 lacks anterior and bears a very small posterior accessory cuspid; presence of both accessory cuspids in the p4; very small talonid without cuspids in the lower carnassial.

Homotherium Fabrini, 1890

Type Species *Machairodus latidens* Owen, 1846.

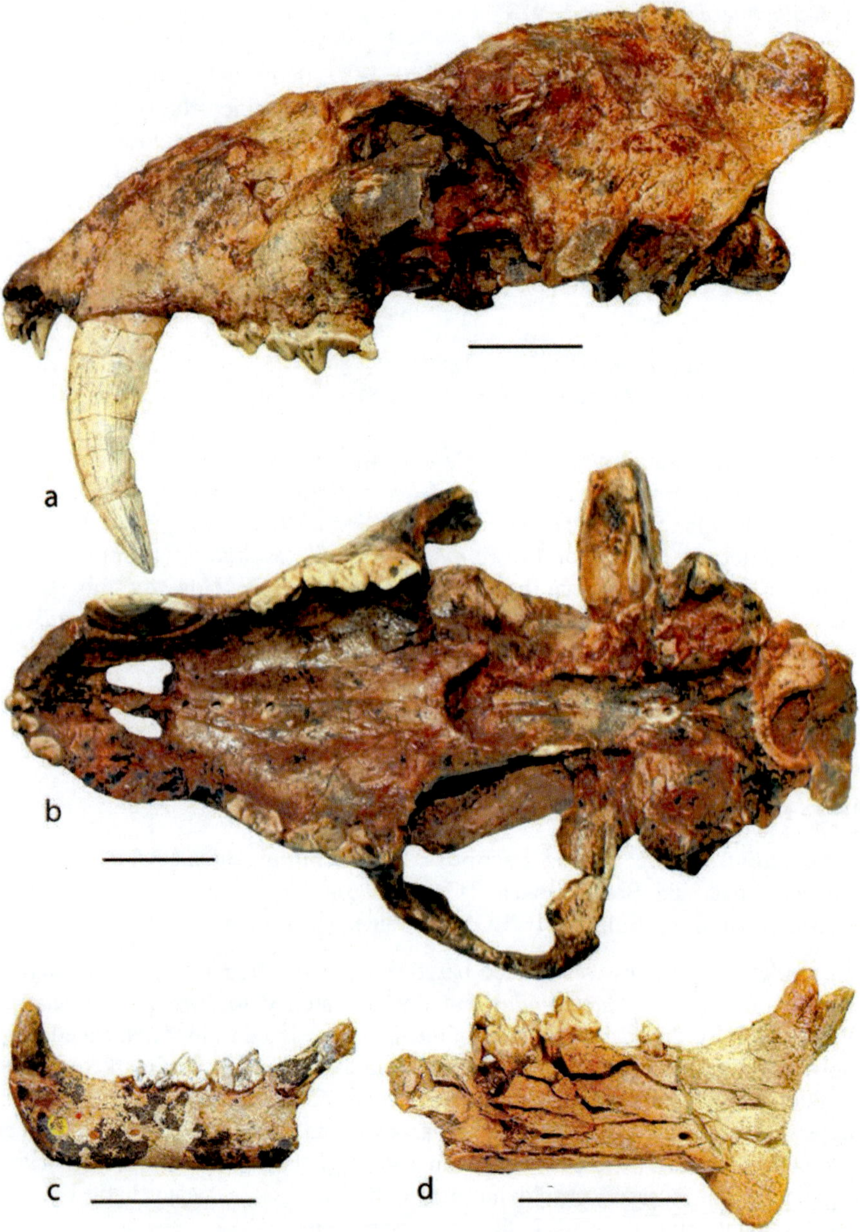

Fig. 3 Selected specimens of Machairodontinae (Felidae) from Greece. **a–b** cranium of *Amphimachairodus giganteus* from Halmyropotamos in **a** lateral, and **b** palatal views, **c** left mandible of *Paramachaerodus orientalis* from Pikermi in buccal view, **d** right mandible of *Megantereon cultridens* from Apollonia in buccal view. Scale bar equals 2 cm

Remarks In the Greek fossil record, there is one lower canine from Tourkobounia 3–5 described as *Homotherium* cf. *crenatidens* (=*H.* cf. *latidens*) by Symeonidis and de Vos (1976) and one upper carnassial and a mandibular fragment from Libakos described as *Homotherium* sp. (Steensma 1988). Some fossils from Sesklon are also related to *Homotherium* and they are referred to as cf. *H. latidens* (Athanassiou 1998).

Homotherium latidens (Owen, 1846)

Nomenclatural and Taxonomical History *Machairodus latidens* in Owen 1846 (new species*);* *Homotherium crenatidens* and *Homotherium nestianus* in Fabrini 1890 (senior synonyms); *Homotherium latidens* in Anton et al. 2005, 2009, 2014 (new combination). See Anton et al. (2014) for an overview.

Type Material No 443 or 103 (holotype), upper canine, housed in Royal College of Surgeons, which was destroyed during bombing in 1941. Any of the other known material (paratypes), housed at different museums, can be chosen as a neotype (Barnett 2014).

Type Locality Kent's cavern, Devonshire, England, Pleistocene.

Distribution The species is known, including its synonymies, from various areas of Eurasia, including Spain, France, Italy, Slovakia, Ukraine, Georgia, Russia, and China. In Greece, it is referred from Milia (Grevena Basin) and Apollonia 1 (Mygdonia Basin).

Remarks The taxon is rare in Greece and the known material is fragmentary and difficult to determine. The oldest occurrence of the species is traced by an isolated upper canine from Milia, in Grevena Basin, dated to the Late/Early Pleistocene, MN16 (Tsoukala et al. 2014). Recently it is recognized in Apollonia 1 with a lower canine, a fragmentary p4, and one McII (Koufos 2018). The main characters of the species (modified from Sardella and Iurino (2012) and Anton et al. (2014) are the serrated upper canine, the absence of the ectoparastyle and the reduced protocone in the upper carnassial, the relatively high and robust mandibular corpus, the deep masseteric fossa with its anterior margin below the middle of the m1, the size of the i3 which is twice the size of the i1, the small, bucco-lingually flattened and serrated on mesial and distal margin lower canine, the small and single-rooted p3, the larger p4 than p3 with small anterior and posterior accessory cuspids, and the elongated and bucco-lingually flattened m1 without talonid.

Megantereon Croizet and Jobert, 1828

Type Species *Ursus cultridens* Cuvier, 1824.

***Megantereon cultridens* (Cuvier, 1824; partim)**

Nomenclatural and Taxonomical History *Ursus cultridens* in Cuvier 1824 (new species); *Megantereon cultridens* in Ficarrelli 1979 (new combination).

Type Specimen IGF 816 (lectotype), right upper canine, housed at the Museum of Natural Sciences, University of Florence (Ficarrelli 1979).

Type Locality Upper Valdarno, Italy, Early Pleistocene (Tasso Faunal Unit).

Distribution The species is known from the Pleistocene of Eurasia reported from Spain, France, Italy, Georgia, and Tajikistan, as well as from Africa. In Greece, it is known from Volax (Drama basin), Apollonia 1 (Mygdonia Basin), and Makinia (southern Greece).

Remarks The relatively scarce fossil material and the limited known evolution of *Megantereon* cause taxonomic problems and the debate continuous during the last decades. The Apollonia *Megantereon* (Fig. 3d) is transferred to the African species *M. whitei* because of its smaller size, like that of Venta Micena (Spain) and Dmanisi (Georgia) (Martínez Navarro and Palmqvist 1996). However, other authors do not agree with this opinion, believing that the small-sized *Megantereon* could be a chronocline of *M. cultridens* or even a new species (Lewis and Werdelin 2010). The Greek material is characterized by large upper canine without serrated mesial and distal margins, small p3 lacking anterior accessory cuspid and displaying a small posterior one, short and narrow p4, large anterior and smaller posterior accessory cuspid in the p4, and absence of talonid in the m1 (Koufos 1992).

Acknowledgments I would like to thank M. J. Salesa for comments that improved this manuscript.

Appendix

List of the Greek localities containing felid fossils. Type localities are marked with bold. Locality numbers refer to the collection numbers of the PaleoBiology Database (PBDB)

Localities[PBDB No]	Age (MN; GPTS in Ma)	Taxon
Apidima-B[32062]	Late Pleistocene	*Lynx lynx*[1] *Panthera pardus*[1]
Apidima-C[32082]	Late Pleistocene	*Lynx lynx*[1] *Felis silvestris*[1] *Panthera pardus*[1]
Apidima-D[32086]	Late Pleistocene	*Lynx lynx*[1]
Loutra Almopias Cave[203847]	Late Pleistocene	*Panthera leo spelaea*[2] *Panthera pardus*[2]
Vraona[183130]	Late Pleistocene	*Felis silvestris*[3] *Lynx lynx*[3] *Panthera pardus vraonensis*[4]

(continued)

Localities[PBDB No]	Age (MN; GPTS in Ma)	Taxon
Petralona Cave[183123]	Late M. Pleistocene	*Felis silvestris*[5] *Panthera leo spelaea*[6]
Apollonia 1[197942]	Epivillafranchian (MNQ20)	*Lynx issiodorensis*[7] *Megantereon cultridens*[7] *Panthera gombaszoegensis*[8] *Homotherium latidens*[8]
Alikes[34782]	?Late Villafranchian	*Panthera gombaszoegensis*[9]
Pyrgos[34655]	?Late Villafranchian	*Lynx issiodorensis*[10]
Libakos[34764]	?Late Villafranchian	*Homotherium* sp.[11]
Gerakarou 1[134617]	Late Villafranchian (MN18)	*Panthera gombaszoegensis*[7]
Marathousa 1[187637]	Middle Pleistocene	*Panthera leo fossilis*?[12]
Sesklon[34614]	Middle Villafranchian (MN17)	cf. *Homotherium latidens*[9]
Volax[34593]	Middle Villafranchian (MN17)	*?Lynx issiodorensis*[13] *Megantereon cultridens*[13]
Tourkobounia 3–5[34592]	Early-Middle Villafranchian (MN16–17)	*Lynx issiodorensis*[14] *Lynx* cf. *issiodorensis*[14] *Homotherium* cf. *latidens*[14]
Milia-5[185859]	Early Villafranchian (MN16; ~3.0–2.5)	*Homotherium latidens*[15, 8]
Milia-1a[185861]	Early Villafranchian (MN16; ~3.0–2.5)	*Homotherium latidens*[15, 8]
Milia-2a[185862]	Early Villafranchian (MN16; ~3.0–2.5)	*Homotherium latidens*[15, 8]
Makinia[34811]	Villafranchian	*Megantereon cultridens*[16]
Volos[34808]	?Villafranchian	*Panthera gombaszoegensis*[10]
Nea Silata[191612]	Miocene/Pliocene (MN13/14)	*Paramachaerodus orientalis*[17]
Samos	Turolian (MN13)	*Pristifelis attica*[18]
Samos Q5[95690]	Late Turolian (MN13; 6.9–6.7)	*Metailurus major*[19] *"Metailurus parvulus"*[19]
Kryopigi[157582]	Middle–late Turolian (MN13; 7.3–6.4)	*"Metailurus parvulus"*[20] *Amphimachairodus giganteus*[20]
Samos A[206461]	Middle Turolian (MN12)	*Metailurus major*[19] *Pristifelis* sp.[19] *"Metailurus parvulus"*[19]
Vathylakkos 3[182750]	Middle Turolian (MN12; ~7.3)	*Pristifelis* sp.[21]
Pikermi[182754]	**Middle Turolian (MN12; 7.3–7.1)**	***"Metailurus parvulus"*[22]** *Metailurus major*[23] ***Pristifelis attica*[24]** ***Amphimachairodus giganteus*[25]** *Paramachaerodus orientalis*[26]
Kerassia 1[195432]	Middle Turolian (MN12)	*"Metailurus parvulus"*[27]
Kerassia 4[195435]	Middle Turolian (MN12)	*Amphimachairodus giganteus*[28]
Chomateres[195562]	Middle Turolian (MN12)	*"Metailurus parvulus"*[29]
Mytilinii 1A[202215]	Middle Turolian (MN12; ~7.1)	*"Metailurus parvulus"*[30]

(continued)

Localities[PBDB No]	Age (MN; GPTS in Ma)	Taxon
Mytilinii 1[202215]	Middle Turolian (MN12; ~7.1)	*Amphimachairodus giganteus*[31]
Halmyropotamos[202213]	?Middle Turolian (MN12?)	*"Metailurus parvulus"*[32] *Metailurus major*[32] *Amphimachairodus giganteus*[32]
Thermopigi[73553]	?Middle Turolian (MN12?)	*Pristifelis attica*[33] *Amphimachairodus* sp.[33] *Paramachaerodus* sp.[33]
Ravin des Zouaves 5[195489]	Early Turolian (MN11; ~8.2)	*Amphimachairodus giganteus*[34]
Ravin X[182745]	?Early Turolian	*Amphimachairodus giganteus*[21]
Ravin de la Pluie[191070]	Late Vallesian (MN10; ~9.3)	*"Metailurus parvulus"*[34]
Antonios[73861]	Early/middle Miocene (MN5)	*Pseudaelurus romieviensis*[35]

GPTS Geomagnetic Polarity Time Scale
[1]Tsoukala (1999), [2]Tsoukala et al. (2006), [3]Symeonidis et al. (1980), [4]Nagel (1999), [5]Sickenberg (1971), [6]Kanellis (1962), [7]Koufos (1992), [8]Koufos (2018), [9]Athanassiou (1998), [10]van der Meulen and van Kolfschoten (1986), [11]Steensma (1988), [12]Sickenberg (1975), [13]Sickenberg (1968), [14]Symeonidis and de Vos (1976), [15]Tsoukala et al. (2014), [16]Symeonidis et al. (1985–86), [17]Koufos (2006), [18]Forsyth Major (1891), [19]Solounias (1981), [20]Lazaridis (2015), [21]Arambourg and Piveteau (1929), [22]Hensel (1862), [23]Roussiakis (2001), [24]Wagner (1857), [25]Wagner (1848), [26]Pilgrim (1931), [27]Roussiakis et al. (2006), [28]Roussiakis and Theodorou (2003), [29]Symeonidis (1978), [30]Koufos (2009), [31]Koufos and Melentis (1982), [32]Melentis (1967), [33]Geraads et al. (2007), [34]Koufos (2012), [35]Koufos (2008)
[a]This study

References

Anton M, Salesa MJ, Galobart A, Tseng ZJ (2014) The Plio-Pleistocene scimitar-toothed felid genus *Homotherium* Fabrini, 1890 (Machairodontinae, Homotheriini): diversity, palaeogeography and taxonomic implications. Quat Sci Rev 96:259–268

Arambourg CJ, Piveteau J (1929) Les Vertébrés du Pontien de Salonique. Ann Paleontol 18:59–138

Argant A, Argant J (2011) The *Panthera gombaszogensis* story: the contribution of the Château breccia (Saône-et-Loire, Burgundy, France). Quat Hors-série 4:247–269

Athanassiou A (1998) Contribution to the study of the fossil mammals of Thessaly. PhD thesis. Gaia 5:1–354

Barnett R (2014) An inventory of British remains of *Homotherium* (Mammalia, Carnivora, Felidae) with special reference to the material from Kent's cavern. Geobios 47:19–29

Baryshnikov GF, Tsoukala E (2010) New analysis of the Pleistocene carnivores from Petralona cave (Macedonia, Greece) based on the collection of the Thessaloniki Aristotle University. Geobios 43:389–402

Carroll RL (1988) Vertebrate paleontology and evolution. WH Freeman, New York

Cherin M, Iurino DA, Sardella R (2013) New well-preserved material of *Lynx issiodorensis valdarnensis* (Felidae, Mammalia) from the early Pleistocene of Pantalla (Central Italy). Boll Soc Paleontol Ital. https://doi.org/10.4435/BSPI.2013.16

Croizet JB, Jobert A (1828) Recherches sur les ossements fossiles du Département du Puy-de-Dôme. Thibaut-Landriot, Paris de Beaumont G (1961) Recherches sur *Felis attica* Wagner eurasiatique avec quelques observations sur les genres *Pseudaelurus* Gervais et *Proailurus* Filhol. Nouv Arch Mus Hist Nat Lyon 6:1–45

de Beaumont G (1975) Recherches sur les félidés (Mammifères, Carnivores) du Pliocène inférieur des sables a *Dinotherium* des environs d'Eppelsheim (Rheihessen). Arch Sci Genève 28:369–405

de Beaumont G (1986) La patte postérieure de *Felis attica* Wagn. (Mammifère, Carnivore) du Turolien de Grèce. Arch Sci 39(3):377–386

de Mequenem R (1924–1925) Contribution à l'étude des fossiles de Maragha. Ann Paleontol 13–14:135–160

Diedrich C (2008) The rediscovery holotypes of the upper Pleistocene spotted hyena *Crocuta crocuta spelaea* (Goldfuss 1823) and the steppe lion *Panthera leo spelaea* (Goldfuss 1810) and taphonomic discussion to the Zoolithen cave hyena den at Geilenreuth (Bavaria South-Germany). Zool J Linn Soc 154:822–831

Etnyre E, Lande J, McKenna A (2011) Felidae (On-line), Animal Diversity Web. http://animal-diversityorg/accounts/Felidae/. Accessed 10 May 2018

Ficarrelli G (1979) The Villafranchian machairodonts of Tuscany. Palaeontogr It 71:17–26

Forsyth Major CI (1891) Considerations nouvelles sur la faune des Vertébres du Miocene supérieur dans l'ile de Samos. C R Acad Sci 113:608–610

Forsyth Major CJ (1894) Le gisement ossifère de Mytilinii et catalogue d'ossements fossiles recueillis à Mytilini, île de Samos, et déposés au Collège Galliard, à Lausanne. Georges Bridel & Co, Lausanne

Gaudry A (1862–1867) Animaux fossils et géologie de l'Attique. Savy, Paris

Geraads D, Tsoukala E, Spassov N (2007) A skull of *Ancylotherium* (Chalicotheriidae, Mammalia) from the late Miocene of Thermopigi (Serres, N.Greece), and the relationships of the genus. J Vertebr Palaeontol 27(2):461–466

Hemmer H, Kahlke R-D, Vekua A (2010) *Panthera onca georgica* ssp. nov. from the early Pleistocene of Dmanisi (republic of Georgia) and the phylogeography of jaguars (Mammalia, Carnivora, Felidae). N J Geol Palaont Abh 257:115–127

Hensel RF (1862) Über die Reste einiger Säugetierarten von Pikermi in der Münchener Sammlung. Monatsber Akad Wiss 27:560–569

Kanellis A (1962) The Petralaona cave. Vuno 7:143–151

Kittl E (1887) Beiträge zur Kennthniss der fossilen Säugethiere von Maragha in Persia. I. Carnivora. Ann Naturhist Hofmus 2:317–338

Koufos GD (1992) The Pleistocene carnivores of the Mygdonia basin (Macedonia, Greece). Ann Paleontol 78:205–257

Koufos GD (2000) Revision of the late Miocene carnivores from the lower Axios valley. Münchner Geowiss Abh A 39:51–92

Koufos GD (2006) Carnivora. In: Koufos GD (ed) The late Miocene vertebrate locality of Perivolaki, Thessaly, Greece. Palaeontographica Abt A 276:39–74

Koufos GD (2008) Carnivores from the early/middle Miocene locality of Antonios (Chalkidiki, Macedonia, Greece). Geobios 41:365–380

Koufos GD (2009) Carnivora. In: Koufos GD, Nagel D (eds) The Late Miocene Mammal Faunas of the Mytilinii Basin, Samos Island, Greece: New Collection. Beit Paläont 31:57–105

Koufos GD (2011) The Miocene carnivore assemblage of Greece. Estud Geol 67:291–320

Koufos GD (2012) New material of Carnivora (Mammalia) from the late Miocene of Axios Valley, Macedonia, Greece. C R Palevol 11:49–64

Koufos GD (2018) New material and revision of the Carnivora, Mammalia from the lower pleistocene locality Apollonia 1, Greece. Quaternary 1:1. https://doi.org/10.3390/quat1010006

Koufos GD, Melentis J (1982) The late Miocene (Turolian) mammalian fauna of Samos island (Greece). Study of the collection of Palaeontological Museum of Mytilinii, Samos. 1. Carnivora. Sci Ann Fac Phys Math Univ Thessaloniki 22:175–193

Koufos GD, Mayda S, Kaya T (2018) New material of late Miocene carnivores from Turkey. Palaeontol Z 92:131–162

Kretzoi M (1938) Die Raubtiere von Gombaszög nebst einer Übersicht der Gesamtfauna (Ein Beitrag zur Stratigraphie des Altquartärs). Ann Mus Nat Hung 31:88–157

Kurtén B (1978) The lynx from Etouaires, *Lynx issiodorensis* (Croizet & Jobert), late Pliocene. Ann Zool Fennici 15:314–322

Kurtén B, Poulianos A (1977) New stratigraphic and faunal material from Petralona cave with special reference to the Carnivora. Anthropos 4:47–130

Kurtén B, Poulianos A (1981) Fossil carnivora of Petralona cave. Status 1980. Anthropos 8:9–56

Lazaridis GT (2015) Study of the Late Miocene vertebrate locality of Kryopigi and other localities of Kassandra Peninsula, Chalkidiki (Greece). Systematics, Taphonomy, Palaeoecology, Biochronology. PhD Thesis, Aristotle University of Thessaloniki

Lewis M, Pacher M, Turner A (2010) The larger carnivora of the west Ranton freshwater bed. Quat Int 228:116–135

Lewis M, Werdelin L (2010) Carnivoran dispersal out of Africa during the early Pleistocene: relevance for hominids? In: Fleagle JG, Shea JJ, Grine FE, Baden AL, Leakey RE (eds) Out of Africa I: Who? When? and Where? Springer, Dordrecht

Martínez Navarro B, Palmqvist P (1996) Presence of the African saber-toothed felid *Megantereon whitei* (Broom, 1937) (Mammalia, Carnivora, Machairodontinae) in Apollonia-1 (Mygdonia Basin, Macedonia, Greece). J Archaeol Sci 23:869–872

Melentis JK (1967) Studien über fossile Vertebraten Griechenlands. 19. Die Pikermifauna von Halmyropotamos (Euböa, Griechenlands). I Teil: Odontologie und Kraniologie. Ann Geol Pays Helleniques 18:283–411

Morales J, Soria D, Montoya P, Pérez SMJ (2003) *Caracal depereti* nov. sp. y *Felis* aff. *silvestris* (Felidae, Mammalia) del Plioceno inferior de Layna (Soria, Espana). Estudios Geol 59:229–247

Nagel D (1999) *Panthera pardus vraonensis* n. ssp., a new leopard from the Pleistocene of Vraona/Greece. N J Geol Palaont Mon 3(3):129–150

O'Brien JSB, Johnson WE (2007) The evolution of cats. Sci Am 2007:68–75

Orlov YA (1968) Fundamentals of paleontology. In: Mammals, vol 13. Israel Program for Scientific Translations, Jerusalem

Pilgrim GE (1913) The correlation of the Siwaliks with mammal horizons of Europe. Rec Geol Surv India 43:264–326

Pilgrim GE (1931) Pontian Carnivora of Europe. British Museum Natural History, London

Roussiakis S (2001) *Metailurus major* Zdansky, 1924 (Carnivora, Mammalia) from the classical locality of Pikermi (Attica, Greece). Ann Paleontol 87(2):119–132

Roussiakis S (2002) Musteloids and feloids (Mammalia, Carnivora) from the late Miocene locality of Pikermi (Attica, Greece). Geobios 35:699–719

Roussiakis SJ, Theodorou GE (2003) Carnivora from the Late Miocene of Kerassia (Northern Euboea, Greece). Deinsea 10:469–497

Roussiakis S, Theodorou G, Iliopoulos G (2006) An almost skeleton of *Metailurus parvulus* (Carnivora, Felidae) from the late Miocene of Kerassia (northern Euboea, Greece). Geobios 39:563–584

Sala B (1990) *Panthera leo fossilis* (v. Reich., 1906) (Felidae) de Isernia la Pineta (Plèistocène moyen inférieur d'Italie). Geobios 23(2):189–194

Salesa MJ, Antón M, Turner A, Alcala L, Montoya P, Morales J (2010) Systematic revision of the late Miocene saber-toothed felid *Paramachaerodus* in Spain. Palaeontology 53:1369–1391

Salesa MJ, Antón M, Morales J, Peigné S (2011) Functional anatomy of the postcranial skeleton of *Styriofelis lorteti* (Carnivora, Felidae, Felinae) from the Middle Miocene (MN 6) locality of Sansan (Gers, France). Estudios Geol 67:223–243

Salesa M, Antón M, Morales J, Peigné S (2012) Systematics and phylogeny of the small felines (Carnivora, Felidae) from the late Miocene of Europe: a new species of Felinae from the Vallesian of Batallones (MN 10, Madrid, Spain). J Syst Palaeontol 10:87–102

Sardella R, Iurino DA (2012) The latest early Pleistocene sabertoothed cat *Homotherium* (Felidae, Mammalia) from Monte Peglia (Umbria, Central Italy). Boll Soc Paleontol Ital 51:15–22
Sickenberg O (1968) Die Unterpleistozäne fauna von Wolaks (Griech.-Mazedonien) II. Carnivoren. Ann Geol Pays Helleniques 19:621–646
Sickenberg O (1971) Revision der Wirbeltier Fauna der Hohle Petralona (Griechisch Mazedonien). Ann Geol Pays Helleniques 23:230–264
Sickenberg O (1975) Eine Säugertierfauna des tieferen Biharius aus dem Becken von Megalopolis (Peloponnes, Griechenland). Ann Geol Pays Helleniques 27:25–73
Solounias N (1981) The Turolian fauna from the island of Samos, Greece. Contrib Vertebr Evol 6:1–232
Spassov N, Geraads D (2014) A new felid from the late Miocene of the Balkans and the contents of the genus *Metailurus* Zdansky, 1924 (Carnivora, Felidae). J Mamm Evol 22:45. https://doi.org/10.1007/s10914-014-9266-5
Steensma KJ (1988) Plio-Pleistozäne Grossäugetiere (Mammalia) aus dem Becken von Kastoria/Grevena südlich von Neapolis-NW Griechenland. PhD thesis, Technical University of Clausthal
Stuart AJ, Lister AM (2011) Extinction chronology of the cave lion *Panthera spelaea*. Quat Sci Rev 30:2329–2340
Symeonidis N (1978) Ein schädel von *Metailurus parvulus* (Hensel) aus Pikermi (Attica, Griechenland). Ann Geol Pays Helleniques 29:698–703
Symeonidis N, de Vos J (1976) Grosssäuger-Funde aus den Pleistozänen spatenfülungen von Turkovunia in Athen. Ann Geol Pays Helleniques 28:135–144
Symeonidis N, Bachmayer F, Zapfe H (1980) Ergebnisse weiterer Grabungen in der Höhle von Vraona (Attica, Griechenland). Ann Geol Pays Helleniques 30:135–144
Symeonidis N, Theodorou GE, Schütt H, Velitzelos E (1985–86) Palaeontological and stratigraphical observations in Achaia and Aitoloakarnania (Western Greece) area. Ann Geol Pays Helleniques 33:329–365
Tsoukala E (1989) Contribution to the study of the Pleistocene fauna of large mammals (Carnivora, Perissodactyla, Artiodactyla) from Petralona cave, Chalkidiki (N. Greece). Sci Ann Aristotle Univ Thessaloniki 1:1–360
Tsoukala E (1999) Quaternary large mammals from the Apidima caves (Lakonia, S. Peloponnese, Greece). Beit Palaont 24:207–229
Tsoukala E, Chatzopoulou K, Rabeder G, Pappa S, Nagel D, Withalm G (2006) Paleontological and stratigraphical research in Loutra Arideas Bear Cave (Almopia Speleopark, Pella, Macedonia, Greece). Sci Ann School Geol (AUTH) 98:41–67
Tsoukala E, van Loghem W, Lazaridis G, Mol D (2014) Carnivores of the early Villafranchian site of Milia, (Grevena, Macedonia, Greece). 6th International Conference on mammoths and their relatives, May 2014. Sci Ann Sch Geol Aristotle Univ Thessaloniki 102:6–7
Turner A, Antón M (1997) The big cats. Columbia University Press, New York
van der Meulen A, van Kolfschoten T (1986) Review of the late Turolian to early Biharian mammal faunas from Greece and Turkey. Mem Soc Geol Ital 31:201–211
Wagner A (1848) Urweltliche Säugetier-Ueberreste aus Griechenland. Abh Bayer Akad Wiss 5(2):155–170
Wagner A (1857) Neue Beiträge zur Kenntniss der fossilen Saugthier-Ueberreste von Pikermi. Abh Bayer Akad Wiss 8:111–158
Weithofer KA (1888) Beiträge zur Kenntniss der Fauna von Pikermi bei Athen. Beit Palaeontol Geol Osterr 6:225–292
Zdansky O (1924) Jungtertiäre Carnivoren Chinas. Paleontologica Sin C 2:38–45

The Fossil Record of Viverrids (Mammalia: Carnivora: Viverridae) in Greece

George D. Koufos

1 Introduction

The family Viverridae includes the genets, civets, and linsangs; its modern representatives consist of 15 genera and 35 species. The family is known from the Old World (Africa, Asia, and southern Europe) and some islands (Madagascar, Indonesia, Philippines). The viverrids are small-to-medium-sized animals with elongated and slender body and relatively short extremities with tetra- or pentadactyl feet. The feet are plantigrade or digitigrade and the claws semiretractile. Their body weight varies from <1 to 15 kg and their length, without the tail, ranges from 30 to 100 cm. The skull of viverrids is elongated, flattened and with elongated muzzle. The premolars are large and the carnassials well developed and trenchants. The P4 is short and broad with strong protocone. The m1 lacks longitudinal blade and has broad trigonid and long talonid. The upper molars are wider than long with three cuspids, and the lower molars have well-developed talonid. The majority of the modern viverrids are arboreal forms but few taxa are scansorial and can climb. They are mainly nocturnal predators and most are carnivorous, feeding on small invertebrates, insects and worms but some taxa include fruits or roots in their diet (Vaughan et al. 2015; Myers 2000; Orlov 1968). The Viverrids are scarce in Greece and known only from two localities; the family is also rare or unknown in the adjacent countries.

G. D. Koufos (✉)
Department of Geology, Laboratory of Geology and Paleontology, Aristotle University of Thessaloniki, Thessaloniki, Greece
e-mail: koufos@geo.auth.gr

E. Vlachos (ed.), *Fossil Vertebrates of Greece Vol. 2*,
https://doi.org/10.1007/978-3-030-68442-6_17

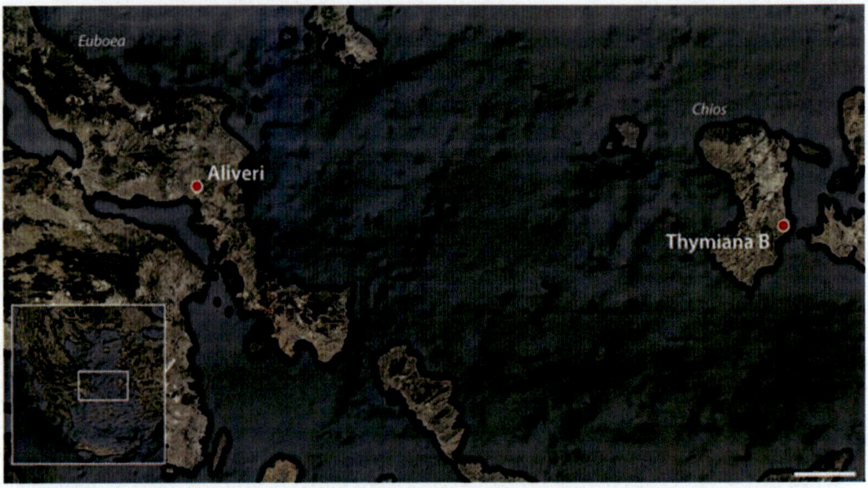

Fig. 1 Map of Greece showing the geographic distribution of the localities with viverrid fossils. See Appendix for more information. Image exported from Google Earth Pro © 2019, map data from US Dept. of State Geographer, SIO, NOAA, U.S. Navy, NGA, GEBCO, image from Landsat/ Copernicus. Scale bar equals 8 km, North faces upward

2 Historical Overview

The Greek fossil record of the viverrids is poor, represented only by two species from the early–middle Miocene that belong to the subfamily Lophocyoninae (Fig. 1 and Appendix). The material is very poor and described by Schmidt-Kittler (1983) and Koufos et al. (1995). The scarcity of the viverrids is possibly due to the limited number of known early–middle Miocene fossiliferous sites. Also, during the late Miocene, the open and dry conditions did not favor the presence of the arboreal viverrids.

3 Phylogenetic Relationships

The family Viverridae *sensu lato* included, until recently, the families Herpestidae and Viverridae *sensu stricto*. Recent molecular studies indicated that the herpestids constitute a separate family, which is considered as the sister group of the family Hyaenidae, whereas Viverridae *sensu stricto* (containing the subfamilies Hemigalinae, Paradoxurinae, Genetinae, and Viverrinae) is considered monophyletic (Veron 2010 and references therein). Some African taxa (*Nandinia*) and oriental linsangs (*Prionodon*) are not real viverrids. The genus *Nandinia* belongs to the separate family Nandiniidae and it is referred as sister group of all modern Felidae (Flynn et al. 2005). The origin of Viverridae and Herpestidae is traced in the late Oligocene at ~25.0 Ma (Werdelin 1996). The early Miocene (>20.0 Ma) *Herpestides* is the oldest known viverrid, traced in France and East Africa (Hunt 1991; Schmidt-Kittler 1987), which, for a long time, was considered as herpestid. During the Neogene/Quaternary the family is known from the tropical/subtropical regions of the Old World, but never moved to the north or migrated to the New World.

4 Systematic Paleontology

Viverridae Gray, 1821

Lophocyoninae Fejfar and Schimt-Kittler, 1987

***Euboictis* Fejfar and Schmidt-Kittler, 1984**

Type Species *Sivanasua aliverensis* Schmidt-Kittler, 1983.

★*Euboictis aliverensis* (Schmidt-Kittler, 1983)

Nomenclatural and Taxonomical History *Sivanasua aliverensis* in Schmidt-Kittler 1983 (new species); *Euboictis aliverensis* in Fejfar and Schmidt-Kittler 1984 (new combination, new genus).

Type Material LGPUT ALI AL-23 (holotype), right P4, housed in the Laboratory of Geology and Paleontology, Aristotle University of Thessaloniki and described and figured by Schmidt-Kittler (1983: p. 303, fig. 1; pl.1, fig. 6).

Type Locality Aliveri, Euboea Island, Greece, early Miocene, MN4.

Distribution It is only known from its type locality.

Remarks The known sample from Aliveri includes a few isolated teeth found during washing sediments for micromammals (Fig. 2a, b). The teeth originally were described as a new species of *Sivanasua*, but soon they were transferred to the new genus *Euboictis*. The species is characterized by a single-rooted and tricuspid P1, a slightly molarized P2 and P3, a paracone-metastyle blade in the P2 and P3 like that of the P4, a short metastyle in the P4, a W-shaped paracone and metacone in the M1 and M2, a continued lingual cingulum in the protocone of the M1 and M2, the absence of conules in the M1 and M2, a relatively molarized p1, the presence of trigonid and short-broad talonid in the lower premolars, an m1 shorter than the m2 with very low trigonid, an m2 talonid longer than trigonid, and an equal-sized hypoconid and hypoconulid in the m2 (Schmidt-Kittler 1983; Fefjar and Schmidt-Kittler 1984).

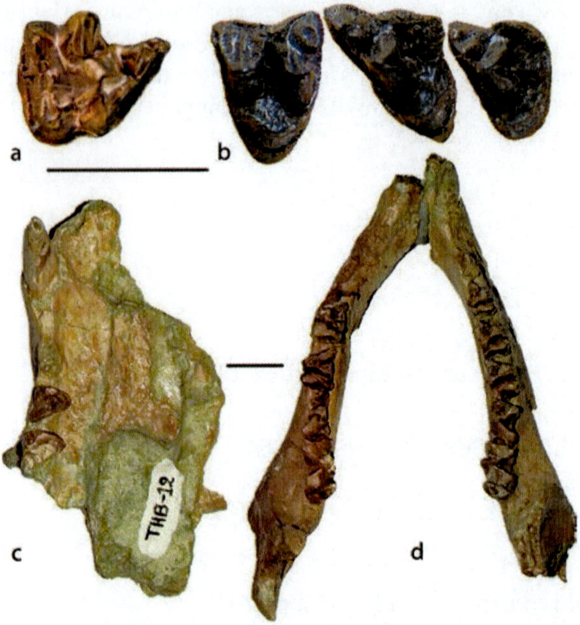

Fig. 2 Selected fossils of Viverridae from Greece. (**a, b**) *Euboictis aliverensis* from the Early Miocene (MN3) of Aliveri, Euboea Island, (**a**) right P4 (holotype) in occlusal view (the tooth has been gilded), (**b**) upper composite dentition (casts) in occlusal view, (**c, d**) *Lophocyon paraskevaidisi*, from the Middle Miocene (MN5) of Thymiana-B, Chios Island, (**c**) right maxilla with P4–M1 in occlusal view, (**d**) mandible (holotype) with the tooth rows in occlusal view. Scale bars equal 1 cm

Lophocyon Fejfar et al., 1987

Type Species *Lophocyon carpathicus* Fejfar et al., 1987.

★*Lophocyon paraskevaidisi* Koufos, de Bonis, Sen, 1995

Type Material LGPUT-THB-1 (holotype), mandible with both tooth rows, housed in the Laboratory of Geology and Paleontology, Aristotle University of Thessaloniki (Koufos et al. 1995: 512, fig. 3; pl. 67, fig. 2).

Type Locality Thymiana-B (THB), Chios Island, Greece, middle Miocene, MN5.

Distribution It is only known from its type locality.

Remarks *Lophocyon paraskevaidisi* was found in the middle fossiliferous horizon of Thymiana (Thymiana-B, THB) in Chios Island and includes some maxillary and mandibular remains (Fig. 2c, d) collected at the beginning of the 1990s. However, the locality Thymiana was found in the 1940s with a quite rich mammal fauna (Paraskevaidis 1940). The molarized P3 and P4, the strongly projected metastyle and shallow groove between it and the parastyle on the P3 and P4, the high-crowned lower teeth, the molarized premolars with broad talonid, and the strong hypoconulid in the m2, are the main characters of *L. paraskevaidisi*. This species differs from *L. carpathicus* in having a shorter size, a higher degree of molarization, and a weaker basal cingulum in the teeth (Koufos et al. 1995).

Appendix

List of the Greek localities containing viverrid fossils. Type localities are marked with bold. Locality numbers refer to the collection numbers of the PaleoBiology Database (PBDB)

Localities[PBDB No]	Age (MN; GPTS)	Taxon
Thymiana B[202728]	Middle Miocene (MN5; ~15.5 Ma)	***Lophocyon paraskevaidisi***[1]
Aliveri[68032]	Early Miocene (MN4)	***Euboictis aliverensis***[2]

MN Mammal Neogene, *GPTS* Geomagnetic Polarity Time Scale
[1]Koufos et al. (1995), [2]Schmidt-Kittler (1983)

References

Fejfar O, Schmidt-Kittler N (1984) *Sivanasua* und *Euboictis* nv. gen.- zwei pflanzenfressende Schleichkatzenvorläufer (Viverridae, Carnivora, Mammalia) im europäischen Untermiozän. Mainz Geowiss Mit 13:49–72

Flynn JJ, Finarelli JA, Zehr S, Hsu J, Nedbal MA (2005) Molecular phylogeny in the Carnivora (Mammalia): assessing the impact of increased sampling on resolving enigmatic relationships. Syst Biol 54:317–337

Hunt RM Jr (1991) Evolution of the aeluroid Carnivora: viverrid affinities of the Miocene carnivoran Herpestides. Am Mus Novit 3023:1–34

Koufos GD, Bonis de L, Sen S (1995) *Lophocyon paraskevaidisi* a new viverrid (Carnivora, Mammalia) from the middle Miocene of Chios Island, Greece. Geobios 28(4):511–523

Myers P (2000) "Viverridae" (On-line), Animal Diversity Web. Accessed March 16, 2021 at https://animaldiversity.org/accounts/Viverridae/. Accessed 27 Mar 2018

Orlov YA (1968) Fundamentals of paleontology, Mammals, vol 13. Israel Program for Scientific Translations, Jerusalem

Paraskevaidis I (1940) Eine obermiocäne fauna von Chios. Neues Jb Miner Geol Palaeont Abh Abt A 83:363–442

Schmidt-Kittler N (1983) The mammals from the lower Miocene of Aliveri (Island of Evia, Greece). On a new species of *Sivanasua* PILGRIM, 1931 (Felidae, Carnivora) and the phylogenetic position of the genus. Proc K Ned Akad Wet 86:301–318

Schmidt-Kittler N (1987) The Carnivora (Fissipedia) from the lower Miocene of East Africa. Palaeontographica Abt A 197:85–126

Vaughan TA, Ryan JM, Czaplewski NJ (2015) Mammalogy, 6th edn. Jones & Bartlett Learning, Burlington

Veron G (2010) Phylogeny of the Viverridae and "Viverrid-like" feliforms. In: Goswami A, Friscia A (eds) Carnivoran evolution: new views on phylogeny, form and function. Cambridge studies in morphology and molecules: new paradigms in evolutionary bio. Cambridge University Press, Cambridge, pp 64–91

Werdelin L (1996) Carnivoran ecomorphology: a phylogenetic perspective. In: Gittleman JL (ed) Carnivore behavior, ecology, and evolution, 2nd volume. Cornell University Press, Ithaca, pp 582–624

The Fossil Record of Percrocutids (Mammalia: Carnivora: Percrocutidae) in Greece

George D. Koufos

1 Introduction

Percrocutidae is an extinct family including some Miocene–Pliocene hyaena-like carnivoran taxa of the Old World, called percrocutoids, which were referred to Hyaenidae for a long time. The genera *Percrocuta, Dinocrocuta, Belbus*, and *Allohyaena* are now included to their own family, the Percrocutidae. Their taxonomy has been a matter of debate and is mainly based on the presence of some derived characters as the enlarged premolars and the absence of M2/m2. Based on the deciduous dentition, Schmidt-Kittler (1976) and Chen and Schmidt-Kittler (1983) proposed that the percrocutoids are phylogenetically distinct from Hyaenidae and share a common ancestry with Stenoplesictinae, a distinct carnivoran family. The discovery of a cranium of *Dinocrocuta gigantea* in China indicated that the morphology of the basicranium and auditory bullae is quite different from that of the true hyaenids. Based on these differences, as well as to those of the deciduous dentition, Werdelin and Solounias (1991) erected the new family Percrocutidae (Qiu et al. 1988).

The family had a great distribution in Eurasia, expanding from Spain to China, and in Africa. In Greece, Percrocutidae were also present, but the known material is scarce; they are traced in Axios Valley, Samos Island, and Chalkidiki Peninsula.

G. D. Koufos (✉)
Department of Geology, Laboratory of Geology and Paleontology,
Aristotle University of Thessaloniki, Thessaloniki, Greece
e-mail: koufos@geo.auth.gr

E. Vlachos (ed.), *Fossil Vertebrates of Greece Vol. 2*,
https://doi.org/10.1007/978-3-030-68442-6_18

2 Historical Overview

The family Percrocutidae was early recognized in Greece after the discovery of the maxillary fragment of a large-sized hyaenid in Axios Valley (Macedonia, Greece), described as *Hyaena salonicae* (Andrews 1918). The specimen, after a long taxonomic wandering, was classified as a separate species of the genus *Dinocrocuta* by Qiu (1987). Quite later, a large-sized hyaenid with percrocutoid features was discovered in the locality Pentalophos 1 of Axios Valley, which was described as *Dinocrocuta gigantea* (Koufos 1995). Some hyaenid specimens from Samos have also a long taxonomical history, referred under various names. Finally, they are classified in the family Percrocutidae, named *Belbus beaumonti* by Werdelin and Solounias (1991). More recently, the genus *Percrocuta* was recognized in the early/middle Miocene locality Antonios, near Thessaloniki (Koufos 2008). In total, three different percrocutoid genera with three different species—two of them are new—have been recognized in the Greek fossil record (Fig. 1 and Appendix).

Fig. 1 Map of Greece showing the geographic distribution of the most important localities with percrocutid fossils (the uncertain locality in Samos is not shown). See Appendix for more information. Image exported from Google Earth Pro © 2019, map data from US Dept. of State Geographer, SIO, NOAA, U.S. Navy, NGA, GEBCO, image from Landsat/Copernicus. Scale bar equals 10 km, North faces upward

3 Phylogenetic Relationships

The Percrocutidae as they share some derived features (enlarged premolars, absence of the second molar in both jaws) with the Hyaenidae *s.s.* were nearly always referred to this family. However, the percrocutids present some derived characters (absence of the metaconid and reduction of the talonid in the m1, reduced protocone in the upper carnassial) distinguishing them from the Hyaenidae *s.s.* Although the monophyly of the Hyaenidae was early questioned (Thenius 1966; Schmidt-Kittler, 1976) the first clear indications originated from the study of a mandible with deciduous dentition from China (Chen and Schmidt-Kittler 1983). The two authors suggested that: (a) the Percrocutidae are phylogenetically separated from the Hyaenidae *s.s.*, having a parallel but independent evolution, and (b) their feloid dp4 could be an indication that they are phylogenetically far from the Hyaenidae *s.s.* and share an ancestry with the "stenoplecictines." The phylogenetic distance from the Hyaenidae *s.s.* was supported by the different cranial morphology of *Dinocrocuta gigantea* described by Qiu et al. (1988). Recently, Xiong (2019) scanned and reconstructed the cranium of *Dinocrocuta gigantea* from China and suggested that the basicranium is like that of the Hyaenidae, questioning the validity of the family Percrocutidae; he suggested that "*it could be an early side-branch of the Hyaenidae*." Besides these evidences, the phylogenetic relationships of the Percrocutidae need more fossils and studies to be clarified.

4 Systematic Paleontology

Percrocutidae Werdelin and Solounias, 1991

***Belbus* Werdelin and Solounias, 1991**

Type Species *Hyaenictis beaumonti* Qiu, 1987.

★*Belbus beaumonti* (Qiu, 1987)

Nomenclatural and Taxonomical History *Hyaena* sp. in Beaumont 1968 (initial identification); *Hyaenictitherium hyaenoides* in Howel and Petter 1980 (new combination); *Hyaenictis beaumonti* in Qiu 1987 (new species); *Belbus beaumonti* in Werdelin and Solounias 1991 (new genus and combination). The taxonomic history of this material is long and complicated, referred under various generic and specific names (Werdelin and Solounias 1991; Turner et al. 2008).

Type Material NHMB-SAM-33 (holotype), left mandibular ramus with c–p3, m1, housed in the Naturhistorisches Museum Basel and described and figured by Beaumont (1968:21–26; pl. 1, figs 1, 2) (Fig. 2a).

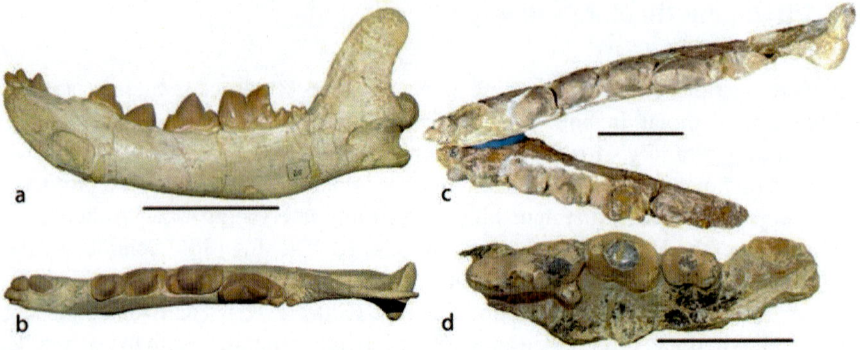

Fig. 2 Selected percrocutids fossils from Greece. (**a, b**) *Belbus beaumonti*, right hemimandible, SMNS-SAM-13118, Samos Loc. Unknown, Greece, in (**a**) lingual and (**b**) occlusal views, (**c**) *Dinocrocuta gigantea*, mandible, LGPUT-PNT-70, Pentalophos 1 (PNT), Axios Valley, Macedonia, Greece, in occlusal view, (**d**) *Dinocrocuta salonicae*, right maxillary fragment with P2–P4, NHML-DVT-M.11413 (holotype), Diavata, Axios Valley, Macedonia, Greece, in occlusal view. Scale bar equals 50 mm

Type Locality Samos Island, Greece, late Miocene, precise locality unknown.

Remarks This taxon is characterized by the presence of the p1 and m2, the absence of the M2, the large and conical P3 and p3 with strongly curved mesial margin, the weak anterior accessory cusps in the premolars, the large posterior accessory cuspid in the p2 and p3, the large m1 with relatively long talonid and strong metaconid, and the dominance of the hypoconid in the m1 talonid (Beaumont 1968; Werdelin and Solounias 1991).

Dinocrocuta Schmidt-Kittler, 1976

Type Species *Hyaena algeriensis* Arambourg, 1959.

Remarks Schmidt-Kittler (1976) erected the subgenus *Percrocuta (Dinocrocuta)* and defined *Hyaena algeriensis* Arambourg, 1959 as its type species; however, Qiu et al. (1988) elevated *Dinocrocuta* to genus level. Besides the various certain species of *Dinocrocuta* found in Greece (see below), there is a right maxillary fragment with dP2–dP4 (LGPUT-XIR-5) from the late Miocene locality Xirochori 1 of Axios Valley. The deciduous teeth differ from those of *Adcrocuta eximia, Percrocuta senyureki,* and *P.* aff. *tungurensis*. Two large-sized hyaenas are recognized in Axios Valley: *D. gigantea* and *D. salonicae* (see below). The deciduous dentition of both species is unknown preventing comparisons and thus the specific determination of the Xirochori specimen is impossible, at the moment; therefore, it is attributed as *Dinocrocuta* sp. (Koufos 2012).

Dinocrocuta gigantea (Schlosser, 1903)

Nomenclatural and Taxonomical History *Hyaena gigantea* in Schlosser 1903 (new species); *Percrocuta (Dinocrocuta) gigantea* in Schmidt-Kittler 1976 (new combination and subgenus); *Dinocrocuta gigantea* in Qiu 1988 (new combination, upgrade to the genus level). The species has a long taxonomical history and it is referred under different generic or sub-generic names (see Howell and Petter 1985; Werdelin and Solounias 1991).

Type Material BSPM-1900 XII. 527 (lectotype), right m1, Shansi, China (Schlosser 1903: 35, fig. 6), by the subsequent designation of Kurtén (1957); housed in the Bayerisches Staadtssamlung für Paläontologie und Geologie, München.

Type Locality Tientsin, Shansi, Tibetfluss (Schlosser 1903), late Miocene.

Distribution The species is known from the late Miocene of Spain, Bulgaria, Turkey, Georgia, China, Mongolia, and from the Greek locality Pentalophos 1.

Remarks Although the wide distribution of *D. gigantea*, the known material is relatively scarce. However, the Chinese material includes some more complete specimens, which allow to see the precise morphology of the taxon. The main morphological characters of the species—modified from Qiu et al. (1988) and Zhang (2005)—are the large-to-very large size, the thick and broad nasal bones, the sharp bending of the cranial roof, the elongated auditory bullae, the short and high cranial proportions, the absence of lingual root in the P3 which is robust and widened mesially, the extremely reduced protocone of the P4, the hypertrophied and strongly oriented lingually p2, the long p4 relative to the p3, and the short m1 relative to the p3. The only known material from Greece is a mandible and two upper canines from Pentalophos 1. The morphology and size of the large-sized hyaenid from the Greek locality Pentalophos 1 (Fig. 2b) allow its attribution to *D. gigantea* (Koufos 1995).

★*Dinocrocuta salonicae* (Andrews, 1918)

Nomenclatural and Taxonomical History *Hyaena salonicae* in Andrews 1918 (new species); *?Crocuta (Percrocuta) salonicae* in Kurtén 1957 (new combination); "*Hyaena salonicae*" in Beaumont 1979 (uncertain attribution); *Allohyaena (Dinocrocuta) salonicae* in Howell and Petter 1985 (new combination); *Dinocrocuta salonicae* in Qiu 1987 (new combination).

Holotype NHML-DVT-M.11413 (holotype), right maxillary fragment with P2–P4, housed in the Natural History Museum of London and described and figured by Andrews (1918: 541) (Fig. 2c, d).

Type Locality Diavata, Axios Valley, Macedonia, Greece. The locality Diavata could correspond to the fossiliferous level of the locality Pentalophos-1 (PNT) but the absence of information for the exact site of *D. salonicae* cannot certify it (Andrews 1918; Koufos 1995).

Remarks Although a certain comparison with the Pentalophos *D. gigantea* is impossible, as there is different material (maxilla vs. mandible), *D. salonicae* seems to be smaller than *D. gigantea* (Koufos 1995). The main characters of the specimen—modified from Pilgrim (1931) and Beaumont (1979)—are the very large size, the large canine, the small P1, the small posterior accessory cusp and the absence of the anterior one in the P2, the absence of anterior and the presence of posterior accessory cusp in the P3, the elongated upper carnassial with extremely strong parastyle, the short protocone, situated behind the mesial margin of the parastyle, and the large M1.

Percrocuta Kretzoi, 1938

Type Species *Crocuta carnifex* Pilgrim, 1932.

Remarks An isolated upper carnassial is only known from Greece, found in the early/middle Miocene locality Antonios. This upper carnassial has similarities with *P. carnifex* and *P. abessalomi*, but the limited material does not allow a definite identification and thus it is attributed to *Percrocuta* sp. (Koufos 2008).

Appendix

List of the Greek localities containing precrocutid fossils. Type localities are marked with bold. Locality numbers refer to the collection numbers of the PaleoBiology Database (PBDB)

Localities[PBDB No]	Age (MN; GPTS in Ma)	Taxon
Samos	Late Miocene (MN11–13)	***Belbus beaumonti***[1]
Xirochori 1[195490]	Late Vallesian (MN10; ~9.6)	*Dinocrocuta* sp.[2]
Diavata[210627]	Late Miocene (MN9–10)	***Dinocrocuta salonicae***[3]
Pentalophos 1[202119]	Early Vallesian (MN9)	*Dinocrocuta gigantea*[4]
Antonios[73861]	Early/middle Miocene (MN4/5)	*Percrocuta* sp.[5]

GPTS Geomagnetic Polarity Time Scale
[1]Qiu (1987), [2]Koufos (2012), [3]Andrews (1918), [4]Koufos (1995), [5]Koufos (2008)

References

Andrews CW (1918) Note on some fossil mammals from Salonica and Imbros. Geol Mag 6:540–543

Chen G, Schmidt Kittler N (1983) The deciduous dentition of *Percrocuta* Kretzoi and the diphyletic origin of the hyaenas (Carnivora, Mammalia). Palaeont Z 57:159–169

de Beaumont G (1968) Une interessante mandibule de Hyaenidae (Carnivora) du Pontien de Samos. Arch Sci Geneve 32:247–250

de Beaumont G (1979) Qu' est-ce que la "*Hyaena salonicae* Andrews" (Mammifere, Carnivore). Arch Sci Geneve 32:247–250

Howell FC, Petter G (1985) Comparative observations on some middle and upper Miocene hyaenids genera: *Percrocuta* Kretzoi, *Allohyaena* Kretzoi, *Adcrocuta* Kretzoi (Mammalia, Carnivora, Hyaenidae). Geobios 18:419–476

Koufos GD (1995) The late Miocene percrocutas of Macedonia (Greece). Palaeovertebrata 24:67–84

Koufos GD (2008) Carnivores from the early/middle Miocene locality of Antonios (Chalkidiki, Macedonia, Greece). Geobios 41:365–380

Koufos GD (2012) New material of Carnivora (Mammalia) from the late Miocene of Axios Valley, Macedonia, Greece. C R Palevol 11:49–64

Kurtén B (1957) *Percrocuta* KRETZOI (Mammalia, Carnivora), a group of Neogene hyenas. Acta Zool Cracoviensia 2(16):375–403

Pilgrim GE (1931) Catalogue of the Pontian Carnivora of Europe in the Department of Geology. The British Museum, Natural History, London

Qiu Z (1987) Die Hyeniden aus dem Ruscinium und Villafranchium Chinas. Münchner Geowiss Abh A 9:1–110

Qiu Z, Xie J, Yan D (1988) Discovery of the skull of *Dinocrocuta gigantea*. Vertebrata Palasiatica 26:128–138

Schlosser M (1903) Die fossilen Säugetiere Chinas nebst einer Odontographie der recenten Antilopen. Abh Bayer Akad Wiss 22:1–221

Schmidt-Kittler N (1976) Carnivores from the Neogene of Asia Minor. Palaeontographica Abt A 155:1–131

Thenius E (1966) Zur Stammesgeschichte der Hyanen. Z Saugetierkunde 31:293–300

Turner A, Anton M, Werdelin L (2008) Taxonomy and evolutionary patterns in the fossil Hyaenidae of Europe. Geobios 41(5):677–687

Werdelin L, Solounias N (1991) The Hyaenidae: taxonomy, systematics and evolution. Fossils Strata 30:1–106

Xiong W-Y (2019) Basicranial morphology of late Miocene *Dinocrocuta gigantea* (Carnivora: Hyaenidae) from Fugu, Shaanxi. Vertebrata Palasiatica 57:274–307

Zhang Z (2005) New materials of *Dinocrocuta* (Percrocutidae, Carnivora) from Lantian, Shaanxi Province, China, and remarks on Chinese Late Miocene biochronology. Geobios 38:685–689

The Fossil Record of Hyaenids (Mammalia: Carnivora: Hyaenidae) in Greece

George D. Koufos

1 Introduction

Hyaenidae is a carnivoran family known with several taxa since the Miocene to the present. Extant hyaenids include four species: the hunting and scavenging large hyaenas *Parahyaena brunnea* (brown hyaena), *Hyaena hyaena* (striped hyaena), and *Crocuta crocuta* (spotted hyaena), as well as the aardwolf *Proteles cristata*, which is feeding on termites. The family is characterized by narrow and low-to-short and broad cranium, moderately elongated and bony auditory meatus, expanded ectotympanic and semi-recumbent to recumbent septum bullae, massive canines without cutting edges or grooves, upper carnassial with elongated parastyle, blade-like metastyle and sometimes reduced protocone, blade-like lower carnassial with metaconid in the primitive forms which decreases and disappears in the later forms and with relatively small talonid, sometimes absence of the P1 and p1, decrease in size and disappear of the M1 and M2, and absence of the m2 in the later forms (Orlov 1968; Myers 2000). In *Proteles*, the teeth are simple and uniform, like flattened cones, except the canines which are strong for protection. The limbs are elongated with the forelegs being longer than the hinder ones; the legs are usually tetradactyls (except in *Proteles*) with nonretractile claws. The hyaenas are generally considered as scavengers but they can hunt and kill large animals, especially if they are hunting as a group. They can cover long distances to find carrions or in other cases, they follow other predators (lions, cheetahs) to find food (Orlov 1968; Myers 2000).

G. D. Koufos (✉)
Department of Geology, Laboratory of Geology and Paleontology,
Aristotle University of Thessaloniki, Thessaloniki, Greece
e-mail: koufos@geo.auth.gr

E. Vlachos (ed.), *Fossil Vertebrates of Greece Vol. 2*,
https://doi.org/10.1007/978-3-030-68442-6_19

2 Historical Overview

The hyaenids were relatively common in the first collections from the various Greek Miocene localities. Some taxa were originally recognized in the first collections from Pikermi and Samos. Several articles on the Greek hyaenids have been published by various Greek and foreign authors as early as the nineteenth century (e.g., Gaudry 1862–1867; Arambourg and Piveteau 1929; Pilgrim 1931; Melentis 1967; Sickenberg 1968; Tsoukala 1989; Roussiakis 1996; Koufos 2000, 2011).

3 Phylogenetic Relationships

The hyaenids originated from the viverrids and arrived in the Old World during the beginning of the Miocene. The early Miocene hyaenids include small-sized and arboreal forms very similar to the modern *Hemigalus derbyanus* ("banded civet"). The genus *Plioviverrops* is one of the earliest hyaenids and occurred in Southern Europe since the early Miocene. It is considered as the ancestral form of the large hyaenids appeared later in the Miocene–Pleistocene (MacDonald 1992). The family reached its maximum diversity in the middle Miocene (~15.0 Ma) with more than 30 species, which are dog-like, nimble, and similar to wolves or jackals (e.g., *Ictitherium viverrinum*). The bone-cracking hyaenids appeared at the end of middle–beginning of late Miocene, and they are common in the late Miocene faunas. The bone-cracking hyaenids survived the different climatic changes from the Miocene to the Pleistocene. *Adcrocuta eximia*, the predominant species in the late Miocene, was replaced in the Pliocene by *Pliocrocuta*, and later by *Pachycrocuta*, which was a large-sized bone-cracking hyaena scavenging on large herbivore carcasses. The decline of the large herbivores at the last Ice Age causes the disappearance of *Pachycrocuta* and its replacement by *Crocuta*, which is still living. The modern aardwolf *Proteles* is directly related with the evolutionary line of *Plioviverrops*, representing a living dog-like hyaenid (MacDonald 1992).

4 Distribution

The family Hyaenidae is widely distributed in the whole Old World; some taxa have a wide distribution and can be used as biostratigraphic markers, e.g., *Adcrocuta eximia*, *Hyanectitherium wongii*, and *Pachycrocuta brevirostris*. In Greece, the hyaenids are well known from the middle Miocene to the Pleistocene with several taxa. They are very common in the late Miocene faunas of Greece with a lot of ictitheriines and hyaenines. The most common species is *Adcrocuta eximia* in the late Miocene and *Crocuta crocuta* in the Pleistocene. They were recognized in several localities all over the country (see Fig. 1 and the Appendix). Besides the referred

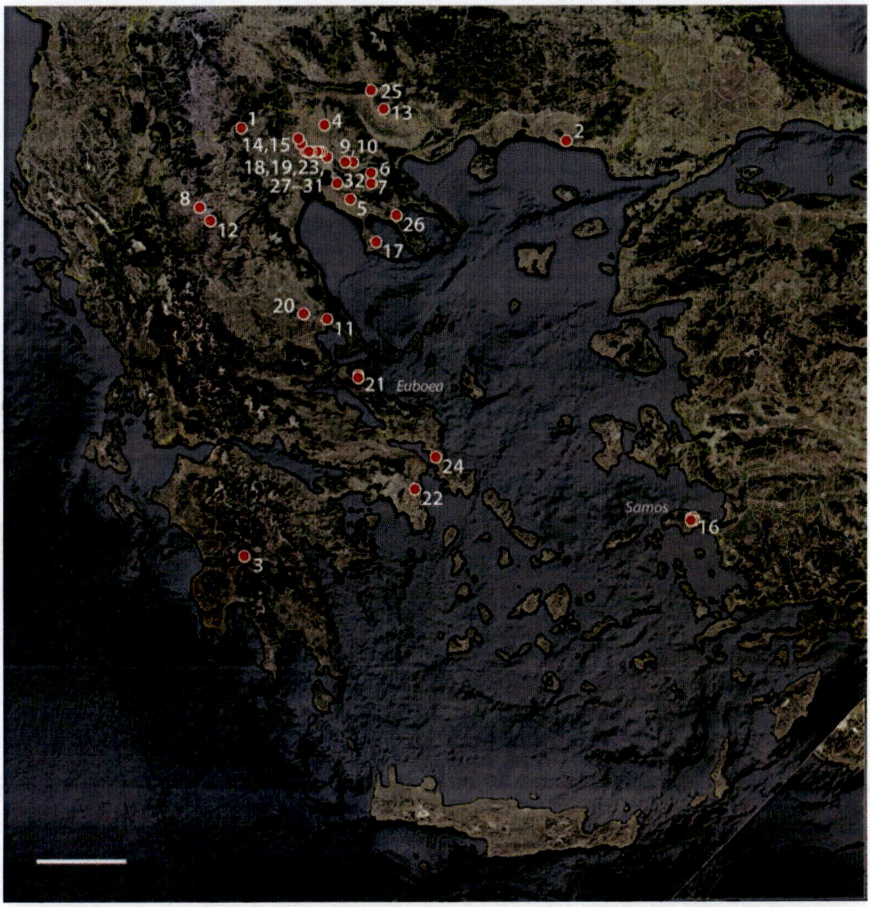

Fig. 1 Map of Greece showing the geographic distribution of the most important localities with hyaenid fossils. **1**, Loutra Almopias Cave; **2**, Maronia Cave; **3**, Megalopolis basin; **4**, Agios Georgios Cave; **5**, Petralona Cave; **6**, Tsiotra Vrissi; **7**, Kalamoto; **8**, Libakos; **9**, Apollonia 1; **10**, Gerakarou 1; **11**, Sesklon; **12**, Dafnero 1; **13**, Maramena; **14**, Dytiko 1; **15**, Dytiko 3; **16**, various localities in Samos Island; **17**, Kryopigi; **18**, Vathylakkos 2; **19**, Vathylakkos 3; **20**, Perivolaki; **21**, Kerassia localities; **22**, Pikermi; **23**, Prochoma 1; **24**, Halmyropotamos; **25**, Thermopigi; **26**, Nikiti 2; **27**, Ravin des Zouaves 5; **28**, Ravin X; **29**, Ravin de la Pluie; **30**, Ravin des Zouaves 1; **31**, Pentalophos 1; **32**, Antonios. See Appendix for more information. Image exported from Google Earth Pro © 2019, map data from US Dept. of State Geographer, SIO, NOAA, U.S. Navy, NGA, GEBCO, image from Landsat/Copernicus. Scale bar equals 80 km, North faces upward

fossiliferous sites in the description of each species, hyaenid remains are referred from some other sites but their material is poor and the specific determinations are doubtful.

5 Systematic Paleontology

Hyaenidae Gray, 1821

Ictitheriinae Trouessart, 1897

Protictitherium Kretzoi, 1938

Type Species *Herpestes crassus* Depéret, 1892.

Remarks A small protictithere is known by two fragments of m1 from the locality Antonios in Chalkidiki. Despite their similarities with *P. crassum,* it cannot be definitely attributed to this species as the material is too scarce and fragmentary and thus it was attributed as *P.* cf. *crassum* (Koufos 2008). The genus is also present in the locality Pentalophos 1; although its similarity with *P. crassum*, it differs from it displaying slightly smaller size, more molarized premolars, long talonid in the m1, higher and stronger metaconid in the m1, and more robust teeth, thus it is referred to as *P.* cf. *crassum* (de Bonis and Koufos 1991). In the material from Ravin de la Pluie, there is an m1 displaying similar morphology and size to that of *P. intermedium* but the limited material cannot allow a definite attribution; therefore, it is identified as *P.* cf. *intermedium* (Koufos 2000).

Protictitherium crassum (Depéret, 1892)

Nomenclatural and Taxonomical History *Herpestes crassus* in Depéret 1892 (new species); *Progenetta crassa* in Forsyth Major 1903 (new combination and mandatory change); *Protictitherium crassum* in Kretzoi 1938 (new combination and mandatory change).

Type Material Specimen number unknown (holotype), some mandibular remains and an isolated P4 (Depéret 1892: p. 31; pl. 1, figs 14–17).

Type Locality La Grive, France, middle Miocene, MN7 + 8.

Distribution The species is recognized in the Greek localities Xirochori 1, and Dytiko 3 in Axios Valley, and Mytilinii 4 in Samos Island. Besides its type locality in France, it is also referred from Spain, Germany, and Turkey.

Remarks The first occurrence of this species in Greece was from Dytiko 3, based on a hemimandible (Koufos 1980) (Fig. 2a), and later in Xirochori 1 and Mytilinii 4 (Koufos 2000, 2009). It is characterized by relatively deep and thin mandibular corpus, large mental foramen below the anterior root of the p2, elongated and narrow symphysis roughly inclined backward, slightly curved ventral margin of the mandibular corpus, small, single rooted and monocuspid p1, elongated and narrow p2 and p3 without anterior and small posterior accessory cuspid, p4 with anterior

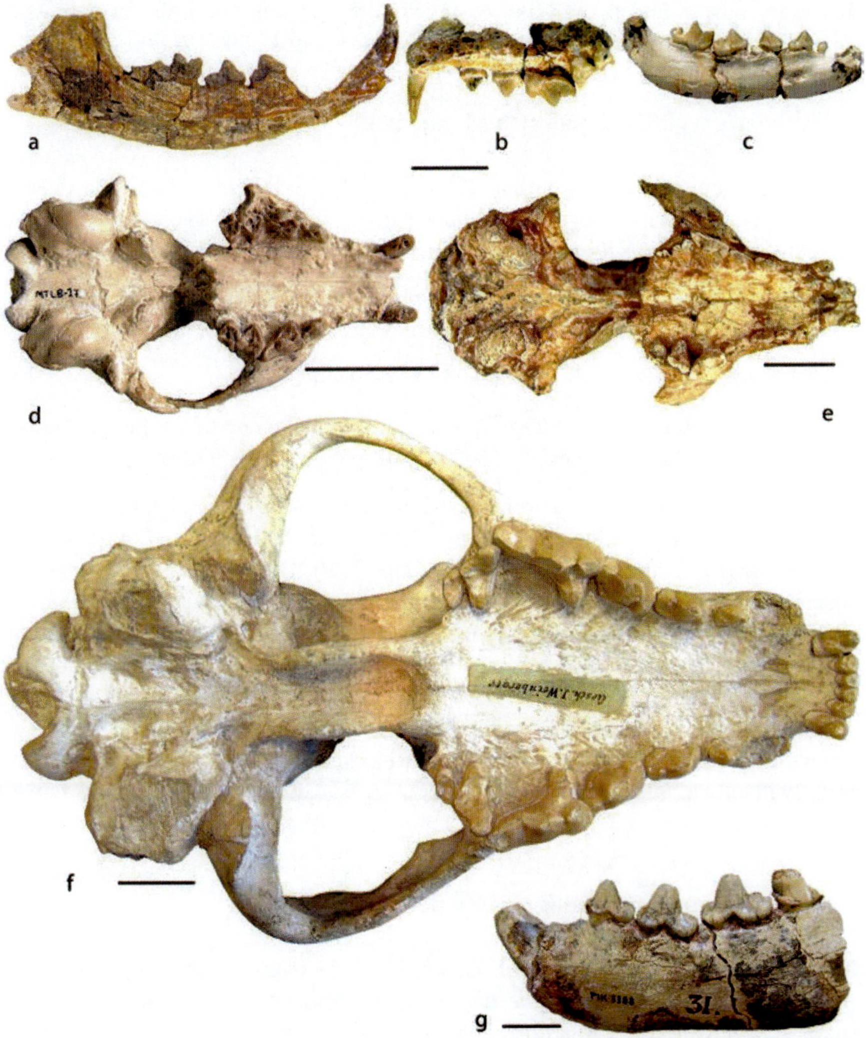

Fig. 2 Selected specimens of Ictitherinae (Hyanidae) from Greece. (**a**) *Protictitherium crassum*, LGPUT-DKO-20, hemimandible with c, p3–m1 from Dytiko 3 in buccal view; (**b**, **c**) *Protictitherium thessalonikensis* from Ravin de la Pluie; (**b**) LGPUT-RPl-69, right maxillary fragment with C–M2 (holotype) in lingual view; and (**c**) LGPUT-RPl-68, left mandibular fragment with c–m1 in buccal view; (**d**) *Plioviverrops orbignyi*, NHMA-MTLB-170, cranium from Mytilinii 1B in palatal view; (**e**) *Ictitherium viverrinum*, MNHNP-PIK-3022, cranium from Pikermi in palatal view; (**f**) *Hyaenictitherium wongii*, NHMW-SAM-A.4746, cranium from an unknown locality of Samos Island in palatal view; (**g**) *Lycyaena chaeretis*, MNHNP-PIK-3383 (holotype), left mandibular fragment with c, p2–p4 from Pikermi in buccal view. Scale bar equals 2 cm

accessory cuspid and talonid-like posterior part, short and wide lower carnassial with clear metaconid and large tricuspid talonid (Koufos 2000, 2009).

Protictitherium gaillardi (Forsyth Major, 1903)

Nomenclatural and Taxonomical History *Herpestes crassus* in Gaillard 1899 (initial identification); *Progenetta gaillardi* in Forsyth Major 1903 (new species); *Protictitherium gaillardi* in Schmidt-Kittler 1976 (new combination).

Type Material Lgr-1352–1353 (holotype) maxilla and associated mandible, housed in Natural History Museum of Lyon (Gaillard 1899: 60; pl. II, figs 1, 3).

Type Locality La Grive, France, middle Miocene, MN7+8.

Distribution This species is also known from the Miocene of Spain and Turkey, and the early/middle Miocene (MN 4/5) locality of Antonios, Greece.

Remarks The type material of this species is characterized by slender upper premolars with developed cingulum, strong protocone in the P4 situated in front of the parastyle and a triangular-shaped M1 (Gaillard 1899). The generic attribution of this species has been subject to several changes until its current referral to *Protictitherium* (Schmidt-Kittler 1976). A maxillary fragment with M1–M2 and a mandibular fragment with p4 are only known from Greece. A number of characters support the attribution of the Greek material to this species, including the strongly projected buccally paracone of M1, the small metacone and the well-developed protocone (higher than the other two cusps on M1), the elongated p3 with a strongly projected distally distal cingulum, the strong posterior accessory cuspid on p4, the high main cuspid, the well-defined metaconid, and the bicuspid talonid on m1 (Koufos 2008).

★*Protictitherium thessalonikensis* Koufos, 2012

Nomenclatural and Taxonomical History *Plioviverrops orbignyi* in Koufos 1980 (initial identification); *Protictitherium* aff. *gaillardi* in de Bonis and Koufos 1991 (new identification); *Protictitherium thessalonikensis* in Koufos 2012 (new species).

Type Material LGPUT-RPl-69 (holotype), right maxillary fragment with C–M2 (Fig. 2b, c).

Type Locality Ravin de la Pluie, Axios Valley, Macedonia, late Miocene, MN10.

Remarks The initial poor material from Ravin de la Pluie did not allow the distinction from the other taxa, though some differences have been observed. During the continuous excavations, the enrichment of the collection allowed the recognition of this new species. The main characters of this species are: small size; low cusps(-ids) in the teeth; protocone of the P4 aligned with the mesial border of the parastyle; large molars, especially M2; slight buccal projection of the paracone in the M1;

strongly molarized p4; strong metaconid; and large talonid with high entoconid on the m1 (Koufos 2012).

Plioviverrops Kretzoi, 1938

Type Species *Viverra orbignyi* Gaudry and Lartet, 1856.

Remarks A few dental remains of a small ictithere are known from the locality Vathylakkos 3 in Axios Valley. The relatively large size, the high and narrow p2 and p3 without accessory cuspids, and the less molarization suggest similarities with *P. guerini* and thus they are identified as *P.* cf. *guerini* (de Bonis and Koufos 1991). New material of *Plioviverrops* from other Greek localities indicates that there is a variation in the size and the presence of the accessory cuspids. Therefore, it is important to reexamine the taxonomy of this species (Koufos 2006, 2009). Some postcranial remains of *Plioviverrops*, known from the fossiliferous site Kerassia-3 in Euboea Island, have similar dimensions to *P. orbignyi* from Pikermi. However, the limited comparative material and the absence of dental remains cannot certify the presence of the species and thus the postcrania is referred to as *Plioviverrops* sp. (Roussiakis and Theodorou 2003).

★*Plioviverrops orbignyi* (Gaudry and Lartet, 1856)

Nomenclatural and Taxonomical History *Viverra orbignyi* in Gaudry and Lartet 1856 (new species); *Thalassictis orbignyi* in Gaudry 1861 (new combination); *Ictitherium orbignyi* in Gaudry 1862–1867 (new combination); *Plioviverrops orbignyi* in Kretzoi 1938 (new combination).

Type Material MNHN-PIK-3022 (holotype), skull associated with the mandible, stored in the Museum national d'Histoire naturelle of Paris and described by Gaudry and Lartet (1856: 273).

Type Locality Pikermi, Attica, Greece, late Miocene, MN12.

Distribution The species is known from the Greek localities Ravin des Zouaves 5, Prochoma 1 and Vathylakkos 2, 3 in Axios Valley, Perivolaki in Thessaly, Mytilinii 1B, Samos Q1, A in Samos Island, and Kryopigi in Chalkidiki. Besides Greece, it is also known from Bulgaria and possibly from Turkey.

Remarks The main characters of the species are: the oval narial opening, the large elliptical orbits, the large auditory bullae, the short and wide palate, the very long postorbital processes, the long-narrow and roughly inclined backward symphysis, the pointed cusps in the teeth, the absence or the rudimentary accessory cusps on the P2 and P3, the elongated upper carnassial with relatively large protocone situated in front of the parastyle's mesial margin, and the long talonid, the high and equal-sized metaconid and protoconid, and the high entoconid of the m1 (Pilgrim 1931; Koufos 2006, 2009) (Fig. 2d).

Ictitherium Wagner, 1848

Type Species *Ictitherium viverrinum* Roth and Wagner, 1854.

Remarks A right mandibular ramus with c, p3–m1 (AMPG-KER-K3/204 from Kerassia 3 in Euboea Island displays larger p3, m1, m3, and more robust p3 than those of *I. viverrinum*, as well as an m2 similar to that of *I. pannonicum*. Besides this similarity, the poor material and its bad preservation cannot allow its definite attribution to this species and thus it is referred to as cf. *I. pannonicum* (Roussiakis and Theodorou 2003). There is also a single p2 from the locality Perivolaki in Thessaly, which has similar morphology and size to *I. viverrinum*, but it is reported as *Ictitherium* sp. because of the poor material (Koufos 2006).

★*Ictitherium viverrinum* Roth and Wagner, 1854

Nomenclatural and Taxonomical History *Galeotherium* sp. in Wagner 1840 (initial identification); *Ictitherium* sp. in Wagner 1848 (new genus); *Ictitherium viverrinum* Roth and Wagner 1854 (new species).

Type Material BSPM-PIK-AS II.66 (holotype), the right mandibular fragment with p2–m1 and the left mandibular fragment with c–p4, housed in the Bayerische Staatssammlung für Paläontologie und Geologie, München, described and figured by Roth and Wagner (1854:49; taf. 8, figs 3, 5).

Type Locality Pikermi, Attica, Greece, late Miocene, MN12.

Distribution Besides the type locality, *I. viverrinum* is widespread in Greece, found in the localities Ravin des Zouaves 5, Prochoma 1, Vathylakkos 3 of Axios Valley, Kryopigi in Chalkidiki, and Q1 and A in Samos Island. The species is also known from Slovakia, Moldova, Ukraine, Turkey, Iran, and China.

Remarks The species is characterized by medium size, infraorbital foramen above the P2, short and relatively wide palate, oval-shaped and hyaenid-like auditory bullae with their anterior border behind the post-glenoid processes, high mandibular corpus with almost straight ventral margin, short angular process, shallow masseteric fossa with its anterior margin situated below the mesial one of the m2, absence of anterior accessory cusp on P2 and P3 but with strong posterior one situated on a distal cingular projection, short and robust P4 with weak protocone aligned mesially with the parastyle, long and narrow M1, long M2 relative to the M1, absence of anterior accessory cuspid in the p2 and p3, small posterior accessory cuspid on a strong distal cingular projection of the p2 and p3, small anterior and large posterior accessory cuspids on the p4, pointed cuspids on the m1, low shearing blade (paraconid + protoconid blade) on the m1, high metaconid relative to the protoconid, and large and oval-shaped m2 (Kurtén 1982; Koufos 2000) (Fig. 2e).

Hyaenictitherium Kretzoi, 1938

Type Species *Ictitheriun hyaenoides* Zdansky, 1924.

Remarks An isolated tooth from the locality Mytilinii 4 of Samos Island is similar to that of *Hyaenictitherium wongii* and therefore it was identified as *H.* cf. *wongii*, indicating its possible presence in the older levels of Samos fossiliferous deposits (Koufos 2009).

Hyaenictitherium wongii (Zdansky, 1924)

Nomenclatural and Taxonomical History *Hyaena hipparionum* in Gervais 1846, 1850 (initial identification); *Ictitherium hipparionum* in Gaudry 1862–1867 (new combination); *Palhyaena hipparionum* in Kittl 1887 (new combination); *Ictitherium wongii* in Zdansky 1924 (new species); *Palhyaena wongii* in Howell and Petter 1980 (new combination); *Thalassictis wongii* in Solounias 1981 (new combination); *Hyaenotherium wongii* in Semenov 1989 (new combination); *Hyaenictitherium wongii* in Andersson and Werdelin 2005 (new combination).

Type Material Although, several skulls and mandibles from China were described by Zdansky (1924:73; taf. 14, figs. 3–6, taf. 15, figs. 1–4, taf. 16, figs. 1, 2), he did not define a holotype. Solounias (1981) defined the cranium and mandible from Locality 109, figured in plate 15 of Zdansky (1924) and housed in the Palaeontological Institute of Uppsala, as the lectotype of the species.

Type Locality Locality 109, Yushe, Shansi, China, MN11–13.

Distribution The species is widely distributed in the whole Eurasia, from Spain to China. In Greece, it is referred from Pikermi in Attica, Ravin des Zouaves 1, Ravin des Zouaves 5, Vathylakos 2, 3 in Axios Valley, and Mytilinii 1A, B, C, Q1, Q4, Q5, and S3 in Samos Island.

Remarks It is a medium-sized ictithere quite common in Eurasia (Fig. 2f). It differs from *Ictitherium* by displaying an elongated metastyle-blade and a large protocone in the P4, reduced molars and relatively short talonid on the m1. The elongated metastyle-blade of the P4, the less high protoconid of the m1, the weak talonid with lower cuspids of the m1, and the smaller size separate it from *Thalassictis*.

Lycyaena Hensel, 1862

Type Species *Hyaena chaeretis* Gaudry, 1861.

★*Lycyaena chaeretis* (Gaudry, 1861)

Nomenclatural and Taxonomical History *Hyaena chaeretis* in Gaudry 1861 (new species); *Lycyaena chaeretis* in Hensel 1862 (new combination).

Type Material MNHNP-PIK-3383 (holotype), mandibular fragment with c–p2 (right) and p2–m1 (left) (Fig. 2g), housed in the Museum national d'Histoire naturelle of Paris, described and figured by Gaudry (1861:534; pl. 11, figs 3–6).

Type Locality Pikermi, Attica, Greece, late Miocene, MN12.

Distribution The taxon is relatively scarce in Eurasia; in Greece, besides the type locality, it is known from the locality A of Samos Island.

Remarks The main characters of the species are: the absence of the anterior and the presence of a strong posterior accessory cusp on the P2 and P3, the elongated and narrow upper carnassial with strong and well separated protocone, the large m1, the well-developed posterior accessory cuspid in the lower premolars situated on a distal cingular projection, the strong anterior accessory cuspid of the p4, and the rudimentary anterior accessory cuspid (like a cingular projection) of the p2 and p3 (Pilgrim 1931) (Fig. 2g).

Hyaeninae Gray, 1821

Hyaenictis Gaudry, 1861

Type Species *Hyaenictis graeca* Gaudry, 1861.

Remarks A right P3 from the locality Ravin de la Pluie of Axios Valley is morphologically and metrically similar to that of *Hyaenictis graeca* but it is referred to as *Hyaenictis* sp. because of the single and not well-preserved tooth (Koufos 2012).

★*Hyaenictis graeca* Gaudry, 1861

Nomenclatural and Taxonomical History *Hyaenictis graeca* in Gaudry 1861 (new species).

Type Material MNHNP-PIK-3002 (lectotype), left mandibular ramus with dP2–dP4 and P4–M1 (Fig. 2b), described and figured by Gaudry (1861:527; pl. 11, figs 1–2), defined by Pilgrim (1931), and housed in the Museum national d'Histoire naturelle of Paris.

Type Locality Pikermi, Attica, Greece, late Miocene, MN12.

Distribution Except the type locality, the species is referred from Spain, where another species is also referred (*H. almerai*). The genus *Hyaenictis* is recognized in Africa (Kenya, Ethiopia and South Africa) by different species.

Remarks Except the type mandible, one maxillary fragment, housed in the Museum national d'Histoire naturelle of Paris, and one maxillary fragment and one mandibular fragment in the Natuhistorisches Museum, Wien (Fig. 3a), belonging possibly to the same individual, are the only known material of *H. graeca*. The main characters of the species are the moderately long slender and shallow mandibular corpus, the large protocone of the upper carnassial which is aligned with the mesial border of the parastyle, the long shearing blade of the P4 with its distal end directed buccally, the large M1, the slender p2 and p3 with strong posterior accessory cuspid and absence of anterior one, the large anterior and posterior cuspids on the p4, the relatively short m1 without metaconid, the short talonid with large entoconid and hypoconid of the m1 (pers. observations). The large protocone of the upper carnassial, as well as the narrow and slender premolars distinguish *H. graeca* from *Adcrocuta eximia*.

Chasmaporthetes Hay, 1921

Type Species *Chasmaporthetes ossifragus* Hay, 1921.

Remarks Some dental remains from the locality Maramena in Serres Basin have been described as *Chasmaporthetes* sp. because they are morphologically closer to this taxon (Schmidt-Kittler 1995:85, fig. 9–13).

★*Chasmaporthetes bonisi* Koufos, 1987

Type Material LGPUT-DTK-126 (holotype), left mandibular ramus with p2–m1, housed in the Laboratory of Geology and Palaeontology, Aristotle University of Thessaloniki, described and figured by Koufos (1987:918; pl. 1, figs 1, 2, 3c).

Type Locality Dytiko 1, Axios Valley, Greece, late Miocene, MN13.

Distribution Besides the type locality, it is known from the locality Ravin des Zouaves 5 of Axios Valley.

Remarks *Chasmaporthetes bonisi* represents a primitive form of hunting hyaenas recognized for the first time in Greece by Koufos (1987) (Fig. 3c). It is characterized by relatively narrow palate, moderately developed P1, large protocone in the P4, well-developed M1, curved and imbricated lower tooth row, oval distal margin in the lower premolars, absence or rudimentary anterior accessory cuspid on the p2 and p3, absence of metaconid on the m1, and small-bicuspid talonid with reduced entoconidon in the m1 (Koufos 1987; de Bonis and Koufos 1994).

Chasmaporthetes lunensis (del Campana, 1914)

Nomenclatural and Taxonomical History *Lycyaena lunensis* in del Campana 1914 (new species); *Hyaena honannesis* in Zdansky 1924 (new species); *Lycaenops lunensis* in Kretzoi 1938 (new genus); *Euryboas bielawski* in Schaub 1941 (new species); *Euryboas lunensis* in Viret 1954 (new combination); *Chasmaporthetes kani* in Galiano and Frailey 1977 (new species); *Chasmaporthetes lunensis* in Kurtén and Werdelin 1988 (new combination).

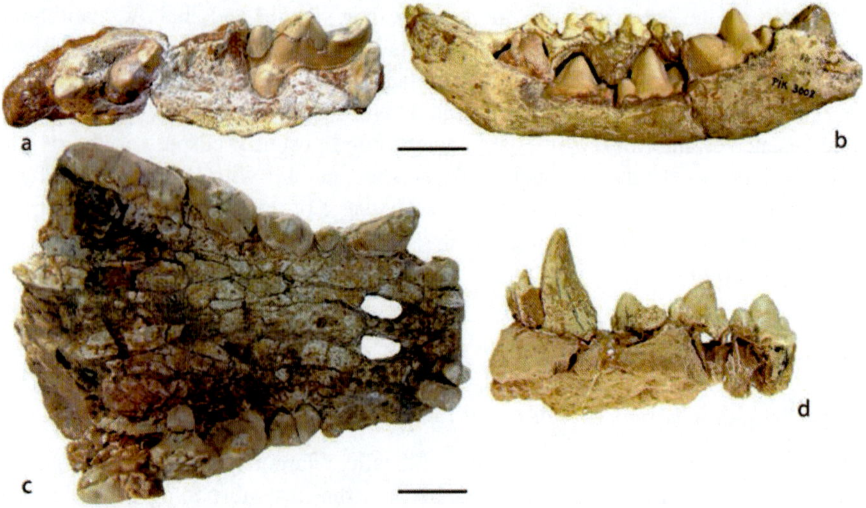

Fig. 3 Selected specimens of Hyaeninae (Hyaenidae) from Greece. (**a, b**) *Hyaenictis graeca* from Pikermi; (**a**) NHMW-PIK- A4715, left maxillary fragment with P3–M1 in lingual view; (**b**) MNHNP-PIK-3002, left mandibular fragment with p2–m2 and dp2–dp4 (lectotype) in buccal view; (**c**) *Chasmaporthetes bonisi* from Ravin des Zouaves 5, LGPUT-RZO-125, maxilla with I2–M1 (right side) and I2, I3, P2–P4 (left side) in palatal view; (**d**) *Chasmaporthetes lunensis* from Dafnero 1, LGPUT-DFN-75, left mandibular fragment with i1–p4 in buccal view. Scale bar equals 2 cm

Type Material IGF 4377 (holotype), partial cranium with the upper dentition, stored in the Museum of Florence, described and figured by del Campana (1914: p. 87; pl. 11).

Type Locality Olivola, Val di Magra, Italy, Early Pleistocene (MNQ17).

Distribution The species is dispersed in Eurasia known from Spain, France, Italy, Ukraine, Russia, Tajikistan, Turkey, China, and Mongolia. In Greece, it is only known from Dafnero 1 in Western Macedonia.

Remarks The known material from Greece includes two mandibular fragments (Fig. 3d) and some isolated teeth, characterized by elongated and slender upper carnassial with strong protocone aligned with the mesial margin of the parastyle, elongated and narrow lower premolars, presence of both accessory cuspids in the lower premolars, m1 short and narrow without or with vestigial metaconid, short bicuspid talonid with a large hypoconid and a small entoconid; all these characters allow its attribution to *C. lunensis* (Koufos 1993).

Adcrocuta Kretzoi, 1938

Type Species *Hyaena eximia* Roth and Wagner, 1854.

★*Adcrocuta eximia* (Roth and Wagner, 1854)

Nomenclatural and Taxonomical History *Hyaena eximia* in Roth and Wagner 1854 (new species); *Crocuta eximia* in Pilgrim 1931 (new combination); *Adcrocuta eximia* in Kretzoi 1938 (new genus).

Type Material BSPM-PIK-105 (holotype), right mandibular ramus with i3–m1 (Fig. 4a, b), stored in the Bayerische Staatssammlung für Paläontologie und Geologie, München (Roth and Wagner 1854: 396; tab. 8, fig. 6).

Type Locality Pikermi, Attica, late Miocene, MN12.

Distribution *Adcrocuta eximia* was widely distributed in Eurasia, from Spain to China. In Greece, besides Pikermi, it is known from Xirochori 1, Ravin de la Pluie, Ravin des Zouaves 1, Ravin des Zouaves 5, Ravin X, Prochoma 1, Vathylakkos 3 in Axios Valley, Nikiti 2 and Kryopigi in Chalkidiki, Mytilinii 1A, Q1, Q4, Q5, A, and S3 in Samos Island, Perivolaki in Thessaly, Thermopigi in Serres Basin, Halmyropotamos and Kerassia 4 in Euboea Island.

Remarks It is a common hyaenid and easily recognizable in the late Miocene mammal faunas of Eurasia. The main characters of the species are: elongated and wide cranium, strong zygomatic arches projecting laterally, large nasal cavity, wide palate, large and flattened laterally auditory bullae with their anterior border at the level of the posterior wall of the post-glenoid fossa, elongated and narrow occipital condyles, large foramen magnum (Fig. 4c), high and thick mandibular corpus with convex ventral margin, relatively low ascending ramus, elongated and narrow mandibular condyle, oval and deep masseteric fossa with its anterior margin below the distal end of the m1, two mental foramens below the p2 and p3 and robust teeth. The P2 and P3 have rather elliptical occlusal outline and lack anterior accessory cusp while the posterior one is large and packed on the main cusp. The upper carnassial is elongated and slender with strongly reduced protocone situated well behind the mesial margin of the parastyle. The M1 is very small and situated vertically to the P4. The p2 is small and oval-shaped without anterior but with large posterior accessory cuspid packed on the main one. The p3 is rectangular more robust than the p2 with only posterior accessory cuspid and the p4 bears anterior and posterior accessory cuspids. The lower carnassial is long and relatively narrow lacking metaconid (in some cases there is a small metaconid) and bearing a small, low, and bicuspid talonid (pers. observations). Two subspecies have been recognized in Greece: *A. e. leptoryncha* de Bonis and Koufos, 1981 in the Vallesian (MN10) and *A. e. eximia* in the Turolian (MN11–13). *A. e. leptoryncha* differs from the nominal subspecies displaying longer snout, narrower palate, less compressed jugal tooth row, and slenderer premolars (de Bonis and Koufos 1981; Koufos 2000).

Pliocrocuta Kretzoi, 1938

Type Species *Hyaena perrieri* Croizet and Jobert, 1828.

Fig. 4 Selected specimens of derived Hyaeninae (Hyaenidae) from Greece. (**a–c**) *Adcrocuta eximia* from Pikermi; (**a**, **b**) BSPM-PIK- AS II105 (holotype), right hemimandible with i3–m1 in (**a**) lingual and (**b**) occlusal views; (**c**) MNHNP-PIK-3000, cranium in palatal view; (**d**) *Pliocrocuta perrieri* from Gerakarou 1, LGPUT-GER-150, right mandibular fragment with i3–m1 in lingual view; (**e, f**) *Pachycrocuta brevirostris* from Gerakarou 1; (**e**) LGPUT-GER-156, left maxillary fragment with DP3–DP4 and erupting P4 in buccal view; (**f**) LGPUT-GER-157, right mandibular fragment with p4–m1 in lingual view. Scale bar equals 2 cm

Pliocrocuta perrieri (Croizet and Jobert, 1828)

Nomenclatural and Taxonomical History *Hyaena perrieri* in Croizet and Jobert 1828 (new species); *Hyaena (Pliohyaena) arvernensis* in Kretzoi 1938 (new combination); *Pachycrocuta perrieri* in Howell and Petter 1980 (new combination); for details about synonymy, see Werdelin and Solounias (1991).

Type Material MNHNP coll. Croizet no. 834 (holotype), mandible, housed in the Museum national d'Histoire naturelle of Paris.

Type Locality Les Étouaires, France, late Pliocene, MN17.

Distribution The species has a wide distribution covering whole Eurasia; it is recognized in Spain, France, Italy, Netherlands, Germany, Austria, Slovakia, Bulgaria, Turkey, Georgia, Ukraine, Russia, Tajikistan, China, and India. In Greece, it is known from Gerakarou 1 in Mygdonia Basin, Petralona Cave in Chalkidiki, and possibly from Sesklon in Thessaly.

Remarks *Pliocrocuta perrieri* is poorly represented in the Greek fossil record, known certainly from Gerakarou 1 (Fig. 4d) and Petralona Cave. Though the species is mentioned from Sesklon (Symeonidis 1992), the material is lost and thus its presence in this locality is questionable (Athanassiou 1998). The main characters of the species are the robust and high mandibular corpus, the presence of a large mental foramen below the p2, the small or rudimentary accessory cuspids on the p2 and p3, the very broad p3, the large accessory cuspids on the p4, and the short lower carnassial relative to the p4 (Koufos 1992). Baryshnikov and Tsoukala (2010) noted that the size of *P. perrieri* is close to that of *Crocuta crocuta* but the previous taxon differs displaying elongated P2, shorter carnassial (smaller metastyle-blade in the upper and enlarged talonid in the lower carnassial), and a well-developed metaconid versus a weak or absent on the m1 of *C. crocuta*.

Pachycrocuta Kretzoi, 1938

Type Species *Hyaena brevirostris* Gervais, 1850.

Pachycrocuta brevirostris (Gervais, 1850)

Nomenclatural and Taxonomical History *Hyaena brevirostris* Aymard in Anonymous 1856 (nomen nudum); *Hyaena brevirostris* in Gervais 1850 (new species); *Pachycrocuta brevirostris* in Kretzoi 1938 (new genus). The species has a long nomenclatural history, referred under various names for the material of the different localities from Eurasia (Werdelin and Solounias 1991). It is also referred with wrong authorship and date for more than 100 years; recently it was corrected and the right name is given (Alba et al. 2015).

Type Material Gervais (1848–1852) mentioned two cranial fragments in the Museum of Le Puy, but he did not define a holotype (specimen number unknown).

Type Locality Sainzelles, Polignac pres de Puy, France, Early Pleistocene, MNQ 20 (Palombo and Valli 2003–04).

Distribution Besides France, where is the type locality of the species, it is recognized in the entire Eurasia from Spain to China. In Greece, it is mentioned from Gerakarou 1, Apollonia 1, Kalamoto, Tsiotra Vrissi in Mygdonia Basin, Libakos in Western Macedonia, and Petralona Cave in Chalkidiki.

Remarks *Pachycrocuta brevirostris*, although present in Greece, its known material is scanty, including a partial cranium of a young individual, some maxillary or mandibular fragments with one or two teeth or with deciduous ones and some isolated teeth (Fig. 4e–f). The species is characterized by large and robust upper carnassial with relatively reduced protocone, elliptical and broader distally p2 with small anterior and large posterior accessory cuspids, elliptical-shaped p3 and p4 with small anterior and larger posterior accessory cuspids, elongated and broad lower carnassial without metaconid and with short talonid in comparison to the trigonid (Koufos 1992; Koufos and Kostopoulos 1997; Tsoukala and Chatzopoulou 2005).

⊙*Crocuta* Kaup, 1828

Type Species *Canis crocuta* Erxleben, 1777.

⊙*Crocuta crocuta* (Erxleben, 1777)—the Spotted Hyena

Nomenclatural and Taxonomical History *Canis crocuta* in Erxleben 1777 (new species); *Crocuta crocuta* in Kaup 1828 (new combination). See Werdelin and Solounias (1991) for information on synonymy.

Distribution The species is dispersed in the whole Old World from the Early Pleistocene-to the present. In Greece, it is traced from Petralona Cave in Chalkidiki.

Remarks *Crocuta crocuta* is the modern spotted hyaena. The main characters modified from the description of the Petralona cave material (Tsoukala 1989; Baryshnikov and Tsoukala 2010) are: short and broad cranium with high occipital crest; triangular occipital bone; convex ventral margin of the mandibular corpus; elongated and relatively narrow symphysis; large masseteric fossa with its anterior margin below the distal end of the m1; the premolars, except the upper carnassial, are situated close to each other and sometimes imbricated; weak posterior accessory cusp on the P2; elongated and strong upper carnassial; well-developed protocone situated in front of the mesial parastyle's margin; p2 and p3 with weak anterior and strong posterior accessory cuspids; p4 with strong anterior accessory cuspid from which starts a crest running across the mesial margin of the main cuspid; posterior accessory cuspid stronger than that of the p3; elongated and relatively narrow lower carnassial; no metaconid; small talonid with tendency to disappear; a small cuspid can only be distinguished on the talonid.

Crocuta crocuta spelaea (Goldfuss, 1823)—the Spotted Cave Hyena

Nomenclatural and Taxonomical History *Hyaena spelaea* in Goldfuss 1823 (new species); *Crocuta spelaea* in Ewer 1954 (new combination); *Crocuta crocuta spelaea* in Cardoso 1993 (new combination). See Sauqué et al. (2017) for further information on synonymy.

Type Material M2609 (holotype), incomplete cranium, housed in the Goldfuss-Museum Bonn (Diedrich 2008: fig. 3).

Type Locality Zoolithen Cave No. D 106, Geilenreuth, Bavaria, Germany, Upper Pleistocene, Weichselian/Würmian (probably OIS 3; Diedrich 2008).

Distribution It is well distributed in Eurasia ranging from Northern China to Iberian Peninsula and British Isles. In Greece, it is known from Agios Georgios Cave, Loutra Almopias Cave, and Haliakmon Basin in northern Greece, Maronia Cave in Thrace, as well as from Ktisti Cave, Ermionis in Peloponnese.

Remarks The main characters of *C. c. spelaea*, as given by Sauqué et al. (2017), are the larger size compared to that of *C. c. crocuta*, the extremely high-crowned P1, the short diastema between C and P1, the absence of diastema between P1 and P2, the disposition of the premolars in the tooth row clearly curved, and the relative shortening of the distal parts of the extremities.

Appendix

List of the Greek localities containing hyaenid fossils. Type localities are marked with bold. Locality numbers refer to the collection numbers of the PaleoBiology Database (PBDB)

Localities[PBDB No]	Age (MN; GPTS in Ma)	Taxon
Loutra Almopias Cave[203847]	Late Pleistocene	*Crocuta crocuta spelaea*[1]
Maronia Cave	Late Pleistocene	*Crocuta crocuta spelaea*[2]
Megalopolis Basin	Late Pleistocene	*Crocuta crocuta spelaea*[3]
Ktisti Cave Ermionis	Late Pleistocene	*Crocuta crocuta spelaea*[4]
Agios Georgios Cave[207192]	Late Pleistocene	*Crocuta crocuta spelaea*[5]
Petralona Cave[183123]	Late M. Pleistocene	*Crocuta crocuta*[6] *Pachycrocuta brevirostris*[7] *Pliocrocuta perrieri*[8]

(continued)

Localities[PBDB No]	Age (MN; GPTS in Ma)	Taxon
Tsiotra Vrissi[192004]	Late Villafranchian (MNQ18; 1.78–1.50)	*Pachycrocuta brevirostris*[9]
Kalamoto[200084]	Late Villafranchian	*Pachycrocuta brevirostris*[10]
Libakos[34764]	?Late Villafranchian	*Pachycrocuta brevirostris*[11]
Haliakmon Basin	Villafranchian	*Crocuta crocuta spelaea*[12]
Apollonia 1[197942]	Epivillafranchian (MN20)	*Pachycrocuta brevirostris*[13]
Gerakarou 1[134617]	Late Villafranchian (MNQ18)	*Pachycrocuta brevirostris*[14] *Pliocrocuta perrieri*[14]
Sesklon[34614]	Middle Villafranchian (MNQ17)	*Pliocrocuta perrieri*[15]
Dafnero 1[34594]	Middle Villafranchian (MNQ17; 2.4–2.3)	*Chasmaporthetes lunensis*[16]
Maramena[32189]	Miocene/Pliocene (MN13/14)	*Chasmaporthetes* sp.[17]
Dytiko 1[32374]	**Late Turolian (MN13; 7.0–6.0)**	***Chasmaporthetes bonisi*[18]**
Dytiko 3[32376]	Late Turolian (MN13; 7.0–6.0)	*Protictitherium crassum*[19]
Samos Q5[95690]	Late Turolian (MN13; 6.9–6.7)	*Adcrocuta eximia*[20] *Hyaenictitherium wongii*[20]
Kryopigi[157582]	Middle–Late Turolian (MN12–13; 7.3–6.4)	*Adcrocuta eximia*[21] *Ictitherium viverrinum*[21] *Plioviverrops orbignyi*[21]
Samos A[206461]	Middle Turolian (MN12)	*Adcrocuta eximia*[20]
Mytilinii 1A[202215]	Middle Turolian (MN12; ~7.1)	*Adcrocuta eximia*[23] *Hyaenictitherium wongii*[23]
Mytilinii 1B[202216]	Middle Turolian (MN12; ~7.1)	*Hyaenictitherium wongii*[23] *Plioviverrops orbignyi*[23]
Mytilinii 1C[202217]	Middle Turolian (MN12; ~7.1)	*Hyaenictitherium wongii*[23]
Samos Q1[95691]	Middle Turolian (MN12; 7.3–7.1)	*Adcrocuta eximia*[20] *Hyaenictitherium wongii*[20] *Ictitherium viverrinum*[20] *Plioviverrops orbignyi*[20]
Samos S3[205351]	Middle Turolian (MN12; ~7.3)	*Hyaenictitherium wongii*[20] *Adcrocuta eximia*[20]
Vathylakkos 2[202703]	Middle Turolian (MN12; ~7.3)	*Hyaenictitherium wongii*[23] *Plioviverrops orbignyi*[23]
Vathylakkos 3[182750]	Middle Turolian (MN12; ~7.3)	*Adcrocuta eximia*[24] *Hyaenictitherium wongii*[24] *Ictitherium viverrinum*[24] *Plioviverrops orbignyi*[24] *Plioviverrops* cf. *guerini*[24]
Perivolaki[194879]	Middle Turolian (MN12; 7.3–7.1)	*Adcrocuta eximia*[26] *Plioviverrops orbignyi*[26] *Ictitherium* sp.[26]
Kerassia 3[195435]	Middle Turolian (MN12)	*Plioviverrops* sp.[27] cf. *Ictitherium pannonicum*[27]
Kerassia 4[195435]	Middle Turolian	*Adcrocuta eximia*[27]

(continued)

Localities[PBDB No]	Age (MN; GPTS in Ma)	Taxon
Pikermi[182754]	**Middle Turolian (MN12; 7.33–7.29)**	***Adcrocuta eximia*[28]** ***Hyaenictis graeca*[29]** ***Ictitherium viverrinum*[28]** ***Lycyaena chaeretis*[30]** *Hyaenictitherium wongii*[31] ***Plioviverrops orbignyi*[32]**
Prochoma 1[202222]	Middle Turolian (MN12; ~7.4)	*Ictitherium viverrinum*[33] *Plioviverrops orbignyi*[24] *Adcrocuta eximia*[33]
Samos A[206461]	Middle Turolian (MN12)	*Lycyaena chaeretis*[20] *Ictitherium viverrinum*[20] *Plioviverrops orbignyi*[20]
Halmyropotamos[202213]	?Middle Turolian (?MN12)	*Adcrocuta eximia*[34]
Thermopigi[73553]	?Middle Turolian (?MN12)	*Adcrocuta eximia*[35]
Samos Q4[95689]	Early Turolian (MN11; ~7.5)	*Adcrocuta eximia*[20] *Hyaenictitherium wongii*[20]
Mytilinii 4[202219]	Late Early Turolian (MN11; ~7.5)	*Protictitherium crassum*[26] *Hyaenictitherium* cf. *wongii*[26]
Nikiti 2[73869]	Early Turolian (MN11; 8.7–8.2)	*Adcrocuta eximia*[36]
Ravin des Zouaves 5[195489]	Early Turolian (MN11; ~8.2)	*Adcrocuta eximia*[23] *Chasmaporthetes bonisi*[37] *Hyaenictitherium wongii*[23] *Ictitherium viverrinum*[33] *Plioviverrops orbignyi*[38]
Ravin X[182745]	?Early Turolian (?MN11)	*Adcrocuta eximia*[22]
Ravin de la Pluie[191070]	**Late Vallesian (MN10; ~9.3)**	*Adcrocuta eximia leptoruncha*[24] ***Protictitherium thessalonikensis*[38]** *?Hyaenictis* sp.[38] *Protictitherium* cf. *intermedium*[33]
Xirochori 1[195490]	Late Vallesian (MN10; ~9.6)	*Adcrocuta eximia*[34], *Protictitherium crassum*[33]
Ravin des Zouaves 1[182746]	Late Vallesian (MN10)	*Adcrocuta eximia*[33] *Hyaenictitherium wongii*[34]
Pentalophos 1[202119]	Early Vallesian (MN9)	*Protictitherium* cf. *crassum*[24] *Dinocrocuta gigantea*[39]
Antonios[73861]	Early/Middle Miocene (MN4/5)	*Protictitherium gaillardi*[40] *Protictitherium* cf. *crassum*[40]

GPTS Geomagnetic Polarity Time Scale
[1]Tsoukala et al. (2006), [2]Pavlides et al. (2005), [3]Melentis (1964), [4]Mitzopoulos (1960), [5]Tsoukala (1991), [6]Kanellis (1962), [7]Baryshnikov and Tsoukala (2010), [8]Sickenberg (1971), [9]Konidaris et al. (2015), [10]Tsoukala and Chatzopoulou (2005), [11]Steensma (1988), [12]Melentis (1966), [13]Koufos and Kostopoulos (1997), [14]Koufos (1992), [15]Athanassiou (1998), [16]Koufos (1993), [17]Schmidt-Kittler (1995), [18]Koufos (1987), [19]Koufos (1980), [20]Solounias (1981), [21]Lazarides (2015), [22]Koufos (2009), [23]de Bonis et al. (1973), [24]Arambourg and Piveteau (1929), [25]de Bonis and Koufos (1991), [26]Koufos (2006), [27]Roussiakis and Theodorou (2003), [28]Roth and Wagner (1854), [29]Gaudry (1846–50), [30]Gaudry (1861), [31]Gaudry (1862–67), [32]Gaydry and Lartet (1856), [33]Koufos (2000), [34]Melentis (1967), [35]Geraads et al. (2007), [36]Koufos (2016), [37]de Bonis and Koufos (1994), [38]Koufos (2012), [39]Koufos (1995); [40]Koufos (2008)

References

Alba DM, Vinuesa V, Madurell-Malapeira J (2015) On the original author and year of description of the extinct hyaenid *Pachycrocuta brevirostris*. Acta Palaeontol Pol 60(3):573–576

Andersson K, Werdelin L (2005) Carnivora from the Late Miocene of Lantian, China. Vertebrata PalAsiatica 10:256–271

Anonymous (1856) La paléontologie de la Haute-Loire offre-t-elle, pour certains époques, des caractères qui la particularisent? In: Congrès Scientifique de France. Tenue au Puy, September 1955. Derache, Paris, pp 261–277

Arambourg CJ, Piveteau J (1929) Les Vertébrés du Pontien de Salonique. Ann Paleont 18:59–138

Athanassiou A (1998) Contribution to the study of the fossil mammals of Thessaly. PhD thesis. Gaia, vol 5, pp 1–354

Baryshnikov GF, Tsoukala E (2010) New analysis of the Pleistocene carnivores from Petralona Cave (Macedonia, Greece) based on the Collection of the Thessaloniki Aristotle University. Geobios 43:389–402

de Bonis L, Koufos GD (1981) A new hyaenid (Carnivora, Mammalia) in the Vallesian (late Miocene) of Northern Greece. Sci Ann Fac Phys Mat Univ Thessaloniki 21:79–94

de Bonis L, Koufos GD (1991) The late Miocene small carnivores of the lower Axios Valley (Macedonia, Greece). Geobios 24(2):361–379

de Bonis L, Koufos GD (1994) Some hyaenids from the late Miocene of Macedonia (Greece) and the phylogeny of hunting hyaenas. Mün Geowiss Abh 26:81–96

de Bonis L, Bouvrain G, Keraudren B, Melentis JK (1973) Premiers resultats des fouilles recentes en Grèce septentrionale (Macédoine). C R Acad Sci Paris 227:1431–1434

del Campana D (1914) La *Lycyaena lunensis* n. sp. dell'ossario pliocenico di Olivola (Val di Magra). Paleontograph Itali 20:87–104

Depéret C (1892) La faune des mammiferes miocènes de la Grive Saint-Alban. Arch Mus Hist Nat Lyon 5:31–96

Diedrich C (2008) The rediscovery holotypes of the Upper Pleistocene spotted hyena *Crocuta Crocuta spelaea* (Goldfuss 1823) and the steppe lion *Panthera leo spelaea* (Goldfuss 1810) and taphonomic discussion to the Zoolithen Cave hyena den at Geilenreuth (Bavaria South-Germany). Zool J Linnean Soc 154:822–831

Gaillard C (1899) Mammifères miocènes nouveaux ou peu connus de La Grive St-Alban. Arch Mus Hist Nat Lyon 7:1–79

Gaudry A (1861) Résultat des fouilles exécutées en Grèce, sous les auspices de l'Académie. C R Acad Sci Paris 50:722–724

Gaudry A (1862–1867) Animaux fossils et géologie de l'Attique. Savy, Paris

Gaudry A, Lartet E (1856) Sur les résultats des recherches paléontologiques entreprises dans l'Attique sous les auspices de l'Académie. C R Acad Sci Paris 43:271–274

Geraads D, Tsoukala E, Spassov N (2007) A skull of *Ancylotherium* (Chalicotheriidae, Mammalia) from the Late Miocene of Thermopigi (Serres, N. Greece) and the relationships of the genus. J Vertebr Paleontol 27(2):461–466

Kanellis A (1962) The Petralaona Cave. Vuno 7:143–151

Kretzoi M (1938) Die Raubtiere von Gombaszög nebst einer Übersicht der Gesamtfauna (Ein Beitrag zur Stratigraphie des Altquartärs). Ann Mus Nat Hung 31:88–157

Konidaris GE, Tourloukis V, Kostopoulos DS, Thompson N, Giusti D, Michailidis D, Koufos GD, Harvati K (2015) Two new vertebrate localities from the Early Pleistocene of Mygdonia Basin (Macedonia, Greece): preliminary results. C R Palevol 14:353–362

Koufos GD (1980) Palaeontological and stratigraphical study of the Neogene continental deposits of the Axios Valley basin. PhD thesis, Aristotle University of Thessaloniki

Koufos GD (1987) *Chasmaporthetes bonisi*, a new hyaenid (Carnivora, Mammalia) from the late Miocene of Macedonia (Greece). Bull Soc Geol Fr 8:913–920

Koufos GD (1992) The Pleistocene carnivores of the Mygdonia basin (Macedonia, Greece). Ann Paleontol 78:205–257

Koufos GD (1993) Late Pliocene carnivores from western Macedonia (Greece). Paläontol Z 67:357–376

Koufos GD (1995) The late Miocene percrocutas (Carnivora, Mammalia) of Macedonia, Greece. Palaeovertebrata 24(1–2):67–84

Koufos GD (2000) Revision of the late Miocene carnivores from the lower Axios valley. Münchner Geowiss Abh A 39:51–92

Koufos GD (2006) Carnivora. Palaeontographica Abt A 276:39–74

Koufos GD (2008) Carnivores from the early/middle Miocene locality of Antonios (Chalkidiki, Macedonia, Greece). Geobios 41(3):365–380

Koufos GD (2009) Carnivora. Beitr Paläont 31:57–105

Koufos GD (2011) The Miocene carnivore assemblage of Greece. Estud Geol 67:291–320

Koufos GD (2012) New material of Carnivora (Mammalia) from the Late Miocene of Axios Valley, Macedonia, Greece. C R Palevol 11:49–64

Koufos GD (2016) Carnivora. Geobios 49(1–2):53–67

Koufos GD, Kostopoulos DS (1997) New Carnivore material from the Plio-Pleistocene of Macedonia, Greece with the description of a new canid. Mün Geowiss Abh 34:33–63

Kurtén B (1982) Status of the fossil hyaenids *Ictitherium viverrinum* and *Thalassictis robusta* (Mammalia) Z Geol Wiss 10:1009–1018

Lazaridis GT (2015) Study of the Late Miocene vertebrate locality of Kryopigi and other localities of Kassandra Peninsula, Chalkidiki (Greece). Systematics, Taphonomy, Palaeoecology, Biochronology. PhD thesis, Aristotle University of Thessaloniki

MacDonald D (1992) The Velvet Claw: a natural history of the carnivores. Parkwest, New York

Melentis JK (1964) Die fossilen rhinocerotiden, hippopotamiden und andere saugetiere aus dem becken von Megalopolis im Peloponnes (Griechenland). Proc Acad Athens 39:388–400

Melentis JK (1967) Studien über fossile Vertebraten Griechenlands. 19. Die Pikermifauna von Halmyropotamos (Euböa, Griechenlands). I Teil: Odontologie und Kraniologie. Ann Geol Pays Helleniques 18:283–411

Mitzopoulos M (1960) Über den ersten Nachweis einer Pleistozänen Hyaene in Griechenland. Ann Géol Pays Hell 11:293–296

Myers P (2000) "Hyaenidae" (On-line), Animal Diversity Web. Accessed March 16, 2021 at https://animaldiversity.org/accounts/Hyaenidae/. Accessed 27 Mar 2018

Orlov YA (1968) Fundamentals of paleontology, vol. 13, Mammals. Israel Program for Scientific Translations, Jerusalem. National Science Foundation, Washington, DC

Palombo MR, Valli AMF (2003–04) Remarks on the biochronology of mammalian faunal complexes from the Pliocene to the middle Pleistocene in France. Geol Romana 37:145–163

Pavlides S, Tsoukala E, Chatzipetros A, Chatzopoulou A, Melfos V, Vasileiadou A, Lazarides G, Vaxevanopoulos M (2005) The Maronia cave in the nummulitic limestone (Thrace, Greece). Geology and palaeontology. In: Proceedings of the 14th international congress of speleology, vol 1, pp 88–90

Pilgrim GE (1931) Pontian Carnivora of Europe. British Museum Natural History, London

Roth J, Wagner A (1854) Die fossilen Knochenüberreste von Pikermi in Griechenland. Abh math–phys Cl Königlich Bayerischen Akad Wiss 7:371–464

Roussiakis S (1996) Contribution to the study of the mammalian findings in the type locality of Pikermi. PhD thesis, University of Athens

Roussiakis S, Theodorou G (2003) Carnivora from the late Miocene of Kerassia (northern Euboea, Greece). Deinsea 10:469–497

Sauqué V, Rabal-Garcés R, Madurell-Malaperia J, Gisbert M, Zamora S, de Torres T, Ortiz JE, Cuenca-Bescós G (2017) Pleistocene cave hyenas in the Iberian Peninsula: new insights from Los Aprendices cave (Moncayo, Zaragoza). Palaeontol Electron 20.1(11A):1–38

Schmidt-Kittler N (1976) Carnivores from the Neogene of Asia Minor. Palaeontographica Abt A 155:1–131

Schmidt-Kittler N (1995) Carnivora, Mammalia. In: Schmidt-Kittler N (ed) The vertebrate locality Maramena (Macedonia, Greece) at the Turolian/Ruscinian boundary (Neogene). Münchner Geowis Abh AbtA 28:75–86

Sickenberg O (1968) Die Unterpleistozäne fauna von Wolaks (Griech.-Mazedonien) II. Carnivoren. Ann Geol Pays Helleniques 19:621–646

Sickenberg O (1971) Revision der Wirbeltierfauna der Höhle Petralona (Griech-Mazedonien). Anls Géol Pays Helleniques 23:230–264

Solounias N (1981) The Turolian fauna from the Island of Samos, Greece. Contrib Vertebr Evol 6:1–232

Steensma KJ (1988) Plio-Pleistozäne Grosssaugetiere (Mammalia) aus dem Becken von Kastoria/Grevena südlich von Neapolis-NW Griechenland. PhD dissertation, Technical University of Clausthal

Symeonidis N (1992) Fossil mammals of Early Pleistocene (Villafranchian) age from Sesklon Basin (Volos). Ann Geol Pays Helleniques 35:1–21

Tsoukala E (1989) Contribution to the study of the Pleistocene fauna of large mammals (Carnivora, Perissodactyla, Artiodactyla) from Petralona Cave, Chalkidiki (N. Greece). Sci Ann Aristotle Univ Thessaloniki 1:1–360

Tsoukala E, Chatzopoulou K (2005) A new Early Pleistocene (Latest Villafranchian) site of large mammals in Kalamoto of the Mygdonia basin (Macedonia, Greece). Preliminary report. Mit Kom für Quart Öster Akad Wissen 14:213–233

Tsoukala E, Chatzopoulou K, Rabeder G, Pappa S, Nagel D, Withalm G (2006) Paleontological and stratigraphical research in Loutra Arideas Bear Cave (Almopia Speleopark, Pella, Macedonia, Greece. Sci Ann, School of Geology, Aristotle Univ, Sp Vol 98:41–67

Werdelin L, Solounias N (1991) The Hyaenidae: taxonomy, systematics and evolution. Fossils Strata 30:1–104

Zdansky O (1924) Jungtertiäre Carnivoren chinas. Paleont Sin C.2:38–45

The Fossil Record of Canids (Mammalia: Carnivora: Canidae) in Greece

George D. Koufos

1 Introduction

The family Canidae is a carnivoran group which includes a variety of mammals, such as wolves, domestic and wild dogs, foxes, jackals, coyotes, and several other modern and fossil dog-like forms. It is the oldest carnivoran family on the fossil record, and its representatives date back to the late Eocene. The canids are widely distributed in all continents except Antarctica; *Canis familiaris* was imported into Australia by humans during prehistoric times and gave the wild dingoes (*Canis familiaris dingo*). They are known by 34 extant species and classified into three subfamilies, the extinct Hesperocyoninae and Borophaginae, and the extant Caninae (Fahey and Myers 2000).

The canids can be defined based on the following characters: elongated and narrow facial region, the length is equal or larger than the breadth at the orbits; the auditory bullae are small, but in some forms they are large; the canines are large and slender and the carnassials large and sectorial; the P4 is longer than the M1 with a small protocone; the m2 is always present and the M3 is present only in the primitive forms; the m2 is moderately elongated and always present, and the m3 is small and sometimes absent; the extremities are plantigrade in the early forms; and digitigrade in the advanced ones (Orlov 1968).

G. D. Koufos (✉)
Department of Geology, Laboratory of Geology and Paleontology,
Aristotle University of Thessaloniki, Thessaloniki, Greece
e-mail: koufos@geo.auth.gr

E. Vlachos (ed.), *Fossil Vertebrates of Greece Vol. 2*,
https://doi.org/10.1007/978-3-030-68442-6_20

2 Historical Overview

In the Greek fossil record, the canids are quite diverse and are represented by at least five genera with twelve species; most of the known material was collected in the last 40–50 years. One of the early found and described canids is a skull from Psychiko, near Athens, which was described as a new species under the name *Canis mitzopouli* by Paraskevaidis (1953) and later transferred to *Nyctereutes negamastoides* by Sickenberg (1968). The known Greek fossil material of canids originates from Pliocene and Pleistocene fossiliferous sites or cave deposits. Most of this material originates from northern Greece and was collected since the 1970s (Sickenberg 1968, 1971, 1972, 1975; Symeonidis et al. 1980, 1985/86; Koufos 1987, 1992, 1993, 2018; Tsoukala 1989, 1992; Koufos and Kostopoulos 1997; Athanassiou 1998; Tsoukala and Chatzopoulou 2005; Tsoukala et al. 2006; Konidaris et al. 2015).

3 Phylogenetic Relationships

The caniform (dog-like) lineage separated from the carnivorans at ~50.0 Ma, and the first known dog-like representative is *Prohesperocyon wilsoni*, found in Texas and dated back at ~40.0 Ma (late Eocene) (Wang and Tedford 2008). Soon after, during the Eocene, the Canidae subdivided into three known subfamilies. The Hesperocyoninae occurred at ~38.0 Ma and disappeared at ~15.0 Ma, the Borophaginae survived during 34.0–2.0 Ma, and the Caninae are still present (Van Valkenburg et al. 2004). All Greek canids belong to the subfamily Caninae and include one new species from Greece.

4 Distribution

The canids arrived in Eurasia through the Beringian landbridge around 8.0 Ma with the genus *Eucyon* (Perini et al. 2010); the earliest European representative of the canids is *Eucyon cipio* found in the late Miocene (Turolian, MN12) of Spain (Rook 2009). The canids then rapidly dispersed in the whole Eurasia with several taxa especially in the Pliocene and Pleistocene. The earliest evidence of the canids in Greece is traced in the early Pliocene localities Megalon Emvolon and Allatini, near Thessaloniki, with the genera *Nyctereutes* and *Eucyon* (Koufos 1997). Most of the Greek canid occurrences belong to the genera *Canis* and *Vulpes*, which are still represented in the extant Greek fauna. The earliest *Vulpes* is known from the Early Pleistocene site Dafnero 1, dated to 2.4–2.3 Ma (Benammi et al. 2020); that of *Canis* originates from Gerakarou 1, dated to ~1.8 Ma (Koufos and Kostopoulos 2016). See Fig. 1 and Appendix for further information.

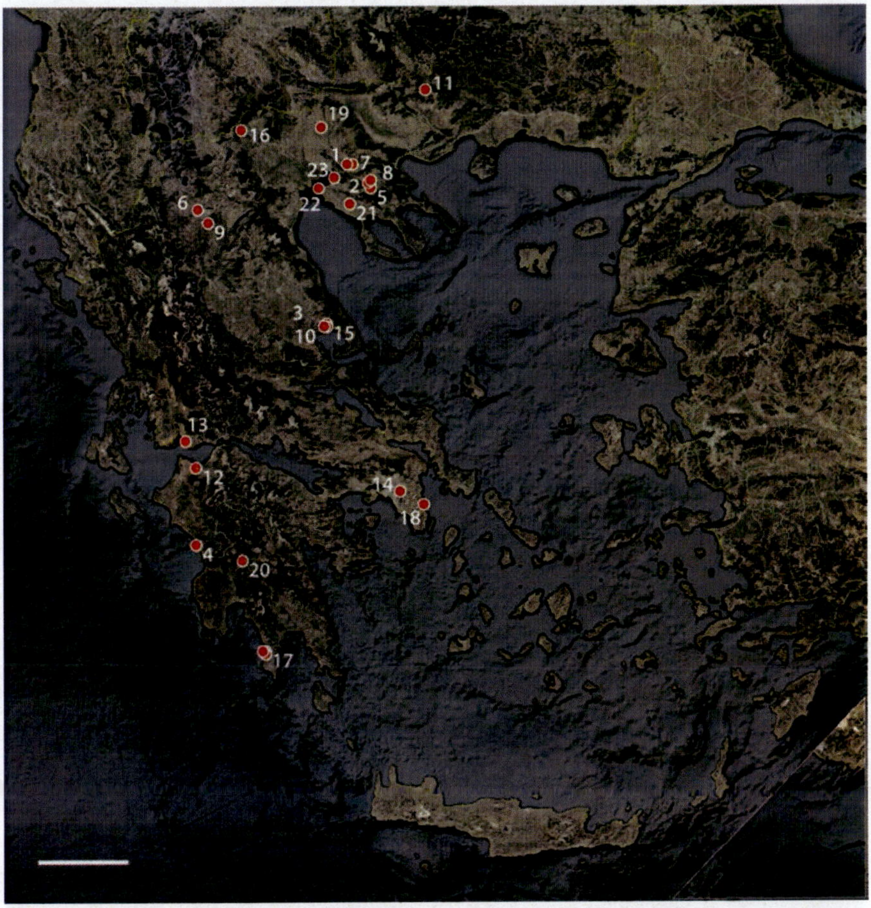

Fig. 1 Map of the occurrences of fossil canids from Greece. See Appendix for further information. **1**, Apollonia 1; **2**, Ravin Voulgarakis; **3**, Alikes; **4**, Kaiafas; **5**, Kalamoto 2; **6**, Libakos; **7**, Gerakarou 1; **8**, Tsiotra Vrissi; **9**, Dafnero 1; **10**, Sesklon; **11**, Volax; **12**, Kastritsi; **13**, Makinia; **14**, Psychiko; **15**, Volos; **16**, Loutra Almopias Cave; **17**, Apidima Caves; **18**, Vraona Cave; **19**, Agios Georgios Cave; **20**, Megalopolis/Marathousa; **21**, Petralona Cave; **22**, Megalo Emvolon; **23**, Allatini. See Appendix for more information. Image exported from Google Earth Pro © 2019, map data from US Dept. of State Geographer, SIO, NOAA, U.S. Navy, NGA, GEBCO, image from Landsat/Copernicus. Scale bar equals 80 km, North faces upward

5 Systematic Paleontology

Canidae Fischer von Waldheim, 1817

***Eucyon* Tedford and Qiu, 1996**

Type Species *Canis(?) davisi* Merriam, 1911.

***Eucyon odessanus* (Odintzov, 1967)**

Nomenclatural and Taxonomical History *Vulpes odessana* in Odintzov 1967 (initial identification); *Eucyon odessanus* in Tedford and Qiu 1996 (new genus and new combination with mandatory spelling change).

Type Material The originally described material from Odessa Catacombs includes numerous cranial and mandibular remains but the holotype is not designated (Odintzov 1967); the list of the type material is given in detail by Rook (1993).

Type Locality Odessa Catacombs, Ukraine, early Pliocene, MN15 (Rook 2009).

Distribution Besides the type locality, the species is known from Romania, Moldova, and Turkey. In Greece, the species is only known from Allatini, near Thessaloniki. This locality does not exist anymore as it was destroyed during the construction of a new road.

Remarks The two mandibular fragments from Allatini have been originally described by Sickenberg (1972) as *Nyctereutes donnezani* (Fig. 2a). Later, Koufos (1997) redescribed them as *E. davisi*, based on the morphological and metrical similarities of these specimens with *E. davisi* and on the reference of Tedford and Qiu (1996) that the two species *E. davisi* and *E. odessanus* could be synonyms, as there are no important differences in their cranial morphology. Besides this identification, it is mentioned that they differ from *N. donnezzani s.s.* in the slightly smaller size and narrower premolars. Rook (2009), considering that *E. odessanus* is valid and different from *E. davisi*, transferred them to *E. odessanus*. The main characters of the taxon are: relatively shallow mandibular corpus with slightly curved inferior margin; elongated symphysis gradually inclined backward; single-rooted p1 with elongated distal cingulum; long and narrow p2 and p3 without accessory cuspids; relatively short lower carnassial with robust metaconid; well-developed hypoconid and entoconid in the m1; and elongated and narrow distally m2 with three cuspids.

***Nyctereutes* Temminck, 1838**

Type Species *Canis viverrinus* Temminck, 1838.

Nyctereutes megamastoides (Pomel, 1842)

Nomenclatural and Taxonomical History *Canis megamastoides* in Pomel 1842 (initial identification); *Nyctereutes megamastoides* in Viret 1954 (new combination); synonym of ★ *Canis mitzopouli* Paraskevaidis, 1953 by Sickenberg 1968 (subjective synonymy).

Type Material Pomel (1842) described a mandibular fragment from Perrier (France) under the name *Canis megamastoides*; Boule (1889) completed the diagnosis of the species based on an almost complete skeleton from Ardé (France) found by Bravard and housed in the Museum Nationale d'Histoire Naturelle in Paris. This specimen was described by Boule (1889), whereas the skull and mandible were figured under the name *Canis borbonicus* by Gervais (1859: pl. 27, fig. 7); this specimen was considered as the type of the species (Viret 1954). However, the original material used to erect the specific name *megamastoides*, given earlier by Pomel (1842), must represent the type of the species.

Type Locality Perrier, Loire Basin, France, Pliocene, MN16 (Viret 1954).

Distribution Besides the type locality, the species found in Spain, France, Italy, Hungary, Romania, Russia, Georgia, and Tajikistan. In Greece, it is known from Dafnero 1 in Grevena Basin, Volax in Drama Basin, Sesklon in Thessaly, and Psychiko in Attica.

Remarks The species is known from several localities of Greece by cranial and mandibular remains. The best-known material is that from Dafnero 1, including two skulls and a mandible. There is also a skull from Psychiko (near Athens) described originally as *Canis mitzopouli* by Paraskevaidis (1953), but it is considered synonym with *N. megamastoides* (Sickenberg 1968; Athanassiou 1998). The Dafnero 1 material is characterized by strong temporal lines connected with the well-developed sagittal crest, rugosity of the parietals, strong postorbital processes directed backward, small, oval, and voluminous auditory bullae with their anterior margin aligned with the posterior surface of the postglenoid fossa, short and flattened paroccipital processes, strong subangular lobe in the mandible, straight ventral margin of the mandibular corpus, short symphysis, anterior margin of masseteric fossa far posteriorly of the m3, elongated and narrow upper premolars (except P4) without accessory cusps, and with strong distal cingular projection, small upper carnassial relative to the molars with small protocone slightly projected in front of the mesial tooth margin, and absence of parastyle, squared upper molars, p2 and p3 without accessory cuspids, short and robust p4 in comparison to the p2 and p3 with strong posterior accessory cuspid, short and wide lower carnassial with almost equal trigonid and talonid, large metaconid and robust hypoconid, accessory cuspid between it and the entoconid, oval-shaped m2 with four cuspids, and small, rounded, and bicuspid m3 (modified from Koufos 1993; Koufos and Kostopoulos 1997).

Nyctereutes tingi Tedford and Qiu, 1991

Type Specimen F:AM 97030 (holotype), left mandibular fragment with i1–m3, (Tedford and Qiu 1991: p. 29, fig. 1F-C).

Type Locality Liujiagou, Yushe Basin, China, Upper Pliocene (Flynn and Qiu 2013).

Distribution Besides China, the species is reported from Bulgaria and Greece. It is only known from a single Greek locality, named Megalon Emvolon (old name "Karaburun" or "Karabouroun") near Thessaloniki.

Remarks Two skulls of this taxon have been described from Megalon Emvolon, representing the only known material of this taxon from Greece and possibly out of China (Fig. 2c). Their morphological characters are similar to those of *N. tingi,* and they are attributed to this taxon (Koufos 1997). The main characters of the Greek sample, modified from Koufos (1997), are: elongated skull with relatively narrow muzzle, elongated nasal bones, exceeding posteriorly the fronto-maxillary suture, presence of a shallow depression in front of the orbits, large infraorbital foramen well pocketed posteriorly, strong postorbital processes bearing the vulpine groove in their dorsal surface, strong and high sagittal crest, strong nuchal crest slightly projected posteriorly, rugose parietal surface, small occipital condyles, rounded foramen magnum, ovoid auditory bullae, elongated and narrow premolars with distal cingular projection, P2 with small posterior accessory cusp, relatively short carnassial with well-developed protocone whose mesial margin exceeds that of the tooth, relatively large subtriangular molars, paracone larger than metacone, presence of metaconule, strong hypocone, and well-developed buccal cingulum.

Vulpes Frisch, 1775

Type Species *Canis vulpes* Linneaus, 1758.

Remarks The genus *Vulpes* is mentioned from the fauna of Kaiafas in Peloponnese and of Marathousa in Megalopolis Basin (van der Meulen and van Kolfschoten 1986; Sickenberg 1975). *Vulpes* cf. *praeglacialis* is also referred from Volos in Thessaly (van der Meulen and van Kolfschoten 1986). The Volos material is undescribed, and the exact locality is unknown; as mentioned, it probably originates from surface collection. Athanassiou (1998) described a right mandibular fragment with p3–m1 from Sesklon in Thessaly as *V.* cf. *alopecoides* because of its slightly larger size than *V. alopecoides*.

Vulpes alopecoides (Del Campana, 1913)

Nomenclatural and Taxonomical History *Canis alopecoides* in Del Campana 1913 (new species); *Vulpes alopecoides* in Stehlin and Dubois 1933 (new combination).

Fig. 2 Selected specimens of *Eucyon* and *Nyctereutes* from Greece. (**a**) Mandible of *Eucyon odessanus* from Allatini in buccal view, (**b**) cranium of *Nyctereutes megamastoides* from Dafnero 1 in palatal view, (**c**) cranium of *Nyctereutes tingi* from Megalo Emvolon in palatal view. Scale bar equals 5 cm

Type Specimen IGF 12110 (holotype), right maxillary fragment with M1 and M2, (Del Campana 1913: pl. 22, fig. 6; Bartolini Lucenti and Madurell-Malapeira 2020: fig. 2).

Type Locality "Il Tasso," Upper Valdarno, Italy, Early Pleistocene (del Campana 1913).

Distribution The species, except Italy, is known from Spain, France, Austria, Germany, Slovakia, Czech Republic, Georgia, Turkey, and possibly Bulgaria. In Greece, it is referred from Dafnero 1 in Grevena Basin, Makinia and Kastritsi in Southern Greece.

Remarks The available material of this species is scarce; it was recognized for the first time in Makinia and Kastritsi (Fig. 3a), and then it was found in Dafnero 1. The main characters of the species are the strong lower canine with distal and mesio-distal crests, the single rooted and buccolingually flattened p1, the elongated and narrow premolars without anterior accessory cuspid, the strong distal cingular projection with a vestigial anterior accessory cuspid in the p3 and a strong posterior accessory cuspid in the p4, the large lower carnassial with strong metaconid, generally present transverse cristid and bicuspid talonid, the presence of a small cuspid in the junction between the trigonid and talonid, the elongated and relatively wide m2 without hypoconid, and the small, rounded, and bicuspid m3 (Koufos 1993; Koufos and Kostopoulos 1997).

Vulpes praeglacialis (Kormos, 1932)

Nomenclatural and Taxonomical History *Alopex praeglacialis* in Kormos 1932 (new species); *Vulpes praeglacialis* in Bonifay 1971 (new combination).

Type Specimen Specimen number unknown (holotype), right mandible with m1–m3 described and figured by Kormos (1932: p. 168; taf.5, fig. 1).

Type Locality Villàny, Kalkberg, Hungary, Early Pleistocene (Kormos 1932; de Bruijn et al. 1992).

Distribution Besides Hungary, the species is known from Spain, France, and Israel. In Greece, the taxon is recognized from Apollonia 1 in Mygdonia Basin and Petralona Cave (group A) in Chalkidiki.

Remarks The material from Apollonia 1 (Fig. 3b, c) includes two mandibular fragments, two upper carnassials, and one upper canine (Koufos 2018); that from Petralona is richer (Baryshnikov and Tsoukala 2010). The species is characterized by sharp and curved distally upper canine with elliptical cross section, upper carnassial with high paracone, trenchant metastyle-blade, small-low and well separated from the paracone protocone with its mesial margin exceeding that of the paracone, shallow mandibular ramus with convex ventral margin after the m1, two small mental foramens below the p2 and p3, oval and deep masseteric fossa ending below the distal margin of the m3, small, monocuspid, and single-rooted p1, elongated and narrow p2 and p3 lacking accessory cuspids, strong distal cingular projection in the p2 and p3, robust p4 relatively to the other premolars bearing posterior accessory cuspid, relatively elongated lower carnassial with small metaconid, short bicuspid talonid, hypoconid larger than the entoconid, and oval m2 with large protoconid and crest-like hypoconid (Koufos 2018).

Vulpes praecorsac Kormos, 1932

Type Specimen Specimen number unknown (holotype), left mandibular fragment with p1–m2 described and figured by Kormos (1932: p. 178; taf. 5, fig. 8).

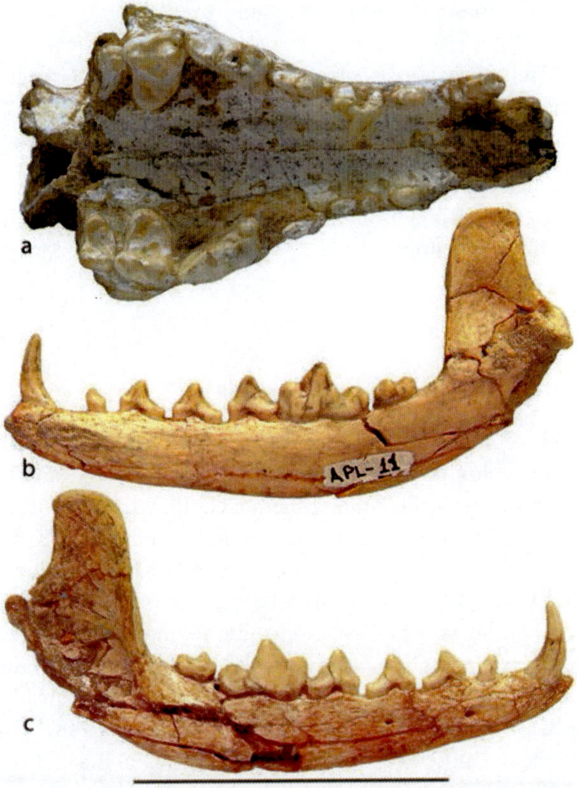

Fig. 3 Selected specimens of *Vulpes* from Greece. (**a**) Cranium of *Vulpes alopecoides* from Kastritsi in ventral view (photo provided by A. Athanassiou), (**b, c**) mandible of *Vulpes praeglacialis* from Apollonia 1 in (**b**) lingual and (**c**) buccal view. Scale bar equals 5 cm

Type Locality Nagyharsányberg, Hungary (= Betfia 2, Romania), Early Pleistocene, MNQ19 (Kormos 1932; van der Meulen and van Kolfschoten 1986).

Distribution Besides Hungary, it is referred from Germany, Ukraine, and Greece. In Greece, it is only known from Volax in Drama Basin.

Remarks The species is only referred from Volax by a badly preserved frontal cranial fragment (Sickenberg 1968); unfortunately, nobody knows where the carnivores of Volax are housed. A great part of Volax material is stored in LGPUT but they are not carnivores, whereas another part (which also did not include carnivores) is housed in the Athens Museum of Geology and Paleontology. Thus, it is impossible now to compare the Volax *Vulpes* with the remaining known material from Greece and to check its identification.

⊙*Vulpes vulpes* (Linneaus, 1758)—the Red Fox

Distribution This species is dispersed all over the world. In Greece, fossil remains have been found from Vraona Cave in Attica, Agios Georgios Cave in Kilkis, Apidima Cave-A, B, C, D in Peloponnese, and Loutra Almopias Cave in Almopia.

Remarks The material from all Greek sites includes several specimens, which are morphologically close to the modern fox, and the observed differences are insufficient for a specific distinction.

Canis Linneaus, 1758

Type Species *Canis familiaris* Linnaeus, 1758.

Remarks Fossil representatives of *Canis* in Greece are referred from Kaiafas in Peloponnese, and Volos in Thessaly. Both provided mainly small mammals, but there are also some large ones; among them a canid, which was attributed, without description, to *C.* cf. *mosbachensis* (van der Meulen and van Kolfschoten 1986). Besides these references, the genus *Canis* is reported from the locality Libakos in Grevena Basin by two different forms, referred to as *Canis* sp. A and *Canis* sp. B. *Canis* sp. A has similar size to *C. etruscus*, but its morphological similarities with both *C. mosbachensis* and *C. senezensis* make it difficult for certain identification. *Canis* sp. B has similar size to *C. arnensis* but the absolute dimensions of the p2 and p4, as well as the premolar/molar ratio differs (Steensma 1988). The genus is also reported from the localities Kalamoto 2 in Mygdonia Basin by a distal radius (Tsoukala and Chatzopoulou 2005) and from Alykes in Thessaly by several postcranial remains, the size of which is like that of *Canis* (Athanassiou 1998).

★*Canis apolloniensis* Koufos and Kostopoulos, 1997

Nomenclatural and Taxonomical History *Canis arnensis* in Koufos 1992 partim (initial identification); *Canis apolloniensis* in Koufos and Kostopoulos 1997 (new species).

Type Specimen LGPUT-APL-530 (holotype), front part of the skull with the mandible (Koufos and Kostopoulos 1997: p. 39, fig. 1, Fig. 4a).

Type Locality Apollonia 1, Mygdonia Basin, Macedonia, Early Pleistocene, MNQ20.

Distribution Known only from its type locality.

Remarks The species was erected on some material from Apollonia 1 (Koufos and Kostopoulos 1997); later, some authors synonymized it with *C. etruscus*, *C. arnensis*, or *C. mosbachensis*. Recently, new material and the revision of the previously published specimens allowed an extensive comparison with these taxa indicating its clear separation from them. *Canis apolloniensis* morphologically is closer to *C. etruscus* and *C. mosbachensis* and can be considered as a plausible ancestor of the last taxon (Koufos 2018). It is characterized by elongated and narrow nasal bones extended to the anterior margin of the orbit, small and oval narial opening, relatively short muzzle, elongated and narrow palate, small, monocuspid, and single-rooted P1 with strong cingulum, elongated and narrow P2 with variably developed posterior accessory cusp and elevated distal cingulum like a secondary posterior cusp, P3 similar to the P2 but larger and with more expressed cusps, relatively short and wide upper carnassial, protocone small, low, well separated from the paracone with its mesial margin aligned with that of the tooth, triangular M1 with paracone larger than metacone, small protocone, crest-like hypocone and presence of a small protoconule, M2 subelliptical similar to the M1, elongated and relatively shallow mandibular corpus, presence of a faint concavity in its inferior margin, two mental foramens below the middle of the p2 and p3, respectively, masseteric fossa oval and deep with its anterior margin below the m3, elongated and narrow p2 with strong distal cingualr projection, p3 similar to p2 but larger and sometimes with a vestigial posterior accessory cuspid, p4 more robust than the other premolars with clear posterior accessory cuspid, short and wide lower carnassial with small metaconid, talonid one-third of tooth's length and bicuspid, m2 oval with large protoconid and without entoconid, and small, rounded m3 with a large protoconid and a small paraconid (Koufos 2018).

Canis arnensis Del Campana, 1913

Type Specimen IGF 867 (lectotype), cranium and mandible of a single individual, IGF 867 described by Del Campana (1913), designated by Torre (1967) and housed in the Museum of the University of Florence.

Type Locality "Il Tasso," Upper Valdarno, Italy, Early Pleistocene, MNQ19–20 (Torre 1967).

Distribution Besides Italy, it is known from Greece, Spain, and France. In Greece, it is known from Gerakarou 1 (Fig. 4b) and Ravin of Voulgarakis in Mygdonia Basin, Alikes in Thessaly, and Petralona Cave in Chalkidiki.

Remarks The species was found for first time in Greece in the locality Gerakarou 1 of Mygdonia Basin (Koufos 1987). Later, it was referred from the locality Apollonia 1, but it recently transferred to *C. apolloniensis* by Koufos (2018). The main features of the taxon are: medium size; relatively short muzzle; short nasal bones, ending at the level of the maxillo-frontal suture; parallel medial walls of tympanic bullae; equal-sized paracone and metacone in the M1; weak posterior and absent anterior accessory cusp in the P2 and P3; robust upper carnassial with small protocone; equally deep protocone and hypocone basins in the M1; crown base of the lower premolars (except p1) at the same level; absence of anterior accessory cuspids in the p3 and p4; weak size difference between the hypoconid and entoco-

Fig. 4 Selected specimens of *Canis* from Greece. (**a**) cranial fragment of *Canis apolloniensis* from Apollonia 1 in ventral view, (**b**) cranium of *Canis arnensis* from Gerakarou 1 in ventral view, (**c**) cranium of *Canis etruscus* from Apollonia 1 in ventral view. Scale bar equals 5 cm

nid of m1; presence of accessory cuspids on the lingual margin of the m1 talonid; equal-sized protoconid and metaconid in the m2; rare presence of entoconid in the m2; and the presence of accessory cuspids on the disto-lingual margin of the m2 (Koufos 1987, 1992; Bartolini Lucenti and Rook 2016).

Canis etruscus Forsyth Major, 1877

Nomenclatural and Taxonomical History *Canis etruscus* in Forsyth Major 1877 (new species); *C. olivolanus* and *C. majori* in Del Campana 1913 (junior synonyms, see Torre 1967).

Type Specimen MCM 47 (lectotype), cranium described by Forsyth Major (1877), designated by Torre (1967), and housed in the Museum of Florence University.

Type Locality Unrecorded locality of Upper Valdarno, Italy, Early Pleistocene, MNQ18–20 (Torre 1967).

Distribution In Eurasia, besides the type locality, it is known from Spain, France, Greece, Romania, Bulgaria, Turkey, Tajikistan, and Kazakhstan. In Greece, it is known from Gerakarou 1, Apollonia 1 (Fig. 4c), Tsiotra Vrissi in Mygdonia Basin, as well as from Alikes in Thessaly.

Remarks *Canis etruscus* has been recorded from four localities, and the known material, although scarce, includes two crania, one from Apollonia and another from Tsiotra Vrissi. The main characters, modified from our material and the emended diagnosis of Cherin et al. (2013), are: medium-size (slightly larger than *C. arnensis*); elongated snout; oval and wide narial opening; nasals end beyond the maxillofrontal suture; long and wide palate; strong sagittal and nuchal crests; narrow upper premolars (except P4); P1 with strong lingual cingulum; P2 and P3 with posterior accessory cusp situated on a distal projection of the distal cingulum; upper carnassial short and slender relative to the other teeth; small protocone aligned with the mesial margin of the tooth; stronger and higher paracone than metacone on M1; high mandibular corpus; strongly curved tooth row; absence of anterior accessory cuspid in the lower premolars; presence of posterior accessory cuspid on the p3 and p4; robust lower carnassial with short and wide talonid and small metaconid; and the hypoconid and entoconid linked by a seldomly sinuous.

⊙*Canis lupus* Linneaus, 1758—The Wolf

Distribution This extant species is widely distributed all over the world. In Greece, fossil remains of the taxon have been found in Vraona Cave in Attica.

Remarks Some postcranial remains of the wolf have been only referred from the Vraona Cave. The species has been also recognized in several paleolithic sites of Greece (not covered herein).

Lycaon lycaonoides (Kretzoi, 1938)

Nomenclatural and Taxonomical History "*Canis*" *gigas* in Kretzoi 1938 (initial identification); *Xenocyon lycaonoides* in Kretzoi 1938 (new species); *Canis (Xenocyon) lycaonoides* in Sotnikova 2001 (new combination; rank change);

Lycaon lycaonoides in Martínez-Navarro and Rook 2003 (new combination). In the meantime, several names were proposed or appeared for this species (see details in Tedford et al. 2009). There are currently two main opinions regarding both the specific identification of this canid material and its generic combination: Tedford et al. (2009) classify all Eurasian Plio-Pleistocene hypercarnivore canids as *Xenocyon lycaonoides*, whereas Martínez-Navarro and Rook (2003) classify them under the genus *Lycaon* represented by three species; *L. falconeri, L. lycaonoides*, and *L. pictus*.

Type Specimen MNM20 (holotype), right M1 (Kretzoi 1938: p. 132; pl. 3, fig. 4).

Type Locality Gombaszög (Hungary), Early Pleistocene.

Distribution Besides Slovakia, the species is known from Spain, France, Germany, Hungary, Montenegro, Russia, Tajikistan, Italy, and Israel. In Greece, it is referred from Apollonia 1 in Mygdonia Basin, and Petralona Cave in Chalkidiki.

Remarks The species is represented in Greece by some maxillary and mandibular remains from Apollonia 1, whereas there is only one m1 from Petralona Cave described under various names (Baryshnikov and Tsoukala 2010; Koufos 2018). A fragmentary skull associated with the mandible found recently in Apollonia 1 allowed the definite specific attribution to *Lycaon lycaonoides* (Koufos 2018). The main characters modified from the last author are: buccolingually flattened upper canine bearing distal and mesiolingual weak crest; single-rooted P1 with strong mesial and distal cingulum; elongated and narrow P2 and P3 lacking anterior accessory cusp and bearing a strong posterior accessory cusp situated in the center of a distal cingular projection; short and robust upper carnassial with a large paracone and a small protocone separated well from the paracone and situated slightly in front of the mesial margin of the tooth; large and triangular-shaped M1 with a large paracone, very small and crest-like hypocone and small entocone; small M2 similar in crown morphology with the M1, lacking hypocone and entocone; mandibular corpus with convex ventral margin; elliptical masseteric fossa with its anterior margin ending below the middle of the m2; small and single-rooted p1; p2 without anterior and posterior accessory cuspids but with a large distal cingular projection; p3 and p4 lack anterior accessory cuspid but they bear a large posterior one; large lower carnassial with small metaconid situated behind the protoconid and relatively short trigonid bearing a large hypoconid; m2 with large protoconid, small and equal-sized metaconid and hypoconid; very small, single-rooted, and round m3.

Acknowledgments Thanks to A. Athanassiou for providing the photo of *Vulpes alopecoides* from Kastritsi. Also, I would like to deeply thank S. Bartolini and G. Iliopoulos for comments that improved this manuscript.

Appendix

List of the Greek localities containing canid fossils. Type localities are marked with bold. Locality numbers refer to the collection numbers of the PaleoBiology Database (PBDB)

Localities[PBDB No]	Age (MN)	Taxa
Apollonia 1[34784]	Epivillafranchian (MNQ20)	***Canis apolloniensis***[1] *Canis etruscus*[1] *Vulpes praeglacialis*[2] *Lycaon lycaonoides*[2]
Ravin Voulgarakis[34783]	Epivillafranchian (MNQ20)	*Canis arnensis*[1]
Alikes[34782]	?Late Villafranchian	*Canis arnensis*[3] *Canis* indet.[3]
Kaiafas[34766]	?Late Villafranchian	*Canis lupus*[4] *Canis lupus* cf. *mosbachensis*[4] *Vulpes* indet.[4]
Kalamoto 2[200084]	?Late Villafranchian	*Canis* sp.[5]
Libakos[34764]	?Late Villafranchian	*Canis* sp. A[6] *Canis* sp. B[6]
Gerakarou 1[34617]	Late Villafranchian (MNQ18)	*Canis arnensis*[7] *Canis etruscus*[7]
Tsiotra Vrissi[192004]	Late Villafranchian (MNQ18; 1.78–1.50)	*Canis etruscus*[8, 9]
Dafnero 1[34594]	Middle Villafranchian (MNQ17; 2.4–2.3)	*Nyctereutes megamastoides*[10, 11] *Vulpes alopecoides*[10]
Sesklon[34614]	Middle Villafranchian (MNQ17)	*Nyctereutes megamastoides*[3] *Vulpes* cf. *alopecoides*[3]
Volax[34593]	Middle Villafranchian (MNQ17)	*Nyctereutes megamastoides*[12] *Vulpes praecorsac*[12]
Kastritsi[34810]	Villafranchian	*Vulpes alopecoides*[13]
Makinia[34811]	Villafranchian	*Vulpes alopecoides*[13]
Psychiko	?Villafranchian	*Nyctereutes megamastoides* **(= *Canis mitzopouli*)**[12,14]
Volos[34808]	?Villafranchian	*Canis lupus*[4] *Canis lupus* cf. *mosbachensis*[4] *Vulpes* cf. *praeglacialis*[4]
Loutra Almopias Cave Ic[203847]	Latest Pleistocene	*Vulpes vulpes*[15]
Apidima-A[32059]	Late Pleistocene	*Vulpes vulpes*[16]
Apidima-B[32062]	Late Pleistocene	*Vulpes vulpes*[16]
Apidima-C[32082]	Late Pleistocene	*Vulpes vulpes*[16]
Apidima-D[32086]	Late Pleistocene	*Vulpes vulpes*[16]
Vraona Cave[183130]	Late Pleistocene	*Vulpes vulpes*[17] *Canis lupus*[17]
Agios Georgios Cave[207192]	Late Pleistocene	*Vulpes vulpes*[18]
Loutra Almopias Cave I[203847]	Late Pleistocene	*Vulpes vulpes*[15]
Loutra Almopias Cave II[203847]	Late Pleistocene	*Vulpes vulpes*[15]
Loutra Almopias Cave III[203847]	Late Pleistocene	*Vulpes vulpes*[15]

(continued)

Localities[PBDB No]	Age (MN)	Taxa
Megalopolis/Marathousa[34765]	Middle Pleistocene	*Vulpes* indet[19]
Petralona Cave[183123]	Early Middle Pleistocene	*Canis arnensis*[20] *Lycaon lycaonoides*[21] *Vulpes praeglacialis*[21]
Megalo Emvolon[36579]	Late Ruscinian (MN15)	*Nyctereutes tingi*[22]
Allatini[182683]	Ruscinian	*Eucyon odessanus*[23]

[1]Koufos and Kostopoulos (1997), [2]Koufos (2018), [3]Athanassiou (1998), [4]van der Meulen and van Kolfschoten (1986), [5]Tsoukala and Chatzopoulou (2005), [6]Steensma (1988), [7]Koufos (1987), [8]Konidaris et al. (2015), [9]Konidaris et al. (2021), [10]Koufos (1993), [11]Benammi et al. (2020) [12]Sickenberg (1968), [13]Symeonidis et al. (1985/86), [14]Paraskevaidis (1953), [15]Tsoukala et al. (2006), [16]Tsoukala (1999), [17]Symeonidis et al. (1980), [18]Tsoukala (1992), [19]Sickenberg (1975), [20]Sickenberg (1971), [21]Tsoukala (1989), [22]Koufos (1997), [23]Sickenberg (1972)

References

Athanassiou A (1998) Contribution to the study of the fossil mammals of Thessaly. PhD thesis. Gaia 5:1–354

Bartolini Lucenti S, Madurell-Malapeira S (2020) Unraveling the fossil record of foxes: an updated review on the Plio-Pleistocene *Vulpes* spp. from Europe. Quat Sci Rev 236:106296 https://doi.org/10.1016/j.quascirev.2020.106296

Bartolini Lucenti S, Rook L (2016) A review on the late Villafranchian medium-sized canid *Canis arnensis* based on the evidence from Poggio Rosso (Tuscany, Italy). Quat Sci Rev 151:58–71

Baryshnikov GF, Tsoukala E (2010) New analysis of the Pleistocene carnivores from Petralona Cave (Macedonia, Greece) based on the Collection of the Thessaloniki Aristotle University. Geobios 43:389–402

Benammi M, Aidona E, Merceron G, Koufos GD, Kostopoulos DS (2020) Magnetostratigraphy and chronology of the lower Pleistocene primate bearing Dafnero fossil site, N. Greece. Quaternary 3:22. https://doi.org/10.3390/quat3030022

Bonifay MF (1971) Carnivores Quaternaires du sudest de La France. Mém Mus Hist Nat Paris C 21:64–153

Boule M (1889) Le *Canis megamastoides* du Pliocene moyen de Perier. Bull Soc Geol Fr 17:321–330

Cherin M, Bertè DF, Rook L, Sardella R (2013) Re-defining *Canis etruscus* (Canidae, Mammalia): a new look into the evolutionary history of Early Pleistocene dogs resulting from the outstanding fossil record from Pantalla (Italy). J Mammal Evol 21(1):95–110

de Bruijn H, Daams R, Daxner-Hock G, Fahlbusch V, Ginsburg L, Mein LP, Morales J (1992) Report of the RCMNS working group on fossil mammals, Reisensburg 1990. Newsl Stratigr 26(2/3):65–117

Del Campana D (1913) I cani pliocenici di Toscana. Palaeontogr Ital 19:189–254

Fahey B, Myers P (2000) Canidae (On-line). Animal diversity web. http://animaldiversity.org/accounts/Canidae/. Accessed 17 May 2018

Flynn LJ, Qiu Z-X (2013) Biostratigraphy and the Yushe Basin. In: Tedford RH, Qiu Z-X, Flynn LR (eds) Late Cenozoic Yushe Basin, Shanxi Province, China: geology and fossil mammals, History geology and magnetostratigraphy, vol I. Springer, Dordrecht, pp 79–82

Forsyth Major CJ (1877) Considerazioni sulle faune dei mammiferi pliocenici e postpliocenici della Toscana. Atti Soc Toscana Sci Nat Mem 3:207–227

Gervais P (1859) Zoologie et Paléontologie françaises. Nouvelles recherches sur les animaux vertébrés dont on trouve les ossements enfouis dans le sol de la France et sur leur comparaison avec les espèces propres aux autres régions du globe. Bertrand, Paris

Konidaris GE, Tourloukis V, Kostopoulos DS, Thompson N, Giusti D, Michailidis D, Koufos GD, Harvati K (2015) Two new vertebrate localities from the early Pleistocene of Mygdonia Basin (Macedonia, Greece): preliminary results. C R Palevol 14:353–362

Konidaris G, Kostopoulos DS, Maron M, Schaller M, Ehlers TA, Aidona E, Marini M, Tourloukis V, Muttoni G, Koufos GD, Harvati K (2021) Dating of the Lower Pleistocene Vertebrate Site of Tsiotra Vryssi (Mygdonia Basin, Greece): Biochronology, Magnetostratigraphy, and Cosmogenic Radionuclides. Quaternary, 4, 1. https://doi.org/10.3390/quat4010001

Kormos T (1932) Die Füchse des ungarischen Oberpliozäns. Folia Zool Hydrobiol 4:167–188

Koufos GD (1987) *Canis arnensis* Del Campana, 1913 from the Villafranchian (Villanyian) of Macedonia (Greece). Paleontol Evol 21:3–10

Koufos GD (1992) The Pleistocene carnivores of the Mygdonia Basin (Macedonia, Greece). Ann Paleontol 78:205–257

Koufos GD (1993) Late Pliocene carnivores from western Macedonia (Greece). Paläontologische Z 67:357–376

Koufos GD (1997) The canids *Eucyon* and *Nyctereutes* from the Ruscinian of Macedonia, Greece. Palaeontol Evol 30–3:39–48

Koufos GD (2018) New material and revision of the Carnivora, Mammalia from the lower Pleistocene Apolonia 1, Greece. In: Palombo M-P (ed) Advances in quaternary studies: the contribution of mammalian fossil record. Quaternary 1: 6; https://doi.org/10.3390/quat1010006

Koufos GD, Kostopoulos DS (1997) New Carnivore material from the Plio-Pleistocene of Macedonia, Greece with the description of a new canid. Mün Geowiss Abh 34:33–63

Koufos GD, Kostopoulos DS (2016) The Plio-Pleistocene large mammal record of Greece: implications for early human dispersals into Europe. In: Harvati K, Roksanic M (eds) Palaeoanthropology of the Balkans and Anatolia: human evolution and its context. Springer, Dordrecht, pp 269–280

Kretzoi M (1938) Die Raubtiere von Gombaszög nebst einer Übersicht der Gesamtfauna (Ein Beitrag zur Stratigraphie des Altquartärs). Ann Mus Nat Hung 31:88–157

Martínez-Navarro B, Rook L (2003) Gradual evolution in the African hunting dog lineage: systematic implications. C R Palevol 2:695–702

Odintzov IA (1967) New species of Pliocene Carnivora, *Vulpes odessana* sp. nov. from the Karst Cave of Odessa. Paleontol Sbornik. L'vov Univ 4:130–137

Orlov YA (1968) Fundamentals of paleontology, vol. 13, Mammals. Israel Program for Scientific Translations, Jerusalem.

Paraskevaidis I (1953) *Canis* und *Equus* aus Attica. Neues Jahrb Mineral Geol Palaeontol Abh Abt A 96:453–496

Perini FA, Russo CAM, Schrago CG (2010) The evolution of South American endemic canids: a history of rapid diversification and morphological parallelism. J Evol Biol 23(2):311–322. https://doi.org/10.1111/j.1420-9101.2009.01901.x

Pomel A (1842) Nouvelle espèce de Chien fossile découverte dans les alluvions volcaniques de l'Auvergne. Bull Soc Geol Fr 14:38–41

Rook L (1993) I cani dell'Eurasia dal Miocene superiore al Pleistocene medio. PhD dissertation, Modena-Bologna-Firenze-Roma "La Sapienza" Universities

Rook L (2009) The wide ranging genus *Eucyon* Tedford & Qiu, 1996 (Mammalia, Carnivora, Canidae, Canini) in the Mio-Pliocene of the Old World. Geodiversitas 31(4):723–741

Sickenberg O (1968) Die Unterpleistozäne fauna von Wolaks (Griech.-Mazedonien) II. Carnivoren. Ann Geol Pays Helleniques 19:621–646

Sickenberg O (1971) Revision der Wirbeltier Fauna der Hohle Petralona (Griechisch Mazedonien). Ann Geol Pays Helleniques 23:230–264

Sickenberg O (1972) Ein unterkiefer des Caniden *Nyctereutes donnezani* (Dep.) aus der umgebung von Saloniki (Griech. Mazedonien) und seine biostratigraphische bedeutung. Ann Nat Mus Wien 76:499–513

Sickenberg O (1975) Eine Säugertierfauna des tieferen Biharius aus dem Becken von Megalopolis (Peloponnes, Griechenland). Ann Geol Pays Helleniques 27:25–73

Sotnikova M (2001) Remains of Canidae from the Lower Pleistocene site of Untermassfeld. Das Pleistozän von Untermassfeld bei Meiningen (Thüringen) 2:607–632

Steensma KJ (1988) Plio-/Pleistozäne Großäugetiere (Mammalia) aus dem Becken von Kastoria/ Grevena, südlich von Neapolis-NW Griechenland. PhD dissertation, Technische Universität Clausthal

Stehlin HG, Dubois A (1933) La grotte de Cotencher, station moustérienne. Mem Soc Paleontol Suisse 52–53:1–292

Symeonidis N, Bachmayer F, Zapfe H (1980) Ergebnisse weiterer Grabungen in der Höhle von Vraona (Attica, Griechenland). Ann Geol Pays Helleniques 30(1):135–144

Symeonidis N, Theodorou GE, Schütt H, Velitzelos E (1985–86). Palaeontological and stratigraphical observations in Achaia and Aitoloakarnania (Western Greece) area. Ann Geol Pays Helleniques 33(1):329–365

Tedford RH, Qiu Z-X (1991) Pliocene *Nyctereutes* (Carnivora: Canidae) from Yushe, Shanxi, with comments on Chinese fossil raccoon-dogs. Vertebrata PalAsiatica 29:176–189

Tedford RH, Qiu Z-X (1996) A new canid genus from the Pliocene of Yushe, Shanxi Province. Vertebrata PalAsiatica 34:27–40

Tedford RH, Wang X, Taylor BE (2009) Phylogenetic systematics of the north American fossil Caninae (Carnivora: Canidae). Bull Am Mus Nat His 325:1–218

Torre D (1967) I cani Villafranchiani della Toscana. Paleontol Ital 33:112–138

Tsoukala E (1989) Contribution to the study of the Pleistocene fauna of large mammals (Carnivora, Perissodactyla, Artiodactyla) from Petralona Cave, Chalkidiki (N. Greece). Sci Ann Aristotle Univ Thessaloniki 1:1–360

Tsoukala E (1992) The Pleistocene large mammals from the Agios Georgios cave, Kilkis (Macedonia, Greece). Geobios 25(3):415–433

Tsoukala E (1999) Quaternary large mammals from the Apidima Caves (Lakonia, S. Peloponnese, Greece). Beit f Palaont 24:207–229

Tsoukala E, Chatzopoulou K (2005) A new early Pleistocene (latest Villafranchian) site with mammals in Kalamotó (Mygdonia Basin, Macedonia, Greece)—preliminary report. Mitt Kommission Quartärforschung Österreichische Akad Wiss 14:213–233

Tsoukala E, Chatzopoulou K, Rabeder G, Pappa S, Nagel D, Withalm G (2006) Paleontological and stratigraphical research in Loutra Arideas bear cave (Almopia speleopark, Pella, Macedonia, Greece). Aristotle Univ Thessaloniki Sci Ann 98:41–67

van der Meulen A, van Kolfschoten T (1986) Review of the late Turolian to early biharian mammal faunas from Greece and Turkey. Mem Soc Geol Ital 31:201–211

van Valkenburgh B, Wang X, Damuth J (2004) Cope's rule, hypercarnivory, and extinction in North American Canids. Science 306(5693):101–104. https://doi.org/10.1126/science.1102417

Viret MJ (1954) Le loess a bancs durcis de Saint-Vallier (Drôme) et sa faune de mammifères. Nouvelles Arch Mus Hist 4:1–197

Wang X, Tedford RH (2008) Dogs, their fossil relatives and Evolutionary history. Columbia University Press, New York

The Fossil Record of Bears (Mammalia: Carnivora: Ursidae) in Greece

Spyridoula Pappa and Evangelia Tsoukala

1 Introduction

The bears (family Ursidae) are the members of the Order Carnivora. All modern bears except *Tremarctos ornatus, U. americanus* (only in North America), and *Ursus maritimus* (in part) inhabit Eurasia. The most extreme example of behavioral flexibility in the Ursidae is the extant brown bear (*U. arctos*), which not only occupies a wider range of habitats in the Palearctic (including Northern Africa) and Nearctic but also demonstrates greater dietary variability than any other species of bear (McLellan et al. 2008). High *intra*specific variability is also very common on extinct species such as cave bears, and their fossil remains vary to a great extent both in size and morphology between populations both geographically and temporally (Grandal d'Anglade 1993; Rabeder et al. 2000; Baryshnikov and Puzachenko 2011).

Major disagreements exist regarding the systematic subdivision of the Ursidae. In the literature, the number of subfamilies recognized within the Ursidae varies from three to five (e.g., McLellan and Reiner 1994; Ginsburg 1999). McLellan and Reiner (1994) discussed five subfamilies: the Hemicyoninae, Agriotheriinae, Tremarctinae, Ursinae, and Ailuropodidae. Ginsburg (1999) recognized three European Neogene subfamilies: the Hemicyoninae, the Phoberocyoninae, and the

S. Pappa (✉)
Curator of Fossil Mammals, Department of Earth Sciences, Natural History Museum, London, UK

Department of Geography, Royal Holloway University of London, Surrey, UK
e-mail: Spyridoula.Pappa@nhm.ac.uk

E. Tsoukala
Laboratory of Geology and Paleontology, Aristotle University of Thessaloniki, Thessaloniki, Greece
e-mail: lilits@geo.auth.gr

E. Vlachos (ed.), *Fossil Vertebrates of Greece Vol. 2*,
https://doi.org/10.1007/978-3-030-68442-6_21

Ursinae. However, most recently, only four subfamilies have been accepted: the Ursinae, Ailuropodinae, Tremarctinae, and the extinct Agriotheriinae (Fulton and Strobeck 2006), a proposal that is followed herein as well.

According to morphological and molecular data (Krause et al. 2008), five extant genera are recognized: *Melursus*, *Ursus, Helarctos, Tremarctos*, and *Ailuropoda.* The family includes a variety of extinct species. The wide morphological diversity of some extinct species has created taxonomic confusion in the past and continues to do so today[1] (Erdbrink 1953; Pacher 2007).

Bears have a heavily built skeleton, so they depend on their strength rather than on speed. They have a plantigrade posture,[2] whereby the entire sole of their posterior feet touches the ground and supports their weight, also allowing them to stand bipedally (Macdonald 2009). This way of walking dictates the morphology of the limb bones, metapodials, and digits. In plantigrade species, the limb bones are shorter and heavier and there is greater lateral mobility at the wrist and ankle than in digitigrades or unguligrades; the metapodials are shorter and more divergent and the digits are arranged fanwise (Ewer 1973). The anterior feet, depending on the species concerned, are not entirely plantigrade but rise partially from the ground (pasemidigitigrade) (Macdonald 2009). In existing forms there is no entepicondylar (supracondyloid) foramen to the humerus (except *Tremarctos*); the true molars are 2/3 (2 upper and 3 lower in each side) and have broad flat tuberculated crowns; the fourth upper premolar, the carnassial tooth, has no third or inner root (Reynolds 1906) apart *Tremarctos* which has three roots (Baryshnikov, pers. comm.).

In Greece, extinct Ursidae has been described from few late Miocene localities (e.g., Pikermi, Samos), whereas from Pliocene onwards ursid remains unearthed from many sites in continental Greece (e.g., Tsoukala 1989, 1992; Koufos et al. 2018; and cited references). During the Middle and Late Pleistocene, Greek ursids were represented by the cave bear group and the brown bear, the latter species has a continuous record from this period until the present day (Tsoukala 1989, 1992; Kostopoulos and Vasileiadou 2006 and cited references). *U. arctos* fossil species has also been discovered at many Paleolithic[3] Greek sites (e.g., Samartzidou 2000; Darlas and Psathi 2008).

[1] A typical example being the extensive list published by Erdbrink (1953), which contains 232 recent and 39 brown bear fossil species and subspecies.

[2] As noted by de Blainville, a bear's skeleton presents certain resemblances to that of man, partly dependent partly on the animal's habbit of sitting on the ischia partly on the plantigrade method of walking (Reynolds 1906).

[3] Around 100,000 BC, encounters between bears and humans led to paintings of bears in the caves of Western Europe and perhaps to bear cults (Bieder 2005). In addition, around 40,000 BC, cult objects and figurines portraying bears originated about this time.

2 Historical Overview

The earliest Ursidae record in the Greek fossiliferous sites is from the middle Turolian (late Miocene) locality Pikermi, where a mandible fragment with the first and the second molars of *Indarctos atticus* species was described by Weithofer (1888). Later, the same species as well as the genus *Ursavus* have been described from Samos Q1 site (Helbing 1932; Bernor et al. 1996). Two species from the genus *Ursavus* were also recognized in two late Miocene localities of Halmyropotamos and Perivolaki (Brunner 1941; Thenius 1947; Koufos 2006). Ursids appear to be absent during latest Miocene–early Pliocene from Greek sites (Kostopoulos and Vasileiadou 2006). After this gap, different members of the family have been described from several Greek sites dated from the late Pliocene to the Late Pleistocene (e.g., Mitzopoulos 1960; Symeonidis and Theodorou 1979; Symeonidis et al. 1980; Tsoukala 1989, 1991; Tsoukala et al. 2006a, b, 2014; Baryshnikov and Tsoukala 2010; Baryshnikov and Tleuderdina 2017). Petralona Cave (Chalkidiki) and Loutra Almopias Cave (North Greece) are identified as key importance sites where the best-preserved material has been discovered. Systematic excavations in the latter site provided thousands of specimens attributed to *Ursus ingressus* Rabeder et al., 2004, a species that has been recorded for the first time in Greece (Rabeder et al. 2006; Tsoukala et al. 2006a). The abundance of milk teeth specimens clearly proves that these bears used the cave as a den (Pappa et al. 2005a, b; Tsoukala et al. 2006a). The Greek fossil record includes four genera and eight certain species (Appendix). From these, *U. arctos* has been also recorded from many Greek archaeological sites (e.g., Samartzidou 2000) and today is still present in Greece representing the southernmost distribution in Europe of the species (Karamanlidis et al. 2015).

3 Phylogenetic Relationships

Evolutionary History and the Fossil Record

Within the Carnivora, the bears form the part of the suborder Caniformia, whose phylogeny has been subject to various revisions. For example, the giant panda genus *Ailuropoda* has alternately been considered as a part of the Ursidae (e.g., Talbot and Shields 1996b) or as a separate family (e.g., Yu et al. 2007). Each family within the Caniformia (except the Ailuridae and Odobenidae, which are today represented by a single species each) includes a range of species with different body shapes, sizes, and habits. For many of these species, plant food makes an important contribution to their diet. This is particularly the case for the Ursidae.

The evolutionary history of the Ursidae family is more recent than that of other carnivores. Ursid-like arctoids can be traced back to the late Eocene (Baryshnikov 2007). The small-bodied *Ursavus* Schlosser, 1899, is considered to be the first genus within the Ursidae (Kurtén 1968) and believed to have evolved from the hemicyonid carnivore genus *Cephalogale* (Erdbrink 1953; Argant and Philippe 1997). *Ursavus* was similar in size to a small bear and possessed characteristic fine vertical wrinkles on the enamel of its teeth (Erdbrink 1953). The most basal species recognized is *Ursavus elmensis* Stehlin, 1917, the remains of which were found during the excavation of a railway tunnel at Elm, between Frankfurt and Fulda (Germany), within Burdigalian (early Miocene) age deposits (Erdbrink 1953). This species lived under relatively stable and subtropical climates in Europe during the early Miocene (McLellan and Reiner 1994). *U. elmensis* had all its premolars, which were adapted for slicing, while its molars show the beginning of the expansion of chewing surface that was to characterize the bear teeth of the later times (Kurtén 1976).

According to the molecular data (see also Sect. 3.2), the radiation of the Ursinae subfamily occurred around 5.39 Ma (Krause et al. 2008; Miller et al. 2012), at the Miocene/Pliocene boundary. This transition was marked by a global drop in temperatures and by increased seasonality (Zachos et al. 2001), with the replacement of forest by open C4 grassland habitats, especially at low latitudes (Cerling et al. 1997).

It is generally accepted that the genus *Ursus* first appeared in the early Pliocene (5.05 Ma) (Krause et al. 2008), in the form of the small-bodied bear *Ursus minimus;* Devèze and Bouillet, 1827 (= *U. arvernensis* Croizet and Jobert, 1828). This species is typical of Ruscinian to early Villafranchian faunal assemblages (Kurtén 1968; Kahlke 1999). The late Pliocene was marked by relatively dry climatic conditions and the widespread development of savanna and steppe grasslands (Kurtén 1968; Stirling and Derocher 1990). *Ursus etruscus* Cuvier, 1823 (*Ursus* aff. *etruscus* sensu Mazza and Rustioni 1992), was initially small and most probably gave rise to the extinct cave bear lineage as well as to the brown bears of Eurasia and north America, and to the extant Asiatic and American black bears (McLellan and Reiner 1994) (Fig. 1). However, it is worth mentioning that Baryshnikov (2007) believes that the extant Asiatic and American black bears separated from *U. minimus*.

Many Pleistocene deposits have proved to be rich in ursid remains, leading to the recognition of multiple new bear species and subspecies, and related phylogenetic studies (e.g., Erdbrink 1953; Kurtén 1968, 1969; Mazza and Rustioni 1994; Argant and Philippe 1997).

U. minimus fossils record shows a progressive increase in size, evolving by the Early Pleistocene (MN17) into *U. etruscus*, although some authors view *U. etruscus* as representing an entirely separate new lineage of perhaps Asiatic origin (Rustioni and Mazza 1993). Early *U. etruscus* were small-bodied (similar to modern *U. thibetanus*), increasing in size until terminal forms approached the size of extant large brown bear (Kurtén 1968; Mazza and Rustioni 1992). In addition, significant changes occurred through time in *U. etruscus* dentition, revealing an increasing importance of herbivory (Kahlke 1999). This is highlighted by the elongation of the molars and the development of a tuberculated occlusal (talon/id field) surface, whereas the carnassial function of the teeth became reduced and the anterior premolars dwindled to small pegs (Kurtén 1968; Mazza and Rustioni 1992). All the premolars, however, were preserved in the Etruscan bear, and it was not until the Middle Pleistocene that the *Ursus* line began to lose these teeth (Erdbrink 1953). These changes in size and morphology in *U. etruscus* may be connected to the

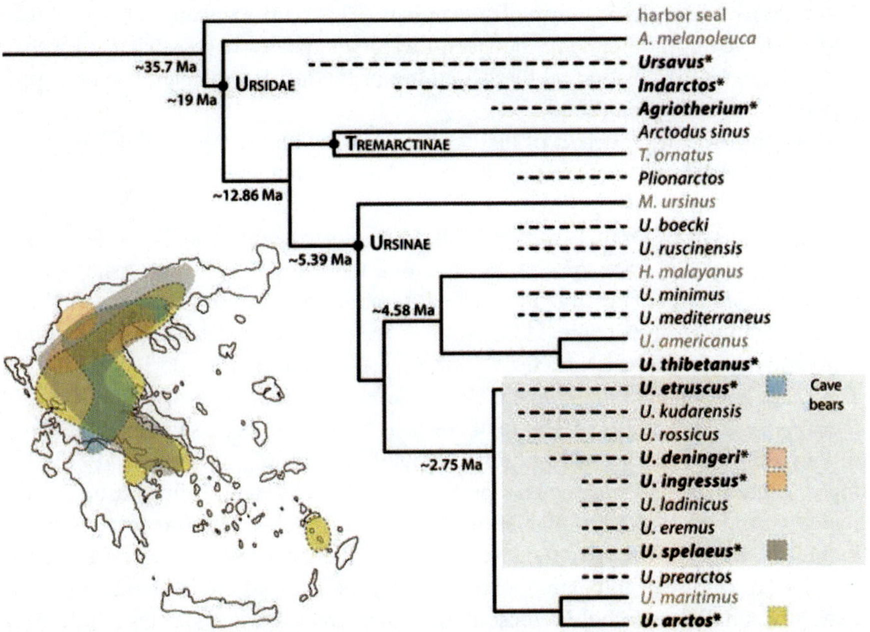

Fig. 1 Simplified phylogeny of extant and fossil bears (based and modified from Krause et al. 2008), marking major events and estimated divergence dates (see text for more information). Cave bears are expanded based on the phyletic tree from Rabeder et al. (2010). Extant taxa are colored gray, whereas extinct bears are colored black. Those found in Greece are marked with boldface and an asterisk, along with the simplified distribution of derived ursids. Map is a schematic distribution of *Ursus* genus in Greece

climatic fluctuations of the Pleistocene and associated paleoenvironmental change. Morphometric analysis of the cheek teeth and metacarpals and metatarsals of the Pleistocene bears of Europe using methods of multivariate and one-dimensional statistics places *U. etruscus* near *U. arctos* (Baryshnikov and Puzachenko 2017; Baryshnikov and Puzachenko 2019, 2020). Despite the commonly held position that *U. etruscus* was the common ancestor of the brown bear, cave bear, and American and Asiatic black bear lineages (Kurtén 1976; Rabeder et al. 2000; Argant 2001) (Fig. 1), Mazza and Rustioni (1994) have proposed that the black bear group (*Ursus minimus–thibetanus*) is ancestral to the other bear groups. Kurtén (1959) proposed that the growth of the Scandinavian and Alpine ice sheets during the Elsterian glaciation (Marine Oxygen Isotope Stage [MIS] 12) caused the eastern and western bear populations to become separated from one other and that the former population gave rise to *U. arctos*, whereas the latter evolved to *Ursus spelaeus* Rosenmüller, 1794 (and not Rosenmüller and Heinroth 1794; see Baryshnikov 1998; Rosendahl and Kempe 2005).

The earliest representatives of spelaeoid bears occur in Europe at approximately the same time as the arctoid ones, with the appearance of *Ursus deningeri* von Reichenau 1904. In Britain, the replacement of *U. deningeri* by *U. spelaeus* has traditionally been placed during the Anglian (Elsterian) glaciation (e.g., Schreve 2001). However, the boundary between these two species is somewhat subjective and a much later transition, at the beginning of the last Interglacial, has been proposed by Rabeder et al. (2000).

The cave bear *U. spelaeus* of the last "Ice Age," with the high-domed skull[4] is a very characteristic extinct species. It is very large with a well-defined sagittal crest and stepped profile of the forehead in lateral view (this is also a clear distinction from the skull of *U. arctos*) (Kurtén 1968, 1976; Torres Pérez-Hidalgo 1988; Mazza and Rustioni 1994). The species had a strong tendency to split into local "races" (Kurtén 1968; Rabeder et al. 2004). The differences between different geographical populations have been expressed in various ways, for example by examination of morphodynamic analysis[5] in cheek teeth (e.g., Grandal d'Anglade 1993; Rabeder 1983, 1989; Rabeder and Tsoukala 1990). Several caves in high mountain areas of the Alps (e.g., in the Totes Gebirge of Austria) were described as yielding "pygmy" or "high-alpine small forms" (Ehrenberg 1929). According to Rabeder (1983), the high plateaux of the Calcareous Alps were occupied by these forms during a relatively warm part of the last glaciation. The dentition of these forms was described by him as developed beyond the level of Middle Pleistocene *U. deningeri* but not to the level of that seen in contemporaneous cave bears from the lowlands. The idea that these small forms could represent more than one cave bear species that lived in Alps between 50 and 40 ka B.P. (Rabeder 1995) was ultimately supported by ancient DNA analyses (Hofreiter et al. 2004a; Rabeder and Hofreiter 2004; Rabeder et al. 2004), and as a result, three new species of cave bears were described within the so-called "*Ursus spelaeus* group" (Rabeder et al. 2004). These are: *Ursus eremus* (holotype from Ramesch bone cave in Totes Gebirge, Austria), *Ursus ladinicus* (holotype from Conturines cave in Italy), and *Ursus ingressus* (holotype from Gamssulzen cave in Totes Gebirge Austria).

A combination of radiometric and DNA analyses has suggested that around 50 ka BP, a very large-bodied bear migrated into the Alps (*U. ingressus*), which had been formerly inhabited by *U. eremus* and probably *U. ladinicus* since 80 ka BP (Hofreiter et al. 2004a; Rabeder and Hofreiter 2004; Rabeder et al. 2004). Interestingly, *U. ingressus* and *U. eremus* apparently coexisted sympatrically for

[4]Among other large mammals, bears have been used in many myths and folklore stories (e.g., as dragons) that ancient Greeks and Romans developed in order to explain the existence of these fossil creatures (Mayor 2000).

[5]Rabeder (1983) developed a statistical method of morphodynamic analysis of the occlusal features of cave bear teeth with the aim of quantifying evolutionary trends, using the fourth upper and fourth lower premolars, and then expanding to consider the rest of the dentition, including molars and incisors (Rabeder 1999). Nagel and Rabeder (1997) further demonstrated that this method of morphodynamic analysis can tested and corroborated by radiometric dating. However, this approach is open to criticism as it is somewhat subjective and by prioritising only certain features of the tooth (e.g., measurements or structure), potentially informative data may have been discarded. Pappa et al. (2005a, b) employed the same method on deciduous bear teeth from Greek sites.

approximately 15 ka without any detectable gene flow between the two populations, leading Hofreiter et al. (2004a) and Rabeder et al. (2004) to propose that a behavioral mechanism existed that prevented these two cave bear species from interbreeding. Currently, it is widely accepted that during the terminal Late Pleistocene, Europe was inhabited by *U. spelaeus* and *U. ingressus* (Rabeder et al. 2004; Rabeder and Hofreiter 2004; Hofreiter et al. 2004a), although debate remains regarding the taxonomical position of these forms, with some authors regarding them as separate species (Rabeder et al. 2004; Rabeder and Hofreiter 2004; Hofreiter et al. 2004a) and others as subspecies of *U. spelaeus* (Baryshnikov and Puzachenko 2011). Recently, nuclear DNA was also obtained from the Middle Pleistocene skull of *U. kudarensis* from Caucasus, confirming the basal position of this species relative to European cave bears (Barlow et al. 2020).

Further revision has been undertaken of cave bears formerly identified as *U. spelaeus* from Greece, with material from Loutra Almopias Cave now re-assigned to *U. ingressus* (Rabeder et al. 2006; Tsoukala et al. 2006a). Baca et al. (2014) recently suggested an earlier appearance of *U. ingressus* in the Sudeten mountains of central Europe, around 80 ka BP, based on the findings from Niedźwiedzia Cave in Poland. This would imply that the immigration of the species in this region occurred earlier than the colonization of the Alps and Swabian Jura.

Molecular Phylogenies of the Ursidae

Key divergence events within the wider Ursidae were investigated by Krause et al. (2008) using mitochondrial DNA (mtDNA). Figure 1 illustrates the resulting simplified ursid phylogeny (with updates where applicable) and highlights the importance of the Miocene/Pliocene boundary in species diversification. According to Krause et al. (2008), the posterior mean of the divergence time between bears and harbor seal was estimated at 36 Ma, and the divergence between the giant panda and the rest of the bears has been calculated between 22.1 and 17.9 Ma (Fig. 1). Further studies, using not only mitochondrial but also nuclear DNA from modern bears, have recently upheld the basal position of the giant panda and the spectacled bear in the ursid evolutionary tree (Talbot and Shields 1996a; Yu et al. 2007). The next divergence event is that of the spectacled bear group from the main bear lineage, which occurred between 15.6 and 12.4 Ma, with a further split between the extant spectacled bear and the extinct American giant short-faced bear, between 7 and 5.3 Ma. Between 5.4 and 4.1 Ma, five modern bear lineages further diversified: the sloth bears, brown bears, American black bears, Asian black bears, and sun bears (Krause et al. 2008). These authors also dated the divergence event between the cave and the brown bears around 3.1 and 2.4 Ma.

MtDNA, in combination with fossil evidence, has also been used to investigate further the relationship between polar bear and brown bear, revealing that they should be considered as sister species (Miller et al. 2012; Hirata et al. 2013). Despite the fact that the two bears are regarded as distinct species and their home ranges do not usually overlap, evidence of interbreeding (exacerbated by climate change bringing the two into contact) is increasingly noted (e.g., Doupe et al. 2006).

As well as wider genus-level definition, the family Ursidae is a very good example in which to investigate complex speciation and rapid evolution of distinct phenotypes (Miller et al. 2012). Taberlet and Bouvet (1994) were the first to identify two main lineages in European brown bears, an eastern and a western one, based on partial MtDNA control region analyses. The eastern lineage is characterized by large populations, whereas the latter comprises only small fragmented populations (Taberlet and Bouvet 1994). Studies on these geographically distinct mtDNA clades were further developed by Leonard et al. (2000), Barnes et al. (2002), and Miller et al. (2006). The eastern lineage is widespread across the Eurasian continent, from northeastern Europe to far eastern Russia, whereas the western lineage comprises two clades, an Iberian one and a Balkan/Italian one (Hirata et al. 2013 and references therein). In addition, in North America, four clades among the extant brown bears have been identified (Talbot and Shields 1996a). As might be expected, the polar bear was not only found to be embedded within the brown bear clade but was also revealed to be closely related to clade 2a from North America, which is currently distributed in the Admiralty, Baranof, and Chichagof island group of southeastern Alaska (Talbot and Shields 1996a; Miller et al. 2012).

Genetic analysis of fossil bear samples has revealed a number of extinct clades and subclades for both ancient brown bears (Barnes et al. 2002) and cave bears (Hofreiter et al. 2004a, b; Baca et al. 2014).

4 Distribution

In Greece, the oldest evidence of ursids is from late Miocene sites with the genera *Indarctos* and *Ursavus*, including sites in continental Greece and one island (Samos Q1). After a gap in the fossil record, the Greek ursids reappeared in Pliocene and Pleistocene with the genus *Ursus*. It is worth mentioning the occurrence of the *Agriotherium* sp. from the early Villafranchian (MN16a) locality of Milia (Grevena, Western Macedonia). The presence of fossil bears is well established with many sites in continental Greece stretching from the northernmost site Volax to the southeast in Tilos. However, most of the archaeological sites with bear remains are gathered in Thessaly (Fig. 2).

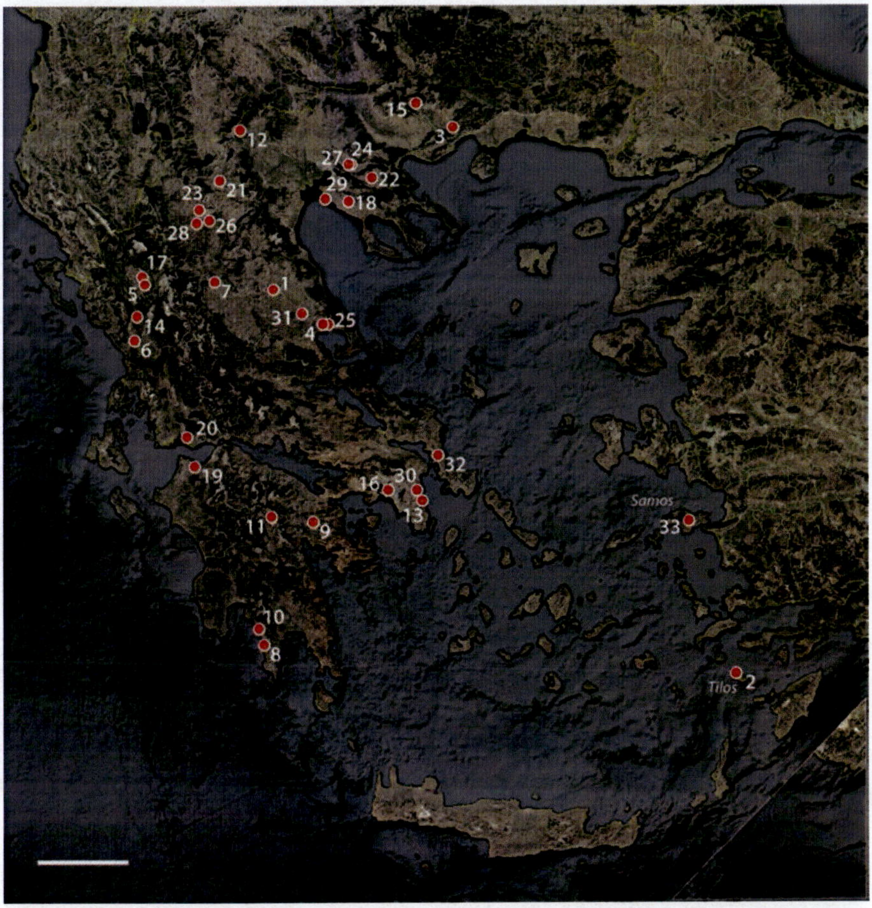

Fig. 2 Map of Greece showing the geographic distribution of the most important localities with bear fossils. **1**, Argissa-Magula; **2**, Charkadio Cave; **3**, Dikili Tash; **4**, Dimini; **5**, Kastritsa; **6**, Asprochaliko; **7**, Theopetra Cave; **8**, Kalamakia Cave; **9**, Klissoura Cave I; **10**, Mavri Spilia; **11**, Kitseli Pothole; **12**, Loutra Almopias Caves; **13**, Vraona Cave; **14**, Kouklesi Cave; **15**, Aggitis Cave; **16**, Korydallos; **17**, Perama Cave; **18**, Petralona Cave; **19**, Kastritsi; **20**, Makinia; **21**, Ptolemaida; **22**, Tsiotra Vrysi; **23**, Livakos; **24**, Vassiloudi; **25**, Sesklon; **26**, Dafnero 1; **27**, Apollonia; **28**, Milia localities; **29**, Epanomi; **30**, Pikermi; **31**, Perivolaki; **32**, Halmyropotamos; **33**, Samos Q1. See Appendix for more information. Image exported from Google Earth Pro © 2019, map data from US Dept. of State Geographer, SIO, NOAA, U.S. Navy, NGA, GEBCO, image from Landsat/Copernicus. Scale bar equals 80 km, North faces upward

5 Systematic Paleontology

The systematic position of the Ursidae is given below, where genera and species from Greece are discussed further in the comments and text and all species in each genus are listed, both fossil and living.

Valid Taxa

Order Carnivora Bowdich, 1821

Suborder Caniformia Kretzoi, 1938

Subfamily Ursavinae Kretzoi, 1945

Tribe Ursavini Hendey, 1980

Genus *Ursavus* Schlosser, 1899

Comments The first remains of *Ursavus* were found in Miocene deposits in Styria (Austria), in Silesia (Poland), and in La Grive-St-Alban (France) (Erdbrink 1953). *Ursavus* genus is known from fossil localities in Greece with two extinct species *U. depereti* and *U. ehrenbergi*.

***Ursavus depereti* Schlosser, 1902**

Type Material Left first and second lower molar described by Schlosser (1902) (p, 149; pl. II, figs. 19, 20, 22 & 23).

Type Locality Melchingen, Germany; Early Vallesian, MN 9.

Distribution The species is known from late Vallesian–early Turolian of central–west Europe (Neogene Old World database, 2020). In Greece, material was collected from late Miocene sites Perivolaki (Thessaly), Pikermi (Attica), and Samos Q1 (Aegean Sea).

Remarks The most characteristic features of the species are its large size, wide, and roughly inclined backward symphysis, strong canine, elongated and narrow p4 with long talonid, and small rounded metaconid (Koufos, 2006: text, figs. 1–2; Koufos, 2011, text, fig. 4). The best specimen so far known from the Greek sites was described by Koufos (2006) from Perivolaki (Thessaly) dated at the beginning of

MN12 (middle Turolian). Additional material referred as *U.* cf. *depereti* has been described by Solounias (1981) and Bernor et al. (1996) from the locality Samos Q1 (eastern Aegean Sea). Both Nagel and Koufos (2009) and Neogene Old World database (2020) included *U.* cf. *depereti* from Samos Q1 in the listed fauna. The species was also present in the famous Pikermi (Attica) late middle Turolian site and is referred as *Ursavus* sp. by Kostopoulos and Vasileiadou (2006) and as *Ursavus* cf. *depereti* in Koufos et al. (2018). We agree with the latter authors.

★*Ursavus ehrenbergi* Brunner, 1941

Type Material Maxilla with P4–M2 left and P1–M2 right, AMPG-HAL-n.

Nomenclatural and Taxonomical History Brunner described first the species based on P4-M2 from Halmyropotamos, Euboea (Greece), which was also included in *Agriarctos* by Kretzoi (1942). Thenius in 1947 reclassified the latter as *Ursavus ehrenbergi* (mentioned also in Abella et al. 2011).

Type Locality Halmyropotamos, Evia Island, middle Turolian, MN12.

Distribution It is only known from its type locality.

Remarks We agree with Kostopoulos and Vasileiadou (2006) that although *U. ehrenbergi* species is valid, its relations with the rest of the species of the genus are uncertain.

Family Ursidae Fischer de Waldheim, 1817

Subfamily Ailuropodinae Grevé, 1894

Tribe Indarctini Abella, Alba et al., 2019

★*Indarctos* Pilgrim, 1913

Comments *Indarctos* includes bears of moderate to large size during the late Miocene distributed in Eurasia, northern Africa, and North America (Petter and Thomas 1986; Hunt 1998). The taxonomy of the genus has been revised from various authors (e.g., Petter and Thomas 1986; Kurtén 1976; Ginsburg 1999). Most researchers recognize a single species in the Turolian of Eurasia, *Indarctos atticus* (Baryshnikov 2002 and references therein). The type locality of this species is near Athens, in Greece.

★*Indarctos atticus* (Weithofer, 1888) ex Dames 1883

Nomenclatural and Taxonomical History The type material initially was described as *Hyaenarctos* sp. (Dames 1883). Dames told to Weithofer that this material was a new species and labeled it as *Hyaenarctos atticus* (original combination) and this was used to describe the holotype (Weithofer 1888). In 1932, Helbing described *Indarctos atticus* from Samos, while Erdbrink (1953) transferred the material from Pikermi to this species. Baryshnikov (2002) considered the Greek material as belonging to a subspecies of *I. punjabiensis atticus*, and this combination was adopted by Albella et al. (2019), whereas Kostopoulos and Vasileiadou (2006) and Koufos et al. (2018) defended a specific-level attribution and referred the material as *I. atticus*. See Roussiakis (2001) for additional information regarding the nomenclatural and the history of the species.

Type Material Left mandible fragment with the first and the second molar described and figured by Weithofer (1888: p. 231; pl. 12, figs. 1–2).

Type Locality Pikermi, Attica, middle Turolian, MN12.

Distribution Besides its type locality, in Greece, this species is also known from Samos Q1. In addition to Greek sites, fossils attributed to this species are known from Ukraine (Baryshnikov 2002), Spain, France, Germany, Hungary, Libya, Iran (Maragheh), Kazakhstan (Kalmakpai) (Baryshnikov and Tleuderdina 2017), and China (Roussiakis 2001 and references therein).

Remarks The most characteristic features of the species are its relatively large size, squarish upper teeth (M2 longer than M1), P4 with moderate protocone and small parastyle, lower first molar with talonid slightly shorter than trigonid and high entoconid, second lower molar with similar talonid as m1 and third lower molar quite rounded (Koufos 2011 and Fig. 3 therein). Postcranial material from the species is quite rare and, however, was discovered in Pikermi site and described by Roussiakis (2001).

Subfamily Agriotheriinae Kretzoi, 1929

Genus *Agriotherium* Wagner, 1837

Remarks One of the best-known representatives of the Agriotheriinae is the "hyaena bear." *Agriotherium insigne* Gervais, 1853, which survived until the earliest Villafranchian (late Pliocene, European Land Mammal Age MN16) in Europe, has been described as a very large ancestral form of bear but with a dentition retaining some dog-like characters (Kurtén 1968; Hendey 1980; Kahlke 1999). *Agriotherium* sp. is only known in Greece from the early Villafranchian locality Milia (Grevena Basin, Western Macedonia) represented by an almost complete calcaneus (Fig. 3) and an imperfect canine (Tsoukala et al. 2014).

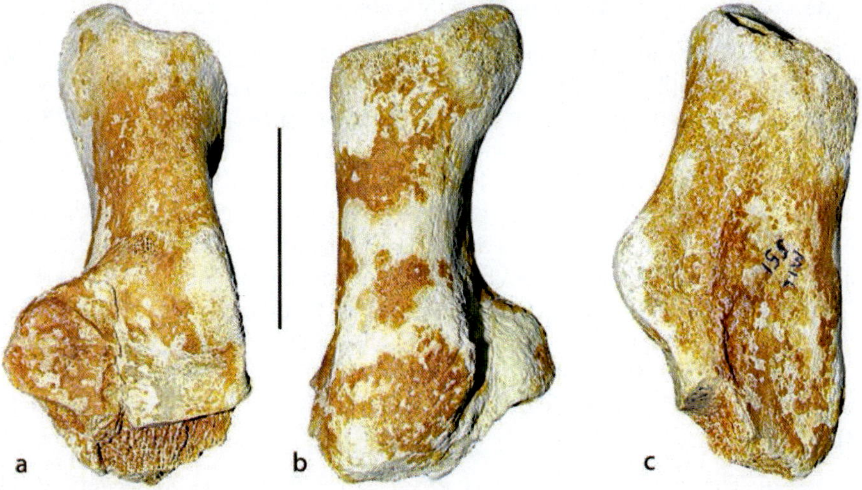

Fig. 3 Left calcaneum (MIL 551) of *Agriotherium* sp. from the Late Pliocene of Milia in (**a**) anterior, (**b**) posterior, and (**c**) lateral views. Scale bar equals 5 cm

Subfamily Ursinae [incorrect Viret, 1955] s. str.

Ursinae Fischer, 1817

Genus *Ursus* Linnaeus, 1758

Comments First appearance of the genus in Europe is at the French site of Montpellier (MN 14), in the early Pliocene (5.05 Ma) (Krause et al. 2008). While in Asia *Ursus* sp. was recorded from Gaozhung Formation (Yushe area, Chine) between 4.5 and 4.2 Ma (Qiulès 2003). *Ursus* genus is known from fossil localities in Greece with the extant species, *U. arctos* and five extinct ones: *U. deningeri*, *U. etruscus*, *U. ingressus*, *U. spelaeus*, and *U. thibetanus*. Additional information per species is provided below. Map in Fig. 1 is only a schematic distribution of the genus in Greece during the last two million years.

Ursus etruscus Cuvier, 1823

Type Material IGF 905, cranium and mandible (lectotype designated by Berzi 1966). *Ursus etruscus* has rather complicated history (see Mazza and Rustioni 1992; Baryshnikov 2007). Mazza and Rustioni (1992) based their amended diagnosis on the male cranium VA 1119 from Naturhistorisches Museum Basel, Switzerland and the female cranium IGF 906 from Museum of Geology and Paleontology of Florence, Italy from Valdarno (Italy).

Type Locality Figline Valdarno, Tuscany, Italy, Early Pleistocene.

Distribution The lectotype is from the Early Pleistocene site of Figline in the Upper Valdarno of Italy, and the species subsequently became widespread in Europe (Mazza and Rustioni 1992; Wagner 2010), including Crimea (Gimranov et al. 2020) and North Africa (Geraads 1997), with some closely related forms reportedly known from Caucasus (Medin et al. 2019), China (Kurtén 1968), and central Asia (Baryshnikov 2007). Baryshnikov (2007) identified four subspecies, which were adopted by Wagner (2010) and not accepted by Medin et al. (2019). In Greece, the species is known from early to late Villafranchian sites of Milia (Western Macedonia), Dafnero 1 (Western Macedonia), Tsiotra Vryssi (Mygdonia Basin), Sesklo (Thessaly), Vassiloudi (Mygdonia Basin), Apollonia 1 (Mygdonia Basin), Epanomi (Thermaikos), and from three Early Pleistocene localities Ptolemaida (West Macedonia), Makyneia (Etoloakarnania, Central Greece), and Kastritsi (Achaia, Peloponnese).

Remarks The first appearance of *U. etruscus* in Greece is in the locality Milia (W. Macedonia) dated to the earliest Villafranchian (Tsoukala et al. 2014). On the root of the lower canine (MIL 510) (Fig. 4a, b) typical gnawing marks, oriented radically below the cervix at the distolingual surface of the tooth, witnessed the presence of the oldest record of *Hystrix refossa* (Lazaridis et al. 2018). The species has also been reported from Dafnero 1 (W. Macedonia) with a single tooth (P4) (Koufos and Kostopoulos 1997). Undoubtedly, the locality Tsiotra Vryssi in the Mygdonia Basin (North, Greece) holds the best-preserved material of the species dated to the late Villafranchian (Koufos et al. 2018). Among cranial, dental, and postcranial remains, an almost complete cranium (TSR-E21-50) was found (Koufos et al. 2018) (Fig. 4c–e). The remains from other Greek sites are limited to a small number of mostly isolated teeth and bones and we agree with their original determinations as *U.* cf. *etruscus* until further remains confirm the presence of the taxon.

In this study, we present an additional material that was found recently, a left lower canine, in a new locality from Epanomi (EPN) (Thermaikos), close to Thessaloniki. The giant tortoise *Titanochelon bacharidisi* (Vlachos et al. 2014) was also found in the same locality within a Pliocene fluvial claystone in the Gonia Formation of western Chalkidiki Peninsula, near Thessaloniki, Greece. In the same horizon, *Gazellospira* cf. *torticornis* (Kostopoulos, pers. commun.) was also identified. The left lower canine EPN 1 is complete and well preserved, slightly worn, of relatively small dimensions but within the range of the size for the etruscoid canines (Table 1 and Fig. 5). Due to the limited material, we propose that the Epanomi ursid is best referred to as *U.* cf. *etruscus*.

Ursus deningeri von Reichenau, 1904

Nomenclatural and Taxonomical History As with the brown bears, the Deninger's bears present considerable variability in morphology and size which has led to the description of numerous subspecies. These are reviewed by Grandal

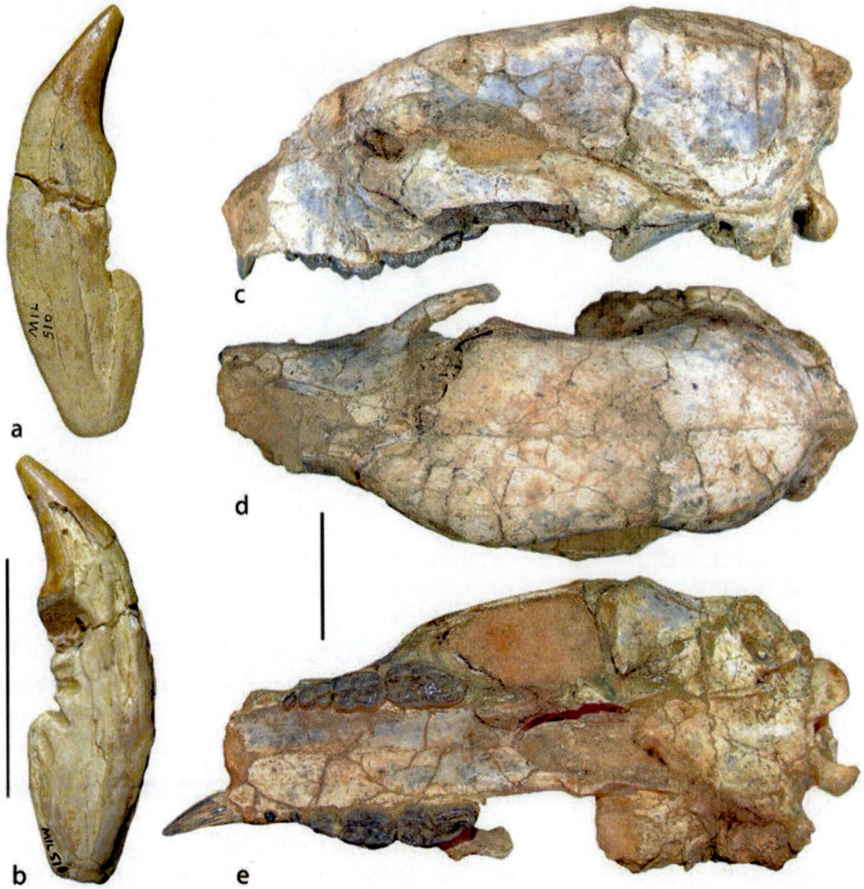

Fig. 4 Selected species of *Ursus etruscus* from Greece. (**a, b**) A lower canine from Milia with gnawing marks (MIL-510) in **a** lingual and (**b**) buccal views. (**c–e**) A skull from Tsiotra Vrysi (TSR - E21-50) in (**c**) left lateral, (**d**) dorsal, and (**e**) palatal views (photographs courtesy of GD Koufos). Scale bars equal 5 cm

Table 1 Descriptive measurements of *Ursus* cf. *etruscus* of a lower canine from Epanomi.

mm	L	B	H	Hcr	Lr	Br
EPN 1	20.31	13.65	74.97	31.43	22.72	13.90

L Length at the base of the crown, *B* breadth at the same point, *H* maximum height of the tooth, *Hcr* height of the crown, *Lr* maximum anteroposterior diameter of the root, *Br* maximum transversal diameter of the root

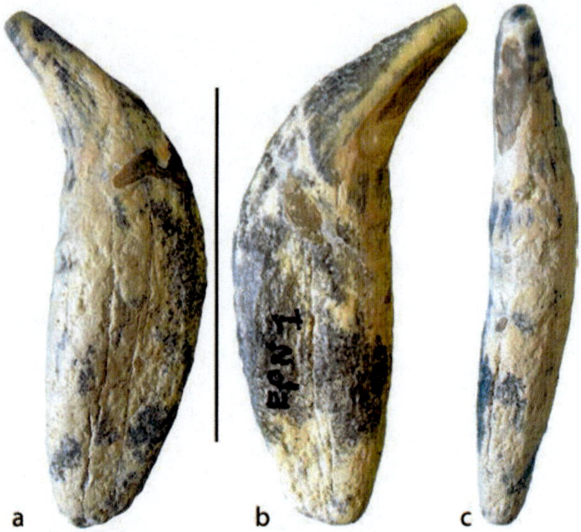

Fig. 5 *Ursus* cf. *etruscus* a lower canine from the Pliocene of Epanomi (Thermaikos) (EPN – 1), in (**a**) lingual, (**b**) labial, and (**c**) distal views. Scale bar equals 5 cm

d'Anglade and Vidal Romani (1997), based on Torres Pérez-Hidalgo (1988) and by Wagner (2010). The nominotypical subspecies *U. deningeri* was described from the locality of Mosbach 2 in Germany by von Reichenau (1904). Early Middle Pleistocene forms such as *Ursus suessenbornensis* Soergel, 1926, have also been widely regarded as an early spelaeoid bear or subspecies of *U. deningeri* (Kurtén 1969; Kahlke 1999; Baryshnikov 2007; Baryshnikov and Puzachenko 2020), although this species has equally been synonymized with *U. arctos* (Mazza and Rustioni 1994; Rabeder et al. 2010). The Süßenborn bear combines some primitive (arctoid) limb proportions with larger body dimensions, which Kahlke (1999) suggested reflect a steppic adaptation during the European early Middle Pleistocene. It is worth mentioning that some researchers have proposed intermediate or transitional taxa such as *U. spelaeus deningeroides* (Mottl 1964; Argant 1991), whereas other authors consider both *U. deningeri* and *U. spelaeus* to be a single species but demonstrating variability through time (e.g., Kurtén 1968; Mazza and Rustioni 1994; Baryshnikov and Puzachenko 2020). Baryshnikov and Foronova (2001) have yet a third opinion, placing both species in the subgenus *Spelearctos* together with *Ursus rossicus* Borissiak, 1930, and the late Early Pleistocene *Ursus savini* Andrews 1922 (= *U. deningeri hundsheimensis* Zapfe, 1946) from West Runton, Norfolk (UK).

Type Material Cranium from Mosbach stored at Mainz Museum described and figured by von Reichenau (1904: p. 226–228; fig. 1; pl. I–III).

Type Locality Mosbach 2, Germany, Early Middle Pleistocene.

Distribution *U. deningeri* is known from many different sites of early Middle Pleistocene age in the Czech Republic, Poland, Hungary, France, Italy, Greece, Spain, Moldova, Turkey, and the UK (e.g., Bishop 1982; Tsoukala 1991; Koenigswald and Heinrich 1999; Argant and Argant 2002; García 2003). A particular site of note for Early *U. deningeri* is Le Vallonet cave in southern France (e.g., Baryshnikov 2007; Argant 2009; Wagner 2010), although the identification of this material was challenged by García (2003), who referred it instead to the arctoid *U. dolinensis* while recognizing some spelaeoid features. Further to the Greek sites, the species was represented in the archaic fauna of the Petralona Cave (Chalkidiki) (Kurtén and Poulianos 1981; Tsoukala 1989, 1991; Baryshnikov and Tsoukala 2010) and to the late Villafranchian site Livakos (Western Macedonia) (this study).

Remarks The Petralona Cave material of the so-called "old collection" is well studied (Tsoukala 1989, 1991; Baryshnikov and Tsoukala 2010). The research during 1960s and so on yielded a great collection of cranial and postcranial material (Sickenberg 1964), and the stalagmite-covered PEC 1002 specimen is one of the best-preserved *U. deningeri* skull; its endocranial traits have been recently studied by Santos et al. (2014) (Fig. 6).

Two specimens, a calcaneus (LIB-179) and the third phalanx (LIB-255) are known from the late Villafranchian Greek site Livakos, Haliakmon basin, in Western Macedonia. This material was initially described as *Ursus* sp. aff. *arctos/deningeri* by Steensma (1988). Later and more recently, both Kostopoulos and Vasileiadou (2006) and Koufos et al. (2018) proposed that it is best to referred as *Ursus* sp. In this study, both specimens were re-examined and it is clear that these belong to *U. deningeri* species. This is based either on morphological or on metrical elements, since it is a robust calcaneus and the distal phalanx is massive with well-developed articular surface, indicating speleoid characters for the Early Pleistocene Libakos bears. The brown bear phalanges are less massive, slenderer, straighter and longer, showing a different lifestyle (Figs. 7 and 8, Table 2).

Ursus spelaeus Rosenmüller, 1794

Nomenclatural and Taxonomical History The cave bear *U. spelaeus* was first described by Rosenmüller in his doctoral thesis, and the often-cited attribution of Rosenmüller and Heinroth, 1794, is therefore no longer valid (Kempe et al. 2005; Rosendahl and Kempe 2005). Kempe et al. (2005) reviewed the taxonomical history the species.

Type Material MB Ma.5017, a skull of an adult male individual, housed in the collection of the Natural History Museum of the Humboldt-University Berlin. In addition, Baryshnikov (2007, 322) designated as a lectotype adult skull MB Ma.5020, specimen at Museum für Naturkunde, Berlin.

Type Locality Zoolithen Cave near Burggeilenreuth, Germany (Rosenmüller 1794).

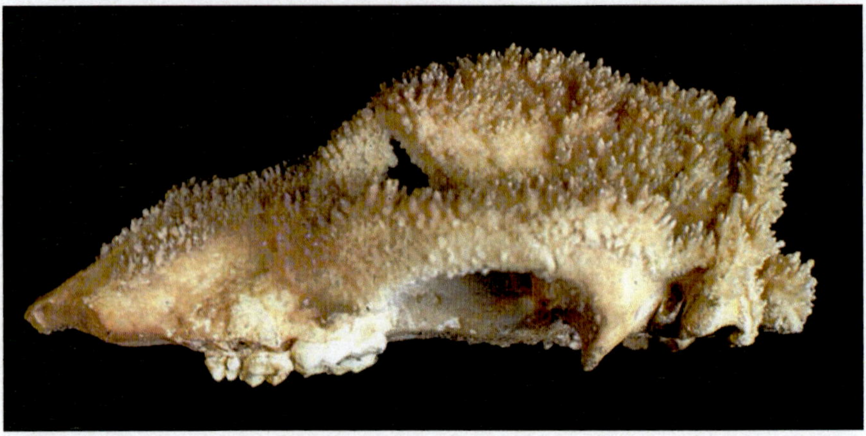

Fig. 6 *Ursus deningeri* skull in lateral view (PEC 1002), from Petralona Cave (Chalkidiki). The skull is partially covered by stalagmitic material

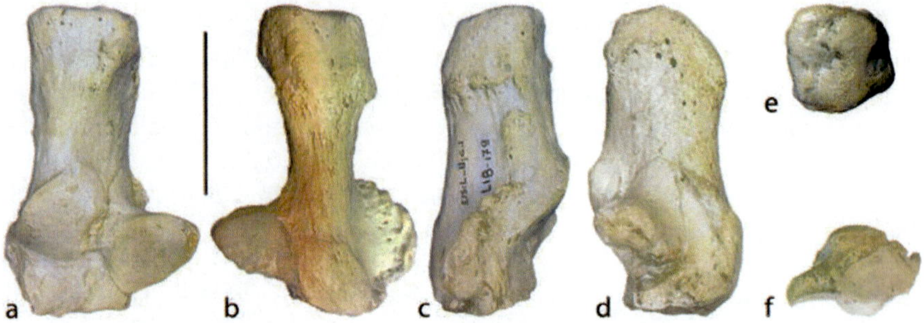

Fig. 7 Right calcaneus (LIB-179) of *Ursus deningeri* from Libakos, in (**a**) anterior, (**b**) posterior, (**c**) lateral, (**d**) mesial, (**e**) proximal, and (**f**) distal views. Scale bar equals 5 cm

Fig. 8 A phalanx of *Ursus deningeri* from Libakos (LIB 255), in (**a, b**) lateral views. Scale bar equals 1 cm

Table 2 Descriptive measurements of *Ursus deningeri* from Libakos, based on the system of measurements of Tsoukala and Grandal d'Anglade (2002)

Specimens	Measurements					
Calcaneus	L	DT	DAP	DTtb	DAPtb	Lmb	Lst
LIB-179	90.55	65.07	49.35	34.15	38.20	45.62	33.00
Phalanx	L	H	DT	Hart	DTart		
LIB-255	35.69	24.85	15.24	13.57	13.87		

art articulation, *DAP* anteroposterior diameter, *DT* transversal diameter, *H* height, *L* length, *Lmb* manubrium, *Lst* sustentaculum tali, *tb* tuber

Distribution The geographical distribution of the cave bear group extends eastwards from northwest Spain across central Europe to the Urals and from Belgium and the Harz region of Germany in the north to Italy and Greece in the south and the Crimea in the southeast. The remains have been found by the thousands in many European caves, such as the celebrated Drachenhöhle near Mixnitz in Styria, Austria (Kahlke 1999). In Greece, the species is recognized from Late Pleistocene sites. The most famous site with well-preserved material is Petralona Cave (Chalkidiki). Additional remains from the spelaeus group have been also reported from the following sites: Aggitis Cave (Eastern Macedonia), Drakos Cave (Kastoria, West Macedonia), Drama (northeastern Macedonia), Karytsa (Southern Greece), Korydallos (Attica), Perama Cave (Ioannina, Epirus, northwestern Greece), and Theopetra, Kalambaka (Thessaly).

Remarks Some characteristic morphological differences between *U. spelaeus* and *U. deningeri* are, for example that the former has a larger body size than the latter (Kurtén 1969), while its cheek teeth show a progressive modification of the occlusal surface during the Pleistocene (Rabeder and Tsoukala 1990; Rabeder 1999). In addition, *U. spelaeus* loses its anterior premolars and the molars become enlarged over time (Kurtén 1969; Kahlke 1999). Recently, studies using CT scanning and 3D modeling of the cranial traits of *U. deningeri* have confirmed that in this species, the stepped forehead does not reach the level of development observed in *U. spelaeus* (García et al. 2006; Santos et al. 2014). The limb bones and, most particularly, the tibia, metapodials, and third phalanges are shorter and more robust in *U. spelaeus* compared to *U. arctos* (Reynolds 1906; Kurtén 1959, 1968, 1976; Kahlke 1999). In Greece, the species has been reported in the famous Petralona Cave, where one of the best-preserved skull with mandible was found (Fig. 9). The remains from the rest Greek sites are limited to a small number of mostly isolated teeth and bones; these Late Pleistocene bears are grouped and referred as the *spelaeus* group until further remains and analysis confirm the presence of the species. The northernmost finding is a fifth metacarpal that was found in the alluvial deposits of Aggitis Cave (Koufos 1981). In Thessaly, two sites are reported with remains of the group, the archaeological site of Theopetra, Kalampaka (Rowley-Conwy and Newton 2000) and Karytsa Cave (Mitzopoulos 1960). The southernmost presence of the *spelaeus*-group is in Korydallos (Paraskevaidis 1961), where remains of an old individual were found in a cave discovered in a local quarry.

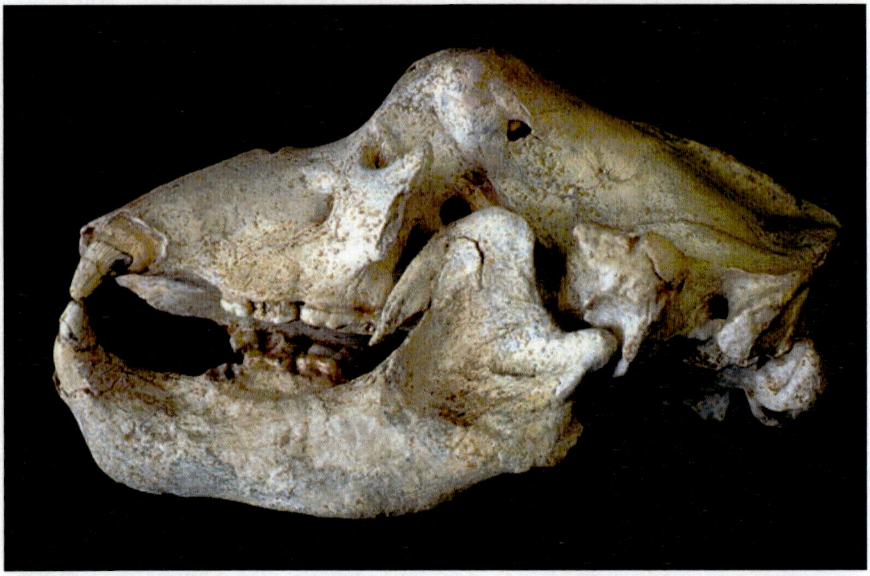

Fig. 9 *Ursus spelaeus* skull and associated mandible in lateral view (PEC 1000), from Petralona Cave (Chalkidiki)

***Ursus ingressus* Rabeder et al., 2004**

Type Material GS 718, a well-preserved skull with both mandibles (Rabeder et al. 2004, pl. 1.1).

Type Locality Gamssulzen Cave in Totes Gebirge Austria (Rabeder et al. 2004).

Nomenclature Remarks Genetically *U. ingressus* is similar to a cave bear from the Urals (Baca et al. 2014; Stiller et al. 2014), which was described as *U. spelaeus kanivetz* Vereshchagin, 1973, which makes it possible to consider the name *ingressus* as the junior synonyms of *kanivetz*. However, morphological differences between bears in Central and Eastern Europe (*ingressus*) and the Urals (*kanivetz*) possibly allow them to be distinguished at a subspecies level (Baryshnikov and Puzachenko 2011, 2017, 2019, 2020).

Distribution This species was present in the central and eastern parts of Europe to the Urals eastwards and to the Crimea southwards (Stiller et al. 2014). Further details regarding the taxonomical position and its distribution have been already discussed in Sect. 3.1 in this chapter. In Greece, the species is recognized from Late Pleistocene deposits in all chambers of Loutra Almopias Cave (North Greece), representing the southernmost occurrence of the species in Europe.

Remarks Loutra Almopias Cave (North Greece) is one of the richest sites in remains of the cave bear *U. ingressus* and was the first site in Greece where this species was described (Tsoukala and Rabeder 2005; Rabeder et al. 2006; Tsoukala et al. 2006a) (Fig. 10). The cave comprises two fossiliferous horizons, one within

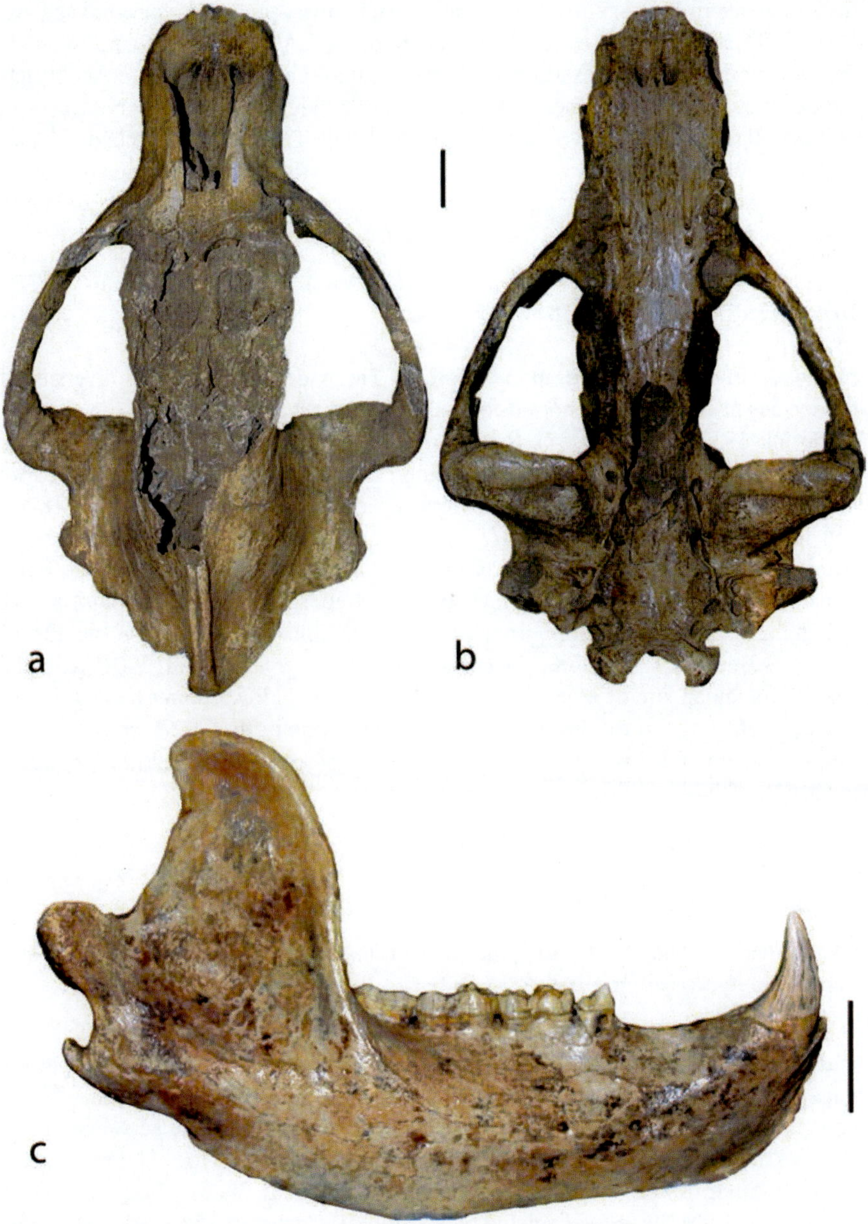

Fig. 10 Select specimens of *Ursus ingressus* from Loutra Almopias Cave (North Greece). (**a**, **b**) skull (LAC 13136) in (**a**) dorsal and (**b**) palatal views, (**c**) right mandible (LAC 998) in buccal view. Scale bars equal 5 cm

the sediments of the cave floor across all five chambers present, referred to as LAC I, II, III, Ib, and Ic. The second (younger) horizon, LAC 1a, is situated 5 m above the cave floor. The first fossiliferous horizon (in all chambers except LAC Ia) has yielded thousands of specimens of *U. ingressus* that are very well preserved, as well as other large mammalian fauna (Tsoukala 1994; Tsoukala and Rabeder 2005; Tsoukala et al. 1998, 2001, 2006a). The abundance of milk teeth remains in LAC is also remarkable, proving that bears used the cave as a den (Pappa et al. 2005a, b; Tsoukala et al. 2006a).

Ursus arctos Linnaeus, 1758

Nomenclatural and Taxonomical History The wide morphological diversity of the species has created taxonomic confusion in the past and continues to do so today (Erdbrink 1953; Pacher 2007). Sixteen subspecies of brown bear are recognized today (Wilson and Reeder 2005), including the grizzly bears in N. America (*U. a. horribilis* Ord, 1815), the Himalayan brown bear (*U. a. isabellinus* Horsfield, 1826), and the Kamchatka brown bear (*U. a. beringianus* Middendorff, 1851), the last being one of the largest forms of brown bear at present. A number of late Early Pleistocene and early Middle Pleistocene arctoid species have been proposed, although, again, the taxonomic position remains controversial. These include *U. prearctos* Boule, 1919; *U. praepriscus* Mottl, 1951; *U. rodei* Musil, 2001 (from the Epivillafranchian site of Untermaßfeld in Germany); and *U. dolinensis* Garcia and Arsuaga, 2001 (from the late Early Pleistocene levels in the Sierra de Atapuerca, northern Spain and from Le Vallonnet, southern France; Rustioni and Mazza 1993, but see Baryshnikov 2007 for a re-identification).

Type Material Unknown. The brown bear was the first species of bear to be described scientifically, by Carl von Linné, under the name *Ursus arctos* (reviewed and cited by Erdbrink 1953).

Type Locality The type locality was most probably in Sweden although there is no known holotype (Miller 1912).

Distribution The extant species is globally the most widely distributed bear species. The largest populations are found in North America and Eurasia at the present day, and the species is relatively abundant in Canada, Alaska, and Russia (Miller et al. 2006; McLellan et al. 2008). Smaller populations are found scattered across Asia, in the Himalayas, Mongolia, India, Japan, and China, in the southern regions of North America and in southern Europe (McLellan et al. 2008). Today, in Greece, the brown bear exists in two isolated populations; the western is in Pindos Mountain and the eastern population is located in the Rodopi Mountain chain (Vlachos et al. 2000). Regarding the fossil brown bear, considerable disagreement remains in the literature regarding the timing of its appearance in Europe. According to a study of material from Bad Deutch-Altenburg in Austria, brown bears were already present in Europe at the end of the Early Pleistocene (Rabeder et al. 2010), whereas other authors do not recognize their existence until the Last Interglacial, for example at

Taubach and Ehringsdorf, Germany and Venice, France (Argant 1996; Koenigswald and Heinrich 1999). Records of *U. arctos* from Heppenloch, Germany (Kurtén 1959) and Lunel-Viel, France (Kurtén 1968), demonstrate that the species was certainly present in Europe by the late Middle Pleistocene. However, although the former has previously been attributed to the Holsteinian interglacial (MIS 11) (e.g., Adam 1975), its age has not been recently reviewed, whereas the latter site is now considered to be possibly of younger, MIS 9, age. It is worth mentioning that the brown bear is absent from uncontested Holsteinian sites such as Steinheim-an-der-Murr in Germany (Kurtén 1959; Schreve and Bridgland 2002) and is also absent from Britain during the equivalent interglacial, the Hoxnian (Schreve 2001). *U. arctos* consequently first makes an appearance in Britain during MIS 9 (Schreve 1997; Schreve 2001) and occurred in both interglacial and glacial stages. Finds of Late Pleistocene brown bears are not only common in Britain (e.g., Reynolds 1906; Currant and Jacobi 2001) but also in southern Europe (e.g., Marra 2003), although remains have been less abundant in central Europe (e.g., Sabol 2001; Guérin 2002). According to Döppes and Pacher (2005), the first evidence of brown bear in the Alpine area after the Last Glacial Maximum is from the southern Alpine site of Grotte Ernesto (Italy) and from northern Alpine site of Neue-Laubenstein—Bärenhöhle (Germany). In Greece, the species is recognized from Late Pleistocene and Holocene deposits including both cave and open localities. The presence of fossil *U. arctos* is rather rare and it has been reported in Petralona Cave (Chalkidiki), Vraona Cave (Attica), Kouklesi Cave (Epirus), Kitseli pothole (Peloponnese), and Loutra Almopias Cave (Z) (North Greece). Additional remains from the species have been also discovered at the following Paleolithic sites: Argissa-Magula (Thessaly), Dikili Tash (Eastern Macedonia), Dimini (Thessaly), Kalamakia cave (Peloponnese), Kassopi (Preveza – Epirus), Klissoura Cave 1 (Southern Greece), Kolominitsa (Peloponnese), Korykeion Antron (Delphi), Loutra Almopias Cave (outside the cave, North Greece), Magula Pefkakia (Thessaly), Mavri Spilia (Peloponnese), and only in the island of Tilos (Dodecanese) at the Charkadio Cave.

Remarks The material of Kouklesi Cave (Epirus) comes from two different potholes and consists of two almost complete skeletons, however, covered with breccia material (Theodorou 1992). Due to their preservation, these specimens as well as the brown bear specimen (maxilla with a broken left canine) from Loutra Almopias Cave (Z) (North Greece) are better to be referred as *U.* cf. *arctos*. In addition, the knowledge of bear remains from two Paleolithic sites Kastritsa (Southern Greece) and Asprochaliko and Louros Valley (Epirus, northwestern Greece) is poor. Bailey et al. (1983) and Kotjabopoulou (2001) included *Ursus* sp. in the faunal assemblage at the rock-shelter of Kastritsa (Southern Greece) (Upper Paleolithic, 23.930–19.000 B.P., Galanidou and Tzedakis, 2001; Galanidou et al. 2000). *Ursus* sp. was also identified from the Middle Paleolithic layers at Asprochaliko, Louros Valley (Epirus, nortrwestern Greece) by Bailey et al. (1983).

Ursus thibetanus Cuvier, 1823

Nomenclatural and Taxonomical History Multiple synonyms for these black bears have been proposed, together with a number of subspecies, of which only two, *U. t. permjak* in the Urals and *U. t. mediterraneus* in the rest of the range in Caucasus and Europe, are considered valid (see Baryshnikov 2007 and Wagner 2010).

Distribution Fossils of *U. thibetanus* are not common but have been recovered from sites in France (Argant 1991; Crégut-Bonnoure 1996), Spain (Torres Pérez-Hidalgo 1988), Greece (Kurtén and Poulianos 1977), and Germany (Kurtén 1975).

Remarks Kurtén and Poulianos (1977) based the description of the species on a single upper molar whose morphology and measurement fit well with that of black bear material from China. The specimen was found in the famous Petralona Cave.

Invalid Taxa

★*Bosdagius felines*

Comments In 1968, Sickenberg described a new ursid genus and species from the locality Volax (Drama Basin, EasternMacedonia) based on a single milk tooth (Sickenberg, 1968: pl. 1; Fig. 1a–c). Spassov (2003) re-examined and described the specimen as a hyaenid tooth. Later Koufos et al. (2018) agreed this designation. Hence, this specimen is no longer representing an ursid.

Acknowledgments We would like to thank Professor emeritus GD Koufos for providing photographs from the Tsiotra Vryssi material. We are also thankful to N Bacharidis that provided specimens from his collection and Prof. D Kostopoulos for useful discussions. We deeply thank G Baryshnikov for comments that improved the manuscript.

Appendix

List of Greek localities with known occurrences of fossil bears, arranged in approximate geochronological order. Locality numbers refer to the collection numbers of the PaleoBiology Database (PBDB)

Localities[PBDB No]	Age	Taxon
Argissa-Magula	Late Pleistocene	*Ursus arctos*[1]
Charkadio Cave[182809]	Late Pleistocene	*Ursus arctos*[2]
Dikili Tash	Late Pleistocene	*Ursus arctos*[1]
Dimini	Late Pleistocene	*Ursus arctos*[1]
Kastritsa[208565]	Late Pleistocene (Upper Paleolithic)	*Ursus* sp.[3,4]
Kolominitsa	Late Pleistocene (Upper Paleolithic)	*Ursus arctos*[5]

(continued)

Localities[PBDB No]	Age	Taxon
Kassopi	Late Pleistocene (Upper Paleolithic)	*Ursus arctos*[6]
Asprochaliko[208564]	Late Pleistocene (Middle–Upper Paleolithic)	*Ursus* sp.[3]
Theopetra Cave	Late Pleistocene (Middle–Upper Paleolithic)	*Ursus spelaeus*-group[7]
Kalamakia cave[184245]	Late Pleistocene (Middle Paleolithic)	*Ursus arctos*[8–10]
Klissoura Cave 1[184243]	Late Pleistocene (Middle Paleolithic)	*Ursus arctos*[11]
Mavri Spilia[208563]	Late Pleistocene (Middle Paleolithic)	*Ursus arctos*[12]
Kitseli pothole[204955]	Late Pleistocene	*Ursus arctos*[13]
Korykeion Antron	Late Pleistocene	*Ursus arctos*[1]
Loutra Almopias Cave (outside the cave)	Late Pleistocene	*Ursus arctos*[14]
Magula Pefkakia	Late Pleistocene	*Ursus arctos*[1]
Vraona Cave[183130]	Late Pleistocene	*Ursus arctos*[15–17]
Kouklesi Cave	Late Pleistocene	*Ursus* cf. *arctos*[18]
Loutra Almopias Cave[203847]	Late Pleistocene	*Ursus ingressus*[14,19–20]
Loutra Almopias Cave Z[208562]	Late Pleistocene	*Ursus* cf. *arctos*[14]
Aggitis Cave[204391]	Late Pleistocene	*Ursus spelaeus*-group[21]
Drakos Cave	Late Pleistocene	*Ursus spelaeus*-group[22–23]
Karytsa	Late Pleistocene	*Ursus spelaeus*-group[24]
Korydallos[182753]	Late Pleistocene	*Ursus spelaeus*-group[25]
Perama Cave[208561]	Late Pleistocene	*Ursus spelaeus*-group[26–28]
Petralona Cave[183123]	Middle Pleistocene	*Ursus deningeri*[29–33]
Petralona Cave[183123] archaic fauna	Late Pleistocene	*Ursus spelaeus*[31,33–34], *Ursus arctos*[31,33–34], *Ursus thibetanus mediterraneus*[31,33–34]
Kastritsi[34810]	Early Pleistocene	*Ursus* cf. *etruscus*[35]
Makinia[34811]	Early Pleistocene	*Ursus* cf. *etruscus*[35]
Ptolemaida	Early Pleistocene	*Ursus* cf. *etruscus*[36]
Tsiotra Vrissi[192004]	Early Pleistocene	*Ursus etruscus*[37]
Livakos[34764]	latest Villafranchian	*Ursus deningeri*[37–39,a]
Vassiloudi[34651]	late Villafranchian	*Ursus* cf. *etruscus*[40–41]
Sesklon[34614]	middle Villafranchian	*Ursus* cf. *etruscus*[37,42]
Dafnero 1[34594]	middle Villafranchian	*Ursus etruscus*[37,41]
Apollonia[34784]	latest Villafranchian	*Ursus* cf. *etruscus*[40–41]
Milia 5[185859]	early Villafranchian	*Agriotherium* sp.[42]
Milia[195290]	early Villafranchian	*Ursus etruscus*[42]
Epanomi	Pliocene	*Ursus* cf. *etruscus*[a]
Pikermi[183123]	middle Turolian (late Miocene)	*Indarctos atticus*[44–44], *Ursavus* cf. *depereti*[43–44]

(continued)

Localities[PBDB No]	Age	Taxon
Perivolaki[194879]	late Miocene	*Ursavus depereti*[45]
Halmyropotamos[202213]	late Miocene	*Ursavus ehrenbergi*[46]
Samos Q1[95691]	late Miocene	*Indarctos atticus*[47–48], *Ursavus* cf. *depereti*[48–49]
NOT VALID		
Volax[34593]	Villafranchian	***Bosdagius felinus***[37,45]

[1]Samartzidou (2000), [2]Kotsakis et al. (1979), [3]Bailey et al. (1983), [4]Kotjabopoulou (2001), [5]Darlas and Psathi (2008), [6]Friedl (1984) (from Theodorou 1992), [7]Rowley-Conwy and Newton (2000), [8]de Lumley and Darlas (1994), [9]Gardeisen et al. (1999), [10]Harvati et al. (2013), [11]Starkovich (2011), [12]Tourloukis et al. (2016), [13]Tsoukala et al. (2006b), [14]Tsoukala et al. (2006a), [15]Symeonidis et al. (1980), [16]Symeonidis and Theodorou (1994), [17]Rabeder (1995), [18]Theodorou (1992), [19]Tsoukala and Rabeder (2005), [20]Rabeder et al. (2006), [21]Koufos (1981), [22]Paraskevaidis and Zervoudakis (1964), [23]Paraskevaidis (1977), [24]Mitzopoulos (1960), [25]Paraskevaidis (1961), [26]Petrochilou (1971), [27]Pavlakis et al. (1993–1994), [28]Pavlakis et al. (1999), [29]Sickenberg (1964), [30]Kurtén and Poulianos (1981), [31]Tsoukala (1989), [32]Rabeder and Tsoukala (1990), [33]Baryshnikov and Tsoukala (2010), [34]Tsoukala (1991), [35]Symeonidis et al. (1985/1986), [36]Stratigopoulos (1986), [37]Koufos et al. (2018), [38]Brunn (1956), [39]Steensma (1988), [40]Koufos and Kostopoulos (1997), [41]Koufos and Kostopoulos (2016), [42]Symeonidis (1992), [41]Koufos and Kostopoulos (2016), [42]Tsoukala et al. (2014), [43]Weithofer (1888), [44]Roussiakis (2001), [45]Koufos (2006), [46]Thenius (1947), [47]Solounias (1981), [48]Bernor et al. (1996), [49]Helbing (1932)
[a]This study

References

Abella J, Montoya P, Morales J (2011) A new species of *Agriarctos* (Ailuropodinae, Ursidae, Carnivora) in the locality of Nombrevilla 2 (Zaragoza, Spain). Estud Geol 67(2):187–191

Adam KD (1975) Die mittelpleistozäne Säugetier-Fauna aus dem Heppenloch bei Gutenberg (Württemberg). Abh zur Karst und Höhlenkunde Reihe D, Paläontol, Zool, München

Argant A (1991) Carnivores quaternaires de Bourgogne. Documents des Laboratoires de Géologie Lyon 115:1–301

Argant A (1996) B. Sous – famílie des Ursínae. In: Guérin C, Patou-Mathis M (eds) Les grand mammiferes Plio-Pléistocenes d' Europe. Masson, Paris, pp 167–177

Argant A (2001) Cave bear ancestors. Cadernos Laboratorio Xeolòxico de Laxe 26:341–348

Argant A (2009) Biochronologie et Grands Mammifères au Pléistocène Moyen et Supérieur en Europe Occidentale: L'Apport des Canidés, des Ursidés et des Carnivores en Général. Quaternaire 20:467–480

Argant A, Argant J (2002) Die Bären von Château (Burgund, Frankreich). In: Rosendahl W, Morgan M, López Correa M (eds) Cave Bear Researches/Höhlen-Bären-Forrschungen. Abh zur Karst und Höhlenkunde, Heft 34, 112 S, München, pp 57–63

Argant A, Philippe M (1997) Les ours et leur évolution. Coll Int "L'Homme et l'Ours", Auberives en Ryans, Isè p 8

Baca M, Mackiewicz P, Stankovic A, Popović D, Stefaniak K, Czarnogórska K, Nadachowski A, Gąsiorowski M, Hercman H, Weglenski P (2014) Ancient DNA and dating of cave bear remains from Niedźwiedzia Cave suggest early appearance of *Ursus ingressus* in Sudetes. Quat Int 339–340:217–223

Bailey GN, Carter P, Gamble C, Higgs HP (1983) Asprochaliko and Kastritsa: further investigations of Paleolithic settlement and economy in Epirus (Northwest Greece). Proc Prehist Soc Greece 49:15–42

Barlow, Axel and Paijmans, Johanna L. A. and Alberti, Federica and Gasparyan, Boris and Bar-Oz, Guy and Pinhasi, Ron and Foronova, Irina and Puzachenko, Andrey Y. and Pacher, Martina and Dalén, Love and Baryshnikov, Gennady and Hofreiter, Michael, Middle Pleistocene Cave Bear Genome Calibrates the Evolutionary History of Palaearctic Bears. Available at SSRN: https://ssrn.com/abstract=3523359 or https://dx.doi.org/10.2139/ssrn.3523359

Barnes I, Matheus P, Shapiro B, Jensen D, Cooper A (2002) Dynamics of Pleistocene population extinctions in Beringian brown bears. Science 295:2267–2270

Baryshnikov G (1998) Cave bears from the Paleolithic of the greater Caucasus. In: Saunders JJ, Styles BW, Baryshnikov GF (eds) Quaternary paleozoology in the northern hemisphere, pp 69–118. Springfield Illinois State Museum Scientific Papers 27

Baryshnikov GF (2002) Late Miocene *Indarctos punjabiensis atticus* (Carnivora, Ursidae) in Ukraine with survey of Indarctos records in the former USSR. Russ J Theriol 1:83–89

Baryshnikov GF (2007) Fauna of Russia and neighbouring countries. Mammals, Ursidae. Nauka, Saint Petersburg

Baryshnikov GF, Foronova I (2001) Pleistocene small cave bear (*Ursus rossicus*) from the South Siberia, Russia. Cadernos Laboratorio Xeoloxico de Laxe 26:373–398

Baryshnikov GF, Puzachenko AY (2011) Craniometrical variability in the cave bears (Carnivora, Ursidae): multivariate comparative analysis. Quat Int 245:350–368

Baryshnikov GF, Puzachenko AY (2017) Morphometric analysis of metacarpal and metatarsal bones of cave bears (Carnivora, Ursidae). Fossil Imprint 73(1–2):7–47

Baryshnikov GF, Puzachenko AY (2019) Morphometry of upper cheek teeth of cave bears (Carnivora, Ursidae). Boreas 48:581–604

Baryshnikov GF, Puzachenko AY (2020) Evolution and morphological variability of cheek teeth in the Kudaro cave bear (*Ursus kudarensis*, Carnivora, Ursidae). Biol Bull 47(7):148–171

Baryshnikov GF, Tleuderdina PA (2017) Late Miocene Indarctos (Carnivora: Ursidae) from the Karabulak formation of the Kalmakpai River (Zaisan depression, Eastern Kazakhstan). Proc Zool Inst RAS 321(1):3–9

Baryshnikov GF, Tsoukala E (2010) New analysis of the Pleistocene carnivores from Petralona Cave (Macedonia, Greece) based on the collection of the Thessaloniki Aristotle University. Geobios 43:389–402

Bernor RL, Solounias N, Swisher CC, van Couvering JA (1996) The correlation of three classical "Pikermian" mammal faunas—Maragheh, Samos, and Pikermi—with the European MN unit system. In: Bernor RL, Fahlbusch V, Mittmann H-W (eds) The evolution of Western Eurasian Neogene mammal faunas. Columbia University Press, New York, pp 137–154

Berzi A (1966) L'orso di Gaville nel Valdarno Superiore. Palaeontographia Italica LX:1–31

Bieder ER (2005) Bear. Reaktion Books Ltd, London

Bishop MJ (1982) The mammal fauna of the early Middle Pleistocene cavern *mull* site of Westbuxy-sub-Mendip, Somerset. Special papers of the Palaeontographical Association 26

Brunn JH (1956) Contribution à l'étude géologique du Pinde septentrional et d'une partie de la Macédoine occidentale. Ann Géol Pays Helléniques 7:1–358

Brunner JH (1941) Ein Bärenrest aus dem Unterpliozän von Euboe. Anz Akad Wiss in Wien Naturwiss Kl 3:24–27

Cerling TE, Harris JM, MacFadden BJ, Leakey MG, Quade J, Eisenmann V, Ehleringer JR (1997) Global vegetation change through the Miocene/Pliocene boundary. Nature 389:153–158

Crégut-Bonnoure E (1996) A review of small Middle Pleistocene bears from France. Acta Zool Cracoviensia 39:89–101

Currant AP, Jacobi R (2001) A formal mammalian biostratigraphy for the Late Pleistocene of Britain. Quat Sci Rev 20:1707–1716

Dames W (1883) Über das Vorkommen von Hyaenarctos in den Pliocän-Ablagerungen von Pikermi bei Athen. Sitzungsber Gesellsch naturf Freunde Berlin 8:132–139

Darlas A, Psathi E (2008) Le Paléolithique supérieur dans la péninsule du Mani (Sud de la Grèce). In: Darlas A, Mihailovic D (eds) The Palaeolithic of the Balkans. British Archaeological Reports International Series 1819, Oxford, pp 51–59 de Lumley H, Darlas A (1994) Fouilles franco-helléniques de la grotte de Kalamakia (Aréopolis, Péloponnèse). Bull Corr Hellénique 118:535–558

Döppes D, Pacher M (2005) Brown bear finds from Caves in the Alpine region. In: Ambros D, Gropp C, Hilpert B, Kaulich B (eds) Neue Forschungen zum Höhlenbären in Europa. Abh Naturhist Gesells Nürnberg 45:91–104

Doupe JP, England JH, Furze M, Paetkau D (2006) Most northerly observation of a grizzly bear (*Ursus arctos*) in Canada. Photographic and DNA evidence from Melville Island, Northwest Territories. Arctic 60:271–276

Ehrenberg K (1929) Die Ergebnisse der Ausgrabungen in der Schreiberwandhöhle am Dachstein. Paläontol Z 11:261–268

Erdbrink DP (1953) A review of fossil and recent bears of the old world. British Museum, Natural History, London

Ewer RF (1973) The carnivores. The world naturalist. Great Britain by Cox and Wyman Ltd., London Fakenham and Reading. Cornell University Press, Ithaca

Fulton TL, Strobeck C (2006) Molecular phylogeny of the Arctoidea (Carnivora): effect of missing data on supertree and supermatrix analyses of multiple gene data sets. Mol Phylogenet Evol 41:165–181

Galanidou N, Tzedakis P (2001) New AMS dates from Upper Palaeolithic Kastritsa. Proc Prehist Soc Greece 67:271–278

Galanidou N, Tzedakis PC, Lawson IT, Frogley MR (2000) A revised chronological and palaeoenvironmental framework for the Kastritsa rock-shelter, Northwest Greece. Antiquity 74:349–355

García N (2003) Revisión de los úrsidos pleistocenos en Europa: relaciones filogené ticas entre los úrsidos del último millón de años; Osos y otros arnívoros de la Sierra de Atapuerca. Fundación Oso de Asturias, Oviedo, pp 414–468

García N, Santos E, Arsuaga JL, Carretero JM (2006) High-resolution X-ray computed tomography applied to the study of some endocranial traits in cave and brown bears. Sci Ann Sch Geology Aristotle Univ Thessaloniki 98:141–146

Gardeisen A, Trantalidou K, Darlas A (1999) Faunal remains from Kalamakia (Peloponnese, Greece). In: Cruz AR, Milliken S, Oosterbeck L, Peretto C (eds) Human population origins in the circum-Mediterranean area: adaptations of the hunter-gatherer groups to environmental modifications. Ark Dent 5:111–120

Geraads D (1997) Carnivores du Pliocène terminal de Ahl al Oughlam (Casablanca, Maroc). Géobios 30:127–164

Gimranov DO, Lavrov AV, Startsev DB, Tarasenko KK, Academician Lopatin AV (2020) First finding of Etruscan bear (Ursus etruscus, Ursidae, Carnivora) in the Crimea (Taurida Cave, Early Pleistocene). Dokl Biol Sci 491:35–38

Ginsburg L (1999) Order Carnivora. In: Rössner G, Heissig K (eds) The Miocene land mammals of Europe. Verlag Dr F. Pfeil, München, pp 109–148

Grandal D'Anglade A (1993) Sexual dimorphism and interpopulation variability in the lower carnassial of the cave bear, *Ursus spelaeus* ROS-HEIN. Cuadernos do Laboratorio Xeoloxico de Laxe 18:231–239

Grandal D'Anglade A, Vidal Romani JR (1997) A population study on the cave bear (*Ursus spelaeus* ROS.-HEIN.) from Cova Eirós (Triacastela, Galicia, Spain). [Étude de populations sur l'ours des cavernes (*Ursus spelaeus* ROS.-HEIN) de Cova Eirós (Triacastela, Galice, Espagne)]. Geobios 30:723–731

Guérin C (2002) Conclusion générale sur la faune du Pléistocéne supérieur de la grotte de Jaurens (commune de Nespouls, Corréze, France). Cahiers sci Mus Hist natur Lyon 1:55–66

Harvati K, Darlas A, Bailey SE, Rein TR, Zaatari SE, Fiorenza L, Kullmer O, Psathi E (2013) New Neanderthal remains from Mani peninsula, Southern Greece: the Kalamakia Middle Paleolithic cave site. J Hum Evol 64:486–499

Helbing H (1932) Über einen *Indarctos*-Schädel aus dem Pontien der Insel Samos. Nebst einen Anhang: *Hyaenarctos* spec. aus dem Pliocaen von Vialette (Haute-Loire). Mém Soc Paléontol Suisse 52:1–18

Hendey QB (1980) *Agriotherium* (Mammalia, Ursidae) from Langebaanweg, South Africa, and relationships of the genus. Ann South African Mus 81:1–109

Hirata D, Mano T, Abramov AV, Baryshnikov GF, Kosintsev PS, Vorobiev AA, Raichev EG, Tsunoda H, Kaneko Y, Murata K, Fukui D, Masuda R (2013) Molecular phylogeography of the brown bear (*Ursus arctos*) in Northeastern Asia based on analyses of complete mitochondrial DNA sequences. Mol Biol Evol 30:1644–1652

Hofreiter M, Rabeder G, Jaenicke-Despres V, Withalm G, Nagel D, Paunovic M, Jambresic G, Pääbo S (2004a) Evidence for reproductive isolation between cave bear populations. Curr Biol 14:40–43

Hofreiter M, Serre D, Rohland N, Rebeder G, Nagel D, Conrad N, Münzel S, Pääbo S (2004b) Lack of phylogeography in European mammals before the last glaciation. Proc Natl Acad Sci U S A 101:12963–12968

Hunt RM (1998) Ursidae. In: Janis C, Scott K, Jacobs L (eds) Evolution of tertiary mammals of North America, Terrestrial carnivores, ungulates and ungulate like mammals, vol 1. Cambridge University Press, Cambridge, pp 174–194

Kahlke RD (1999) The history of the origin, evolution and dispersal of the Late Pleistocene *Mammuthus-Coelodonta* faunal complex in Eurasia (large mammals). Fenske Companies, Rapid City

Karamanlidis AA, de Gabriel HM, Krambokoukis L, Gimenez O (2015) Evidence of a large carnivore population recovery: counting bears in Greece. J Nat Conserv 27:10–17

Kempe S, Rosendahl W, Döppes D (2005) The making of the cave bear – Die wissenschaftliche Entdeckung des, *Ursus spelaeus'*. In: Nagel D (eds) Festschrift Gernot Rabeder. Mitt Kommission Quartärfrschung der Österreichischen Akad Wissen, vol 14, pp 89–106

Koenigswald W, Heinrich WD (1999) Mittelpleistozäne Säugetierfaunen aus Mitteleuropa – der Versuch einer biostratigraphischen Zuordnung. Kaupia 9:53–112

Kostopoulos DS, Vasileiadou K (2006) The Greek Late Neogene-quaternary ursids in relation to palaeogeography and palaeoenvironment. Sci Ann Sch Geology Aristotle Univ 98:285–292

Kotjabopoulou E (2001) Patterned fragments and fragments of patterns: upper palaeolithic rockshelter faunas from Epirus Northwestern Greece. PhD thesis, University of Cambridge

Koufos GD (1981) A new Late Pleistocene (Würmian) mammal locality from the basin of Drama (N. Greece). Sci Ann Aristotle Univ Thessaloniki 21:129–148

Koufos GD (2006) 4. Carnivora. In: Koufos GD (ed) The late Miocene vertebrate locality of Perivolaki, Thessaly, Greece. Palaeontographica Abt A Band 276 Lieferung 1–6:39–74

Koufos GD (2011) The Miocene carnivore assemblage of Greece. Estud Geol 67:291–320

Koufos GD, Kostopoulos D (1997) New carnivore material from the Plio-Pleistocene of Macedonia (Greece) with the description of a new canid. Müncher Geowiss Abh A 34:33–63

Koufos GD, Kostopoulos DS (2016) The Plio-Pleistocene large mammal record of Greece: implications for early human dispersals into Europe. In: Harvati K, Roksandic M (eds) Human evolution in the southern Balkans. Springer, Dordrecht, pp 269–280

Koufos GD, Konidaris GE, Harvati K (2018) Revisiting *Ursus etruscus* (Carnivora, Mammalia) from early Pleistocene of Greece with description of new material. Quat Int 497:222–239

Krause J, Unger T, Nocon A, Malaspinas A-S, Kolokotronis SO, Stiller M, Soibelzon L, Spriggs H, Dear PH, Briggs AW, Bray SCE, O'Brien SJ, Rabeder G, Matheus P, Cooper A, Slatkin M, Pääbo S, Hofreiter M (2008) Mitochondrial genomes reveal an explosive radiation of extinct and extant bears near the Miocene-Pliocene boundary. BMC Evol Biol 8:220

Kurtén B (1959) On the bears of the Holsteinian interglacial. Stockholm Contrib Geol 2:73–102

Kurtén B (1968) Pleistocene mammals of Europe. Aldine Publishing Company, Chicago

Kurtén B (1969) Cave bears. Stud Spelaeol 2:13–24

Kurtén B (1975) Fossile Reste von Hyänen und Bären (Carnivora) aus den Travertinen von Weimar-Ehringstdorf. III. Internationales Paläontologisches Kolloquium 1968. Das Pleistozän von Weimar- Ehringstdorf. Teil II. Abh Zentralen Geol Inst 23:465–484

Kurtén B (1976) The cave bear story. Life and death of a vanished animal. Columbia University Press, New York

Kurtén B, Poulianos A (1977) New stratigraphic and faunal material from Petralona cave with special reference to the Carnivora. Anthropos 4:47–130

Kurtén B, Poulianos A (1981) Fossil Carnivora of Petralona cave, status of 1980. Anthropos 8:9–56

Lazaridis G, Tsoukala E, Maul LC (2018) The earliest *Hystrix refossa*. A new early Villafranchian record from Milia (Grevena, Macedonia, Greece). Hytrix It J Mammal 30:12–18

Leonard JA, Wayne RK, Cooper A (2000) Population genetics of ice ages brown bears. Proc Natl Acad Sci U S A 97:11651–11654

Macdonald DW (2009) The encyclopedia of mammals. OUP, Oxford

Marra AC (2003) *Ursus arctos* from selected Pleistocene sites of eastern Sicily. Boll Soc Paleontol Italiana 42:145–150

Mayor A (2000) The first fossil hunters: paleontology in Greek and Roman times. Princeton Univertity Press, Princeton

Mazza P, Rustioni M (1992) Morphometric revision of the Eurasian species *Ursus etruscus* Cuvier. Palaeontograph Ital 79:101–146

Mazza P, Rustioni M (1994) On the phylogeny of Eurasian bears. Palaeontogr Abt A 230:1–38

McLellan B, Reiner DC (1994) A review of bear evolution. Int Conf Bear Res Manage 9:85–96

McLellan BN, Servheen C, Huber D (2008) Ursus arctos. wwwiucnredlistorg. Accessed 21 Jun 2019

Medin T, Martínez-Navarro B, Madurell-Malapeira J, Figueirido B, Kopaliani G, Rivals F, Kiladze G, Palmqvist P, Lordkipanidze D (2019) The bears from Dmanisi and the first dispersal of early *Homo* out of Africa. Sci Rep 9(1):1–16

Miller GS (1912) Catalogue of the mammals of Western Europe (Europe exclusive of Russia) in the collection of the British Museum. BMNH, London, pp 284–303

Miller CR, Waits LP, Joyce P (2006) Phylogeography and mitochondrial diversity of extirpated brown bear (*Ursus arctos*) populations in the contiguous United States and Mexico. Mol Ecol 15:4477–4485

Miller W, Schuster SC, Welch AJ, Ratan A, Bedoya-Reina OC, Zhao F, Kim HL, Burhans RC, Drautz DI, Wittekindt NE, Tomsho LP, Ibarra-Laclette E, Herrera-Estrella L, Peacock E, Farley S, Sage GK, Rode K, Obbard M, Montiel MR, Bachmann L, Ingólfsson Ó, Aars J, Mailund T, Wiig Ø, Talbot SL, Lindqvist C (2012) Polar and brown bear genomes reveal ancient admixture and demographic footprints of past climate change. Proc Natl Acad Sci U S A 109:2382–2390

Mitzopoulos M (1960) Über das Vorkommen von *Ursus spelaeus* in Agrapha-Gebirge (Griechenland). Ann Géol Pays Helléniques 11:150–153

Mottl M (1964) Bärenphylogenese in Südost-Österreich. Mitt Mus Bergbau, Geol Technik am Landesmus Joanneum 26:1–55

Nagel D, Koufos GD (2009) The late Miocene mammal faunas of the Mytilinii Basin, Samos Island, Greece: new collection. 15. Carnivore guild structure. Beitr Paläontol 31:391–396

Nagel D, Rabeder G (1997) Chronologie de L' ours des caverns en Autriche "MAN and BEAR". Et Rec ArchéologUniv Liège 100:265–287

Neogene Old World database (2020). www.helsinki.fi/science/now/database.htm. Accessed 21 Jun 2019

Pacher M (2007) The type specimen of *Ursus priscus* Goldfuss, 1818 and the uncertain status of late Pleistocene brown bears. N Jb Geol Paläontol Abh Stuttgart 245:331–339

Pappa S, Tsoukala E, Lazaridis G, Rabeder G (2005a) Milk teeth of quaternary carnivores from northern Greek caves-preliminary report. N Forschungen Höhlenbären Europa Abh Band 45:169–182

Pappa S, Tsoukala E, Lazaridis G, Rabeder G (2005b) Cave bear milk teeth from Loutra Arideas (Macedonia, Greece). Preliminary report. In: Proceedings of the 14th international congress of speleology, vol 1, pp 106–113

Paraskevaidis I (1961) Neue Funde Quartärer Wirbeltierreste in Attika. Ann Géol Pays Helléniques 12:149–152

Paraskevaidis I (1977) Säugetierreste aus Griechenland. In: Proceedings of the VI colloquium on the Geology of the Aegean Region, vol 2, pp 1143–1154

Paraskevaidis I, Zervoudakis TZ (1964) The cave-bear from the Drakos-cave, Kastoria. Bull Speleolog Soc Greece 7:198–199

Pavlakis P, Fistani A, Symeonidis N (1993–1994) New evidence of Middle Pleistocene mammal fauna from the Perama Cave (Ioannina, Greece). Bull Speleolog Soc Greece 21:340–360

Pavlakis P, Fistani A, Symeonidis N (1999) New evidence of Middle Pleistocene mammal fauna from the Perama Cave (Ioannina, Greece). In: Proceedings of the ICOPAG conference

Petrochilou A (1971) The cave of Perama in Ioannina. In: Int Speleological Meeting Greek Speleological Soc, pp 60–62

Petter G, Thomas H (1986) Les Agriotheriinae (Mammalia, Carnivora) néogènes de l'ancien monde presence du genre Indarctos dans la faune de Menacer (ex-Marceau), Algérie. Geobios 19(5):573–586

Quilès J (2003) Les Ursidae du Pléistocène moyen et supérieur en Midi méditerranéen: Apports paléontologiques, biochronologiques et archéozoologique [PhD Thesis]. Paris: Museum National d'Histoire Naturelle. p.322

Rabeder G (1983) Neues vom Höhlenbären. Zur Morphogenetik der Backenzähne. Die Höhle 2(34):67–85

Rabeder G (1989) Modus und Geschwindigkeit der Höhlenbären-Evolution. Schriftenr Verbr naturwiss Kenntnisse Wien 127:105–126

Rabeder G (1995) Die Gamssulzenhöhle imToten Gebirge. Mitt Komm Quartärforsch Österr Akad Wiss 9:1–133

Rabeder G (1999) Die Evolution des Höhlenbärengebisses. Mitt Quartärkomm Österr Akad Wiss 11:1–102

Rabeder G, Hofreiter M (2004) Der neue Stammbaum der alpinen Höhlenbären. Die Höhle 55:1–19

Rabeder G, Tsoukala E (1990) Morphodynamic analysis of some cave-bear teeth from Petralona cave (Chalkidiki, N. Greece). Beit Paläontol 16:103–109

Rabeder G, Nagel D, Pacher M (2000) Der Höhlenbär. Thorbecke Verlag, Stuttgart. Species

Rabeder G, Hofreiter M, Nagel D, Withalm G (2004) New taxa of alpine cave bears (Ursidae, Carnivora). Cahiers sci Mus Lyon Hors 2:49–67

Rabeder G, Tsoukala E, Kavcik N (2006) Chronological and systematic position of cave bears from Loutra Arideas (Pella, Macedonia, Greece). Sci Ann Sch Geol Aristotle Univ Thessaloniki 98:69–73

Rabeder G, Pacher M, Withalm G (2010) Early Pleistocene bear remains from Deutsch-Altenburg (Lower Austria). Mitt Quartärkomm Österr Akad Wiss Wien 17:1–135

Reynolds SH (1906) A monograph on the British Mammalia. Vol. II. British Pleistocene Hyaenidae, Ursidae, Canidae and Mustelidae. Mon Palaeontogr Soc London 2:1–35

Rosendahl W, Kempe S (2005) *Ursus spelaeus* Rosenmüller 1794 and not Rosenmüller and Heinroth-Johann Christian Rosenmüller, his life and the *Ursus spelaeus*. In: Ambros D, Gropp C, Hilpert B, Kaulich B (eds) Neue Forschungen zum Höhlenbären in Europa. Abh Naturhist Gesell Nürnberg 45:191–198

Roussiakis S (2001) Postcranial remains of *Indarctos atticus* (Ursidae, Mammalia) from the classical locality of Pikermi (Attica, Greece), with a description of the front limb. Senckenb Lethaea 81(2):347–358

Rowley-Conwy P, Newton S (2000) Late Paleolithic and Mesolothic animal bones from Theopetra cave. Prelim. Report. In: Kyparissi-Apostolika N (ed) Theopetra cave, 12 years of excavation and research. Proceedings of the international science conference, Trikala, pp 129–134

Rustioni M, Mazza P (1993) The genus *Ursus* in Euarsia: dispersal events and stratigraphical significance. Riv It Paleontol Stratigr 98:487–494

Sabol M (2001) Fossil Brown bears of Slovakia. Los osos pardos fósiles de Eslovaquia Cadernos Lab Xeolóxico de Laxe 26:311–316

Samartzidou E (2000) Study of zooarchaelogical material from the early Neolithic sites of Axos, Giannitsa, Western Macedonia, MA thesis, Aristotle University of Thessaloniki

Santos E, Garcia N, Carretero JM, Arsuaga JL, Tsoukala E (2014) Endocranial traits of the Sima de los Huesos (Atapuerca, Spain) and Petralona (Chalkidiki, Greece) Middle Pleistocene ursids. Phylogenetic and biochronological implications. Ann Paléontol 100(4):297–309

Schlosser M (1902) Beiträge zur Kenntniss der Saugerthierreste auf den Süddeutschen Bohnherzen. Geol Pälaontol Abh Iena NF 5(3):115–258

Schreve DC (1997) Mammalian biostratigraphy of the later Middle Pleistocene in Britain. PhD thesis, University of London

Schreve DC (2001) Differentiation of the British late Middle Pleistocene interglacials: the evidence from mammalian biostratigraphy. Quat Sci Rev 20:1693–1705

Schreve DC, Bridgland DR (2002) Correlation of English and German Middle Pleistocene fluvial sequences based on mammalian biostratigraphy. Geol Mijnb 81:357–373

Sickenberg O (1964) Die Saugerierfauna der Höhle Petralona bei Thessaloniki (vor. Mit.). Geol Geoph Res Inst Geol Subsurf Res 9:1–16

Sickenberg O (1968) Die Unterpleistozäne Fauna von Wolaks (Griech., Mazedonien). II. Die Carnivoren. Ann Géol Pays Helleniques 19:621–646

Solounias N (1981) The Turolian fauna from the island of Samos, Greece. Contrib Vertebr Evol 6:1–232

Spassov N (2003) The Plio-Pleistocene vertebrate fauna in South-Eastern Europe and the megafaunal migratory waves from the east to Europe. Rev Paleobiologique Geneve 22:197–229

Starkovich B (2011) Trends in Subsistence from the Middle Paleolithic through Mesolithic at Klissoura Cave 1 (Peloponnese, Greece). PhD thesis, University of Arizona

Steensma KJ (1988) Plio-Pleistozäne Groβsäugetiere (Mammalia) aus dem Becken von Kastoria/Grevena, sudlich von Neapolis-NW-Griechenland. Disserstation, Math-Naturwissen Fak Techn Univ Clausthal

Stiller M, Molak M, Prost S, Rabeder G, Baryshnikov G, Rosendahl W, Münzel S, Bocherens H, Granda-d'Anglade A, Hipert B, Germonpré M, Stasyk O, Pinhasi R, Tintori A, Rohland N, Mohandesan E, Ho Y.W. S, Hofreiter M, Knapp M (2014) Mitochondrial DNA diversity and evolution of the Pleistocene cave bear complex. Quat Int 339–340:224–231

Stirling I, Derocher AE (1990) Factors affecting the evolution and behavioral ecology of the modern bears. In: Bears: international conference on bear research and management, vol 8, pp 189–204

Stratigopoulos S (1986) On the fossil remains of *Ursus etruscus* and *Hippopotamus amphivius major* in Ptolemais basin. Bull Geol Soc Greece 18:183–197

Symeonidis N, Theodorou G (1979) Ein neuer Nachweis des Höhlenbären in Nord-Griechenland. Ann Géol Pays Helléniques 29:704–715

Symeonidis N, Bachmayer F, Zapfe H (1980) Ergebnisse weiterer Grabungen in der Höhle von Vraona (Attika, Griechenland). Ann Géol Pays Helléniques 30:291–299

Symeonidis N, Theodorou G, Schütt H, Velitzelos E (1985/1986) Paleontological and stratigraphical observations in the area of Achaia and Etoloakarnania (W. Greece). Ann Géol Pays Helléniques 33:329–338

Taberlet P, Bouvet J (1994) Mitochondrial DNA polymorphism, phylogeography, and conservation genetics of the brown bear *Ursus arctos* in Europe. Proc R Soc Lond B 255:195–200

Talbot SL, Shields GF (1996a) Phylogeography of brown bears (Ursus arctos) of Alaska and paraphyly within the Ursidae. Mol Phylogenet Evol 5:477–494

Talbot SL, Shields GF (1996b) A phylogeny of the bears (Ursidae) inferred from complete sequences of three mitochondrial genes. Mol Phylogenet Evol 5:567–575

Thenius E (1947) *Ursavus ehrenbergi* aus dem Pont von Euböa (Griechenland). Sitzber Akad Wissen Wien 156:225–249

Theodorou G (1992) On *Ursus* cf. *arctos* from the upper Pleistocene of Kouklesi cave (Ioannina, W. Greece). Ann Géol Pays Helléniques 30:291–299

Torres Pérez-Hidalgo T (1988) Osos (Mammalia, Carnivora, Ursidae) del Pleistoceno Ibérico: I. Filogenia; Distribución estratigráfica y geográfica. Estudio anatómico y métrico del cráneo. Bol Geol Minero 99:3–46

Tourloukis V, Thompson N, Garefalakis C, Karkanas P, Konidaris GE, Panagopoulou E, Harvati K (2016) New Middle Palaeolithic sites from the Mani Peninsula, Southern Greece. J Field Archaeol 41:68–83

Tsoukala E (1989) Contribution to the study of the Pleistocene fauna of large mammals (Carnivora, Perissodactyla, Artiodactyla) from Petralona cave (Chalkidiki, N. Greece). Thesis, Sci Ann, Scholl of Geology, Thessaloniki University

Tsoukala E (1991) Contribution to the study of the Pleistocene fauna of large mammals (Carnivora, Perissodactyla, Artiodactyla) from Petralona cave (Chalkidiki, N. Greece). Preliminary report. C R Acad Sci Paris 312:331–336

Tsoukala E (1992) Quaternary faunas of Greece. Courier Forsch Inst Seckenberg 153:79–92

Tsoukala E (1994) Bärenreste aus Loutraki (Mazedonien, Griechenland)-"*Ursus spelaeus*". In: 2nd Höhlenbären symposium

Tsoukala E, Grandal d'Anglade A (2002) System of measurements on ursid skeleton. Etud Rec Archéol Univ Liège:265–287

Tsoukala E, Rabeder G (2005) Cave bears and late Pleistocene associated faunal remains from Loutra Arideas (Pella, Macedonia, Greece). 15 years of research. Naturhist Gesell Nürnberg 45:225–236

Tsoukala E, Rabeder G, Verginis S (1998) *Ursus spelaeus* and Late Pleistocene associated faunal remains from Loutraki (Pella, Macedonia, Greece)-excavations of 1999. Cadernos Lab Xeolóxico de Laxe 26:441–446

Tsoukala E, Rabeder G, Verginis S (2001) *Ursus spelaeus* and associated faunal remains from Loutraki (Pella, Macedonia, Greece)-excavations of 1996. Geological and geomorphological approach. In: 4th international Höhlenbären symposium

Tsoukala E, Chatzopoulou K, Rabeder G, Pappa S, Nagel D, Withalm G (2006a) Paleontological and stratigraphical research in Loutra Arideas bear cave (Almopia Speleopark, Pella, Macedonia, Greece). Sci Ann Sch Geol Aristotle Univ 98:41–67

Tsoukala E, Bartsiokas A, Chatzopoulou K, Lazaridis G (2006b) Quaternary mammalian remains from the Kitseli Pothole (Alea, Nemea, Peloponnessus). Sci Ann Sch Geol Aristotle Univ 98:273–284

Tsoukala E, van Logheın W, Lazaridis G, Mol D (2014) Carnivores of the early Villafranchian siteof Milia, (Grevena, Macedonia, Greece). Sci Ann Sch Geol Aristotle Univ 102:6–7

Vlachos CG, Bakaloudis DE, Dimitriou M, Kritikou K, Chouvardas D (2000) Seasonal food habits of the European brown bear (*Ursus arctos*) in the Pindos Mountains, Western Greece. Folia Zool 49(1):19–25

Vlachos E, Tsoukala E, Corsini J (2014) *Cheirogaster bacharidisi*, sp. nov., a new species of a giant tortoise from the Pliocene of Thessaloniki (Macedonia, Greece). J Vertebr Paleontol 34(3):560–575

Wagner J (2010) Pliocene to early Middle Pleistocene ursine bears in Europe: a taxonomic overview. J Nat Mus (Prague) Nat Hist Ser 179(20):197–215

Weithofer A (1888) Beiträge zur kenntniss der fauna von Pikermi bei Athen. Beit Palaeontol Öster Ungarns 6:225–292

Wilson DE, Reeder DM (2005) Mammal species of the world. A taxonomic and geographic reference, 3rd ed. Johns Hopkins University Press, Baltimore

Yu L, Li YW, Ryder OA, Zhang YP (2007) Analysis of complete mitochondrial genome sequences increases phylogenetic resolution of bears (Ursidae), a mammalian family that experienced rapid speciation. BMC Evol Biol 7:198

Zachos J, Pagani M, Sloan L, Thomas E, Billups K (2001) Trends, rhythms, and aberrations in global climate 65 Ma to present. Science 292:686–693

The Fossil Record of Ailurids (Mammalia: Carnivora: Ailuridae) in Greece

George D. Koufos

1 Introduction

The sole living member of the family Ailuridae is the red panda (*Ailurus fulgens*), although the fossil record contains several extinct ailurid taxa (*Amphictis*, *Actiocyon*, *Alopecocyon*, *Simocyon*, *Protursus*, *Magerictis*, *Pristinailurus*, and *Parailurus*). The systematic position of the red panda has been discussed for a long time; it was considered as a relative of the American racoon (*Procyon lotor*) and thus belonging to the Procyonidae Gray, 1825. Recent molecular studies indicated that it belongs to a separate family, the Ailuridae Gray, 1843, originated from the trichotomy of the superfamily Musteloidea (Flynn et al. 2000). Today, all extinct taxa related to the red panda are included to the family Ailuridae, which consists of three subfamilies Amphictinae, Simocyoninae, and Ailurinae (Peigné et al. 2005; Kullmer et al. 2008; Salesa et al. 2011; Spassov and Geraads 2011). The species of the genus *Simocyon* belong to the subfamily Simocyoninae together with *Actiocyon, Alopecocyon*, and *Protursus*. In the family Ailuridae, the posterior border of the palate is situated behind the distal border of the upper tooth row, the external auditory canal is long and ossified, and the M1 cingulum is continuous around the protocone base (Ginsburg 1999). The main features of Simocyoninae are the tendency to hypercarnivory, the sharp upper and lower carnassials, the reduced protocone of the upper carnassial, and the high p4 and m1 (Ginsburg 1999).

G. D. Koufos (✉)
Department of Geology, Laboratory of Geology and Paleontology, Aristotle University of Thessaloniki, Thessaloniki, Greece
e-mail: koufos@geo.auth.gr

E. Vlachos (ed.), *Fossil Vertebrates of Greece Vol. 2*,
https://doi.org/10.1007/978-3-030-68442-6_22

2 Historical Overview

The ailurids are well distributed in Eurasia. In Greece they represented by the single genus *Simocyon*, which was originally described from Pikermi (Roth and Wagner 1854; Roussiakis 1967; Fig. 1 and Appendix). *Simocyon* is quite rare in Greece and the known material is scarce. Besides Pikermi, the ailurids with the genus *Simocyon* were also recognized in Halmyropotamos (Melentis 1967). They are still absent from the rich late Miocene collections of Axios Valley and Samos Island (Fig. 1).

3 Phylogenetic Relationships

The Late Paleocene and Eocene Miacidae, lived from 55.0 to 34.0 Ma, are considered as the earliest ancestors of the ailurid red panda (Wesley-Hunt and Flynn 2005); during late Eocene, miacids gave two groups, the Hesperocyonidae (early dogs) and the primitive ursids, represented by the genus *Mustelavus* (Wang et al. 2005). The latter genus is near the dichotomy of Ailuridae + Procyonidae and Mustelidae. The ailurids originated in Europe with *Amphictis* during late Oligocene–early Miocene; *Amphictis* is a primitive mustelid, which appears to be the closest ancestor of the ailurids (Wang et al. 2005; Peigné et al. 2005). Wang et al. (2005) considered *Simocyon* as the sister group of extant *Ailurus*. The phylogenetic relationships of Ailuridae and other related families are given in Salesa et al. (2006, 2011).

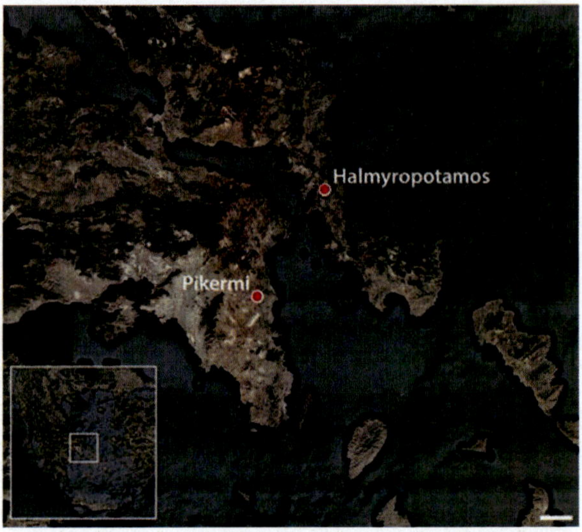

Fig. 1 Map of Greece showing the geographic distribution of the localities with ailurid fossils. See Appendix for more information. Image exported from Google Earth Pro © 2019, map data from US Dept. of State Geographer, SIO, NOAA, U.S. Navy, NGA, GEBCO, image from Landsat/ Copernicus. Scale bar equals 10 km, North faces upward

4 Systematic Paleontology

Ailuridae Gray, 1843

Simocyoninae Dawkins, 1868

***Simocyon* Wagner, 1858**

Type Species *Gulo primigenius* Roth and Wagner, 1854

★*Simocyon primigenius* (Roth and Wagner, 1854)

Nomenclatural and Taxonomical History *Gulo primigenius* in Roth and Wagner 1854 (new species); *Simocyon primigenius* in Pilgrim 1931 (new combination). In the meantime, it was referred with several generic and specific names (e.g., *Canis lupus primigenius* in Roth and Wagner 1854; *Pseudocyon robustus* in Wagner 1857; *Metarctos diaphorus* in Gaudry 1861).

Type Specimen BSPM-PIK-AS II.53 (holotype), right mandibular ramus with c–m1, (Roth and Wagner 1854: taf. 8, figs 1, 2) (Fig. 2c).

Type Locality Pikermi (PIK), Attica, Greece, Late Miocene, MN12.

Distribution Besides the type locality, the species is recognized in the late Miocene locality Halmyropotamos in Evia Island (Fig. 2a, b). It is also referred from Spain, Bulgaria, Moldova, and China; there is also a reference for its presence in Africa (Howell and Garcia 2007).

Remarks The taxonomy of the species is debated, as it is classified to the family Procyonidae, Ailuridae, or Canidae. The main characters of the species, modified from Pilgrim (1931) and Roussiakis (2002), are the moderately high frontal region, the low occiput, the short muzzle, the strong and wide zygomatic arches, the high sagittal crest, the small auditory bullae, the short mandible, the deep mandibular corpus, the mesial margin of the masseteric fossa below the posterior border of the m2, the small P1 situated near the canine, the absence of the P2 and P3, the high paracone and trenchant metacone in upper carnassial, the low protocone situated considerably behind the mesial margin of the tooth, the broader than long M1 with subtriangular occlusal outline, the crescentic protocone, the absence of accessory cusps, the sub-rectangular M2 with metacone much smaller than paracone, the occasionally absent p1, the absence of the p2 and p3, the small posterior accessory cuspid in the p4, the variable presence of the anterior one, the stout m1 with moderately strong metaconid, the large hypoconid forming a cutting crest mesio-distally directed, the larger m2 than m1, and the presence of a crenulated cingulum around the talonid of the m2.

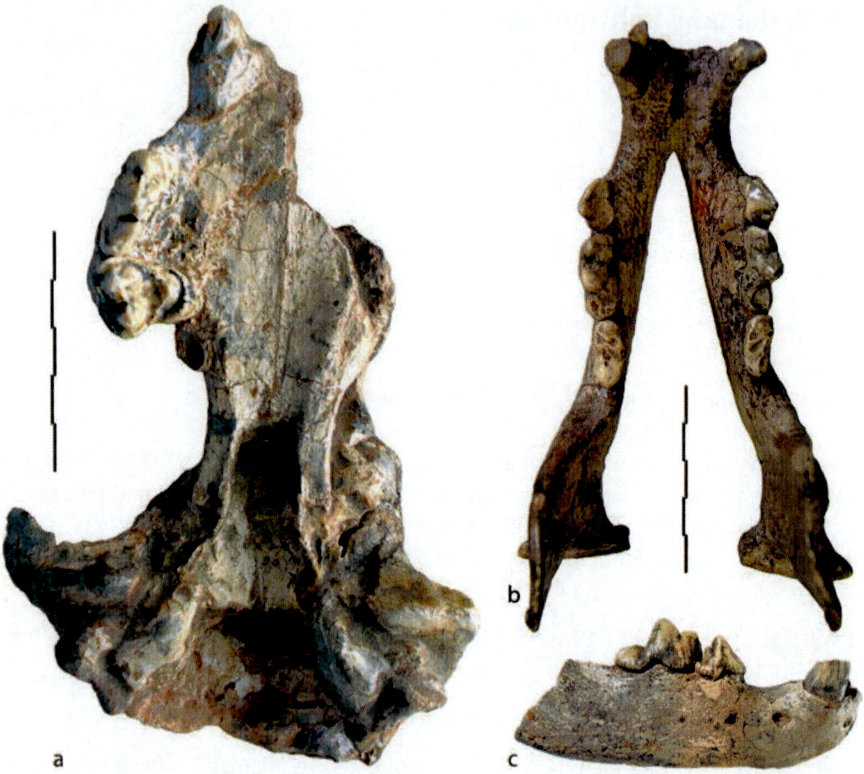

Fig. 2 Specimens of *Simocyon primigenius* from the late Miocene of Greece; (**a**) AMPG-HAL-1967/8, cranial fragment from Halmyropotamos, Evia Island, Greece in occlusal view; (**b**) MNHN-PIK-3020, mandible from Pikermi, Attica, Greece in occlusal view; (**c**) BSPM-AS II 53 (holotype), mandibular fragment with c–m1 from Pikermi, Attica, Greece in buccal view. Scale bars equal 50 mm

Appendix

List of the Greek localities containing ailurid fossils. Type localities are marked with bold. Locality numbers refer to the collection numbers of the PaleoBiology Database (PBDB)

Localities[PBDB No]	Age (MN; GPTS in Ma)	Taxon
Pikermi[182754]	Middle Turolian (MN12; 7.3–7.1)	***Simocyon primigenius***[1]
Halmyropotamos[202213]	?Middle Turolian (MN12)	*Simocyon primigenius*[2]

MN Mammal Neogene, *GPTS* Geomagnetic Polarity Time Scale
[1]Roth and Wagner (1854), [2]Melentis (1967)

References

Flynn JJ, Nedbal MA, Dragoo JW, Honeycutt RL (2000) Whence the red panda? Mol Phylogenet Evol 17(2):190–199

Gaudry A (1861) Note sur les carnassiers fossils de Pikermi (Gréce). Bull Soc Geol Fr 18:527–538

Ginsburg L (1999) Order Carnivora. In: Rössner G, Heissig K (eds) The Miocene land mammals of Europe. Verlag Dr Friedrich Pfeil, Munchen, pp 109–148

Howell FC, Garcia N (2007) Carnivora (mammalia) from Lemudong'o (Late Miocene: Narok district, Kenya). Kirtlandia 56:121–139

Kullmer O, Morlo M, Sommer J, Lutz H, Engel T, Forman M, Holzförster F (2008) The second specimen of *Simocyon diaphorus* (Kaup, 1832) (Mammalia, Carnivora, Ailuridae) from the type–locality Eppelsheim (early late Miocene, Germany). J Vertebr Paleontol 28(3):928–932

Melentis JK (1967) Studien über fossile Vertebraten Griechenlands. 19. Die Pikermifauna von Halmyropotamos (Euböa-Griechenland). 1. Teil: Odontologie und Kraniologie. Ann Geol Pays Helléniques 19:285–411

Peigné S, Salesa MS, Antón M, Morales J (2005) Ailurid carnivoran mammal *Simocyon* from the late Miocene of Spain and the systematics of the genus. Acta Paleontol Polonica 50(2):219–238

Pilgrim GE (1931) Catalogue of the Pontian Carnivora of Europe in the Department of Geology. The British Museum, Natural History, London

Roth J, Wagner A (1854) Die fossilen Knochenüberreste von Pikermi in Griechenland. Abh math–phys Cl Königlich Bayerischen Akad Wiss 7:371–464

Roussiakis SJ (2002) Musteloids and feloids (Mammalia, Carnivora) from the late Miocene locality of Pikermi (Attica, Greece). Geobios 35:699–719

Salesa MJ, Antón M, Peigné S, Morales J (2006) Evidence of a false thumb in a fossil carnivore clarifies the evolution of pandas. PNAS 103:379–382

Salesa MJ, Peigné S, Antón M, Morales J (2011) Evolution of the family Ailuridae: origins and Old-World fossil record. In: Glatston AR (ed) Red panda: biology and conservation of the first panda. Academic Press, London, pp 27–41

Spassov N, Geraads D (2011) A skull of *Simocyon primigenius* (Roth & Wagner, 1854) (Carnivora, Ailuridae) from the late Miocene of Karaslari (Republic of Macedonia), with remarks on the systematics and evolution of the genus. Neues Jahrb Mineral Geol Palaeontol Abh Abt A 262(2):151–161

Wagner A (1857) Neue Beitraege zur Kenntniss der fossilen S.ugethier-.berreste von Pikermi. Abh bayer Akad Wissensch Mathem naturwiss Klasse Munchen 8:111–158

Wang X, McKenna MC, Dashzeveg D (2005) *Amphicticeps* and *Amphicynodon* (Arctoidea, Carnivora) from Hsanda Gol Formation, Central Mongolia and phylogeny of basal arctoids with comments on zoogeography. Am Mus Novit 3983:1–57

Wesley-Hunt GD, Flynn JJ (2005) Phylogeny of the Carnivora: basal relationships among the Carnivoramorphans, and assessment of the position of "Miacoidea" relative to Carnivora. J Syst Palaeontol 3:1–28

The Fossil Record of Mephitids (Mammalia: Carnivora: Mephitidae) in Greece

George D. Koufos

1 Introduction

The family Mephitidae includes the skunks and stink badgers. For a long time, the skunks have been classified within Mustelidae as the separate subfamily Mephitinae and stink badgers within the Mustelinae together with the badgers. Recent molecular evidence indicates that the skunks and stink badgers share a common ancestor and should be separated from the mustelids. Thus, the subfamily Mephitinae was upgraded to the family level (Wund 2005). The extant mephitids include 13 species; the skunks inhabit the New World and the stink badgers (*Mydaus*) occur in Indonesia and Philippines (Koepfli et al. 2008; Wund 2005). The mephitids display a relatively short palate, which does not project beyond the posterior margin of the cheek teeth or near it, slightly inflated auditory bullae, a moderately elongated upper carnassial with low protocone, an M1 that is shorter or longer than the P4, an m1 with metaconid and short and basin-like talonid that sometimes has marginal cusplets, and a clear m2 (Orlov 1968). Fossil mephitids are distributed in Eurasia where they are known by several taxa from Spain to China. In Greece, the family is known from continental Greece and Samos Island with two late Miocene species (Fig. 1 and Appendix).

G. D. Koufos (✉)
Department of Geology, Laboratory of Geology and Paleontology, Aristotle University of Thessaloniki, Thessaloniki, Greece
e-mail: koufos@geo.auth.gr

E. Vlachos (ed.), *Fossil Vertebrates of Greece Vol. 2*,
https://doi.org/10.1007/978-3-030-68442-6_23

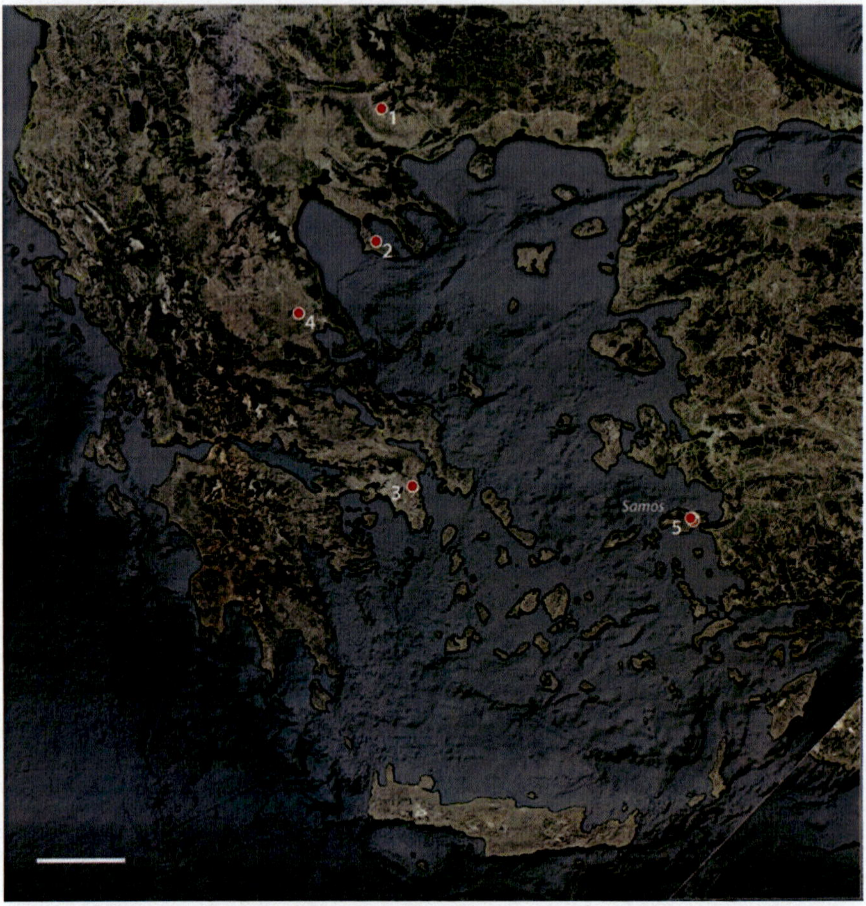

Fig. 1 Map of the occurrences of fossil mephitids from Greece. See Appendix for further information. **1**, Maramena; **2**, Kryopigi; **3**, Pikermi; **4**, Perivolaki; **5**, various localities and quarries in Samos Island. See Appendix for more information. Image exported from Google Earth Pro © 2019, map data from US Dept. of State Geographer, SIO, NOAA, U.S. Navy, NGA, GEBCO, image from Landsat/Copernicus. Scale bar equals 80 km, North faces upward

2 Historical Overview

The Greek fossil record includes few reports of mephitids belonging to the late Miocene genus *Promephitis*, which is considered as the only certain fossil Mephitidae in the Old World (Geraads and Spassov 2016 and references therein). It was originally described from Pikermi by Gaudry (1861a, b) as the new species *Promephitis lartetii*; another new species, *Promephitis majori*, was later described from an unknown locality of Samos Island by Pilgrim (1933). Recently, *Promephitis* has been also discovered in Central and North Greece (Koufos 2006; Lazaridis 2015).

3 Phylogenetic Relationships

The position of the previously recognized subfamily Mephitinae within arctoids has been a matter of debate, as molecular analyses suggest that they occupy a basal position within Arctoidea outside Mustelidae, whereas analyses of morphological characters suggest a placement within Mustelidae (see Wang and Qiu 2004 and references therein). The position of *Promephitis* within Mephitinae has been recently discussed in the parsimony analysis of Geraads and Spassov (2016), who suggest that the extinct genus occupies a basal position within Mephitidae, as a sister to the clade of *Mydaus* and Mephitini.

4 Systematic Paleontology

Mephitidae Bonaparte, 1845

★***Promephitis* Gaudry,** 1861a

Type Species *Promephitis lartetii* Gaudry, 1861a.

Remarks *Promephitis lartetii* is reported from the localities Q1, Q2, Q6, and A in Samos Island without description and reference of the material (Solounias 1981). However, these references are questionable as there is another *Promephitis* species from Samos (see below) and the material needs revision; thus, the Samos sample is referred to as *Promephitis* sp. in this chapter. A mandibular fragment with the m1 is described from the latest Miocene locality of Kryopigi in Chalkidiki as cf. *Promephitis lartetii* by Lazaridis (2015), whereas a single tooth from the Miocene/Pliocene locality of Maramena in Serres Basin is referred to as *Promephitis* sp. (Schmidt-Kittler 1995).

★***Promephitis lartetii* Gaudry,** 1861a

Type Material MNHN-PIK-3019 (holotype), cranium with the mandible, housed in the Muséum national d'Histoire naturelle, Paris, and described by Gaudry (1861a, b).

Type Locality Pikermi, Attica, late Miocene, MN12.

Distribution Besides its type locality, it is also reported from Perivolaki in Thessaly.

Remarks The holotype of *P. lartetii* cannot provide much information because the cranium is tightly connected to the mandible and both specimens are dorsoventrally compressed and deformed. However, Pilgrim (1931) refers that the M1 has a convex

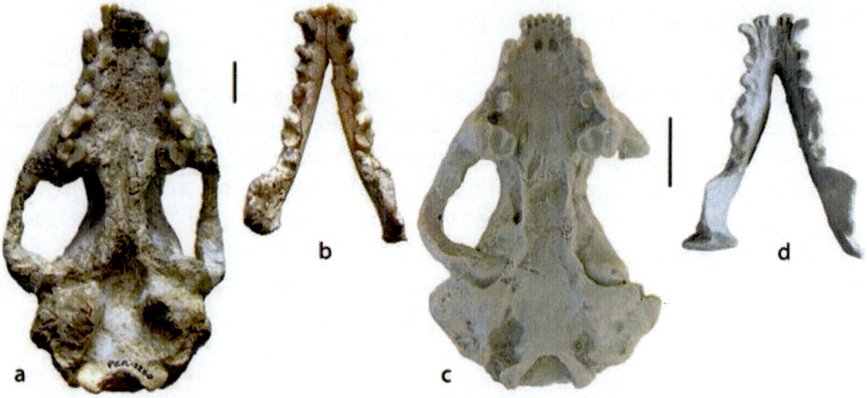

Fig. 2 Selected cranial specimens of *Promephitis* from Greece. (**a, b**) LGPUT PER 1280, cranium and associated mandible of *Promephitis lartetii* from Perivolaki in (**a**) palatal view of the cranium, and (**b**) occlusal view of the mandible. (**c, d**) AMNH-SAM-20585, cast of the holotype cranium and associated mandible of *Promephitis majori* from Samos in (**c**) palatal view of the cranium, and (**d**) occlusal view of the mandible. Scale bars equal 10 mm

lingual and straight buccal side and the m1 bears a metaconid situated somewhat in front of the protoconid. Based on the Perivolaki material (Fig. 2a, b), *P. lartetii* displays a wide nasal cavity, a strong post-orbital process, a strong sagittal crest, a short and relatively wide palate, small and deep choanae with their anterior border well behind the distal margin of the M1, a short and broad P4 with large protocone and without parastyle, a large M1 with the paracone larger than the metacone and a large crest-like protocone, a robust p4 without accessory cuspids, an elongated m1 with relatively wide talonid, and a rounded m2 (Koufos 2006).

★*Promephitis majori* Pilgrim, 1933

Type Material AMNH-SAM-20585 (holotype), cranium and associated mandible described and figured by Pilgrim (1933: p. 2, figs 1–6) (Fig. 2c, d; cast of the specimen) and housed in the American Museum of Natural History, New York.

Type Locality Adrianos ravine, Samos Island (precise locality unknown), Greece, late Miocene, MN12.

Distribution It is known only from the type locality.

Remarks It differs from *P. lartetii* in its smaller size, stronger post-orbital processes, less prominent occipital condyles, and smaller transverse diameter of M1 (Pilgrim 1933).

Acknowledgments I would like to deeply thank M. Cherin and G. Iliopoulos for comments that improved this manuscript.

Appendix

List of the Greek localities containing mephitid fossils. Type localities are marked with bold. Locality numbers refer to the collection numbers of the PaleoBiology Database (PBDB)

Localities[PBDB No]	Age (MN; GPTS in Ma)	Taxon
Maramena[32189]	Miocene/Pliocene (MN14/15)	*Promephitis* sp.[1]
Kryopigi[157582]	Middle–Late Turolian (MN12–13; 7.3–6.4)	cf. *Promephitis lartetii*[2]
Pikermi[182754]	Middle Turolian (MN12; 7.3–7.1)	***Promephitis lartetii***[3]
Perivolaki[194879]	Middle Turolian (MN12; 7.3–7.1)	*Promephitis lartetii*[4]
Samos Q1[95691]	Middle Turolian (MN12; ~7.1)	*Promephitis* sp.[5,a]
Samos Q2[206460]	Middle Turolian (MN12; ~7.1)	*Promephitis* sp.[5,a]
Samos A[206461]	Middle Turolian (MN12; ~7.1)	*Promephitis* sp.[5,a]
Samos Q6[206462]	Early Turolian (MN11)	*Promephitis* sp.[5,a]
[b]**Samos[206463]**	Turolian (MN11–13)	***Promephitis majori***[6]

GPTS Geomagnetic Polarity Time Scale
[1]Schmidt-Kittler (1995), [2]Lazaridis (2015), [3]Gaudry (1861a, b), [4]Koufos (2006), [5]Solounias (1981), [6]Pilgrim (1933)
[a]This study
[b]Precise locality unknown

References

Bonaparte CLJL (1845) Catalogo methodico dei mammiferi Europei. Giacomo Pirola, Milano

Gaudry A (1861a) Résultats des fouilles enteprises en Grèce sous les auspices de l'Académie des Sciences. C R Acad Sci Paris 52:722–724

Gaudry A (1861b) Note sur les carnassiers fossils de Pikermi (Gréce). Bull Soc Geol Fr 18:527–538

Geraads D, Spassov N (2016) Musteloid carnivores from the upper Miocene of South-Western Bulgaria, and the phylogeny of the Mephitidae. Geodiversitas 38(4):543–558

Koepfli KP, Deere KA, Slater GS, Begg C, Begg K, Grassman L, Lucherini M, Veron G, Wayne RK (2008) Multigene phylogeny of the Mustelidae: resolving relationships, tempo and biogeographic history of a mammalian adaptive radiation. BMC Biol 6:10. https://doi.org/10.1186/1741-7007-6-10

Koufos GD (2006) Carnivora. In: Koufos GD (ed) The late Miocene vertebrate locality of Perivolaki, Thessaly, Greece. Palaeontographica Abt A 276:39–74

Lazaridis G (2015) Study of the late Miocene vertebrate locality of Kryopigi and other localities of Kassandra Peninsula, Chalkidiki (Greece). Systematics, taphonomy, paleoecology, biochronology. PhD dissertation, School of Geology, Aristotle University, Thessaloniki, Greece

Orlov YA (1968) Fundamentals of Paleontology, vol 13, mammals. Israel Program for Scientific Translations, Jerusalem

Pilgrim GE (1931) Pontian Carnivora of Europe. British Museum Natural History, London

Pilgrim GE (1933) A fossil skunk from Samos. Am Mus Novit 633:1–15

Schmidt-Kittler N (1995) Carnivora, mammalia. In: Schmidt-Kittler N (ed) The vertebrate locality Maramena (Macedonia, Greece) at the Turolian/Ruscinian boundary (Neogene). Münchner Geowis Abh Abt A 28:75–86

Solounias N (1981) The Turolian fauna from the island of Samos, Greece. Contrib Vertebr Evol 6:1–232

Wang X, Qiu Z (2004) Late Miocene *Promephitis* (Carnivora, Mephitidae) from China. J Vertebr Paleontol 24(3):721–731

Wund M (2005) Mephitidae. In: Animal Diversity Web. http://animaldiversity.org/accounts/Mephitidae/. Accessed 27 Apr 2018

The Fossil Record of Mustelids (Mammalia: Carnivora: Mustelidae) in Greece

George D. Koufos

1 Introduction

The mustelids (family Mustelidae) include a variety of extant and extinct carnivorans such as weasels, badgers, otters, minks, wolverines, and related forms. The skunks were traditionally referred as a subfamily of the Mustelidae, named Mephitinae, but today they constitute the separate family Mephitidae—they are treated elsewhere in this volume. Extant mustelids include 22 genera and 56 species dispersed all over the World, except Australia, Antarctica, and some isolated oceanic islands. Despite the great diversification of the mustelids, they all share some common characters like the elongated and slender body, the short skull, the short facial region, the inflated or flattened auditory bullae, the short legs with five digits, the non-retractile claws (the claws are strong in the digging forms), the upper carnassial bearing a variably developed protocone, situated in the middle or in the forepart of the tooth, the lingual length of the M1 is longer than the labial one, the very small M2 and m2 (if present), and the absence of the M3 and m3. The oldest mustelids are dated to the Oligocene or middle Miocene of the Old World, but their ancestry is not clear (Orlov 1968; Wund 2018).

G. D. Koufos (✉)
Department of Geology, Laboratory of Geology and Paleontology,
Aristotle University of Thessaloniki, Thessaloniki, Greece
e-mail: koufos@geo.auth.gr

E. Vlachos (ed.), *Fossil Vertebrates of Greece Vol. 2*,
https://doi.org/10.1007/978-3-030-68442-6_24

2 Historical Overview

The family Mustelidae was recognized early in the first collections from Pikermi and Samos Island (e.g., Gaudry 1861a, b, 1862–67; Weithofer 1888; Forsyth Major 1894). Later, different members of the family have been described from several Neogene–Quaternary fossiliferous sites (e.g., Pilgrim 1931; Symeonidis and Sondaar 1975; Koufos 1982, 2006, 2009; Tsoukala 1992, 1999). The known material from Greece is scarce and sometimes does not allow definite attributions; however, there are some well-preserved cranial and mandibular remains, providing clear morphological evidence and thus certain determinations. The Greek fossil record includes twelve genera, 17 certain and three possible species among which there are six certain and two possible new species.

3 Phylogenetic Relationships

The origin and phylogenetic relationships of the mustelids are not clear, and they are still debated. The earliest fossil mustelids are known from the Oligocene or middle Miocene and there is a discussion concerning which of these early forms can be considered as ancestral for the crown mustelids (Wund 2018). The study of the nucleotide sequences in the extant taxa indicated that the Musteloidea *s.l.* includes three major lineages, Ailuridae, Mephitidae, and the monophyletic group of Procyonidae + Mustelidae *s.s.* (excluding the skunks) with unclear interrelationships for this trichotomy (Flynn et al. 2000). Based on the morphological characters of the primitive arctoids, Wang et al. (2005) suggested that the late Eocene miacids gave rise to two major lineages, that is the Hesperocyonidae (early dogs) and the Ursida represented by *Mustelavus*, a genus which is close to the clade formed by Ailuridae and Procyonidae + Mustelidae. Genetic analyses indicate that the mustelids are separated into four primary clades and three monotypic lineages; two main diversification events, due to major paleoenvironmental and biotic changes during the Neogene, have been recognized: one in the Miocene giving rise to the extant clades and another during the Pliocene from which the high extant diversity of the family originated (Koepfli et al. 2008).

4 Distribution

In Greece, the mustelids are well-distributed from the Miocene onwards, including several localities in Continental Greece and its islands (including Crete; see Fig. 1 and Appendix). The oldest evidence of the mustelids is traced in the early/middle Miocene with the genus *Proputorius*. During the late Miocene, several species have

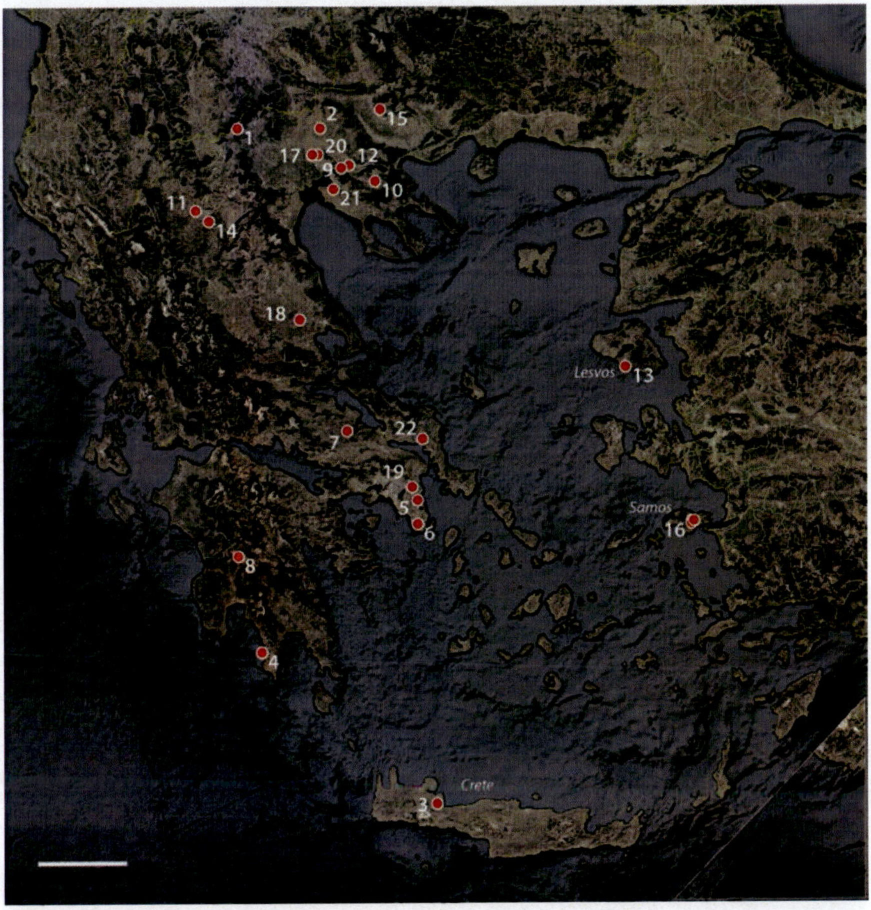

Fig. 1 Map of the occurrences of fossil mustelids from Greece. **1**, Loutra Almopias Cave; **2**, Agios Georgios Cave; **3**, Liko Cave; **4**, Apidima Cave C; **5**, Vraona Cave; **6**, Kitsos Cave; **7**, Sarakenos Cave; **8**, Marathousa 1; **9**, Apollonia 1; **10**, Ravin Voulgarakis; **11**, Libakos; **12**, Gerakarou 1; **13**, Vatera F; **14**, Dafnero 1; **15**, Maramena; **16**, various localities in Samos Island (see Appendix); **17**, Vathylakkos 3; **18**, Perivolaki; **19**, Pikermi; **20**, Ravin de la Pluie; **21**, Antonios; **22**, Aliveri. See Appendix for more information. Image exported from Google Earth Pro © 2019, map data from US Dept. of State Geographer, SIO, NOAA, U.S. Navy, NGA, GEBCO, image from Landsat/ Copernicus. Scale bar equals 80 km, North faces upward

been recognized in various fossiliferous sites. In the Pliocene and Pleistocene of Greece, the known mustelid species are fewer than the Neogene ones. Most of the Late Pleistocene mustelids from Greece are also encountered in the extant fauna of the country (e.g., *Meles meles*, *Martes foina*, *Martes martes, Lutra lutra, Mustella putorius, Mustella nivalis*, *Vormela peregusna*).

5 Systematic Paleontology

Mustelidae Fischer von Waldheim, 1817

Lutrinae Bonaparte, 1838

***Enhydriodon* Falconer, 1868**

Type Species *Enhydriodon sivalensis* Falconer, 1868.

★*?Enhydriodon latipes* Pilgrim, 1931

Type Material NHML-PIK-M.9002 (holotype), left hind foot described and figured by Pilgrim (1931: p. 56, figs 6–27).

Type Locality Pikermi, Attica, Greece, late Miocene, MN12.

Distribution It is only known from its type locality.

Remarks The single known specimen of this species is a series of foot bones that cannot provide information about its distinction from the other species, which are known by cranial and mandibular remains. Most of the Pikermi bones probably belong to the same individual (Pilgrim 1931). Recently, more postcranial remains from Pikermi have been described and attributed to as *E. ?latipes*, but without additional information of its differences from the known species of the genus (Roussiakis 2002). For the moment, it is considered as a distinct species, but certainly more material, especially cranial, is necessary to allow a more precise identification.

***Lutra* Brisson, 1762**

Type Species *Mustella lutra* Linnaeus, 1758.

"*Lutra affinis*" Gervais, 1859

Type Material A hemimandible with m1 from France is referred as the holotype of this species, but it is not figured (Gervais 1859). The specimen is probably lost, and the material of Maramena, Greece, housed in the Museum of Paleontology and Geology of the University of Athens, is proposed as the neotype for the species (Ginsburg 1999: p. 147). The materials from Maramena (Greece) and Venta del Moro (Spain) are significantly different in size (Willemsen 1992; Cherin et al. 2016). Thus, the species is doubtful, and the Greek material could be referred to as "*Lutra affinis*" until a review of the taxon is carried out.

Type Locality Sables de Montpellier, France, Pliocene.

Fig. 2 Selected specimens of Lutrinae (Mustelidae) from Greece. (**a**, **b**) *Lutra affinis* from Maramena, the proposed neotype for the species. Left mandible in (**a**) buccal view and (**b**) lingual view. Scale bar equals 1 cm

Distribution Besides its type locality, the species is recognized from the late Miocene of Spain and from the Greek Miocene/Pliocene locality of Maramena in Serres Basin.

Remarks The Maramena material (Fig. 2) is scanty and morphologically close to the modern *L. lutra* but its size is smaller and closer to *L. affinis,* thus it is attributed to the last species (Schmidt-Kittler 1995).

Lutra simplicidens Thenius, 1965

Type Material Hdsh. VIII/36 (holotype), left mandible with p2, p3, and m1, housed in the Palaeontological Institute of the University of Vienna (Willemsen 1992).

Type Locality Hundsheim bed VIII, near Deutsch-Altenburg, Austria (Willemsen 1992).

Distribution Besides its type locality, the species is also recognized in England, Germany, and Russia. In Greece, it is known from the Middle Pleistocene of Marathousa 1, in Megalopolis Basin.

Remarks The single known mandibular fragment with p3–m1 from Greece is characterized by trigonid broader than talonid, shallow talonid basin, large metaconid, and absence of hypoconulid. These morphological features separate it from *L. lutra* and allow its attribution to *L. simplicidens* (Konidaris et al. 2018). According to the phylogenetical analysis of Cherin et al. (2016), it should be moved to another genus and until the discovery of complete cranial material should be referred to as *"Lutra" simplicidens.*

Lutrogale Gray, 1865

Type Species *Lutra perspicillata* Geoffroy, 1826.

★*Lutrogale cretensis* (Symeonidis and Sondaar, 1975)

Remarks See Lyras et al. (this volume).

Mustelinae Fischer von Waldheim, 1817

Martes Pinel, 1792

Type Species *Martes domestica* Pinel, 1792.

★*Martes woodwardi* Pilgrim, 1931

Nomenclatural and Taxonomical History *Martes woodwardi* in Pilgrim 1931 (new species); *Pilgrimeles woodwardi* in Ginsburg 1999 (new combination and genus). This species has been recently attributed to its original combination (Roussiakis 2002), an opinion followed herein.

Type Material NHML-PIK-M.2031 (holotype), right mandibular fragment with p4–m1, described and figured by Pilgrim (1931: p. 40; pl. 2, figs 2, 2a; Fig. 2a) and housed in the Natural History Museum of London.

Type Locality Pikermi, Attica, Greece, late Miocene, MN12.

Distribution It is only known from its type locality.

Remarks There is a debate concerning the taxonomy of the Pikermi material (Fig. 3a). Long after its original description, it was included in the new genus *Pilgrimeles* (Ginsburg 1999), and as a member of Melinae. Other authors argued on the validity of the diagnostic characters of the new genus (Roussiakis 2002) and reclassified the Pikermi material to the genus *Martes*. The species is characterized by large-size, m1 with relatively large metaconid, elongated and strongly concave talonid in the m1, and elongated m2 (Pilgrim 1931). Roussiakis (2002) refers to the same species one m1 with small metaconid and a rounded basin-like talonid, situated slightly more labially to the metaconid–protoconid junction.

★*Martes lefkonensis* Schmidt-Kittler, 1995

Type Material AMPG-MAR-MA401 (holotype), right mandibular fragment with p4–m1, housed in the Palaeontological and Geological Museum, University of Athens (Schmidt-Kittler 1995: p. 76; pl. 1, fig. 1; Fig. 3b).

Type Locality Maramena, Serres Basin; Miocene/Pliocene boundary, MN13/14.

Distribution It is only known from its type locality.

Remarks The Maramena material was described as a new *Martes* with small-size, relatively short m1 (shorter than that of the modern species), steeply sloping mesially lingual border of the m1 talonid, stronger metaconid than the modern species,

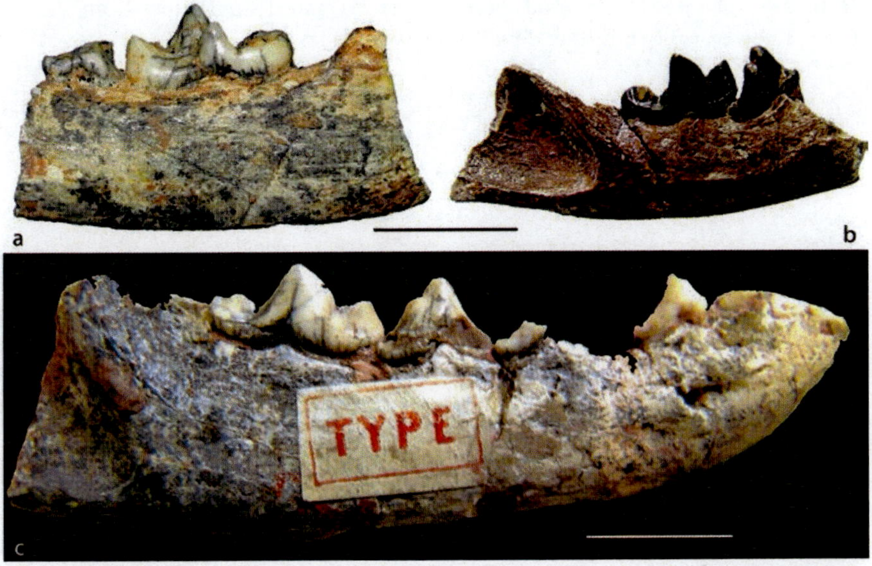

Fig. 3 Selected specimens of Mustelinae (Mustelidae) from Greece. (**a**) Right mandibular fragment (holotype) of *Martes woodwardi* from Pikermi in lingual view, (**b**) right mandible (holotype) of *Martes lefkonensis* from Maramena in buccal view, (**c**) left mandible (holotype) of ?*Sinictis pentelici* from Pikermi in buccal view. Scale bar equals 1 cm

and well-developed basal cingulum all around the protocone of the M1 (Schmidt-Kittler 1995).

⊙*Martes foina* (Erxleben, 1977)—the Beech Marten

Distribution In Greece, it is reported from the Late Pleistocene deposits of Apidima Cave C, in south Peloponnese. Besides Greece, it is also known from France and Czech Republic. The extant distribution of this species is much more extensive.

Remarks The known Greek material includes some isolated upper canines and two mandibular fragments. The lower carnassial has an elongated trigonid, a well-defined metaconid, and a narrow talonid with well-developed hypoconid and hypoconulid (Tsoukala 1999).

⊙*Martes martes* (Linneaus, 1758)—the European Pine Marten

Distribution The extant taxon is well distributed in Europe and Western Asia; however, there are indications for its presence in Africa and North America as well. Fossil remains of the taxon are reported from the Early Pleistocene of Europe (NOW 2019). In Greece, it is only known from the Late Pleistocene deposits of Vraona Cave, Attica.

Remarks The known material includes one m1 and a single ulna from Vraona Cave, referred without further description by Rabeder (1995).

Sinictis Zdansky, 1924

Type Species *Sinictis dolichognathus* Zdansky, 1924.

★?*Sinictis pentelici* (Gaudry, 1861)

Nomenclatural and Taxonomical History *Mustela pentelici* in Gaudry 1861a, b (new species); *Martes* (*Mustela*) *pentelici* in Schlosser 1888 (new combination); *Sinictis* (*?*) *pentelici* in Pilgrim 1931 (new combination).

Type Material MNHN-PIK-3260 (holotype), left mandibular fragment with p2–m1, described and figured by Gaudry (1861b: p. 527; pl.10, figs 4, 5; Fig. 3c) housed in the Museum national d'histoire Naturelle of Paris.

Type Locality Pikermi, Attica, Greece, late Miocene, MN12.

Distribution It is only known from the type locality.

Remarks The species was reported under various names since a long time, but ultimately Pilgrim (1931) described it as a possible member of the genus *Sinictis*. The species is characterized by relatively high mandibular corpus, single mental foramen below the p2, anterior margin of the masseteric fossa below the m2, small p1, elongated and narrow p2 without anterior accessory cuspid and with strong distal cingular projection, p4 with strong mesial cingular projection and a small posterior accessory cuspid packed on the main one, small metaconid, talonid with hypoconid and absence of entoconid, and small and rounded m2 (pers. observations).

Mustela Linnaeus, 1758

Type Species *Mustela erminea* Linnaeus, 1758.

Remarks The genus *Mustela* has been traced from the latest Early Pleistocene locality Ravin Voulgarakis in Mygdonia Basin and from the Middle Pleistocene of Marathousa 1 in Megalopolis Basin. The single known specimen from Ravin Voulgarakis has some morphological similarities with *Mustela stromeri,* but it is smaller and displays a large metaconid in m1. On the other hand, it is smaller and has longer talonid in the m1 than *Mustella eversmanni* (Koufos and Kostopoulos 1997). The material from Marathousa 1 includes a tibia, which falls into the range of variation of *Mustela* and thus it is reported as *Mustela* sp. (Konidaris et al. 2018).

⊙*Mustela nivalis* Linnaeus, 1766—the Least Weasel

Distribution The Least Weasel is widely distributed, covering nearly the entire Europe (except Ireland and Iceland), North Africa, and a major part of Asia and North America. In Greece, it is mentioned from the Late Pleistocene of Vraona Cave in Attica, and Kitsos and Sarakenos Caves in Viotia.

Remarks The taxon is referred without further description from Vraona Cave (Bachmayer et al. 1988/89; Rabeder 1995). Recently it was recognized in Kitsos and Sarakenos Caves by some mandibular fragments characterized by an elongated and quite narrow lower carnassial, weakly broadened crown, quite long and broad talonid, and the absence of third root (Marcizak 2016).

⊙*Mustela putorius* Linneaus, 1758—the European Polecat

Distribution The European Polecat is widely distributed in Europe and North America but rare in Asia and absent in Africa. In Greece, it is known only from the Late Pleistocene deposits of Agios Georgios Cave in Kilkis.

Remarks The sole known specimen is a mandibular fragment with c and p3–m1 (SGK-209, housed in the Cave's exhibition), which is attributed to the subspecies *Mu. p. robusta*. The main characters are: single-rooted p2; larger protoconid than paraconid in m1; elongated and ridge-like talonid in m1; and single-rooted m2 (Tsoukala 1992).

Proputorius Filhol, 1891

Type Species *Proputorius sansaniensis* Filhol, 1891.

Remarks This genus is recognized in the early/middle Miocene locality Antonios in Chalkidiki from a mandibular fragment with p3–p4 and one p4, which were described as *Proputorius* cf. *sansaniensis* by Koufos (2008).

Baranogale Kormos, 1934

Type Species *Baranogale helbingi* Kormos, 1934.

Remarks This genus is recognized in Greece only by a mandibular fragment from the middle Villafranchian of Dafnero 1. This specimen shows the characters of *Baranogale*, but it is larger than *B. helbingi* and smaller than *B. balcanica*, and closer to *B. adroveri*; therefore, and considering the limited material, it is attributed to *Baranogale* aff. *B. helbingi* by Koufos and Kostopoulos (1997).

Mellivorinae Gray, 1865

Eomellivora Zdansky, 1924

Type Species *Eomellivora wimani* Zdansky, 1924.

Eomellivora piveteaui Ozansoy, 1965

Nomenclatural and Taxonomical History *Eomellivora piveteaui* in Ozansoy 1965 (new species); *Eomellivora wimani piveteaui* in Wolsan and Semenov 1996 (referral to the subspecies level). Lately, this species group name is used again at its original sense (Valenciano et al. 2015).

Type Material MNHN-SIN-1004 (lectotype), right mandibular fragment with i2–p4 and m1 trigonid, described and figured by Ozansoy (1965: pl. 2, fig. 1), housed in the Museum national d'histoire Naturelle of Paris. Wolsan and Semenov (1996) proposed this specimen as the lectotype of the species.

Type Locality Yassiören, Middle Sinap, Turkey, late Miocene, MN9–10.

Distribution Besides its type locality, the species is known from Spain and Moldova. In Greece, it is only recognized from the Late Miocene (Vallesian) locality Ravin de la Pluie in Axios Valley.

Remarks The only known material from Greece is a single M1, originally attributed to *E. wimani* by Koufos (2012). At that time, the type material was referred to the subspecies *E. wimani piveteaui* Wolsan and Semenov, 1996, but the single tooth from Greece does not allow a subspecific identification. Recently, a skull and a mandible from Spain confirmed the specific value of *E. piveteaui*, and the Greek material was transferred to this species (Valenciano et al. 2015). The emended diagnosis for the species is: buccolingually compressed hypoconid, small trenchant hypoconulid in line with the protoconid-hypoconid in the m1; m2 trigonid larger than the talonid, metaconid absent, sharp paraconid, protoconid, and hypoconid situated along the mesiodistal axis (Valenciano et al. 2015).

Guloninae Gray, 1825

Plesiogulo Zdansky, 1924

Type Species *Lutra brachygnatha* Schlosser, 1903.

Remarks Two humeral fragments from Pikermi, Attica, are described as *Plesiogulo* sp. by Symeonidis (1975).

Plesiogulo crassa Kurtén, 1970

Nomenclatural and Taxonomical History *Plesiogulo brachygnathus* in Zdansky 1924 partim (initial identification); *Plesiogulo brachygnathus* forme *crassa* in Teilhard and Leroy 1945 (new identification); *Plesiogulo crassa* in Kurtén 1970 (new species).

Type Material Front part of a cranium and associated mandible (holotype; number unknown), figured by Teilhard and Leroy (1945: figs 9, 10B) and housed in the Palaeontological Institute, University of Uppsala.

Type Locality Yushe, Shansi, China, late Miocene, MN11–13.

Distribution Besides its type locality, *P. crassa* is known from the late Miocene of Italy and Kazakhstan. In Greece, it is known from Vathylakkos-3 in Axios Valley, and Perivolaki in Thessaly.

Remarks The species was unknown in the fossil record of Europe until the beginning of the 1970s, when it was recognized in the Axios Valley; later, a cranium and the associated mandible were also found in Perivolaki (Fig. 4). The main characters of the Greek material are the relatively long cranium with short face, the oval and wide narial opening, the high sagittal crest, the large triangular-shaped and flattened bullae, the relatively short and narrow palate with narrow choanae, the robust P4 with rounded protocone well separated by a deep constriction, the elongated and abruptly inclined posteriorly symphysis, the deep and oval-shaped masseteric fossa,

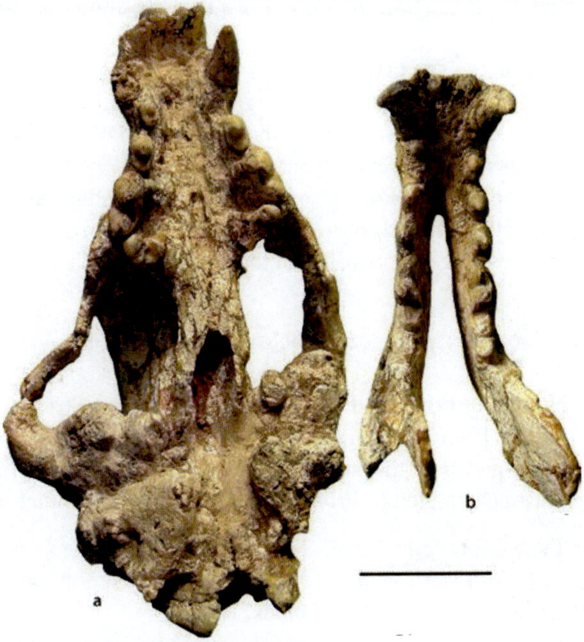

Fig. 4 Selected specimens of Guloninae (Mustelidae) from Greece. (**a**, **b**) *Plesiogulo crassa* from Perivolaki. (**a**) Cranium in palatal view, and (**b**) the associated mandible in occlusal view. Scale bar equals 5 cm

the long and narrow m1, the elongated basin-like talonid in the m1, and the small metaconid support its attribution to this species (Koufos 1982, 2006).

Melinae Bonaparte, 1838

Promeles Zittel, 1893

Type Species *Mustela palaeattica* Weithofer, 1888.

★*Promeles palaeattica* (Weithofer, 1888)

Nomenclatural and Taxonomical History *Mustela palaeattica* in Weithhofer 1888 (new species); *Meles? (Mustela) palaeattica* in Schlosser 1888 (new combination); *Meles palaeatticus* in Schlosser 1890 (new combination); *Promeles palaeattica* in Zittel 1893 (new combination and genus).

Type Material IPUW-PIK without a number (holotype), cranium and associated mandible described and figured by Weithofer (1888: p. 226; taf. 1, figs 1–11); it is housed in the Palaeontological Institute of Vienna University.

Type Locality Pikermi, Attica, Greece, late Miocene, MN12.

Distribution Besides its type locality, the species is known from Perivolaki in Thessaly, and it is mentioned from Samos Q1 and A without description. Outside Greece, it has been found in Europe (Italy and Germany) and Asia (Iran).

Remarks The taxon is rare in Greece, recognized only in the late Miocene localities Pikermi and Perivolaki (Fig. 5a). The main characters of the species are the relatively short and wide snout, the oval shaped nasal cavity with almost vertical lateral margins, the flattened frontals, the absence of the parastyle in the P4, the strong protocone in the P4, the strong metaconule and lingual cingulum in the M1, the long m1 with elongated basin-like talonid, the strong metaconid in the m1 connected to the crest-like entoconid, and the small, rounded, and bicuspid m2 (Koufos 2006).

★*Promeles macedonica* Schmidt-Kittler, 1995

Nomenclatural and Taxonomical History *Promeles macedonicus* in Schmidt-Kittler 1995 (initial identification, new species). The word "meles" is feminine in Latin and thus the name "*Promeles*" as well. Therefore, the name of the species must be changed to *Promeles macedonica* for gender agreement.

Type Material AMPG-MAR-MA406 (holotype), right P4–M1, (Schmidt-Kittler 1995: p. 77, fig. 3; pl. 1, figs 2–4); it is housed in the Museum of Paleontology and Geology, University of Athens.

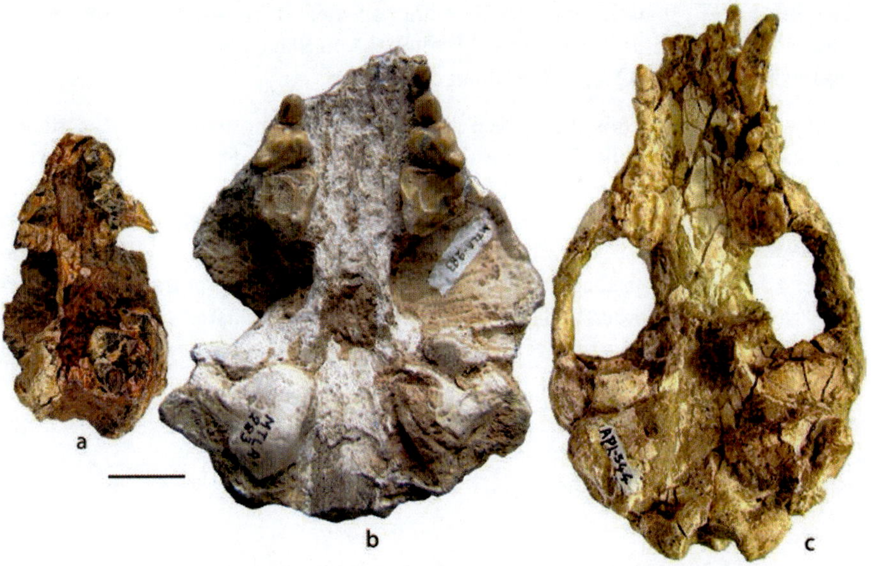

Fig. 5 Selected specimens of Melinae (Mustelidae) from Greece. (**a**) Cranium of *Promeles palaeattica* from Pikermi in palatal view, (**b**) cranium of *Parataxidea maraghana* from Mytilinii 1A, Samos Island in palatal view, (**c**) cranium of *Meles dimitrius* from Apollonia 1 in palatal view. Scale bar equals 2 cm

Type Locality Maramena, Serres Basin; Miocene/Pliocene boundary, MN-13/14.

Distribution It is only known from its type locality.

Remarks Similar to *Promeles palaeattica*, from which it differs, displaying a slightly concave distal border in M1, a more expressed parastyle angle in the M1, a longer trigonid and a shorter talonid in the m1 (Schmidt-Kittler 1995).

Parataxidea Zdansky, 1924

Type Species *Meles maraghana* Kittl, 1887.

Parataxidea maraghana (Kittl, 1887)

Nomenclatural and Taxonomical History *Meles maraghana* in Kittl 1887 (new species); *Parataxidea maraghana* in Zdansky 1924 (new combination).

Type Material NHMW-MRG without a number (holotype), right maxillary fragment with C–M1 (Kittl 1887: p. 337; taf. 15, fig. 4).

Type Locality Middle Maragheh, Iran, late Miocene, MN11–12.

Distribution Besides its type locality, this species is known from Greece and Turkey. In Greece, it is known from Mytilinii-1A in Samos Island, and it is further reported from Samos Q1 and A without description.

Remarks In Greece, this species is only known from Samos where it was certainly described from the locality Mytilinii 1A (Koufos 2009; Fig. 5b). Its main characters are the short and wide muzzle, the oval narial opening, the narrow frontals and rounded orbits, the absence of sagittal crest, the elongated and narrow palate with relatively wide choanae, the robust P2 and P3 with strong cingulum, the short and wide P4 with strong protocone, the large M1 with long lingual mesiodistal diameter and straight mesial border, and the large and robust m1 with narrow talonid (Koufos 2009); all these characters fit well the morphology of *Parataxidea maraghana*.

Meles Brisson, 1762

Type Species *Ursus meles* Linnaeus, 1758.

Meles thorali Viret, 1950

Type Material The species was described from Saint-Vallier by Viret (1950) without definition of holotype; the type material includes cranial and mandibular remains.

Type Locality Saint-Vallier, France, Early Pleistocene, MN17.

Distribution Besides its type locality, this species is known from Spain, Greece, and Bulgaria. In Greece, it is known only from Vatera-F in Lesvos Island.

Remarks According to the revision of the European sample of *Meles thorali*, the following diagnostic features distinguish this taxon from *Meles meles*: small infraorbital foramen situated above the P4; equal-sized paracone and metacone in the M1; more distolingually situated metaconule relative to the metacone thus creating a notch on the buccal crown margin; longer trigonid in the m1 relative to the talonid; on average shorter m1 and larger P4 although showing a considerable overlap (Madurell-Malapeira et al. 2011). Pending detailed analysis, the Vatera-F material is attributed to *M. thorali* due to the similar dental dimensions (de Vos et al. 2002).

★*Meles dimitrius* Koufos, 1992

Type Material LGPUT-GER-161 (holotype), left mandibular ramus with c–m2, housed in the Laboratory of Geology and Paleontology, University of Thessaloniki (Koufos, 1992: p. 15; pl. 4, fig. 3).

Type Locality Gerakarou 1, Mygdonia Basin, Early Pleistocene, MNQ18.

Distribution This species is also known from Apollonia 1 in Mygdonia Basin (Greece).

Remarks The material of *Meles* from Gerakarou 1 and Apollonia 1 (Fig. 5c) has been recently transferred to *M. thorali* and *M. meles atavus*, respectively (Madurell-Malapeira et al. 2011). The description of some new material and the revision of the old collection indicated that there are significant morphological differences from both taxa (Koufos 2018). The Greek material differs from *M. thorali* in displaying tympanic bullae on line with the postglenoid processes, overall smaller and more robust P4, shorter P4 relative to the buccal length of the M1, smaller metacone than paracone in the M1, smaller buccal length of M1, overall smaller m1, and longer talonid relative to the trigonid. The Greek sample also differs from *M. m. atavus* in having wider and more robust upper carnassial, more triangular occlusal outline in the P4 with an angular lingual margin, relatively large cusp in the lingual angular margin of the P4, small cuspule in the mesial margin between the paracone and the cusp of the lingual margin, a labial incision between the metacone and metaconule, well-developed postprotocrista reaching the lingual crown margin of the M1, and short buccal length of M1 relative to the carnassial length (Koufos 2018). Taking in mind the great morphological variation of the modern badgers, the unresolved problem of the conspecificity between European and Asian badgers, and the very few fossil materials, I think that it is better to keep, at the moment, *M. dimitrius* as a separate species.

⊙*Meles meles* (Linneaus, 1758)—the European Badger

Distribution Today, it is widespread in Eurasia and present in Africa. This species is known in Greece from Apidima Cave-C in Peloponnese, Loutra Almopias Cave in NW Greece, and Vraona Cave in Attica.

Remarks This taxon is given in the faunal lists of Vraona and Apidima Caves without description. Among the found material in there is a maxillary fragment with P2–P4 from Loutra Almopias Cave, the carnassial of which is characterized by large paracone, prominent blade, weak metacone, and well-developed basal cingulum (Tsoukala et al. 2006).

Galictinae Reig, 1957

Pannonictis Kormos, 1931

Type Species *Pannonictis pliocaenica* Kormos, 1931.

Remarks Steensma (1988) described one upper and two lower canines from Libakos in Grevena Basin as *Enhydrictis ardea*. However, the morphology and dimensions of these teeth are close to *Pan. nestii* and thus they are referred to as *Pannonictis* sp. (Koufos 2014).

Carnivora *incertae sedis*

Palaeogale von Meyer, 1846

Type Species *Palaeogale minuta* (Gervais, 1848–1852).

Remarks The genus *Palaeogale* is often associated with mustelids, but its dentition preserves both mustelid and feliform features and thus should be considered as *incertae sedis* within Carnivora (Baskin 1998; Hayes 2000). It is only reported from the Early Miocene locality Aliveri without description and reference of the material (Schmidt-Kittler 1983).

Acknowledgments I would like to deeply thank M. Cherin and G. Iliopoulos for comments that improved this manuscript.

Appendix

List of the Greek localities containing mustelid fossils. Type localities are marked with bold. Locality numbers refer to the collection numbers of the PaleoBiology Database (PBDB)

Localities[PBDB No]	Age (MN; GPTS in Ma)	Taxon
Loutra Almopias Cave[203847]	Late Pleistocene	*Meles meles*[1]
Agios Georgios Cave[207192]	Late Pleistocene	*Mustela putorius*[2]
Liko Cave, Crete[92902]	**Late Pleistocene**	***Lutrogale cretensis*[3]**
Apidima Cave C[32082]	Late Pleistocene	*Martes foina*[2] *Meles meles*[2]
Vraona Cave[183130]	Late Pleistocene	*Martes martes*[4] *Mustela nivalis*[5] *Meles meles*[5]
Kitsos Cave	Late Pleistocene	*Mustela nivalis*[6]
Sarakenos Cave[203845]	Late Pleistocene	*Mustela nivalis*[6]
Marathousa 1[187637]	Middle Pleistocene (0.42–0.56)	*Lutra simplicidens*[7] *Mustela* sp.[7]
Apollonia 1[34784]	Epivillafranchian (MN20)	*Meles dimitrius*[8]
Ravin Voulgarakis[34783]	Epivillafranchian (MNQ 20)	*Mustela* sp.[9]
Libakos[34764]	?Late Villafranchian	*Pannonictis* sp.[10]
Gerakarou 1[134617]	**Late Villafranchian (MNQ18)**	***Meles dimitrius*[8]**
Vatera F[183341]	Middle Villafranchian (MN17)	*Meles thorali*[11]
Dafnero 1[34594]	Middle Villafranchian (MN17)	*Baranogale* cf. *helbingi*[9]
Maramena[32189]	Miocene/Pliocene (MN13/14)	***Promeles macedonica*[12]** *Lutra affinis*[12] ***Martes lefkonensis*[12]**
Samos A[206461]	Middle Turolian (MN12)	*Promeles palaeattica*[13], *Parataxidea maraghana*[13]
Mytilinii 1A[202215]	Middle Turolian (MN12; ~7.1)	*Parataxidea maraghana*[14]

(continued)

Localities[PBDB No]	Age (MN; GPTS in Ma)	Taxon
Samos Q1[95691]	Middle Turolian (MN12; 7.3–7.1)	*Parataxidea maraghana*[13] *Promeles palaeattica*[13]
Vathylakkos 3[182750]	Middle Turolian (MN12; ~7.3)	*Plesiogulo crassa*[15]
Perivolaki[194879]	Middle Turolian (MN12; 7.3–7.1)	*Plesiogulo crassa*[16] *Promeles palaeattica*[16]
Pikermi[182754]	**Middle Turolian (MN12; 7.3–7.1)**	***Promeles palaeattica*[17]** ***Martes woodwardi*[18]** **?*Enhydriodon latipes*[18]** **?*Sinictis pentelici*[19, 20]** *Plesiogulo* indet[21]
Ravin de la Pluie[191070]	Late Vallesian (MN10; ~9.3)	*Eomellivora piveteaui*[22]
Antonios[73861]	Early/Middle Miocene (MN4/5)	*Proputorius* cf. *sansaniensis*[23]
Carnivora *incertae sedis*		
Aliveri	Early Miocene (MN4)	*Palaeogale* sp.[24]

GPTS Geomagnetic Polarity Time Scale

[1]Tsoukala et al. (2006), [2]Tsoukala (1992), [3]Symeonidis and Sondaar (1975), [4]Rabeder (1995), [5]Bachmayer et al. (1988/89), [6]Marcizak (2016), [7]Konidaris et al. (2018), [8]Koufos (1992), [9]Koufos and Kostopoulos (1997), [10]Steensma (1988), [11]de Vos et al. (2002), [12]Schmidt-Kittler (1995), [13]Solounias (1981), [14]Koufos (2009), [15]Koufos (1980), [16]Koufos (2006), [17]Weithofer (1888), [18]Pilgrim (1931), [19]Gaudry (1861a), [20]Gaudry (1861b), [21]Symeonidis (1975), [22]Koufos (2012), [23]Koufos (2008), [24]Schmidt-Kittler (1983)

References

Bachmayer F, Malez V, Symeonidis N, Theodorou G, Zapfe H (1988–89) Die Ausgrabund in der Höhle von Vraona (Attica), in Jahre 1985 Österr Akad Wiss Math Naturwiss Kl Abt 1 197:287–307

Baskin JA (1998) Mustelidae. In: Janis CM, Scott KM, Jacobs LL (eds) Evolution of tertiary mammals of North America: terrestrial carnivores, ungulates, and ungulate like mammals. Cambridge University Press, Cambridge, pp 152–173

Cherin M, Iurino AA, Willemsen G, Carnevale G (2016) A new otter from the early Pleistocene of Pantalla (Italy), with remarks on the evolutionary history of Mediterranean Quaternary Lutrinae(Carnivora, Mustelidae). Quarter Sci Rev 135:92–102

de Vos J, van der Made J, Athanassiou A, Lyras G, Sondaar PY, Dermitzakis MD (2002) Preliminary note on the late Pliocene fauna from Vatera (Lesvos, Greece). Ann Geol Pays Helleniques 39:37–70

Flynn JJ, Nedbal MA, Dragoo JW, Honeycutt RL (2000) Whence the red panda? Mol Phylogenet Evol 17(2):190–199

Forsyth Major CJ (1894) Le gisement ossifère de Mytilinii et catalogue d'ossements fossiles recueillis à Mytilini, île de Samos, et déposés au Collège Galliard, à Lausanne. Georges Bridel & Co, Lausanne

Gaudry A (1861a) Résultats des fouilles enteprises en Grèce sous les auspices de l'Académie des Sciences. C R Acad Sci Paris 52:722–724

Gaudry A (1861b) Note sur les carnassiers fossils de Pikermi (Gréce). Bull Soc Geol Fr 18:527–538

Gaudry A (1862–67) Animaux fossiles et géologie de l'Attique. Savy, Paris

Gervais P (1859) Zoologie et Paléontologie françaises. Nouvelles recherches sur les animaux vertébrés dont on trouve les ossements enfouis dans le sol de la France et sur leur comparaison avec les espèces propres aux autres régions du globe. Bertrand, Paris

Ginsburg L (1999) Order Carnivora. In: Rössner G, Heissig K (eds) The Miocene land mammals of Europe. Verlag Dr Friedrich Pfeil, Munchen, pp 109–148

Hayes FG (2000) Brooksville 2 local fauna (Arikareean, latest Oligocene) Hernando County, Florida. Bull Florida Mus Nat Hist 43(1):1–47

Kittl E (1887) Beiträge zur Kennthniss der fossilen Säugethiere von Maragha in Persia. I. Carnivora. Ann Naturhist Hofmuseums 2:317–338

Koepfli K-P, Deere KA, Slater GJ, Begg S, Begg K, Grassman L, Lucherini M, Veron G, Wayne RK (2008) Multigene phylogeny of the Mustelidae: resolving relationships, tempo and biogeographic history of a mammalian adaptive radiation. BMC Biol 6(10):1–22. https://doi.org/10.1186/741-7007-6-10

Konidaris GE, Athanassiou A, Tourloukis V, Thompson N, Giusti D, Panagopoulou E, Harvati K (2018) The skeleton of a straight-tusked elephant (*Palaeoloxodon antiquus*) and other large mammals from the Middle Pleistocene butchering locality Marathousa 1 (Megalopolis Basin, Greece): preliminary results. Quat Int 497:65–84

Koufos GD (1980) Palaeontological and stratigraphical study of the Neogene continental deposits of the Axios Valley basin. PhD thesis, Aristotle University of Thessaloniki

Koufos GD (1982) *Plesiogulo crassa* from the upper Miocene (lower Turolian) of Northern Greece. Ann Zool Fenn 19:193–197

Koufos GD (1992) The Pleistocene carnivores of the Mygdonia basin (Macedonia, Greece). Ann Paleontol 78:205–257

Koufos GD (2006) Carnivora. In: Koufos GD (ed) The late Miocene vertebrate locality of Perivolaki, Thessaly, Greece. Palaeontographica Abt A 276:39–74

Koufos GD (2008) Carnivores from the early/middle Miocene locality of Antonios (Chalkidiki, Macedonia, Greece). Geobios 41:365–380

Koufos GD (2009) Carnivora. In: The late Miocene mammal faunas of the Mytilinii Basin, Samos Island, Greece: new collection (Koufos, G. D. & Nagel, D., eds). Beit Palaeontol 31:57–105

Koufos GD (2012) New material of Carnivora (Mammalia) from the late Miocene of Axios Valley, Macedonia, Greece. C R Palevol 11:49–64

Koufos GD (2014) The Villafranchian carnivoran guild of Greece: implications for the fauna, biochronology and paleoecology. Integr Zool 9:444–460

Koufos GD (2018) New material and revision of the Carnivora, Mammalia from the lower Pleistocene Apolonia 1, Greece. In: M-P Palombo (ed) Advances in quaternary studies: the contribution of mammalian fossil record. Quaternary 1(6):1–38. https://doi.org/10.3390/quat1010006

Koufos GD, Kostopoulos DS (1997) New Carnivore material from the Plio-Pleistocene of Macedonia, Greece with the description of a new canid. Münchener Geowiss Abh 34:33–63

Lyras GA, Athanassiou A, van der Geer AAE (this volume) The fossil record of insular endemic mammals from Greece. In: Vlachos E (ed) The fossil vertebrates of Greece Vol. 2—laurasiatherians, artiodactyles, perissodactyles, carnivorans, and island endemics. Springer – Nature Publishing Group, Cham

Madurell-Malapeira J, Alba DM, Marmi J, Aurell J, Moyà-Solà S (2011) The taxonomic status of European Plio-Pleistocene badgers. J Vertebr Paleontol 31(4):885–894

Marcizak A (2016) The common weasel *Mustela nivalis* L., 1766 from Sarakenos Cave (Greece) and the difficulty to distinguish small mustelid species in the fossil record. Acta Zool Cracoviensia 59(1):25–35

Orlov YA (1968) Fundamentals of paleontology, vol. 13, Mammals. Israel Program for Scientific Translations, Jerusalem

Ozansoy F (1965) Étude des gisements continentaux et des mammifères du Cénozoïque de Turquie. Mém Soc Geol Fr 44:1–92

Pilgrim GE (1931) Catalogue of the Pontian Carnivora of Europe in the Department of Geology. The British Museum, Natural History, London

Rabeder G (1995) Jungpleistozäne und Fruhholozäne Säugetierreste aus der Höhle von Vraona auf Attica, Griechenland. Ann Geol Pays Helleniques 36:47–58

Roussiakis S (2002) Musteloids and feloids (Mammalia, Carnivora) from the late Miocene locality of Pikermi (Attica, Greece). Geobios 35:699–719

Schmidt-Kittler N (1983) The mammals from the lower Miocene of Aliveri (island of Evia, Greece). On a new species of *Sivanasua* PILGRIM, 1931 (Felidae, Carnivora) and the phylogenetic position of the genus. Proc K Ned Akad Wet 86:301–318

Schmidt-Kittler N (1995) Carnivora, Mammalia. In: Schmidt-Kittler N (ed) The vertebrate locality Maramena (Macedonia, Greece) at the Turolian/Ruscinian boundary (Neogene). Münchner Geowis Abh Abt A 28:75–86

Solounias N (1981) The Turolian fauna from the island of Samos, Greece. Contrib Vertebr Evol 6:1–232

Steensma KJ (1988) Plio-/Pleistozäne Großäugetiere (Mammalia) aus dem Becken von Kastoria/Grevena, südlich von Neapolis-NW Griechenland. PhD dissertation, Technische Universität Clausthal

Symeonidis N (1975) Ein grosser Mustelidae aus Pikermi. Ann Geol Pays Helleniques 26:314–319

Symeonides N, Sondaar PY (1975) A new otter from the Pleistocene of Crete. Ann Géol Pays Hellenique 27:11–24

Teilhard de Chardin P, Leroy P (1945) Les Mustelidés de Chine. Institute of Geology, Pékin, pp 4181–4185

Tsoukala E (1992) The Pleistocene large mammals from the Agios Georgios Cave, Kilkis, (Macedonia, N. Greece). Geobios 25(3):415–433

Tsoukala E (1999) Quaternary large mammals from the Apidima Caves (Lakonia, S. Peloponnese, Greece). Beit Palaeontol 24:207–229

Tsoukala E, Chatzopoulou K, Rabeder G, Pappa S, Nagel D, Withalm G (2006) Paleontological and stratigraphical research in Loutra Arideas bear cave (Almopia speleopark, Pella, Macedonia, Greece). Aristotle Univ Thessaloniki Sci Ann 98:41–67

Valenciano A, Abella J, Sanisidro O, Hartstone-Rose A, Álvarez-Sierra MA, Morales J (2015) Complete description of the skull and mandible of the giant mustelid *Eomellivora piveteaui* Ozansoy, 1965 (Mammalia, Carnivora, Mustelidae), from Batallones (MN10), late Miocene (Madrid, Spain). J Vert Paleontol 35(4):e934570

Viret MJ (1950) *Meles thorali* n. sp. du loess Villafranchien de Sait-Vallier. Eclogae Geol Helv 43(2):274–287

Wang X, McKenna MC, Dashzeveg D (2005) *Amphicticeps* and *Amphicynodon* (Arctoidea, Carnivora) from Hsanda Gol Formation, Central Mongolia and phylogeny of basal Arctoids with comments on zoogeography. Am Mus Novit 3983:1–57

Weithofer A (1888) Beiträge zur kenntniss der fauna von Pikermi bei Athen. Beit Palaeontol Öster Ungarns 6:225–292

Willemsen GF (1992) A revision of the Pliocene and Quaternary Lutrinae from Europe. Scr Geol 101:1–115

Wolsan M, Semenov YA (1996) A revision of the late Miocene mustelid carnivoran *Eomellivora*. Acta Zool Cracov 39:593–604

Wund M (2018) Mustelidae. In: Animal diversity web. http://animaldiversity.org/accounts/Mustelidae/. Accessed 27 Apr 2018

The Fossil Record of Insular Endemic Mammals from Greece

George A. Lyras, Athanassios Athanassiou, and Alexandra A. E. van der Geer

1 Introduction

The islands of the Aegean Sea (Fig. 1) constitute an extensive archipelago situated between mainland Greece to the West and North and Turkey to the East. During the Miocene, this area was part of the mainland that later became fragmented because of large-scale extensional tectonic movements in combination with eustatic sea-level fluctuations. These fluctuations are a direct result of alternating abrupt climatic shifts particularly occurring since the Middle Pleistocene. As a result, dramatic changes in the paleogeography of the Aegean took place, because of the great differences between the highest and lowest sea-level stands (approximately +10 m and −120 m, respectively, with regard to the present-day sea level; Grant et al. 2014; Spratt and Lisiecki 2016). Thus, during glacial periods, when much of the surface water was captured as ice in the vast glaciers, the sea level remained low, and wide, previously submerged areas were exposed, enlarging existing land masses and narrowing the sea straits. Warm, interglacial climate had the opposite effect, increasing an island's isolation from the mainland. During the Pleistocene, particularly during glacial periods, several Aegean islands were colonized by continental vertebrates.

G. A. Lyras (✉)
Department of Geology and Geoenvironment, National and Kapodistrian University of Athens, Athens, Greece
e-mail: glyras@geol.uoa.gr

A. Athanassiou
Ephorate of Palaeoanthropology–Speleology, Hellenic Ministry of Culture and Sports, Athens, Greece
e-mail: aathanas@geol.uoa.gr

A. A. E. van der Geer
Naturalis Biodiversity Center, Leiden, The Netherlands
e-mail: alexandra.vandergeer@naturalis.nl

E. Vlachos (ed.), *Fossil Vertebrates of Greece Vol. 2*,
https://doi.org/10.1007/978-3-030-68442-6_25

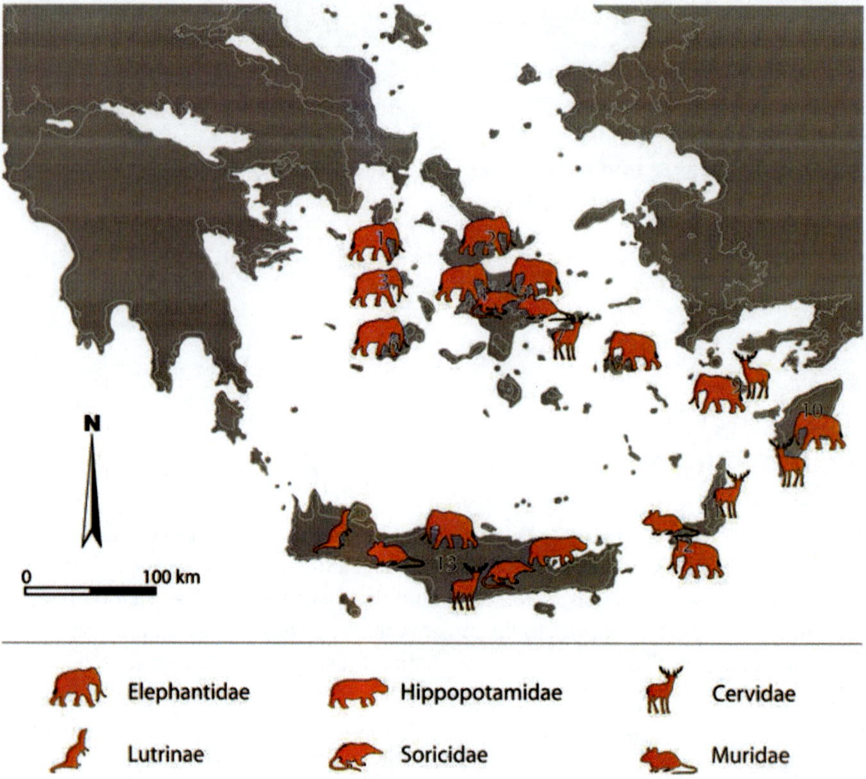

Fig. 1 Map of the southern Aegean Sea indicating the geographic position of the present-day islands with Pleistocene insular mammals discussed in text, and the composition of endemic mammal faunas, which evolved on certain paleo-islands during the Pleistocene: **1** Kýthnos, **2** Delos, **3** Sériphos, **4** Paros, **5** Naxos, **6** Milos, **7** Amorgós, **8** Astypálaea, **9** Tilos, **10** Rhodes, **11** Kárpathos, **12** Kassos, **13** Crete. The gray area represents the approximate extent of the exposed land during a sea-level drop of 100 m (based on maps of the Hellenic Navy Hydrographic Service). The present-day shoreline is indicated by a white line (based on a public-domain topographic map from Wikimedia Commons). Animal figures source: http://phylopic.org/

Island size is directly related with the number of species found on the island, with larger islands tending to contain more species (known as the species–area relationship). Consequently, most Aegean islands, being small, were inhabited by only a limited number of species. Crete, being the largest island of Greece, supported the highest level of mammalian biodiversity of all Aegean islands. A pervasive pattern exhibited across island mammals worldwide is the general trend for gigantism in smaller-bodied species and dwarfism in larger-bodied species, also known as the "island rule" (Lomolino 2005; Lomolino et al. 2012, 2013). This body-size evolution is influenced by ecological release from mammalian competitors or predators, shifts in resource subsidies, immigrant selection, and thermoregulation (Lomolino et al. 2012). Nearly all insular taxa of the Aegean followed this trend of body size evolution. An exception were the deer of Crete and Kárpathos, which were likely the result of sympatric speciation (de Vos and van der Geer 2002).

This chapter presents the endemic mammals that underwent evolutionary changes during the Pleistocene to adapt to the life on an isolated island. Endemic mammals are restricted to the Aegean islands. The proximity of the Ionian islands to the mainland did not constitute adequate isolation, and thus recurrent immigrations did not permit the populations to evolve into endemic forms. The same likely holds for Kýthera, south of Peloponnese, although the island's fossil mammals have frequently been characterized as endemic (Athanassiou et al. 2019). Since the insular taxa of the Aegean were endemic to each island (or paleo-island, consisting of two or more present-day islands), they are presented here first geographically arranged and then taxonomically. Taxa of family or higher rank are explained only once, in their first appearance. For information regarding the endemic marginated tortoises of Crete and Tilos, see Vlachos (volume 1); for the mainland taxa on present-day islands, which were too close to or part of the mainland during the geological time these taxa lived (e.g., Miocene of Samos and Crete, Plio-Pleistocene of Rhodes, Pleistocene of Kýthera, etc.), see the respective chapters on each taxon. Transliterated local geographic names are typed accented when necessary, in order to help with their correct pronunciation.

Abbreviations: ka, thousand years before present; MIS, Marine Isotope Stage; M/m, P/p, D/d, upper/lower molars, premolars, deciduous teeth, respectively; AMPG, Museum of Paleontology and Geology, National and Kapodistrian University of Athens, Greece; NHMUK, Natural History Museum of London, Great Britain; FAUU, Faculty of Geosciences, Utrecht University, the Netherlands; MGGC, Museo Geologico "Giovanni Capellini", Bologna, Italy; MNHW, Museum of Natural History of Vienna, Austria; MPUR, Palaeontological Museum, Department of Earth Sciences, University of Rome "La Sapienza", Italy; NCB, Naturalis Biodiversity Center, Leiden, the Netherlands; NHMH, Natural History Museum of Crete, Heráklion, Greece; RGM, Rijksmuseum voor Geologie en Mineralogie, which although now part of NCB, its abbreviation is still in use for the fossil vertebrate collection.

2 Cyclades

Historical Overview

Fossils of insular mammals have been discovered on Delos, Paros, Naxos, Milos, Sériphos, Kýthnos, and Amorgós. The first fossil mammal of the Cyclades was discovered on Delos. Cayeux (1908) reported the discovery of an isolated upper molar found near the Apollo temple in the deposits of the Inopós River and referred the find to the continental species *Elephas antiquus* (now *Palaeoloxodon antiquus*). Vaufrey (1929) recognized that the specimen belongs to a dwarf form, similar to the

Maltese species *Elephas mnaidriensis*. In a revision of the same specimen, Sen et al. (2014) followed Vaufrey's opinion, but refrained from referring it to a species (*Palaeoloxodon* sp.). Concurrently with Vaufrey, Georgalas (1929) reported in an encyclopedia entry the presence of dwarf elephant fossils on Paros, unfortunately without any further information about the finds, except for a vague geographical indication of the locality, "close to Antiképhalos hill." Much later, a third elephant fossil was discovered on the adjacent island of Naxos, in the Trypití Ravine, by a villager and presented to a school teacher of Apeíranthos. The latter brought the specimen, an upper jaw of an adult dwarf elephant bearing both third molars, to the AMPG, where it was subsequently described by Mitzopoulos (1961) as *Palaeoloxodon antiquus melitensis*. The same specimen was re-examined by van der Geer et al. (2014a), who attributed it to a new species of endemic elephant, *P. lomolinoi*. Papp (1953) reported the presence of dwarf elephant fossils on Milos and Sériphos, without giving any further information on the finds and the find spots. Kuss (1973) recorded the presence of deer and small mammal fossils on Amorgós. He suggested that the deer resembled the Cretan deer, without, however, providing a description. Honea (1975) reported a partial elephant tusk from the north-west area of the island Kýthnos (the exact location is not defined). According to the author, the find was at the base of a calc-tufa cemented breccia, associated with unidentified fossil bones, charcoal, and lithic artifacts, and its age was estimated to much in excess of 9 ka.

In addition to the finds mentioned above, there is an ambiguous report by Paraskevaidis (1956), who described horse molars as *Equus caballus*, which originated from tuffs on Milos. Dermitzakis and Sondaar (1979) suggested that the molars might be subrecent.

The sporadic fossils from the Cyclades have been included in several revisions, such as those by Sondaar (1971), Sondaar and Boekschoten (1967), Dermitzakis and Sondaar (1979), Kotsakis et al. (1980), Sen (2017), and Athanassiou et al. (2019). The elephant material from Naxos is curated at AMPG, that from Delos at the local Archaeological Museum, while the small mammals from Naxos are stored at FAUU. At present, there are no data on the whereabouts of the rest of the findings.

Paleogeography and Endemism

The Cycladic islands are located in the central Aegean Sea. They are formed on a partly submerged plateau, the Cyclades Plateau, which is characterized by maximum water depth of 250 m and by numerous outcropping islands (Le Pichon and Angelier 1981; Kapsimalis et al. 2009). This plateau formed a large paleo-island during cold stages of the Pleistocene (Fig. 1), whose size and geography was greatly influenced by the changes of the sea level. During the Last Glacial Maximum, a single large island with extensive flat areas was subaerially exposed extending to an area of approximately 7600 km^2 (Kapsimalis et al. 2009). This mega-island extended over several present-day islands of the Eastern Cyclades, including Naxos, Paros, and Delos. However, not all islands of the Eastern Cyclades were included. Amorgós, Anáphi, and several smaller islands remained separate, as well as the closely located island of Astypálaea, which geographically belongs to the Dodecanese. During the Late Pleistocene, even during the lowest sea stands of the Last Glacial Maximum, the islands of the Western Cyclades, including Milos, Sériphos, and Ký́thnos, were separated from the mega-island by a marine strait, approximately 14 km wide (Kapsimalis et al. 2009).

The paleogeography of the Cyclades during the Middle Pleistocene is less clear. According to Lykousis (2009), during the Middle Pleistocene glacial periods, the Cyclades Plateau formed land connections with the mainland. According to this scheme, during the major low sea-level periods of MIS 12, 10, and 8, a continuous, elongated landmass was exposed in this area, which connected central Greece with Anatolia. Since MIS 6, the Cyclades Plateau was separated from Anatolia, but, during low sea-level periods, it was connected to mainland Greece. Following the paleogeographic reconstructions of Lykousis (2009), the paleo-islands of the Cyclades were separated from the mainland only during the Late Pleistocene. However, this model (Lykousis, 2009) is in conflict with the view that the seaways between Euboea–Andros and Sounion–Kea were already open during the Early Pleistocene (Anastasakis et al. 2006). This implies a significantly more isolated condition of the Cycladic Plateau throughout the Pleistocene. Although the Pleistocene mammal-bearing localities of Cyclades are very few, and with poor faunal content, their endemic character is pronounced. This is clear from the exclusive presence of elephants or deer, both large mammals with good swimming capabilities. Apparently, the sea strait between the mainland (Attica, Euboea) and the Cycladic paleo-islands was broad enough to act as an insurmountable barrier for most mammals. It should be noted, however, that the Naxos rock mouse hardly differs from the continental species. On the other hand, the available finds are inadequate for drawing well-supported conclusions regarding its endemicity.

Systematic Paleontology

Eulipotyphla Waddel, Okada and Hasegawa, 1999

Soricidae Fischer von Waldheim, 1817

⊙*Crocidura* Wagler, 1832

Distribution Found on Naxos, near the Trypití Ravine.

Remarks Van der Geer et al. (2014a) depicted a left mandible and a left maxillary of *Crocidura* sp., but did not provide any description.

Rodentia Bowdich, 1821

Muridae Illiger, 1811

⊙*Apodemus* Kaup, 1829

Type Species *Mus agrarius* Pallas, 1771.

⊙*Apodemus* cf. *mystacinus* (Danford and Alston, 1877)

Type Material MCZ 14887, skin and skull of a male.

Type Locality Zebil, Bulgar Dagh, Adana Province, Turkey.

Taxonomic and Nomenclatural History *Apodemus* sp. in Sondaar (1971). This identification was cited in later publications, such as Dermitzakis and Sondaar (1979) and Kotsakis et al. (1980). Recently, van der Geer et al. (2014a) referred the sample to *Apodemus* cf. *mystacinus*.

Distribution Found on Naxos, near the Trypití Ravine.

Remarks The presence of *Apodemus* on Naxos was first noted by Sondaar (1971). The available specimens were figured and described by van der Geer et al. (2014a). The size of the Naxos *Apodemus* agrees well with the measurements of *A. mystacinus*, which, according to Masseti (2012), still occurs on Naxos.

Proboscidea Illiger, 1811

Elephantidae Gray, 1821

Type Genus *Elephas* Linnaeus, 1758.

Remarks Two of the elephantid occurrences in Cyclades, those from Delos and Naxos, are referred to the genus *Palaeoloxodon*. Additional fossil elephant remains have been reported from four other Cycladic islands (Sériphos, Milos, Paros, and Kýthnos), but their taxonomic affinities are unknown, because they were not figured or described, they lack any provenance data, and their current whereabouts remain unknown. In the following section, we present the elephant finds from Delos and Naxos, which have been described in the literature.

***Palaeoloxodon* Matsumoto, 1924**

Type Species *Elephas namadicus naumanni* Makiyama, 1924.

Included Taxa The genus is known from two single finds, a molar from Delos referred to *Palaeoloxodon* sp. and a maxilla from Naxos referred to *P. lomolinoi*.

Remarks A very common elephant genus in the Middle and Late Pleistocene of Europe, represented by the species *P. antiquus* (Falconer and Cautley, 1847), which is considered as the parent taxon of most endemic proboscidean species of the Mediterranean islands. The two occurrences of the genus in Cyclades may belong to different species, despite the fact that Delos and Naxos were parts of the same mega-island during most of the Pleistocene. Athanassiou et al. (2019) postulated that the two finds belonged to geochronologically distant populations.

Palaeoloxodon sp. (from Delos)

Remarks The elephant molar from Delos was initially identified as a D4 or M1 of the continental species *E. antiquus* (see Cayeux 1908). Vaufrey (1929) corrected its anatomical position to M3, realizing that the find belongs to an endemic elephant. The same author provided a drawing of the occlusal surface. Recently, Sen et al. (2014) described in detail and figured a cast of the specimen, estimating that it is about 60% the size of an M3 of the continental *P. antiquus* in linear dimensions. This places the Delos elephant dimensionally between *P. tiliensis* from Tilos and *P. creutzburgi* from Crete (see Sects. 4 and 7), possibly close to *P.* aff. *creutzburgi* from Kassos (Sen et al. 2014; Sect. 6).

★*Palaeoloxodon lomolinoi* van der Geer et al. 2014

Taxonomic and Nomenclatural History The Naxos find was initially referred to *Palaeoloxodon antiquus melitensis* (Falconer, 1862), a name assigned to dwarf elephant material from Malta (Mitzopoulos 1961). Sondaar (1971), who apparently disagreed with such a name sharing between unrelated populations (see also Sondaar and Boekschoten 1967), referred to it as an unnamed dwarf elephant. Kotsakis et al. (1980) attributed it to the size-group of *P. mnaidriensis*, despite the much larger size of the latter species. More accurately, Caloi et al. (1996) regarded it as an *E. antiquus*-derived elephant similar in size to *E. melitensis*. Alcover et al. (1998) cited it as *Elephas* unnamed species B. Palombo (2001) repeated the identification as an elephant probably belonging to the palaeoloxodontine line, similar in size to *E. melitensis*, while later, she (Palombo 2004) regarded it as an endemic species smaller than *E. mnaidriensis*. In the first edition of their book on the evolution of island mammals, van der Geer et al. (2010) considered it as undescribed species and referred to it as *Elephas* sp. Finally, van der Geer et al. (2014a) described it as a new species of the genus *Palaeoloxodon*, *P. lomolinoi*.

Type Material AMPG(V)999 (holotype), an isolated maxilla preserving both third molars (M3) (Fig. 2; van der Geer et al. 2010, pl. 6e; 2014a, fig. 5).

Type Locality Trypití Ravine, Naxos, Greece; probably Late Pleistocene.

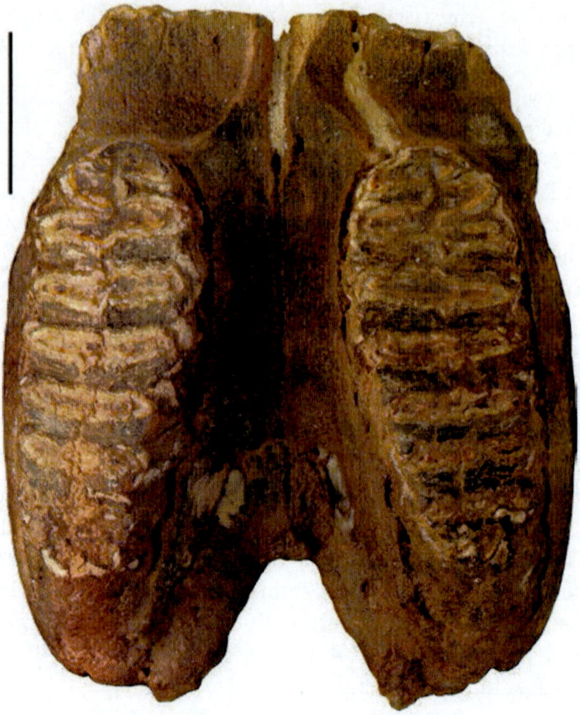

Fig. 2 The holotype AMPG(V)999 of *Palaeoloxodon lomolinoi* van der Geer et al., 2014; ventral (occlusal) view (rostral end is at the top). The graphical scale equals 5 cm

Remarks The Naxos elephant had a body mass of approximately 8% of the mass of its mainland ancestor (van der Geer et al. 2014a). It thus had a similar body size as the dwarf elephants from Tilos (*P. tiliensis*) and Luparello, Sicily (*P. "melitensis"*). Based on this degree of dwarfism, the Naxos elephant was originally attributed to *P. melitensis* (see Mitzopoulos 1961). However, size alone is not sufficient to distinguish the insular elephants, as it is the result of parallel evolution, not common ancestry. According to the current view on the taxonomy of insular dwarf elephants, every paleo-island harbored its own endemic species as a result of a distinct colonization event and subsequent evolutionary process (Athanassiou et al. 2019).

Artiodactyla Owen, 1848

Cervidae Goldfuß, 1820

Type Genus *Cervus* Linnaeus, 1758.

Remarks Kuss (1973) reported the presence of fossil deer on Amorgós. According to him they resemble the Cretan Pleistocene deer. Kuss provided no description of the material, nor any locality data, making it impossible to evaluate his determination. Furthermore, the whereabouts of these fossils are unknown. The deer findings on Amorgós may belong to a species of deer that was endemic to Cyclades, or even just to Amorgós alone, which always constituted a separate island (Fig. 1). Its similarity to the Cretan deer could be the result of parallel evolution, as in the case of the dwarf elephants.

3 Astypálaea

Systematic Paleontology

Palaeoloxodon Matsumoto, 1924

Remarks In the late 1990s, a partial elephant tusk was discovered during construction works at Sýgkairo, at the center of Astypálaea, a Dodecanesian island situated east of Cyclades. The tusk was excavated by a team of the Ephorate of Palaeoanthropology–Speleology (Ministry of Culture). It was recently described and figured by Athanassiou et al. (2019), who referred it to *Palaeoloxodon* sp. This taxonomic identification was based on the weak curvature and torsion of the specimen, which is consistent with the morphology of this genus. However, the find offers no clues about the size of this elephant, because tusk size is known to vary greatly among individuals of the same population.

4 Tilos

Tilos is a small island of the Dodecanese (SE Aegean), characterized by rugged relief. It hosted two successive endemic faunas, an older dominated by deer and a more recent dominated by elephants.

Historical Overview

The first fossil mammals on Tilos were discovered in 1971 by Prof. Nikolaos Symeonidis, in the deposits of Charkadió Cave. Symeonidis soon carried out the first excavation in the cave (April 1972, on behalf of the University of Athens), recovering dental as well as postcranial remains of dwarf elephants (Fig. 3). In the following years, several research expeditions were made by teams of the University of Athens, often in collaboration with the Natural History Museum of Vienna. The systematic excavations in the cave yielded several thousands of fossils, mainly of

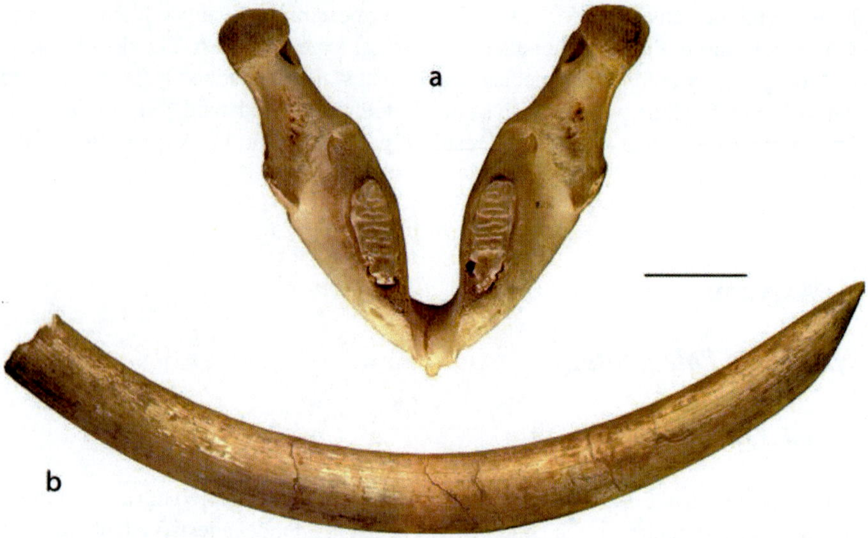

Fig. 3 *Palaeoloxodon tiliensis* Theodorou, Symeonidis and Stathopoulou, 2007; (**a**) the mandible AMPG-1907 in occlusal view; (**b**) the tusk AMPG-V-1914 in lateral view. The graphical scale equals 10 cm

elephants, but also of deer in the deeper layers, as well as of tortoises (Symeonidis 1973; Symeonidis et al. 1973; Bachmayer and Symeonidis 1975b; Bachmayer et al. 1976, 1984; Theodorou 1983; Theodorou and Symeonidis 1994, 2001; Theodorou et al. 2007; Mitsopoulou et al. 2015). The material from Tilos is curated at AMPG and MNHW. The paleontological research in the cave is still in progress.

Paleogeography and Endemism

Tílos is located close to the Anatolian mainland. Its present-day minimal distance from it is 21 km. Given the steep coasts on both sides of the sea strait and its considerable depth (over 400 m), this distance likely remained about the same (>20 km) even during the Last Glacial Maximum. Another possible migration route to the island is that via the volcanic islands Nísyros and Gyalí. Nísyros is located 11 km to the north-west of Tilos (about 9 km during glacial periods). This route would require an additional 5 km crossing from Kos Island (a peninsula of Asia Minor during low sea-level stands) to Gyalí and Nísyros. A third option would be a migration from Rhodes, from where an endemic mammal fauna is also known (see Sect. 5). This option requires a swimming distance during low sea-level stands of about 30 km, or about 15 km via Chálki. Contrary to the small changes in the width of the sea straits between Tilos and the mainland, the island's area (now a mere 61.5 km^2) was considerably larger during low sea-level stands. During the Late Pleistocene, the surface area was up to 30% larger than the present one, based on the 125-m isobath (van der Geer et al. 2017). As noted by Theodorou (1988), the marine regressions caused wide lowland areas to emerge, which made the island's terrain more hospitable to migrant populations of large herbivores. According to the stratigraphic data collected at Charkadió (Bachmayer et al. 1976, 1984), Tilos witnessed two migration events, one by deer at about 140 ka and a later one by elephants at about 50 ka. The deer became extinct long before the arrival of the elephants, and their existence on the island seems to have been short (judging from their stratigraphic representation in the Charkadió sequence). Based on the current knowledge, it is hypothesized that these two distinct migration events occurred during a cold stage, when the larger area of the island offered better chances for survival.

Systematic Paleontology

★*Palaeoloxodon tiliensis* (Theodorou, Symeonidis and Stathopoulou, 2007)

Taxonomic and Nomenclatural History The elephant sample from Charkadió was initially (Symeonidis 1973) referred to two distinct subspecies, originally defined on material from Malta, the smaller *Palaeoloxodon antiquus falconeri* (Busk, 1867) and the larger *P. antiquus melitensis* (Falconer, 1862). In the same year, the identification was changed in part, by referring the material to *P. a. falconeri* and *P. a. mnaidriensis* Adams, 1874 (Symeonidis et al. 1973). This taxonomic scheme, based on typological identification, was followed for the next decade, considering that the larger-bodied taxon (*P. a. mnaidriensis*) was stratigraphically older (Bachmayer et al. 1976). Kotsakis et al. (1980) were the first to recognize the distinctive character of the Tilos elephant sample and referred to it as *Palaeoloxodon* sp., followed later by Caloi and Palombo (1994) and Palombo (2001). The subsequent recovery of large-sized fossils close to the cave's surface, as well as the statistical study of the whole excavated sample, led Theodorou (1983, 1984) to challenge the two-taxa model and correctly attribute the observed differences to sexual dimorphism, which is known to be substantial in elephant species. Theodorou (1983, 1984) also recognized the taxonomic uniqueness of the Tilos sample, but refrained from coining a new name for it, referring to it provisionally as *P. a. falconeri*. A new species was erected for the Tilos dwarf elephant much later, classified in the extant genus *Elephas* (*E. tiliensis*; Theodorou et al. 2007).

Type Material The species was defined on a large number of syntypes (2 dental and 104 postcranial specimens), curated in AMPG. The hypodigm consists of more than 15,000 specimens curated in AMPG and MNHW, which belong to at least 45 individuals (Theodorou et al. 2007).

Type Locality Charkadió Cave, Tilos Island; Late Pleistocene—early Holocene.

Distribution Its geochronologic distribution extends through the latest Pleistocene and Holocene (~50–4 ka), based on ^{14}C and U/Th dates of 45–4.4 ka (Bachmayer et al. 1976, 1984). An increased fossil density, which may imply a population peak, is observed at 18–17 ka (Theodorou et al. 2007).

Remarks According to Theodorou et al. (2007), *P. tiliensis* probably had a maximal withers height of 1.8–1.9 m. Lomolino et al. (2013) estimate its body mass to approximately 9% of the mass of its mainland ancestor, *P. antiquus*.

The causes for the extinction of *P. tiliensis* are unknown. Theodorou (1988) discussed the potential roles of the diminishing living area due to the rapid sea level rise at the very end of the Pleistocene and the early Holocene, active volcanism, and the human newcomers. Extinction through human agency, based on the presence of lithic artifacts and broken tusks believed to be knapped (Bachmayer and Symeonidis 1975a), is the least supported. The artifacts were found only in disturbed surface layers (Bachmayer et al. 1976), while the alleged tusk tools morphologically resemble naturally broken tusk tips (compare, e.g., Haynes 1988, fig. 15; Haynes and Klimowicz 2015, p. 136, fig. 1).

Cervidae Goldfuß, 1820

Remarks Bachmayer et al. (1976) reported on the discovery of deer fossils at the lower levels of the excavation at Charkadió, within layers that did not include elephants. No antler specimens were recovered, hampering the taxonomic identification. Bachmayer et al. (1976) recognized two size groups and referred to these as Cervidae I and II, respectively, acknowledging the possibility of significant sexual dimorphism within a single species. A deer bone was U/Th-dated at 140 ± 11 ka (Bachmayer et al. 1984). Apart from Charkadió Cave, deer fossils have also been discovered at an open-air site elsewhere on the island (dubbed "Tilos 2" in Theodorou et al. 2007), but these have not yet been studied. The deer settlement on Tilos seems to have been brief, probably terminated because of a volcanic eruption of nearby Nísyros (Theodorou 1988).

5 Rhodes

Historical Overview

Citations and descriptions of fossil mammal localities on Rhodes in the older literature (e.g., de Bruijn et al. 1970; Sondaar 1971; Kuss 1973) refer to faunas of a continental, balanced type. (Nevertheless, a deer sample from Kritinía, similar in size to red deer, described by Kuss 1975b, is regarded by Dermitzakis and Sondaar 1979 as endemic.) A discovery of an unambiguously endemic assemblage was first reported by Marinos and Symeonidis (1973), who identified the presence of dwarf elephant remains in the debris of an illegal excavation carried out by treasure hunters in a cave on the hill Erimókastro. The collected finds (mostly postcranial bones) were studied by Symeonidis et al. (1974). No other excavation has been made in the cave since 1973, except for a very brief one in 1993 (by the University of Athens) which yielded an almost complete tusk. An additional indirect indication of the presence of elephants on the island was provided by Milàn et al. (2007), who interpreted certain sedimentary structures in eolianites of Late Pleistocene to early Holocene age as elephant ichnofossils. The elephant material is curated at AMPG.

Paleogeography and Endemism

Rhodes is currently separated from Asia Minor by an 18-km-wide and more than 300-m-deep sea strait. It is estimated that during the glacial periods of the Pleistocene, the swimming distance from the mainland would be about 15–16 km. This is well within the capabilities of elephants, which have been documented on the island, and deer, the presence of which as endemic forms is unconfirmed.

Systematic Paleontology

Palaeoloxodon Matsumoto, 1924

Taxonomic and Nomenclatural History The elephant remains from Erimókastro were preliminarily reported as belonging to the Maltese taxon *P. antiquus mnaidriensis* Adams, 1874 (Marinos and Symeonidis 1973). The same attribution was published the following year by Symeonidis et al. (1974), emphasizing the relatively large size of the fossil remains. Indeed, the sample belongs to a medium-sized endemic species of the genus *Palaeoloxodon*, as evidenced by the weakly curved and untwisted tusk excavated in 1993, but it is unrelated to the similar-sized population from Malta. Following the current taxonomic practice, according to which each insular migratory event is considered evolutionary unique, and thus the resulting endemic population should be named individually, the Rhodian elephant should have its own name. Nevertheless, due to the inadequacy of available material, it is referred to as *Palaeoloxodon* sp. (see also Athanassiou et al. 2019).

6 Kárpathos and Kassos

Historical Overview

The majority of fossil mammal findings from Kassos and Kárpathos belong to deer. The Kárpathos findspot was discovered in 1963 by R. Kinzelbach and others in the extremely consolidated filling of a cave, close to the island's capital. Kuss (1967) and Sondaar and Boekschoten (1967) were the first authors to describe deer fossils from this cave. Later, Sondaar collected murid rodents belonging to the genus *Mus* (personal communication of David Mayhew in Dermitzakis and Sondaar 1979) and additional deer fossils from other caves, while Kuss (1975b) mentioned the presence of a tortoise in association with the deer finds. Shortly after the first discoveries on Kárpathos, Kuss collected the first fossil deer specimens from Kassos (Kuss 1969), associated with a bird bone. Recently, Sen et al. (2014) reported the discovery of a single elephant molar on Kassos, which they referred to as *Palaeoloxodon* aff. *creutzburgi*, assuming a relation to the Cretan elephant.

Paleogeography and Endemism

Kárpathos and Kassos are separated today only by a shallow (maximum 60 m deep) and narrow (6 km wide) sea strait. For the major part of the Middle and Late Pleistocene the two islands were united (Fig. 1) hosting a shared fauna. This is indicated by the deer fossils, which are very similar on the two islands. The provenance of the faunas is a yet unresolved problem. The Kárpathos and Kassos deer may have come from Crete (Kuss 1967, 1969; Sondaar and Boekschoten 1967), in which case they should be included in the Cretan genus *Candiacervus* (see Sect. 7). Another option is that they migrated from Rhodes, and the similarities to the Cretan deer are the result of parallel evolution. Both ways of migration require the crossing of long overseas distances up to 40 km, even in time periods of very low sea level.

Systematic Paleontology

***Palaeoloxodon* aff. *creutzburgi* (Kuss, 1965)**

Remarks There is only a single elephant find on Kassos, an isolated third lower molar (m3), found in a loose isolated block on a mountain slope close to the northwestern coast of the island (Sen et al. 2014). This m3 is morphologically and dimensionally similar to the Cretan *P. creutzburgi*, as reflected in the taxonomic attribution by Sen et al. (2014). The considerable distance between Crete and Kassos makes it, though, unlikely that their elephant populations were conspecific. Probably, these populations represent distinct migratory and evolutionary processes. If so, the Kassos elephant apparently needs to be named as a separate species. As in the case of the Rhodes dwarf elephant, the currently available material is inadequate for this purpose. The fossil molar is curated at the Archaeological Museum of Kassos.

Cervidae Goldfuß, 1820

★"*Candiacervus*" Kuss, 1975

Type Species *Anoglochis cretensis* Simonelli, 1907.

Remarks *Candiacervus* is a single-island taxon that lived during the Pleistocene on Crete (see Sect. 7). Kuss (1975b) applied the same name to the Pleistocene deer taxa of Kárpathos and Kassos. However, these latter deer differ from *Candiacervus* in having more robust metapodials and certain differences in antler morphology, including a rugose surface (van der Geer et al. 2010), thus a phylogenetic connection to the Cretan deer seems to be not well founded. Already Kuss (1969) remarked that their antlers are more complex, and explained this as an evolution toward a higher complexity relative to their Cretan ancestors. Also in the case of Crete, he assumed an evolution from simple, unbranched antlers in older deposits to more complex antlers in younger deposits. Van der Geer (2018) dismissed this scenario, in which case the argument of Kuss (1969) of derivation from a Cretan complex form does not hold anymore. In case of different phylogenetic decent, the taxonomic use of *Candiacervus* should be restricted to the Cretan endemic deer alone (see also Sondaar et al. 1996, who first argued that the Karpathos–Kassos deer should be classified in a separate genus). Since no phylogenetic analysis has yet been carried out, it is not possible to assign the Kárpathos–Kassos material to an existing genus or, alternatively, name a new one. Therefore, we provisionally continue to use the name *Candiacervus* here, listed within quotation marks, pending a future revision of the material.

★"*Candiacervus*" *cerigensis* Kuss, 1975

Taxonomic and Nomenclatural History The deer from Kárpathos were initially referred by Kuss (1967) to the Cretan species *Megaceros cretensis* (Simonelli 1907), considering it particularly similar to the sample from the locality Gridá Avláki. Sondaar (1971) followed this use and referred the Kárpathos sample to *Megaloceros cretensis*. Kuss (1969) also described deer specimens from Kassos and referred them to *Cervus cretensis*. In his review of the Pleistocene insular mammals of the Eastern Mediterranean, Kuss (1973) listed the Kárpathos and Kassos deer as Cervidae ex aff. *Cervus cretensis*. In 1975, the same author (Kuss 1975b) coined two new species names for this material, both classified in his newly erected Cretan genus *Candiacervus*; *C. cerigensis* and *C. pigadiensis*. Both are present, according to the author, on both Kassos and Kárpathos. The attribution of the material to two species was soon questioned by Dermitzakis and Sondaar (1979), who regarded Kuss' relevant arguments as inadequate. Petronio (1990) expressed a similar opinion, considering the environment of the Karpathos–Kassos paleo-island as insufficient to support two similar cervid species. On the contrary, de Vos (2006) followed Kuss' taxonomy, even adding that the Kassos deer might constitute a third, still unnamed, distinct species.

Type Material The holotype is a left five-tined antler, figured in Kuss (1975b, pl. III, fig. f). The hypodigm is curated at NHMH and NCB.

Type Locality Kandília Cave, on the East coast of Kárpathos, east of Pigádia.

Distribution Pleistocene of Kárpathos and Kassos.

Remarks This is a small deer species with a short skull, and very short and robust metapodials. The antlers are complex, five tined in adult individuals.

★*"Candiacervus" pigadiensis* Kuss, 1975

Taxonomic and Nomenclatural History As for *"C." cerigensis*.

Type Material The holotype is a right tibia, figured in Kuss (1975b, pl. IV, fig. 1). The hypodigm is curated at NHMH and NCB.

Type Locality Kandília Cave, on the East coast of Kárpathos, east of the island's capital town Pigádia.

Distribution Pleistocene of Kárpathos and Kassos.

Remarks About 20% larger than the mean *"C." cerigensis* (based on the holotype tibia, the only postcranial specimen referred by Kuss 1975b to this species), with antlers bearing at most four tines, very similar in ontogenetic development and morphology to those of *"C." cerigensis*. The antlers of *"C." pigadensis* may form a three-tined crone at their distal end.

7 Crete

Crete is by far the largest island in the Aegean, characterized by a rugged relief with high mountains and steep ravines. During the Pleistocene, the island was inhabited by two successive endemic faunas. Over 100 localities with endemic mammals have been discovered on it (Fig. 4; Lax 1996; Iliopoulos et al. 2010).

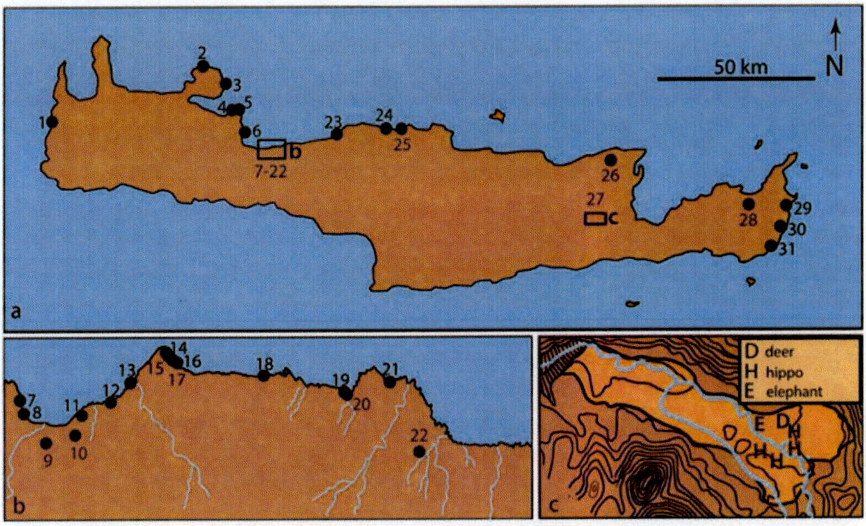

Fig. 4 The most important Pleistocene localities of Crete. (**a**) Map of Crete. (**b**) Map of Réthymnon area. (**c**) Map of Katharó Plateau. Localities: **1** Sphinári, **2** Stavrós localities, **3** Cape Maléka localities, **4** Koutalás Cave, **5** Vámos Cave, **6** Liko Cave, **7** Geráni Cave 2, **8** Geráni Cave 4, **9** Geráni Cave 3, **10** Geráni Cave 5, **11** Geráni Cave 6, **12** Bate Cave, **13** Zourída Cave, **14** Mávro Mourí 1, **15** Mávro Mourí 2, **16** Mávro Mourí 3, **17** Mávro Mourí 4, **18** Simonelli Cave, **19** Koumbés II = Gridá Avláki, **20** Koumbés B, **21** Koumbés III, **22** Réthymnon fissure, **23** Skaléta, **24** Balí localities, **25** Kaló Choráfi, **26** Mílatos caves, **27** Katharó localities, **28** Sitía 1, **29** Karoúmbes localities, **30** Káto Zákros, **31** Xeros. Several localities appear in the literature with alternative spellings. For example, Sourida instead of Zourída, Charoumbes or Kharoumbes instead of Karoúmbes, Mawro Muri and Mavromouri instead of Mávro Mourí, Koumpes and Gumbes instead of Koumbés, and Melekas instead of Maléka. All of them are different transliterations to the Roman alphabet by the various authors that worked on Crete. See Lax (1996) for a complete list of alternative spellings

Historical Overview

The first who reported fossil vertebrates from Crete was the British traveler Richard Pococke, who in 1745 described a fossiliferous cave on a peninsula of Chania. In the mid nineteenth century, the British navy officer and geologist Thomas Abel Spratt explored Crete (Spratt 1865). He discovered several fossiliferous deposits and sent some fossils to Hugh Falconer of the British Natural History Museum in London for identification. Falconer thought these were the remains of a goat, a roebuck, or a stag, as well as of a small mouse *Myoxus*, but did not publish his findings (van der Geer et al. 2010). In 1869, the French geologist and botanist Félix Victor Raulin visited Crete and also reported the presence of mammalian fossils. In 1893, the Italian geologist Vittorio Simonelli discovered fossils of deer and elephants in coastal caves near Réthymnon. Simonelli (1894, 1907) described the elephant fossils as the mainland species *Elephas priscus* and recognized the unique character of the deer, which he described as *Anoglochis cretensis* (in Simonelli 1907). The British paleontologist and archaeozoologist, Dorothea Bate, explored Crete a few years later in 1904 (for her travel accounts, see Bate 1905). In the years thereafter, she published several descriptions of the material she had collected: the pygmy mammoth as *Elephas creticus* (Bate 1907; now *Mammuthus creticus*), the Cretan rats *Mus catreus* (Bate 1912; now *Kritimys catreus*) and *Rattus kiridus* (Bate 1942; now *Kritimys kiridus*), and the Cretan mouse, *Mus minotaurus* (Bate 1942). In 1966, Boekschoten and Sondaar named the Cretan dwarf hippopotamus as *Hippopotamus creutzburgi*, mainly based on the extensive material collected five decades before by Bate.

The search for fossil vertebrates on Crete peaked during the late 1960s and throughout 1970s. During that period, three main parties explored and systematically excavated sites on Crete. The first was led by S. Kuss of the University of Freiburg, Germany, the second by A. Malatesta of Sapienza University of Rome, Italy, and the third by P. Sondaar of the University of Utrecht, the Netherlands. These teams excavated numerous sites on Crete, together producing an extensive amount of fossils, which are now curated at NHMH, MPUR, FAUU, NCB, and AMPG, respectively. In the following years, several researchers used this material to answer various questions about the Cretan endemic mammals. The biostratigraphy of Crete was established by Mayhew, who recognized two main biozones, consisting of three and two subzones, respectively, based on the occurrence of fossil murids (Mayhew 1977). Much research has been done on the taxonomy and phylogeny of the Cretan deer, for which Kuss established the endemic genus *Candiacervus*, to which he added a new species, *C. rethymnensis*, in the same contribution (Kuss 1975b). In the years thereafter, six more species were described: *C. ropalophorus*, *C. major*, *C. dorothensis*, *C. devosi*, *C. listeri*, and *C. reumeri* (de Vos 1984; Capasso Barbato and Petronio 1986; Capasso Barbato 1992; van der Geer 2018). In total, six size groups and eight nominal deer species are recognized on Crete, making *Candiacervus* one of the most interesting taxa of Mediterranean endemics regarding evolutionary radiations (de Vos 1996, 2000). Other taxa, such as the hippopotamus, the otter, and the proboscideans were studied in detail as well (e.g., Kuss 1975a; Symeonidis and Sondaar 1976; Spaan 1996; Poulakakis et al. 2002; Herridge and Lister 2012).

Apart from biostratigraphy and taxonomy, the fossil mammal material collected in previous excavations formed the basis for works covering various aspects of their paleobiology, including their paleopathology (Dermitzakis et al. 2006; Palombo and Zedda 2016; Lyras et al. 2016, 2019), functional morphology (e.g., Palombo and Petronio 1989; Caloi and Palombo 1996; Spaan 1996), paleoecology (e.g., van der Geer et al. 2006b, 2013, 2014b; van den Hoek Ostende et al. 2017), paleoneurology (Angelelli 1980), and genomics (Poulakakis et al. 2006).

Paleogeography and Endemism

Crete emerged *de novo* from the sea after the Early Pleistocene marine regression and has remained since then isolated from the mainland, receiving its species by immigration. Only a few mammalian species, however, managed to cross the sea and successfully colonize Crete. Isolated from the rest of the world, they evolved into new species, unique to Crete. Therefore, although Crete was poor in species composition (low high-order diversity), it had a high level of endemism (100%). Two key geographical features that influence the nature of the island biotas are geographic isolation and island size (Lomolino 2000). In the case of Pleistocene Crete, both features were strongly influenced by sea-level changes. For example, the shortest distance between western Crete and the mainland today is about 90 km but at low sea levels only the strait of Antikythera (SE of Peloponnesus) remains considerably broad, being about 20 km wide. The surface area of Crete today is 8336 km^2, but increases by 20% when the sea level drops 120 m, as it did during the Last Glacial Maximum (van der Geer et al. 2017). Island size is directly related with the number of species found on the island, with larger islands tending to contain more species (known as the species–area relationship). Crete follows this general pattern and supported the highest level of mammalian biodiversity of all Greek islands, being at the same time the largest island of the Aegean.

Based on the stratigraphic occurrence of two different lineages of endemic murids, two main Pleistocene biozones can be discerned on Crete: the *Kritimys* biozone (Early–early Middle Pleistocene), and the, younger, *Mus* biozone (late Middle–Late Pleistocene) (Dermitzakis and de Vos 1987; Mayhew 1996). Both biozones are characterized by typical island faunas lacking large terrestrial predators and consisting only of taxa capable of long-distance overseas dispersal (Sondaar 1977). The oldest fauna consisted exclusively of a lineage of giant mice (*Kritimys catreus* and the younger chronospecies *K. kiridus*), a dwarf mammoth (*Mammuthus creticus*), and a dwarf hippopotamus (*Hippopotamus creutzburgi*). The younger biozone consisted of a lineage of giant mice (*Mus bateae* and the younger chronospecies *M. minotaurus*), a dwarf elephant (*Palaeoloxodon creutzburgi*), eight species of deer (*Candiacervus* spp.), an otter (*Lutrogale cretensis*), and the still living Cretan shrew (*Crocidura zimmermanni*).

Most Cretan species follow the "island rule." For example, the Cretan dwarf mammoth (*M. creticus*), weighing approximately 300 kg, had evolved a body mass reduction to a mere 4% of that of its mainland relative, *M. meridionalis* (see Lomolino et al. 2013). In contrast, the Cretan giant rat (*K. catreus*), weighing just over 0.5 kg, was 6.7 times larger than its mainland relative, *Praomys* (see van den Hoek Ostende et al. 2017).

However, not all Cretan mammals evolved toward either a smaller or a larger form. The Cretan deer, represented by no less than eight species, seem to have embarked on a different evolutionary path. Their shoulder heights vary between 40 cm for the smallest species (*C. ropalophorus*) and 165 cm for the largest species (*C. major*). Most of these species were contemporaneous, and likely the result of sympatric speciation followed by an adaptive radiation to occupy all possible empty herbivore niches ranging from dense forest to jagged rocks (de Vos and van der Geer 2002). Today, all Pleistocene endemic species of Crete are extinct, except for the Cretan shrew, currently listed as vulnerable by the IUCN Red List of Threatened Species. All other mammals living today on Crete derived from Holocene colonizers or anthropogenic introductions, such as the Cretan feral goat, locally known as kri-kri or agrími (*Capra aegagrus cretica*) and the Cretan spiny mouse (*Acomys minous*) (Kouvari and van der Geer 2018).

Systematic Paleontology

Soricidae Fischer von Waldheim, 1817

★⊙*Crocidura zimmermanni* Wettstein, 1953

Taxonomic and Nomenclatural History The presence of fossil shrews was initially noticed by Bate (1905). Reumer (1986) demonstrated that the Pleistocene shrew and the living *Crocidura zimmermanni* are one and the same species. The living form was initially recognized as a subspecies of the greater white-toothed shrew, *Crocidura russula zimmermanni* in Wettstein (1953), but was later elevated to species level (*Crocidura zimmermanni* in Vesmanis and Kahmann 1978).

Type Material NMHW-5510 (holotype), a female skull with mandible (extant specimen).

Type Locality Nida Plateau, Ida Mountain, Crete.

Distribution Besides its type locality and mountainous highlands of Crete, this species also occurs in Pleistocene deposits at Mílatos 2, Stavrós micro, Stavrós Cave, Réthymnon fissure, and Liko Cave.

Remarks The shrew *Crocidura zimmermanni* exhibited only minor changes in body size through time (van der Geer et al. 2013).

Muridae Illiger, 1811

★*Kritimys* Kuss and Misonne, 1968

Type Species *Mus catreus* Bate, 1912.

★*Kritimys catreus* (Bate, 1912)

Taxonomic and Nomenclatural History The species was originally named *Mus catreus*. Kuss and Misonne (1968) transferred it to the new genus *Kritimys*.

Type Material NHMUK PV M10472 (holotype), a right mandible with m1–m2, figured in Bate (1912, fig. 1–1a).

Type Locality Sphinári, Chaniá; probably Early–Middle Pleistocene.

Distribution Besides its type locality, this species also occurs in several other sites of Crete such as Stavrós, Cape Maléka 2, Balí 2, Mílatos I, Mílatos III, Karoúmbes A, Karoúmbes 4, and Xeros.

Remarks *Kritimys catreus* presents a clear trend toward gigantism. However, its gradual increase in body size shows fluctuations. Furthermore, after the genus *Mus* colonized Crete, *Kritimys* exhibited a significant decrease in size just before its disappearance from the fossil record (van der Geer et al. 2013).

★*Kritimys kiridus* (Bate, 1942)

Taxonomic and Nomenclatural History Bate (1912) described an upper jaw from Cape Maléka 1 as belonging to a large rat (*Epimys*). Much later, she created a new species for it (*Rattus kiridus* in Bate 1942). The species was transferred to the genus *Kritimys* by Kuss and Misonne (1968).

Type Material NHMUK PV M10555 (holotype) maxilla with M1–M3, figured in Bate (1942, fig. 1a, b).

Type Locality Cape Maléka 1, Chaniá; probably Early–Middle Pleistocene.

Distribution Besides its type locality, this species also occurs in Cape Maléka 3, and as *Kritimys* aff. *kiridus* in Sitía I.

Remarks There are two chronospecies of *Kritimys*: the older *K. kiridus* and the younger *K. catreus*, both showing a clear tendency toward gigantism, with the largest reaching a body mass of about half a kilogram (van den Hoek Ostende et al. 2017). The earliest record of *Kritimys* is the material of *Kritimys* aff. *kiridus* from Sitía 1 (Mayhew 1977). According to Mayhew (1977), the size and morphology of a single murid m2 from Rhodes is similar to that of *Kritimys* aff. *kiridus* from Sitía.

Mus Linnaeus, 1758

Type Species *Mus musculus* Linnaeus, 1758.

Remarks There are two chronospecies of Cretan *Mus*: the older *M. bateae* and the younger *M. minotaurus*, which may be part of a single lineage derived from an ancestral population resembling *M. musculus* (Mayhew 1977, 1996). During the evolutionary history of the Cretan *Mus*, there is an overall tendency toward increase in body size.

★*Mus minotaurus* Bate, 1942

Taxonomic and Nomenclatural History Initially Bate (1912) considered the material she collected as belonging to a spiny mouse (*Acomys*). Later she assigned it to a new mouse species (*Mus minotaurus* in Bate 1942).

Type Material NHMUK PV M15950 (holotype), maxilla with M1–M3, figured in Bate (1942, fig. 2a, b).

Type Locality "Cave deposit in a small limestone ridge between Canea and Suda," Chaniá. Probably Late Pleistocene to Holocene.

Distribution Besides its type locality, this species also occurs in several other sites of Crete such as Liko Cave, Stavrós micro, Mílatos 2, upper layer of Mílatos III, Mílatos IV, Skaléta, Simonelli Cave, Koumbés B, Réthymnon fissure, Geráni Cave 2, Zourída, and Mávro Mourí 4.

Remarks This is one of the larger species in the genus. The molars from the various localities show a variation in size (Mayhew 1977). These differences correspond to temporal changes in body size and may match the climate fluctuations, where the largest specimens occurred during the intermediate cold and dry phases (van der Geer et al., 2013).

★*Mus bateae* Mayhew, 1977

Taxonomic and Nomenclatural History A second endemic mouse species was described from Stavrós Cave (*Mus bateae* Mayhew, 1977).

Type Material FAUU no 266 (holotype), upper right M1, figured by Mayhew (1977, fig. 7).

Type Locality Stavrós micro, Akrotiri, Chaniá; probably late Middle Pleistocene.

Distribution Known only from its type locality.

Remarks The mouse *Mus bateae* is the oldest chronospecies of Cretan mouse. It occurs only at the Stavrós locality together with the last *Kritimys* species. The species is of medium size (intermediate between *Mus musculus* and *Mus minotaurus*), its maxillae and mandibles have sizes similar to those of the extant *M. musculus*, but it shows longer and wider molars with strongly developed cusps. *Mus bateae* eventually evolved into the large *Mus minotaurus*.

Apodemus Kaup, 1829

Remarks A single lower jaw of an *Apodemus* was collected in Sitía I together with *Kritimys* aff. *kiridus* teeth (Mayhew 1977). Nevertheless, according to the same author "*the recognition of this genus as part of the Pleistocene fauna of Crete should await confirmation.*"

Gliridae Muirhead in Brewster, 1819

Remarks Dormice are represented on Crete by a single molar of uncertain genus-level affinities (Mayhew 1977), coming from Stavrós micro, Akrotíri, Crete. It may represent the evidence of a short-lived establishment or it might be the result of a sampling error (Mayhew, personal communication).

Carnivora Bowdich, 1821

Mustelidae Fischer von Waldheim, 1817

Lutrogale Gray, 1865

Type Species *Lutra perspicillata* Geoffroy Saint Hilaire, 1826.

★*Lutrogale cretensis* (Symeonidis and Sondaar, 1976)

Taxonomic and Nomenclatural History The new genus and species *Isolalutra cretensis* was erected by Symeonidis and Sondaar (1976). Willemsen (1992) transferred the species to the genus *Lutrogale*, based on similarities with the extant smooth-coated otter of Asia (*L. perspicillata*).

Type Material AMPG(V) 942 (holotype), a nearly complete skeleton, curated at AMPG (Fig. 5).

Type Locality Liko Cave, Likotinará, Chaniá; probably late Middle Pleistocene–Late Pleistocene.

Remarks No further remains apart from the holotype have been described. The holotype is a nearly complete skeleton found in articulation in situ at the cave Liko. The skull and mandible were described by Symeonidis and Sondaar (1976), while the postcranial skeleton was described by Willemsen (1980). According to recent phylogenetic analysis (Cherin et al. 2016), *L. cretensis* is the sister taxon of the extant *L. perspicillata.* These two species have been hypothesized to share a common ancestor with Pleistocene otters from Italy, namely *Sardolutra ichnusae* (Sardinia), *Lutraeximia trinacriae* (Sicily), and *Lutraeximia umbra* (central Italy) (Cherin 2017). This common ancestor may be related to the widespread European fossil species "*Lutra*" *simplicidens*, which is also reported in Greece (see Konidaris et al. 2018). However, this hypothesis cannot be currently verified by phylogenetic analysis due to the lack of complete cranial material of "*L.*" *simplicidens* (see Cherin 2017). If valid, *Lutraeximia* and *Sardolutra* should then be considered junior synonyms of *Lutrogale.*

Fig. 5 The Cretan otter *Lutrogale cretensis* (AMPG(V) 942), from the late Middle Pleistocene–Late Pleistocene of Liko Cave. (**a**) Mounted skeleton, (**b**) ventral view of skull, (**c**) right lateral view of skull. Scale bar equals 20 cm for (**a**), and 5 cm for (**b**, **c**)

Elephantidae Gray, 1821

Mammuthus Brookes, 1828

Type Species *Elephas primigenius* Blumenbach, 1799.

★*Mammuthus creticus* (Bate, 1907)

Taxonomic and Nomenclatural History The species was originally referred by Bate (1907) to the genus *Elephas*, while acknowledging morphological similarities to *M. meridionalis*. Osborn (1942) transferred the species to *Palaeoloxodon* and Kuss (1965) to *Loxodonta*. Later, Kuss (1966) synonymised it with the Sicilian dwarf elephant, "*Hesperoloxodon*" *antiquus falconeri*. A mammoth ancestry was proposed by Mol et al. (1996), which was followed by van der Geer et al. (2006a), referring to it as *Mammuthus creticus*. Herridge and Lister (2012) reached the same conclusion, based on a thorough revision and phylogenetic analysis.

Type Material No holotype was designated originally. The lectotype, fixed by Osborn (1942), is a lower third molar (NHMUK PV M9381; Herridge and Lister 2012: fig. 1b).

Type Locality Cape Maléka 1, Chaniá; probably Early–Middle Pleistocene.

Distribution Known only from its type locality.

Remarks The Cretan mammoth, with a body mass of approximately 300 kg, was the smallest mammoth that ever lived (Herridge and Lister 2012). Although Bate (1907) acknowledged its morphological similarity to *M. meridionalis*, for many years it was considered as a palaeoloxodontine elephant and the final product of the insular anagenetic evolution toward a miniature body size from the mainland *P. antiquus* to the dwarf "*P.*" *creticus* with the small *P. creutzburgi* as intermediate stage (e.g., Bate 1907; Symeonidis and Theodorou 1983). However, subsequent studies demonstrated that this dwarf proboscidean is in fact a mammoth (Mol et al. 1996; Poulakakis et al. 2006; Herridge and Lister 2012). In addition, *M. creticus* is restricted to the *Kritimys* biozone and is therefore stratigraphically older than the Cretan dwarf elephant, to which it is, therefore, unrelated (Mayhew 1977; Dermitzakis and de Vos 1987).

Palaeoloxodon Matsumoto, 1924

★*Palaeoloxodon creutzburgi* (Kuss, 1965)

(= ★*Elephas chaniensis* Symeonidis, Theodorou and Giannopoulos, 2000)

Taxonomic and Nomenclatural History Since its discovery and for nearly a century, straight-tusk elephant remains on Crete were attributed either to *Palaeoloxodon* sp. or to *P. antiquus* (and its synonyms): *Elephas priscus* in Simonelli (1894), *E. antiquus* in Bate (1907), *Hesperoloxodon antiquus* in Osborn (1942), *E. antiquus* in Kuss (1966); ?*Elephas* sp. in Kuss and Misonne (1968), *Palaeoloxodon* sp. in Melentis (1974), *P. antiquus* in Kotsakis et al. (1979), *Elephas* n. sp. in Symeonidis and Theodorou (1983), *P. antiquus* n. ssp. in Symeonidis and Theodorou (1983), *P.* cf. *antiquus* in Mol et al. (1996). Vaufrey (1929) used the name of the Maltese middle-sized dwarf elephant *E. mnaidriensis*. Kuss (1965) described the material from Kaló Choráfi as *Loxodonta creutzburgi*. Kuss believed that there were five proboscidean species on Crete. He considered that the large-sized elephants he found in Panagía I cave belonged to *E. antiquus* (= *P. antiquus*). His new species *creutzburgi* was referred to by later authors either as a species or as a subspecies of *Palaeoloxodon* or *Elephas*: *E. creutzburgi* in Belluomini and Delitala (1983), *E. (P.) creutzburgi* in Palombo and Petronio (1989), *E. antiquus creutzburgi* in Poulakakis et al. (2002), and *P. creutzburgi* in Lomolino et al. (2013). A new species of a dwarf elephant from Vamos Cave, *E. chaniensis*, was described more recently (Symeonidis et al. 2000). This species was initially thought as larger than *P. creutzburgi*, but

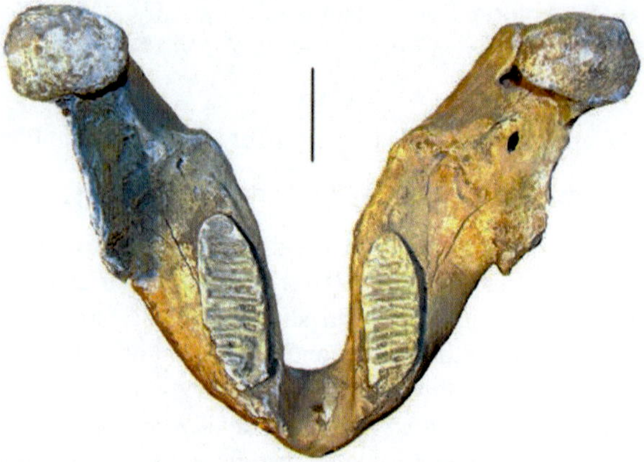

Fig. 6 Occlusal view of the mandible of *Palaeoloxodon creutzburgi* from an undefined cave at Koumbés (AMPG). Scale bar equals 10 cm

comparisons by Athanassiou et al. (2019) showed that there are specimens of the former species that are comparable in size with the latter. Thus, the elephant remains from Vamos Cave were considered as a large-sized *P. creutzburgi*, and *E. chaniensis* as a junior synonym of *P. creutzburgi* (Athanassiou et al. 2019).

Type Material NHMH nr. 22/1963 (holotype), a left M1 (Kuss 1965, pl. III).

Type Locality Kaló Choráfi, near Sísses, Réthymnon; probably late Middle–Late Pleistocene.

Distribution Besides its type locality, this species also occurs in Kaló Choráfi, Skaléta, Vamos, Katharó, Simonelli Cave, Zourída, Koumbés II (Gridá Avláki), Koumbés III, and Karoúmbes 3.

Remarks The Cretan dwarf elephant (Fig. 6) had a body mass of approximately three metric tonnes, and thus less than half the body mass of its mainland relative *P. antiquus* (see Lomolino et al. 2013). The linear dimensions of its dental and skeletal elements are approximately 20% smaller than those of *P. antiquus*. Consequently, several authors considered the Cretan elephant as conspecific with *P. antiquus* or at most a subspecies. The Cretan elephant was significantly larger than the stratigraphically older Cretan mammoth, *Mammuthus creticus* (see Athanassiou et al. 2019). The presence of deer may have blocked its evolution toward extreme dwarfism, preventing it from evolving similar pygmy proportions as the geologically older mammoth did, according to Palombo (2007).

Hippopotamidae Gray, 1821

Hippopotamus Linnaeus, 1758

Type Species *Hippopotamus amphibius* Linnaeus, 1758.

Comments Many of the hippopotamid occurrences from Crete (e.g., from Siteia I, Stavrós Cave, Mílatos cave I, Karoúmbes I, Karoúmbes V, Phangrómouro I) have not been identified beyond the genus level.

★*Hippopotamus creutzburgi* Boekschoten and Sondaar, 1966

Taxonomic and Nomenclatural History For more than a century after its discovery, the names of hippopotamus species from other islands have been applied to the Cretan form. Initially, it was assigned to *Hippopotamus medius* by Owen (1845). He used the name *medius* that had been applied by Desmarest (1822) to fossils that Cuvier had considered a middle-sized hippo. In reality, those fossils belonged to a sea cow, now known as *Metaxytherium medium*. De Blainville (1847) assigned the Cretan hippo to the extant species *H. amphibius*, while Spratt (1865) applied the name *H. minor* and von Zittel (1925) the name *H. minutus*. Both latter authors used names referring to the Cypriot hippo. Falconer (1868) applied the name *H. pentlandi* of the Sicilian dwarf hippo. Bate (1905) referred to it as *Hippopotamus* sp. ("pigmy"). Depéret (1921), on his turn, used the name *H. melitensis* of the Maltese hippo. To make matters worse, Thenius and Hoffer (1960) used the name *H. cretensis,* but without any prior taxonomical act (*nomen nudum*). The Cretan hippo was finally properly named *H. creutzburgi* by Boekschoten and Sondaar (1966).

Type Material NHMUK PV-M10286 (holotype), a fragmentary upper jaw, with P4–M3 left and right (Fig. 7; Boekschoten and Sondaar 1966, pl. I, fig. 1).

Type Locality Katharó Plateau, Lassíthi; Middle Pleistocene.

Fig. 7 Holotype of *Hippopotamus creutzburgi* (NHMUK-M10286) from the Middle Pleistocene of Katharó Plateau. Maxilla with left and right P4–M3, in occlusal view (rostral end is on the left). Scale bar equals 5 cm

Distribution Besides its type locality in the Katharó Plateau, this species has also been reported from Koutalás Cave, Mastabás (Heráklion), Alogarás, Katharádes, Xerókambos, Phangrómouro I, Sitía I, Karoúmbes I, IV, V, Mílatos, Káto Zákros, and Stavrós Cave (Schlager 1996; Spaan 1996; Iliopoulos et al. 2010). The material of the two latter localities has been referred by Kuss (1975a) to a distinct subspecies (see below).

Remarks The Cretan hippo (Fig. 7) is an insular offshoot of *H. antiquus* (see Boekschoten and Sondaar 1966). Its body weight was slightly less than 400 kg, which is 12% of that of its mainland relative (Lomolino et al. 2013). The most significant site with Cretan hippos is at Katharó, an upland basin high in the mountains above Kritsá village with several localities with hippo fossils. The stratigraphically oldest hippo fossil was found in Sitía I. Although the specimen is only a rib fragment, it was classified as hippo by Spaan (1996).

★*Hippopotamus creutzburgi parvus* Kuss, 1975

Taxonomic and Nomenclatural History The subspecies was named by Kuss (1975a) who had the opinion that the material from Káto Zákros and Stavrós is significantly dimensionally smaller than the material from Katharó. Van der Geer et al. (2010, p. 52) considered this difference too small for a recognition of a distinct subspecies.

Type Material NHMH KZ 68/4a (holotype), a mandible with p2–p4, m2–m3, alveolus of m1 (right) and p4, m2–m3 alveoli for dm1–p1 and incisors (left), figured in Kuss (1975a, pl. 1, figs 1–2).

Type Locality Káto Zákros, Lassíthi; probably Middle Pleistocene.

Distribution Besides Káto Zákros, it has been also identified in Stavrós Cave.

Remarks Kuss (1975a) noticed differences in size between hippos from different sites, upon which he based his subspecies for the Cretan hippo, the small *H. c. parvus* from Káto Zákros and Stavrós. This material is smaller sized than the hippo from Katharó (Kuss 1975a). The dimensional overlap in the molar measurements between the two forms is small, but the *parvus* sample remains scanty and of unknown variation. A direct consequence of naming the subspecies *H. c. parvus* is that the type series from Katharo should be retained for the subspecies *H. c. creutzburgi.* Indeed, several authors (e.g., Dermitzakis and de Vos 1987; Spaan 1996; de Vos 1996) refer to the material from Katharo as belonging to the subspecies *H. c. creutzburgi.*

Cervidae Goldfuß, 1820

Remarks An antler fragment from Katharó Plateau, Lassíthi (?Middle Pleistocene; NHMH collection) has been mentioned in the literature as a form close to

Megaloceros (Dermitzakis et al. 2007) or as dwarf form of the giant deer (van der Geer et al. 2011). Also, Dermitzakis et al. (2007) suggest that this taxon is ancestral to Late Pleistocene *Candiacervus*, although without giving diagnostic features. Based on all available evidence, we consider this occurrence as an indeterminate Cervidae.

★*Candiacervus* Kuss, 1975

Type Species *Anoglochis cretensis* Simonelli, 1907.

Remarks *Candiacervus* is a single-island taxon that lived during the Pleistocene on Crete. There were eight species of *Candiacervus* living on the island during the late Middle and Late Pleistocene, varying in size from dwarf to giant forms (Fig. 8).

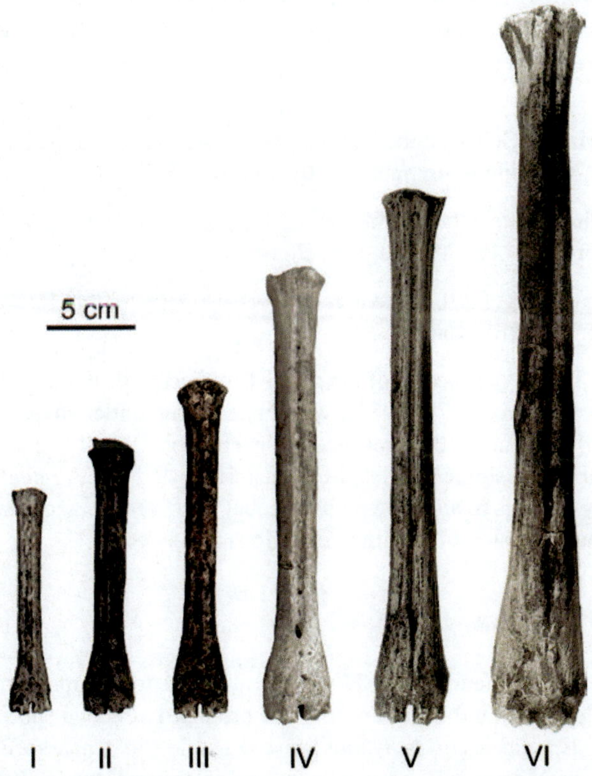

Fig. 8 Metatarsals of the six different size groups of the Cretan deer *Candiacervus*, in dorsal view. Roman numbers indicate the size groups of de Vos (1979). Size group I: *C. ropalophorus*; size group II: *C. devosi*, *C. listeri* and *C. reumeri*; size group III: *C. cretensis*; size group IV: *C. rethymnensis*; size group V: *C. dorothensis*; size group VI: *C. major*. Note that size group II includes three species, which cannot be distinguished on postcranial elements alone. AMPG (sizes I–IV) and MPUR (sizes V–VI)

According to van der Geer (2018), its antlers are with generally strongly compressed beams with longitudinal grooves and without pearling, they are with at most four tines (two anterior, one posterior, one subbasal tine), but mostly a single anterior (basal) tine and often one posterior (back) tine, no crown is developed, and a rudimentary palmation may be present. The overall antler morphology of *Candiacervus* suggests a phylogenetic placement close to *Dama* (see van der Geer 2018).

★*Candiacervus cretensis* (Simonelli, 1907)

Taxonomic and Nomenclatural History Simonelli classified the Cretan deer under the genus *Anoglochis* (*Anoglochis cretensis* in Simonelli 1907). In his view, the Cretan deer was a Miocene relict. The species was later classified under various genera: *Cervus* (*Eucladoceros*) *creticus* in Vaufrey (1929), *Nesoleipoceros cretensis* in Radulesco and Samson (1967), *Megaloceros cretensis* in Sondaar and Boekschoten (1967), and *Praemegaceros cretensis* in Kurtén (1968, p. 164). In 1975, Kuss proposed a new genus to accommodate the Cretan deer (*Candiacervus cretensis* in Kuss 1975b) after Candia, the Venetian name of the city of Heráklion, and the island in general.

Type Material MGGC number unknown, metacarpal III-IV (lectotype, fixed by de Vos 1984), figured by Simonelli (1907, figs 24–25).

Type Locality Crete Island, unknown site, possibly a cave near Gridá Avláki, Rethymon; probably late Middle–Late Pleistocene.

Distribution Except for the type locality, it is also known from Réthymnon fissure, Liko Cave, and Mávro Mourí 4.

Remarks It is the type species of the genus *Candiacervus*. It is a middle-sized deer (size class 3 of de Vos 1979, 1984), with undescribed antler shape and size. Until Kuss (1975b) recognized the presence of several species, all Cretan deer material was assigned to this single species. Most material listed as *C. cretensis* in older literature likely belongs to a different species, but to solve this taxonomic issue properly, a thorough revision of existing collections is needed.

★*Candiacervus rethymnensis* Kuss, 1975

Taxonomic and Nomenclatural History Kuss noted the large variability in size and morphology within the Cretan deer and proposed a second species for a larger morphotype, *Candiacervus rethymnensis* in Kuss 1975b. Capasso Barbato (1990) erected the new subgenus *Leptocervus* and classified the species as *Cervus* (*Leptocervus*) *rethymnensis*. Later she transferred the subgenus and species to the genus *Pseudodama* (*Pseudodama* (*Leptocervus*) *rethymnensis* in Capasso Barbato 1995). De Vos (1979, 1996) and van der Geer (2018) retained the classification of Kuss as *Candiacervus rethymnensis*.

Type Material NHMH number unknown (holotype), right metacarpal III-IV, figured by Kuss (1975b, pl. 6 fig. o).

Type Locality Mávro Mourí 4, Réthymnon; probably late Middle–Late Pleistocene.

Distribution Except for the type locality, it is also known from Liko Cave, Koumbés B, and Gridá Avláki.

Remarks A middle-sized deer of the genus *Candiacervus* (size class 4 in de Vos 1979, 1984), larger than *C. cretensis* (Kuss 1975b). Limb bones are about twice as long as those of size class 2 (*C. devosi, C. listeri*, and *C. reumeri*). The basal tine of the antler may have been trifurcated. No further information on antler shape and size is known.

★*Candiacervus ropalophorus* de Vos, 1979

Taxonomic and Nomenclatural History Initially, the material of this species was included in *Candiacervus cretensis*. In 1979, de Vos coined the new name *Candiacervus ropalophorus* for the smallest size group (his size class 1). Later, Capasso Barbato reduced *Candiacervus* to subgenus of *Megaloceros* and soon after of *Megaceroides*: *Megaceros* (*Candiacervus*) *ropalophorus* in Capasso Barbato (1992) and *Megaceroides* (*Candiacervus*) *ropalophorus* in Capasso Barbato (1995). Later authors (e.g., van der Geer 2018) followed the original classification of de Vos (1979).

Type Material RGM 437854 (holotype), male skull, figured in de Vos (1984, pl. 10); the original type series contained also RGM 438460 (paratype), antler.

Type Locality Geráni Cave 4, Réthymnon; probably late Middle–Late Pleistocene.

Distribution Geráni Cave 4, Geráni Cave 2 (upper layer), Geráni Cave 5, Geráni Cave 6, Mávro Mourí 4 (layer c), Zourída, Réthymnon fissure, and Kaló Choráfi.

Remarks This is the smallest species of *Candiacervus* with a shoulder height of about 40 cm and a body mass of about 22 kg on average (van der Geer et al. 2013). Limb bones were proportionally more massive than those of the fallow deer (*Dama dama*). The antler is very long (up to 77 cm; may exceed total head-rump length) with a single tine and a characteristic bludgeon-shaped end (Fig. 9). The antler beam is either straight or gently curved.

★*Candiacervus major* (Capasso Barbato and Petronio, 1986)

Taxonomic and Nomenclatural History The presence of very large-sized deer on Crete was initially reported by Kotsakis et al. (1976) as "Cervo taglia grande" (= large-sized deer). De Vos (1979) distinguished six size groups within the genus *Candiacervus,* which he considered as taxonomic units. He attributed the large-sized deer to *Candiacervus* sp. VI. A new species, *Cervus major*, was erected by

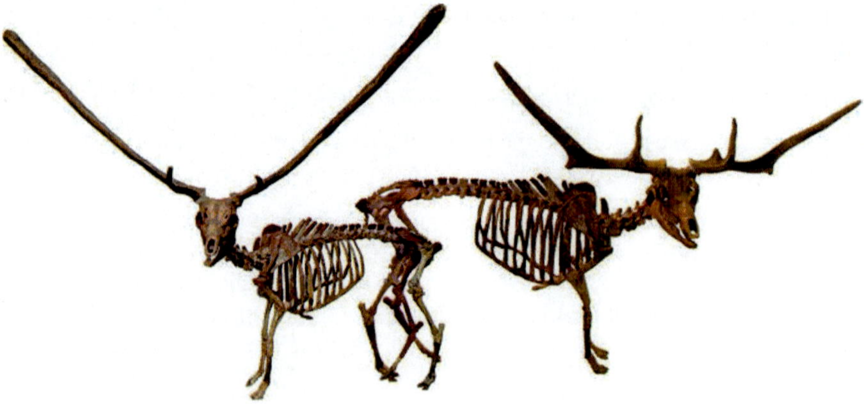

Fig. 9 Mounted skeletons of *Candiacervus ropalophorus* (left) and *Candiacervus devosi* (right) (AMPG collection). Shoulder height of *C. devosi* is approximately 50 cm

Capasso Barbato and Petronio (1986) for the material reported by Kotsakis et al. (1976) as large-sized deer. The species was considered as belonging to the subgenus *Leptocervus*, either of red deer (*Cervus* (*Leptocervus*) *major* in Capasso Barbato 1990) or of fallow deer (*Pseudodama* (*Leptocervus*) *major* in Capasso Barbato 1995). Caloi and Palombo (1996) expressed uncertainty about the generic attribution, and listed it as ?*Pseudodama* (*Leptocervus*) *major*. De Vos (2000), finally, placed the species under *Candiacervus*.

Type Material MPUR nr 30 (holotype), metatarsal, figured in Capasso Barbato and Petronio (1986, pl. 3).

Type Locality Bate Cave, Réthymnon; probably late Middle–Late Pleistocene.

Distribution Except for the type locality, it is also known from Liko Cave.

Remarks This is the largest species of *Candiacervus* with withers height of about 1.65 m (de Vos 1979) and an average body mass of about 316 kg (van der Geer et al. 2013). No antler or dental material has been described. The metatarsal is about four times the length of that of *C. ropalophorus*, which is the smallest species (see above; de Vos 1979).

★*Candiacervus dorothensis* (Capasso Barbato, 1992)

Taxonomic and Nomenclatural History The species was initially referred to as "Cervo taglia media" (= medium-sized deer) in Kotsakis et al. (1976). De Vos (1979) placed it in his size group *Candiacervus* sp. V. The species name *Cervus* (*Leptocervus*) *dorothensis* was formally coined by Capasso Barbato (1992), although the same author had already mentioned it in earlier publications (e.g., Capasso Barbato 1990). Later, Capasso Barbato (1995) transferred it to *Pseudodama* (*Leptocervus*). De Vos (2000) moved the species to *Candiacervus*.

Type Material MPUR nr 25 (holotype), right radius, figured in Capasso Barbato (1992, fig. 4).

Type Locality Bate Cave, Réthymnon; probably late Middle–Late Pleistocene.

Distribution Only known from the type locality.

Remarks A large-sized species of *Candiacervus*, only smaller than *C. major*. No antler or dental material has been described.

★*Candiacervus devosi* van der Geer, 2018

Taxonomic and Nomenclatural History There are three species of small-sized Cretan deer that are larger than *Candiacervus ropalophorus* (Fig. 9) and differ from each other in skull and antler morphology. De Vos (1979) referred to them as *Candiacervus* sp. IIa, *C.* sp. IIb, and *C.* sp. IIc. Capasso Barbato (1990) and later Caloi and Palombo (1996) assigned the sizes I and II of de Vos to a single species, *Megaceros* (*Candiacervus*) *ropalophorus*, ignoring differences in body size and antler morphology. In 2018, van der Geer coined the name *Candiacervus devosi* for de Vos' "*Candiacervus* sp. IIb."

Type Material AMPG(V)1735 (holotype), male skull, figured in de Vos (1984, pl. 13); the type series contains also AMPG(V)1733 (paratype), a right antler.

Type Locality Liko Cave, Likotinará, Chaniá; probably late Middle–Late Pleistocene.

Distribution Except for the type locality, it is known from Mávro Mourí 4.

Remarks A small-sized species of *Candiacervus*. Antler with a double first tine, a very small back tine and a very long, upward pointing last segment, tapering into a sharp pointed end (Fig. 9). The two tines making up the first tine are generally connected by webbing. Specimens from a deeper layer of the same cave may show a rudimentary sub-basal tine or distal bifurcation of the first tine.

★*Candiacervus listeri* van der Geer, 2018

Taxonomic and Nomenclatural History The species was initially referred to as *Candiacervus* sp. IIa by de Vos (1979). Capasso Barbato (1990) assigned it to *Megaceros* (*Candiacervus*) *ropalophorus*. The name *Candiacervus listeri* was coined by van der Geer (2018).

Type Material AMPG(V)1734 (holotype), male skull (Fig. 10; de Vos 1984, pl. 12); the original type series contains AMPG(V)1726 (paratype), a partial skull with both antlers.

Type Locality Liko Cave, Likotinará, Chaniá; probably late Middle–Late Pleistocene.

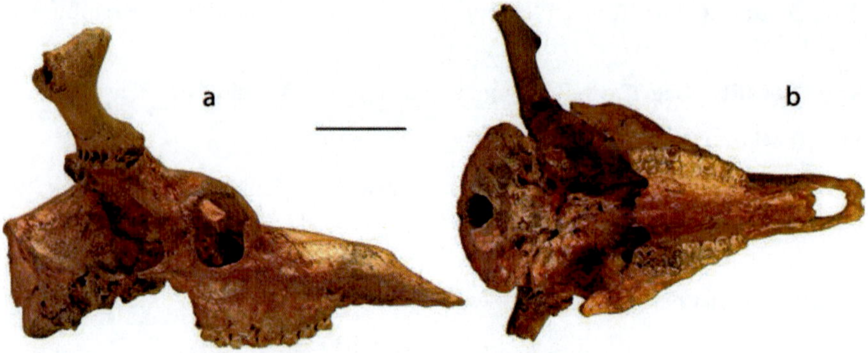

Fig. 10 Holotype skull of *Candiacervus listeri* (AMPG(V)1734) from the late Middle–Late Pleistocene of Liko Cave, in (**a**) lateral, and (**b**) ventral views. Scale bar equals 5 cm

Distribution Except the type locality, it is also known from Geráni Cave 2, Geráni Cave 4, Gridá Avláki, Koumbés B, Kaló Choráfi, Mávro Mourí 3, Mávro Mourí 4, Peristéri 2, and Sifanos.

Remarks A small-sized species of *Candiacervus* with very long antlers (up to 70 cm) with a single tine, gradually becoming thinner toward the end. The main beam is either straight or gently curved. A back tine, as well as a serrated posterior border, may be present.

★*Candiacervus reumeri* van der Geer, 2018

Taxonomic and Nomenclatural History The species was initially listed as *Candiacervus* sp. IIc in de Vos (1979). Malatesta (1980) referred the corresponding material from Simonelli cave to *Praemegaceros cretensis*. Capasso Barbato (1990) and Caloi and Palombo (1996) did not distinguish it from smallest-sized *Candiacervus* and listed it as *Megaceros* (*Candiacervus*) *ropalophorus*. The species was named *Candiacervus reumeri* by van der Geer (2018).

Type Material AMPG(V) 1736 (holotype), male skull, figured in de Vos (1984, pl. 14); the type series contains also RGM 442702 (paratype), a left antler figured in de Vos (1984, pl. 8).

Type Locality Liko Cave, Likotinará, Chaniá; probably late Middle–Late Pleistocene.

Distribution Except the type locality, it is also known from Peristéri 2. *Candiacervus* cf. *reumeri* occurs in Simonelli Cave, Kaló Choráfi, and Mávro Mourí 3.

Remarks A small-sized species of *Candiacervus* with a moderately long antler with two anterior tines and a back tine. The beam segments have a straight or minimally concave posterior border. The first three segments are in a line, whereas the last segment bends upward. A rudimentary sub-basal tine or a very small second basal tine may be present, generally in specimens from deeper layers of the same cave. Specimens from Simonelli Cave lacking the second basal tine but with overall similar shape and size may belong to this species (van der Geer 2018).

Acknowledgments We thank John de Vos, Michael Dermitzakis, and the late Paul Sondaar for introducing us to the study of island mammals. We also thank George Iliopoulos, Ross MacPhee, Mark Lomolino, Victoria Herridge, and the late Dave Mayhew for discussions we had with them. We are much indebted to Hans Brinkerink for his invaluable help in the field. We are also thankful to the reviewers of the original manuscript, John de Vos, Rafaelle Sardella, and Marco Cherin, for their corrections and comments.

References

Alcover JA, Campillo X, Macias M, Sans A (1998) Mammal species of the world: additional data on insular mammals. Am Mus Nov 3248:1–29

Anastasakis G, Piper DJW, Dermitzakis MD, Karakitsios V (2006) Upper Cenozoic stratigraphy and paleogeographic evolution of Myrtoon and adjacent basins, Aegean Sea, Greece. Mar Petrol Geol 23:353–369

Angelelli F (1980) Endocranial morphology of the dwarf deer of Crete. Quad Accad Naz Lincei 249:101–109

Athanassiou A, van der Geer AAE, Lyras GA (2019) Pleistocene insular Proboscidea of the Eastern Mediterranean: a review and update. Quat Sci Rev 218:306–321

Bachmayer F, Symeonidis NK (1975a) Eigenartige Abspaltungen von Stosszähnen der Zwergelefanten aus der Höhle 'Charkadio' auf der Insel Tilos – Artefakte? Ann Géol Pays Hellén 26:320–323

Bachmayer F, Symeonidis NK (1975b) Schilkrötenreste (*Testudo marginata* Schoepff) aus der Höhle 'Charkadio' auf der Insel Tilos (Dodekanes, Griechenland). Ann Géol Pays Hellén 26:324–327

Bachmayer F, Symeonidis N, Seemann R, Zapfe H (1976) Die Ausgrabungen in der Zwergelefantenhöhle „Charkadio" auf der Insel Tilos (Dodekanes, Griechenland) in den Jahren 1974 und 1975. Ann Naturhist Mus Wien 80:113–144

Bachmayer F, Symeonidis N, Zapfe H (1984) Die Ausgrabungen in der Zwergelefantenhöhle der Insel Tilos (Dodekanes, Griechenland) im Jahr 1983. Sitzungsber Österr Akad Wis – Math-Naturwis Kl 193:321–328

Bate DMA (1905) Four and a half months in Crete in search of Pleistocene mammalian remains. Geol Mag 2(5):193–202

Bate DMA (1907) On elephant remains from Crete, with description of *Elephas creticus* n. sp. Proc Zool Soc London 1907:238–250

Bate DMA (1912) On a new species of mouse and other rodent remains from Crete. Geol Mag 9:4–6

Bate DMA (1942) New Pleistocene Murinae from Crete. Ann Mag Nat Hist 9(49):41–49

Belluomini G, Delitala L (1983) Datazione di resti ossei e denti del Pleistocene superiore e dell'Olocene dell'area mediterranea con il metodo della racemizzazione degli aminoacidi. Geogr Fis Dinamica Quat 6:21–30

Boekschoten GJ, Sondaar PY (1966) The Pleistocene of the Katharo Basin (Crete) and its *Hippopotamus*. Bijdr Dierk 36(8):17–44

Caloi L, Palombo MR (1994) Functional aspect and ecological implications in Pleistocene endemic herbivores of Mediterranean islands. Hist Biol 8:151–172

Caloi L, Palombo MR (1996) Functional aspects and ecological implications in hippopotami and cervids of Crete. In: Reese DS (ed) Pleistocene and Holocene fauna of Crete and its first settlers, Monographs in world archaeology, vol 28. Prehistory Press, Madison, pp 125–151

Caloi L, Kotsakis T, Palombo MR, Petronio C (1996) The Pleistocene dwarf elephants of Mediterranean islands. In: Shoshani J, Tassy P (eds) The Proboscidea: evolution and paleoecology of elephants and their relatives. Oxford University Press, New York, pp 234–239

Capasso Barbato L (1990) Les Cervidés endémiques de Crète. Quaternaire 1:265–270

Capasso Barbato L (1992) Nuova specie di cervide del pleistocene di Creta. Atti Accad Naz Lincei Cl Sci Fis Mat Nat Rend 1(7):183–220

Capasso Barbato L (1995) Un tentativo di analisi cladistica applicata ad un cervide endemico del Pleistocene di Creta. Geol Rom 31:243–248

Capasso Barbato L, Petronio C (1986) *Cervus major* n. sp. of Bate Cave (Rethymnon, Crete). Atti Accad Naz Lincei Cl Sci Fis Mat Nat Rend 18(2):59–100

Cayeux L (1908) Découverte de l'*Elephas antiquus* à l'île de Délos (Cyclades). CR Hebd Séances Acad Sci 147:1089–1090

Cherin M (2017) New material of *Lutra simplicidens* (Carnivora, Mustelidae, Lutrinae), a key taxon for understanding the evolution of European otters. Riv It Paleontol Strat 123(3):433–441

Cherin M, Iurino DA, Willemsen G, Carnevale G (2016) A new otter from the Early Pleistocene of Pantalla (Italy), with remarks on the evolutionary history of Mediterranean Quaternary Lutrinae (Carnivora, Mustelidae). Quat Sci Rev 135:92–102

de Blainville HMD (1847) Sur les Hippopotames (Buffon) (*Hippopotamus* L.) et les cochons (Buffon) (*Sus* L.). In: Ostéographie ou description iconographique comparée du squelette et du système dentaire des mammifères récents et fossiles pour servir de base à la zoologie et à la géologie, vol 22. J.B. Baillière, Paris, pp 1–248

de Bruijn H, Dawson MR, Mein P (1970) Upper Pliocene Rodentia, Lagomorpha and Insectivora (Mammalia) from the isle of Rhodes (Greece). Proc K Ned Akad Wet 73(5):535–584

de Vos J (1979) The endemic Pleistocene deer of Crete. Proc K Ned Akad Wet 82(1):59–90

de Vos J (1984) The endemic Pleistocene deer of Crete. Verh K Ned Akad Wet 31:1–100

de Vos J (1996) Taxonomy, ancestry and speciation of the endemic Pleistocene deer of Crete compared with the taxonomy ancestry and speciation of Darwin's finches, Monographs in world archaeology, vol 28. Prehistory Press, Madison, pp 111–124

de Vos J (2000) Pleistocene deer fauna in Crete: its adaptation, radiation and extinction. Tropics 10:125–134

de Vos J (2006) Notes about the parallels in evolution of the Pleistocene cervids from Greece (Crete, Kassos and Karpathos), Japan (the Ryukyu-islands) and Philippines (Masbate). Hell J Geosci 41:127–140

de Vos J, van der Geer AAE (2002) Major patterns and processes in biodiversity: taxonomic diversity on islands explained in terms of sympatric speciation. In: Waldren B, Ensenyat JA (eds) World Islands in prehistory, international insular investigations. BAR international series, vol 1095. Archaeopress, Oxford, pp 395–405

Depéret C (1921) La phylogénie des Hippopotames. C R Soc Geol Fr 21:163–165

Dermitzakis MD, de Vos J (1987) Faunal succession and the evolution of mammals in Crete during the Pleistocene. N Jb Geol Pal Abh 173(3):377–408

Dermitzakis MD, Sondaar PY (1979) The importance of fossil mammals in reconstructing paleogeography with special reference to the Pleistocene Aegean Archipelago. Ann Géol Pays Hellén 29:808–840

Dermitzakis MD, van der Geer AAE, Lyras GA (2006) Paleopathological observations on a population of fossil deer from the Late Pleistocene of Crete. In: Kalofourtis A, Papadopoulos N, Spiliopoulou C, Maravelias K, Chatziioannou A (eds) Volume in honour of Professor Andonis Koutselinis. Parisianou Editions, Athens pp. 43–51

Dermitzakis M, Iliopoulos G, van der Geer AAE, Lyras GA (2007) The rise and fall of the Cretan deer. Quat Int 167–168(Suppl 1):97

Desmarest AG (1822) Mammalogie, ou description des espèces de mammifères. Agasse, Paris

Falconer H (1868) On the fossil remains of *Elephas melitensis*, an extinct pygmy species of elephant; and other Mammalia, &c. from the ossiferous caves of Malta. In: Murchison C (ed) Palæontolgical memoires and notes of the late Hugh Falconer, A.M., M.D., with a biographical sketch of the author, vol 2, pp 292–307

Georgalas GC (1929) [Elephas (palaeozoology)]. Megáli Hellinikí Egkyklopaédia 9:945–946 (in Greek)

Grant KM, Rohling EJ, Bronk Ramsey C, Cheng H, Edwards RL, Florindo F, Heslop D, Marra F, Roberts AP, Tamisiea ME, Williams F (2014) Sea-level variability over five glacial cycles. Nat Commun 5:1–9

Haynes G (1988) Longitudinal studies of African elephant death and bone deposits. J Archaeol Sci 15:131–157

Haynes G, Klimowicz J (2015) A preliminary review of bone and teeth abnormalities seen in recent *Loxodonta* and extinct *Mammuthus* and *Mammut*, and suggested implications. Quat Int 379:135–146

Herridge VL, Lister AM (2012) Extreme insular dwarfism evolved in a mammoth. Proc R Soc London B 279(1741):3193–3200

Honea K (1975) Prehistoric remains on the island of Kythnos. Am J Archaeol 79:277–279

Iliopoulos G, Eikamp H, Fassoulas C (2010) A new late Pleistocene mammal locality from Western Crete. Bul Geol Soc Greece 43:918–926

Kapsimalis V, Pavlopoulos K, Panagiotopoulos I, Drakopoulou P, Vandarakis D, Sakelariou D, Anagnostou C (2009) Geoarchaeological challenges in the Cyclades continental shelf (Aegean Sea). Z Geomorphol 53:169–190

Konidaris GE, Athanassiou A, Tourloukis V, Thompson N, Giusti D, Panagopoulou E, Harvati K (2018) The skeleton of a straight-tusked elephant (*Palaeoloxodon antiquus*) and other large mammals from the Middle Pleistocene butchering locality Marathousa 1 (Megalopolis Basin, Greece): preliminary results. Quat Int 497:65–84

Kotsakis T, Melentis J, Petronio C, Sirna G (1976) Seconda spedizione paleontologica Lincea nell'isola di Creta. Quad Accad Naz Lincei 223:3–10

Kotsakis T, Petronio C, Sirna G (1980) The Quaternary vertebrates of the Aegean islands: palaeogeographical implications. Ann Géol Pays Hellén 30:31–64

Kouvari M, van der Geer AAE (2018) Biogeography of extinction: the demise of insular mammals from the Late Pleistocene till today. Palaeogeogr Palaeoclimatol Palaeoecol 505:295–304

Kurtén B (1968) Pleistocene mammals of Europe. The World Naturalist. Weidenfeld and Nicolson, London

Kuss SE (1965) Eine pleistozäne Säugetierfauna der Insel Kreta. Ber Naturf Ges Freiburg i Br 55(2):271–348

Kuss SE (1966) Beiträge zur Pleistozän-Fauna der Insel Kreta. I. Die von Bate 1904 gesammelten Elefanten- und Cerviden-Reste. Ber Naturf Ges Freiburg i Br 56(2):169–181

Kuss SE (1967) Pleistozäne Säugetierfunde auf den ostmediterranen Inseln Kythera und Karpathos. Ber Naturf Ges Freiburg i Br 57:207–216

Kuss SE (1969) Die erste pleistozäne Säugetierfauna der Insel Kasos (Griechenland). Ber Naturf Ges Freiburg i Br 59:169–177

Kuss SE (1973) Die pleistozänen Säugetierfaunen der ostmediterranen Inseln: Ihr Alter und ihre Herkunft. Ber Naturf Ges Freiburg i Br 63:49–71

Kuss SE (1975a) *Hippopotamus creutzburgi parvus* n. spp., ein pleistozänes Zwergflußpferd von der Insel Kreta. Ber Naturf Ges Freiburg i Br 65(1–2):5–23

Kuss SE (1975b) Die pleistozänen Hirsche der ostmediterranen Inseln Kreta, Kasos, Karpathos und Rhodos (Griechenland). Ber Naturf Ges Freiburg i Br 65(1–2):25–79

Kuss SE, Misonne X (1968) Pleistozäne Muriden der Insel Kreta. N Jb Geol Pal Abh 132(1):55–69

Lax EM (1996) A gazetteer of Cretan paleontological localities. In: Reese DS (ed) Pleistocene and Holocene fauna of Crete and its first settlers, Monographs in world archaeology, vol 28. Prehistory Press, Madison, pp 1–32

Le Pichon X, Angelier J (1981) The Aegean Sea. Phil Trans R Soc London 300:357–372

Lomolino MV (2000) Ecology's most general, yet protean pattern: the species–area relationship. J Biogeogr 27:17–26
Lomolino MV (2005) Body size evolution in insular vertebrates: generality of the island rule. J Biogeogr 32:1683–1699
Lomolino MV, Sax DF, Palombo MR, van der Geer AAE (2012) Of mice and mammoths: evaluations of causal explanations for body size evolution in insular mammals. J Biogeogr 39:842–854
Lomolino MV, van der Geer AAE, Lyras GA, Palombo MR, Sax D, Rozzi R (2013) Of mice and mammoths: generality and antiquity of the island rule. J Biogeogr 40:1427–1439
Lykousis V (2009) Sea-level changes and shelf break prograding sequences during the last 400 ka in the Aegean margins: subsidence rates and palaeogeographic implications. Cont Shelf Res 29:2037–2044
Lyras GA, Giannakopoulou A, Lillis T, Veis A, Papadopoulos GC (2016) Bone lesions in a Late Pleistocene assemblage of the insular deer *Candiacervus* sp. II from Liko cave (Crete, Greece). Int J Paleopathol 14(1):36–45
Lyras GA, Giannakopoulou A, Lillis TD, van der Geer AAE (2019) Paradise lost: evidence for a devastating metabolic bone disease in an insular Pleistocene deer. Int J Paleopathol 24:213–226
Malatesta A (1980) Dwarf deer and other Late Pleistocene fauna of the Simonelli Cave in Crete. Quad Accad Naz Lincei 249:3–97
Marinos G, Symeonidis N (1973) Erstmalige Funde von Zwergelefanten auf der Insel Rhodos. Anz Math-Naturwiss Kl Österr Akad Wiss 110:129–131
Masseti M (2012) Atlas of terrestrial mammals of the Ionian and Aegean islands. De Gruyter, Berlin
Mayhew DF (1977) The endemic Pleistocene murids of Crete I–II. Proc K Ned Akad Wet 80(3):182–214
Mayhew DF (1996) The extinct murids of Crete. In: Reese DS (ed) Pleistocene and Holocene fauna of Crete and its first settlers, Monographs in world archaeology, vol 28. Prehistory Press, Madison, pp 167–171
Melentis JK (1974) Paläontologische Ausgrabungen in den Höhlen des Gebietes von Rhethymnon, Kreta. Sci Ann Faculty Phys Math Univ Thessaloniki 14:17–24
Milàn J, Bromley RG, Titschack J, Theodorou G (2007) A diverse vertebrate ichnofauna from a Quaternary eolian oolite, Rhodes, Greece. In: Bromley RG, Buatois LA, Mángano G, Genise JF, Melchor RN (eds) Sediment–organism interactions: a multifaceted ichnology, SEPM special publications, vol 88. Tulsa, SEPM Society for Sedimentary Geology, pp 333–343
Mitsopoulou V, Michailidis D, Theodorou E, Isidorou S, Roussiakis S, Vasilopoulos T, Polydoras S, Kaisarlis G, Spitas V, Stathopoulou E, Provatidis C, Theodorou G (2015) Digitizing, modelling and 3D printing of skeletal digital models of *Palaeoloxodon tiliensis* (Tilos, Dodecanese, Greece). Quat Int 379:4–13
Mitzopoulos MK (1961) Über einen pleistozänen Zwergelefanten von der Insel Naxos (Kykladen). Prakt Akad Ath 36:332–340
Mol D, de Vos J, van den Bergh GD, Sondaar PY (1996) The taxonomy and ancestry of the fossil elephants of Crete. Faunal turnover and comparison with proboscidean faunas of Indonesian islands. In: Reese DS (ed) Pleistocene and Holocene fauna of Crete and its first settlers, Monographs in world archaeology, vol, vol 28. Prehistory Press, Madison, pp 81–98
Osborn HE (1942) Proboscidea – a monograph of the discovery, evolution, migration and extinction of the mastodonts and elephants of the World, Vol II: Stegodontoidea and Elephantoidea. American Museum Press, New York
Owen R (1845) Descriptive and illustrated catalogue of the fossil organic remains of Mammalia and Aves contained in the Museum of the Royal College of Surgeons of England. R. and J. Tayler, London
Palombo MR (2001) Endemic elephants of the Mediterranean Islands: knowledge, problems and perspectives. In: Cavarretta G, Gioia P, Mussi M, Palombo MR (eds) Proceedings of the 1st international congress "The World of Elephants", Rome, 2001, pp 486–491

Palombo MR (2004) Dwarfing in insular mammals: the endemic elephants of mediterranean islands. Zona Arqueol 4:354–371

Palombo MR (2007) How can endemic proboscideans help us understand the "island rule"? A case study of Mediterranean islands. Quat Int 169–170:105–124

Palombo MR, Petronio C (1989) Morphostructural characters of *Elephas (Palaeoloxodon) creutzburgi* (Kuss, 1965) femur from Simonelli Cave (Rethymnon, Crete). Hystrix 1:95–105

Palombo MR, Zedda M (2016) Surviving in a predator-free environment: hints from a bone remodelling process in a dwarf Pleistocene deer from Crete. C R Palevol 15(1–2):245–254

Papp A (1953) Die paläogeologische Entstehung der Ägäis nach dem derzeitigen Stand unserer Kenntnisse. In: Wettstein O (ed) Herpetologia aegaea. Sitzungsber Österr Akad Wiss – Math-Naturwiss Kl 162:815–818

Paraskevaidis I (1956) Observations sur le Quaternaire de la Grèce. In: Actes de IV Congrès International du Quaternaire, Rome–Pise, 1953, pp 167–178

Petronio C (1990) Les cervidés endémiques des îles méditerranéennes. Quaternaire 1:259–264

Pococke R (1745) A description of the East and some other countries. Bowyer, London

Poulakakis N, Mylonas M, Lymberakis P, Fassoulas C (2002) Origin and taxonomy of the fossil elephants of the island of Crete (Greece): problems and perspectives. Palaeogeogr Palaeoclimatol Palaeoecol 186(1–2):163–183

Poulakakis N, Parmakelis A, Lymberakis P, Mylonas M, Zouros E, Reese DS, Glaberman S, Caccone A (2006) Ancient DNA forces reconsideration of evolutionary history of Mediterranean pygmy elephantids. Biol Lett 2(3):451–454

Radulesco C, Samson P (1967) Sur un nouveau cerf mégacérin du Pléistocène moyen de la dépression de Brasov (Roumanie). Geol Rom 6:317–344

Raulin FV (1869) Description physique de l'île de Crète. Bertrand, Paris

Reumer JWF (1986) Notes on the Soricidae (Insectivora, Mammalia) from Crete. I. The Pleistocene species *Crocidura zimmermann*. Bonn zool Beitr 37(3):161–171

Schlager N (1996) Phangromouro I and II: two recently discovered paleontological sites, with a note on Hogarth's Zakros sealings nos. 92 and 139. In: Reese DS (ed) Pleistocene and Holocene fauna of Crete and its first settlers, Monographs in world archaeology, vol 28. Prehistory Press, Madison, pp 33–45

Sen S (2017) A review of the Pleistocene dwarfed elephants from the Aegean islands, and their paleogeographic context. Fossil Impr 73:76–92

Sen S, Barrier E, Crété X (2014) Late Pleistocene dwarf elephants from the Aegean islands of Kassos and Dilos, Greece. Ann Zool Fenn 51:27–42

Simonelli V (1894) Appunti sopra i terreni neogenici e quaternari dell'Isola di Candia. Atti Accad Naz Lincei Cl Sci Fis 2:265–268

Simonelli V (1907) Mammiferi quaternari dell'isola di Candia. Mem R Acad Sci Inst Bologna, Cl Sc Fis 4:455–471

Sondaar PY (1971) Paleozoogeography of the Pleistocene mammals from the Aegean. In: Strind A (ed) Evolution in the Aegean. Opera Botan 30:65–70

Sondaar PY (1977) Insularity and its effect on mammal evolution. In: Hecht MK, Goody PC, Hecht BM (eds) Major patterns in vertebrate evolution. Plenum, New York, pp 671–707

Sondaar PY, Boekschoten GJ (1967) Quaternary mammals in the South Aegean island arc, with notes on other fossil mammals from the coastal regions of the Mediterranean. Proc Konin Nederl Akad Wet 70:556–576

Sondaar PY, Dermitzakis MD, de Vos J (1996) The palaeogeography and faunal evolution of the land mammals of Crete. In: Reese DS (ed) Pleistocene and Holocene fauna of Crete and its first settlers, Monographs in world archaeology, vol 28. Prehistory Press, Madison, pp 61–67

Spaan A (1996) *Hippopotamus creutzburgi*: the case of the Cretan *Hippopotamus*. In: Reese DS (ed) Pleistocene and Holocene fauna of Crete and its first settlers, Monographs in world archaeology, vol 28. Prehistory Press, Madison, pp 99–111

Spratt TAB (1865) Travels and researches in Crete. John van Voorst, London

Spratt RM, Lisiecki LE (2016) A Late Pleistocene sea level stack. Clim Past 12:1079–1092

Symeonidis NK (1973) Die Entdeckung von Zwergelefanten in der Höhle „Charkadio" auf der Insel Tilos (Dodekanes, Griechenland). Ann Géol Pays Hellén 24:445–461

Symeonidis NK, Sondaar PY (1976) A new otter from the Pleistocene of Crete. Ann Geol Pays Hellén 27:11–24

Symeonidis NK, Theodorou GE (1983) [New findings of fossil elephants on Crete Island]. Ann Geol Pays Hellén 31:113–129 (in Greek)

Symeonidis N, Bachmayer F, Zapfe H (1973) Grabungen in der Zwergelefanten-Höhle „Charkadio" auf der Insel Tilos (Dodekanes, Griechenland). Ann Naturhist Mus Wien 77:133–139

Symeonidis N, Bachmayer F, Zapfe H (1974) Entdeckung von Zwergelefanten auf der Insel Rhodos (Ausgrabungen 1973). Ann Naturhist Mus Wien 78:193–202

Symeonidis NK, Theodorou GE, Giannopoulos B (2000) [The new species *Elephas chaniensis* from the submerged Pleistocene deposits of Vamos Cave at Chania, Crete]. Bull Soc Spéléol Gr 22:95–108 (in Greek)

Thenius E, Hoffer H (1960) Stammesgeschichte der Säugetiere. Springer-Verlag, Berlin

Theodorou GE (1983) [The fossil dwarf elephants of the cave "Charkadio" (Tilos island, Dodecanese)]. PhD thesis, National and Kapodistrian University of Athens (in Greek)

Theodorou GE (1984) Die fossilen Zwergelefanten der Höhle ‚Charkadio' auf der Insel Tilos (Dodekanes, Griechenland). Anz Math-Naturwiss Kl Österr Akad Wiss 120:83–85

Theodorou GE (1988) Environmental factors affecting the evolution of island endemics: the Tilos example from Greece. Modern Geol 13:183–188

Theodorou GE, Symeonidis NK (1994) [Palaeontological excavations in Charkadio cave during the decade 1984–1994]. Bull Soc Spél Gr 21:253–262 (in Greek)

Theodorou GE, Symeonidis NK (2001) The excavations of the last ten years at Charkadio cave on Tilos Island, Dodekanese, Greece. In: Cavarretta G, Gioia P, Mussi M, Palombo MR (eds) Proceedings of the 1st international congress "The World of Elephants", Rome, 2001, pp 514–518

Theodorou G, Symeonidis N, Stathopoulou E (2007) *Elephas tiliensis* n. sp. from Tilos island (Dodecanese, Greece). Hell J Geosci 42:19–32

van den Hoek Ostende LW, van der Geer AAE, Wijngaarden CL (2017) Why are there no giants at the dwarves feet? Insular micromammals in the eastern Mediterranean. Quat Int 445:269–278

van der Geer AAE (2018) Uniformity in variety: antler morphology and evolution in a predator-free environment. Palaeontol Electron 21(1):1–31

van der Geer AAE, Dermitzakis M, de Vos J (2006a) Crete before the Cretans: the reign of dwarfs. Pharos 13:121–132

van der Geer AAE, de Vos J, Lyras GA, Dermitzakis M (2006b) New data on the Pleistocene Cretan deer *Candiacervus* sp. II (Mammalia, Cervinae). Cour Forschungsinst Senck 256:131–137

van der Geer AAE, Lyras GA, de Vos J, Dermitzakis M (2010) Evolution of island mammals: adaptation and extinction of placental mammals on islands. Wiley-Blackwell, Oxford

van der Geer AAE, Iliopoulos G, Lyras GA, de Vos J (2011) The Middle Pleistocene deer of the Katharo basin (Crete, Greece) and its importance for a better understanding of the evolutionary history of the insular fauna of Crete. In: 9th annual meeting of the European Association of Vertebrate Palaeontologists, Heraklion, Crete, Greece, June 2011. Program and Abstracts, p 55

van der Geer AAE, Lyras GA, Lomolino M, Palombo MR, Sax D (2013) Body size evolution of palaeo-insular mammals: temporal variations and interspecific interactions. J Biogeogr 40:1440–1450

van der Geer AAE, Lyras GA, van den Hoek Ostende LW, de Vos J, Drinia H (2014a) A dwarf elephant and a rock mouse on Naxos (Cyclades, Greece) with a revision of the palaeozoogeography of the Cycladic Islands (Greece) during the Pleistocene. Palaeogeogr Palaeoclimatol Palaeoecol 404:133–144

van der Geer AAE, Lyras GA, MacPhee RDE, Lomolino MV, Drinia H (2014b) Mortality in a predator-free insular environment: the dwarf deer of Crete. Am Mus Novit 3807:1–26

van der Geer AAE, Lomolino MV, Lyras GA (2017) "Island Life" before man: biogeography of palaeo-insular mammals. J Biogeogr 44(5):995–1006

Vaufrey R (1929) Les éléphants nains des îles méditerranéennes, et la question des isthmes pléistocènes. Arch Inst Paléontol Hum Mem 6:1–220

Vesmanis I, Kahmann H (1978) Morphometrische Untersuchungen an Wimperspitzmäusen (Crocidura) 4. Bemerkungen über die Typusreihe der kretaischen *Crocidura russula zimmermanni* Wettstein, 1953 im Vergleich mit *Crocidura gueldenstaedti caneae* (Miller, 1909). Säugetierkdl Mitt 26:214–222

von Zittel KA (1925) Text book of paleontology. III Mammalia. MacMillan, London

Wettstein O (1953) Die Insectivora von Kreta. Z Säugetierkd 17:4–13

Willemsen GF (1980) Comparative study of the functional morphology of some Lutrinae, especially *Lutra lutra*, *Lutrogale perspicillata* and the Pleistocene *Isolalutra cretensis*. Proc Konin Nederl Akad Wet (Ser B) 83:289–326

Willemsen GF (1992) A revision of the Pliocene and Quaternary Lutrinae from Europe. Scr Geol 101:1–115

Index

E. Vlachos (ed.), *Fossil Vertebrates of Greece Vol. 2*,
https://doi.org/10.1007/978-3-030-68442-6

B

C

E